This book

Metallurgical Processes for the Year 2000 and Beyond

Metallurgical Processes for the Year 2000 and Beyond

Proceedings of the International Symposium on Metallurgical Processes for the Year 2000 and Beyond, held at the TMS 1989 Annual Meeting, Las Vegas, Nevada, February 27-March 3, 1989 and sponsored by the TMS Physical Chemistry Committee and the Pyrometallurgy Committee. The symposium was co-sponsored by the following:

Deutsche Gesellschaft fur Metallkunde Ev
Finnish Association of Mining and Metallurgical Engineers
The Institute of Metals (U.K.)
The Iron and Steel Institute of Japan
Iron and Steel Society (USA)
Jernkontoret (Sweden)
The Mining and Metallurgical Institute of Japan
Verein Deutscher Eisenhuttenleute (West Germany)

Edited by

H.Y. Sohn
University of Utah
Salt Lake City, Utah

and

E.S. Geskin
New Jersey Institute of Technology
Newark, New Jersey

A Publication of
TMS
Minerals • Metals • Materials

A Publication of The Minerals, Metals & Materials Society
420 Commonwealth Drive
Warrendale, Pennsylvania 15086
(412) 776-9024

Printed in the United States of America
Library of Congress Catalog Number 88-63681
ISBN Number 0-87339-084-9

TMS
Minerals • Metals • Materials

Symposium Co-Chairmen

H.Y. Sohn and E.S. Geskin

Advisory Organizing Committee

Robert W. Bartlett
University of Idaho
Moscow, Idaho

Tasuku Fuwa
Nippon Steel Corporation
Kawasaki, Japan

Suk Joong Im
Korea Mining and Smelting Co.
Seoul, Korea

Roland A. Kammel
Technische Universitat
Berlin, West Germany

Klaus W. Lange
Institute fur Eisenhutten
Aachen, West Germany

Frank Lawson
Monash University
Clayton, Victoria
Australia

Helmet Maczek
Berzelius Mettalhutten-GmbH
Duisburg, West Germany

Juho Makinen
Outokumpu Oy
Harjavalta, Finland

Takeshi Nagano
Mitsubishi Metal Corporation
Tokyo, Japan

David G.C. Robertson
University of Missouri-Rolla
Rolla, Missouri

Jef R. Roos
Katholieke Universiteit
Leuven, Belgium

Yogesh Sahai
Ohio State University
Columbus, Ohio

Klaus Schwerdtfeger
Technische Universitat
Clausthal, West Germany

Peter D. Southwick
Inland Steel Company
East Chicago, Indiana

Julian Szekely
Massachusetts Institute of Technology
Cambridge, Massachusetts

Akira Yazawa
Tohoku University
Sendai, Japan

Shoukun Wei
Beijing University of Iron and
Steel Technology
People's Republic of China

Noel A. Warner
University of Birmingham
Birmingham, Great Britain

Session Chairmen

Plenary Session

H.Y. Sohn
University of Utah
Salt Lake City, Utah

E.S. Geskin
New Jersey Institute of Technology
Newark, New Jersey

Metals and Materials Processing

Session I

R.H. Nafziger
U.S. Bureau of Mines
Albany, Oregon

S. Asai
Nagoya University
Nagoya, Japan

Session II

J.G. Byrne
University of Utah
Salt Lake City, Utah

R.D. Pehlke
University of Michigan
Ann Arbor, Michigan

Session III

H.Y. Sohn
University of Utah
Salt Lake City, Utah

D.R. Morris
University of New Brunswick
Fredericton, New Burnswick

Non-Ferrous Processing

Session I

M.G. King
ASARCO
Salt Lake City, Utah

M.T. Hepworth
University of Minnesota
Minneapolis, Minnesota

Session II

A.E. Morris
University of Missouri-Rolla
Rolla, Missouri

T. DebRoy
Pennsylvania State University
University Park, Pennsylvania

Biological and Aqueous Processing

M.K. McCarter
University of Utah
Salt Lake City, Utah

T.D. Chatwin
Resource Recovery and
Conservation Consultants
Salt Lake City, Utah

Electrolytic Processing

J.B. Hiskey
University of Arizona
Tucson, Arizona

S.K. Kim
University of Ulsan
Ulsan, Korea

Ironmaking and Steelmaking

Session I

P.C. Chaubal
Inland Steel Company
East Chicago, Indiana

R.I.L. Guthrie
McGill University
Montreal, Canada

Session II

D.G.C. Robertson
University of Missouri-Rolla
Rolla, Missouri

E.S. Geskin
New Jersey Institute of Technology
Newark, New Jersey

Raw Materials

R.G. Bautista
University of Nevada
Reno, Nevada

N. El-Kaddah
University of Alabama
University, Alabama

Preface

The eighties have been rollercoaster years for metallurgists. In the beginning a good start. Things looked promising. Then all of a sudden a real downturn. Drastic drops in production and the closing of mills, plants and mines, causing mass layoffs and unemployment. Especially hard hit was the research and development in minerals industries. And the verdict of the experts was that the time of smokestack industry in this country is over: "High-tech" is all we need. Then again a change. Shutdowns stopped. Metallurgical industry is experiencing an upswing. Things seem to have quieted down.

But the questions still remain: Are the experts right after all? Is metallurgy in this country ever going to see the good days again? The answer is sought not only in the corporate boardroom but also in the living room.

Metallurgical industry is going to stay, but changes must be made. Old facilities and processes must be modernized and replaced to adopt those based on the most advanced technical principles. Technology developments and their adoption will determine the future of metallurgy and how much longer the iron age will last.

Technical progress enables us to utilize an everwidening range of chemical and physical conditions for producing, shaping and treating metals. At the same time, the increasing costs of energy, raw materials, labor, and environmental control impose severe restrictions on the metallurgical processes. If previously the process selection was mainly based on technological considerations, it now has become an optimization problem having many constraints. There are many routes for converting raw materials to desired products; there are different products to meet the demands of society. If conventionally the search for a new technology was based on intuition and trial-and-error, the same is no longer possible. The metallurgical community must obtain a method to solve these problems in a greatly more systematic way and through an integrated effort by enlisting the help of communities with different expertises.

It is quite clear that the development of a healthy future industry must be based on the critical analysis of current practices and, based on it, an analytical determination of the future direction. It is with this in mind that this international symposium was organized. We hope that the symposium will provide a forum for fruitful exchange of information and views as well as an opportunity to seriously contemplate the future of a discipline which is important to all of us.

We wish to thank all the contributing authors for their participation and the members of the Advisory Organizing Committee for their valuable help and encouragement. Special thanks go to Marlene Karl, Judy Parker and

Judi Heiles at TMS Headquarters and the secretarial staff at the University of Utah and the New Jersey Institute of Technology for their assistance.

H.Y. Sohn
University of Utah
Salt Lake City, Utah

E.S. Geskin
New Jersey Institute of Technology
Newark, New Jersey

November 1988

Table of Contents

Organizing Committee....................v

Session Chairmen....................vi

Preface....................ix

Plenary Papers

Influences on Future Primary Metal Processes....................3
R.W. Bartlett

Some Directions for Metal Processing....................9
M.C. Flemings and S.B Brown

Recent Studies on Electromagnetic Processing of Materials....................17
S. Asai

Materials Processing in Space in the Space Station Era (Abstract Only)....................37A
R.S. Sokolowski

Future Trends of Non-Ferrous Metals Smelting and Refining....................37B
T. Nagano

Metals and Materials Processing

Ultra Fine Particle Deposition (Abstract Only)....................57
I.S. Min, S.J. Im and D. Kim

Development of Electromagnetic Atomization Process....................59
K. Sassa, T. Kozuka and S. Asai

Processing and Oxidation Protection of Carbon/ Carbon Composites (Abstract Only)....................69
G.R. St.Pierre, G. Holcomb and R.A. Rapp

Advanced Metal Matrix Composites Through XD™ Processing (Abstract Only)....................71
L. Christodoulou, J.M. Brupbacker and D.C. Nagle

Intelligent Realtime Carbonization Control....................73
W.J. Pardee and M.A. Shaff

Manufacturing of Metal-Ceramics Composites by Mashy-State Processings 83
M. Kiuchi and S. Sugiyama

Synthesis of Nonequilibrium Materials by Laser Surface Modification (Abstract Only) 101
J. Mazumder and A. Kar

Consolidation and Annealing Studies of Rapidly Solidified 304 Stainless Steel 103
W. Kim, I.-O. Shim, J.G. Byrne, J.-J. Kim and J.E. Flinn

Continuous Casting of Steel in the 21st Century 115
R.D. Pehlke

The Levitation Casting Process 129
H.R. Lowry and A.S. Klein

Inverse Solidification Analysis—A Valuable Tool for Computer Aided Process Engineering 139
K.L. Schwaha, H. Holl, H.W. Engl and T. Langthaler

Production of Advanced Materials by the Ohno Continuous Casting Process 155
A. Ohno

Electromagnetic Flow Control in Metals Processing Operations (Abstract Only) 165
J. Szekely

The Design, Operation and Characterization of a Plasma Reactor for the Generation of Value Added Products (Abstract Only) 167
P.R. Taylor and S. Pirzada

Computer Aided Design of Hybrid Plasma Reactors for Use in Materials Synthesis 169
J.W. McKelliget and N. El-Kaddah

Knowledge Based Control of Materials Processing: Challenges and Opportunities for the Third Millennium 183
D. Apelian and A. Meystel

Waterjet Cutting—Emerging Metal Shaping Technology 203
E.S. Geskin, W.L. Chen and A. Vora

Technology Necessary to Develop a Truly Automated Manufacturing Facility 215
R.C. Progelhof and W.F. Ranson

Technological Impact of Customer Requirements in Aluminum Metal Working Processes 225
R.E. Fanning

Non-Ferrous Processing

Control of Potential Difference in Continuous Smelting System 241
A. Yazawa

Mitsubishi Process—Prospects to the Future and Adaptability to Varying Conditions 253
T. Shibasaki, K. Kanamori and S. Kamio

Technologies for Low Cost Retrofitting of the ER&S Copper Smelter 273
P.G. Cooper

The Development Trends of the Outokumpu Flash Smelting Process for the Year 2000 (Abstract Only) 285
P. Hanniala and J. Sulanto

KLS Process Development for Lead Smelting 287
N. Wakamatsu, Y. Maeda and S. Suzuki

Semi-Pilot Plant Test for Injection Smelting of Zinc Calcine 301
S. Goto, M. Nishikawa and M. Fujikawa

Fundamental Studies on Nickel Metallurgy in China 323
H. Jiang, S. Yang and B. Guo

Separation of Various Elements in Crude Metal or Alloy During Vacuum Distillation 335
D. Yongnian

The New Concept in Zinc Electrowinning Operation 347
K. Kaneko, H. Ohba and T. Kimura

Computer-Aided Analysis and Simulation of High-Temperature Processes (Abstract Only) 361
I. Barin, G. Eriksson and F. Sauert

On-Line Measurements at Elevated Temperatures in Metallic Solutions, Mattes and Molten Salts Using Solid Electrolytes 363
D.J. Fray

A Manufacture Process of New Type Tungstate and Hydrogen Reduction of Tungsten Oxide Thereof 383
R.-J. Cao and X.-H. Tang

Physico-Chemical Fundamentals of Melting Non-Ferrous Concentrates With the Production of Metal (Matte) & Ferrite-Calcium Slags (Abstract Only) 391
A.I. Okunev

Biological and Aqueous Processing

Flotation of Microorganisms—Implications in the Removal of Metal Ions from Aqueous Streams 395
R.W. Smith, Z. Yang and R.A. Wharton, Jr.

Biotechnology Applied to Raw Materials Processing.......................... 411
S.K. Kawatra and T.C. Eisele

Bioaccumulation of Metals from Solution: New Technology for Recovery, Recycling and Processing 427
G.J. Olson, F.E. Brinckman, T.K. Trout and D. Johnsonbaugh

Bio-Leaching of Sulfide Ores.. 439
L.C. Thompson

Model for Bacterial Leaching of Copper Ores Containing a Variety of Sulfides 451
B.C. Paul, H.Y. Sohn and M.K. McCarter

Process Consideration for the Fabrication of pH Microelectronic Sensors.. 465
M. Moinpour, P.W. Cheung, E. Liao, C.Y. Aw and D.J. Brown

Electrolytic Processing

Potential for Fused Salt Electrolysis for Metal Winning and Refining 493
D.J. Fray

The Reduction of Alumina Beyond the Year 2000 Overview of Existing and New Processes.. 517
A.F. Saavedra, C.J. McMinn and N.E. Richards

Future Technological Developments for Aluminium Smelting.. 535
K. Grjotheim and B. Welch

Lunar Production of Aluminum, Silicon and Oxygen........................ 551
R. Keller

The Cathode Process of Aluminum Chloride Electrolysis.. 563
Y.J. Zhang and R. Tunold

Electrometallurgy of Silicon 571
G.M. Rao

Mathematic Models of Lost Currents in Bipolar Cells.. 583
Z. Guangwen, D. Shuzhen, T. Qiuzhan and W. Tan

Ironmaking and Steelmaking

Oxygen Steelmaking in the Future 603
R.D. Pehlke

The Application of the Second Law Technique for the Prediction of Trends in Steel Making Technology 613
E.S. Geskin

An Integrated Steel Plant for the Year 2000 623
I.F. Hughes

The AISI Program for Direct Steelmaking 637
E. Aukrust

Steel in the Year 2000 647
W.T. Hogan

The Consteel Process for Continuous Feeding-Preheating-Melting and Refining Steel in the Electric Furnace 655
J.A. Vallomy

Counter-Current Contacting in Metal Refining—A Reality in the Year 2000? (Abstract Only) 671
D.G.C. Robertson

Analysis of Bath Smelting Processes for Producing Iron 673
R.J. Furehan, K. Ito and B. Ozturk

Coal-Based Ironmaking Via Melt Circulation 699
N.A. Warner

Plasma Technology for Metal and Alloy Production Present Status and Future Potential 721
S.O. Santén and J. Feinman

Stainless Steelmaking Process by Direct Use of the Smelting Reduction of Chromium Ore in BOF 733
S. Nishioka, K. Yamada, T. Takaoka, Y. Kikuchi, Y. Kawai, A. Ozeki and M. Yamaga

Acoustic Sensors for Process Control in the Year 2000 745
N.D.G. Mountford, S. Dawson, I.D. Sommerville and A. McLean

Direct Reduced Iron: An Advantageous Charge Material for Induction Furnace (Abstract Only) 761
K. Sadrnezhaad

Raw Materials

The Influence of Solid State Imperfections in Mineral and Metal Processing 765
G. Simkovich and F.F. Aplan

Some Milling Practices and Technological Innovations on Beneficiation of Antimony Sulfide Ores in China 783
H.-K. Hu

Oxygen Production Technologies 791
K.J. Murphy, A.P. Odorski and A.R. Smith

Recent "Advances" in Sulfuric Acid Technology—Review and Analysis (Abstract Only) 807
L.J. Friedman

Processing of Complex Sulphides 809
J.M. Figueiredo, M.C. Coelho, A.R. Silva and R.M. Guedes

Some Promising Techniques for Complex Iron Ore Metallurgical Processing 825
L.I. Leontjev, N.A. Vatolin and S.V. Shavrin

Author Index 841

Plenary Papers

INFLUENCES ON FUTURE PRIMARY METAL PROCESSES

Robert W. Bartlett
College of Mines and Earth Resources
University of Idaho
Moscow, Idaho 83843

ABSTRACT

Economic and other factors, mostly outside of metallurgy, that strongly influence applied research, development and especially successful commercial adoption of new processes for producing primary metals and other commodity materials are reviewed. These factors, which can be both positive and negative in their influence, are often too lightly regarded in the research planning stage. They include continuing imperatives that have long influenced new process development as well as more recent trends. Both are discussed. Nearly all new materials eventually become commodities--products made to standard specifications with no producer control over price. Success requires being a low cost producer on a world wide basis. Composites and other complex material assemblages are unlikely to escape this fate as they mature.

Metallurgical Processes for the Year 2000 and Beyond
Edited by H.Y. Sohn and E.S. Geskin
The Minerals, Metals & Materials Society, 1988

INTRODUCTION

The adoption of new processes, beginning with research and development, for primary metals and other commodity materials is particularly difficult because each new process must supplant an existing process that is already being used commercially. There are many requirements besides technical innovation and cost superiority that must be met by the new process. Consequently, few innovations make it all the way from the research laboratory to commercial success.

This paper discusses some of the factors that influence success in developing new processes for primary metals and commodity materials, based on the authors experience as a research metallurgist and industrial research manager. It begins with some of the factors, that are always present and constrain successful process research and development, and proceeds to current trends that are expected to continue long enough to influence processes in the year 2000. Many of these trends are contradictory in their effect on new process development, and the future remains uncertain.

The prevailing importance of these trends will vary with the material, process concept and other factors intrinsic to each case. Nevertheless, an awareness of these influences can help to avoid some of the R&D pitfalls and dead ends. Such awareness is deemed to be a necessary condition to avoid failure when commercial adoption of a new process is the goal.

ECONOMIC AND TECHNICAL VERITIES

A number of ever present factors, mostly economic, add to the technical difficulties and constrain a successful R&D program and commercial implementation:

- Fierce materials competition and substitution will remain with continued uncertain effects on demand for current primary metals and commodity materials.

- Primary metal and commodity materials firms cannot control their product price and they must be low cost producers on a worldwide basis, if they are to survive inevitable periodic price recessions. New processes that do not conform to this requirement should not be adopted.

- New plants require a marginal price equal to their manufacturing costs **PLUS** capital amortization. Existing plants must only meet manufacturing costs. A new process usually cannot supplant an old process when demand is static or shrinking. Therefore new process adoption usually only occurs during growth in demand and price elevation. Obsolescent processes are eliminated by recessions, when their product prices fall below their manufacturing costs and plant deterioration or full abandonment prevent reopening during a period of increased product price. The author learned these lessons the hard way during the 1980's.

- The process "learning curve" works against first adoption of new processes but benefits them thereafter. A high cost of entry will continue to deter or prevent new process adoption in some cases where large economies of scale and the learning curve of entrenched processes are dominant. Probable examples are the iron blast furnace and the Hall-Heroult aluminum reduction process.

- Cost of capital greatly influences new process adoption with low interest rates required to justify investment. This has been and remains a major advantage for Japan, where lower interest rates are caused by their high savings rate and low military spending.

CAPITAL, FINANCING AND MONEY EXCHANGE TRENDS

The high dollar exchange value over the 1980's nearly ruined the American primary metals industry and hindered adoption of new processes. However, the recently cheapened dollar will likely be sustained at low exchange rates because of the high net USA international debt. Also, many metal price increases (dollars) haven't fully reflected the dollar decline---this bodes well for domestic producers.

Overexpansion of production capacity in third world countries using debt capital has stopped and should not begin again because of their debt repayment problems. Banks and governments are unlikely to repeat this mistake before the year 2000. Eventually, this will provide a better balance of supply with demand than occurred through the 1980's, and hopefully better prices.

Although there will be much fluctuation, historically high average interest rates are expected to continue, because of our international debt and low savings rate. A drastic change in tax policy could alter this forecast, with its negative implications.

BUSINESS AND LABOR TRENDS

Both negative and positive influencing trends on new process development exist relative to the situation in past years. The trends with negative impact on adoption of new processes follow:

- o New product research is believed by most professional business managers to have a much greater payoff than cost-lowering process research. This discourages process R&D expenditures.
- o The Federal Government places very little of their huge materials R&D budget in improved processing for primary metals.
- o There has been a marked decline and frequent disappearance of large corporate R&D Labs in the primary metals industry.
- o New technology is readily and quickly transferred anywhere in the World regardless of patents. This has been accelerated by the information and air travel revolutions. The newest technology is implemented in new plants regardless of location, including less developed countries. Examples are steel in Korea, copper in Chile and ferroalloys in South Africa. This reduces incentives for R&D expenditures in the USA.

The trends with mostly positive impact on adoption of new processes for primary metals and commodity materials follow:

- o There is disaggregation of vertically integrated, large industries---which creates intrepreneuring opportunities. Entrepreneurs will usually take more risks on new processes than corporate bureaucracies. However, entrepreneurs in this industrial sector usually spend less on R&D.
- o With relaxation of anti-trust laws there will be more cooperative R&D in the future, including multicompany consortia to do R&D projects, and Govt/industry coop programs, e.g., MITI in Japan, and the DOE "Steel Initiative" in the USA. More international cooperation will occur on R&D, process development (large pilot plants) and financing of first plants.
- o Future materials shortages will lead to long term product marketing contracts to finance large new capital projects. This has already been successfully pioneered in developing new domestic coal mines.
- o Wage rates have become relatively unimportant for new or modernized plants due to their high capital cost. Plants are now being built at a cost of $1,000,000 per employee. Compare the amortization of this capital with the wages of one worker. Capital costs in third world countries often involve greater infrastructure and higher capital costs. Hence, wage rates are less of a USA disadvantage than before. Also, American wages in the primary metals sector are declining relative to other domestic wages and foreign wages. Labor skill and dependability are critical, both individually and collectively (avoiding

strikes and other work stoppages). Hence, we may still be disadvantaged against Japan and a few other countries with respect to education and employee attitudes.

TECHNOLOGY TRENDS

An expanding diversity of new materials and their supporting raw materials demands will continue. This will create opportunities for new processes. A current example is the interest in thallium for superconductors. We may turn to extracting thallium from western USA gold mill tailings, which contain a significant amount of this element.

New materials and semi-fabricated product markets will influence primary materials process R&D and selection to provide more value, customer service, shape forming, etc. in fewer steps and at lower total cost during primary processing. Primary metal producers will need to match or incorporate new downstream processing steps. An example is to provide prealloyed titanium powder for using in near net shape fabrication by hot isostatic pressing. This would compete favorably with machining VAR alloy billets. More specialized materials and niche markets will occur, differentiated by product, customer, service, etc.

New processes will be adopted to improve quality and yield. Examples are thin section continuous casting, vacuum and inert gas processing, etc. Product (material) quality and yield from existing processes will also improve through better process monitoring, using improved sensors and instrumentation, and better control through more extensive use of on-line computers.

More computerized optimization and decision making will be used to provide quality improvements and flexibility in both products and production runs, and in making other operating choices. Use of artificial intelligence and computers as low cost on-line process operators may permit an economic resurgence in batch processing. This approach is already being used for some pharmaceutical materials.

Large economies of scale will not dominate as much as they have in the past. Advantages of smaller scale will occur through:

- -Satisfying capital limitations,
- -Providing flexibility in products and turndown of production rates,
- -Becoming closer and more responsive to markets or feedstocks (e.g. ministeel plants--markets and scrap),
- -Producing minor metals and specialty materials where quantities are small.

New processes will be strongly driven by capital reduction and simplified process flow sheets. Comparative preference examples are:

- -Pneumatic flow > fluid bed flow > fixed (moving) bed flow,
- -Coal based rather than coke for pyrometallurgical reduction, and
- -Solution mining and direct reduction from solution > multisteps in separate plants of: excavation mining, beneficiation,smelting and refining.

High intensity processes will be favored because they provide compact equipment, short residence time, fast kinetics, fast mass/heat transport and low capital cost. Examples may be cyclone smelting and arc plasma processing---or perhaps plasma cyclone smelting.

Innovations in specialty equipment invented for one material will be adopted for several other materials. Examples are column flotation, continuous belt thin section casting machines, and solvent extraction.

New technologies will be combined with each other and with existing processes to simplify production of materials. An example is combining alloy reduction (synthesis) with gas atomization on discharge from the reduction reactor to produce rapidly solidified alloy powders.

Electricity cogeneration has become a major factor in process selection. Metal reduction may be a "topping step" in a combined cycle system with metal production cost relatively insensitive to fuel (reductant) usage, provided a low-cost fuel such as surface mined coal is used. Any excess fuel passes through the reduction step and is "purchased" by the electricity step at no cost to the metal process.

Low grade, bulk minable, near surface ore deposits will continue to be the dominant ores, especially in developed countries where rich ores are depleted, and the available ores will strongly influence new process development. This will mandate processes with very low costs per ton of rock and will lead to noninvasive (in situ) processes when possible.

For many new materials, the material and the component are inseparable in processing and manufacture---for example composites and integrated circuits. Most of the value in new materials is at the component stage. Primary materials firms have usually not been successful in moving to the component stage of new materials. End manufacturers are integrating back to become new materials producers.

ENERGY, ENVIRONMENT, RECYCLING AND GOVERNMENT

Energy cost and conservation will again become important in process selection by the year 2000 because the present oil glut is expected to disappear by then. Methanol, derived from natural gas, will become an important transportation fuel. Coal and electricity will become even more important for pyrometallurgical processes.

Environmental mitigation will continue to drive much new process R&D and process selection. A reduction in fluid volumes being processed will usually significantly reduce environmental control costs. Use of more oxygen enrichment will lower gas volumes in smelting operations. In some cases environmental control costs equal or exceed the primary processing costs. Recycling process research and adoption will accelerate. Although the high cost of collecting dispersed waste materials works against recycling, society (government) may decide to financially incentivize collection and recycle through tipping fees or other penalties.

Strategic (defense) considerations may effect domestic materials processing through increased stockpiling and by subsidizing domestic plants and technologies that are not internationally cost competitive.

POSSIBLE NEW PROCESS TECHNOLOGIES–YEAR 2000

New extraction technologies by year 2000 may include the following:

- -Forced air injection in copper dump leaching (present extraction is only 20 to 30%).
- -Use of supercritical fluids as selective extractants
- -Large volume, subsurface pressure processing using the natural hydrostatic head for pressure.
- -Metal by-product recovery from geothermal brines
- -Borate solution mining--a pilot plant has been run
- -Selective sorption of very dilute heavy metals (e.g. tungsten and molybdenum) from brines.
- -Continuous thin section shape casting of steel and nonferrous metals.
- -Increased role of biomaterials in both metal extraction and selective sorption of metals.
- -On-line computer sorting and separation of mixed industrial and municipal waste materials.
- -Selective organic extractants for nearly all metals of commercial interest.
- -Manufacture of low-performance, high-tonnage composite structural materials from partially separated waste.
- -Large scale commercial adoption of centrifugal (gravity enhancing) processes in combination with flotation and solvent extraction separations.

SOME DIRECTIONS FOR METAL PROCESSING

M. C. Flemings and S. B. Brown

Department of Materials Science and Engineering
Massachusetts Institute of Technology
Cambridge, MA

Abstract

Metallurgy, as all of material science and engineering, rests on the intellectual foundations of processing-structure-property-performance relationships. Today it is widely recognized that processing has been too long neglected and presents exciting opportunities for exploitation. New processes, many of them already in the laboratory, will be exploited in the year 2000 and beyond, for near net shape forming and for achieving new metallurgical structures and composite structures. Process modelling will alter the way we think about process and product design. Processes themselves will increasingly be automated and controlled by computers, altering the nature of the workplace and the tasks of the metallurgical engineer of the future.

Metallurgical Processes for the Year 2000 and Beyond
Edited by H.Y. Sohn and E.S. Geskin
The Minerals, Metals & Materials Society, 1988

Introduction

Innovations in metal processing is hardly a new phenomenon, with the field thousands of years old. Nonetheless, the next few decades offer particularly important opportunities to improve our ability to process these materials. We appreciate today the variety of processing paths available to us, including vapor-, plasma- and electro-deposition, multiple phase and casting technologies, and non-equilibrium processes, and how these affect structure at the atomic level and at higher levels of aggregation. We understand the basic physical mechanisms involved far better than even before and have remarkable new abilities to predict, control, and simulate the kinetics underlying processes, and the processes themselves.

The following sections describe these new opportunities in greater detail. First, we examine developments in the modelling of metal processing. We then consider some implications associated with process control and automation. Finally we present what we consider to be particularly promising metal processing technologies, and conclude with a few summary comments.

Processing Modelling

The advent of the computer has changed the way we think about processing today, and certainly the way we think about what it will be in the year 2000 and beyond. Many of the governing equations and models for materials behavior in processing were developed decades, and in some cases centuries before computers existed. Unfortunately the nature of problems associated with these models, due to either their large size (thousands of variables), or nonlinear character (necessitating iterative solution methods) prevented the solution of all but limited classes of problems.

All of this changed with the application of numerical analysis, including techniques such as finite difference and finite element methods. These techniques permitted the solution of complex boundary-valued problems involving arbitrary boundary conditions and highly non-linear material and geometric behavior. Computers applied these techniques efficiently and quickly. The upshot was that problems in thermal analysis, fluid flow, electromagnetism, and solid deformation have suddenly become solvable. Processes which were understood only qualitatively now could be analyzed in rigorous, quantitative manner, and process parameters could be varied, evaluated, and optimized.

There are two consequences of this new power to analyze. First, process simulation is no longer the domain of the applied mathematician. Engineers can now design a part and evaluate its operating stresses without having to solve the governing equations from scratch. Designs and the design process therefore become more efficient, since engineers have the means to evaluate their ideas before production. Second, the applied mathematician can address problems of previously considered impossible complexity. Subtleties of material processing such as turbulence or nonlinear heat transfer can be explored, and processing effects can be explained and controlled. Entirely new processes can be designed and simulated before construction.

This is not to say that computers provide the only advances in process modelling. An important contribution to the modelling effort is, and will continue to be, new experimental methods. Advances in optical pyrometry, high temperature measurement, high vacuum technology, computer-aided-tomography, sensor technology, and laser spectroscopy all have provided new insights into the process kinetics and constitutive behavior. The new experimental data resulting from these methods provide both better insight into the details of processes and better information to calibrate the models simulating those processes. New, different, and more sensitive sensors will continue to be developed and used in this way.

The advance of process simulation will most likely occur along several fronts. One front of progress is that process simulation will become a regular part of process design. Simulation techniques are not yet part of the everyday design process. We expect to see the nonlinear models which are now used in the research laboratory become part of the manufacturing activity. Die and mold design, selection and sizing of equipment, and process optimization (e.g. for casting or forging) will all be accomplished using simulation techniques. The sophistication of these techniques will increase as computer memory sizes and processor speeds continue to increase.

Another front will be marked improvement in process models. One such improvement is internal variable constitutive models, where the process model simultaneously tracks the evolution of material internal structure with the process. One example of such a model is constitutive models for hot working, which predict the effects of hardening and recovery on the internal structure of an alloy while accurately simulating its flow behavior. These internal variable models, although requiring substantial computer support, allow a very accurate representation of the process, which in turn permits more accurate optimization of the process. One other advantage of these models is that they have the potential of indicating non-intuitive processing paths which would not be obvious given the myriad of process parameters.

On still another front, we should expect better control of material processing given more accurate models. Any control system requires a model of the system to provide an "ideal" or target output. Incorporation of more sophisticated process models will permit more accurate and stable control of the actual manufacturing processes, improving both material quality and tolerances. This is an important implication of improved material process modeling, for we frequently assume that the benefits of modelling occur at the "front-end" of the manufacturing process. Improved models, however, have the potential of influencing every step in the production process.

It is difficult to say exactly what industry or process will be most influenced by our new simulation capability. No industry, traditional or emerging, will be unaffected, and there is no apparent limit to the changes which may occur as a result.

Looking beyond process simulations of the type we can undertake today, we can envision the day when computer simulations will link the materials, design, processing, and performance aspects of a material manufacturing operation. Consider as example, the design of a turbine disk for an aircraft engine. Under ordinary conditions, the designer starts by attempting to achieve given performance specifications in a way that satisfied a few basic constraints, perhaps minimization of weight in the present circumstance. Fabrication engineers are then asked to see whether the part can be produced. If problems exist, or if the cost appears excessive, some design modifications may be required. Finally, maintenance and inspection specialists are consulted, but it is usually too late by this stage in the procedure for their needs to have a major impact on design. This serial approach emphasizes mechanical aspects of the design at the expense of production and maintenance considerations.

A far better approach would be to incorporate all or most of these considerations into computer simulations carried out during the initial stages of the design process. To start, one might examine possible processing paths in order to optimize metallurgical microstructures for different properties in different regions of the component. In the example of the turbine disk, the microstructure could be optimized for creep strength at the rim and for low cycle fatigue and ultimate tensile strength in the bore. The same detailed simulations might also address issues of technical feasibility and economics. Models which relate microstructural properties to processing paths might also be used to examine manufacturing options and even to optimize for ease of maintenance.

It seems quite likely that such an integrated approach to materials design eventually can lead to small but significant changes in technology that, in turn, will produce large improvements in performance and cost over the lifetimes of products. It even seems possible, because we are dealing with complex problems in systems analysis, that the results of these integrated simulations occasionally will turn out to be very different from what was expected. When that happens, the technological impact is apt to be very great indeed. The limiting factors in this program are the availability of analytic and numerical models, and the availability of the specially trained scientists and engineers who are needed to develop these models and bring them to bear on technology.

Process Control and Automation

The Bureau of Labor Statistics has estimated that by the year 2000, the number of jobs in manufacturing in the U.S. will have decreased slightly, to 18.2 million workers from the 19 million workers today. Within manufacturing however, the nature of the work will change and the nature of the jobs available will be different. There are expected to be 165,000 more jobs for engineers, 23,000 more jobs for computer and other scientists, and 70,000 more jobs for engineering and other technicians. At the same time, there will be hundreds of thousands fewer jobs for less skilled workers. This, of course, is a result of the "smart machines" and automated, flexible processes we are already seeing sweep the field.

Over the last decade, we have discovered advantages of introducing process automation that were not widely considered initially. To properly control a complex process, one must understand the process itself in considerable detail, closely standardize and control incoming raw materials or other inputs to the process, and then determine and control, with appropriate feedback, the important variables. Failure to carry out these steps properly has resulted in many failed attempts at process automation, but properly carrying them out has resulted in processes of greatly improved reliability and consistency, and lower reject rate. The success in process improvement in these cases has, of course, been only partly due to the computers themselves.

A second important lesson we have learned is that since computers and machines can do some things better, and most things faster, than human beings, we do not have the same boundary conditions for processes we had when it was necessary to carry them out using human brains or muscles. The result has been development, in some cases, of radically different processes, and usually processes which proceed at a much more rapid rate than would be possible with humans alone.

New Processes

In the decades ahead, we can expect that vapor-solid processes for metals will continue their rapid development - for coatings, for interconnects in VLSI manufacture and for manufacture of thin films for other uses (e.g. superconductivity, capacitors). A host of important processes are available today to produce metal films in this way, broadly characterized as physical or chemical deposition processes. While one tends to think first of these processes as applying very small components, this will not necessarily be the case in the future. Already, heavy protective coatings are applied to turbine blades by this process, and active development programs are underway to produce steel sheet in high volume production with decorative and protective coatings produced by plasma deposition processes.

In the year 2000 and beyond, we can expect solidification processing to retain and increase its industrial importance. This is because it is in many cases the shortest, most economic route from starting raw material to near net shape. In other cases, unique microstructures can be obtained by solidification.

In continuous casting, there is intense worldwide interest in developing new near net shape casting methods for strip, rod, and simple shapes such as "T's" and "U's". Nowhere is the interest more intense than in making steel strip of a thickness such that the hot mill can be eliminated. In such an operation, liquid metal would be cast directly to a strip of thickness such that it could be cold rolled to a thickness suitable, for example, for automotive plate. One method of strip casting being studied by many companies is the twin roll process in which liquid metal is introduced in the gap between two rolls and solidifies largely or completely within that gap. Many other processes are being examined as well. In all of these processes, problems to be overcome include those of metal flow and distribution, surface finish, internal soundness, and cost.

Net shape casting of discrete parts has undergone a dramatic revolution over the past few decades, and that revolution continues. Permanent mold casting of aluminum alloys has been perfected and automated to a point beyond the dreams of the foundrymen of a few decades ago. Investment casting, once the province of the jeweler and the dentist, has become a worldwide casting process of great commercial importance, and a multi-billion dollar business. Investment cast superalloys, usually vacuum cast, comprise an important component of the business, but there is an important and growing segment of investment castings produced by air melting for a wide range of applications; e.g. automotive components and golf clubs. Innovations in materials and methods will continue to drive the cost of these castings downwards to make them more competitive with machine parts and with forgings.

Pressure die-casting has for many years been the process of choice for producing small, high-volume components of zinc and aluminum. Important developments in this field over the coming years will be new methods of metal handling (e.g. electro-magnetic transport) to more efficiently and economically bring the liquid metal to the die, techniques to more economically die cast alloys with melting points above that of aluminum alloys, and methods to use the hot chamber die casting process for aluminum alloys.

A process which can be viewed as something of a hybrid between casting and forging, termed "semi-solid forming" or "Rheocasting" was initially developed nearly 20 years ago and is now beginning to be exploited commercially in this country and in Europe. It is also the subject of large development programs throughout the world - including a MITI sponsored research company entitled "Rheotec". In this process, metal is vigorously stirred during its initial solidification in order to break up the dendritic structure. The stirring process produces a semi-solid "slurry" which can then be formed directly using conventional casting processes if the fraction solid is low enough. At higher fractions solid where the slurry becomes quite thick, it can be formed by conventional metalworking processes. If desired, the slurry can be fully solidified in, for example, some billet form. The billet can than be reheated to the semi-solid state and then formed as described above.

Rapid solidification processing will continue to develop, mature, and find more applications. We see two important commercial applications today in amorphous sheets for transformers and in the new strong magnetic alloys, the neodymium iron boron alloys. In addition, specialty super-hard materials are being made by the process and we can expect other applications to come with time.

Metal matrix composites is an important area for future growth in metal processing. Because of inherent economies, the most important processes of the future for these components are likely to be those involving liquid metal. Interest today focuses most strongly on two such processes. One is the "mixing in" to a liquid or semi-solid metal of ceramic particulates or fibers, and then casting or otherwise forming the resultant slurry. The second important method is infiltration of a liquid metal into a fibrous or particulate preform. A third process, solidification of "in-situ composites" has received much study over the last decade or two. In this case, a eutectic or near-eutectic alloy is directionally solidified, thus aligning the two or more phases that may be present. Such alloys have the potential of being outstanding high strength materials, and this process will no doubt one day become commercial.

Powder processing, as another near net shape process, and as a way for producing composite materials, will continue to grow. There is much interest today in applying the injection molding process to metals and this process could become of significant importance commercially, but much work yet needs to be done on binder burnout, shrinkage and distortion, and other processing aspects.

One promising application for powder metals is the production of forging preforms. Powder technology permits preform geometries optimized to a given die shape. Subsequent forging then acts to achieve final dimensions and eliminate residual porosity.

Metal working, including continuous processes such as rolling and discontinuous processes like forging, will continue to develop. Many of the simulation techniques mentioned earlier in this paper will permit the optimization of working operations through the reduction of waste, acceleration of design, and coupling of structure to the deformation process. An excellent example of this is the recent progress made by the modelling community in the production of deformation textures. Anisotropy may become a readily controllable design option, made possible by our improved ability to use basic physics to represent complex deformation processes.

Electrolytic processing has an important segment of the broader electrochemical industry and includes metal production, plating, semiconductor processing, and chemical production. New materials, new processes, and process modelling have had a dramatic influence on the course of the industry in recent years and promise to further alter its character in the years to come. We can expect in the future to see electroprocessing as a route for electrosynthesis of advanced materials. Codeposition is one method. For example, the ternary magnetic alloy neodymium-iron-boron can be produced from codeposition from a molten salt. Electrolysis is a highly non-equilibrium process and has the capability of generating metastable structures in the form of coatings, epitaxial layers or powders. Compositionally modulated structures are also readily produced. Electrodeposition is usually carried out at or near room temperature in aqueous and organic electrolytes or at elevated temperatures in molten salts. We will see the development in the future of low temperature electrolytic processing in cryogenic liquids. Cryogenic electroprocessing offers a new window of opportunity in a temperature region where kinetic processes occur at quite different rates than at room or elevated temperatures, providing new opportunities for achieving metastable structures.

With respect to forming of fully solid materials, we can expect continuing major developments in metal removal methods, both by new removal processes including laser machining and electrolytic machining, but also as a result of new abrasive and cutting tool materials. New ceramic cutting tools and abrasives produced by the sol gel route, new chemically modified diamond structures, improved cubic boron nitride materials and others promise a continuing rapid revolution in our ability to remove metal quickly and efficiently.

The advent of complementary classes of materials will assist metals processing by improving process equipment. New refractory materials have improved the quality and effective lifetime of process equipment, and will continue to do so. Die alloys and coatings will improve the yield and wear resistance of forming and casting equipment. Although not necessarily the most publicized achievements, these incremental improvements in the manufacturing capability of processing equipment may provide the greatest economic benefit to the current metals industry.

Concluding Remarks

The year 2000 is, to the dismay of many of us, not very far away. It is possible to see quite clearly the directions we will travel in metal processing over that time period, although the speed of travel is certainly more difficult to predict. Surely, many new processes now in or emerging from the laboratory, will find their way into commercial utilization. Process modelling, computer and experimental, will play an increasing role in process as well as product development. Computers will be used increasingly to carry out the processes themselves. The nature of work and of the workplace in metal processing facilities will change, even more rapidly than it has already. Generally, our plants will be a better place to work, but a higher level of education will be needed throughout our work force, with more engineers, and with workers who are comfortable with technology.

RECENT STUDIES ON ELECTROMAGNETIC PROCESSING OF MATERIALS

Shigeo Asai

Department of Iron & Steel Engineering
Faculty of Engineering, Nagoya University
Furo-cho, Chikusa-ku, Nagoya 464
Japan

Abstract

Electromagnetic processing of materials is newly proposed to be considered as one of the engineering fields where functions of electromagnetism are utilized for processing electrically conducting materials, such as liquid metals, electrolytes and plasmas. The background requiring this processing and the engineering characteristics supporting it are described. The applicable functions of electromagnetism to material processing are revealed from the view point of magnetohydrodynamics. The existing processes and the proposed ones are classified into utilized functions and engineering applications are discussed.

Metallurgical Processes for the Year 2000 and Beyond
Edited by H.Y. Sohn and E.S. Geskin
The Minerals, Metals & Materials Society, 1988

Introduction

Electric energy has been used in the metallurgical field for the processes of melting, refining and solidification. In these cases both the magnetic field which is induced by an imposed electric current, and the electric current are applied on electrically conducting materials such as molten metals. Thus, it turns out that the magnetohydrodynamic (MHD) phenomena have unconsciously been seen in metallurgical processes using electric current. The academic backbone studying MHD phenomena is the Magnetohydrodynamics which is being developed in the fields of plasma physics, astrophysics, fusion research and nuclear energy.

The first attempt where magnetohydrodynamics was intentionally applied on metallurgy was a Symposium of IUTAM (The International Union of Theoretical and Applied Mechanics) in titled as "Metallurgical Applications of Magnetohydrodynamics" at Cambridge, UK, 1982 [1]. This timely event stimulated a few metallurgists to recognize the importance of MHD and was a good warning to the metal societies in the world. Especially in Japan, the Steelmaking Investigative Subcommittee of Research Planning Committee in the ISIJ (The Iron and Steel Institute of Japan) discussed the application of MHD to metallurgy and selected it as one of the four major innovative subjects for future steelmaking technology in Japan [2]. Indeed, the term of "Electromagnetic Metallurgy" itself was firstly coined by this subcommittee. Upon the request of a report from the subcommittee, the ISIJ has established a committee on "Electromagnetic Metallurgy" at 1985, under the author's chairmanship.

For material processing by use of electric and magnetic energies, an intentional application of "Electromagnetic Processing of Materials" is indispensable which is newly defined as an engineering field combined with magnetohydrodynamics and material processing.

Background and Characteristics

The background where the study of electromagnetic processing of materials has been demanded and the characteristics of electromagnetic processing are described here.

Background

Increase of electric energy consumption in metal industries.

A strong demand for high grade materials is accompanied by an increase in the consumption of electric energy. Especially, in the process of steel-making the number of treatments to refine steel are increasing in the requisition for clean steel. Therefore, time needed for refining tends to be prolonged and addition of thermal energy to molten steel is recognized to be an important technical subject. This thermal energy is mainly supplied by electric energy.

Development of the technology connected with this field.

These days, the technology connected with electricity and magnetism is drastically developed. For instance, the recently produced high energy permanent magnet (for example Fe-Nd-B alloy) has easily given us a strong magnetic field at a cheap cost. Also, super conducting magnets will be used in the related industries in near future on the conjecture that the critical temperature of super conducting materials keeps increasing. Moreover, an electrode made of ZrB_2, (melting temperature: 3000℃ and electric resistivity: $10^{-5}\Omega\cdot m$), has been developed. This has opened the gate to impose electric current directly on metals with high melting point such as molten steel and seems to bring a large impact on the process metallurgy. Until now the electric current induced by alternating magnetic field has been mainly adopted in metal industries (for example, linear type electromagnetic stirring, electromagnetic pump, etc.) instead of directly imposing electric current on molten metal. From now on, by direct imposition of current we could expect alternative methods for transport and mixing of molten metal with a more efficient use of electric power.

Characteristics

The characteristics of electromagnetic processing of materials are given as follows:

High density, cleanliness and controllability of electric energy.

Making use of electric current and magnetic field is the cheapest and the most convenient method to impose high density energy on metals. This energy source is extremely clean except the contamination from electrode. Furthermore, the controlling technique of the regarding electric current and magnetic field has so much advanced these days.

Effective use of electric energy.

In the metal refining field, electric energy has been mainly used as the heat source. The values of ordinary mixing power employed in steelmaking processes (1W/ton~500W/ton) are corresponding to the heating rate from about (1℃/one week) to (1℃/30 min.). On the other hand, the heating rate required as heat addition in steelmaking processes is about 1~6 ℃/min. Then, it can be recognized that the amount of mixing power is extremely small in comparison with that for heat addition. Thus, a part of electric energy should be converted into kinetic energy through a way to add some functions of electromagnetism such as mixing, splashing, etc.(the details will be mentioned in later parts) as shown in Fig.1. The energy used for functions of electromagnetism is not wasted but is finally converted into thermal energy.

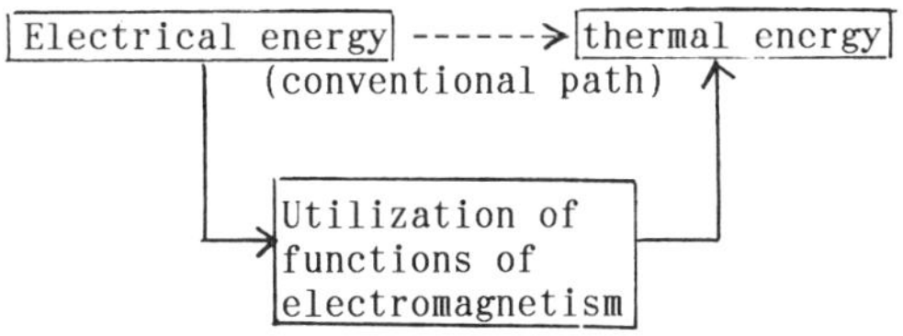

Figure 1 - The recommended energy path (——) and the conventional one (----).

Application of magnetohydrodynamics to metallurgical problems.

Magnetohydrodynamics, the study of the dynamics of electrically conducting fluid such as plasma, the term of which was coined at 1942 by Alfvén, of course, can cover the dynamics of molten metal. The comparison between plasma and liquid metal of fluidic characteristics is as follows:

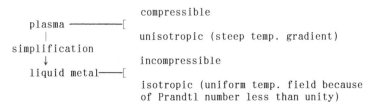

plasma —— [compressible; unisotropic (steep temp. gradient)

| simplification ↓

liquid metal —— [incompressible; isotropic (uniform temp. field because of Prandtl number less than unity)

The vast knowledge stored in magnetohydrodynamics is applicable to solve the problems of liquid metal processing. The advantage of this application is the simplified procedure from a compressible and unisotropic fluid to an incompressible and isotropic one.

Governing Equations for Motion of Electrically Conducting Fluid

In order to analyze fluid motion driven by electromagnetic force, Maxwell's equations and Ohm's law governing electric and magnetic fields and Navier-Stokes equation governing fluid flow field have to be solved simultaneously.

The basic equations under the Magnetohydrodynamic Approximations may now be written as follows:
Maxwell's equations are

$$\nabla \times \mathbf{E} = -\partial \mathbf{B}/\partial t \tag{1}$$
$$\nabla \times \mathbf{B} = \mu \mathbf{J} \tag{2}$$
$$\nabla \cdot \mathbf{B} = 0 \tag{3}$$

Ohm's law in the case accompanied with fluid motion is

$$\mathbf{J} = \sigma (\mathbf{E} + \mathbf{V} \times \mathbf{B}) \tag{4}$$

where, **E** is the intensity of electric field, **J** is the electric current density, **B** is the magnetic flux density, μ is the magnetic permeability, σ is the electric conductivity and **V** is the velocity of fluid.

The electromagnetic force (Lorentz force) driving fluid is obtained by substituting **J** and **E** which are obtained by solving Eqs. (1)～(4), into Eq.(5).

$$\mathbf{f} = \mathbf{J} \times \mathbf{B} \tag{5}$$

On the other hand, the equation of continuity of the incompressible fluid is shown by Eq.(6)

$$\nabla \cdot \mathbf{V} = 0 \tag{6}$$

The equation of motion of the Newtonian incompressible fluid, so called Navior-Stokes equation, is written as Eq.(7).

$$\rho \{\partial \mathbf{V}/\partial t) + (\mathbf{V} \cdot \nabla)\mathbf{V}\} = -\nabla p + \mu_f \nabla^2 \mathbf{V} + \mathbf{f} \tag{7}$$

where, ρ is the density of fluid and μ_f is the viscosity coefficient of fluid.

The flow field, **V**, can be obtained by substituting the body force evaluated from Eq.(5) into the third term in the right hand side of Eq.(7) and by simultaneously solving Eqs.(6) and (7). In this procedure, however, **V** has to be determined before evaluating **E** and **B** since **V** also appeares in the second term in the right hand side of Eq.(4). That is, in the system driving the electrically conducting fluid by electric and magnetic fields, the velocity field and the electric and magnetic fields are coupling and influence each other.

In general, as the velocity of fluid and the vessel size are comparatively small in metallurgical processes, it can be considered in many cases that the electric and magnetic fields affect the velocity field but the velocity field has little influence on them (this means that the magnetic Reynolds number, $Rm = \mu \sigma UL << 1$, where, U is the characteristic velocity and L is the characteristic length).

Functions of electromagnetism applied to materials processing

Function of shape control

As being indicated in Fig.2, let us take the coordinate where x-z plain corresponds to the surface of molten metal and y-axis is taken toward the normal outerward direction from an origin on the surface (the right-hand rule). The electric current passing through the wire which is set along x-direction above the surface of molten metal (presuming that the width of the wire is larger than that of molten metal) induces a uniform magnetic flux in the z-direction, B_z, based on Eq.(2). The current in the metal induced by B_z can be obtained by use of Eq.(2) as follows:

$$J_x = (1/\mu)\cdot(\partial B_z/\partial y) \tag{8}$$

$$J_y = -(1/\mu)\cdot(\partial B_z/\partial x) \tag{9}$$

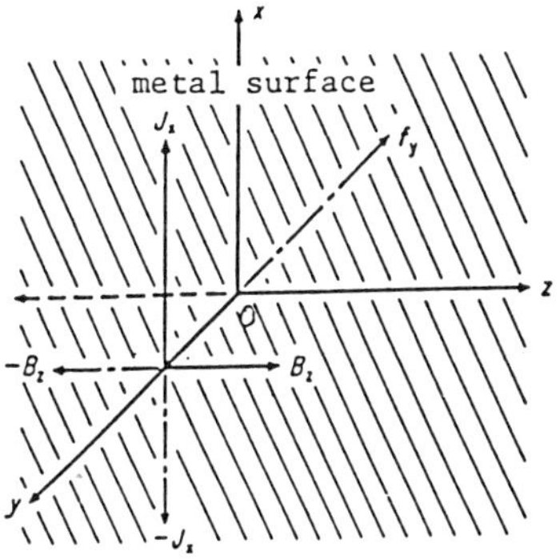

Figure 2 - Principle of the function of shape control.

Assuming a uniform B_z along the x-direction, $J_Y=0$ is. Thus, the electromagnetic body force induced by J_x and B_z is written as Eq.(10) on the basis of Eq.(5).

$$f_y = -J_x\cdot B_z = -(1/\mu)\cdot(\partial B_z/\partial y)\cdot B_z$$

$$= -(1/2\mu)\cdot(\partial B_z^2/\partial y) \tag{10}$$

The magnetic pressure can be evaluated by integrating the body force, f_Y, from $y=-\infty$ (the position of $y=-\infty$ indicates a deeper point in the metal phase than the electromagnetic skin depth, δ_m) to y=0 (surface). This pressure acts on the surface of molten metal.

$$P_m = \int_{-\infty}^{0} f_y\, dy = -(1/2\mu)B_{zo}^2 \tag{11}$$

where, the boundary conditions of $B=B_{zo}$ at y=0 and B=0 at $y=-\infty$ were adopted. The negative sign of P_m in Eq.(11) indicates that P_m presses the molten metal from its surface toward the inner part. That is, it means that the electric current passing through the wire set along the surface of molten metal always acts to push the molten metal. Based on this principle, it is possible to control the shape of molten metal by use of coil current.

Let us show a more general explanation of the magnetic pressure, P_m. By substituting Eq.(2) into Eq.(5) and using the vector identities, Eq.(12) can be derived

$$\mathbf{f} = \mathbf{J}\times\mathbf{B} = (1/\mu)\cdot(\nabla\times\mathbf{B})\times\mathbf{B}$$

$$= \underbrace{(\mathbf{B}\cdot\nabla)\mathbf{B}/\mu}_{f_1} - \underbrace{\nabla\mathbf{B}^2/2\mu}_{f_2} \qquad (12)$$

It does not necessarily hold that the rotation $(=\nabla\times)$ of $f_1=(\mathbf{B}\cdot\nabla)\mathbf{B}/\mu$ in Eq.(12) is zero, $(\nabla\times(\mathbf{B}\cdot\nabla)\mathbf{B}/\mu\neq0)$, but that of $f_2=\nabla B^2/2\mu$ is always zero $(\nabla\times(\nabla B^2/2\mu)=0)$. That is, f_1 has a possibility to be a rotational force which drives the fluid motion, but f_2 is always irrotational without any contribution to the fluid flow. The ratio of f_2 to f_1 is given as Eq.(13) [3].

$$|f_2/f_1| \sim L/\delta_m \qquad (13)$$

where, $\delta_m=(2/\mu\sigma\omega)^{1/2}$ is the electromagnetic skin depth and ω is the angular velocity of alternation. With decrease of δ_m accompanied by increase in ω, $|f_2/f_1|$ becomes large. This indicates that the electromagnetic force works mainly as a irrotational force. That is, in order to control the shape of molten metals, high frequency is favorable and for the aim of mixing which will be mentioned later, low frequency is suitable. Another type of shape control is a process utilizing the Lorents force, $\mathbf{f}=\mathbf{J}\times\mathbf{B}$ (Eq.(5)), which can be caused by the imposed direct magnetic field, $\mathbf{B}$, and the current, $\mathbf{J}=\sigma\mathbf{V}\times\mathbf{B}$, induced by fluid motion, $\mathbf{V}$, under the magnetic field.

Function of fluid driving

By imposing the direct electric current density J_x (x-direction) and the direct magnetic flux density B_Y (y-direction) perpendicular to each other, the electromagnetic body force f_z (z-direction) is induced on the basis of Eq.(5).

$$f_z = J_x B_y \qquad (14)$$

This can be extended to the case imposing alternating electric current ($J_x = \sqrt{2}J_e\sin(\omega t)$) and alternating magnetic field ($B_Y=\sqrt{2}B_e\sin(\omega t-\beta)$) by taking the average value of the body force over one period.

$$f_z = \frac{1}{T}\int_0^T \sqrt{2}J_e\sin(\omega t)\cdot\sqrt{2}B_e\sin(\omega t-\beta)dt$$

$$= J_e B_e \cos\beta \qquad (15)$$

where, β is the phase angle between the current and the field, T is a period of alternation and J_e and B_e are the effective values of alternation.

There is an other way to drive the electrically conducting fluid where the alternating travelling magnetic field is imposed instead of the direct imposition of current. By applying the vector operator, rotation $(=\nabla\times)$, on Eq.(2) and eliminating $\mathbf{E}$ and $\mathbf{J}$ by use of Eqs.(1) and (4), Eq.(16) is obtained.

$$\nabla\times\nabla\times\mathbf{B} = \mu\nabla\times\mathbf{J} = \mu\nabla\times\sigma(\mathbf{E}+\mathbf{v}\times\mathbf{B})$$

$$= -\mu\sigma(\partial\mathbf{B}/\partial t) + \mu\sigma\nabla\times(\mathbf{v}\times\mathbf{B}) \qquad (16)$$

By using the vector identity, $\nabla\times\nabla\times\mathbf{B} = \nabla(\nabla\cdot\mathbf{B}) - \nabla^2\mathbf{B}$, and Eq.(3),

Eq.(17) is derived from Eq.(16).

$$\nabla^2 \mathbf{B} + \mu\sigma\nabla\times(\mathbf{v}\times\mathbf{B}) = \mu\sigma\partial\mathbf{B}/\partial t \tag{17}$$

Let us explain the function of fluid driving by use of alternating travelling magnetic field in an example of electromagnetic stirring in continuous casting machine. In order to drive molten metal toward the casting direction as shown in Fig.3, the alternating travelling magnetic flux, $\mathbf{B}(\mathbf{x},t)=\mu\mathbf{h}(\mathbf{x})\cdot\exp(j\omega t)$ has to be imposed on the surface of slab. Equation (18) is obtained by substituting the $\mathbf{B}(x,t)$ into Eq.(17).

$$\nabla^2 \mathbf{h} + \mu\sigma\nabla\times(\mathbf{V}\times\mathbf{h}) = j\mu\sigma\omega\mathbf{h} \tag{18}$$

By assuming that $\mathbf{h}$ distributes in the x-direction (the direction of slab thickness) and travels in the z-direction by the propagating constant, γ, $\mathbf{h}$ can be written as Eq.(19).

$$\mathbf{h} = \begin{cases} h_x = \hat{h}_x(x)\cdot\exp(-j\gamma z) \\ h_Y = 0 \\ h_z = \hat{h}_z(x)\cdot\exp(-j\gamma z) \end{cases} \tag{19}$$

As the second term of the left hand side of Eq.(17) can be eliminated by the assumption that the magnetic Reynolds number $Rm=\mu\sigma UL$ is less than 1, the z-component of Eq.(17) is given as Eq.(20).

$$d^2\hat{h}_z(x)/dx^2 - \beta^2\hat{h}_z(x) = 0\ ,\quad \beta^2 = \gamma^2 + j\mu\sigma\omega \tag{20}$$

Eq.(20) is solved under the boundary conditions of Eqs.(21) and (22). And B_z is obtained as Eq.(23).

$$\hat{h}_z = h_o \quad \text{at} \quad x = x_o \tag{21}$$

$$\partial\hat{h}_z/\partial x = 0 \quad \text{at} \quad x = 0 \tag{22}$$

$$B_z = \mu h_o \frac{\cosh(\beta x)}{\cosh(\beta x_o)}\cdot\exp\{j(\omega t-\gamma z)\} \tag{23}$$

Equation (3) is described as Eq.(24).

$$\partial B/\partial x + \partial B/\partial z = 0 \tag{24}$$

By substituting Eq.(23) into Eq.(24) and solving the resulting equation under the condition of Eq.(25), Eq.(26) is obtained.

$$B_x = 0 \quad \text{(symmetry)} \quad \text{at} \quad x = 0 \tag{25}$$

$$B_x = j\gamma\mu h_o \frac{\sinh(\beta x)}{\beta\cosh(\beta x_o)}\cdot\exp\{j(\omega t-\gamma z)\} \tag{26}$$

By substituting Eqs.(23) and (26) into Eq.(2), the electric current density is obtained as Eq.(27).

$$\mathbf{J} = \begin{cases} J_x = 0 \\ J_Y = h_o(\gamma^2-\beta^2)\dfrac{\sinh(\beta x)}{\beta\cosh(\beta x_o)}\cdot\exp\{j(\omega t-\gamma z)\} \\ J_z = 0 \end{cases} \tag{27}$$

Finally by substituting Eqs.(23),(26) and (27) into Eq.(5) and taking the real part of the resulting equation, **f** is derived as Eq.(28)

$$\mathbf{f} = \begin{cases} f_x = \dfrac{h_o^2 \mu^2 \omega \sigma}{2\sqrt{\gamma^4 + (\mu \omega \sigma)^2}} \cdot \dfrac{-\beta_i \sinh(2\beta_r x) + \beta_r \sin(2\beta_i x)}{\cosh(2\beta_r x_o) + \cos(2\beta_i x_o)} \\ f_Y = 0 \\ f_z = \dfrac{h_o^2 \mu^2 \omega \sigma \gamma}{2\sqrt{\gamma^4 + (\mu \omega \sigma)^2}} \cdot \dfrac{\cosh(2\beta_r x) - \cos(2\beta_i x)}{\cosh(2\beta_r x_o) + \cos(2\beta_i x_o)} \end{cases} \quad (28)$$

where, β_r and β_i are corresponding to the real part and the imaginary part of β. As shown in Eq.(28), the alternating magnetic field travelling toward the z-direction has the force acting along the x-direction (compressive force) and the force acting along the z-direction (driving force). Thus, Eq.(28) indicates the function of fluid driving (mixing).

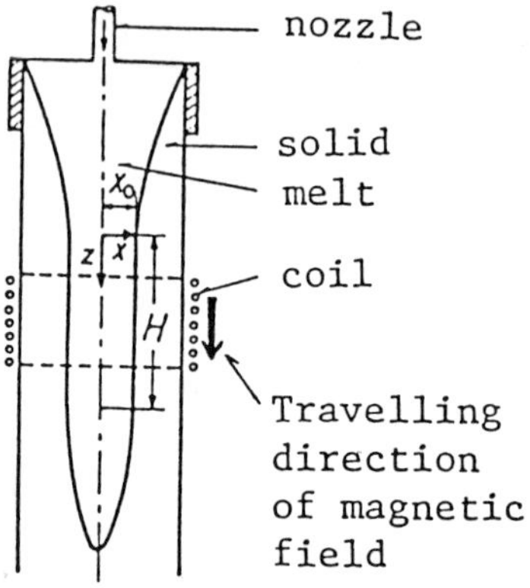

Figure 3 - Explanation of the function of fluid driving by travelling magnetic field in an example of electromagnetic stirring in continuous casting.

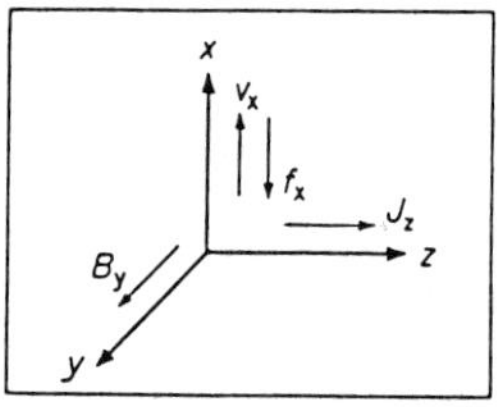

Figure 4 - Principle of flow suppression.

Function of flow suppression

The current is induced in the z-direction based on Eq.(4) if the fluid flow of the x-direction is caused under the imposition of the magnetic field of the y-direction, as shown in Fig.4.

$$\mathbf{J} = \sigma \mathbf{V} \times \mathbf{B} = (0,\ 0,\ \sigma V_x B_Y) \quad (29)$$

Substituting $J_z = \sigma V_x B_Y$ into Eq.(5) yealds f_x.

$$\mathbf{f} = \mathbf{J} \times \mathbf{B} = (-J_z B_Y, 0,\ 0) = (-\sigma V_x B_Y^2,\ 0,\ 0) \quad (30)$$

This electromagnetic body force, $f_x = -\sigma V_x B_Y^2$, indicates the suppression force which is exerted to the opposite direction of V_x as implied by the negative sign in Eq.(30). Following the same procedure, the body force $f_z = -\sigma V_z B_Y^2$ is derived under the exsistance of B_Y and V_z. That is, it can be understood that fluid flow of any direction except that of the imposing magnetic field can be suppressed by imposing DC magnetic field. However, it should be noticed that f_x can not be induced in a case where the electric circuit is not established in accordance with the electrical boundary condition [4].

Function of levitation (gravity shift)

An electrically conducting material can be levitated by balancing the gravity force with the electromagnetic force given by Eq.(5)

$$\mathbf{J}\times\mathbf{B} = \rho\mathbf{g} \qquad (31)$$

Equation (31) indicates the function of levitation.

By imposing direct current and direct magnetic field, it is possible to change the acceleration of gravity, **g**, to **g**' as given in Eq.(32)

$$\rho\mathbf{g}' = \rho\{\mathbf{g} - (\mathbf{J}\times\mathbf{B})/\rho\} \qquad (32)$$

Equation (32) shows the function of gravity shift.

Function of splashing

When electromagnetic force induced by direct electric current and direct magnetic field is much larger than the gravity force or the adhesion force due to surface tension

$$|\mathbf{J}\times\mathbf{B}| > \max\{|\rho\mathbf{g}|,\ 6\kappa/a^2\} \qquad (33)$$

where, κ is the surface tension and a is the radius of molten metal.

Equation (33) expresses the function of splashing.

Function of heat generation

When the current **J**, passes through an electrically conducting material, Joule heat is generated as given in Eq.(34).

$$q = |\mathbf{J}|^2/\sigma \qquad (34)$$

Equation (34) indicates the function of heat generation. Based on Eqs.(1) and (4), it can be understood that the current **J** is decomposed into three sources;

1) The first term is the imposed current
2) The second term is the induced current by alternating magnetic field
3) The third term is the induced current by motion of the electrically conducting fluid in magnetic field

$$\mathbf{J} = \underbrace{J_0}_{\text{first term}} + \underbrace{\sigma(-\partial\mathbf{A}/\partial t)}_{\text{second term}} + \underbrace{\sigma\mathbf{V}\times\mathbf{B}}_{\text{third term}} \qquad (35)$$

where **A** is the vector potential of magnetic field (defined as $\mathbf{B}=\nabla\times\mathbf{A}$, $\nabla\cdot\mathbf{A}=0$)

On the other hand, Joule heat generated by alternating magnetic field is given as Eq.(36)

$$q = \frac{|J|^2}{\sigma} = \frac{h_0^2}{\sigma\,\delta_m^2}\cdot\exp(2y/\delta_m) \qquad (36)$$

From Eq.(36), the heat generation rate is increased and concentrated in the vicinity of surface with increase of ω. Equation (36) expresses the function of heat generation by electric current.

Function of velocity detecting

The principle is based on Lenz's law given in Eq.(37), where the electric field is induced by motion of the electrically conducting fluid in the magnetic field.

$$\mathbf{E} = -\mathbf{V} \times \mathbf{B} \tag{37}$$

Based on Eq.(37), the velocity is evaluated by measuring **E** under a given value of **B**.

Combined function

By using a number of the functions of electromagnetism mentioned above, new combined functions such as the refining function and the function of controlling solidification structure are born. The refining function is composed of the functions of fluid driving, levitation, splashing and heat generation. The functions of fluid driving, flow suppression and levitation are adopted in the function of controlling solidification structure.

The overview of electromagnetic processing of materials and the classification based on the functions

Figure 5 reveals the overview of electromagnetic processing of materials as a tree. The roots indicate the academic background supporting this engineering field as follows.

electromagnetic processing of materials {
- plasma engineering
- process metallurgy —— { transport phenomena; thermodynamics }
- magnetohydrodynamics —— { electromagnetism; fluidmechanics }

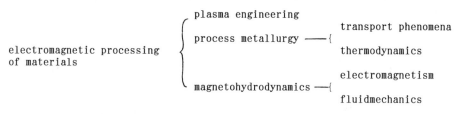

The branches predict the functions mentioned in the previous chapter and the leaves in each branch show processes and technologies belonging to the corresponding function. The processes and the technologies are classified by each function and explained in the followings.

Function of shape control

The magnetic pressure, $P_m = -B_{z0}^2/2\mu$, given in Eq.(11), is used in processes such as vertical electromagnetic casting [5], cold crucible [6], levitation melting, making a thin film of molten metal [7] and plastic deformation.

The cold crucible is composed of a segmented copper mould put in a coil as shown in Fig.6. This water cooled cupper mould concentrates the magnetic field enough to levitate and melt a relatively large amount of metal (say, kg unit), without any contact with the mould. Growing a single crystal of ThO_2(melting point 3400℃) was conducted by use of the cold crucible [8]. We can expect several applications such as melting of chemically active metals, treatment of radioactive materials and production of uranium fuel bar by making use of the advantages of the cold crucible. Moreover, the direct process of melting and casting of Ti and Zr from scrap can be considered by use of the continuous casting type of cold crucible instead of consumable electrode method.

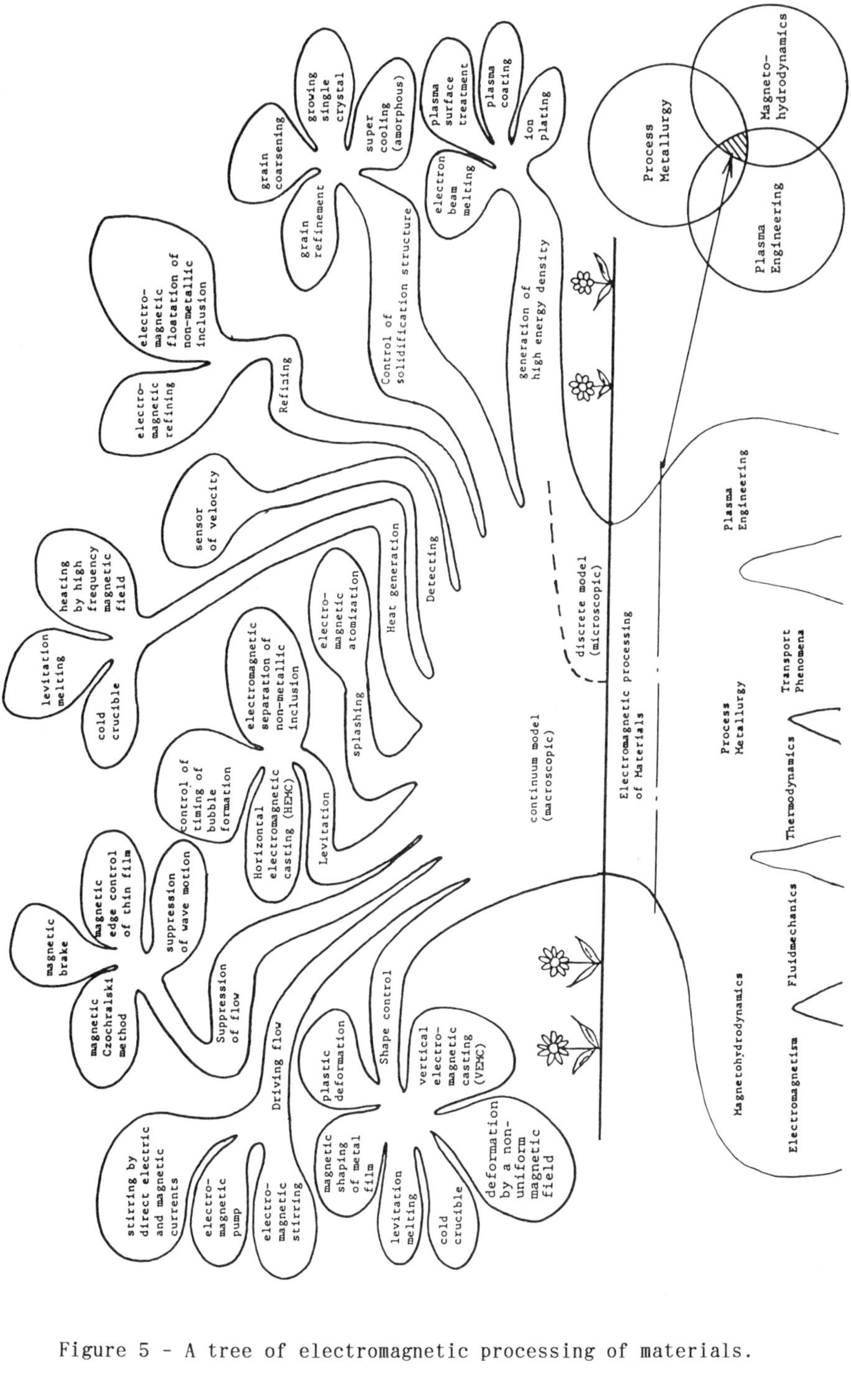

Figure 5 - A tree of electromagnetic processing of materials.

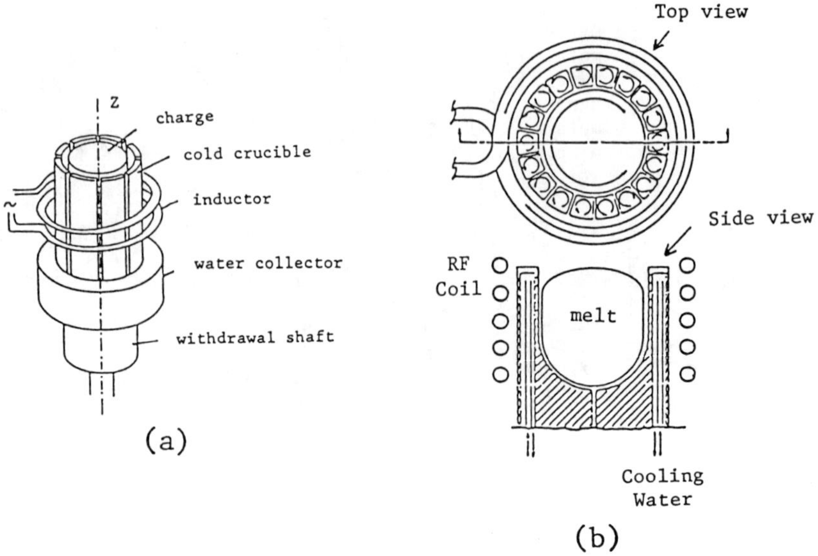

Figure 6 - Schematic view of cold crucible
(a) continuous casting type, (b) batch type [6].

Another type of shape control was proposed [9] by the use of a non-uniform direct magnetic field. The shape of the injected molten metal in the non-uniform magnetic field is deformed as shown in Fig.7. This principle of shape control can be applied to make a thin film of molten metal in a process such as direct production of thin plates from molten metal.

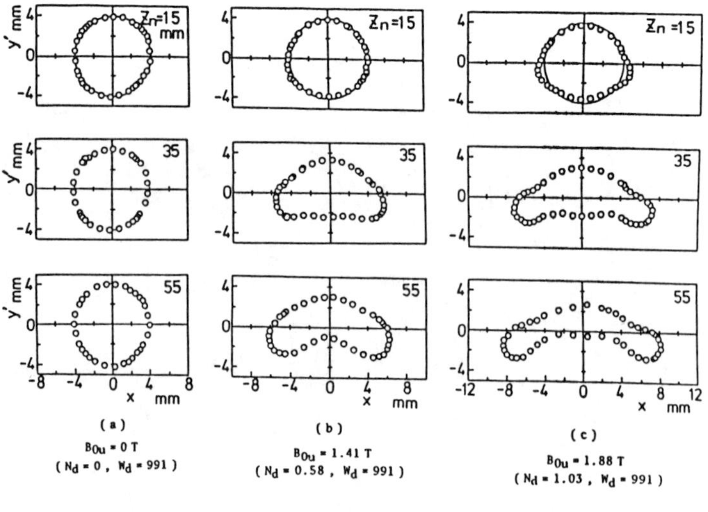

Figure 7 - Cross sectional shape of liquid jet [9]
(Nd=Stewart number, Wd=Weber number) $d = 8mm, w_0 = 2.10 m/s$

Function of fluid driving

The travelling magnetic field induces a body force as given in Eq.(28) which drives molten metal along the travelling direction of magnetic field and pinches it perpendicularly. The combined imposition of direct electric current and direct magnetic field induces a body force as shown in Eq.(5) which drives molten metal to a direction perpendicular to those of current and field. The typical processes which employ the function of fluid driving are electromagnetic stirring of continuous casting and electromagnetic pump.

Function of flow (wave) suppression

The Lorentz force, $\mathbf{f}=(\sigma \mathbf{V}\times \mathbf{B})\times \mathbf{B}$, which is given in Eq.(30), is used to suppress fluid flow motion. This force can suppress all of the fluid motions except the flow parallel to the direction of the imposed magnetic field. In the magnetic Czochralsky method [10] the natural convection due to the temperature difference is suppressed to control the oxygen content in a single crystal of silicon. In order to reduce non-metallic inclusions in steel and to get a good surface quality of slab, direct magnetic field is applied to the liquid pool of continuous casting. It reduces the velocity of nozzle flow together with a reduction in wave motion at the meniscus. This refers to a magnetic brake [11].

In the production of thin sheets by a twin roll direct caster, a saw shape edge often appears on both edges of sheets. To get a smooth edge shape, direct magnetic field is applied to a pair of composite rolls made of magnetic and non-magnetic materials, such as plain steel and stainless steel as shown in Fig.8. A strong magnetic field appears at the plain steel part between the rolls and reduces the velocity of molten metal along the axis of the rolls. The quick reduction of this velocity at the vicinity of both edges gives comparatively smooth edges [12].

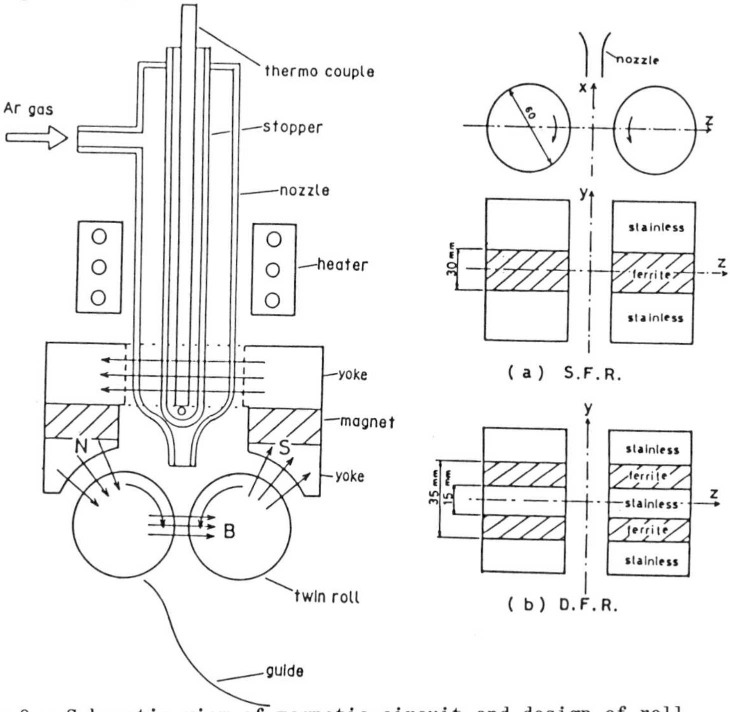

Figure 8 - Schematic view of magnetic circuit and design of roll.

The wave suppression is considered to be a crucial element technology in metallurgical processes. For instance the wave motion appearing at pouring flow and metal pool in the twin-roll process, meniscus of conventional continuous casting and free surface of molten metal held in vertical electromagnetic casting is strongly desired to be suppressed. The effect of a direct magnetic field on the wave suppression can be classified into the following three cases:
① magnetic field is imposed horizontally to the wave motion, where the gradient of magnetic field is indispensable [4], ② the direction of magnetic field is parallel to the wave vector [13,14] and ③ the direction of magnetic field is perpendicular to the wave vector [15,16].

Especially, the wave motion existing at the meniscus of conventional continuous casting of steel causes irregular oscillation marks on the surface. Thus, the suppression of wave motion at the meniscus is a crucial problem for getting a slab with good surface quality. The nice quality of slab allows sending slab directly to the rolling stage without passing through the stage of surface treatment. This can save a large amount of thermal energy for reheating them. The effect of magnetic field on the appearance of oscillation marks was studied on a metal with low melting point by Hayashida et al. [17]. As shown in Photo.1, the direct magnetic field has the function of changing irregular oscillation marks to regular ones. The magnetic effect was found to be obvious in the low frequency oscillation range.

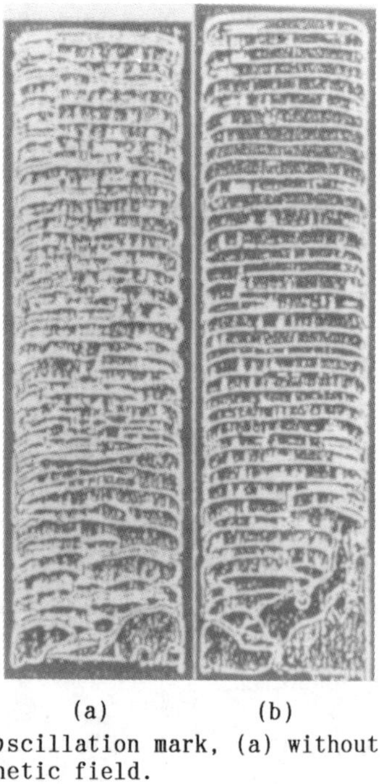

(a) (b)

Photo. 1 - Overview of oscillation mark, (a) without magnetic field and (b) with magnetic field.

Function of levitation (gravity shift)

On the basis of the principle of levitation given in Eq.(31), a new process of horizontal electromagnetic casting (HEMC), so called as mouldless casting was developed [18,19]. It not only enables the elimination of surface defects of slab caused by contacting with mould wall, but also intends the near net shape casting. The additional advantage of this process is the favorable application to the casting of heavy metals at less investment than for the ordinary electromagnetic casting making use of high frequency magnetic field. Figure 9 shows a schematic view of the HEMC apparatus. A magnified view of the cast rod and thin plate and a cross-section of the thin plate are given in Photo.2. The surface appearance of products is very smooth without any defects and the cross sectional shape slightly deforms from the slit shape of nozzle due to surface tension.

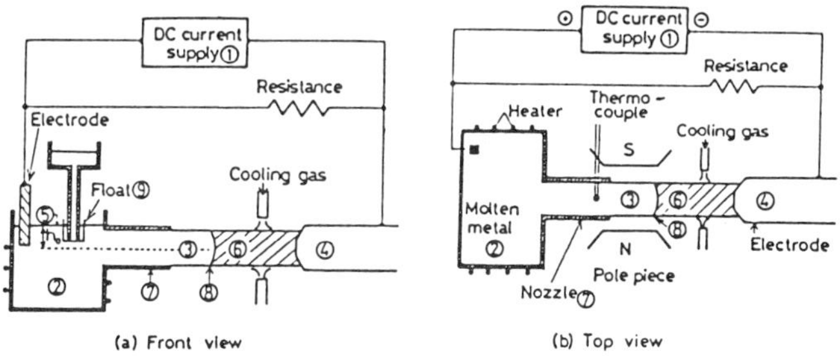

Figure 9 - Schematic view of the experimental appararus of horizontal electromagnetic casting.

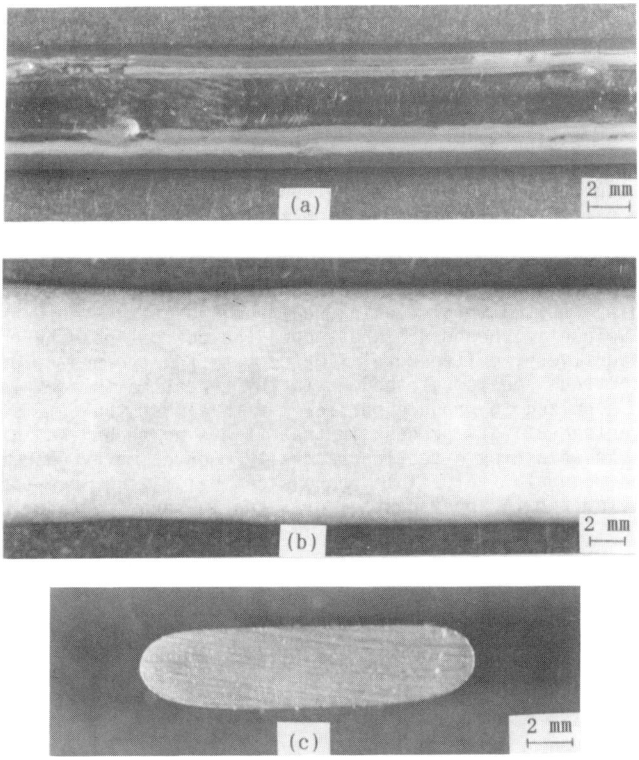

Photo. 2 - Magnified view of the cast (a) rod, (b) thin plate and (c) cross-section of the thin plate.

By imposing alternating electric current and direct magnetic field, the electromagnetic body forces which are exerted upon the molten metal in the directions along the gravity force and against that are induced periodically with the same switching frequency as that of the imposed alternating current. Following this principle, frequency of bubble formation can be controlled as shown in Fig.10. Bubbles are formed randomly when no electric current and no magnetic field are applied, but the frequency of bubble formation can be precisely controlled by their application [20].

Another example is the electromagnetic floatation of non-metallic inclusions where electromagnetic force induced by direct electric current and direct magnetic field is exerted upon the molten metal in the same direction as that of the gravity force. This force yields an apparent density heavier than that of molten metal and accelerates the separation rate of non-metallic inclusions without induction of electromagnetic force [21].

B (T)	I (A)	f (Hz)	probe signal
0	0	-	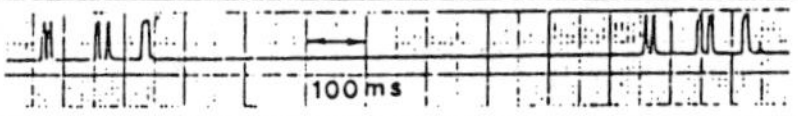
0.6	100	40	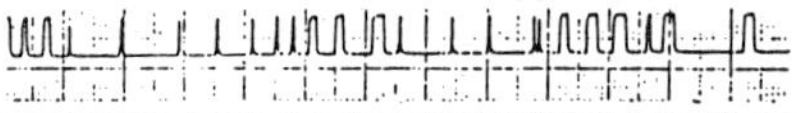
0.6	200	40	

Figure 10 - Formation of bubbles under various electric and magnetic conditions.

Function of splashing

By making use of the principle given in Eq.(33), a new atomization process, so called electromagnetic atomization, has been developed [22]. A schematic view of the experimental apparatus is shown in Fig.11. When electric voltage and magnetic field, perpendicular to each other, are imposed between a nozzle and an electrode, molten metal is splashed at the moment when the molten metal flowing out from the nozzle contacts with the electrode. Splashing phenomenon shuts down the current passing through the metal. Consecutively, molten metal flowing out from the nozzle contacts with the electrode and is splashed again. By repeating this cycle, molten metal is so atomized to produce particles with almost the same size. Another advantage of this process is that it has no mechanical driving device and the splashing direction is easily controlled by changing the direction of magnetic field. This process can be of great use for designing a new type atomizer in spray casting or a gun in spray coating.

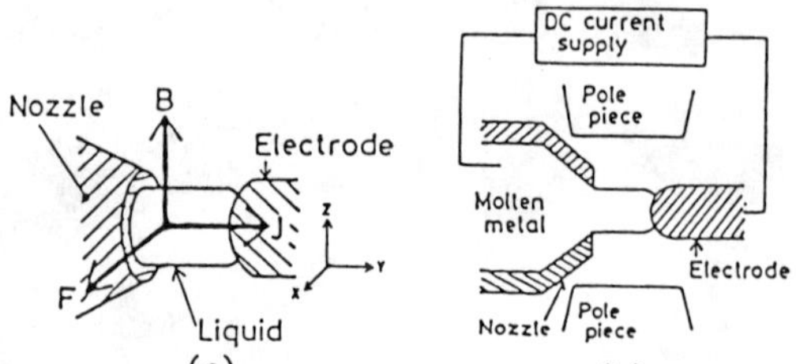

Figure 11 - Electromagnetic atomization process. (a)principle, (b)schematic view of apparatus

Function of heat generation

There are two methods for heating up the molten metal by making use of Joule heat (q=J /2σ). One is the direct imposition of electric current on molten metal and the other is the indirect method where electric current is induced in molten metal by imposing high frequency magnetic field. The latter has mainly been employed in metal industries regardless of less efficient conversion from electric to thermal energy because of no need for electrode. The former is not practical for heating since it needs a current supply with high current capacity under low voltage due to high electric conductivity of metal. However, as the electrode with high electric conductivity and high melting point has recently been developed, for instance ZrB , the direct imposition of current should be noticed as an important future technology to develop alternative figures for electromagnetic stirring and pumping [23].

Function of velocity detecting

A velocity sensor was proposed by Vivès [24]. The principle of this sensor is based on Lenz's law given in Eq.(37), where the velocity of molten metal is detected by observing **E** between two parallel lead wires set on both sides of a strong and small permanent magnet as shown in Fig.12. This sensor is sensible enough to observe fluctuation of turbulent flow. However, it is not usable over the Curie temperature of the magnet. Thus, it is strongly desired to develop a sensor being able to work in high temperature range.

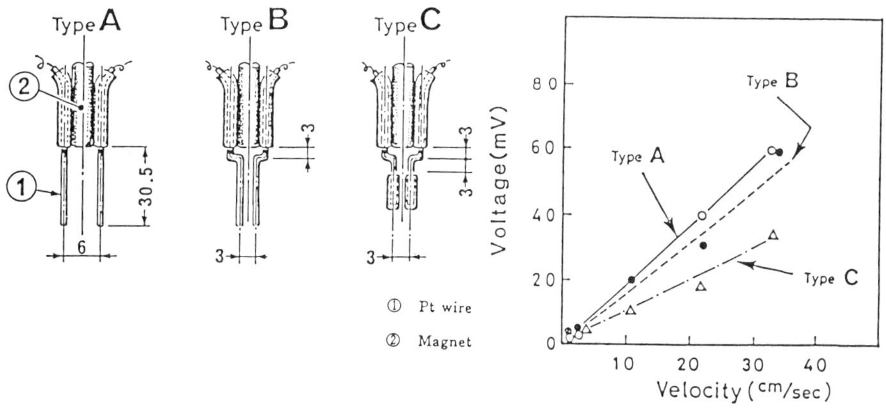

Figure 12 - Magnetic sensors used in experiment and relation between velocity and voltage [25].

Combined function

(I)Refining function.

An electromagnetic refining process is proposed where the driving force of molten metal caused by gas lifting in RH degassing process is replaced by electromagnetic force [26]. The advantage of this process is that the splashing of molten metal accompanied with arc spark in a vacuum chamber could accelerate mass transfer enormously and the capacity of gas ejector system could be reduced significantly due to the elimination of argon gas injection.

The electromagnetic separation of non-metallic inclusions described before, also belongs to this function [21].

(II)Control of solidification structure.

As a direct magnetic field suppresses natural convection in the vicinity of solidifying front by the induced body force, $\mathbf{f}=\mathbf{J}\times\mathbf{B}=-\sigma(\mathbf{V}\times\mathbf{B})\times\mathbf{B}$, nuclei multiplicated on liquid-solid interface have no chance to be transported into the bulk liquid. This leads to a columnar structure [27]. On the other hand stirring by the function of fluid driving causes multiplication due to breaking of dendrite tips and transports nuclei into bulk liquid so that the solidified structure tends towards an equiaxed one in general [27].

Recently levitation technique was applied to get a large amount of undercooling as much as 200K in steel [28]. This seems to be one of the promising methods for obtaining a bulk amorphous metal.

Lielausis et al. [29] presented an interesting paper regarding the direct effect of electric current and magnetic field on solidifying phenomenon. This paper seems to promise the finding of a new function of electromagnetism connected with solidification of metal.

Conclusion

The electromagnetic processing of materials has been newly defined as one of the promising engineering fields where the functions of electromagnetism are utilized for processing electrically conducting materials such as liquid metals, plasmas and electrolytes. The background requiring this processing and the engineering characteristics supporting it were described. The existing processes and the proposed ones making use of electric current and magnetic field were classified on the basis of the functions of electromagnetism and were discussed from the view point of theoretical principles and engineering applications.

The electromagnetic processing of materials is located in a small valley between the two big mountains well known as "Metallurgy" and "Electromagnetism". These mountains have preserved very useful mines for a long time and supplied unfathomable profundity of benefits to humankind. However, nobody knows whether the immense vein of the mines found in the two mountains reaches to the underground in the valley or not. Sometimes one have found useful ores there by chance. In fact, most of metallurgists have unconsciously made use of the phenomena which should belong to electromagnetic processing of materials.

Let MHD and metallurgical engineers and scientists cooperate to find new functions which should exist on the boundary between each field of magnetohydrodynamics, process metallurgy and plasma engineering. The discovery of new functions and those applications are indispensable for developing the electromagnetic processing of materials as a new engineering field.

Acknowledgement

I would like to express my deep thanks to the Iron and Steel Institute of Japan for the establishment of the committee on "Electromagnetic Metallurgy". This committee is the earliest establisment of a research organization in this field in the world. It has strongly stimulated the field of the electromagnetic processing of materials in Japan. Thanks are also given to the Ministry of Education of Japan for financial supports of the researches in this field (No.61850125 and No.61470056).

References

[1] H.K.Moffatt and M.R.E.Procter, Metallurgical Application of Magnetohydrodynamics, Proceedings of a Symposium of the IUTAM, (The Metal Society, 1984)

[2] K.Kawakami, 'Avenues to Innovative Steelmaking Technologies in Japan', Trans. ISIJ, 24(1984), 754

[3] M.Garnier, 'La Levitation Electromagnetique', GIS MADYLAM, (1987), 6

[4] T.Kozuka, S.Asai and I.Muchi, 'Effect of Gradient of Transverse Direct Magnetic Field on Suppression of Wave Motion J. of The Iron and Steel Institute of Japan, (1988) (in contribution)

[5] Z.N.Getselev, U.S.patent 3467166

[6] T.F.Ciszek, 'Some Application of Cold Crucible Technology for Silicon Photovolatic Material Film', J. Electrochem. Soc., 132(1985), 963

[7] T.Kozuka, S. Asai and I.Muchi, 'Effect of Applying Electromagnetic Force on Falling Behavior of Molten Metal Film', J. of The Iron and Steel Institute of Japan, 73(1987), 828

[8] C.C.Herrick and R.G.Behrens, 'Growth of Large Uraninite and Thorianite Single Crystal from the Melt Using a Cold-Crucible Technique' J. of Crystal Growth, 51(1981), 183

[9] S.Oshima, R. Yamane, Y.Mochimaru and Y.Matsuoka, 'The Shape of Liquid Metal Jets under a Non-Uniform Magnetic Field 1st Report and 2nd Report', Trans. of the Japan Society of Mechanical Engineers (Series B), 52(1986), 2888, 2897

[10] K.Hoshi, T.Suzuki, T.Okubo and N.Isawa, 'CZ Silicon Crystal Grown in Transverse Magnetic Fields', Extended Abstracts electrochem,Soc. Spring Meeting', (Electrochem.Soc., Pennington, NJ) 80-1(1980), 811

[11] J.Nagai, K.Suzuki, S.Kojima and S.Kollberg, 'Steel Flow Control in a High-Speed Continuous Slab Caster Using an Elecrtomagnetic Brake', Iron and Steel Engineer, 61(1984), 41

[12] T.Yuhara, T.Kozuka, S.Asai and I.Muchi, 'Effect of Direct Magnetic Field on the Edge Shape of Thin Plate Cast by Twin Roll Process', CAMP-ISIJ, 1(1984), 389

[13] M.Kinoshita, T.Kozuka, S.Asai and I.Muchi, 'Suppression of Surface Wave of Molten Metal by DC Magnetic Field Parallel to the Direction of Wave Progagation', CAMP-ISIJ, 1(1988), 387

[14] Y.Kishida and K.Takeda, 'Boundary Effect of Suppression of Free-Surface-Wave in a Magnetic Field', CAMP-ISIJ, 1(1988), 386

[15] T.Kozuka, S.Asai and I.Muchi, 'Effect of Direct Magnetic Field Imporsed Vertically to Surface on Suppression of Wave Motion', J. of The Iron and Steel Institute of Japan, 74(1988), (in print)

[16] T.Muguruma, S.Kobayashi, 'Suppression of Surface Waves in Molten Metal by DC Magnetic Field', J. of The Iron and Steel Institute of Japan, 73(1987), S1448

[17] M.Hayashida, T.Ohno, H.Ono and K.Tsutsumi, 'The Simulation of Relationship between Surface Waves in Mold and the Quality of Strands' Surfaces Using Sn-Pb Alloy', J. of The Iron and Steel Institute of Japan, 73(1987), S686

[18] S.Asai, T.Kozuka and I.Muchi, 'Process Development and Stability Analysis of Horizonal Electromagnetic Casting Method', J. of The Iron and Steel Institute of Japan, 72(1986), 2218

[19] T.Kozuka, S.Asai and I.Muchi, 'Horizontal Electromagnetic Casting Process of Thin Plate and Its Stability Analysis', J. of The Iron and Steel Institute of Japan, 74(1988), 1793

[20] K.Takeda, M.Nakamura, H.Ohno, K.Kuwabara and T.Ohashi, 'Development of Electromagnetic Valve for Continuous Casting Tundish', J. of The Iron and Steel Institute of Japan, 73(1987), S1449
[21] P.Marty and A.Alemany, 'Theoretical and Experimental Aspects of Electromagnetic Separation', Proceedings of a Symposium of the IUTAM, (The metal Society 1984), 245
[22] K.Sassa, N.Agata, T.Kozuka and S.Asai, 'Atomization of Molten Metal by Use of Electromagnetic Force', CAMP-ISIJ, 1(1988), 390
[23] K.Wada, E.Takeuchi, K.Ando, S.Kitamine, H.Mori and I.Noda, 'Direct Electric Charging to Molten Steel', J. of The Iron and Steel Institute of Japan, 73(1987), S687
[24] Ch.Vivès and R.Ricou, 'Experimental Study of Continuous Electromagnetic Casting of Aluminum Alloys', Met. Trans., 16B(1985), 377
[25] K.Hosotani, H.Nakato, K.Saito, M.Oguchi, H.Okuda and A.Sugano, 'Velocity Measurement of Liquid Metal Flow', J. of The Iron and Steel Institute of Japan, 73(1987), S688
[26] S.Asai, Report of the Committee on Electromagnetic Metallurgy in ISIJ, (1986)
[27] S.Asai, K.Yasui and I.Muchi, 'Effects of Electromagnetic Forces on Solidified Structure of Metal', Trans. ISIJ, 18(1978), 754
[28] M.Nakamura, A.Ozeki and K.Mori, 'Solidification Behavior of Undercooled Molten Steel by Levitation Method', CAMP-ISIJ, 1(1988), 264
[29] O.Lielausis, A.Mikelsons, E. Shcherbinin and Yu.Gelfgat, 'Electric Currents in Molten Metals and Their Interaction with a Magnetic Field', Proceedings of a symposium of the IUTAM, (The Metal Society,1984), 234

MATERIALS PROCESSING IN SPACE

IN THE SPACE STATION ERA

Robert S. Sokolowski
NASA
Washington, DC

ABSTRACT

The future of Materials Processing in Space (MPS) in the Space Station Era is being assured today by a strong coordinated science program which looks to solve these problems, the answers to which serve to underpin the technical foundation upon which future metallurgical and materials processes will be built. The critical link between the product and process development activities that accompany this science endeavor is the "nuts and bolts" hardware that will ultimately operate in the space environment. Ground-based facilities including drop facilities and aircraft provide excellent product development and process concept study tools. Shuttle flights with Spacelab and commercially developed space facilities represent the evolutionary path to Space Station processing facilities. Today the most promising areas of MPS are crystal growth, containerless, and quasi-containerless processing. The unique features of these processes will be highlighted. Finally a more visionary perspective of MPS will be presented that incorporates the limits of today's knowledge with the economic "necessity" of tomorrow.

Metallurgical Processes for the Year 2000 and Beyond
Edited by H.Y. Sohn and E.S. Geskin
The Minerals, Metals & Materials Society, 1988

FUTURE TRENDS OF NON-FERROUS METALS SMELTING AND REFINING

Takeshi Nagano

President, Mitsubishi Metal Corporation
5-2, 1-chome, Ohte-machi,
Chiyoda-ku, Tokyo 100, Japan

Abstract

Prior to discussing future trends of non-ferrous metals smelting and refining, the past developments are looked back. Social and economic environments in Japan over the past two decades are reviewed. Technological developments to comply with these environments are explained, taking for examples the cases of the Mitsubishi continuous smelting and converting process and Onahama's jumbo tank system for copper tankhouse. After considering the factors which will influence the improvements in non-ferrous processing, future trends are forecasted generally for copper, zinc, lead, aluminum and rare earth metals. Recent development works in Japan for the future are also introduced. In the discussed area, it appears that the emergence of new innovative processes would be difficult. The progress will be made rather toward the sophistication of the existing technologies, utilizing the advancement of the related technologies such as computers, electronics, etc.

Metallurgical Processes for the Year 2000 and Beyond
Edited by H.Y. Sohn and E.S. Geskin
The Minerals, Metals & Materials Society, 1988

Introduction

This paper discusses the future trends of non-ferrous metals smelting and refining on the occasion of the international symposium, "The Metallurgical Processes for the Year 2000 and Beyond".

It is considered that the improvement of technology is greatly influenced by the social and economic environments which surround it. From this standpoint, review was made of Japanese case over the past two decades. Various new technologies were developed also in the non-ferrous metal world to comply with the requirements under the above environments. The examples are described in the cases of the Mitsubishi continuous smelting and converting process and Onahama's jumbo tank system for copper electro-refining.

After foreseeing the major factors which will affect the technological developments in the future, the trends are forecasted for copper, zinc, lead, aluminum and rare earth metals generally. Some development works in Japan, which will give suggestions for future trends, are also introduced in some detail.

Review of Social and Economic Environments in Japan

Environmental Issues

The rapid growth in industry and the economy since the early 1960s in Japan brought about various pollution problems, such as increase of asthma or bronchitis patients around industrial complex, Minamata disease and so on. The rapid growth of industry at that time is symbolically shown in Figure 1 on the production of steel. Starting from the promulgation of Environment Basic Law in 1967, various anti-pollution laws have been enacted. Through the improvement of pollution control in the industrial world and due to people's concern for environmental matters, the ambient air and water quality has been significantly recovered over the years. As for SOx concentration in the ambient air, which is the concern of non-ferrous metals industry, it has been within the regulated standard almost all over the country these days. The decrease of SOx emissions in Japan is illustrated in Figure 2.

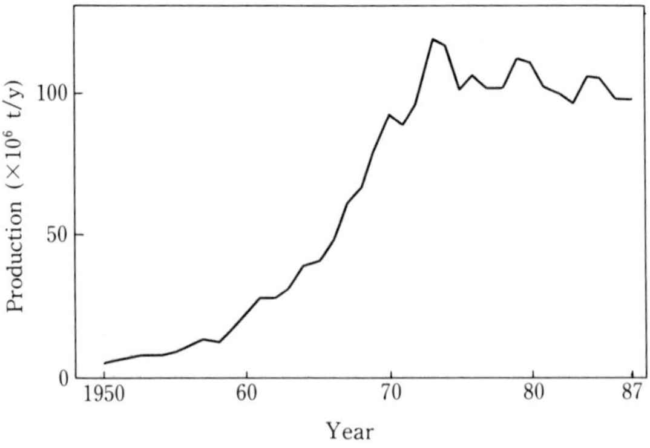

Figure 1 - Production of Steel (1)

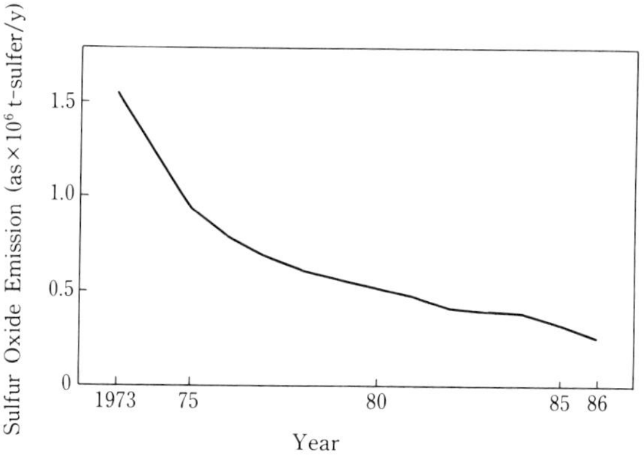

Figure 2 - Emissions of Sulfer Oxide (2)

Labor Cost

The economic growth over the past 20 years contributed to the increase of employment and corporate profits. This has caused the rising wages. The changes of the average annual earnings of Japanese employees are shown in Figure 3. The rising salaries and wages have made firms, especially labor-intensive firms concentrate their efforts on manpower savings. The recent appreciation of the yen has stressed further the hike of salaries and wages in dollar terms as given in Figure 3 and it is promoting the transfer of production facilities overseas in some industrial fields. In any case constant rises of wages have forced the industry to make efforts for manpower savings.

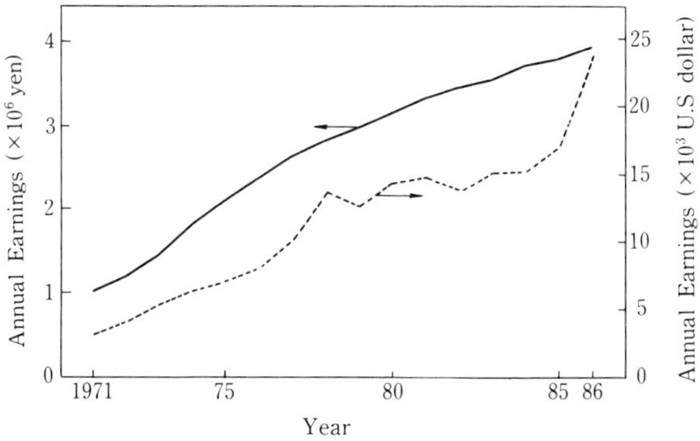

Figure 3 - Annual Earnings (3)

Energy Cost

Saving energy cost is basically a theme for engineers in energy consuming non-ferrous industry. The marked rise of energy cost since oil crisis in 1973 enforced the production company to save further the energy consumption. Figure 4 shows the rise of energy prices over the past two decades. The rise of electric power price is characteristic of Japan, where reliance on oil is high as an energy source of electric power. The situation has made non-ferrous industry concentrate their efforts on reduction of fuels and electric power consumption. Even with outstanding technology most of the aluminum refineries in Japan closed because of the hike of power price, with only 40 thousand tonnes production remaining in 1987, down from the peak of 1.19 million tonnes in 1979.

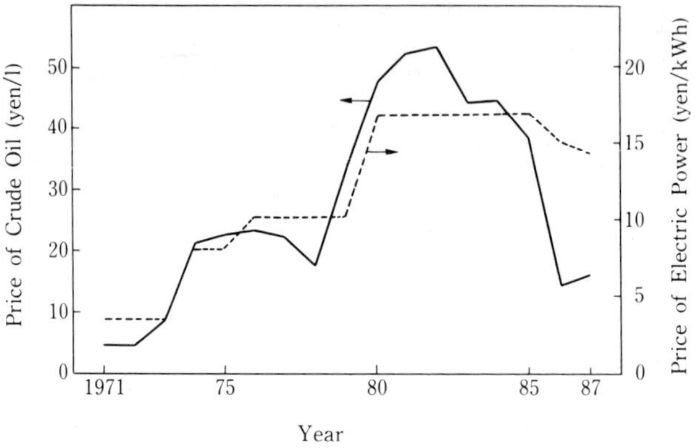

Figure 4 - Price of Energies (4)

Appreciation of Yen

Yen rate had been fixed as 360 yen for a dollar since 1949. However, following its appreciation to 308 yen in December, 1971, it had been floated since February 1972. Later, since 1977 the revaluation had started by the favorable economy backed up mainly by the increase of exports to the U.S. The sharp appreciation had started since September, 1985 when the G5 conference by finance ministers recognized the need for correction of international trade imbalance. The changes of yen rate are shown in Figure 5.

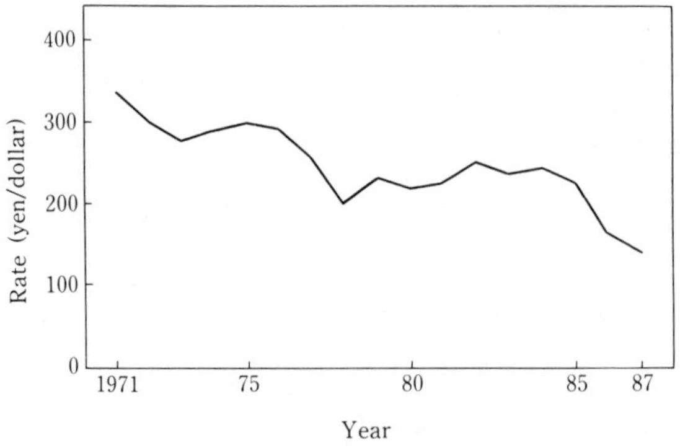

Figure 5 - Yen Rate against U.S. Dollar

Development of Implementing Technology

Under the above mentioned social and economic environments, various development works had been carried out in the non-ferrous metals industry. As an example, technical developments performed by Mitsubishi Metal Corporation are described in this section with respect to copper smelting and refining.

Copper Smelting

The Mitsubishi continuous smelting and converting process was developed against the background mentioned above. The technical cores of the process are summarized as follows (5).

Firstly, it employs injection smelting, where the raw materials are injected through top blowing lances with oxygen enriched air into molten bath. The concentrates and fluxes, with pulverized coal as supplementary fuel,are caught by melts in the solid state, melt, then react with well dispersed oxygen. The procedure brings about instantaneous smelting, low mechanically carried-over dust, and high oxygen utilization efficiency.

Secondly, it employs a multi-furnace system. To produce continuously blister copper and low-copper slag, it was a theoretically necessary choice to utilize two furnaces for reaction, one under the low oxygen-potential and the other under the high oxygen-potential. This enabled the process to produce low sulfer and low impurity blister copper.

Thirdly, it employs unique lime slag of $CaO-FeO-Fe_2O_3-Cu_2O$ system as converting furnace (C-furnace) slag. Compared with the conventional silicate slag, this slag has enough fluidity under the high oxygen potential for blister making, with operating temperature of 1150 to 1250 C. Under such conditions the silicate slag requires 40-50% copper content only for melting, while the lime slag contains 10 to 15% copper and allows the low rate of copper return to the smelting furnace (S-furnace), producing good fluidity slag and good quality blister.

Fourthly, because of continuous operation and oxygen enrichment, off-gases from S and C-furnaces are small and constant in volume and high in SO_2 concentration. This enables the boilers, electrostatic precipitators and acid plant to be compact and steady in operation.

Lastly, since the molten products are transported in the closed launders, the fugitive gases are quite small in volume and easy to handle, compared with the process with the conventional P.S. converter.

Because of the above features, the process implements the requirements described earlier, i.e. energy and manpower savings and environmentally free from pollution.

Table I shows the recent operating data. Regarding the matte grade, it used to be 65%, however recent improvements resulted in a rise to 68-70%, maintaining copper concentration in slag at around 0.6%. Operating campaign period has been two years so far, with 95% on-line time, however it will be extended to three years in the future. The water cooled copper jacket and improvement of the brickwork technique have contributed to the extension of campaign period.

Table I. Recent Operating Data (May, 1988)

Copper Concs. Treated	(t/m)	26,868
Anode Copper Produced	(t/m)	8,667
Major Assay		
Copper in Concs.	(%)	32.2
Matte Grade	(%)	69.5
Copper in S-f'ce Slag	(%)	0.6
Operating Conditions at S-f'ce		
Feed Rate	(t/h)	38.0
Oxygen in Lance Blast	(%)	40.0
Fuel required		
*Coal mixed with Concs.	(kg/h)	1400
*Oil for Burner	(l/h)	0
Operating Conditions at C-f'ce		
Anode Copper Production Rate	(t/h)	12.3
Oxygen in Lance Blast	(%)	31.0
Fuel required	(l/h)	0

Copper Refining

In 1975 the Onahama Smelter and Refinery of Onahama Smelting and Refining Co. Ltd. (a subsidiary of Mitsubishi Metal Corporation) started to construct a new copper refinery of 8000 tpm (No.3 tankhouse) in addition to the existing No.1 and No.2 tankhouses of 10,000 tpm capacity together. The No.3 tankhouse has been designed with several innovative concepts to meet the requirements mentioned previously (6).

Design Principles The principles placed when designing are as follows;

(1) Marked labour savings - Labor savings were aimed not only at reducing cost but also at freeing the operators from warm and humid working conditions.

(2) Savings on the interest for inventory - The conventional tankhouse has much inventory copper (gold and silver as well) in the cell, mostly as anode, cathode and electrolyte. This means expending much inventory interest.

(3) Employment of parallel flow of electrolyte with electrodes - For the purpose of uniform temperature, composition and additives in the electrolyte, parallel flow of electrolyte was regarded as ideal and this method was employed.

Measures To realize the above objectives the following measures were taken;

(1) Jumbo size cell- Twenty conventional cells were grouped in one jumbo cell which is illustrated in Figure 6. The electrolyte flows parallel to the electrode as illustrated. The dimensions of electrode

and cell are shown on Table II.

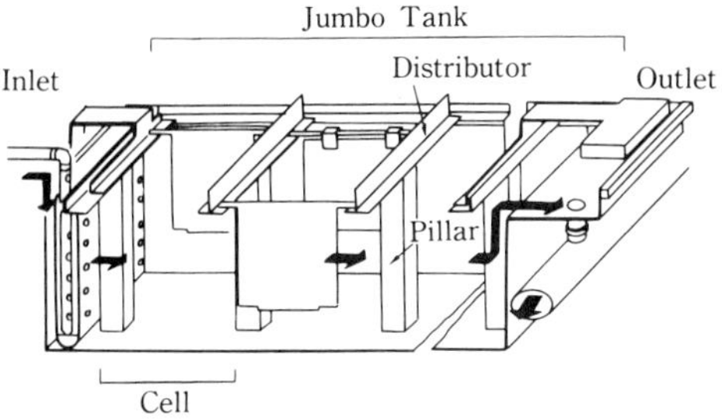

Figure 6 - Jumbo Tank

Table II. Dimensions of Electrode and Cell

		Conventional system	Jumbo tank system
Anode			
weight	(kgs)	330	144
thickness	(mm)	40	15
length	(mm)	980	980
width	(mm)	960	960
life	(days)	28,30	12,13,14
scrap ratio	(%)	15	26
Cathode			
weight	(kgs)	140	110
length	(mm)	1,000	1,000
width	(mm)	1,000	1,000
life	(days)	14,15	12,13,14
Cell			
length	(mm)	4,800	4,130
width	(mm)	1,200	28,130
depth	(mm)	1,270	1,320
numbers		728	30

(2) Thin and light-weight anode - The thickness of anode was cut to almost half as compared with the standard anode for one cropping of cathode. Since the conventional casting such as turn-table type is hard to cast such a thin anode with the accurate dimension, the Hazelett casting machine was employed after investigation.

(3) Innovative electrode handling system - For handling electrodes between cells and receiving or shipping area, an innovative system comprising a charging bridge, carrier cars and hoists was developed and employed. The charging bridge feeds (or pulls) a set of anodes of 51 plates and cathodes of 50 plates into (or from) a cell. It runs over the cells on the rails with basement on the ground. The

carrier cars are circulating, with a set of electrodes (51 anodes and 50 cathodes), among receiving area of anodes and starting sheets, the charging bridge and shipping area of anode scrap and electrolytic copper. Hoists serve for transferring anodes (starting sheets) from anode spacing machines (starting sheet assembly machines) to carrier cars, and also transferring anode scrap (cathode) from carrier cars to an anode scrap washing machine (cathode washing machine). The arrangement of the above handling system is illustrated on Figure 7.

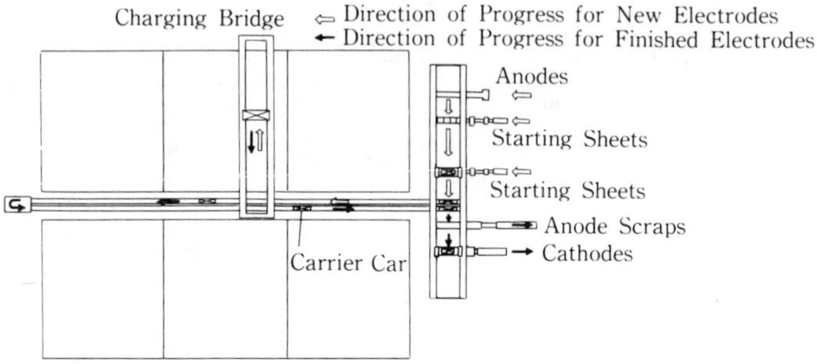

Figure 7 - Arrangement of Electrodes Handling System

<u>Result</u> The innovative concept introduced into the construction of No.3 tankhouse at Onahama brought about the outcome as expected. The labor force of the new system is shown on Table III. The decrease of copper inventory is presented on Table IV, and the recent operating data on Table V.

<u>Table III. Labor Force</u>

Anode Conveyer	1
Starting Sheet Assembly Machine	1
Hoists and Carrier Cars	1
Charging Bridge	1
Cathode and Anode Scrap Conveyers	1
Electrolyte Control	1
Total	6

Table IV. Copper Inventory in Process

	Conventional system	Jumbo tank system
Production Capacity (t/m)	12,000	8,000
Copper Inventory (t)	11,320	4,290
Unit Copper Inventory (kg/t of cathode)	943	536

Table V. Recent Operating Data (Apr.'87-Mar.'88)

		Conventional system	Jumbo tank system
Production	(t/m)	9,398	5,622
Operating Cells	(No.)	744	24
Electric Power			
current density	(A/m^2)		
day time		128	100
night time		219	200
current efficiency	(%)	98.3	95.3
Energy Consumption			
electric power (electrolysis and general)	(kWh/t)	247	254
steam	(kg/t)	336	323

Future Trends

After forecasting environments surrounding the industry, future trends for copper, zinc, lead, aluminum and rare earth metals are reviewed generally, with introduction of some development works in Japan.

Social Environments, Raw Materials and Related Technologies

Social Environments The price of oil is decreasing in the past couple of years due mostly to less demand and financial situation of oil producing countries, however, it would increase again in the long term, because of lower possibility of development of big oil wells and prospective increase of the mining cost. The salaries and wages would also continue to rise worldwidely, depending on the economic growth.

Hence, the efforts for saving energy and manpower will be continued in the future.

Efforts will also be made for the improvement of environmental issues. The environmental concern from global aspects will be further stressed in such issues as destruction of ozone in strato-sphere and ambient temperature increase by carbon dioxide. Improvement of workplace environment will also be promoted further.

Raw Materials Treatment of non-ferrous recycling materials will become more important. Attention is paid to the demand of gold. Figure 8 shows the changes of demand of gold in Japan. The figures also shows quantity of gold consumed for plating and electronic instruments. The recycling procedure in the non-ferrous processing will play an important role, when the gold used in this area is scrapped.

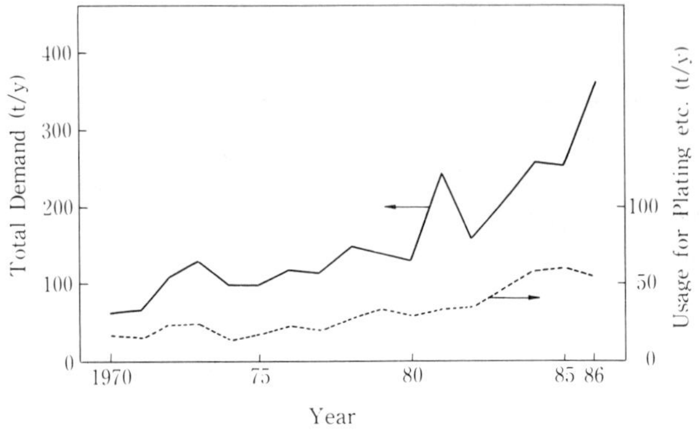

Figure 8 - Total Demand of Gold and the Usage for Plating and Electronic Instruments (7)

Related Technologies Computers are getting more compact and efficient following the advancement of further highly integrated LSI. The utilization of computers will proceed more and more and total control in the plant, comprising control of process itself, raw materials, products, purchasing of supplies, plant maintenance, finance and administration, will rather easily be realized. The intelligent robot will widely be utilized in the industrial world, by the improvement of sensors and computers.

Copper

As regards smelting, considering the long time period which is required for the development of a new smelting process and the current situation of the development, it would be hard to expect the emergence of a new process. The Outokumpu/Inco flash, Noranda and Mitsubishi processes would continue their status as standard technology into the early 21st century.

The rapid economic growth is expected in Southwest Asia including the People's Republic of China which will lead to increased demand for copper in the area. Thus, construction of a new copper smelter will be foreseen and in that case the continuous copper smelting and converting process would be the most reasonable choice because of the superiority in terms of construction cost, savings on energy and manpower and environmental control.

As to electro-refining, the trends will be toward further sophistication of the existing electrolytic process in which high quality copper is produced, and the savings on energy and manpower will be achieved further.

As it was realized recently in Kidd Creek, modification of Onahama's jumbo tank system, in which permanent blanks of stainless steel are used instead of the conventional starting sheet, appear to provide a good example for an ideal copper tankhouse to be built in the future. In any case the electrode and other materials handling system will further be automated by the improvement of the sensor and the computer and the tankhouse will be operated almost fully automatically with a very few people in the future.

The recent production increase of solvent extraction-electrowon copper is remarkable. The process is applied for the ore dumped in the past for heap leaching, low grade ore by in-situ leaching or ore below a cut-off grade. By the progress of solvent extraction technology and by the introduction of permanent blanks, the quality of cathode is considerably improving - such quality as they can be used for the wire rod. Thus the process will be more popular in the future, although in the limited areas as described above.

Future Developments of the Mitsubishi Continuous Process

(1) Productivity of the Furnace Productivity of the furnace has improved at both Naoshima and Kidd Creek, as seen in Table VI. And in both cases further improvement is expected (8)(9). The increased productivity of the furnace has been achieved by using more tonnage of oxygen.

Table VI. Improvement of Furnace Productivity

	Naoshima	Kidd Creek
Furnace Dimension		
S-f'ce (inner dia. : m)	8.25	10.3
C-f'ce (inner dia. : m)	6.71	8.16
Concentrates Throughput		
designed (t/h)	24	35
operating (t/h)	40	60
tested (t/h)	50	-

The limitation of throughput for the S-furnace with a given dimension will be possibly determined by ;

* kinetics of related metallurgical reactions, and

* thermally autogenous point.

In the Mitsubishi process the metallurgical reactions proceed in a narrow space beneath lances and complete almost instantaneously. Hence, it is considered that the rates of reactions are very large and cannot be a limitation factor. Therefore the throughput, at which the reaction proceeds autogenously, i.e. without supplemental fuel, is the limitation for the assumed furnace.

Considering the above in mind and assuming furnaces with the dimensions in Table VII, which are similar in size to the ones in Kidd Creek, and operating conditions which are based on the actual operation at Naoshima, the maximum throughput was calculated (10). The calculation was carried out for three types of concentrates as shown in Table VII with adjustment of matte grade respectively. The result is summarized in Table VIII. For calculating annual production, operating days are assumed as 340 days a year and on-line time when operating as 95%. Annual capacity is 160,000 tonnes copper for the concentrates of 30% Cu, 200,000 tonnes for 35% Cu and 240,000 tonnes for 40% Cu respectively.

Table VII. Design Parameters for Model Calculation

Design parameters		S-furnace	C-furnace	
Furnace size	(m)	10	8	
Heat loss	(Mcal/h)	4,400	5,000	
Total lance blow	(Nm^3/h)	30,000	22,000	
Oxygen enrichment	($\%O_2$)	50	35	
Concentrate grade				
		Type 1	Type 2	Type 3
Cu	(%)	30.0	35.0	40.0
Fe	(%)	24.7	22.8	20.8
S	(%)	30.1	28.8	27.5

Table VIII. Result of Model Calculation

Type of concentrate	Type 1 30%Cu	Type 2 35%Cu	Type 3 40%Cu
Matte grade (%Cu)	67	71	75
Feed rate (t/h)	70	75	80
Annual copper production ($10^3t/y$)	160	200	240

(2) Treatment of Anode Scrap and Secondary Raw Materials At Naoshima, anode scrap and No.2 copper scrap are treated in the reverberatory furnace process currently. However, the treatment is required in any case at the stand-alone type continuous smelter. In that case they will be processed in the C-furnace utilizing surplus heat. In the C-furnace, heat surplus is realized easily by controlling the usage of tonnage oxygen. Actually melting of various shapes of No.2 scrap and also shredded anode scrap had been successfully tested. No problems were found for melting. In the future, if required, melting of anode scrap and No.2 scrap can be carried out in the C-furnace by specially designed feeding equipment.

As to the secondary raw materials of rather low grade, they are currently processed in the S-furnace by mixing with concentrates or feeding directly to the furnace after crushing.

(3) Treatment of High Grade Copper Concentrates In some big mines now operating or under development, they produce high grade copper concentrates. The copper content in this case is 40-45% with 20-30% S, compared with ordinary concentrates of 26-28% Cu and 30-32% S.

The reaction heat generated, when high grade copper concentrates are treated, is lower as compared with that generated from the treatment of ordinary concentrates. Hence, if matte grade is kept at the standard value of 68% for high grade concentrates, fuel requirement at the S-furnace increases. To confirm if the fuel for the S-furnace is reduced and also if copper output is increased by utilizing the room for lance blast at the C-furnace obtained from raised matte grade, test operations on high matte grade were conducted (10). As a result, reduced fuel requirement and increased production were proved. In the test, the relation between matte grade and copper content in slag was also investigated. The results obtained are given in Figure 9. As shown in this figure, slag loss increases sharply as matte grade reaches around 73%. However, if matte grade is held below 72%, slag loss stays in the range from 0.7% to 0.8%. Rather low slag loss for high matte grade is one of the advantages with the Mitsubishi process.

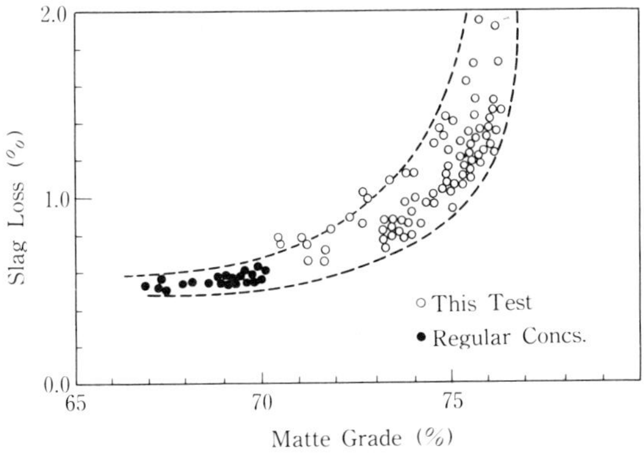

Figure 9 - Relation between Slag Loss and Matte Grade

Zinc

The electrolytic and the Imperial Smelting (I.S.) processes are two major streams for zinc extraction. With various improvements, these will continue the current standing into the next century.

The electrolytic process has the feature of the production of high quality zinc. The improvements will be carried out for further automation and mechanization in the process and also savings in the electric power consumption. One of the examples for a significant improvement in power consumption is described later. Improvements in the treatment of primary leach residues will continue particularly in solving a difficult problem in the disposal of the final residues, in the jarosite, goethite and hematite processes.

Direct leaching of zinc sulfide concentrates which replaces roasting in the electrolytic process is also noticeable. The process is operated in Trail, Cominco Metals Ltd. and Kidd Creek, Falconbridge Ltd. A couple of refineries are reportedly planning to introduce it. The distinctive feature of recovering sulfur as elemental form, in addition to the elimination of residue treatment, will continue to give this process an important status in terms of sulfer fixation.

The I.S. process has its characteristics in the treatment of complex lead-zinc concentrates, the low production cost especially for PW grade zinc and production of salable or disposable slag. Further development will be made for the improved productivity and economics.

Recently a new process has been developed in Japan. The purpose of the development is to switch energy source from electricity to coke. The process was proposed by Prof. Sakichi Goto and developed by the Committee of Joint Research Organization of Injection Metallurgy (11). It comprises roasting and the successive injection smelting of the calcine. The calcine is fed to the smelting furnace with pulverized coke through a lance. Zinc volatilizes, then is collected by lead splash condenser as in the I.S. process and slag is tapped out of the furnace. Although the process is still on a semi-pilot scale and has problems such as lower condensation efficiency to be solved, it appears to provide one solution to modify the electrolytic process where the power cost is extremely high compared with coke.

New Concept for Power Savings in Electrowinning Recently a new concept was developed for remarkable power savings in the electrowinning operation by Akita Zinc Refinery of Mitsubishi Metal Corporation (12).

For the reduction of power consumption in the electrowinning, two approaches are considered. One is the improvement of current efficiency and the other is the decrease of the cell voltage. The former is over 90% in the ordinary zinc refineries and it will be hard to improve it economically by further electrolyte purification. On the other hand the decrease of cell voltage is expected, because it is actually almost 3.4V (under current density of 500 A/m^2) for the theoretical value of 2.67V. The difference is mainly contributed by the resistance at electrolyte and the surface layer of anode. The following process was developed for the reduction of cell voltage.

The surface to surface distance was reduced from the present 30 mm to 15 mm to reduce the electrolyte resistance. The narrower spacing causes difficulty when pulling up cathode, however, the problem was solved by simultaneous pulling out of anode and cathode. Further, the surface of anode is cleaned every time it is pulled out to remove MnO_2 crust on the surface. The narrower spacing and frequent cleaning of anode surface decreased the cell voltage substantially and brought about a 10% reduction in power consumption for a tonne of zinc deposited.

The same current efficiency and purity of cathode are maintained in spite of the narrower spacing. This is achieved by the constant spacing by the frame attached to the anode as shown in Figure 10, which prevents electrodes from electric short circuit. The anode and cathode are handled as a unit as illustrated in Figure 11. The specially designed handling machine pulls out anodes and cathodes simultaneously, enlarges spacing, transfers, cleans anodes, washes and strips cathodes and comes back to the cell for charging. These movements are carried out automatically.

Potential savings on construction cost from decreased space occupancy by the reduction of electrodes distance, are also the feature of the new concept.

It is now commercialized in Mitsubishi's Akita Zinc Refinery after a year and half semi-commercial test.

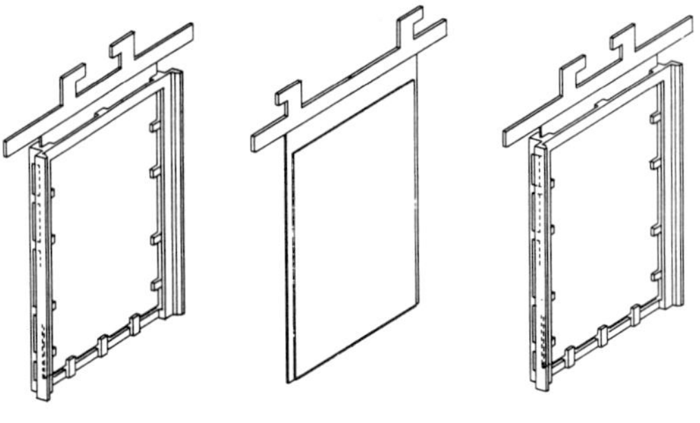

Figure 10 - Anodes installed with Frame

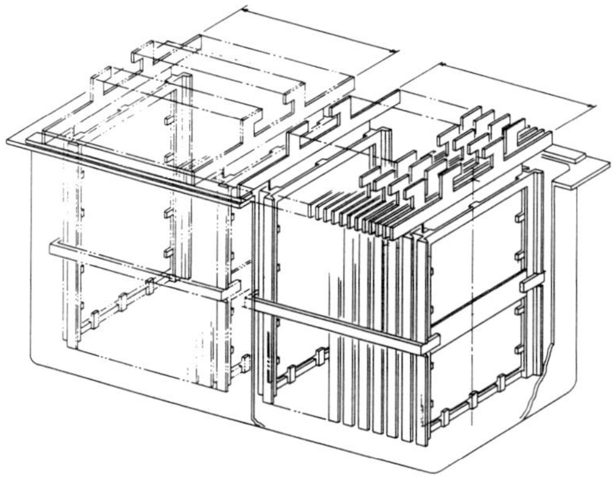

Figure 11 - Electrodes Units in a Cell

Lead

Various direct smelting processes have been proposed and developed in recent years for the replacement of conventional sintering and blast furnacing procedure.

Among them is the Kivcet process that has been commercialized at last. QSL commercial plants are now under construction. In February, 1987 at Portovesme Smelter of Nuova Samim, the Kivcet process of 600 t of concentrates per day was brought into operation. It is reported that QSL plants are under construction in the province of Gansu in the People's Republic of China and in Trail of Cominco Metals Ltd., Canada (13). Of these the one in Trail is said to commence its commercial operation from September 1989 with the capacity of 160 thousand tonnes of lead per year.

Both processes employ a single furnace with a partition wall and oxidize lead concentrates in one compartment and reduce slag in the other compartment with crude lead of low sulfur and discard slag as the molten products from the furnace. In the Kivcet process smelting is carried out by the flash furnace method and reduction by electrothermic process with reduction of zinc as well which is recovered as zinc or zinc oxide. In the QSL process, the pelletized concentrates are oxidized in the smelting zone by submerged injection technique and the reduction of slag also by submerged injection of coal. Both processes use tonnage oxygen for oxidation and produce high SO_2 concentration off-gas.

Thus both processes achieve considerable cost savings on energy and manpower and improvement of pollution control and workplace conditions.

Further progress and improvement are expected for both processes after their commercialization. These processes will be the major candidates as the standard procedure of lead smelting in the 21st century.

Aluminum

The production of aluminum has been carried out almost exclusively by the Bayer/Hall-Héroult process for more than a century since its invention. Various new processes have been proposed and challenged, however the process still dominates in the production of aluminum because of its reasonableness and perfectness.

One of the challenges when using blast furnace was intensively investigated in Japan, where extremely expensive power price forced almost all the aluminum refining capacity to close. The research had lasted almost ten years. It was started by Mitsui Aluminum Co, Ltd. in 1977 and later was succeeded by a research team jointly organized by major Japanese aluminum production and heavy industry companies. The process comprises reduction of bauxite by coke in the blast furnace with production of iron, silicon and aluminum alloy of 20-25% aluminum, extraction of aluminum from the alloy by molten lead and separation of aluminum from lead. The purity of aluminum obtained was better than that by the Hall-Héroult process, under the production rate of 500 kgs/d. The project terminated, finding the problems for commercialization in rather low recovery rate of aluminum (60-65%) and low grade of aluminum which requires costly refining.

The challenge for establishing an alternative process will continue in the future, however it would take quite a long time to find out an innovative technology to replace the existing process. Hence, efforts will be made continuously for the improvement on saving energy and manpower in the Bayer/Hall-Héroult process.

Rare Earth Metals

Rare earth metals are now spotlighted with the recent discovery of high temperature super-conductivity materials, originating in the invention of the characteristics for La-Ba-Cu-O system by Drs. Bednorz and Müller.

Because of the chemical similarity among the elements in the group, it had been hard to separate them each other and they were used as mixtures such as misch metal. However, improvement of the recent solvent extraction technology enabled the sharp isolation of each element. Various uses for the elements have now been found. The usage of yttrium and europium oxide as fluophor of the cathode ray tube for color TV is well-known. High purity lanthanum oxide is used for the additives of new functional glasses. Several rare earth oxides are useful for engineering ceramics. The major usage of rare earth metals is in the magnet ; Sm-Co, Nd-Fe-B and Pr-Fe-B-Cu magnets. The use of LaNi alloy for alkaline battery electrode and terbium and dysprosium for the magneto-optic disc will also increase in the future.

Although sales amount of rare earth metals is very small compared with that of common metals, the growth rate in the future will be large. Technological efforts will be made for the production of high purity metals and halides and the research of new functional materials will also be continued in the future.

Conclusion

On the occasion of the international symposium, " The Metallurgical Processes for the Year 2000 and Beyond ", future trends were reviewed for some common metals and rare earth metals as well. In conclusion, in the discussed area, the processes developed in this century would be succeeded to the 21st century. And it appears that the possibility of emerging new technologies replacing existing ones is small. However, the existing technologies will be further sophisticated and competition among them will intensify. The trends for technology sophistication will be in the savings on energy and manpower, in which case technologies related to computers, electronics, etc. will play an important role.

References

1. Ichiro Yano, Nihon Kokusei Zue (Tokyo, Japan; Kokuseisha Co. Ltd., 1988), Appendix I

2. Handbook of Revised Pollution-related Health Damage Compensation System (Tokyo, Japan; Energy Journal Co. Ltd., 1988)

3. Monthly Statistical Survey on Labor, Ministry of Labor

4. Oil price from Monthly Bulletin of Import and Export, Ministry of Finance. Power price from typical sample for industrial use in a certain electric power company.

5. J.C. Yannopoulos, and J.C. Agarwal, Extractive Metallurgy of Copper (New York, NY: The Metallurgical Society of AIME, 1976) 439-457

6. ibid. 588-608

7. Bulletin of Japan Mining Industry Association Aug. Sept. 1978, and Aug. Issue from 1979 to 1987.

8. M. Goto et al., "Improvements in High Intensity Operation of the Mitsubishi Process at Naoshima Smelter" (Paper Presented at 115 AIME Annual Meeting, New Orleans, LA, March 3 - 6, 1986)

9. C. Díaz, C. Landolt, and A.A. Luraschi, Copper 87 vol.4 Pyrometallurgy of Copper (Santiago, Chile: Facultad de Ciencias Fisicas Matematicas, Universidad de Chile, 1988), 123-137

10. T. Shibasaki, K. Kanamori, and S. Kamio, "Mitsubishi Process - Prospects to the Future and Adaptability to Varying Conditions" (Paper to be presented at the 118th AIME Annual Meeting, Las Vegas, Nevada, Feb.28 - Mar.3, 1989)

11. S. Goto, M. Nishikawa, and M. Fujikawa, "Semi-pilot Plant Test for Injection Smelting of Zinc Calcine" (Paper to be presented at the 118th AIME Annual Meeting, Las Vegas, Nevada, Feb.28 - Mar.3, 1989)

12. K. Kaneko, H. Ohba, and T. Kimura, "The New Concept in Zinc Electrowinning Operation" (Paper to be presented at the 118th AIME Annual Meeting, Las Vegas, Nevada, Feb.28 - Mar.3, 1989)

13. "Lurgi 1987", Brochure by Lurgi GmbH

Metals and Materials Processing

ULTRA FINE PARTICLE DEPOSITION

I.S. Min, Suk J. Im, Daesoo Kim

Research Center, Korea Mining & Smelting Co.
327-23 Garibong Dong
Seoul, Korea

Abstract

UFP are defined by atoms or their clusters having diameter in the range of 10Å ~ 0.1 μm. Due to a ratio of very large surface area to volume, they have peculiar properties such as extra high surface energy. Recently UFP deposition technique may find its new applications to various areas such as conductive/resistive film formation. In this method, gas flow with UFP jet through the nozzle which is considered to be a free molecular flow is maintained at a very high speed (~100m/s). Using this technique, mixture film of Fe and Ni, and Ag/Ag-Alloy etc. were obtained, and their mechanical and electro-magnetic properties were determined. The experimental results are pending. The test results and discussion will be presented in the main paper.

Metallurgical Processes for the Year 2000 and Beyond
Edited by H.Y. Sohn and E.S. Geskin
The Minerals, Metals & Materials Society, 1988

DEVELOPMENT OF ELECTROMAGNETIC ATOMIZATION PROCESS

K. Sassa, T. Kozuka and S. Asai

Department of Iron & Steel Engineering
Faculty of Engineering, Nagoya University
Nagoya, Japan

Abstract

A new atomization process is developed. The principle of this process is completely different from that of conventional processes. This process uses electromagnetic energy induced by direct electric current and direct magnetic field instead of kinetic energy produced by fluid impingement or disk rotation.

Experimental work of this process was carried out using pure tin, and it was found that the crucial factor for a stable operation is the contact state of molten metal with the electrode. In order to control the condition, a delay time controlling unit was developed and combined with this process. It is noticed that this controller works well to produce uniformly sized particles.

A linear relation between the impulse and the momentum of atomized particles is found by use of experimental data and is explained by theoretical analysis.

Metallurgical Processes for the Year 2000 and Beyond
Edited by H.Y. Sohn and E.S. Geskin
The Minerals, Metals & Materials Society, 1988

Introduction

In atomization processes prevailing in metal powder technology, kinetic energy is currently employed for atomizing molten metal. That is, molten metal is broken into small particles by other fluid (e.g. water, argon gas, etc) or by impinged upon a rotating solid disk. This splashing process is always accompanied with uncertain factors for shape and size of the particles. The diameter distribution of products, therefore, comparatively spreads over a wide range. From another standpoint, the restriction of atmosphere is inevitable in some of these processes. As a pulverizing technology, the process which can produce uniformly sized particles and permits the choice of atmosphere is strongly desired.

On the other hand, a great attention has been paid to the spray casting process as a new near net shape casting. In this process, a small size spraying device is favorable.

In this paper, a new atomization process is proposed, which atomizes molten metal by electromagnetic energy instead of kinetic energy produced by fluid impingement or disk rotation in the conventional atomization processes. This process can produce more uniformly sized particles with a comparatively less contamination from atmosphere and a smaller size device.

Principle of electromagnetic atomization process

Figure 1 shows the basic principle of electromagnetic atomization process, which is well known as Fleming's left hand rule. An electric circuit with direct current supply is shorted when the molten metal flown from the horizontal nozzle contacts with the electrode facing to the nozzle. Magnetic field, **B** is imposed vertically so as to be perpendicular to electric current, **J**.

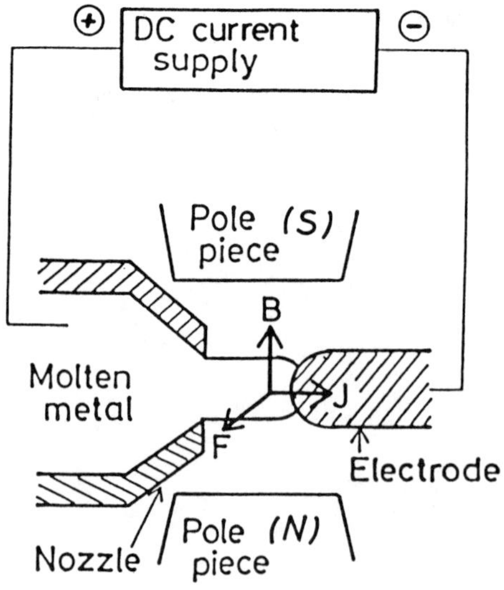

Figure 1 - Schematic view of electromagnetic atomization.

Atomizing process is devided into two stages as follows:

Stage I. Molten metal flows from the nozzle, but does not contact with the electrode.

Stage II. After the molten metal contacted with the electrode, the molten metal between the nozzle and the electrode is splashed to be particles. And electric current shuts down.

Stage I and II occur sequentially and are repeated so that the molten metal can be atomized.

Experiment

Figure 2 shows the schematic view of experimental apparatus. This apparatus is composed of ① direct current supply, ② supplying system of molten metal, ③ electrode, ④ permanent magnet, ⑤ measurement system of electric current and voltage between the electrode and the nozzle, and ⑥ delay time controlling unit which will be mentioned later in detail.

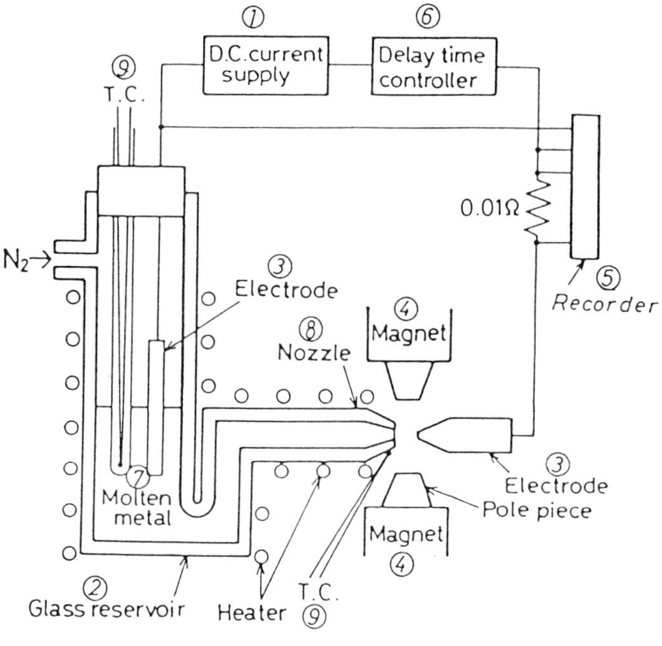

Figure 2 - Schematic view of experimental apparatus.

Pure tin (300g weight) is melt in a reservoir made of pyrex glass. Temperatures of the metal bath ⑦ and the tip of the nozzle ⑧ are monitored by thermo-couples ⑨ and kept at a preset value by the controlling unit and the heater. The positive electrode is submerged into the metal bath and negative one is set facing to the nozzle. After the voltage is set at a given value, molten metal is driven to flow out from the nozzle by nitrogen gas pressure. The flow rate is controlled by the flowing rate of nitrogen gas. Pole pieces of the magnet are placed above and under the gap between the nozzle and the electrode, so that the magnetic field is imposed on the gap vertically. The above piece is south pole and the under one is north pole. The length of the gap is controlled by a manipulator.

The results of prior experiments indicated that the contact state of molten metal with the electrode was a crucial factor for stabilizing the operation and getting particles of uniform size. So the delay time controlling unit is developed and combined into the electric circuit. The function of this unit is the control of the delay time which is the time interval between the moment at which the molten metal contacts with the electrode and the moment at which the electric current is imposed by using this controlling unit. By introducing this unit, an atomizing process composed of two stages is modified as follows:

Stage I. Molten metal flows from the nozzle, but does not contact with the electrode.

Stage I-II. Molten metal contacts with the electrode, but electric current is not imposed by the function of delay time controlling unit.

Stage II. When the preset time of the controlling unit is reached, the molten metal between the nozzle and the electrode is splashed.

This new stage I-II is required for stabilizing the operation and getting particles of uniform size. In the stage I-II, molten metal is able to contact with the electrode more firmly than in the case without the controlling unit.

The transitional behavior of electric current and electric voltage in the case with the delay time controlling unit are schematically shown in Fig. 3. When the current is passing, molten metal suffers the electromagnetic force for pulverizing. At the point A, molten metal in the gap between the nozzle and the electrode splashes out. Simultaneously, electric current shuts down and electric voltage appears. Then, molten metal continuously flows from the nozzle to charge in the gap (stage I). At the point B, molten metal contacts with the electrode and the voltage falls down to zero. However, current is not imposed at this moment. So, the volume of molten metal in the gap between the nozzle and the electrode continuously increases so as to establish a better contact state (stage I-II). After the period, t_d (delay time), is elasped (at the point C), current is imposed on molten metal so that it is splashed out from the gap during the period, t_s (stage II). At the point D, molten metal departs from the nozzle and the electrode, so electric current shuts down again.

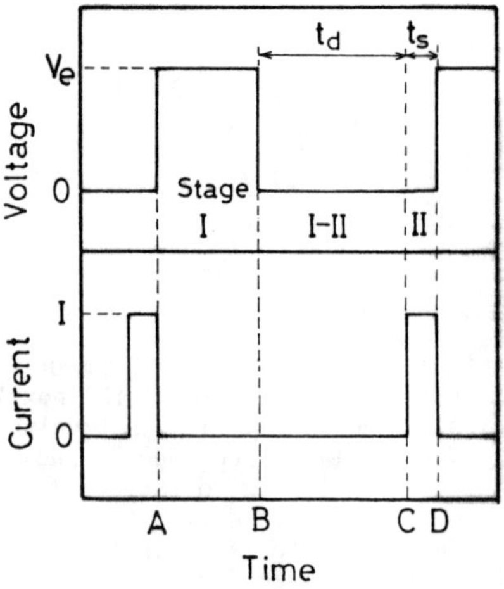

Figure 3 - Schematic patterns of voltage and current indicating the function of the delay time controlling unit.

In the experiments, splashing behavior is observed by a high speed VTR system. The imposed current and its duration are measured for calculating the impulse of particles, and initial velocity of particles is measured for the momentum of particles. Experimental conditions are shown in Table 1.

Table 1 - Experimental conditions

Electric current, I	5~40 A
Electric voltage, V_e	2~8.5 V
Imposed magnetic field, B	1.3 T
Nozzle diameter, d_n	0.5 mm
Distance between the nozzle and the electrode, l	0.5 mm
Temperature at nozzle tip, T	523~573 ℃
Delay time, t_d	0~100 ms
Molten metal flow rate, q	$(0.07\sim1.42)\times10^{-6}$ m^3/s

Results

Figure 4 shows the observed transitional behavior of the electric current and electric voltage in the case without the delay time controlling unit. In this case, stage I-II does not exist. The cyclic period of stage I and II is measured to be as short as 0.7ms. Furthermore, it is noticed that the pattern of current is irregular.

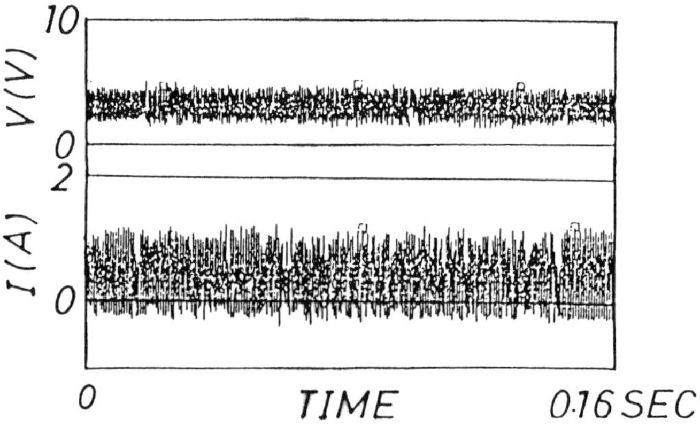

Figure 4 - Observed transitional change of electric voltege and current in the case without delay time controlling unit.

On the other hand, Fig.5 predicts the current behavior in the case with delay time controlling unit. In this case, the delay time is set to be 20ms. The interval in which each particle is splashed is about 27ms. The interval and the value of the current peak appear in very uniform and regular pattern.

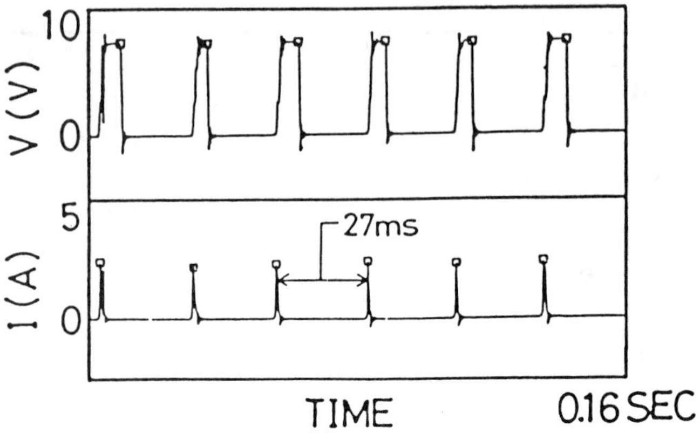

Figure 5 - Observed transitional change of electric voltage and current under the condition of 20ms delay time.

Photo. 1 shows the splashing behavior during pulverizing and SEM view of the obtained particles in the case without the delay time controlling unit. Figure 6 shows the size distribution of particles in that case. It is clear that atomized particles are comparatively irregular in size and shape.

Photo. 2 I) shows the trajections of particles in the case with 20ms delay time. In this case, particles are splashed horizontally in regular way as shown by the trajectories of a particle seen in this photo. Particles are solidified in the air to be spherical shape as shown in Photo. 2 II).

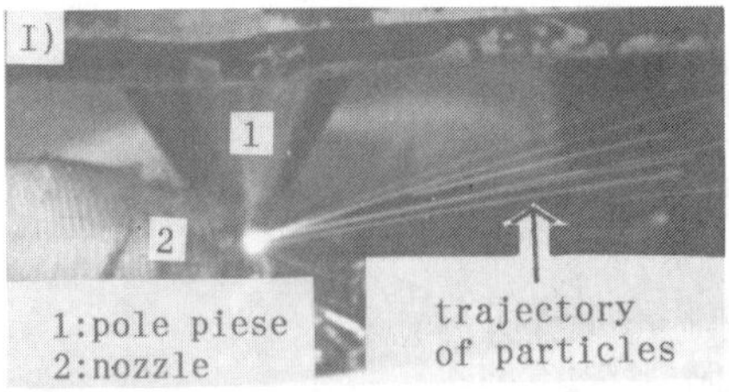

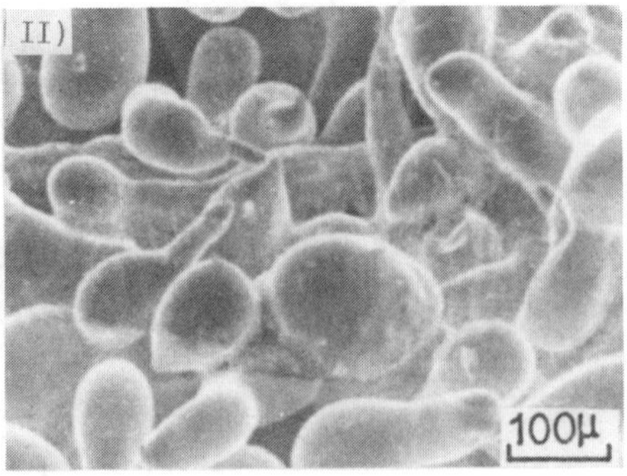

Photo. 1 - I) Trajectories of a particle in the electromagnetic atomization process in the case without delay time controlling unit. (shutter speed: 1/30 s)

II) SEM view of particles.

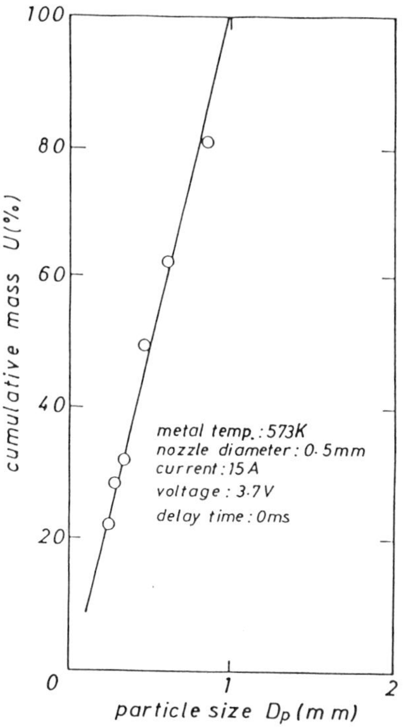

Figure 6 - Relationship between cumulative mass and particle size.

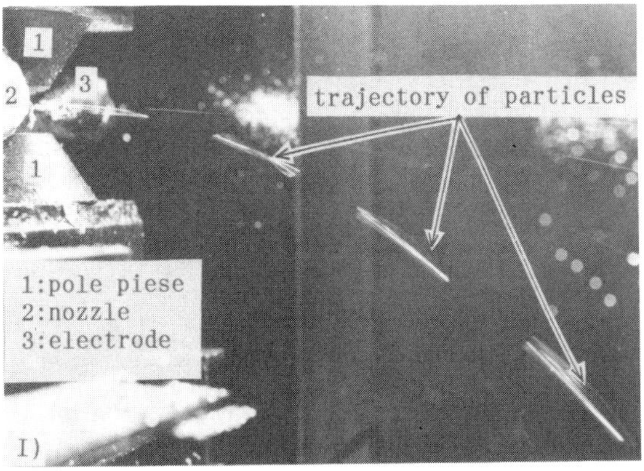

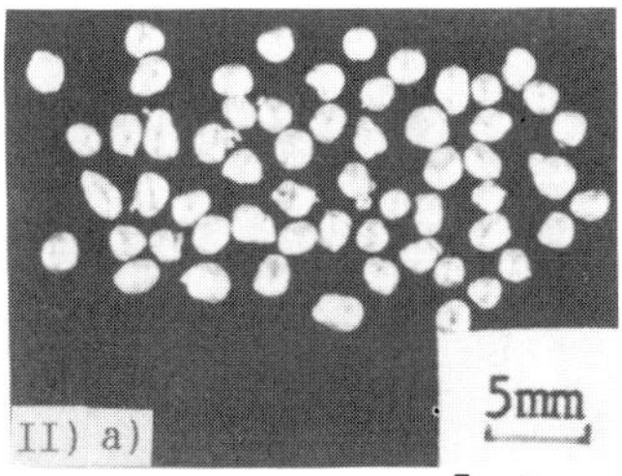

$q = 3.43 \times 10^{-7}\ m^3/s$

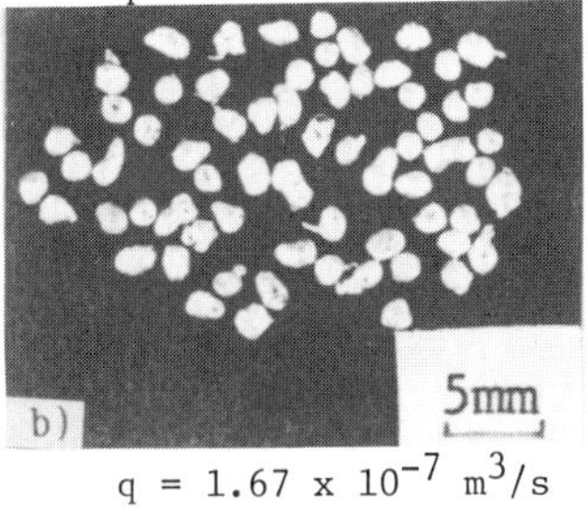

$q = 1.67 \times 10^{-7}\ m^3/s$

Photo. 2 - I) Trajectories of a particle in the electromagnetic process under the condition of 20ms delay time. (shutter speed: 1/30 s)

II) Micrograph of particles.

Discussion

In this chapter, the relation between the impulse and the momentum of particles is discussed. The impulse is given by the product of the electromagnetic force and the current imposing time, t_S. The momentum is calculated from the initial velocity and the mass of a spherical particle.

The relation between the impulse and the momentum is given by;

$$mv_0 = VJBt_S \quad (1)$$

Equation (2) is derived from Equation (1).

$$\rho v_0 = JBts \quad (2)$$

Where, v_0 is the initial velocity of spherical particles, which can be evaluated from the falling distance, h, and the horizontal distance between the place where particles are collected and the tip of the nozzle, x, as given by Equations (3) and (4), respectively.

$$h = (1/2)gt^2 \quad (3)$$

$$x = v_0 t \quad (4)$$

and v_0 is given by;

$$v_0 = x\sqrt{g/2h} \quad (5)$$

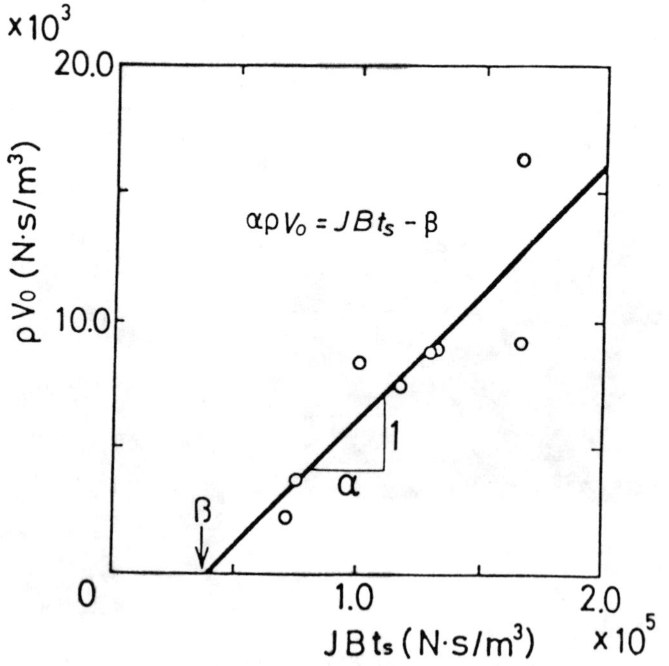

Figure 7 - Relationship between momentum and impulse of particles.

According to Eq.(2), experimental data are plotted in Fig.7. The value of J and ts are measured from Fig.5. It is noticed that a linear relation between the impulse and the momentum is observed as is theoretically predicted (Eq.(2)) Thus;

$$\alpha \rho v_0 = JBt_s - \beta \qquad (6)$$

The following values of α and β are calculated from Fig. 7.

$$\alpha = 0.1, \qquad \beta = 0.38 \times 105 \ N\cdot s/m^3$$

The reciprocal value of α implies the number of particles produced from the molten metal existing between the nozzle and the electrode per an imposition of current. β indicates the required force to overcome the clinging force. It is found that this atomization process strongly relies on the contact state of the molten metal with the electrode and the surface tension of molten metal.

Conclusion

A new atomization process making use of electromagnetic energy instead of kinetic energy was proposed, and the test apparatus was built up to indicate the reality of this process. The test experiments have proved the followings.

1. The contact state of molten metal with the electrode is the important factor for stabilizing the operation and producing uniformly sized particles.

2. Delay time controlling unit is useful for getting a better contact state of molten metal with the electrode.

3. A linear relation exists between the impulse and the momentum of a particle.

Nomenclature

$\mathbf{B}$,B	: magnetic flux density	[T]
D_p	: diameter of a particle	[mm]
d_n	: inner diameter of a nozzle	[mm]
g	: acceleration of gravity	[m/s^2]
h	: falling distance in the vertical direction	[m]
I	: imposed electric current	[A]
$\mathbf{J}$,J	: electric current density	[A/m^2]
l	: length between nozzle and electrode	[mm]
m	: mass of a particle	[kg]
q	: flow rate of molten metal	[m^3/s]
T	: temperature of molten metal	[K]
t	: time	[s]
t_d	: delay time	[s]
t_s	: current imposing time	[s]
U	: cumulative frequency	[%]
v_0	: initial velocity in the horizontal direction	[m/s]
V	: volume of molten metal between nozzle and electrode	[m^3]
V_e	: imposed electric voltage	[V]
ρ	: density of molten metal	[kg/m^3]

PROCESSING AND OXIDATION PROTECTION OF

CARBON/CARBON COMPOSITES

George R. St.Pierre, Gordon Holcomb, and
Robert A. Rapp
The Ohio State University
Department of Metallurgical Engineering
143 Fontana Lab
116 W. 19th Avenue
Columbus, OH 43210

Abstract

Carbon/Carbon composites offer promise as high-temperature structural materials. A primary limitation on their application is the need to provide protection against oxidation in hot oxidizing environments. Silicon carbide conversion coatings have been shown to provide extended protection at temperatures of 1600°C. SiC coated C/C components have been used in several aerospace applications. In the present paper, the preparation of carbon fibers and their treatment during subsequent processing into coated components are discussed with particular emphasis on the processing variables that can result in improved oxidation resistance. In particular, the influence of substrate inhibitors and selective sealants for cracks in coated C/C components are analysed.

Metallurgical Processes for the Year 2000 and Beyond
Edited by H.Y. Sohn and E.S. Geskin
The Minerals, Metals & Materials Society, 1988

ADVANCED METAL MATRIX COMPOSITES THROUGH XD™ PROCESSING

L. Christodoulou, J.M. Brupbacher and D.C. Nagle

Martin Marietta Laboratories
1450 S. Rolling Road
Baltimore, MD 21227

Abstract

The XD™ * approach to producing metal matrix composites employs an in-situ precipitation technique for the formation of the reinforcement directly in the matrix of interest. Key features include: thermodynamically stable reinforcement; clean particle/matrix interfaces, single-crystal reinforcement; size, chemistry, and shape control; uniform distribution; compatibility with conventional ingot techniques; and grain refinement.

Using XD™ technology it is possible to simultaneously introduce a number of different reinforcements in a given matrix, thus allowing for designing the microstructure to meet a specific material need. Thus, XD™ technology is a very useful technique for achieving "designer" or "engineered" microstructures. XD™ technology has been applied to a number of systems including Al, NiAl, TiAl and Cu. Materials property and processing benefits will be discussed.

* XD™ is a trademark of Martin Marietta Corporation.

Metallurgical Processes for the Year 2000 and Beyond
Edited by H.Y. Sohn and E.S. Geskin
The Minerals, Metals & Materials Society, 1988

INTELLIGENT REALTIME CARBONIZATION CONTROL

William J. Pardee and Michael A. Shaff

Rockwell International 1049 Camino dos Rios, Thousand Oaks, CA 91360.

Deeper materials science knowledge from mathematical models and in situ material property sensors will be exploited in next generation factory process controllers to improve productivity and quality. These intelligent control systems will accelerate productivity growth and materials science application through the ease with which they can be modified to accept new materials knowledge and sensors. Carbonization of resin matrix carbon fiber composites, specifically, is a violent solid state chemical reaction with large stresses due to gas pressure, matrix shrinkage, and differences in thermal expansion. The matrix strength falls sharply, then rises again during the process. At least two competing reaction paths are important; the first cleaves methylene bonds, reduces toughness, density and stiffness; the second reforms those bonds, ultimately resulting in aromatic ring condensation that increases density, strength, and modulus. Since these reactions have different temperature dependencies, one can expect to shift the process towards the second path by changing the heating rate, but that change must be accomplished without producing macroscopic interlaminar crack growth. To accomplish this, we employ gas sensors to infer matrix chemical state and acoustic emission to identify microstructural changes and to recognize brittle fracture. The latter does not prevent delamination, but is important in testing models for the causes of delamination and in improving the processing cycle. This paper describes a control strategy to accelerate the process while still lowering the risk of failure.

Research supported by DARPA and the Office of Naval Research under contract N00014-87-C-0724.

Metallurgical Processes for the Year 2000 and Beyond
Edited by H.Y. Sohn and E.S. Geskin
The Minerals, Metals & Materials Society, 1988

Introduction

Deeper materials science knowledge from mathematical models and in situ material property sensors will be exploited in next generation factory process controllers to improve productivity and quality. These intelligent control systems will be more easily modified to accept new materials knowledge and sensors, thus accelerating productivity growth. Such a control strategy for carbonization control is described below.

Carbon-carbon composites are time consuming and costly to manufacture. Processing methods have been developed empirically and are usually very conservative because the process often fails when any change is made. Apparently minor changes in materials or geometry can produce failures that are difficult to diagnose and correct. We will describe below several essentially unavoidable sources of variability. These are best dealt with by applying materials science to adapt (*i.e.*, closed loop control on *materials properties*) the process in real time to reduce the effect of unavoidable model idealizations, material variability and environmental noise. The most critical issue in control of the carbonization process control is avoiding catastrophic delamination. The strategy developed below makes use of a priori materials knowledge and real time sensors to accelerate the now very slow process without destroying the material, and without requiring unreasonable resources to determine the statistical properties of the material.

The entire manufacturing process, itself, is summarized briefly in Section 1. The control strategy is described first assuming idealized sensors and actuators to make the system integration goals clear. The idealized sensors and actuators are then replaced by combinations of real sensors, models, and a priori materials knowledge of cause and effect. The top level control system is being implemented using machine problem solving methods from artificial intelligence. Those methods encode materials knowledge in a disciplined, modular, human intelligible format with resultant ease of maintenance and growth. The machine intelligence tools include interactive explanation of the materials science based decisions. The resulting approach to manufacturing process control brings more materials science to the factory floor, and should improve technology transition from materials science to engineering and manufacturing.

Resin Matrix Carbon-Carbon Composite Manufacturing

A typical resin matrix carbon-carbon composite is manufactured from layers of woven carbon fibers preimpregnated with a phenolic resin (prepreg). The prepreg can be woven in a variety of two dimensional patterns. Processing requirements depend on the weave, because different weaves exhibit different resistance to delamination. Layers of the fabric are overlaid on a tool. The resulting layers of fabric are cured under heat and pressure to make a hard, stiff, slightly cracked plastic component. Typical cure is performed at 325°F (160°C) and 200 psi.

Much higher temperatures are required in the next (first carbonization) step, so it is normally performed in a separate retort in an inert gas (usually nitrogen) at atmospheric pressure. The temperature is increased slowly in a series of ramps and holds to about 1500°F (800°C). Non-carbon atoms (hydrogen, oxygen) of the plastic matrix are converted to a variety of volatile products and removed. Approximately 50% of the matrix mass is lost in this process. Gas evolution is intermittently high. Important species include[1] H_2, CO, CO_2, CH_4, C_2H_6, C_3H_8 (from the decomposition of isopropyl alcohol) and H_2O. The component first expands as it is heated and then shrinks as the plastic matrix is converted to carbon. The matrix shrinkage is considerable, ~ 20% in linear dimension. Because the fibers constrain *component* shrinkage, the reaction can produce considerable thermophysical stress which can result in catastrophic damage to the component during the process or make it too weak to survive subsequent processing. Thermal expansion also produces thermophysical stresses which can destroy the component. As the reaction proceeds, the strength of the matrix decreases to a small fraction of its intial value, reaching a minimum at about the time weight loss is complete, then *increases* again substantially as the component is heated further. Satisfactory completion of the carbonization process requires *both* elimination of most of the non-carbon atoms and achieving adequate strength to survive subsequent processing.

Cracking is extensive and multifaceted during carbonization. Some crack formation is beneficial by reducing long range stresses and providing channels for escaping gases that may otherwise reach pressures high enough to cause delamination. Crack formation is also essential to allow channels for later reimpregnation of the component. Unsuitable crack formation, however, can easily destroy the component. Carbonization of a large component may take two weeks in standard manufacturing practice. The resulting component is porous and has a dense network of fine cracks.

After most of the matrix has been converted to carbon, the component is cooled and transferred to another retort to be heated to much higher temperatures intended to produce the desired crack structure and a somewhat graphitic microstructure. This step is called graphitization, although the graphitized matrix achieved from a resin matrix has much less of the graphite-like long range order obtained from pitch. Graphitization temperatures may be as high as 4500°F (2500°C) and the environment is nitrogen at atomospheric pressure. Little further gas evolution occurs. Because the chemical changes are less drastic, the component is somewhat stronger, and stresses are lower, failure is uncommon.

After graphitization, the component is reimpregnated (usually with a different resin or pitch) and the carbonization and graphitization steps repeated. This cycle is often repeated a third time to achieve desired density or strength. The first carbonization step has the greatest mass loss with concomitantly high gas evolution and potentially high internal pressures, the greatest shrinkage and the lowest strength. First carb has, not surprisingly, the highest failure rate, particularly when developing a process for a component with a new size, shape, chemistry or fiber architecture.

Control Goals and Strategy

The primary manufacturing control system goal should be to complete the carbonization process as rapidly as possible while keeping the risk of delamination low. This is a goal (to complete the process) and a constraint (acceptable risk). The process completion goal can be broken into subgoals: 1) remove by heating (nearly all) non-carbon elements of the matrix; 2) achieve acceptable matrix strength by further heating if necessary (the strength first decreases and then increases), and 3) cool to near room temperature.

The IPM control strategy can be most easily grasped with the conceptual aid of idealized sensors. Imagine three idealized sensors that measure, respectively, completion of carbonization, material strength, and risk-of-failure. Imagine further that we have heating and cooling actuators that allow very rapid change of material temperature. With these idealized sensors and actuators, the process goals can be achieved by a fairly simple strategy. The system would start heating the component slowly and increase the heating rate until either the heater's limit was reached or the risk of failure rose to the limit (say 4%), then adjust the heating rate to keep the risk of failure constant until the desired strength and carbonization states were reached. It would then begin to cool slowly and then more rapidly until the risk of failure limit or the cooling capacity limit was reached. This simple strategy can be refined to achieve higher speed and more complex goals, but let us first examine how it can be implemented.

The carbon-carbon IPM control system approaches this ideal strategy by inferring the risk of failure using models, multiple sensors, and heuristic knowledge. It compensates for unavoidable thermal lag by predicting the future risk of failure and planning its future control actions. Because mathematical models for the physical and chemical processes are too complex for optimal inversion, they must be augmented by heuristic reasoning to plan a control path that meets the risk and speed objectives. Real time replanning makes it also possible to choose actions that will improve final material properties.

There are two critical issues: 1) can we avoid delamination if we know the risk of failure is significant? and, 2) can the system actually infer something like a risk of failure from a combination of real sensors, models and heuristics? The answer to the first is that we certainly can in many cases. Process-induced delaminations (distinguished from possible foreign objects or wrinkles introduced by previous steps) originate from continuously changing values of gas pressure, chemical change induced shrinkage, temperature (thermal stress), modulus and matrix strength. Since reaction rates are directly and strongly influenced by temperature, and these, in turn, drastically effect gas evolution, it is clear that intelligent control of temperature and the rate of change of temperature can control the risk of delamination due to gas pressure or thermal stresses. The answer to the second question is also a qualified yes. We will describe in the following how models and sensors can be combined to obtain something like a "risk-of-failure" sensor.

It is less clear at this time whether shrinkage stress or matrix strength can be controlled, although process chemistry and mechanics arguments suggest that it is possible. In some cases, however, the best that an intelligent control system may be able to do is to recognize that the process has failed, terminate in a way that minimizes losses, and perhaps modify subsequent process cycles. The acoustic sensor research described earlier has demonstrated the ability to detect delaminations as they happen, so this capability is within reach.

Materials Process Origin of Risk

We denote by $F_{strength}(\sigma)$ the probability that the strength of the composite is less than σ, and by $p(\sigma)$ the probability that the peak stress is between σ and $\sigma + d\sigma$, related to a probability density $f_{stress}(\sigma)$ by $p(\sigma) = f_{stress}(\sigma)d\sigma$. The probability of failure p_F is related to the overlap of the stress and strength probability distributions, and is given by

$$p_F(t) = \int_0^{\infty} f_{stress}(t, \sigma) F_{strength}(t, \sigma) d\sigma \tag{1}$$

Both distributions change significantly during the process, and the system must have some knowledge of each.

The similarity of the risk issues to those encountered in structural design suggests simplifications. Engineering design practice deals with uncertain strength and variable loads by designing so that the maximum stress σ_{max} is less than some fraction 1/n of the nominal (expected or average) strength S,

$$\sigma_{max} \leq S/n \ . \tag{2}$$

The number n is called the *safety factor*. Some simple probability analyses can show that, since existing carbonization practices succeed most of the time in spite of substantial variability in stresses and strengths, the effective safety factor is now considerably greater than 1 during much of the process. This excess safety factor offers the opportunity for considerable economies. A somewhat more sophisticated approach, also borrowed from engineering design practice, uses the *safety index*, defined by

$$\beta = \frac{\langle (S - \sigma_{max}) \rangle}{\sqrt{\left[\mathrm{var}\, S + \mathrm{var}\, \sigma_{max} \right]}} \tag{3}$$

The quantity in angle brackets is the expected or mean value of strength minus maximum stress. The denominator is the standard deviation of the numerator assuming statistical independence. The safety index, therefore, is the expected value of the failure variable (strength minus stress) measured in units of its own standard deviation. A safety index of three means three standard deviations from failure. This is a more useful measure of risk than the safety factor because it recognizes that risk increases when the uncertainty is large even if the mean values don't change. It is much easier to obtain, however, than the actual probability of failure. Control based on estimated safety *index* offers a practical, semi-quantitative way to substantially increase processing speed without increasing failures.

The strength and stress both change significantly during the process. The strength distribution moves first down[2,3] (by roughly a factor of 10) and then up (by perhaps a factor of four). Stiffness also decreases and then increases.

We expect from experience with the fracture of other brittle materials that the strength distribution can be represented by a Weibull distribution, with both median and width changing as the process proceeds. Fig. 1 illustrates hypothetical stress (normal with standard deviation 25% of mean) and cumulative strength (Weibull with standard deviation 12% of mean) distributions with a safety factor of 3.

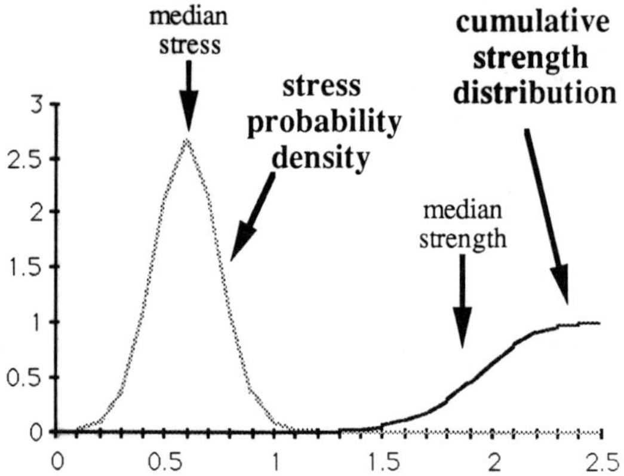

Fig. 1. Qualitative illustration of stress and strength probability distributions at a point during the process.

Carbonization is done very conservatively, now, with large but unknown safety factors during most of the process. These excess safety factors offer significant opportunities for increased speed if the safety index is estimated from sensors and *a priori* knowledge.

The stress arises from 1) gas pressure, 2) matrix shrinkage due to carbonization, and 3) thermal expansion and contraction constrained by the fibers. The most probable stress is, therefore, increasing and decreasing as temperatures and reactions rates change. The contributions to the stress from these sources are additive, but during different phases of the process often one or more of these stresses is negligible. This is a potentially significant simplification. The shape of the distribution also changes as the reaction proceeds. The qualitative effects of the controllable process parameters (temperature, heating rate, duration) on the stress and strength distributions are discussed below.

The statistical or probabilistic character of the *strength* of a composite is familiar. It arises from a variety of microstructural features, such as the size, orientation, and spacing of microcracks, and variations in layup, starting materials, and oven heating that contribute to inhomogeneous carbonization.

Stresses during carbonization have been less studied, and less is known about their variability. Several physical examples suggest that variability is likely to be substantial and unavoidable. Variations in resin density, for example, are common and certainly influence the rate of gas production, just as variations in crack density and morphology influence gas escape. Similar conditions produce statistical variability in shrinkage and thermal stresses. *Process* actions to set temperatures, heating rates and hold times can increase or decrease the variability as well as the mean value of the stress. One expects, for example, that rapid heating increases the *variability* as well as the most probable value of gas pressure induced stress, while a long temperature hold probably decreases both.

Inferring Safety Index

Failure of a brittle material can be usefully separated into three issues: the toughness of the material, the size, shape, and orientation of the most significant defect, and the stress. We have planned a series of experiments to measure these as a function of reaction progress.

The effective matrix toughness changes dramatically during the process. We must, therefore, relate toughness to a state description that can be inferred from sensor data and whose subsequent values can be modeled. A good choice of state description can have very high benefit in improving the reliability of the control system's estimates of toughness. In integrating chemical and mechanical modeling we have exploited the polymer chemists' qualitative knowledge of the relationship between sensed gases, bond type, and bond stiffness and strength. It is likely, in particular, that the observed sharp decrease in matrix strength in the processing regime between about 400°C and 550°C is a consequence of the breaking of methylene bonds to form methane and oxides of carbon, while an alternative reaction path that produces hydrogen appears to strengthen (and stiffen) the matrix. If the measured toughness were parametrized in terms of a single variable of weight loss, it would exhibit a deep minimum, but the location of the minimum would depend on temperature/time history because the changes in path would change the ratio of the two reactions (one of which weakens and the other of which strengthens) for the same weight loss.

The practical consequence is that using one variable would lead to significant scatter in the predicted strength. That is, the toughness is a function of more than one variable. Parametrizing the toughness in terms of a wisely chosen combination of two variables, such as methane and hydrogen evolved, should give much less scatter. (Not just any two variables will do. For example, the present knowledge of the reaction chemistry and its relation to mechanics suggests that the use of methane and carbon-oxides would *not* reduce scatter very well.) Thus our substrategy is to measure the toughness as a function of methane and hydrogen and subsequently use methane and hydrogen evolution to estimate the toughness. *Future* toughness will be predicted by predicting the state from a process model[4] (which also will predict gas evolution), and using this to calculate the strength by the same empirical relation. If we can infer a more detailed state description, such as one that distinguishes aromatic hydrogen[5] from hydrogen in methylene bonds, we can considerable reduce the scatter in toughness predictions, thereby enabling us to perform the process faster and more reliably.

The second failure issue is the size of the largest defect. The largest defects will be estimated from experimental measurements of strength *in situ* with known effective toughness. Those experiments will determine whether the defect distribution changes in important ways during the process. Qualitative studies[6] have shown that the nature of the crack distribution changes considerably by, for example, fiber matrix interface cracking, but there is no evidence yet bearing on changes in the population of potential delamination inducing defects.

The third failure factor is the internal stress. Gas pressure can be calculated from process models and knowledge of the permeability, which will also be measured and parametrized. Shrinkage and thermal stresses can be predicted by extensions of current process models. We are planning experiments to check the model predictions of long range (essentially thermophysical) stresses by monitoring the force on an asymmetrical cantilever beam. Transferring such measurements from the initial laboratory setup to a subseqent process control situation requires, again, a wise choice of independent variables which will make predictions more reliable. The best choice is still an open issue.

Inferring Process Completion

Reaction completion can be inferred from any of several sources: a) time at temperature; b) matrix electrical conductivity (eddy current); c) cumulative and current gas evolution; or d) some combination of the above. Cooldown failures are only due to thermal stress, so temperature distribution monitoring is the most important indicator of risk of failure. When combined with the asymmetrical cantilever beam experiments, temperature measurements seem likely to be a reliable indicator of cooling stresses. The parametrized defect size experiments earlier produce a strength value that can be used to estimate end-process strength, as well as in process. Thus both removal of non-carbon materials and adequacy of strength can be reasonably inferred from the available sensor data and models.

Control Strategy Summary

This completes a sketch of the domain knowledge available for controlling carbonization. It appears to be a complete picture. If the chemical kinetic models are good enough to describe and predict state (especially the fractions of hydroxyl groups, methylene bonds, aromatic carbon, and aromatic hydrogen in the matrix), it is likely that the system will have a basis for material property control as well as accelerating the process and reducing the failure rate. The use of a knowledge based (artificial intelligence) control architecture offers a practical method to improve factory floor productivity on a continuing basis by easy additional of improved sensors and models.

References

1 Kwang Chung, T. Liao, and I. Goldberg, Research on Intelligent Processing of Carbon-Carbon Composites under DARPA/ONR contract N00014-87-C-0724, Rockwell International, First Annual Report, October, 1988.
2 F.I. Clayton, AFWAL/ML contract F33615-83-C-5020, 1987.
3 Z. Lausevic and S. Marinkovic, "Mechanical Properties and Chemistry of Carbonization of Phenol Formaldehyde Resin", Carbon, 24, pp575-580, 1986.
4 W.C. Loomis and R. b. Dirling, Jr., "Development of a Process Environment Model for 2D Carbon-Carbon Composites", Supported by AFWAL/MLBC under F33615-83-C-5042, Ms. Frances Abrams, AFML Project Engineer, and subsequently by NASA/MSFC contract NAS8-36294, Ms. Louise Semmel, NASA Project Engineer.
5 It appears that an infrared reflectance sensor could recognize aromatic carbon directly.
6 James Bulau, Research on Intelligent Processing of Carbon-Carbon Composites under DARPA/ONR contract N00014-87-C-0724, Rockwell International, First Annual Report, October, 1988.

MANUFACTURING OF METAL-CERAMICS COMPOSITES

BY MASHY-STATE PROCESSINGS

M. Kiuchi and S. Sugiyama

Institute of Industrial Science
University of Tokyo
7-22-1, Roppongi, Minato-ku, Tokyo, Japan 106

Abstract

New metal working processes, such as the mashy-state mixing, mashy-state forging, mashy-state rolling and mashy-state extrusion are developed and applied to manufacture the so-called particle reinforced metals, fiber reinforced metals, particle reinforced cladding metals and so forth. These metal-ceramics composites are made of metal sheets, metal powders, ceramics particles and ceramics fibers. The mashy-state processes are also used to work those composites into bars, wires, tubes, plates, sheets, machine parts and other products. The working conditions necessary to manufacture sound composite products are clarified and the mechanical properties of the products are investigated. Through the investigation, it becomes clear that the mashy-state processes are useful and effective to manufacture metal-ceramics composites. It is also known that these processes may be applied in various ways to development of new metal matrix compounds.

Metallurgical Processes for the Year 2000 and Beyond
Edited by H.Y. Sohn and E.S. Geskin
The Minerals, Metals & Materials Society, 1988

Introduction

The so-called metal-ceramics composites and processes to manufacture them are now being widely investigated in order to comply with increasing demands for new materials with novel and excellent mechanical properties and functions. The metal-ceramics composites are expected to have superior characteristics with respect to the elastic modulus, hardness, anti-wear resistance, high temperature strength and other mechanical properties to those of conventional metallic materials.

The present authors have been investigating into new metal working processes which are based upon the characteristics of metals and their alloys in the mashy-state. In the mashy-state, the metal (or alloy) includes both of the solid and liquid components. According to the fact that the partial melting takes place and the liquid component usually exists at grain boundaries, the metal in the mashy-state has quite different mechanical properties from those of the metal in the solid-state. Because of that the bonding force among grains is very weak due to the partial melting occurred at grain boundaries, the mashy metal has distinctively low flow-stress and high deformability. The mashy metal splits into grains or particles and sticks to others very easily. It is so soft as to be stirred up just like a lump of sherbet, but it is sufficiently viscous to be mixed up with other materials, such as ceramics particles and ceramics short-cut fibers.

By making use of these characteristics, new metal working processes, such as the mashy-state extrusion, mashy-state forging and mashy-state rolling have been developed. Through a series of investigations, it has become clear that these processes are useful for not only manufacturing metallic bars, wires, tubes and sheets, but also manufacturing various types of metal-ceramics composites. These processes are also useful to work those composites into bars, wires, tubes, sheets, machine parts and other near-net shape products /1/~/8/.

In this paper, the basic concepts of those processes, their characteristics and the mechanical properties of the manufactured products are explained.

Mashy-State Extrusion of Wires, Bars and Tubes

By utilizing the mechanical properties of the mashy metal, especially its distinctively low flow stress, the extrusion of wires, bars and tubes from billets heated up to the mashy-state can be performed very naturally.

The out-line of the mashy-state extrusion is as follows. The billet is heated up to the mashy-state by the furnace. When the billet is heated, if the weight percent of liquid component is less than 20~30%, that means the solid fraction φ is higher than 70~80%, the billet is able to keep its initial shape and dimension in the furnace. When the liquid component is included more than 20~30% in the billet, it may deform under the effect of the gravity force. Therefore, a kind of container is necessary to hold the billet. After the scheduled mashy-state of the billet is attained, it is transfered to the extrusion press. The billet is inserted to the container of the press and is extruded into the required product. During the process, the billet should be protected from cooling and should be kept at the scheduled mashy-state. In order to keep the mashy-state, the container, die and punch need to be preheated. The necessary pre-heating temperature depends upon the heat capacity of the billet, container, die, punch, ram and other tools.

The characteristics of the mashy-state extrusion are summerized as follows.

(1) Because of the distinctively low flow stress of the billet, the extrusion pressure is very low comparing with that for the conventional hot extrusion. The extrusion pressure for the mashy-state extrusion is about 1/4~1/5 of that for the hot extrusion.

(2) Due to the low extrusion pressure, very high reduction, that is very high extrusion ratio, is attainable through single pass extrusion.

(3) The liquid component of the billet works as a lubricant in the container and die.

(4) Owing to the lubrication effect of the liquid component and the low extrusion pressure, products with complicated cross-section and thin wall-thickness can be extruded rather easily.

(5) Materials with poor deformability can be extruded easily owing to the liquid components in mashy billets.

Fig. 1 shows some examples of extruded wires, bars and tubes of Al-7075. Their surfaces and internal structures are sound and no defects are observed. The products manufactured by the mashy-state extrusion are usually rather soft and their yield stresses are lower than those of hot-extruded products. On the other hand, they have adequate deformability comparing with hot-extruded products.

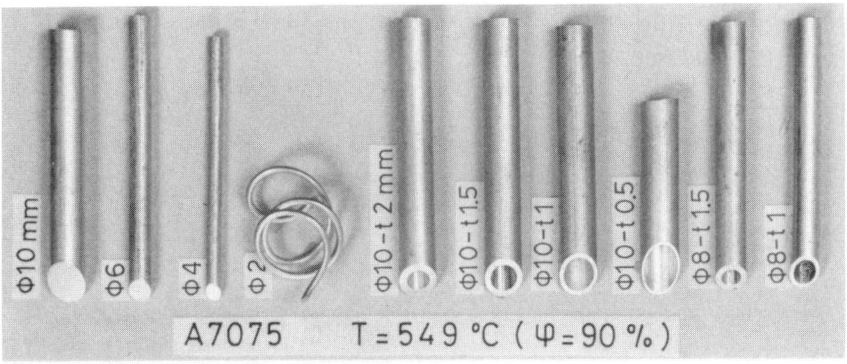

Fig. 1 Some examples of wires, bars and tubes manufactured by the mashy-state extrusion

Mashy-State Rolling of Sheet Metal

Fig. 2 shows a schematic illustration of the mashy-state rolling. Before the rolling, the mashy sheet consists of the solid and liquid components. At the entrance of roll-gap, the liquid component in the mashy sheet is apt to flow to the top and bottom surfaces under the effect of the pressure induced at the roll gap. But the liquid component does not flow out of the mashy sheet. It is cooled by the roll and solidifies and pulled into the roll-gap. On the other hand, the solid component, that consists of grains, is compressed, deformed and drawn into the roll-gap. At the roll-gap, each grain is deformed and elongated. Before the mashy sheet comes to the exit of roll-gap, the liquid component usually finishes its solidification. In some cases, however, the rolled sheet may include the liquid component due to uncompleteness of solidification, but the rate of the residual liquid component should be less than a few weight percent.

Fig. 3 shows some examples of internal structures observed in Al-alloy sheets manufactured by the mashy-state rolling. The solid fraction φ of the mashy sheet before rolling, thickness reduction, rolling speed and other working conditions are shown in the figure. The roll surface temperature is kept at the room temperature before the rolling, therefore, the mashy sheet is rolled and cooled simultaneously under rather high cooling rate.

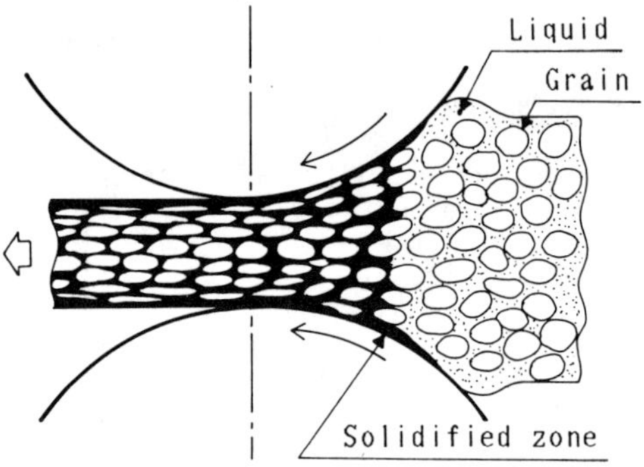

Fig. 2 Schematic illustration of the mashy-state rolling

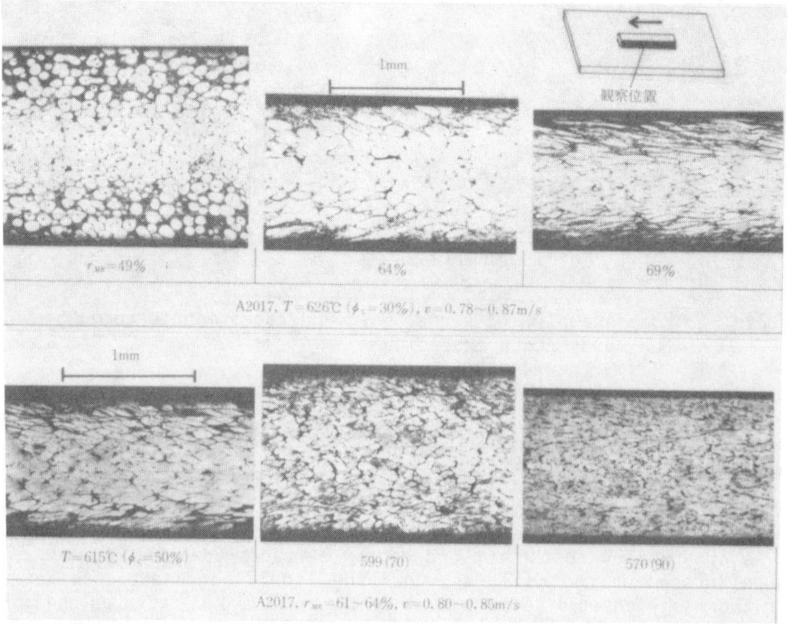

Fig. 3 Internal structures of sheets manufactured of the mashy -state rolling

From the figure, the following results are known. When the mashy sheet includes adequate amount of the liquid component before the rolling and the thickness reduction given to the mashy sheet is not so high, the solid component, that is the solid grain, does not deform at the roll-gap. Almost all grains keep their initial geometry during the mashy-state rolling. It looks like that the grains are wafting in the liquid and they go through the roll-gap like the so-called slurrey without any deformation. But, of course, the liquid component solidifies during it passes through the roll-gap.

When the thickness reduction or the solid fraction of the mashy sheet is high enough, the grains in the mashy sheet are not able to pass the roll-gap without contacting with each other and deforming themselves. The grains restrict each other and are forced to deform and elongate to pass through the roll-gap. Thus each grain is elongated by the rolls and a kind of fiber structure is formed in the rolled sheet. As the thickness reduction and the solid fraction of the mashy sheet increases, the elongation of the grains is more and more promoted and a typical fiber structure is observed in the rolled sheet.

It should be noticed that the rolled sheet has not always homogenious internal structure. When the mashy sheet includes adequate amount of the liquid component before the rolling, the liquid component comes to the surface zone and solidifies and consequently the grains gather to the inner zone. Therefore, sometimes, the rolled sheet has two different structures at the surface zone and the inner zone, respectively. This kind of inhomogenuity gives distributions of mechanical properties in the thickness direction.

Process to Manufacture Particle Reinforced Metals (PRM)

A schematic diagram of the process to manufacture the so-called particle reinforced metals (PRM) is shown in Fig. 4. Each step of the process is based upon the utilization of the above mentioned characteristics of the mashy metal. The outline of the process is as follows.

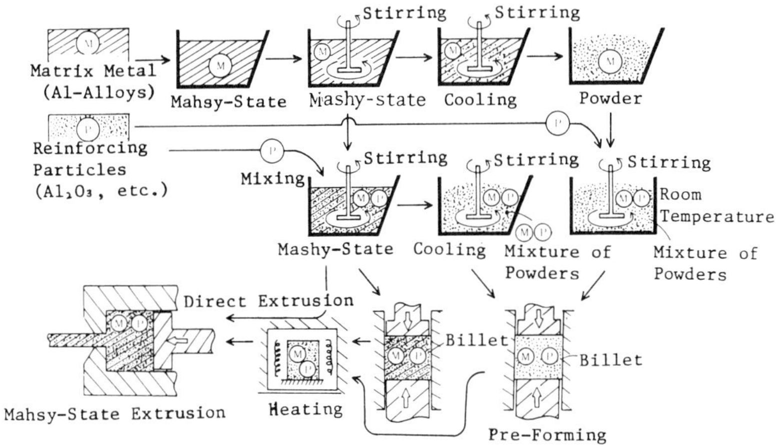

Fig. 4 Schematic diagram of the mashy-state processing to manufacture particle reinforced metals and products

(1) The metal matrix is heated up to the aimed temperature which corresponds to the scheduled mashy-state. At the temperature, the solid and liquid components co-exist in the metal matrix and the weight rate of the solid component to the total weight of metal matrix, that is defined as the solid fraction φ, is kept constant. When the solid fraction φ is less than the critical value, it becomes possible to stir up the metal matrix. In this state, the reinforcing particles, that are the employed ceramics particles, are mixed up with the metal matrix.

(2) The mashy mixture of the metal matrix and the reinforcing particles is transfered directly to the working machine, such as the extrusion press, forging press and rolling mill.

(3) According to circumstances, the mashy mixture is cooled to the solid mixture. Then it is preformed into the composite billet. The composite billet is reheated to the mashy-state and supplied to the secondary mashy-state processes.

(4) The mashy mixture or the reheated mashy composite billet is worked into bars, wires, tubes, sheets or machine parts by the mashy-state extrusion, mashy-state forging or mashy-state rolling.

(5) The particle reinforced metal, which is usually very hard and brittle and has poor deformability in the solid-state, has distinctively low flow stress and high deformability in the mashy-state, therefore, it can be worked into required products very easily.

(6) Instead of the mixture made in the mashy-state, the mixture of the metal powder and the reinforcing particles is also capable of being supplied to the secondary mashy-state working processes. In this case, the mixture is preformed into the billet and the billet is heated up to the aimed mashy-state. Then it is transfered to the working machine and formed into the product.

Mashy-State Extrusion of Particle Reinforced Metals (PRM)

The mashy-state extrusion of particle reinforced metals are carried out as follows. In the following cases, aluminum alloys, such as A-7075, A-5056, A-5052, A-2011 and others, are employed as the metal matrix and Al_2O_3, SiC and SiN particles are used as the reinforcing particle. The billets are preformed out of the mixture of metal matrix and ceramics particles made in the mashy-state or the mixture of metal powder and ceramics particles. The diameter of billet is ϕ40mm and the diameters of extruded bars are ϕ10, ϕ8, ϕ6 and ϕ4mm. The solid fraction φ, that is the weight percent of the solid component, of the metal matrix of billet in the container is varied in the range from 0% to 90%. The volume fraction denoted by V_p of the reinforcing particles in the billet is varied in the range from 0% to 50%.

Fig. 5 shows some extruded bars and wires of the particle reinforced metal. The surface of each bar or wire is very smooth and no defect is observed on it. Needless to say, the suitable geometry of the die, the appropriate solid fraction φ of the metal matrix in the billet and the appropriate temperature of the die are essentially necessary to get sound extruded products. For instance, the suitable solid fraction φ for the billet with high V_p is lower than that for the billet with low V_p. When V_p is fixed, the appropriate solid fraction φ for the billet including small reinforcing particles is lower than that for the billet including large reinforcing particles.

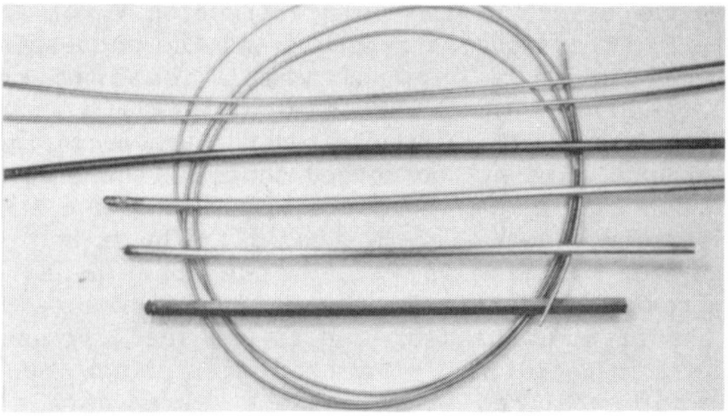

Fig. 5 Some examples of the extruded bars and wires of particle reinforced metals

These results mean that the liquid component in the billet, that is the liquid component included in the mashy mixture, plays a dominant role to promote smooth flow of the mashy mixture through the die. In other words, the liquid component is indispensable for smooth metel flow and smooth relative slip between metal matrix and ceramics particles in the die.

Fig. 6 shows a typical internal structure of the extruded composite bar. Its composition is ⟨(A-5056)+Al_2O_3⟩. In the pictures, the reinforcing particles fixed in the metal matrix are observed clearly. No cavity between metal matrix and reinforcing particle is observed.

Fig. 7 shows the flow-stress of the extruded composite bar of ⟨(A-7075)+Al_2O_3⟩, measured by the uni-axial compression test. The abscissa shows the tested temperature. The ordinate shows the compressive flow-stress at 10% strain. Each curve in the figure corresponds to the varied volume fraction V_p of Al_2O_3 particles. It is clear that the flow-stress of the composite increases according to the increase in V_p and the flow-stress of the composite with V_p=30% is almost twice of that of the metal matrix at every tested temperature.

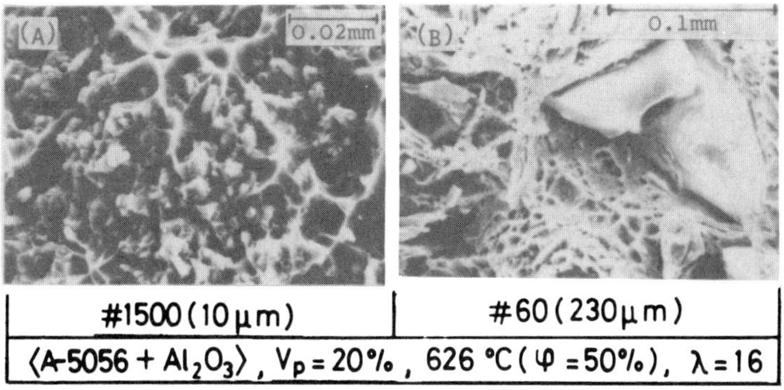

Fig. 6 Internal structures of the extruded bars of particle reinforced metals

Fig. 8 shows the effect of the volume fraction V_p of Al_2O_3 particles on the elongation E_ℓ, the impact strength E and the anti-wear resistance M of the extruded composite bars of which compositions are ⟨(A-7075)+Al_2O_3⟩, ⟨(A-5056)+Al_2O_3⟩ and ⟨(A-2011)+Al_2O_3⟩, respectively. Here, the anti-wear resistance is evaluated through measuring the specific amount of wear per unit time by the method shown in the figure. From these results, it is known that as V_p increases, the elongation E_ℓ decreases and it becomes almost zero when V_p reaches to 30%. The impact strength also decreases due to the increase in V_p. On the other hand, the anti-wear resistance is improved distinctively through the increase in V_p of Al_2O_3 particles. In the figure, the decrease in the specific amount of wear M due to the increase in V_p of Al_2O_3 particles is shown clearly. When V_p gets to 20%, the anti-wear resistance of the tested particle reinforced aluminum alloys becomes almost equal to that of stainless steel SUS316 or high carbon steel S45C.

The results explained in this chapter indicate the fact that the particle reinforced metals, even though they are very hard and brittle in the solid-state and the cold or hot forming of them is very difficult, become very soft and formable in the mashy-state and are able to be formed into various products without any difficulty. This is one of the important advantages of the mashy-state processes.

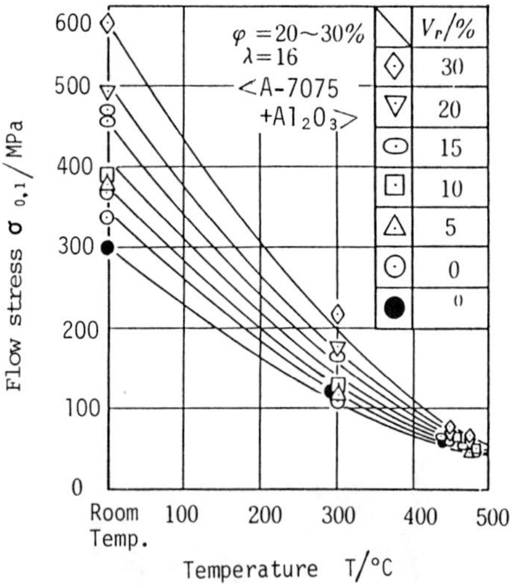

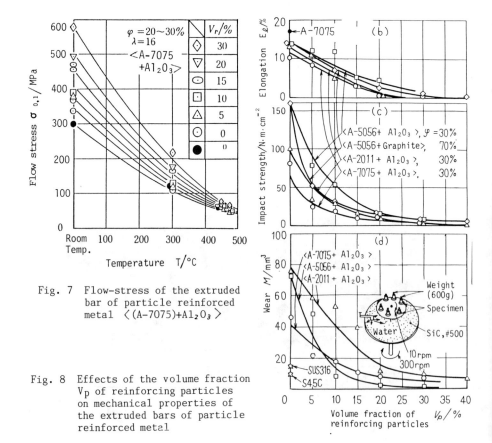

Fig. 7 Flow-stress of the extruded bar of particle reinforced metal ⟨(A-7075)+Al_2O_3⟩

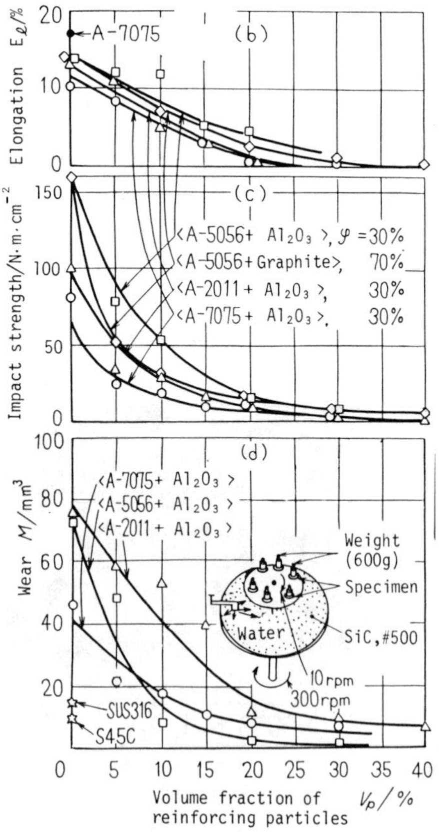

Fig. 8 Effects of the volume fraction V_p of reinforcing particles on mechanical properties of the extruded bars of particle reinforced metal

Process to Manufacture Particle Reinforced Cladding Metals

Here, the particle reinforced cladding metal means the metal matrix compound with two layers made of the base metal and the particle reinforced metal laminated on it.

A schematic illustration of the process to manufacture the particle reinforced cladding metals by means of the mashy-state bonding and forging is shown in Fig. 9. The outline of the process is as follows.

(1) The mashy mixture of metal matrix and reinforcing particles made by the mashy-state mixing is preformed into plates or sheets by the mashy-state forging. The preliminary bonding between metal matrix and reinforcing particles is performed in this preforming process.

(2) According to circumstances, the mixture of metal powder and reinforcing particles is employed and it is also preformed into plates or sheets by the hot or cold forging. In this case, the preforming is carried out just for the compaction of the mixture.

(3) The preformed composite plate or sheet, that is the particle reinforced metal, is heated up to the mashy-state and laminated on the heated base metal by the mashy-state forging or upsetting. When the lamination is carried out, in order to attain satisfactory bonding between base metal and composite plate or sheet, the base metal should be heated up to the appropriate temperature and, if necessary, it should be also heated up to the mashy-state.

(4) In the modified process, the mixture of metal powder and reinforcing particles is directly piled up on the surface of the base metal and they are subjected together to the heating and afterwards to the mashy-state forging or upsetting. In this case, the bonding between metal matrix and reinforcing particle and the bonding between base metal and composite lamina are performed at the same time.

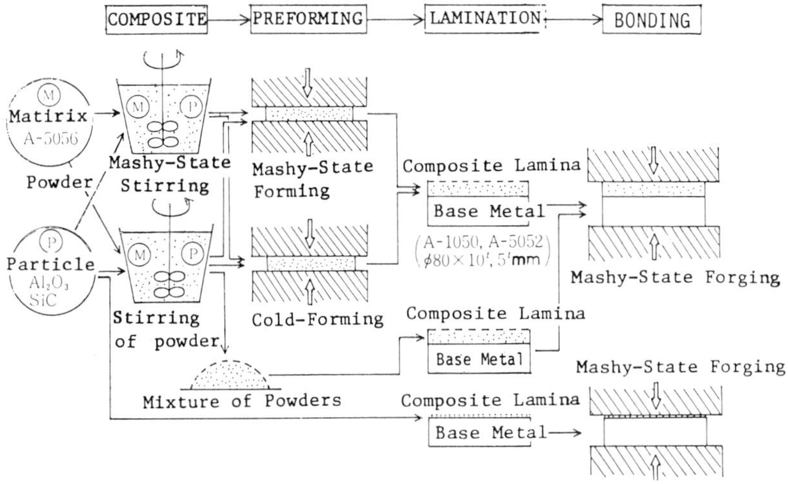

Fig. 9 Schematic illustration of the mashy-state processing to manufacture particle reinforced cladding metals

(5) In another modified process, the reinforcing particles are directly embedded to the surface layer of the base metal by the mashy-state forging or upsetting. In this process, the base metal should be heated up to the mashy-state and the reinforcing particles piled up on the base metal are pressed by the tool to penetrate into the liquid component of the base metal. After solidification of the liquid component, the reinforcing particles are fixed in the surface layer of the base metal. Thus the surface layer is made into the composite lamina, that is the particle reinforced metal.

(6) In the particle reinforced cladding metals made by these processes, the base metals keep their original ductility, therefore, the manufactured cladding metals usually have good deformability. They are capable of being formed into products with various shapes by conventional hot and/or cold plastic working processes.

Particle Reinforced Cladding Metals Manufactured by Mashy-State Forging

Fig. 10 shows a product of particle reinforced cladding metal manufactured by the mashy-state forging. The external appearance of its cross-section shows two layers. The upper black layer is the composite lamina, that is the particle reinforced metal, and the lower grey layer is the base metal. The metal matrix of composite lamina is A-5056, the reinforcing particle is Al_2O_3(#1500) and the base metal is A-5052. The mashy-state forging conditions are shown in the figure.

Fig. 11 shows the effects of the forging temperature T and the forging pressure p on the internal structure of the forged cladding metal. Here, the metal matrix is A-5056 and the base metal is A-5052. The reinforcing particle is Al_2O_3(#1500). The mashy-state forging conditions are shown in the figure.

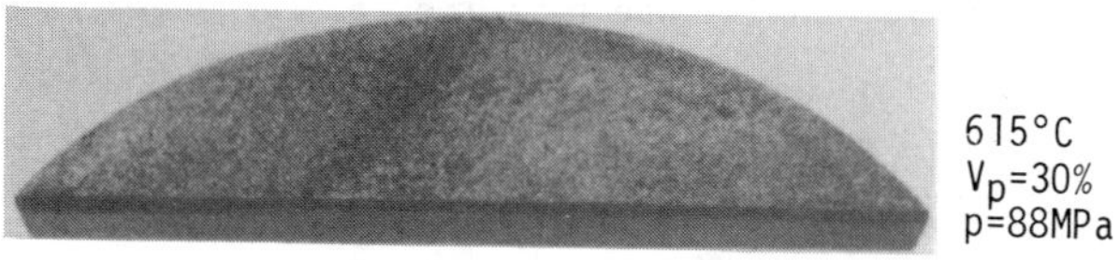

Fig. 10 A product of particle reinforced cladding metal manufactured by the mashy-state forging

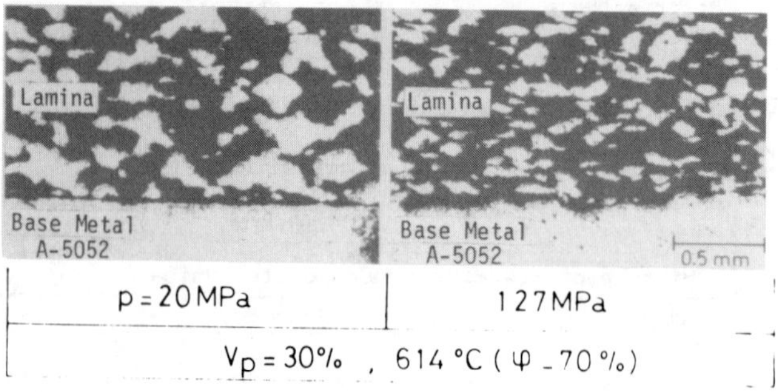

Fig. 11 Effects of the forging temperature and the forging pressure on the internal structure of the forged particle reinforced cladding metals

In the mashy-state forging, as mentioned previously, the reinforcing particles are wrapped and fixed by the liquid component. Therefore, a sufficient amount of the liquid component and high enough hydrostatic pressure are essentially necessary to wrap and fix all of the reinforcing particles firmly. In this sense, the internal stractures of composite laminae shown at the right hand of the figure are better than those shown at the left hand.

In the internal structures shown at the right hand of the figure, it is found that at the boundary between composite lamina and base metal, the liquid components of the metal matrix and the base metal penetrate into each other and solidifies. Thus the very firm bonding between composite lamina and base metal is attained. Because of the mechanism of bonding based on mutual penetration of the liquid components, a sufficient amount of the liquid component is necessary to get enough bonding strength between composite lamina and base metal.

As mentioned above, the forging pressure p is another important factor to get sound internal structure of the product. The high enough pressure is indispensable to make the liquid component fill up every gap around reinforcing particles in the composite lamina and to make both of the liquid components of the metal matrix and base metal penetrate into each other at the boundary between them.

Fig. 12 shows the internal structure of the surface layer of base metal to which Al_2O_3 particles are directly embedded by the mashy-state forging. Here, the base metal is A-5052. As shown in the figure, Al_2O_3 particles are fixed firmly by the base metal. The surface layer has been made into the particle reinforced composite lamina and it includes Al_2O_3 particles with high volume fraction.

The mechanical properties of the composite lamina cladded on the base metal and the particle reinforced surface layer of the base metal are same with those of the particle reinforced metal explained in the previous

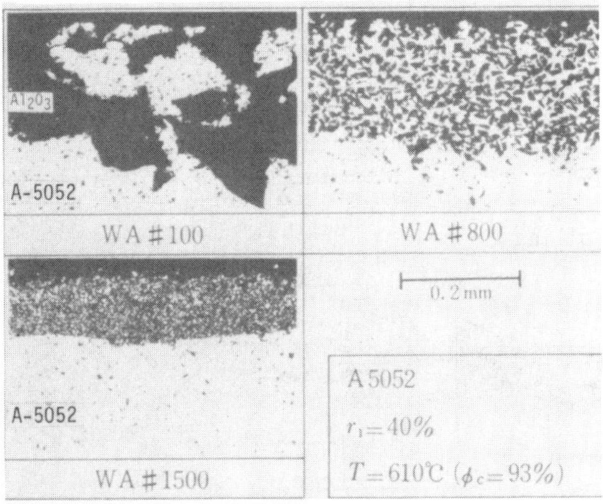

Fig. 12 Internal structure of the surface layer of the metal matrix reinforced by directly embedded Al_2O_3 particles

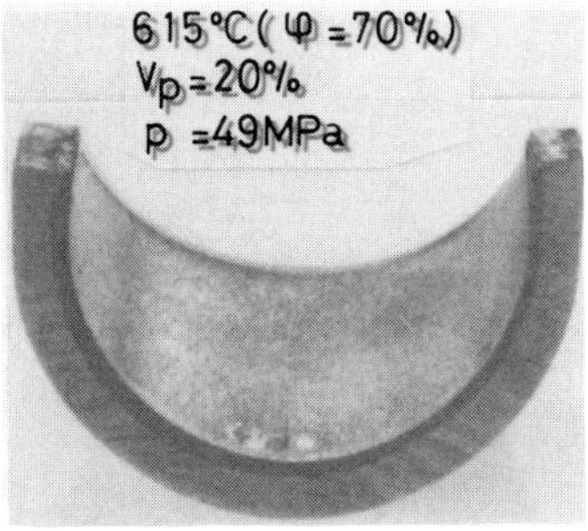

Fig. 13 A product of the particle reinforced cladding metal

chapter. Their hardness H_V naturally increases through the increase in the volume fraction V_p of Al_2O_3 particles. H_V also increases according to the increase in the forging pressure p. This is due to the fact that when the liquid component wraps and fixes Al_2O_3 particles, the high forging pressure promotes firm contact between liquid component of metal matrix and surfaces of Al_2O_3 particles and consequently, the bonding strength between them is improved.

Fig. 13 shows a product made of a particle reinforced cladding metal (a flat plate) by means of the cold-bending. The base metal and the metal matrix are A-5056. The reinforcing particle is Al_2O_3. It is clear that the cold-bending has been performed successfully, even though the composite lamina is very hard and brittle. This cold deformability is one of important and advantageous mechanical properties of the particle reinforced cladding metals, manufactured by the present process.

Particle Reinforced Composite Sheets Manufactured by Mashy-State Rolling

A schematic illustration of the process to manufacture particle reinforced composite sheets is shown in Fig. 14. In this process, at the beginning, the mixture of metal powder (metal matrix) and ceramics particles (reinforcing particles) is piled up uniformly on the mother sheet. Then they are heated up together to the aimed temperature. At the temperature, the raw composite lamina, that is the mixture of metal powder and ceramics particles, includes the liquid component due to partial melting of the metal powder.

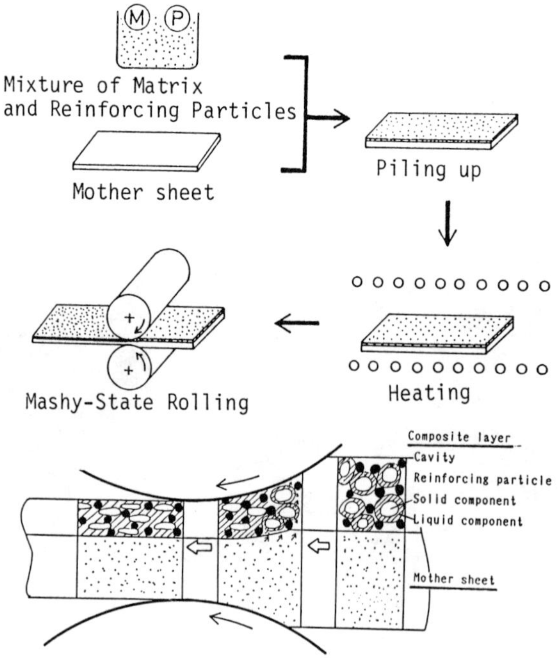

Fig. 14 Schematic illustration of the mashy-state rolling to manufacture particle reinforced composite sheets

According to circumstances, the mother sheet also includes the liquid component due to partial melting of itself. The solid fraction φ of the raw composite lamina or the mother sheet naturally depends on the temperature. The appropriate temperature should be chosen so that the following mashy-state rolling can be performed stably.

The heated mother sheet and raw composite lamina are rolled and bonded into an unified composite cladding sheet. In order to get high bonding strength between composite lamina and mother sheet and to get the sound internal structure of the composite lamina, the mashy-state rolling should be carried out by using rolls which have suitable diameters and appropriate surface temperature. The appropriate rolling speed and the suitable thickness reduction are also necessary.

In the modified process, the reinforcing particles are directly embedded into the surface layer of mother sheet by the mashy-state rolling. In this process, the mother sheet should be heated up to the mashy-state and the reinforcing particles are embedded into the liquid component of the mother sheet by the roll. By the direct embedding, the surface layer of the mother sheet is made into the particle reinforced composite lamina which usually has higher volume fraction of reinforcing particles than that of the composite lamina made of metal powder and ceramics particles.

The internal structures and mechanical properties of composite sheets manufactured by the mashy-state rolling are similar to those of particle reinforced cladding metals made by the mashy-state forging. The surface of mother sheet is reinforced by the composite lamina with respect to hardness, anti-wear resistance and high temperature strength. On the other hand, the mother sheet reinforces the composite lamina with respect to toughness or impact strength and it saves deformability of the composite sheet.

Fig. 15 shows some examples of internal structures of composite sheets manufactured by the mashy-state rolling. In these cases, the mother sheet is stainless steel (SUS304). The matrix of the composite lamina is the mixture of cast iron powder and deoxidized iron powder. The reinforcing particle is Al_2O_3 and its average size is varied for each case. The volume fraction V_p of Al_2O_3 particles in the composite lamina, the thickness reduction r and the rolling temperature T are shown in the figure. From the figure, it is known that the reinforcing particles are fixed firmly in the composite lamina. No cavity is observed.

In order to get sound internal structure of the composite lamina, sufficient amount of the liquid component and high enough hydrostatic pressure are necessary at the roll-gap. The amount of the liquid component and the hydrostatic pressure necessary for the satisfactory bonding between matrix and reinforcing particle and the sound bonding between composite lamina and mother sheet depend upon the size of employed reinforcing particle, its volume fraction V_p, thickness of composite lamina, composition of metal matrix, composition of mother sheet, diameter of roll and rolling speed.

Fig. 16 shows the mashy-state rolling conditions necessary to get sound internal structure of the composite lamina and satisfactory bonding strength between mother sheet and composite lamina. These conditions are for the cases when the mother sheet is SUS304, the matrix is the mixture of cast iron and deoxidized iron and the reinforcing particle is Al_2O_3. The abscissa shows the volume fraction V_p of Al_2O_3 in the composite lamina. The ordinate shows the thickness reduction r necessary to get sound internal structure and bonding strength.

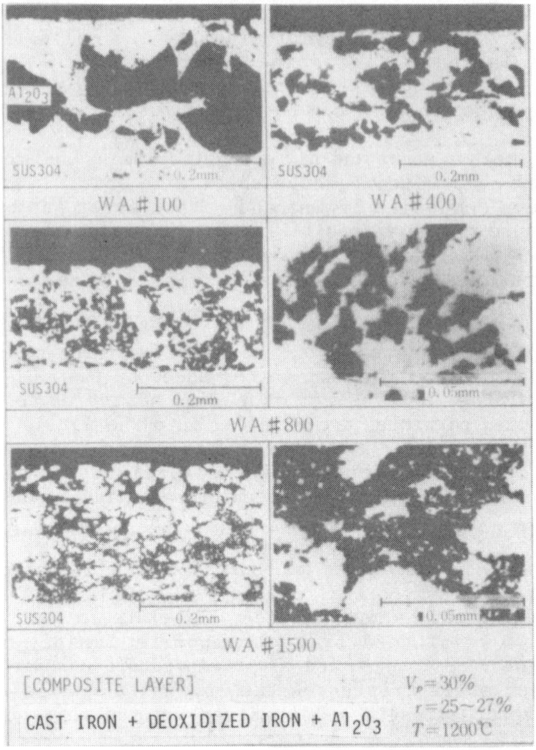

Fig. 15 Internal structures of the composite sheets manufactured by the mashy-state rolling

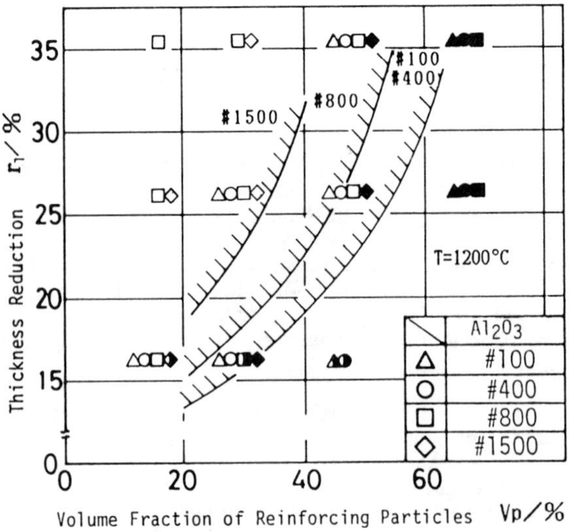

Fig. 16 Mashy-state rolling conditions necessary to manufacture sound composite sheets

From these results, it is known that the composite sheet which has high volume fraction of small reinforcing particles needs larger thickness reduction, that means higher hydrostatic pressure at the roll gap, than those for the composite sheet with low volume fraction of large reinforcing particles. Similar results necessary to get sound composite sheets are obtained for various combinations of mother sheets and composite laminae.

Fig. 17 shows some of composite sheets manufactured by this process and the welded pipes made of these sheets. The mother sheet is SUS304. The matrix of composite lamina is cast iron and deoxidized iron. The reinforcing particle is Al_2O_3. The internal structure of composite lamina on the inside surface of pipe is also shown in the figure. The composite sheets were formed into the pipes by the so-called U-O bending process. After the bending, the both edges were welded. In the bending process, no trouble such as fracture of the composite lamina took place. The outside and inside surfaces of pipes are very smooth and their qualities are satisfactory.

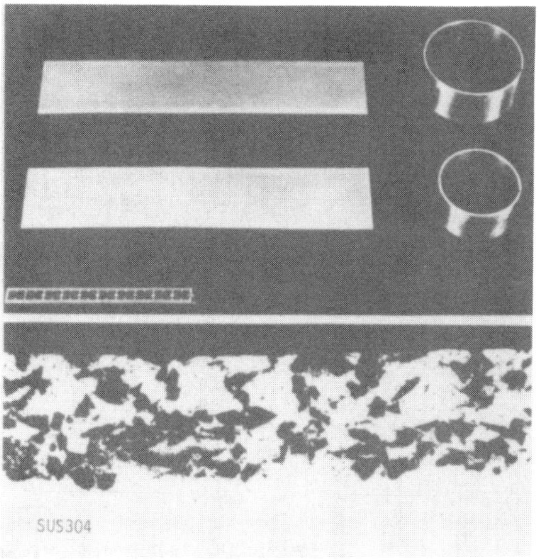

♯100 V_p 30% r 25~27% T 1200°C

[COMPOSITE LAYER] :

CAST IRON + DEOXIDIZED IRON + Al_2O_3

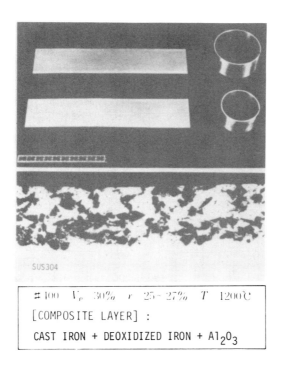

Fig. 17 Composite sheets manufactured by the mashy-state rolling and the welded pipes made of them

Process to Manufacture Fiber Reinforced Metals (FRM)

In this chapter, the mashy-state process, which aims to manufacture metal matrix compounds reinforced by the so-called short-cut ceramics fibers, is explained.

The outline of the process is shown in Fig. 18. In the following cases, the aspect ratio of the employed ceramics fibers is in the range from 100 to 300. The fibers of which aspect ratio is larger than 80$\sim$100 are usually tangled. Therefore, at the first step, the ceramics fibers should be separated in a suitable viscous solvent. In order to perform separation, the ceramics fibers are put into the solvent which is being stirred. Through continuation of stirring, the ceramics fibers are separated one by one. The viscosity of the solvent and the speed of stirring should be adjusted so as to promote effective separation and to prevent breakage of the fibers. The appropriate viscosity and speed of stirring depends upon the dimension and property of the employed ceramics fibers.

After the separation is finished, the metal powder used for metal matrix is added to the same solvent and mixed with the separated ceramics fibers through stirring. Thus the mixture of metal matrix and reinforcing fibers is obtained. The mixture is a kind of slurry and is able to flow easily.

Then the mixture is extruded through a specially designed die into bars. The extruded bar is naturally very soft. If the solvent is excessive, it deforms under the effect of the gravity force. Therefore, according to circumstances, a kind of container is necessary to hold the extruded bar. Through the extrusion, the ceramics fibers in the bar are oriented in the longitudinal direction. In order to attain the regular orientation and to prevent breakage of the fibers, the suitable geometry of die and the appropriate extrusion speed are necessary.

At the next step, the extruded bars are heated in order to get rid of the solvent. Through the heating, the solvent evaporates and the raw composite bars consisting of the metal powder and the oriented ceramics fibers are obtained. The raw composite bars thus obtained are dry and porous.

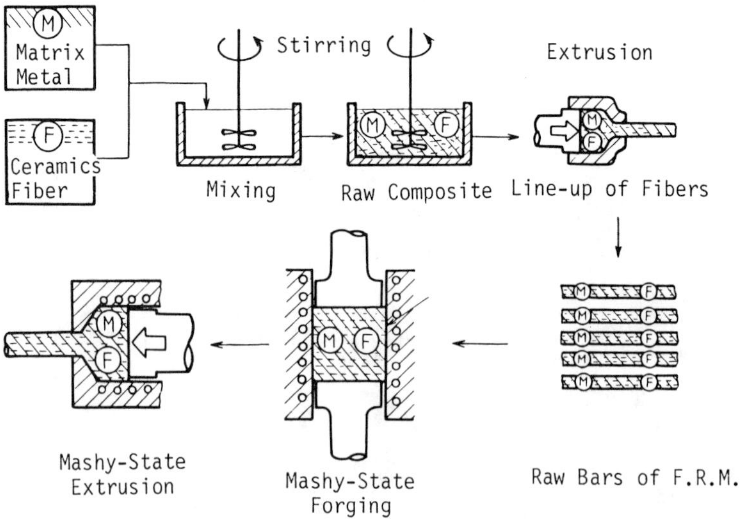

Fig. 18 Schematic diagram of the mashy-state processing to manufacture fiber reinforced metals and products

After getting rid of the solvent, appropriate number of raw composite bars are put together in a die and heated up to the mashy-state and forged into an unified fiber reinforced composite product.

In the mashy-state forging, the partial melting of metal powder, that is the metal matrix, takes place and the liquid component flows and occupies cavities among metal matrix and ceramics fibers completely. Through solidification of the liquid component, the ceramics fibers are fixed firmly in the forged composite product. If dies with appropriate shapes and dimensions corresponding to required products are used, various kind of machine parts of the fiber reinforced metal are obtained.

When wires, bars and tubes of the fiber reinforced metal are requested, the mashy-state extrusion is carried out by employing composite billets made by the above-mentioned mashy-state forging. In this case, the forged composite billet is heated up to the mashy-state and transfered to the extrusion press and extruded. It should be noticed that the forged composite billets are necessary to get sound products. The forged billet has no cavity in it and the full density is essentially necessary to keep the orientation of reinforcing fibers and to prevent breakage of fibers. In addition, the liquid component of metal matrix in the billet plays very important role to attain smooth flow of the fibers. The liquid component reduces the shear resistance between metal matrix and reinforcing fiber and is effective to prevent breakage of the fibers.

Fig. 19 shows a forged billet and extruded bars of the fiber reinforced metal. The metal matrix is A-5056 and the reinforcing fibers are SiC short-cut fibers of which diameter is $15\mu m$ and aspect ratio is around 200. The density of the billet is about 100%. The surfaces of the extruded bars are very smooth and no defect is observed.

Fig. 20 shows an example of internal structures observed on cross-sections of the extruded bars. It is clear that all fibers are oriented in the longitudinal direction and there are no cavities between metal matrix and reinforcing fiber.

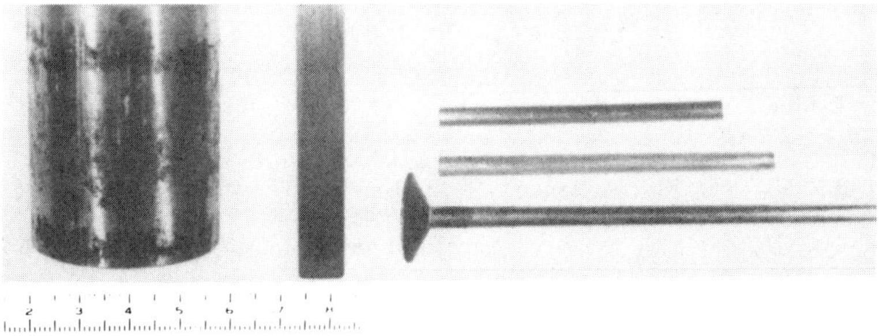

Fig. 19 Billets and extruded bars of the fiber reinforced metal
<(A-5056)+SiC>

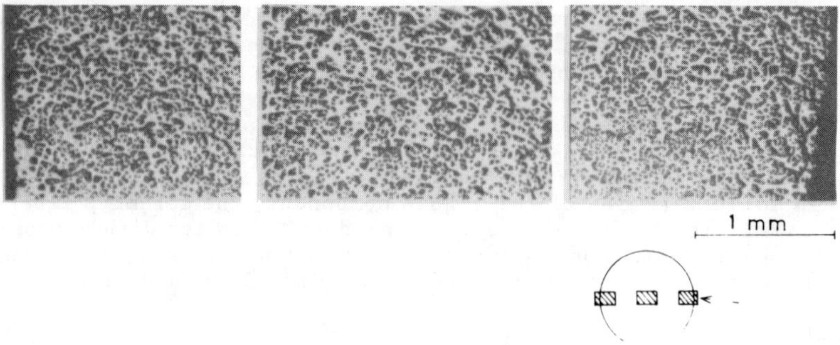

Fig. 20 Internal structure of a extruded fiber reinforced bar 〈(A-5056)+SiC〉

Conclusion

By utilizing the characteristics of metals in the mashy-state, the new processes to manufacture various kind of metal-ceramics composites have been developed. The particle reinforced metals, the particle reinforced cladding metals, the particle reinforced composite sheets, the fiber reinforced metals, and wires, bars and tubes made of them have been successfully manufactured. The necessary working conditions to get sound products and the properties of those products have been systematically investigated. Through a series of investigations, it has become clear that the mashy-state processings are very useful and effective for development of various types of metal matrix compounds.

References

/1/ M. KIUCHI, et. al., Proc. of the 20th International MTDR Conference (1979), 71

/1'/ M. KIUCHI, et. al., Proc. of the 20th International MTDR Conference (1979), 79

/2/ M. KIUCHI, et. al., J. of Japan Society for Technology of Plasticity (JSTP), 23-230 (1982), 915

/3/ M. KIUCHI, et. al., J. of JSTP, 24-272 (1983), 974

/4/ M. KIUCHI, et. al., J. of JSTP, 24-274 (1983), 1113

/5/ M. KIUCHI, et. al., Mechanical Behavior of Materials-IV, Pergamon Press (1984), 1013

/6/ M. KIUCHI, et. al., Annals of the CIRP, Vol. 36-1 (1987), 173

/7/ M. KIUCHI, et. al., Advanced Technology of Plasticity. Vol II (1987), 753

/8/ M. KIUCHI, et. al., Proc. of the 14th NAMRC (1986), 359

SYNTHESIS OF NONEQUILIBRIUM MATERIALS BY

LASER SURFACE MODIFICATION

J. Mazumder and A. Kar

Laser Aided Materials Process Lab
Department Mechanical & Industrial Engineering
University of Illinois
1206 W. Green Street
Urbana, IL 61801

Abstract

Inherent rapid cooling in laser processing often produces novel materials due to nonequilibrium partitioning during solidification. This paper reports theoretical and experimental investigations on in situ alloy formation with extended solid solution.

A mathematical model for determining the composition of metastable alloys and the effects of various process parameters on the composition of extended solid solution is presented. To model the nonequilibrium partitioning of solute in concentrated solution, rate equations and conservation of mass equations are used and an expression for nonequilibrium partition coefficients is derived for concentrated solution, which is realistic for laser alloying and cladding. Effect of rapid cooling on solute segregation is presented by deriving an expression for nonequilibrium partition coefficient.

Experiments were performed with Ni-Al and Ni-Hf systems to obtain the composition of solute in the extended solid solution produced during laser cladding.

Metallurgical Processes for the Year 2000 and Beyond
Edited by H.Y. Sohn and E.S. Geskin
The Minerals, Metals & Materials Society, 1988

CONSOLIDATION AND ANNEALING STUDIES OF

RAPIDLY SOLIDIFIED 304 STAINLESS STEEL

W. Kim, In-Ok Shim, and J. G. Byrne
Department of Metallurgy and Metallurgical Engineering
University of Utah
Salt Lake City, Utah 84112-1183

Jong-Jip Kim
Korean Standards Research Institute
P.O. Box 3
Taedok Science Town
Taejon Chungnam 300-31
Republic of Korea

John E. Flinn
INEL EGG Idaho Inc.
P.O. Box 1625
Idaho Falls, Idaho 83415

Abstract

Two rapid solidification processes for 304 stainless steel were studied: centrifugal atomization (CA) in a helium atmosphere and atomization by exposing argon-containing melt to vacuum (VGA). Positron annihilation and x-ray line broadening measurements of explosively consolidated monoliths were used to clarify microhardness measurements as a function of annealing temperature. Interesting differences in positron response between CA and VGA materials are ascribed to more carbide formation in the CA material than in the VGA material in the 600°C range. In addition, hot isostatic pressing of the CA material was examined as to the relations between mechanical behavior, pore shape, and relative density. A variable stress concentration factor is introduced into the theory to allow for changing pore shape as relative density increases.

Metallurgical Processes for the Year 2000 and Beyond
Edited by H.Y. Sohn and E.S. Geskin
The Minerals, Metals & Materials Society, 1988

Introduction

There are many potential advantages of rapid solidification processing (RSP), such as a high degree of matrix supersaturation, finely dispersed microstructures, improved chemical homogeneity, and retention of metastable phases, which have been studied extensively. To consolidate rapidly solidified material which is in powder form into useful shapes often requires subjecting the material to elevated temperature for extended periods of time under mechanical forces, as in hot isostatic pressing or hot extrusion. In such cases, some of the characteristics of the rapid-cooling step can be strongly affected. An alternative is dynamic consolidation (1), for example by explosive forming, where the time frame at least is extremely short.

In the present work, effects of the consolidation processes of hot isostatic pressing (HIP) and explosive forming in a die are examined. The particular material is 304 stainless steel. The characteristics of both RS and VGA 304 stainless-steel powders were described by Wright, Flinn, and Korth (2) earlier.

Experimental Details

Commercially obtained CA and VGA 304 stainless-steel powders were explosively consolidated and isochronally annealed at INEL/EGG at temperatures from 200 to 1200°C. X-ray particle size measurements were performed at the University of Utah on samples annealed up to 600°C using Mn filtered Fe K_α x-radiation. Step scans of the f.c.c. (220) peak at intervals of 0.04° 2θ with dwell times of 10 s per step were used. A single-peak Fourier analysis was used to separate particle size and microstrain contributions (3,4). Vickers diamond pyramid hardness measurements were made of all samples. Positron Doppler broadening measurements were made as described elsewhere (5) using ^{22}Na positrons. Briefly, this entails measuring the number and energies of the gamma rays which accompany positron annihilation in a sample. This is done in a multichannel analyzer. The positron parameter used to describe changes is called the peak-to-wings (P/W) parameter and is the ratio of the area in the central 21 channels (56 eV per channel, centered at 511 keV) of the annihilation peak, divided by the sum of the two wing regions, each 50 channels wide with gaps of 30 channels on each side of the central 21 channels of the peak region. A sharpening of the distribution of these energies is indicative of increasing positron trapping sites in the sample. This in turn indicates increasing defect density in the sample. The converse is then true for a broadening of the distribution.

A different experimental direction consisted of the exploration of the hot isostatic pressing of 304 CA stainless-steel powder in order to develop knowledge of the relation between tensile strength and porosity as a function of HIP temperature, pressure and time. HIP samples of powders of average particle sizes of each of 30 μm and 100 μm were produced at INEL at pressure of 5 and 15 ksi for times between 1.0 and 10 hours at temperatures of 900 and 1100°C. The powders were placed into stainless-steel tubes, which were evacuated and sealed prior to being heated by resistance heaters in a chamber in which the pressure was applied to the tube via compressed argon gas. The stainless-steel sheath was machined away, and specimens such as for tensile or density measurements were machined at the University of Utah. For density determination, samples were weighed in air and in water in a pycnometer using a balance accurate to 0.001 gm. Tensile tests were performed on a 10,000-lb Instron machine at a cross-head speed of 0.02 in per min. Optical microscopy was used to study pore shape and neck size (interparticle contact length) and scanning electron microscopy to observe fracture morphology.

Results and Discussion

Annealing of Explosively Consolidated R.S. Powders and X-Ray Particle Size

Figure 1 shows the positron annihilation and x-ray particle size results for the annealing of explosively consolidated VGA and CA powders. The VGA samples gave a classical recovery-recrystallizaiton curve shape for the P/W parameter. The decrease in P/W, which begins at 200°C, is accompanied by an increase in x-ray particle size which would be appropriate for recovery-recrystallization. For the CA samples, which were atomized in a helium atmosphere, the positron P/W values initially fall more steeply than for the VGA values; however, they subsequently show a maximum at 600°C. This maximum evidently was caused by carbide precipitation which was observed by TEM in this temperature range (6). This precipitation would not be expected to affect the x-ray particle size, which is seen in Fig. 1 to remain constant, but it could very well account for the microhardness maximum seen for the CA material in Fig. 2 between 400 and 600°C. The VGA material also shows a microhardness at about 500°C, but at a lower hardness level. Both hardness peaks may be caused by carbide precipitation; however, important distinctions should be made between the two materials to better understand the differences in all of the observations. Firstly, the CA alloy has a much higher carbon content (0.05 wt pct) than does the VGA alloy (0.01 wt pct). Secondly, the CA

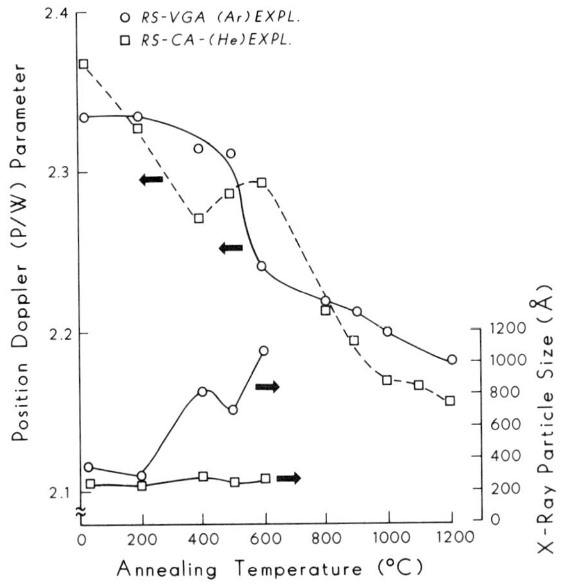

Fig. 1. Positron peak-to-wings (P/W) parameter values (arbitrary units) and x-ray particle size (Å) versus annealing temperature (°C) for the indicated rapid solidification (RS) and explosively consolidated powders: CA for centrifugally atomized in helium and VGA for vacuum gas atomized in argon.

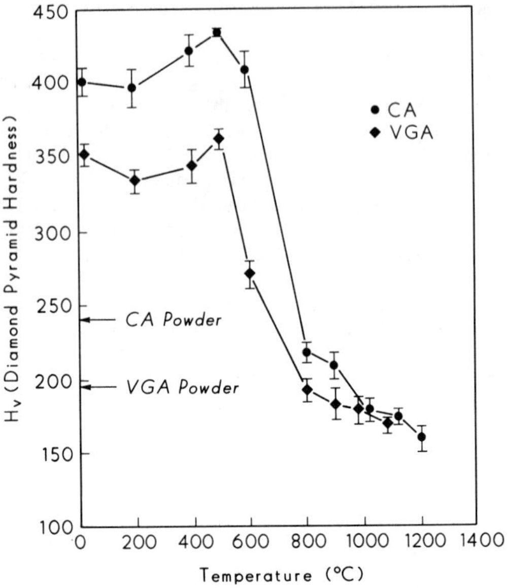

Fig. 2. Influence of annealing temperature on the microhardness of explosively consolidated type 304 stainless-steel powders. Bars indicate one standard deviation.

alloy had a much higher cooling rate than did the VGA alloy. Thirdly, the amount of helium contained in the CA material was 8 atomic ppm, compared with less than 1 atomic ppm of argon in the VGA material. TEM observations report (7) that helium voids apparently serve as heterogeneous nucleation sites for carbides. This observation could be quite important in the present context. For example, the higher cooling rate of the CA powder would have given a higher initial vacancy concentration, and this combined with the higher trapped helium content and higher carbon content than for the VGA alloy would all work cooperatively to facilitate carbide precipitation in the CA alloy when heated to the 600°C temperature range. With this in mind, one can rationalize the maximum in positron annihilation response in the CA alloy versus the lack of such a response in the VGA alloy in Fig. 1. The increase in x-ray particle size in the weakly precipitating VGA alloy in Fig. 1 can be interpreted as showing that there was less resistance in the VGA alloy to enlargement of the subgrain size during annealing than in the CA alloy, where much more precipitation is present. The subgrain size is within a factor of 2 of the x-ray particle size (8).

The above data, together with the independent TEM observations mentioned (6,7), seem to indicate a gradual hardening due to carbide precipitation superimposed on the recovery range of both CA and VGA powders explosively consolidated and then annealed. The precipitation is so much less in the VGA material as to be undetected by the positron Doppler technique.

Experiments and Theory on Hot Isostatically Pressed CA Powder

Connecting the many processing variables such as powder shape, size, compaction technique, and sintering technique with mechanical properties is an almost overwhelmingly complex problem. However, attempts at porosity/tensile strength correlations have seemed reasonable in the past. At least fifteen experimentally based empirical or theoretical expressions exist in the literature relating tensile strength to density, but a listing of these (9) is beyond the scope of the present paper. Predictions made from simplified treatments of this unwieldy problem usually deviate from experimental data and often the discrepancy is attributed to the effect of pore size. German (10) for instance related strength to interparticle neck geometry, assuming spherical powders, and Eudier (11) considered fractional load-bearing area with a constant stress concentration factor included. Several other studies (12-15) have attempted to predict fracture strain from growth and coalescence of voids with varying degrees of success. It became obvious from our own data and a thorough literature survey (8) that much more is involved in improving mechanical properties than just decreasing total porosity, since large changes in mechanical properties result due to processing changes even at a constant density (9).

The purpose of this part of our study was to identify the trade-offs between pore fraction and pore morphology in achieving desired levels of strength and ductility. We also wanted to acquire an understanding of the underlying mechanisms of deformation and fracture responsible for those trade-offs. This understanding involved a rather large mathematical development whose details (9) will need to be published in a separate paper; however, we

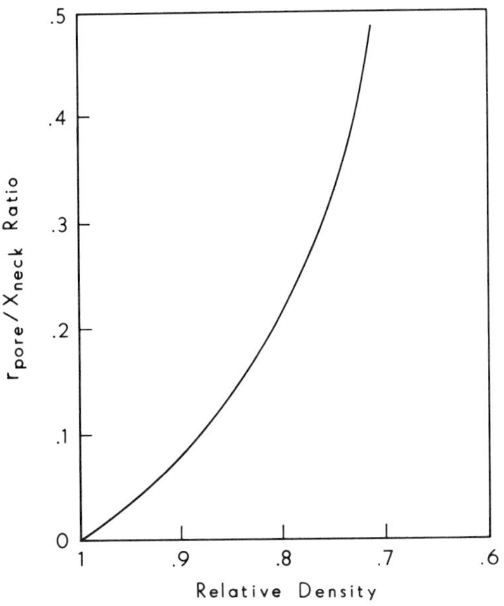

Fig. 3. The ratio of pore radius to neck radius (r_{pore}/X_{neck}) versus relative density.

will present here the basic ideas quantitatively insofar as is possible and also present some comparisons of data with predictions.

First-stage sintering is usually modeled as a set of regularly packed spheres of equal size deforming at their contacts (17). Here, we extend this model to the third stage, eliminating for convenience the usual second stage in which the porosity is modeled as a continuous network of cylindrical pores along grain edges (18). The third stage models particle shapes as tetrakaidecahedra with spherical voids at each corner.

Peterson (19) studied the stress concentration effect of ellipsoidal cavities on the general stress state, and one of the main new contributions of Kim (9) was to develop a relationship between elliptical cavity geometry and relative density. This involved consideration of the neck radius and its (orthogonal) curvature as particles join during hot isostatic pressing. Figure 3 shows the resulting theoretical dependence of the ratio of pore radius, r_{pore}, over neck radius, X, on relative density. This ratio approaches infinity as the relative density approaches its initial (low) value of 0.64 for HIP. Ultimately a relationship was developed (9) for relative strength, i.e., actual strength, σ_{TS}, over theoretical strength, σ^o_{TS}, for full density, as a function of porosity, P, pore shape, a variable stress concentration factor for pores, K_t, and one adjustable constant, A, which decreases as material becomes more ductile. The final equation is

$$\frac{\sigma_{TS}}{\sigma^o_{TS}} = \frac{1 - P}{1 + A(K_t - 1)(\frac{r}{X})^3} \tag{1}$$

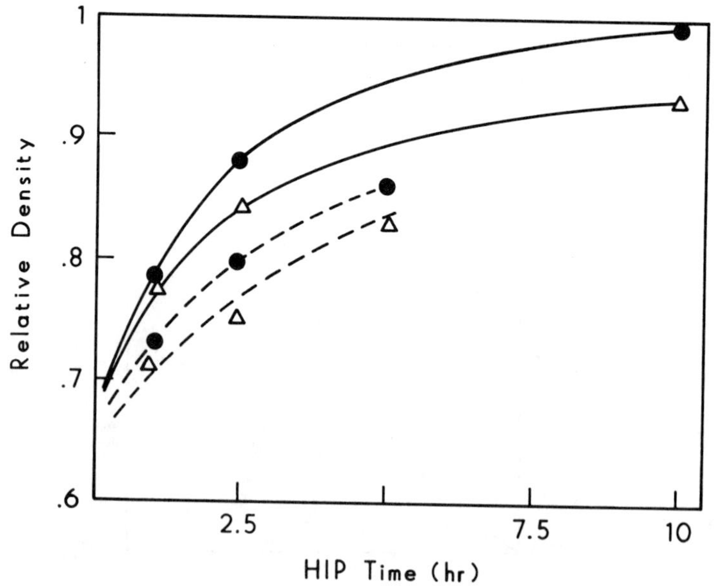

Fig. 4. Relative density versus HIP time for CA 304 stainless-steel powders. Solid circles and open triangles are for 30 μm and 100 μm powder, respectively; solid and dashed curves are for 15 ksi and 5 ksi respectively. HIP temperature was 900°C.

Figure 4 shows how relative density increased with HIP time at 900°C for 30 and 100 μm powders at 15 ksi. Figure 5 shows how the relative density changed with HIP temperature. Figure 6 shows how the neck radius varies with relative density. The straight lines are shown merely to locate the extreme theoretical values of neck size at initial (0.64) and full relative densities according to Aino Helle's equation for neck radius (16). Our data points for ths 100 μm and 30 μm powders lie higher than predicted from Helle's theory, and the curvatures of the two data point sets are positive (concave upward) rather than the opposite, which would be expected from Helle's equation, which is:

$$X = \frac{1}{\sqrt{3}} \left(\frac{D - D_o}{1 - D_o}\right)^{1/2} R \qquad (2)$$

In this, X is neck radius, D is instantaneous relative density, D_o is initial relative density (0.64), and R is particle radius (50 and 15 μm).

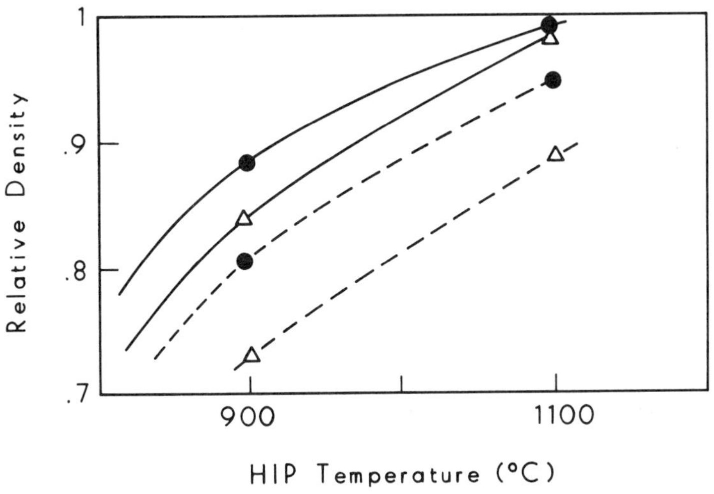

Fig. 5. Relative density versus HIP temperature for CA 304 stainless-steel powder; solid circles and open triangles are for 30 μm and 100 μm powders respectively. Solid curves and dashed curves are for 15 ksi and 5 ksi respectively. HIP time was 2.5 hours.

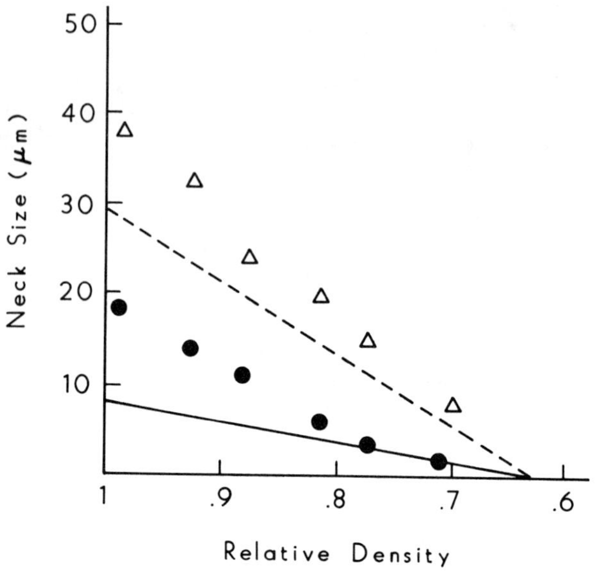

Fig. 6. Neck size, X, versus relative density. Solid circles and open triangles are for 30 μm and 100 μm powders respectively. Dotted and solid line intercepts on the vertical axis are neck size values predicted by Helle's Eq. 2. Value of relative density of 0.64 is the initial value in the HIP model.

Let us now present some plots to give the reader at least some feeling for the ability of the theory to fit the data. Figure 7 shows relative strength versus relative density for 30 μm powder HIPed at 900°C and 5 ksi. The theory of Haynes (20,21), who considered elastic stress concentration concepts and extended them to plastic cases, is shown by the dashed and dotted curve. The completely dashed curve shows the minimum relative strength, which is given by

$$\sigma_{relative\ (min)} = \frac{\sigma_{TS\ min}}{\sigma^{o}_{TS}} = \frac{(1 - P)}{K_t} \qquad (3)$$

in which σ^{o}_{TS} is the theoretical strength of fully dense material, P is porosity, K_t is the theoretical stress concentration factor, a function of pore shape and relative density (9), and $\sigma_{TS\ min}$ is the minimum tensile strength

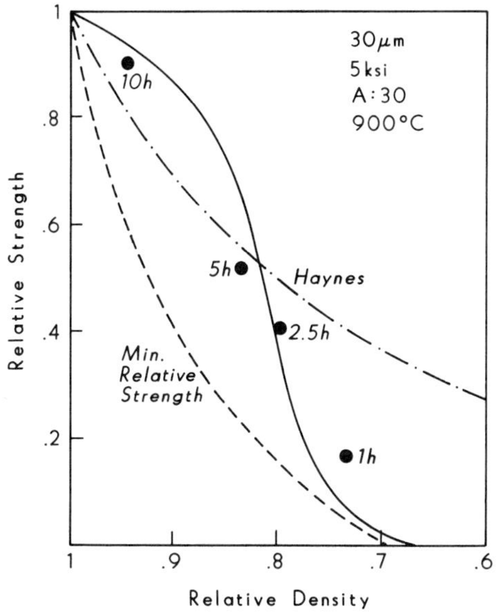

Fig. 7. Relative strength versus relative density for HIP at 5 ksi, 900°C, and 30 μm powder. Solid line is current theory, dashed and dotted line is theory of Haynes, and dashed line is minimum relative strength. HIP times are indicated on figure.

which is obtained from a reduction of the maximum tensile strength by considering the effect of K_t. The solid curve is the present theoretical prediction and the points are our experimental data for various HIP times. The fit is certainly promising, relative to previously available theory. Figure 8 shows relative strength versus relative density for theoretical curves and data points for the 30 μm powder HIP'ed at 900°C and two pressures. Again, the agreement is the best to date. As in the previous figure, one notices that at lower relative densities the actual data are higher in relative strength than the theory would predict. Figure 9 shows again the same coordinates as in Figs. 7 and 8 but treats 100 μm powder HIPed at 1100°C at two pressure for various times. Again, one can say the fit improves as the density increases.

Many microstructural observations were made and showed that larger particle size reduces the volume of plastic deformation as a result of large stress concentrations at the necks due to fewer contacts between particles. Pore morphology determines whether macroscopic specimen necking instability will precede fracture (spherical pores which coalesce slowly) or the converse (angular pores which coalesce very rapidly).

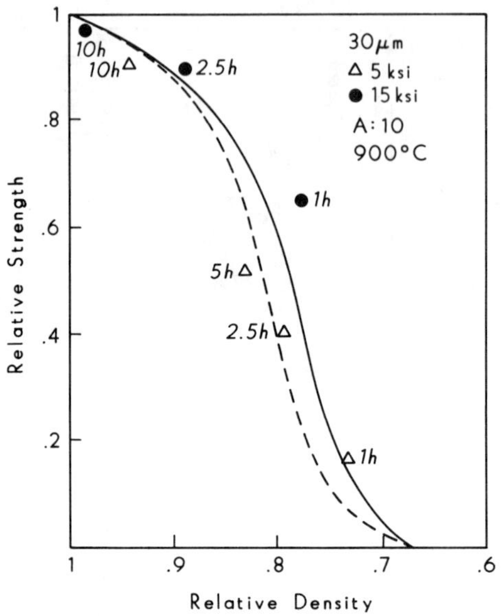

Fig. 8. Relative strength versus relative density. Solid and dashed lines are current theory for 15 ksi and 5 ksi, respectively. Solid circles and open triangles are for 15 ksi and 5 ksi respectively. HIP temperature 900°C.

Conclusions

Combining the techniques of positron annihilation Doppler broadening, microhardness, and x-ray line broadening was helpful in delineating the various mechanisms occurring during the annealing of explosively consolidated RS powder, both centrifugally atomized and vacuum gas atomized. The CA material had carbide precipitation superimposed in its recovery-recrystallization stages. This precipitation is believed to be assisted in nucleating by the presence of helium bubbles trapped during the atomization process.

Hot isostatic pressing of CA powder produced mechanical properties which can be understood better by theory which considers the idea of a variable stress concentration factor which is a function of density. The ductility of HIPed CA 304 stainless-steel powder is a stronger function of HIP temperature at the same relative density than the HIP pressure or particle diameter. The pore shape plays a strong role in determining ductility at constant pore fraction. Angular pores localize stress and result in rapid void coalescence. Spherical pores allow more uniform plastic flow and slower void coalescence than for angular voids, and hence promote higher ductility.

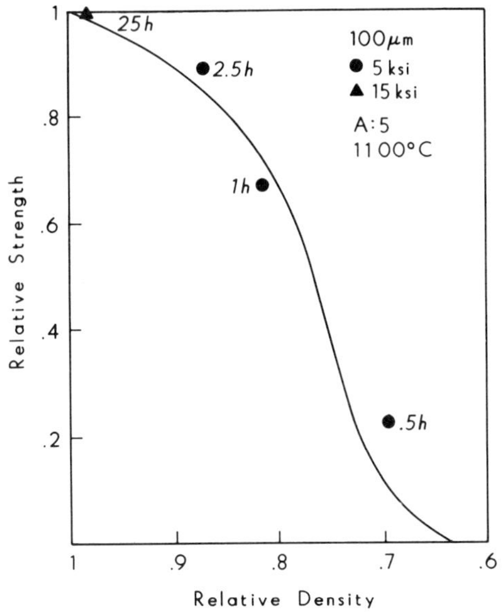

Fig. 9. Relative strength versus relative density. Solid line is current theory. Solid circles and solid triangle are for 5 ksi and 15 ksi respectively. HIP temperature is 1100°C.

It is hoped that the two parts of our research on R.S. powders excite interest on the part of readers in this type of material processing, which should have great potential in the near net shape manufacturing techniques of the future.

Acknowledgement

The authors appreciate the financial, material, and technical support of the U.S. Bureau of Mines via INEL/EGG.

References

1. John E. Flinn, Rapid Solidification Technology for Reduced Consumption of Strategic Materials (Park Ridge, NJ, Noyes Pub., 1984).

2. Richard N. Wright, John E. Flinn, and Garry E. Korth, Rapidly Solidified Alloys and their Mechanical and Magnetic Properties (Pittsburgh, PA, Materials Research Society, 1986), 437-440.

3. A. Gangulee, J. Appl. Cryst. 7 (1974) 434-439.

4. J. Mignot and D. Rondot, Acta Metall., 23 (1975) 1321-1324.

5. Po-We Kao, S. Panchanadeeswaran, and J. G. Byrne, Metall. Trans., 13A (1982) 1177-1180.

6. John E. Flinn, private communication with authors, May 1988.

7. John E. Flinn and T. Kelly, paper presented at the TMS Annual Meeting, Phoenix, AZ, February 1988.

8. D. E. Mikkola and J. B. Cohen, Local Atomic Arrangements Studied by X-Ray Diffraction (New York, NY, Gordan and Breach Science Publishers, 1966), 289.

9. W. Kim, "Mechanical Properties of Hot Isostatically Consolidated Rapidly Solidified Stainless Steel Powder" (Ph.D. Thesis, University of Utah, 1988), 3.

10. R. M. German, Intl. Jour. of Powder Met. and Powder Tech., 13 (1977) 259.

11. M. Eudier, Second European Symposium on Powder Metallurgy, Stuttgart, 1968.

12. F. A. McClintock, Ductility (Metals Park, OH, Amer. Soc. for Metals, 1968), 255.

13. A. Needleman, Jour. of Appl. Mechanics, 39 (1972), 964.

14. R. F. Thomason, Jour. of the Inst. of Metals, 96 (1968) 360.

15. J. Gurland and J. Plateau, Trans. ASM, 56 (1963) 442.

16. A. Helle, Ph.D. Thesis, Lulea University, 1986.

17. D. S. Wilkinson, Ph.D. Thesis, University of Cambridge, 1977.

18. E. Arzt, M. F. Ashby, and K. E. Easterling, Met. Trans. 14A (1983) 211.

19. R. E. Peterson, Non-Destructive Testing (1960) 193.

20. R. Haynes, Powder Metallurgy, 14 (1971) 64.

21. R. Haynes, Powder Metallurgy, 14 (1971) 71.

CONTINUOUS CASTING OF STEEL IN THE 21st CENTURY

Robert D. Pehlke
The University of Michigan
Materials Science and Engineering
2300 Hayward Street
2158 Dow Building
Ann Arbor, Michigan 48109-2136

Abstract

Current continuous casting processes are going through a rapid evolution of advances and new developments. At the same time, intense efforts are being directed to the development of radical new technologies such as continuous casting of thin slabs and strip steel.

Traditionally, the casting processes for steel have been anticipated by the development and implementation of processes in the non-ferrous industry. This trend is reviewed and what it may portend for steel is described.

Prepared for the International Symposium for the Year 2000 and Beyond, AIME Annual Meeting, February, 1989, Las Vegas, Nevada.

Metallurgical Processes for the Year 2000 and Beyond
Edited by H.Y. Sohn and E.S. Geskin
The Minerals, Metals & Materials Society, 1988

Introduction

The continuous casting of steel has been a widely adopted process in the production of steel. The concepts for this processing go back into the early and mid 1800's; however, commerical development was delayed for a variety of reasons until the mid twentieth century.

Beginning in the early 1960's, adoption of continuous casting was initiated and continues at this point in time to be accelerated. In Japan, more than 90% of the steel produced is continuously cast (1). More than 50% of the steel produced in the United States is continuously cast and this fraction continues to rise (2). Numerous examples can be cited of installations costing well over $200,000,000 which have provided a payoff in less than a year and a half. As this process has continued to be applied and its technology to increase, computer control systems have been installed. In particular, in the larger plants producing slabs, sophisticated computer systems involving the integration and networking of mini-computers, micro-computers, and mainframe including back-up systems at all levels are being utilized. Pehlke (3), and Harabuchi and Pehlke (4) have described the structure of these systems.

The emphasis on this processing is based on the fact that continuous casting provides an improved yield where from the caster system 96% of the steel produced ends up in shippable product, whereas in the previous ingot casting case, this was limited to 86.4% on the average. In addition, product uniformity is insured. There is a substantially higher quality of product and the energy savings are substantial because the blooming or slabbing mill is eliminated and the soaking pits are not involved which result in an energy savings of about 3.2 million BTUs per ton (3).

The future of continuous casting will involve further adoption to the 90% of steel production level. Two areas where additional growth will occur are in direct charging/direct rolling and in computer control of the process. New more radical technologies which will evolve in this technology early in the coming century are near net shape casting techniques for thin slabs and steel strip.

A case can be made for anticipation of casting processes for steel by developments in non-ferrous casting technologies. Certainly processes which have succeeded for aluminum and for copper and brass alloys have been redesigned for higher temperatures and applied, at least at the pilot plant level, to casting of steel.

Direct Charging/Direct Rolling

The integration of the casting and rolling processes offer further substantial energy savings in the steelmaking process. This "concatenation" has been achieved in a few instances. The future holds this operating accomplishment. Tsubakihara has outlined the evolution of technologies which have made this possible and the quality assurance requirements to gain wide adoption (5). Operating considerations have been detailed by Harabuchi and Pehlke (6).

Direct linking of continuous casting and hot rolling, either by direct charging or in-line rolling, will be widely adopted in the 21st century. Continuous casting installations with in-line hot rolling of slabs and of billets have been constructed. However, the accomplishment of this technology with assured quality and maximum productivity has required the development and implementation of a number of technologies. These technologies are for manufacturing a defect-free strand, for ensuring rollable temperatures when delivered for hot rolling, for on-line width changing for slabs, and as discussed below overall control of steelmaking operations. The rapid adoption of this technology is exemplified in Japan where a major fraction of continuously cast product is hot charged or direct rolled.

Computer Control of Continuous Casting

The installation of these computer systems offer opportunities for a substantial extension of control strategies. Whereas at the present time, guidelines have been established for monitoring and control of the continuous casting system, direct dynamic control, utilizing on-line sensor inputs can now be implemented. One particular area which is critical to process operation and quality control of product is the status of the temperature distribution in the product being cast, including not only the shell but its internal liquid core. The ability to monitor and operate the system on the basis of the thermal status of the product could markedly enhance productivity, improve product quality and minimize machine maintenance. An overall optimization of the process could result.

A large number of process input variables which are now being monitored including tundish temperatures, product surface temperatures, withdrawal forces, roll pressures, strand deformations and other secondary variables can be utilized as input to a control function for purposes of optimizing strand operation and productivity. This control function can be directly integrated into the overall computer control system.

The benefits achieved utilizing current integrated computer controlled manufacturing by continuous casting have been extensive. Production throughout the plant has been optimized by a supervisory scheduling system. The product has become of more consistent and higher quality with an increased yield and reduced costs.

Integrated and networked computer control systems have been installed in a number of major slab casting installations for steel. These systems involve a computer architecture which includes mainframe computers with back-up, several mini-computers and a number of programmed logic controllers, digital controllers and micro-processor based controlled operator stations. This computer hardware supports a control system hierarchy which includes a database for the entire production system. Management of the melting function and its correlation with caster scheduling is implemented at a lower level, along with data collection and reporting and tracking of the slabs in the system. At the lower level(s) workstations for operator guidance and the implementation of supervisory set-point control have been installed along with data acquisition/reduction and reporting functions. This lower level also involves variable monitoring with alarm systems and local control, such as for mold level, and certain control and sequencing functions such as tundish movement and ladle turret rotation.

This control structure has provided a dramatic improvement in plant productivity and quality. The productivity has been achieved through improved scheduling and extended sequence casting. Product quality has been improved through process optimization, improved process control, and a substantial upgrade of overall product flow, monitoring, and integrated supervision of downstream functions such as slab cutting, marking and direct charging to the reheat furnaces at the hot strip mill. The success of these systems is perhaps best illustrated by the outstanding improvements achieved in productivity, as for example, at the Gary Works of U.S. Steel Corporation of USX.

The latent power of these computer controlled installations and the ability to incorporate and integrate other control functions represent a resource of great potential for improvement in the overall productivity of the steelmaking operation. In particular, the implementation of direct monitoring, simulation and control of the behavior of the cast strand itself is now at hand.

The next steps will be retrofitting of existing slab casters and an adaptation to the EAF-Caster production system of the mini-mill of computer control and production scheduling systems. These systems, supported by additional sensor technologies and computing power, will improve quality and production efficiencies in nearly all continuous casting facilities.

Advances in Sensor Technology

The assurance of quality in the high production environment of continuous casting and the key inputs to the computer integrated manufacturing system require additional sensors. These sensors will find application in off-line and on-line product inspection and equipment diagnosis. For example, Tsubakihara has listed a number of sensors vital to monitoring continuous casting quality, as presented in Table I (5). While many of these sensors are widely used, and others are being adopted, a number are in the development stage and represent the future of sensor technology in continuous casting of steel.

Sensor	Purpose	System	Diffusion degree
Ladle slag flow-out detection	To decrease inclusion	Judgment by electromagnetic method or judgment by nozzle vibration	
Molten steel level in the mold	To control the level of molten steel and to improve quality	Eddy current (feedback amplification), electromagnetic coupling, magnetic flux balance, radioisotope, thermocouple, or optical method	◎
Molten powder layer thickness in the mold	To monitor the powder flow-in conditions and the feeding of powder	Two-frequency eddy current	
Solidified shell thickness in the mold	To measure solidifying and cooling conditions	Electromagnetic ultrasonic	
Cast shell contact conditions to the mold	To monitor the powder flow-in conditions	Ultrasonic	
Cast shell friction conditions with the mold	To predict breakout and surface flaws	Mold vibration acceleration and transmission function	
Temperature distribution on the mold	To predict breakout and to measure the heat removal amount	Thermocouples mounted inside of the mold	○
Shorter side taper of the mold	To control the mold width change and the cooling process	Differential transformer or electromagnetic displacement meter	○
Powder film thickness just below mold	To monitor the powder flow-in conditions	Multiple-wavelength radiation pyrometer	
Cast surface temperature distribution just below the mold	To predict breakout and to monitor casting conditions	Optical fiber two-color pyrometer	
Cast piece shorter side shape just below the mold	To predict breakout and to monitor casting conditions	Water flow type ultrasonic distance meter or differential transformer	
Cast surface temperature distribution at the secondary cooling zone	To control secondary cooling and to improve quality	Two-color pyrometer and radiation pyrometer equipped with brush	◎
Solidified shell thickness at the secondary cooling zone	To control secondary cooling and to determine the solidification completion position	Electromagnetic ultrasonic	
Roll reaction force at the secondary cooling zone	To determine the solidification completion position	Load cell	
Hot cast piece thickness	To measure cast piece bulging and to calculate the weight of unit length	Differential transformer or laser distance meter	
Hot cast piece width (shape)	To measure cast piece bulging and to calculate the weight of unit length	Differential transformer or laser distance meter	○
Hot cast piece surface flaw	Cast piece handling, conditioning, and operation feedback	High pressure mercury lamp and camera, laser scan method, or temperature distribution	○
Hot cast piece subsurface defect	Cast piece handling, conditioning, and operation feedback	Eddy current flaw detection	
Hot cast piece internal crack	Cast piece handling, conditioning, and operation feedback	Electromagnetic ultrasonic crack detection	

◎: generally diffused; ○: now being diffused; No marks: now used for trial.

Table I - Sensors to measure the quality at the continuous casting process.

Direct Casting - Current Status and Long Term Perspectives

The near-net-shape casting of steel has in recent years become a subject of renewed interest. The background for this processing goes back into the 1800's. However, the development has been delayed for a variety of causes. Even in the 1930's when direct casting of non-ferrous metals into strip was first applied commercially, difficulties in control of liquid steel prevented serious consideration for steel manufacturing.

However, in the 1960's, a research program on strip casting was initiated at Jones & Laughlin Steel Corporation which resulted in the pilot scale test of an "Inside the Ring" process extending into the early 1970's. This process produced a 0.25-inch-thick and 13.5-inch-wide strip at about 100 feet per minute. In the early 1960's, programs were undertaken to develop thin slab casting of steel based on the Hazelett twin-belt caster which was already being used with success commercially for non-ferrous metals, including aluminum and copper. Subsequent to this period, as reported by Dancy (7), none of the ambitious programs of the 1960's led to commerical application. Several reasons for this lack of continuity included the economic status within large integrated steel companies and the potential for the technology.

Within the last two years, two key symposia-related publications have appeared which present the results of development work in this technology.(8,9) More recently, a survey of world activity in near-net-shape casting research and development reported 22 major projects underway. (10)

The direct casting processes would produce a casting close to the final product dimensions of the next conventional processing step. For example, in conventional continuous casting of low carbon, aluminum-killed steels for sheet applications, a slab, 8 to 10 inches thick is cast and subsequently rolled on a hot strip mill to thicknesses ranging from 0.05 to 0.5 inches. A direct casting process for sheet steels would be capable of producing a sheet or thin slab close to the final hot rolled dimensions. The objective would be to produce material similar in physical dimensions and properties to conventional hot rolled material, and in fact to eliminate hot rolling as a processing step.

Direct casting would allow established companies to upgrade outmoded ingot casting and rolling operations at a fraction of the cost of current technologies. Furthermore, the lower investment costs would allow new companies to enter the strip and plate markets.

The major areas in direct casting technology are shown schematically in Table II which presents a flow sheet for an integrated strip producer with three direct casting alternatives: thin slab casting, strip casting and spray forming (11).

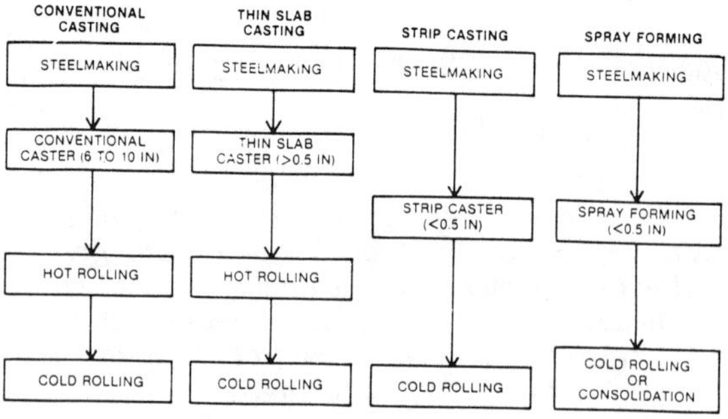

Table II - Casting Process Steps

In thin slab casting, a slab 1 to 2 inches thick is produced, and although hot rolling is not eliminated, the amount of reduction necessary to produce hot band is markedly reduced with a corresponding reduction in investment and operating costs. Thin slab technology is currently a reality, and thin slab machines are either at the pilot stage or under construction (12). For example, SMS-Concast has developed a simple, low-head casting machine which can produce a slab at a rate of 240 inches per minute in widths up to 64 inches and sheet bar thickness down to 1.5 inches.(13) Nucor Steel has announced the adoption of this technology for an 800,000 ton per year hot band mill using an in-line 4-stand continuous strip mill.

In strip casting, a strip less than 0.2 inches thick would be cast which could eliminate the need for a hot strip mill. Strip casting machines also could be very small, inexpensive and cost efficient. (14). A number of problems remain to be solved before commercialization of this process, as discussed below. The three primary techniques used in strip casting are single-roll processes, twin-roll processes and electromagnetic levitation casting. A summary of current activity in this field is presented in Table III (11).

Country	Company	Caster Type
United States	BSC-Armco-Inland-Weirton	Twin roll
	Armco	Single roll
	Allegheny-Ludlum	Single roll
	Argonne National Labs	Electromagnetic levitation
	United Technologies	Single roll
	LTV	Drum in drum
Japan	Nippon Steel	Twin roll
	Kawasaki Steel	Twin roll
	Nippon Kokan	Twin roll
	Nippon Metals	Twin roll
	Kobe	Twin roll
	Nippon Yakin	Twin roll
France	IRSID	Twin roll
Italy	CSM	Twin roll
	Danieli	Thin strip
Austria	Voest-Alpine	Single roll
		Twin roll
Switzerland	Concast	Single roll
Germany	Mannesman-Battelle	Single substrate?

Table III - Alternative Casting Technique Participants - Thin Strip

Spray forming, a process where liquid metal is atomized and sprayed onto a substrate to form a layer, is a third major new casting technique. Casting rates can be high and the product from spray casting has a very fine microstructure with minimum segregation. This process eliminates conventional casting and hot rolling processes and is a true near-net-shape casting process. This technology allows casting of a wide variety of shapes and pre-forms and has been used in the manufacture of rings, tubes, small billets and pipes for both ferrous and non ferrous metals (15-17).

In the longer term, strip casting, which is not as far advanced as the other two technologies, would appear to have the greatest incentive for commercialization. The process could permit the design of facilities with one, or, at most two in-line rolling stands, the overall cost savings, including the reduced cost of capital and a decreased operating cost, might be as much as $100 per ton (7) This advantage fits in well where replacement of capacity is to be considered or new facilities on a smaller scale are being considered.

Thin Strip Casting

Following work on rapid solidification processing which had been started by Duwez at Cal Tech (18) interest in strip casting was renewed at the end of the 1970's and single-roll and twin-roll development work started again. As noted above, Jones & Laughlin Steel Corporation had started work in the mid 1960's on strip casting. A decade later they successfully produced a considerable amount of strip from a pilot drum-in-drum caster (19). At that point in time, the major problems to be overcome to advance this technology were related to metal feeding, temperature loss, strip irregularities and edge control.

In September, 1984, Allegheny Ludlum announced the successful development of a single-roll planar flow casting process for type 304 stainless steels with thicknesses up to 0.1 inches and widths up to 20 inches (20,21).

In rapid solidification, one of the preferred fabrication methods was by chill block casting which involved pouring the stream of liquid metal onto a rotating chilled mold to form a thin rivet of metal (22). Chill block casters are variations of the single-roll and twin-roll casters. In recent developments, liquid metal handling and distribution systems have been refined to develop the melt drag and planar flow casting processes that have subsequently been developed for strip steel production on single rolls as illustrated in Figure 1 (11). A typical configuration of a twin-roll caster is shown in Figure 2 (23).

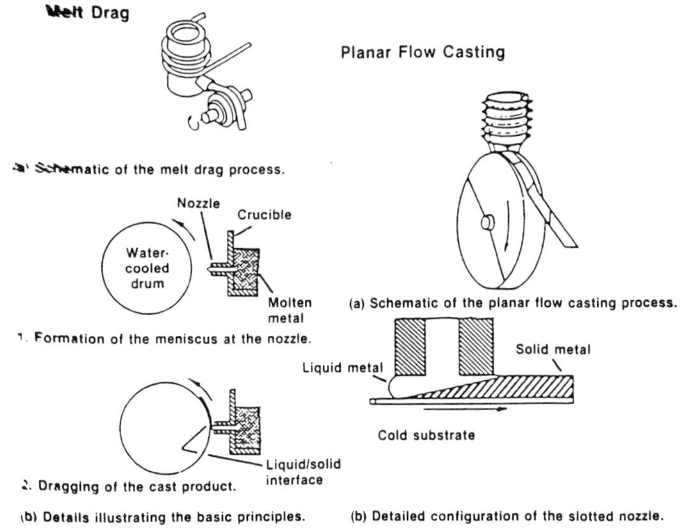

Figure 1 - Melt Drag and Planar Flow Casting

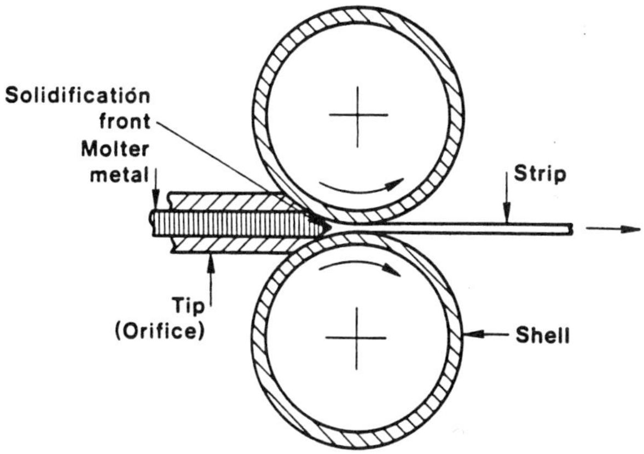

Figure 2 - Twin Roll Casting Schematic

At this point in time, the problems to be solved in order to commercialize direct strip casting are (12,24):

(1) Integration into current production facilities.

(2) Edge containment which relates to pool stability in single-roll casting and edge dam design in twin-roll casting. Electromagnetic containment could play a role in solving these problems.

(3) Strip quality - the strip must be homogenous, consistent and free of defects. Steel flow and distribution must be uniform and cleanliness at the highest level, since inspection and conditioning of this extended surface product would be impossible.

(4) Reoxidation must be eliminated, and inclusion or nozzle blockage elements must be at the lowest possible level because of small nozzle sizes.

(5) Strip width should be extended to conventional levels (34 to 80 inches or more).

(6) Process control and equipment costs - It appears at present that roll costs can be high due to limited life, and coiling and handling costs will be significant unless yield losses are absolutely minimized. Costs relating to assurance of quality could limit broad application of the technology.

At present, single-roll processes seem nearer commercialization; however, thickness constraints, due to single sided solidification and problems with upper side solidification structure, refractory compatability, reoxidation and width extension may limit application of the process. Progress has been made on control of single-roll casters, but development of twin-roll machines appears at this time to be no further than it was when Hazlett abandoned the process a number of years ago (24).

The limitations in direct strip casting for the manufacture of sheet metal products fall in the ability to identify, monitor and control relevant process parameters. The development of a suitable process control strategy must allow production of high quality strip on a reproducible basis and within narrow product specifications. It is necessary that an understanding of process dynamics be developed in order to utilize fully computing capabilities which are available to interpret sensory inputs and to utilize this information for control of direct strip casting in real time.

These challenges should be answered in the next several years, although Dancy challenges the industry by tying these accomplishments to "sufficient continuous funding....from users-the steel producers" (7). In the absence of intensive effort now, we are indeed looking toward the year 2000 and beyond.

Anticipation In Non-Ferrous Technologies

In 1933, S. Junghans proposed the oscillating mold, and the first plant for commercial casting of brass in a vertical machine using an open ended oscillating mold was built in Germany. In 1935, a plant was put in operation at Scovill Manufacturing Company for continuous production of brass plates using casting rolls. These technologies anticipated several vertical casting processes for steel. In the Soviet Union and in North America following World War II, several continuous casting installations for steel were brought on stream in several pilot plants and in a few commercial plant installations.

Subsequently, several non-ferrous casting technologies have evolved into ferrous casting processes. One noteable example is horizontal continuous casting which was adopted from non-ferrous and iron casting to steel casting at General Motors by incorporating three key aspects: a boron nitride break ring, intermittent product withdrawal from the fixed mold, and staged, controlled cooling in the mold (25,26).

The Hazlett process using twin belts has been successful in producing thin slabs of non-ferrous metals including aluminum and copper, and has recently been commercialized for production of copper anodes (7). Efforts continue to utilize this technology for production of thin slabs of steel.

Non-ferrous casting processes have been modified and adopted for casting of steel in many instances. This area of metals processing technology has anticipated many steel production processes.

Conclusions

The processes for continuous casting of steel are undergoing rapid advances and evolution. The linking of continuous casting and hot rolling of steel is being utilized for a wide variety of continuously cast steel products.

Computer control of the entire steelmaking process with the center being the caster is rapidly improving the quality and productivity of steel manufacturing systems. Integrated and networked computer control systems will be widely adopted in the future, including in the mini-mill sector.

Sensors have created the technology underlying computer management of continuous casting and the interrelated steelmaking and hot rolling processes. Substantial further development of monitoring technologies is required, and the coming century can anticipate dramatic advances in sensor technologies.

Direct casting processes for steel will be developed in the future. Thin slab casting is being realized, and strip casting of steel is receiving substantial development effort. This technology should come to fruition after the year 2000.

New developments in continuous casting of steel have often been anticipated by advances in non-ferrous casting processes. A view of continuous casting of steel to the year 2000 and beyond should include consideration of current non-ferrous processes.

References

1. *Continuous Casting Machines for Steel- WORLD SURVEY, 14th Edition, as of January 1, 1988, Concast Service Union AG, Zurich, 1988.*

2. *"Continuous Caster Roundup - U.S.A.", Iron and Steelmaker, 14, 1987, pp. 17-22.*

3. *R.D. Pehlke, "Computer Control of Continuous Casting of Steel", In preparation.*

4. *T.B. Harabuchi and R.D. Pehlke, Volume Four - Continuous Casting - Design and Operations, "Process Control Computer Systems", Section 3.4, 1988, pp. 120-127, Iron and Steel Society - AIME, Warrendale, PA.*

5. *O. Tsubakihara, "Technologies That Have Made Direct Concatenation of Continuous Casting and Hot Rolling Possible", Transactions of the Iron and Steel Institute of Japan, 27, No. 2, 1987, pp. 81-102.*

6. *T.B. Harabuchi and R.D. Pehlke, Volume Four - Continuous Casting - Design and Operations, Chapter Three - Operations, pp. 79-129, 1988, Iron and Steel Society - AIME, Warrendale, PA.*

7. *T.E. Dancy, "Strip Casting - Where Do We Go From Here", Iron and Steelmaker, 14, No. 12, 1987, pp. 25-28.*

8. *Near Net Shape Casting, 1987, Iron and Steel Society - AIME, Warrendale, PA.*

9. *Proceedings of International Symposium on Near-Net-Shape Casting of Strip, 1987, Canadian Institute of Mining and Metallurgy, Toronto, Ontario.*

10. *R. Wolfgang, W. Kapellner and R. Steffen, Stahl u. Eisen, 108, No. 9, 1988, pp. 409-417.*

11. *A.W.Cramb, "New Steel Casting Processes for Thin Slabs and Strip - A Historical Perspective", Iron and Steelmaker, 15, No. 7, 1988, pp. 45-60.*

12. *M. Cygler and M. Wolf, "Continuous Strip and Thin Slab Casting of Steel - An Overview", Iron and Steelmaker, 13, No. 8, 1986, pp. 27-33.*

13. *"New SMS Technology for Thin Slab Casting", Iron and Steel Engineer, May 1987, pp. 51-52.*

14. *K. Schwaha, W. Egger, H. Rametsteiner, R. Landerl, A. Niedermayr and F. Hirschmanner, "Strip Casting of Low-Carbon Steel at VOEST-ALPINE", Paper No. 220 of Reference 9.*

15. *A.R.E. Singer, "The Principles of Spray Rolling of Metals", Metals and Materials Science, 1970, pp. 246-257.*

16. *B.A. Rickinson, F.A. Kirk and D.R.G. Davies, "CSD - A Novel Process for Particle Metallurgy Products", Powder Metallurgy, 24, No. 1, 1981, pp. 1-7.*

17. *R.W. Evans, A.G. Leatham and R.G. Brooks, "The Osprey Preform Process", Powder Metallurgy, 28, No. 1, pp. 13-20.*

18. *P. Duwez, "Structure and Properties of Alloys Rapidly Quenched from the Liquid State", ASM Transactions, 60, 1967, pp. 607-633.*

19. *E. Aukrust, E.A. Mizakar and C.C. Gerding, "Recent Progress on a Unique Strip Casting Process", Presented at 4th Advanced Technology Symposium on Near-net-shape Casting, Myrtle Beach, SC, May 1987, ISS-AIME, Warrendale, PA.*

20. *R.K. Pitler, Proceedings of the N.J. Grant Symposium on Processing and Properties of Advanced High Temperature Alloys, Cambridge, MA, June 1985.*

21. *D.C. Dean, "The Physical Metallurgy of Strip Cast Type 304 Stainless Steel", Iron and Steelmaker, 15, No. 3, 1988, pp. 12-15.*

22. *S. Kavesh, "Principles of Fabrication", Metallic Glasses, 1976, pp. 36-73, ASM International, Metals Park, OH.*

23. *R.J. O'Malley and M.E. Karabin, "A Theoretical and Experimental Study of the Roll Casting Process", Paper No. 189 of Reference 3.*

24. *A.W. Cramb, "New Developments in the Continuous Casting of Steel", Iron and Steelmaker, 15, No. 8, pp. 65-68.*

25. *F.J. Webbere, R.G. Williams and R. McNitt, "Steel Scrap Reclamation using Horizontal Strand Casting", GM Research Publication GMR-11, October 20, 1971.*

26. *W.G. Patton, "GM Casts In-Plant Scrap into In-Plant Steel", Iron Age, (December 9, 1971), pp. 53-55.*

The Levitation Casting Process

Hugh R. Lowry and Arthur S. Klein

Electrical Distribution & Control
Technology Department, GE Company
Bridgeport, CT, USA

Abstract

The Levitation Casting Process is a new continuous casting method that uses an electromagnetic levitation field, instead of a mold, to support and contain a column of solidifying metal. By counteracting gravitational forces and hydrostatic pressure on the metal column, this process eliminates friction and adhesion at the interface between the solidifying metal and the cooled walls of the casting chamber. The process inherently provides high casting speeds, excellent dimensional control and smooth, continuous emergence of solidified product from the top of the casting chamber.

Of major importance is that the electromagnetic field produces an intense stirring of the liquid metal both before and during solidification. As a result of this stirring, the cast product has a homogeneous, equiaxed, fine grain structure generally suitable for immediate drawing or other forming operations without the need for hot rolling or other expensive additional processing.

This paper reviews the fundamentals of the Levitation Casting Process, results obtained in casting various pure metals and alloys, and the production sized system at Lowell, MA, that uses this process.

Metallurgical Processes for the Year 2000 and Beyond
Edited by H.Y. Sohn and E.S. Geskin
The Minerals, Metals & Materials Society, 1988

Introduction

The Levitation Casting process was specifically developed to meet several requirements including:

- Economic at low casting rates
- Usable with pure metals and alloys
- High, uniform product quality suitable for fine wire drawing or cold rolling into flat strip and tape
- Easy cast product shape and size changes
- Process inherently stable and easily automated

In meeting these criteria it was also desired to retain the advantages of General Electric's well known Dip Forming continuous casting process; namely the inherent "fail-safe" properties of upward casting, solidifying metal by means other than pressure contact with a cooled mold, and use of a sealed furnace, casting chamber and cooling tower, filled with a protective atmosphere, to prevent oxidation of the molten metal or cast product.

Levitation Casting Process Principle

The dictionary defines the verb levitate as "to cause to rise and float in the air" and our early experiments were directed toward investigating means of levitating a column of liquid metal by an electromagnetic field. Rather than simply use conventional electromagnetic levitation apparatus, we chose to use an unconventional means ------- namely an upwardly moving electromagnetic field.

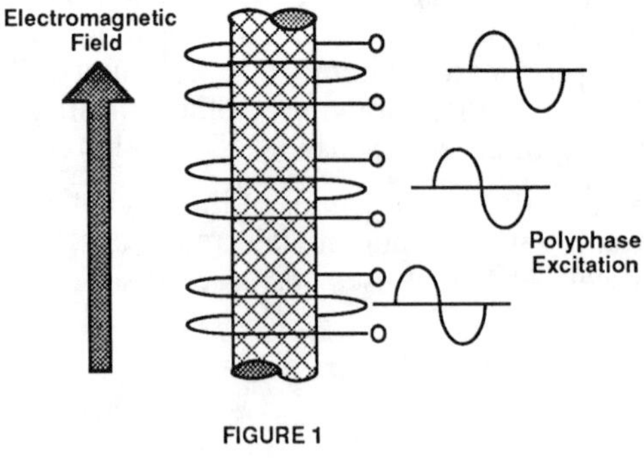

FIGURE 1

A cylindrical, upwardly moving field can be created by passing polyphase electric current through a series of coils as illustrated by Fig. 1.

The field's upward velocity is directly proportional to the product of the excitation frequency and the effective wave length of the coil. In our equipment the levitation field (not the molten metal !) moves upward at several hundred metres per second.

This moving field induces currents in a conductor placed within the field and, as illustrated by Figure 2, there are two effects caused by the interaction of these induced currents with the field.

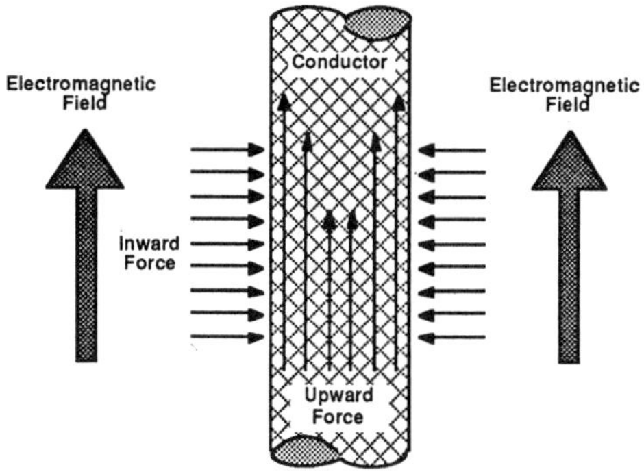

FIGURE 2

First there is an upward force which is strongest at the outside of the conductor and weakens (due to skin effect) toward the center. The second effect is an inward radial pressure perpendicular to the upward force. The magnitude of these forces can be approximated by a solution of Maxwell's equations with several simplifying assumptions. Computer simulation by finite element techniques gives results within about ± 5% of measured values and also provides information on the spatial distribution of these forces.

When liquid metal is introduced into the bottom of such a field the result is illustrated by Fig. 3.

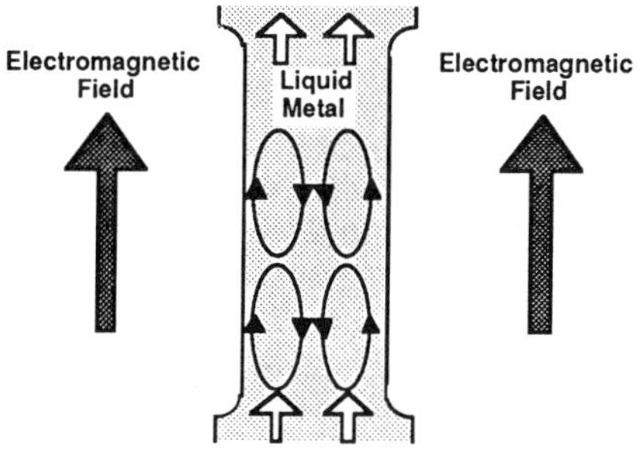

FIGURE 3

In this deliberately exaggerated drawing it will be seen that the liquid is "pinched" inward by the radial pressure of the field so the upward moving column is reduced in diameter.

The upward or levitation force produced by the field induces an upward motion of the liquid metal at the outside of the column (where the upward force is greatest) and a down flow near the center where the levitation force diminishes.

Those somewhat familiar with electromagnetic pumps used in nuclear reactors and other applications will recognize that this levitation apparatus is far different from conventional electromagnetic pumps and is, in reality, a very inefficient way to "pump" metal.

Great pains are taken in electromagnetic pumps to prevent "pinching" of the metal column from taking place since it reduces the flow rate and causes other problems. Additionally, an electromagnetic pump is specifically designed to produce a uniform flow profile across the diameter of the metal to prevent undesired local circulating currents that create turbulence and waste energy.

In the Levitation Casting process, however, both pinching and violent stirring are desired so the levitation coil and its associated polyphase power supply are designed to enhance these effects.

When cooled walls are added to the configuration of Figure 3 the upper portion of the levitated metal column will solidify as illustrated in Figure 4.

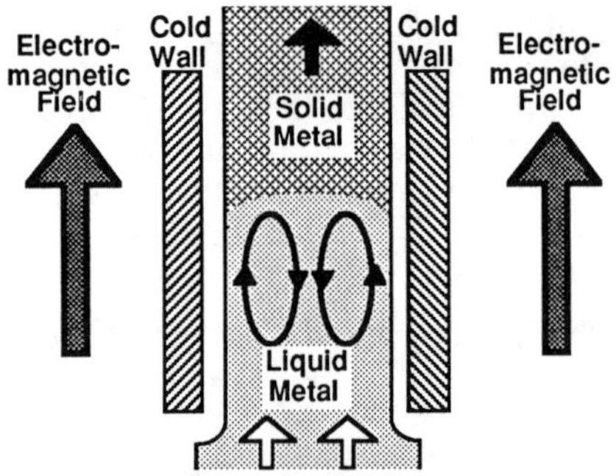

FIGURE 4

It is well known that the electrical conductivity of solidified metal is several times higher than the conductivity of the same metal when molten. Since the levitation force is directly proportional to the conductivity of the metal on which it acts, it is obvious that the upward force on the solidified rod will be several times greater than the upward force on the liquid metal.

From a practical standpoint, this means the solidifying rod is fully supported and pushed upward out of the levitation field. It is not necessary to exert a strong tensile force on the hot rod to pull it out of the casting chamber. This, of course, minimizes any tendency for the very hot, still relatively weak rod to develop cracks or other casting defects.

If liquid metal is continuously supplied to the bottom of the casting chamber and a means provided to take away the solidified rod, the result is a unique method of continuous casting. Such a GELEC* continuous casting apparatus is illustrated schematically by Figure 5.

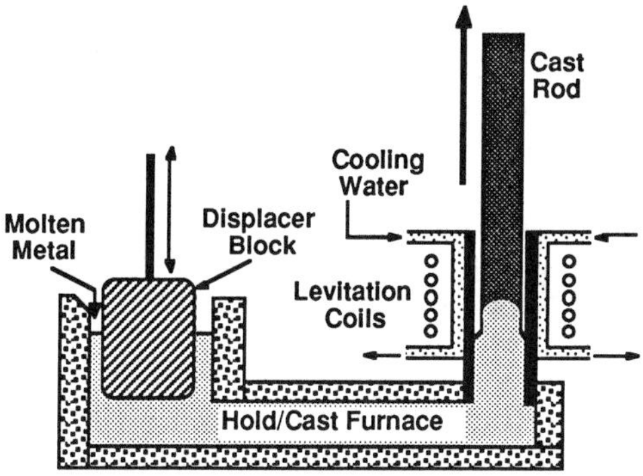

FIGURE 5

As shown in Figure 5, the liquid metal underpours from the large holding furnace on the left up into the levitation coils on the right. A displacer block in the holding furnace moves up and down to keep the metal level in the levitation coils approximately constant and to compensate for periodic refillings of the holding furnace from a ladle or tiltable melting furnace.

In Figures 3, 4 and 5 the amount of "pinching" of the liquid metal appears to be relatively large in comparison to the diameter of the column. In reality, an appreciable gap between the liquid metal and the cooled wall of the casting chamber would greatly impede heat removal from the liquid metal and hence the rate at which solidified products could be produced. For this reason, the strength of the levitation field is set so that the gap between the liquid metal and the wall of the casting chamber is about 25 μm (0.001 inches). A gap of this magnitude is sufficient to prevent friction and adhesion between the solidified metal and the walls of the casting chamber but does not appreciably affect the heat flow rate.

Grain Structure

One consequence of the electromagnetic stirring illustrated in Figures 3 and 4 is that alloys are kept well mixed while molten and during solidification. This, of course, is very desirable since it prevents variations in composition across the diameter of the cast product. In one experiment, an aluminum based material was successfully cast with a homogeneous structure even though it contained a high percentage of essentially solid particles that tend to settle out in more normal casting procedures. This continual, violent stirring until solidification takes place makes feasible the continuous casting of alloys previously thought to be "impossible" (or at least very difficult) to cast continuously by other methods.

** A Trademark of General Electric Company, USA*

An even more interesting and useful consequence of counteracting the force of gravity on the liquid metal and stirring it violently is that large dendrites are broken before they can grow into large grains. It has been hypothesised that a newly solidified dendrite will experience a strong increase in upward force (due to its high conductivity) in comparison to the upward force on the surrounding liquid metal. This tends to break the the still weak dendrites thus forming fragments that act as nucleation sites for new dendrites and creating a multiplicity of small grains.

In any case, the structure of rod made by the Levitation Casting process is fine grained, equiaxed and homogeneous --- essentially the same as the structure obtained by tandem hot rolling. In alloys, the grain size is about 25 to 50 μm. In pure metals, such as low oxygen copper, the grains are roughly 200 to 300 μm. This difference is felt due to the fact that a pure metal has a single and unique temperature for the transition between liquid and solid whereas in alloys there is a temperature band for the transition thus giving more opportunity for dendrites to be broken before the structure is completely solid.

The remarkable grain structure of rod made by the Levitation Casting process was totally unexpected. The original intent of levitating the molten and solidifying metal was to prevent pressure contact (and hence friction) with the walls of the casting chamber. The serendipitous result of also creating a small equiaxed grain structure means that the cast structure is satisfactory for immediate use in wire drawing and other applications without additional processing. The ability to go directly from molten metal to a usable "as cast" structure and eliminate hot rolling or heat treating is an obvious economic benefit.

Laboratory Results

The initial development work on this process was done in our lab in Bridgeport, CT, using a 25 kg capacity resistance heated furnace connected through a short tube to the levitation coil and heat exchanger assembly mounted directly above. The molten metal was moved up into the levitation coil by gas pressure. The levitation coil/heat exchanger assembly was only 19 cm high but rod speeds of 60 to 70 mm per second were achieved when casting 12 mm low oxygen copper rod. This corresponds to a casting rate of about 250 kg per hour but, of course, the small furnace only permitted runs of 5 to 10 minutes duration.

In addition to copper, many other alloys were successfully cast in this research unit including beryllium copper, leaded beryllium copper, cadmium copper, silver copper, cadmium silver, copper-nickel-silicon and various aluminum alloys. The alloys all exhibited the fine grain equiaxed structure previously described.

A length of low oxygen copper GELEC® continuously cast rod was successfully drawn by our wire plant from 14 mm down to 0.13 mm (5 mils) through three machines without any intermediate annealing. A 100:1 reduction in diameter (corresponding to a 99.99% reduction in area) without annealing is considered to be a good test of rod quality and uniformity.

Additionally, a short length of "as cast" 9.5 mm (0.375 inch) low oxygen copper GELEC rod was cold rolled down into tape 0.135 mm thick and 19 mm wide (0.005 by 0.75 inches) without annealing. There was no edge cracking or surface defects of the type seen in copper with nonuniform grain structure. Some of the alloy rods were also drawn or otherwise processed successfully without additional steps to refine their grain structure. This leads to the conclusion that intense electromagnetic stirring during solidification can minimize or eliminate much of the processing now done to "as cast" materials before cold working.

Another interesting aspect of the Levitation Casting process is the inherent size uniformity of the cast rod. Typical diameter variations are $\pm$ 75 μm ($\pm$ 3 mils) or less. The surface, however, has a slightly rippled appearance since solidification takes place while the metal is suspended in the field and not in contact with the walls of the casting chamber. This slight rippling on the surface does not seem to have any effect on drawability, cold rolling or other properties.

Production Equipment

Because the levitation coil/heat exchanger assembly (i.e.the casting chamber) worked so successfully in the research size system, it was decided to bypass the usual scale-up or pilot production steps in this project. Instead, we moved immediately into installing equipment that could continuously supply molten metal to the casting chamber and take away solidified product at rates of up to one ton per hour. This equipment was designed and built by the Inductotherm Corporation under the terms of an R & D and patent license agreement from GE.

A prototype holding/casting furnace, complete with computer controlled metal level subsystem, was first installed and tested in our Bridgeport lab. An improved version of this hold/cast furnace and the rest of the equipment necessary for full production system was then built by Inductotherm and installed in a cable factory in Lowell, MA.

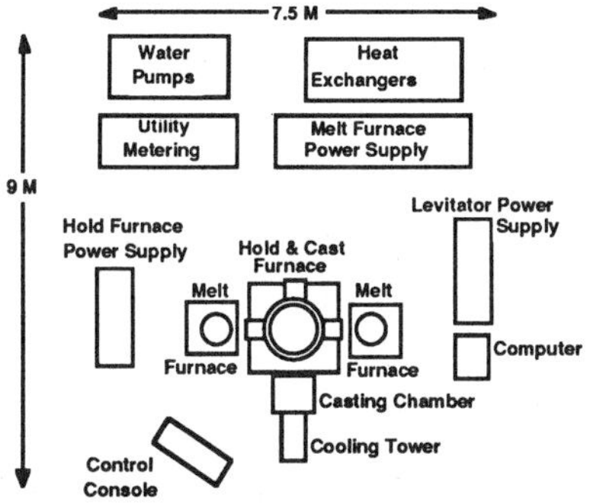

FIGURE 6

The layout of this production system is shown by Figure 6.

Two 350 kg Dura-Line® coreless induction melting furnaces from Inductotherm have been installed. One melt furnace is adequate for normal operations but, under some circumstances, it may be desirable to melt a charge in one furnace while the other is being loaded and then energize the second furnace after the first is dumped into the hold/cast furnace

In addition to new copper, we have successfully used briquetted factory scrap as input material in the system. The melt furnaces and the hold/cast furnace have tight fitting lids and nitrogen gas is fed in to prevent oxidation. Fumes from the top of the molten metal are ducted away to an outside "bag house" where particulate matter is removed. With this type of sealed furnace, plus the stirring inherent in induction furnaces, it has been found that a small amount of graphite added to the melt furnace will quickly reduce the oxygen content of scrap copper to a few parts per million.

When a charge is completely melted it is poured into the holding/casting furnace which has a capacity of about 700 kg of copper. This special Inductotherm furnace is also of the "coreless" type in which the metal is contained in a refractory crucible and heated by induction coils outside the crucible.

Figure 7 is a simplified side view of the production sized holding/casting furnace indicating the relative size of various components.

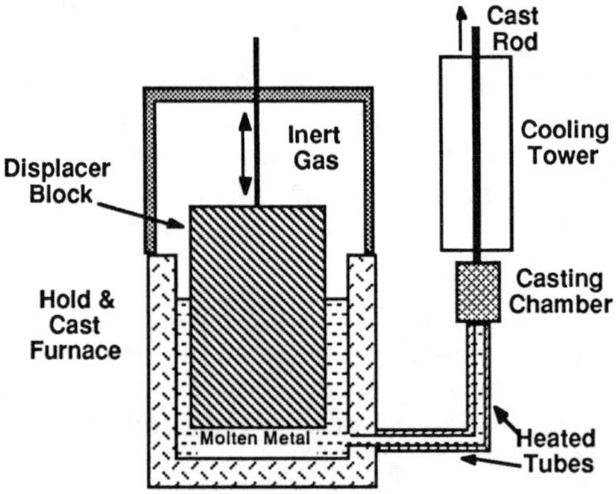

Figure 7

As illustrated in the figure, the displacer block is moved up and down to maintain a more or less constant metal level in the casting chamber to compensate for periodic large additions of molten metal and the gradual depletion of metal when casting. When it is desired to stop casting, such as on weekends or when changing alloys, the remaining metal in the holding/casting furnace can be dumped out and the power turned off.

The molten metal from the holding/casting furnace underpours from the bottom of furnace through a heated tube and then moves upward into the levitation coil/heat exchanger assembly through another heated tube. The levitation coil and heat exchanger assembly (the casting chamber) is held in place by wing nuts so it can be replaced in about 10 minutes when maintenance is needed or another size or shape of product is to be made. When such a change is made, the holding/casting furnace is not emptied or turned off. The metal level in the furnace is simply lowered from the normal "cast" level by raising the displacer block.

After solidified rod emerges from the top of the levitation coil/heat exchanger assembly, it enters a sealed cooling tower filled with an inert atmosphere and is cooled by water sprays. The sensing coil of an eddy current flaw detector is also located in this cooling chamber so that the rod quality can be continuously monitored and recorded when desired.

The rod is cooled to about 80 °C before it leaves the sealed cooling tower so that formation of surface oxides is prevented. Synchronized pinch rolls in the cooling tower and top drive support the weight of the cool rod and guide it to a conventional type of coiler designed to handle 2500 kg coils.

All temperatures, rod speed, metal level, water flow, levitation coil current, startup sequence and other process variables are monitored and controlled by a GE Series Six programmable controller. (Some process variables can also be adjusted manually during a run). This PC is also programmed with logic and interlocking commands to display setup errors such as doors left open, cooling water valves not turned on, etc. Casting can not be started until all equipment is in the correct configuration and critical temperatures are at the desired set point. Conversely, if a serious problem occurs during casting, the PC will detect it and immediately shut the system down. As a result of this monitoring and control by the PC, it is anticipated that only one operator will be necessary to operate the system.

The electrical power consumed by the total GELEC system when ready to cast (furnaces full and at proper temperature) is 145 kw. Of this, 65 kw is needed to operate the air exhaust and cooling water circulation units. When casting and coiling rod, the power consumption rises to about 250 kw.

Melting of new copper is done when initially filling the hold/cast furnace and periodically thereafter when replenishment is necessary. The electrical energy needed to melt 350 kg of cold copper scrap and pour it into the hold/cast furnace is about 140 to 150 kw-hr.

When desired, the furnaces can be emptied and cooled down completely in about 18 hours after which all electrical power to the system can be turned off.

Expected Applications

The near term applications of the Levitation Casting process will be to make low oxygen copper and copper alloy rod of 8 to 14 mm diameter. The ability to go directly from molten metal to a usable rod size provides considerable cost savings over manufacturing rod by billet casting followed by extrusion or bar casting followed by hot rolling. The recycling of bare wire scrap into rod, as will be done at the Lowell plant, also provides significant cost savings over current procedures for scrap recovery.

Longer term, the process will be further developed in two directions. One will be to apply Levitation Casting to rod manufacturing using high melting point alloys including steel. This will require use of high temperature refractory materials in the furnaces and casting chamber plus appropriate changes in the power supply for the levitation coils to compensate for the higher electrical resistivity of these alloys.

The second type of development will be in casting of different shapes including tubes. The configuration of forces produced by the levitation field results in a tendency to cast hollow cylinders and this can be augmented by various mechanical and electrical arrangements. Initial experiments along these lines have given encouraging results.

References

H. R. Lowry and R. T. Frost , "General Electric Levitation Casting (GELEC™) Process", presented at TMS-AIME Fall Meeting, Detroit , MI, September 1984.

H. R. Lowry and R. T. Frost, US Patents 4,414,285; 4,662,431; 4,709,749; 4,719,965; and 4,770,724; Japanese Patent 1,348,818 and corresponding issued patents and pending applications in other countries covering the GELEC process, apparatus and product.

A. S. Klein and H. R. Lowry, "The GELEC™ Process" presented at the International Wire & Machinery Association Conference, Scanwire 85, Copenhagen, Denmark, 13-16 May 1985

H. R. Lowry, "Levitation Casting™ Process for Producing Rod and Other Shapes", presented at the 10th ASM Conference on "Advances in the Production of Tubes, Bars and Shapes", Orlando, Florida, 9-10 October 1986

H. R. Lowry, "Levitation Casting Process for the Production of Tube", presented at the International Tube Association Conference, Cherry Hill N.J., 21-24 September 1987

INVERSE SOLIDIFICATION ANALYSIS -
A VALUABLE TOOL FOR COMPUTER AIDED PROCESS ENGINEERING

K.L. Schwaha *, H. Holl *, H.W. Engl **,
and T. Langthaler **

* VOEST-ALPINE Industrieanlagenbau,
** University of Linz, Department of Mathematics,
Linz, Austria

Abstract

The trend towards near net shape casting requires the development of more complex design and control algorithms for continuous casting machines in particular with respect to the solidification process.

Compared to conventional continuous casting the reduction in casting thickness allows to influence more directly the solidification rate by control variables like surface cooling and/or casting speed.

In order to enable the determination of these control variables for a given solidification profile, an algorithm (inverse solidification analysis) has been developed based on nonlinear constrained optimization. The application of the model is illustrated by practical examples ranging from conventional continuous casting to advanced thin slab casting.

Metallurgical Processes for the Year 2000 and Beyond
Edited by H.Y. Sohn and E.S. Geskin
The Minerals, Metals & Materials Society, 1988

Introduction

Process development in the Iron and Steel Industry is directed towards the realization of continuous process sequences. Since the direct linkage of processes imposes some restrictions onto the individual process steps, more complex algorithms are needed for design of equipment and control of operation. In particular the recent trend to reduce casting thickness, i.e. to cast thin slabs, has made the direct coupling of continuous casting and rolling absolutely necessary. The prospective development in strip production is schematically shown in fig. 1.

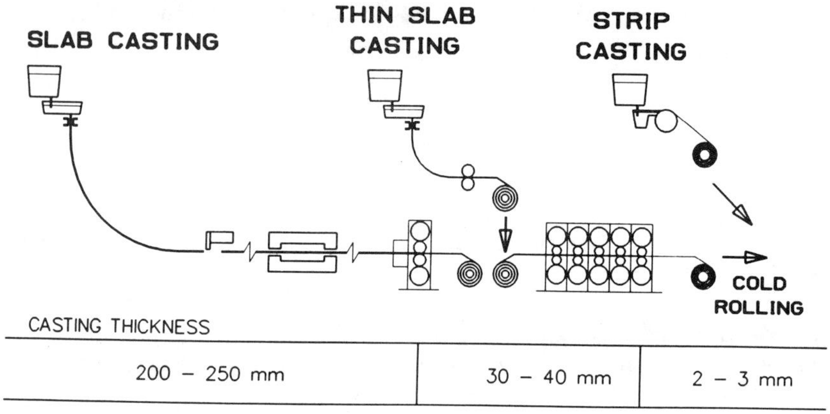

Figure 1 - Comparison of process routes for strip production

Direct hot rolling of continuously cast slabs or strips does not only require to match the output and input of the two subsequent processes with respect to quantity, but also to optimize exiting and feeding temperature of the slabs even in cases of disturbances. In consequence, the control strategies for cooling in the caster have to be adapted to this demand.

In previous studies (1) it has been shown that the most important factor in producing high temperature slabs is the utilization of the latent heat of solidification which means that the end of the liquid pool is positioned rather close to the end of the strand containment. With conventional cooling strategies this can only be achieved by keeping the speed constant over a whole casting sequence because the solidification is mainly determined by the solidification time. On the other hand constant casting speed means a variation of casting rate depending on width, which causes problems in the coordination with the BOF which is run on a constant tap-to-tap time schedule.

From these considerations it becomes obvious that cooling control has to be designed and operated to provide a "dead" band of permissible casting speed variations without changing the location of the crater end. This means that the shell growth curve is defined as a function of strand length rather than of solidification time. Although the solidification process is routinely investigated by a straightforward thermal analysis (2,3,4,5), there is little information available on solving the inverse problem which means to derive the cooling conditions from a defined shell growth curve. Some relevant references are given in (6).

In the present paper the technological basis, the mathematical procedure, and the application of the inverse solidification analysis shall be outlined and discussed going from conventional continuous casting to new processes like thin slab casting.

Solidification control in continuous casting

In continuous casting of a given steel grade at a given strand thickness the solidification is influenced by the operational parameters casting speed, cooling intensity, and superheat of steel in the mold. The first task in analysing the operational limits of solidification control was to investigate the sensitivity of these individual parameters.

For this purpose a parameter study has been carried out using conventional direct solidification analysis i.e. solving the Fourier differential equation by a proven numerical method e.g. the implicit finite difference method. The procedure of direct solidification analysis is illustrated in fig. 2.

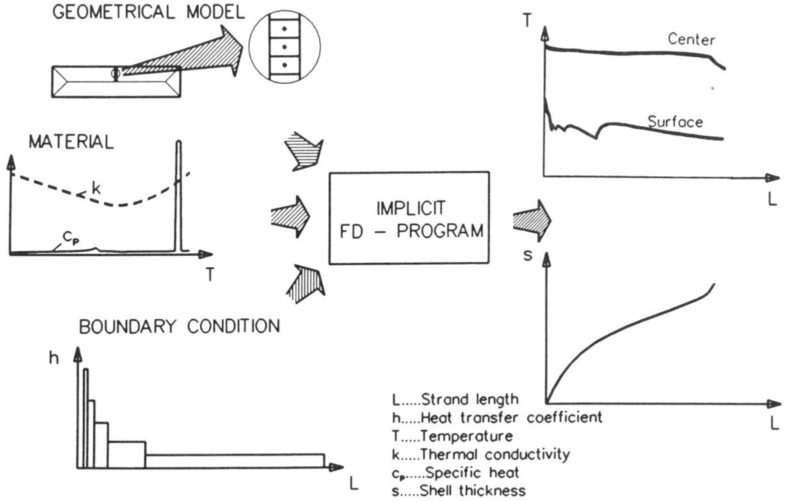

Figure 2 - Scheme for direct analysis of the solidification process

Since the analysis was restricted to slab casting with symmetric cooling on both faces a one-dimensional model representing only one half of the strand is seen to be adequate within the practically relevant accuracy. The material properties were taken temperature dependent. The boundary conditions in the mold were defined as a function of the surface temperature, in the strand support the cooling was simulated by a step-like function of the heat transfer coefficient.

The parameter studies concentrated on evaluating the influence of the above mentioned operational parameters regarding

- the local solidification rate just below the mold and
- the overall metallurgical length.

In the first analysis directed towards the local soldificiation rate the initial shell thickness s_o has been varied according to the casting speeds in the range of 1.0 to 3.0 m/min for an initial shell thickness of 10 to 20 mm, heat transfer coefficients were varied between 500 and 2000 W/m^2K, and the superheat has been varied from 0 to 50 ^oC. From the resulting transient temperature fields the local soldification rate for an initial shell thickness of 15 mm was determined at different time steps (fig. 3).

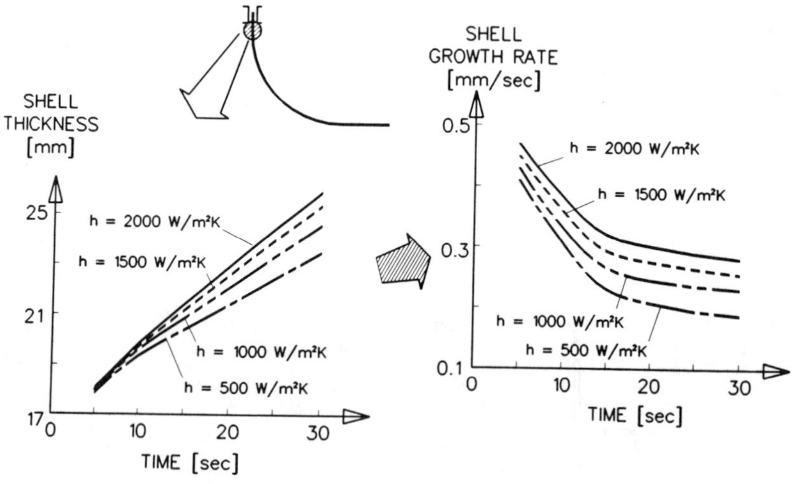

Figure 3 - Determination of local shell growth rate depending on cooling intensity

The functional relationships derived by the above mentioned procedure are visualized in fig. 4 in the form of three-dimensional surfaces. As can be learned from the dependence on the initial shell thickness the casting speed has the dominating influence on the solidification rate.

In order to determine the contribution of each parameter quantitatively the tangents of the three-dimensional surfaces have been calculated and the partial slopes are used as an indicator for the influence of each parameter which leads to the following rating:

$$\frac{\partial s}{\partial v} : \frac{\partial s}{\partial h} : \frac{\partial s}{\partial T_{super}} = 11 : 3 : 1 \qquad (1)$$

(The symbols for all equations are defined in table 1)

The same procedure has been applied with respect to the metallurgical length assuming a ratio of 1 : 3 of available heat transfer representing the turn-down ratio of 1 : 10 of an air/water cooling zone and a variation of the casting speed form 1 to 2 m/min. The variational ratio is found to be rather similar with:

$$\frac{\partial L_{met}}{\partial v} : \frac{\partial L_{met}}{\partial h} : \frac{\partial L_{met}}{\partial T_{super}} = 10 : 3 : 1 \qquad (2)$$

Table I <u>Definition of symbols used in the paper if not specified otherwise</u>

s	shell thickness
v	casting speed
h	heat transfer coefficient
x	length coordinate
k_E	solidification coefficient
t	time
k	thermal conductivity
ρ	density
Q	latent heat of solidification
L	strand length
L_{met}	metallurgical length i.e. crater end
T	temperature
T_l	liquidus temperature
T_{super}	superheat temperature
T_s	surface temperature
T_{en}	ambient temperature
s_o	initial shell thickness
c	effective heat capacity

The results of the numerical calculations have also been compared to the ones coming from analytical formulas which are frequently used for calculating the solidification. With some rough simplifications the analytic solution of the classical Stefan problem (7) leads to the following equations:

$$s = k_E \sqrt{t} \tag{3}$$

$$k_E = \sqrt{\frac{2k}{\varrho Q}(T_1 - T_s)} \tag{4}$$

The functional relationship derived by this analytic approach is shown in fig. 5 for the specified assumptions of material properties and operating parameters.

Influence of control variables on local shell growth

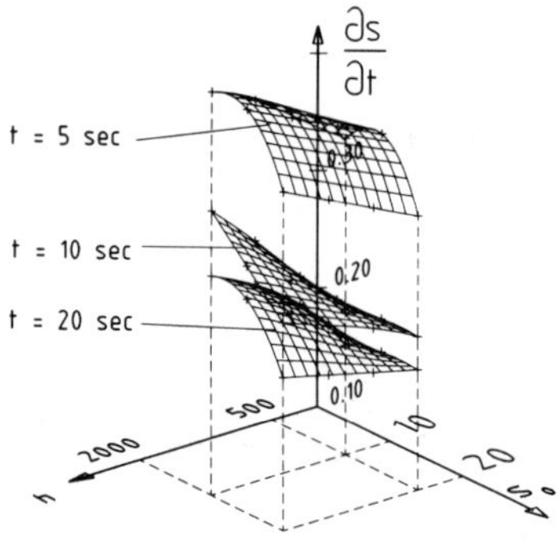

Figure 4 - Numerical results

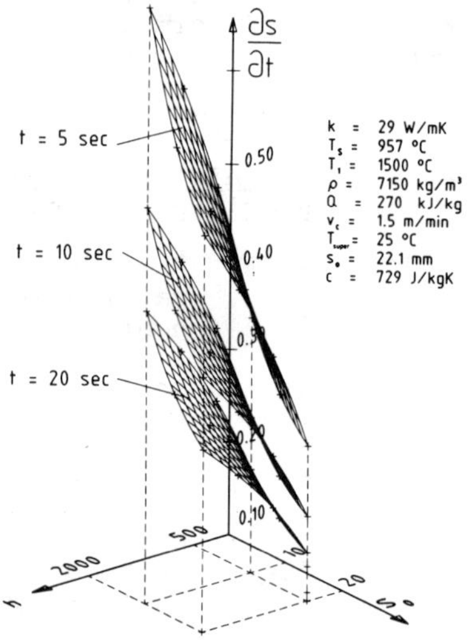

Figure 5 - Analytical results

When calculating the three partial derivatives of the shell thickness the following ratios result

$$\frac{\partial s}{\partial v} : \frac{\partial s}{\partial h} : \frac{\partial s}{\partial T_{super}} = 9 : 8.6 : 1 \qquad (5)$$

The equivalent influence of casting speed and heat transfer coefficient can also be derived directly from equ. (3) and (4).

$$k\,(T_1 - T_s) = s\,h\,(T_s - T_{en}) \qquad (6)$$

$$s = f\sqrt{h}\sqrt{\frac{L}{v}} \qquad (7)$$

According to this relation a change of v by a multiplication factor results in the same change of h to hold the shell thickness constant.

From the comparison of the numerical and the analytical results it is seen that the simple formulas may lead to wrong conclusions when analysing the dependence of local solidification behaviour on operating parameters. An explanation will be given in the next section. Nevertheless, the simple formulas can be taken for rough estimates of the metallurgical length as long as similar systems are compared because the solidification coefficients are derived from specific empirical data.

Regarding the original problem of optimizing direct rolling, the operating conditions have to be derived from a given solid shell profile. As the solidification in slab casting is influenced by many variables (e.g. the water flow rates in the individual zones) the direct solution of this inverse problem by parameter studies of conventional solidification analysis would be rather problematic and time consuming. Consequently, a new efficient algorithm has been developed to directly provide the answers for the solidification control adaptable to direct rolling of slabs or strips.

Mathematical Method for Inverse Solidification Analysis

In the "direct" solidification analysis the transient temperature field and the solidification front are calculated with all other quantities, especially the heat transfer function, being given (see fig.2). In contrast, the inverse problem means to compute the heat transfer function from a given solidification profile with all other quantities being given. Actually, for the practical use it is sufficient to compute a function such that the resulting solidification front is "sufficiently close" to the given one ("control problem").

Since the latter two problems are inverse problems, their nature regarding well-posedness has to be clarified. According to Hadamard, a problem is called "well-posed" if for all admissible data, a unique solution exists and that solution depends continuously on the data, where the concept of continuity used has to be such that it is meaningful for the actual problem to be solved. Otherwise, a problem is called ill-posed; see (8) for many aspects of ill-posed problems and "regularization methods" for their numerical treatment.

Before doing any numerical work, it is crucial to find out if one deals with a well-posed or an ill-posed problem since the numerical treatment of an ill-posed problem without taking the ill-posedness into account usually leads to severe numerical instability. For the present problem results on uniqueness and stability can be found in (9).

The relevant equations for the solidification analysis are specified below (cf.(6,10)). As explained in (6), these equations are derived from an enthalpy formulation with a gradual change in enthalpy along a solidification interval in contrast to the classical Stefan problem. This is one reason for the fact that equations (3) and (4) lead to unsatisfactory results, the other one being that the material parameters are temperature dependent.

$$\frac{\partial}{\partial x}\left[k(T(x,t))\frac{\partial T(x,t)}{\partial x}\right] = \varrho c(T(x,t))\frac{\partial T(x,t)}{\partial t} \quad \text{for } (x,t)\in\left[0,\tfrac{d}{2}\right]\times[0,t] \qquad (8)$$

$$k(T(0,t))\frac{\partial T(0,t)}{\partial x} = h(t)\left[T(0,t)-T_W\right] \quad \text{for } t\in[0,t_E] \qquad (9)$$

$$\frac{\partial T}{\partial x}\left(\tfrac{d}{2},t\right) = 0 \quad \text{for } t\in[0,t_E] \qquad (10)$$

$$T(x,0) = T_O(x) \quad \text{for } x\in\left[0,\tfrac{d}{2}\right] \qquad (11)$$

$$T(s(t),t) = T_1 \quad \text{for } t\in[0,t_E] \qquad (12)$$

$[0,d]$	Cross section of the strand in x-direction, cooled by spray cooling at x=0
T_W	Spray water temperature
t_W, t_E	Times when the strand leaves the secondary cooling zone and is completly solid
T_O	Temperature profil at time t=0

The numerical treatment of the problem is done by taking advantage of the special structure of the heat transfer function h(x), i.e. the existence of locally defined cooling zones, in which the spray cooling can be regulated independently, but within one zone it has to be constant. Hence, there are only finitely many, say n, degrees of freedom, a fact that is likely to regularize the problem (i.e. approximate an ill-posed problem by a well-posed one). Therefore the n "heat transfer coefficients" $h_1, \ldots\ldots, h_n$ and possibly the casting speed v are the control variables which can vary between prescribed ranges, whereas the superheat has been neglected as control variable because of its small influence.

It is precisely this observation that makes the problem easily numerically tractable when looking for a heat transfer function with infinitely many degrees of freedom, one would have to construct a regularization algorithm in order to avoid numerical instability. It is the "built-in regularization" of h(x) being described by n parameters that prevents numerical trouble; hence it is essential to use this.

For each choice of these heat transfer coefficients the deviation between the resulting and the desired solidification front is measured by a weighted sum of squares at finitely many points. The aim is to minimize this objective function by choosing the heat transfer coefficients appropriately within their prescribed bounds.

This is a linearly constrained nonlinear optimization problem. Each evaluation of the objective functional involves solving the direct problem specified above, i.e. solving a system of nonlinear partial differential equations, which is very time consuming. Thus, it is crucial for any numerical algorithm to use as few evaluations of the objective functional as possible.

For solving this constrained optimization problem, Rosen's gradient projection method with line search has been chosen (11). Fig. 6 shows a three-dimensional interpretation of Rosen's gradient projection method using a surface of second order.

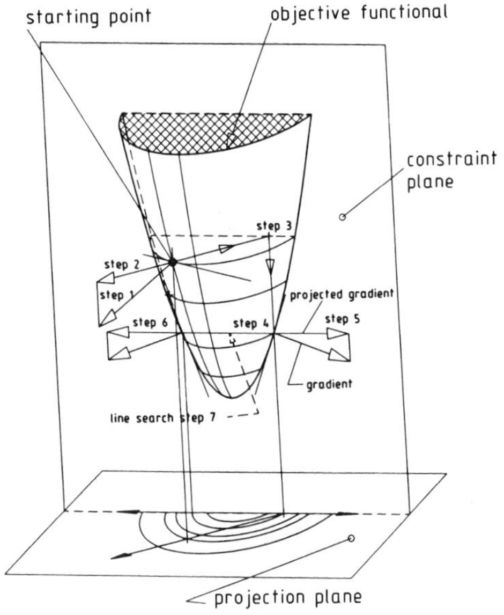

Figure 6 - Schematics of gradient projection method

The gradient of the objective functional was approximated by forward differences. The choice of the line search turned out to be crucial, since the objective function appeared to have quite flat minima, which seems to reflect the underlying ill-posedness of the inverse problem. Details are given in (10) and a flowchart of the used program is visualized in fig. 7.

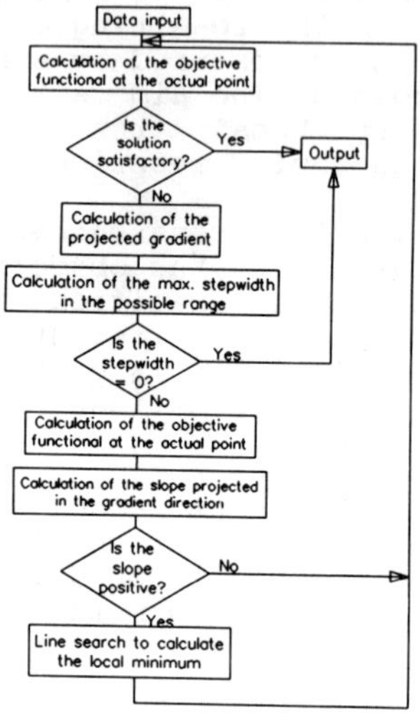

Figure 7 - Simplified flow diagram of the inverse solidification analysis

In an explicitely given constrained optimization problem with even only a few variables, naive approaches like evaluating the objective functional for a few parameter configurations and using some kind of interpolation will not very likely lead to a satisfactory solution in reasonable time. This can be expected much less for the optimization problem under consideration whose objective functional involves the solution of a nonlinear partial differential equation.

From the many efficient algorithms for nonlinear constrained optimization Rosen's method has been chosen as being the simplest one. Due to the complicated structure of the objective functional, it cannot be taken for granted that the algorithm arrives at a global minimum, but only at a local one. However, it is only needed that the objective functional becomes small enough, so that the prescribed and the achieved solidification fronts are close enough for practical purposes; this can be achieved by the present alorithm, as the practical examples show.

Application of inverse solidification analysis

The inverse solidification program has been implemented on a VAX-11/780 so that interactive use is possible. In the following the versatility of the program for computer aided engineering shall be demonstrated with particular emphasis on the new casting processes.

Analysis for conventional slab casting

Starting from the input data like the solidification profile, the initial and limit values of heat transfer, the material properties, and the model geometry, the inverse algorithm performs several forward calculations in order to determine the gradients necessary for the application of Rosen's method. The solution is found by minimizing the weighted squares deviations. The iterative procedure is illustrated in fig. 8 by the deviations after each iteration step and by the achieved approximation.

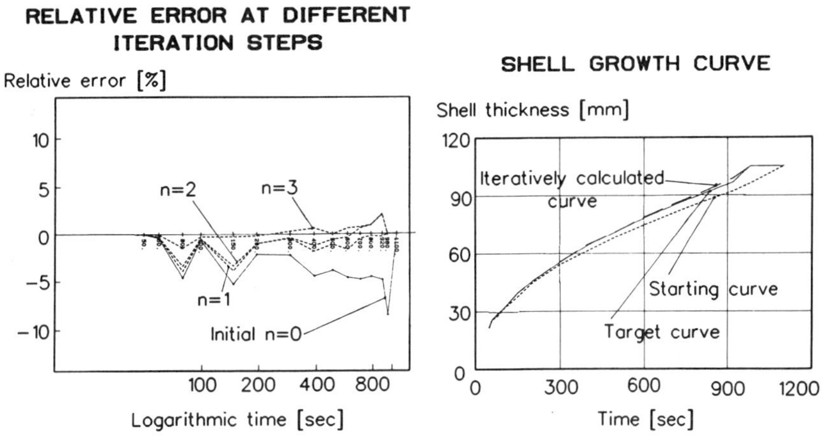

Figure 8 - Illustration of the inverse analysis

The computation time for this example is typically 170 seconds. The resulting heat transfer function is shown in fig. 9 together with the relative deviation of the heat transfer function, which has been calculated for conventional cooling.

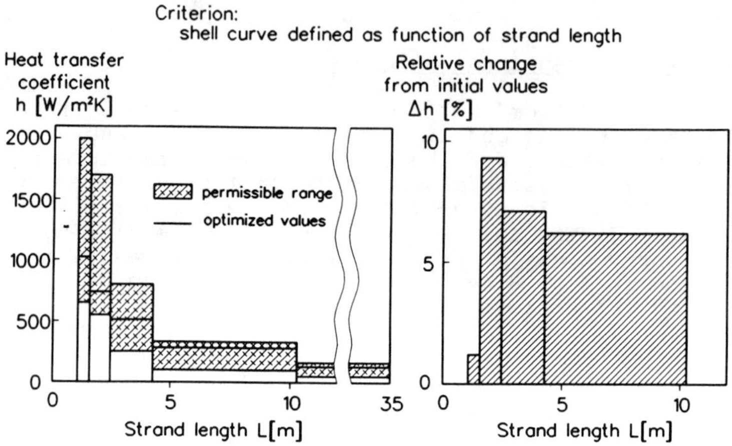

Figure 9 - Optimized heat transfer function by inverse solidification analysis

Analysis for conventional and advanced thin slab casting

After having established and tested the new calculation procedure in comparison to experimental data, the new tool of computer aided engineering has been applied to investigate specific problems coming from new processes. For their economic success the integration with the up- and downstream processes has to be optimized regarding lay-out and logistics of operation.

One important aspect is the possibility of compensating speed variations in the caster due to delays in liquid steel supply or to operational problems in the finishing stands. Taking the other condition of minimum variation in the temperature of the thin slabs into account the solution has to come from adaptations in the strand cooling. For this purpose the inverse solidification algorithm has been used with the casting speed to be the free parameter floating around a base curve. Different casting thicknesses have been investigated assuming identical through-put for the base curve.

Fig. 10 shows the tolerable speed variations as a function of casting thickness when applying the new cooling strategy derived from inverse solidification analysis.

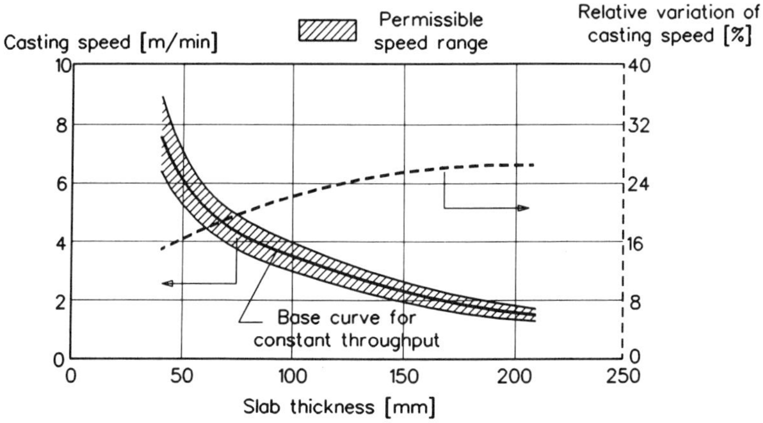

Figure 10 - Control range of casting speeds depending on thickness

Considering the relative values, the variational band is being reduced with decreasing casting thickness. This result is explained by the fact that with the decrease of thickness the speed is increased proportionally and that at higher speeds a longer section with indirect cooling, i.e. contact cooling in the mold and by cooling plates, becomes necessary so that less time is available to directly influence the solidification process.

The importance of exactly positioning the end of the liquid pool becomes also evident from the consequences on machine length. As thin slab casting requires rather high operating speeds, speed variations lead to significant changes in the metallurgical length when applying conventional cooling strategy as shown in fig. 11. In contrast, the new cooling strategy keeps the machine length within narrow limits despite substantial variations in speed.

The further decrease in casting thickness as proposed for Advanced Thin Slab casting ATS (12) leads to casting speeds of approximately 15 m/min which makes the use of a mold with moving cooling walls inevitable and which drastically reduce the time for controlling the solidification.

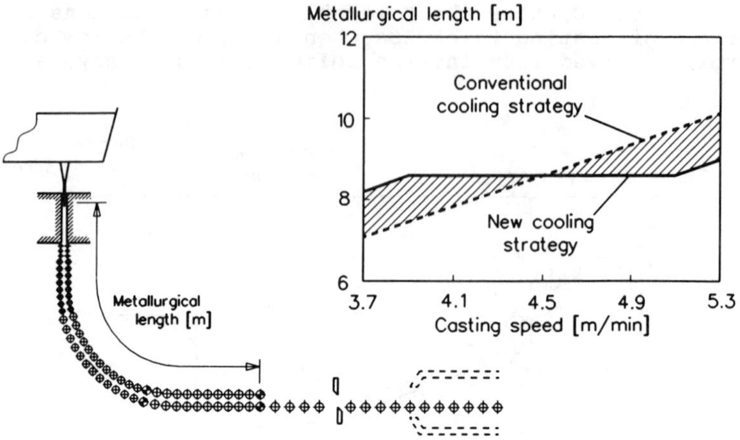

Figure 11 - Optimization of machine length by inverse solidification analysis for conventional thin slab casting

One aspect of the fundamental investigations to determine the proper mold length was the analysis of possible solidification control. Due to the nature of the process the heat transfer by direct contact cannot be influenced very much because at high heat transfer rates (i.e. intimate contact) a gap forms instantly between strand and cooling surface. In consequence, for the required close control of metallurgical length the secondary cooling by direct water sprays has to be made sufficiently long. The results of these investigations are shown in fig. 12 using again the amount of permissible speed variations to be the relevant parameter for the evaluation.

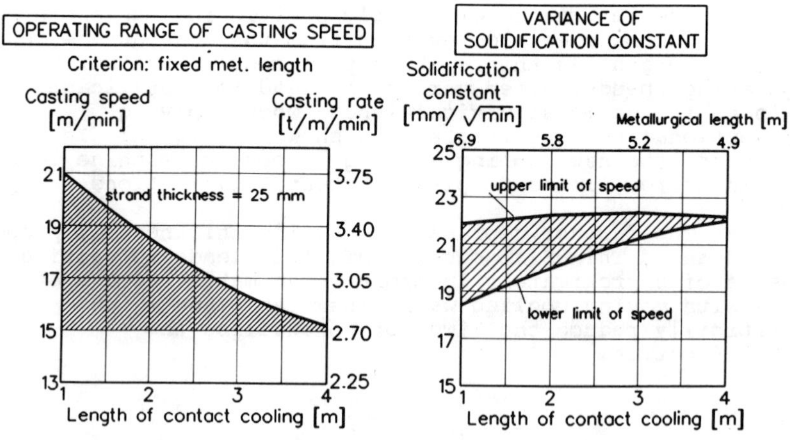

Figure 12 - Investigation of contact cooling length for advanced thin slab casting

In combination with calculations of the strand support which will be necessary below the mold these relationships are the basis to define the fundamental data for the design of the ATS caster.

Summary and Outlook

The development of an efficient algorithm for inverse solidification analysis allowed to use it as a tool to design continuous casting machines and to specify the limits of the new casting processes. A new cooling strategy has been derived which is primarily aiming at a locally fixed solidification front, so that the enthalpy of slabs at the exit of the caster is kept almost constant within a defined range of casting speeds by optimum use of the latent heat of solidification.

The use of the model can also be extended to investigate the formation of the crystalline structure by defining the local solidification rate as input parameter and thereby to locate the transition from dendritic to equiaxed structure. This type of investigations is planned for future research in combination with metallurgical data.

Acknowledgement

The authors gratefully acknowledge the great assistance by Mr. Bruno Lindorfer in the realization and implementation of the inverse solidification programme.

T. Langthaler has been supported by the Austrian Fonds zur Förderung der wissenschaftlichen Forschung (project S32/03).

References

1. G. Holleis, H. Bumberger, T. Fastner, F. Hirschmanner, K. Schwaha, Prerequisites for production of cc semis for direct rolling and hot charging, Continuous Casting Conf. 85, London, May 1985, paper 46.

2. K.L. Schwaha, T. Fastner, G. Holleis, E. Misera, Strand cooling and internal quality of continuously cast slabs, 2nd Voest Alpine Continuous Casting Conference 1981, paper 16.

3. H. Wiesinger, K. Schwaha, O. Kriegner, Mathematical Analysis of the Solidification Process in High-Speed continuous casting at various cooling conditions, Proc. of the 2nd Process Technology Conf. Vol. 2, Chicago, Feb. 1981, 86-94.

4. J.K. Brimacombe, Design of cc machines based on a heat-flow analysis: state-of-the-art review, Can. Met. Quart. Vol 15, No 2, (1976) 163-175.

5. A.Etienne, Modelling Solidification Structure to assess the internal quality of continuously cast product, Preprints of the 4th International Conference Continuous Casting, Brussels 1988, 597 - 608.

6. H.W. Engl and Th. Langthaler, "Control of the solidification front by secondary cooling in continuous casting of steel" in H.W. Engl, H. Wacker, W. Zulehner (eds.), Case Studies in Industrial Mathematics (Teubner, Stuttgart (1988)),51-77.

7. H. S. Carslaw & J. C. Jäger, Conduction of heat in solids (Oxford at the Clarendon Press (1959))

8. H.W. Engl, C.W. Groetsch (eds.), Inverse and Ill-Posed Problems (Academic Press, Boston (1987))

9. H.W. Engl, Th. Langthaler, and P. Manselli, "On an inverse problem for a nonlinear heat equation connected with continuous casting of steel" in K.H. Hoffmann, W. Krabs (eds.), Optimal control of Partial Differential Equations II (Birkhäuser, Basel (1987)), 67-89.

10. H.W. Engl and Th. Langthaler, "Numerical solution of an inverse problem connected with continuous casting of steel", Zeitschrift für Operations Research 29 (1985), B185-B199.

11. J.B. Rosen, "The gradient projection method for nonlinear programming (part I: linear constraints)," SIAM Journ. Appl. Math. 8 (1960), 181-217.

12. K.L. Schwaha, W. Egger, F. Hirschmanner, F. Landerl, A. Niedermayr, H. Rametsteiner, Strip Casting of Low Carbon Steel at VOEST ALPINE, Int. Symposium on Near-Net-Shape Casting of Strip of CIM, Toronto May 1987, paper 220.

PRODUCTION OF ADVANCED MATERIALS

BY THE OHNO CONTINUOUS CASTING PROCESS

A. Ohno
Department of Metallurgical Engineering, Chiba Institute of Technology, Narashino, Chiba-ken, Japan
and
Department of Metallurgy and Materials Science, University of Toronto, Toronto, Ontario, Canada

Abstract

A new near net shape casting process, OHNO Continuous Casting (OCC), has been recently developed. In this process, molten metal is fed into a heated mold maintained at a temperature above the solidification temperature of the metal being cast. Solidification begins from the central core of the casting rather than the wall of the mold. With the heated mold, the formation of crystals on the mold wall is prevented and solidification occurs out of contact with the mold surface. This produces near net shaped products with smooth, mirror-like surfaces and unidirectional grain structures free from cavities, porosity and segregation. Any crystals formed at the beginning of solidification decrease in number as solidification progresses and eventually a single crystal product is obtained. This process is applicable to any metal alloys, converting them into new materials with superior properties and quality.

Metallurgical Processes for the Year 2000 and Beyond
Edited by H.Y. Sohn and E.S. Geskin
The Minerals, Metals & Materials Society, 1988

Introduction

As a result of growing technical demands for more advanced and diversified materials, the OHNO Continuous Casting (OCC) technology has found many new applications in industry. Considerable interest in the process has been shown in Japan and other countries because of its distinct ability to:

produce a single crystal, (1)
produce a remarkably clean cast surface without a witness mark,
provide extremely good ductility,
and reduce casting defects and segregation.

This paper provides an overview of the principle of the OCC process and describes the remarkable characteristics of the material produced by this method.

The Principle of the OCC Process

The principle of the OCC process is very simple. It uses a heated mold, the temperature of which is held just above the solidification temperature of the metal being cast. This prevents nucleation on the mold walls and allows solidification to take place at the mold outlet only. The heat is extracted through the product being cast by a cooling device located near the mold exit. A schematic illustration of the general principle of the OCC process is shown in Figure 1.

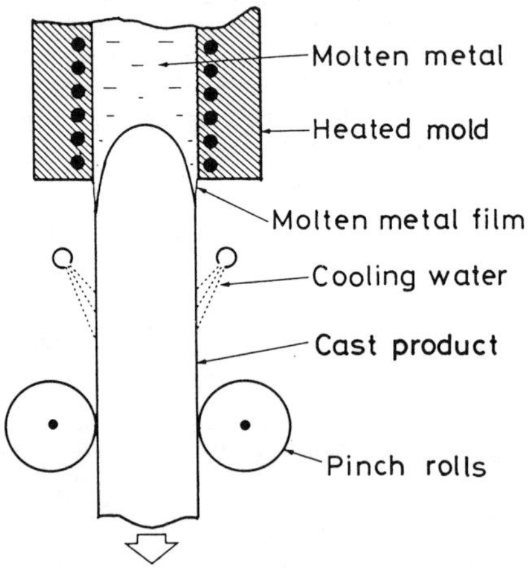

Figure 1 - Schematic illustration of the OCC Process.

Since the mold is heated, crystals are formed only at the edge of the dummy bar at the start of casting, and, as the dummy bar travels away from the mold exit, solidification progresses without contacting the mold wall. This produces an unidirectionally solidified structure which is very clean cast and having a mirror-like surface. As casting proceeds,a crystal with the most favourable growth condition survives, eliminating all other crystals. A schematic drawing of the mode of solidification at the start of casting and the formation of a single crystal is shown in Figure 2. A photograph of the macrostructure of a Sn plate showing the reduction in the number of crystals and subsequent formation of a single crystal is shown in Figure 3.

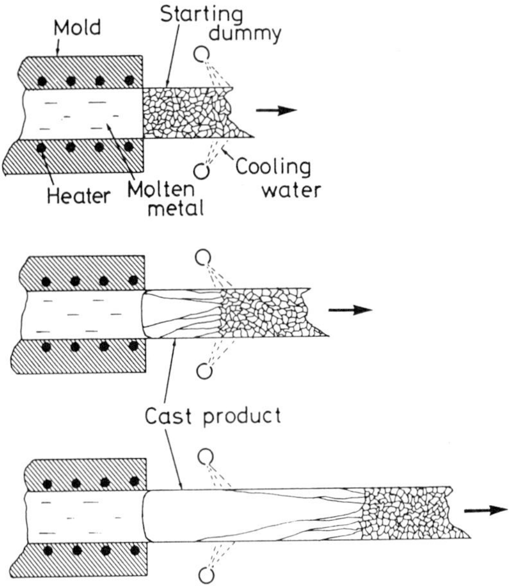

Figure 2 - Schematic illustration of the formation of a single crystal casting.

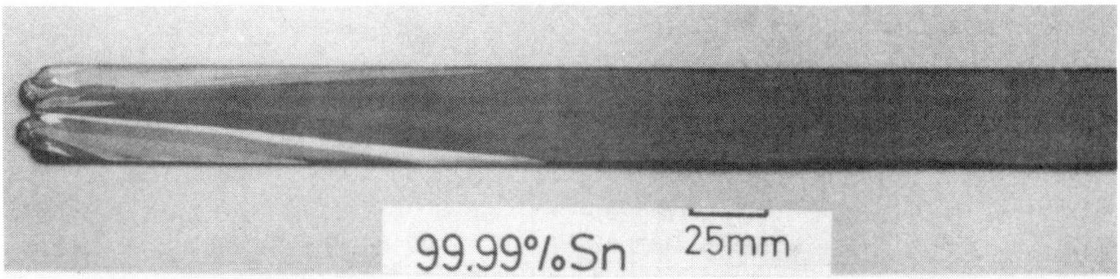

Figure 3 - The formation of a single crystal of Tin.

The Integrated Melting and Casting Facility of the OCC Process

The principle of the OCC process can be applied to horizontal, vertical and rotor-type casting methods.

A detailed diagram and photograph of a horizontal OCC apparatus for experimental use are shown in Figures 4 and 5 respectively.

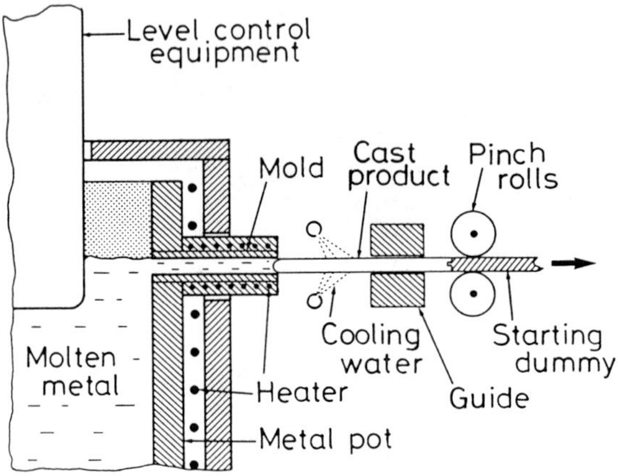

Figure 4 - Schematic illustration of horizontal OCC apparatus.

Figure 5 - Actual casting apparatus for copper with a casting capacity of about 10 Kg.

The OCC method is well suited for the casting of very small wires, plates and tubes. The apparatus consists of a melting crucible placed in an electric furnace, a heated mold, a cooling device, pinch rolls and a level control device to supply molten metal to the mold.

To start up casting, the dummy bar is set at the mold exit and the level control ramrod is lowered into the molten metal to raise the metal surface level to the desired height, insuring the delivery of a constant supply of liquid metal to the mold. As soon as the molten metal comes into contact with the dummy bar, water is introduced into the cooling device and the dummy bar is pulled by rotating pinch rolls to initiate casting. The withdrawal and ramrod controls are set to the desired speeds, taking into consideration such parameters as cooling capacity, mold temperature and melt temperature.

Remarkable Characteristics of the OCC Process

The OCC Process has applicability to a wide range of existing metals and alloys, being able to convert them into new materials with superior properties and qualities. Significant advantages of the OCC process are: (2)

improved workability
improved corrosion resistance
improved fatigue resistance
reduced segregation and uniform structure

Workability of OCC Materials

It is well known that the eutectic composition of Sn-Zn alloy material has very limited workability. However, the same material processed by the OCC method exhibits extraordinary workability. A detailed comparison of the workability of Sn-Zn alloys processed by the OCC method and the conventional method has been outlined by G. Motoyasu et al. , of the Chiba Institute of Technology in Japan.(3) The effect of the amount of reduction on the rate of breakage of Sn-Zn alloy wires during drawing operations is shown in Figure 6.

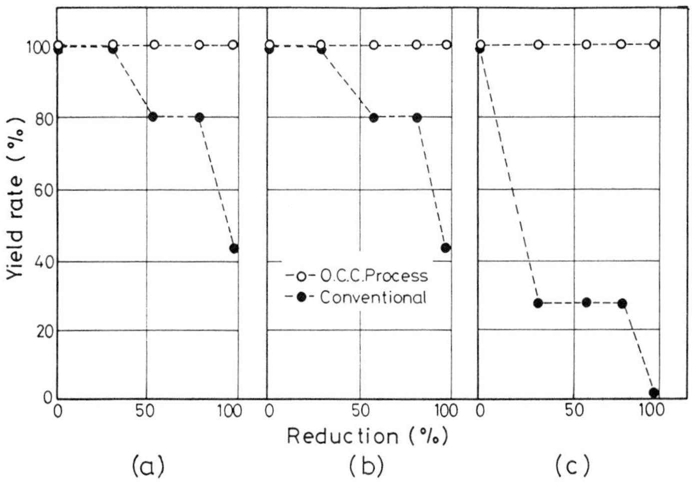

Figure 6 - The effect of the amount of reduction on the breakage rate of Sn-Zn alloy wires.
(a) Sn-6.5% Zn (b) Sn-9% Zn (c) Sn-30% Zn

The OCC processed materials possess excellent workability as is apparent if the yield values are compared Figure 6. Samples of a single crystal Sn-30% Zn alloy wire 4 mm in diameter cast by the OCC process, and 0.1 mm wire, which was drawn down to its very fine size from 4 mm cast wire without an intermediate process annealing are shown in Figure 7.

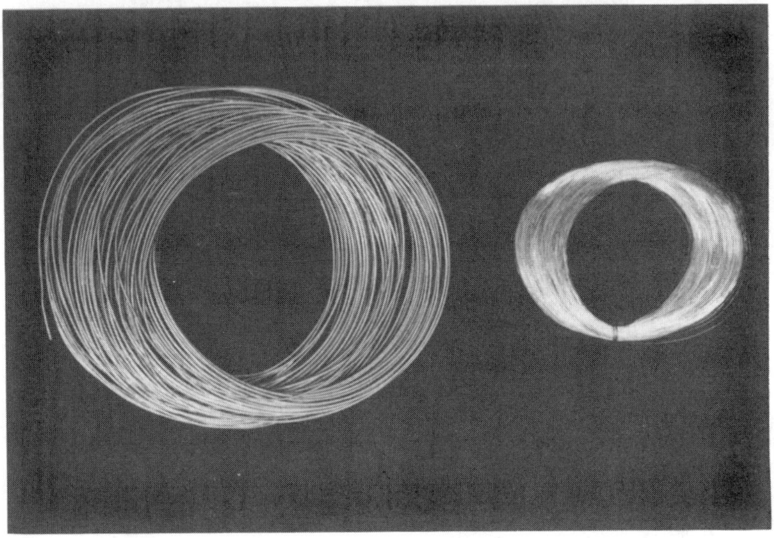

(a) (b)

Figure 7 - (a) Single crystal of Sn-30% Zn cast wire, 4 mm in diameter.
(b) 0.1 mm wire of same alloy drawn down directly from 4 mm cast wire.

Pure magnesium also has workability limitations. With use of the OCC process, this material also shows remarkably improved properties, i,e., a pure magnesium plate, 5 mm in thickness, cast by the OCC process can be rolled down to a foil 0.05 mm in thickness. This foil can be deep formed in the shape of a speaker diaphragm. Photographs of pure magnesium foil and a speaker diaphragm are shown in Figure 8 and Figure 9 respectively.

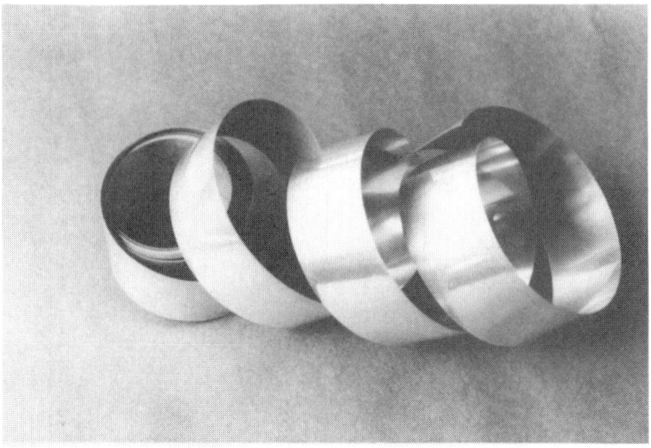

Figure 8 - Pure magnesium foil 0.05 mm in thickness.

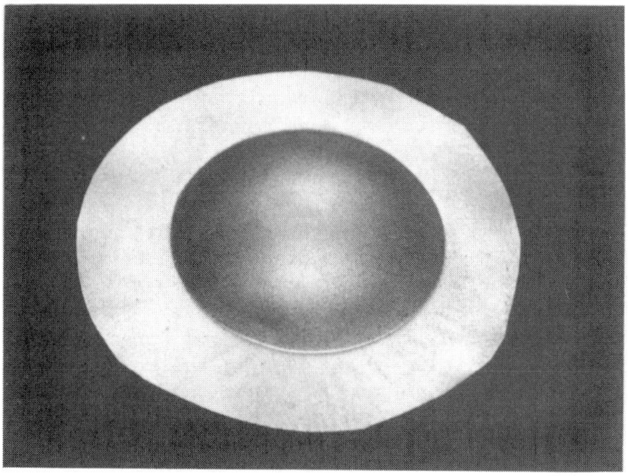

Figure 9 - Pure magnesium speaker diaphragm formed from foil of 0.05 mm in thickness.

Corrosion Resistance

The production of a single crystal material makes a great contribution to its anti-corrosion properties. Pure magnesium, 99.98% in purity solidified in the cold mold, and a single crystal material of the same purity processed by the OCC method were subjected to the salt splay test using a 5% salt solution. The result of the test is shown in Figure 10 where the resistance to corrosion is expressed in weight loss.. It is apparent in the figure that the corrosion rate for a single crystal material produced by the OCC process is significantly reduced.

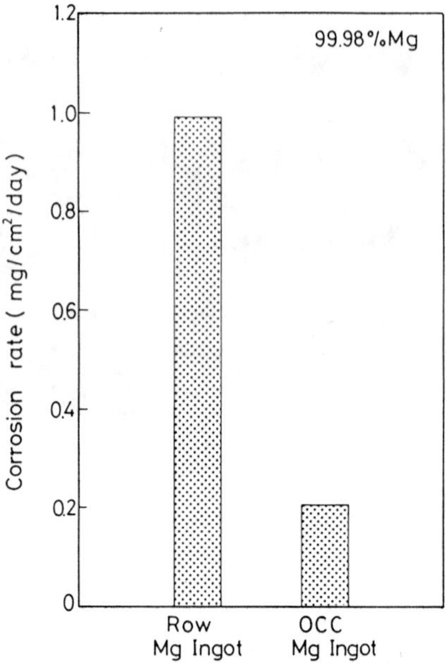

Figure 10 - The result of 5% salt solution splay test on pure magnesium.

Reduced Segregation and Uniform Structure

Because of the distinct mode of solidification of the OCC process, whereby solidification first takes place at the central core, no segregation will be observed in the central position of the casting. This results in the creation of a more uniform structure. Figure 11-a shows the uneven distribution of Al in a 6 mm wire extruded from a 160 mm billet of AZ 31 magnesium alloy cast in the cold mold. In contrast, Figure 11-b shows a very uniform distribution of Al in 6 mm wire of same alloy cast directly from the melt by the OCC process.

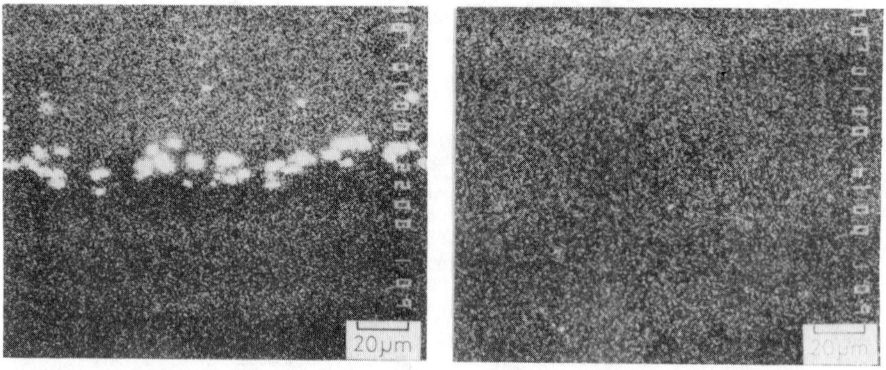

(a) (b)

Figure 11 - Examination of Segregation of Al in AZ 31 alloy by EPMA. (a) Cast conventionally extruded (b) Directly cast

Fatigue Resistance of the OCC Material

It is known that when metal materials are placed under or subjected to the cyclic-stress condition, fatigue fracture occurs. The condition of the material is, of course, critical to its potential for fracture. Therefore it is of great importance to attempt to improve its properties by eliminating cavities, grain boundaries and inhomogeneity which often lead to sudden immature failure in the cast product. The OCC process can improve fatigue resistance significantly by eliminating various defects inherent in the conventional cast product.

An Al-4% Cu wire, 6 mm in diameter and solidified in cold metal mold and a material of same composition produced by the OCC method were subjected to a 90° cyclic bending test, the result of which is shown in Figure 12.

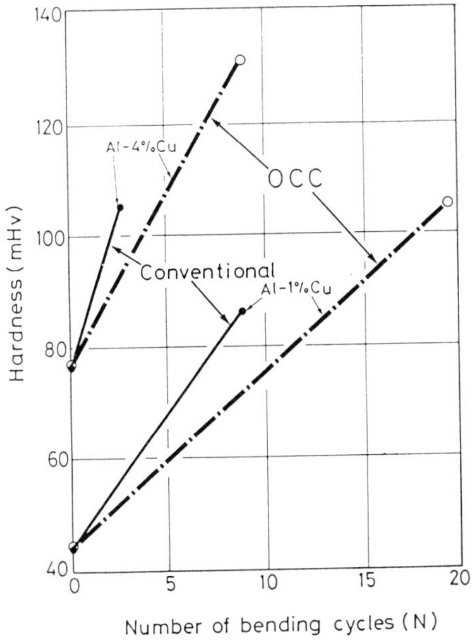

Figure 12 - Hardness result obtained by 90° cyclic bending.

It can be seen that the conventional material fractured quite rapidly, usually upon application of bending 3 to 5 times. However, the OCC material withstood much more stress, and required up to 30 bendings before the material fractured. As well, the hardness value for the OCC material increased to VH 130 compared to VH 100 for conventional material. This demonstrates the potential of the OCC process and indicates that it may be possible to use this method for the production of many new structural materials. It is for this reason that the OCC process is attracting attention as a potential method for the production of future new and improved high quality materials.

Acknowledgments

Author would like to express his gratitude to the Osaka Fuji Co. Ltd. for their cooperation and assistance in obtaining invaluable information.

References

1. A. Ohno, "Continuous Casting of Single Crystal Ingot by the OCC Process", Journal of Metals, Jan. (1986) p. 14.

2. A. Ohno, "The Continuous Casting Process with a heated mold-OCC Process", Bulletin of the Japan Institute of Metals, Sept. (1984) p. 773.

3. G. Motoyasu, T. Motegi, A. Ohno, "Structures and Workability of Unidirectionally Solidified Sn-Zn alloy Ingots obtained by a New Continuous Casting Process", J. Japan, Inst. Metals, Dec. (1987) p. 935.

ELECTROMAGNETIC FLOW CONTROL IN METALS PROCESSING OPE

Julian Szekely

Dept. of Materials Science & Engineering
Massachusetts Institute of Technology
Cambridge, MA 02139

Abstract

A review is given of the current metals processing operations, where electromagnetic forces play an important part in affecting the performance of the system. These include arc furnaces, plasmas, the Hall Cell, electromagnetic stirring and others. These considerations will be extended to discuss what new frontiers may be breached by the intelligent application of electromagnetic phenomena. The topics discussed will include magnetic flow control, levitation melting and cold crucible melting and refining.

Metallurgical Processes for the Year 2000 and Beyond
Edited by H.Y. Sohn and E.S. Geskin
The Minerals, Metals & Materials Society, 1988

THE DESIGN, OPERATION AND CHARACTERIZATION OF A PLASMA REACTOR FOR THE GENERATION OF VALUE ADDED PRODUCTS

P. R. Taylor and S. Pirzada

Dept. of Metallurgical & Mining Engineering
College of Mines & Earth Resources
University of Idaho
Moscow, Idaho 83843

Abstract

A reactor was designed and constructed to perform experiments on the generation of value added products using a plasma heat source. The engineering drawings were developed to include multiple sample ports for temperature and velocity measurements and view ports were designed to allow the Physics group to make optical measurements. Two methods for feeding solids into the system were developed and installed. Preliminary experiments were performed to characterize the system at various flow rates and power settings. Experiments were performed using zinc metal and oxygen to form zinc oxide. The particles that were generated were characterized chemically using atomic absorption and x-ray diffraction; and physically using scanning electron microscopy. The particles appear to be of value added size; but there is a tendency for the fine particles to adhere to one another forming agglomerates. Additional experimental work will continue the evaluation of the major parameters.

Metallurgical Processes for the Year 2000 and Beyond
Edited by H.Y. Sohn and E.S. Geskin
The Minerals, Metals & Materials Society, 1988

COMPUTER AIDED DESIGN OF HYBRID PLASMA REACTORS FOR USE IN MATERIALS SYNTHESIS

J. W. McKelliget* and N. El-Kaddah**

*Dept. of Mechanical and Energy Engineering
University of Lowell, Lowell, MA 01854

**Dept. of Metallurgical and Materials Engineering
The University of Alabama, Tuscaloosa, AL 35487-9644

Abstract

A mathematical model for the analysis and design of hybrid plasma reactors for advanced materials synthesis is presented. The model is based upon a solution of the electromagnetic vector potential equation and is capable of predicting the two-dimensional velocity, temperature, and electromagnetic fields as well as the reaction kinetics inside the reactor for any axi-symmetric coil configuration. The model is used as a computer aided design tool for the design of hybrid plasma reactors consisting of a conventional DC torch augmented by an RF induction coil and it is demonstrated that the hybrid system possesses superior characteristics for materials synthesis over conventional DC or RF systems. The coupling between the RF and DC components is found to affect both the temperature field and the flow field and to have a significant effect on the reaction kinetics and on materials recovery. The model is used to study the thermal decomposition of silicon tetrachloride as a function of reactor design and operating parameters and it is demonstrated that through a CAD approach it is possible to significantly improve the operation of materials synthesis systems.

Metallurgical Processes for the Year 2000 and Beyond
Edited by H.Y. Sohn and E.S. Geskin
The Minerals, Metals & Materials Society, 1988

Introduction

In recent years there has been growing interest in the use of the inductively coupled plasma for chemical synthesis as a way of meeting the increasing demand for high quality, high performance materials for use in critical applications/1/. Although the basic feasibility of this approach has been demonstrated in the laboratory/2,3/ there still exists a number of operational problems that adversely affect system performance for some applications. In particular, the recirculation and back flow effects that arise from the electromagnetic forces in the plasma can result in the repulsion of the reactants from the hot plasma core and their subsequent deposition on the walls of the reactor. Another operational problem arises from the fact that the energy dissipated in the plasma has to compensate for the heat losses to the environment by conduction, convection, and radiation in order to maintain the ionization of the plasma and, hence, the stability of the plasma may be affected by the presence of the reactants and their products.

Recently it has been suggested/4/ that some of these problems may be alleviated by the use of a low power DC torch augmented by an RF induction coil, the so called hybrid plasma. In theory the presence of the DC component at the center of the reactor counteracts the electromagnetically driven recirculating flow while, at the same time, providing a constant stream of ionized gas which can act to improve plasma stability. By its very nature the hybrid reactor is an extremely complicated system involving the mutual interaction of electromagnetic fields, fluid flow fields, temperature fields, and reaction kinetics and the key to the successful exploitation of these systems is to develop a quantitative description of these phenomena through an analysis of the underlying physical laws. It is the contention of the authors that in the twenty first century a very important element in the design, construction, and operation of complex materials processing systems will involve a computer aided design methodology based upon the application of a sophisticated mathematical model of the system.

In this paper the power of the CAD approach is demonstrated using a model reaction system based on the thermal decomposition of silicon tetrachloride and the predicted performance of the hybrid torch is compared to that of conventional RF and DC systems.

Model Formulation

In developing realistic models of RF plasma reactors for materials synthesis it is necessary to address the following aspects of the system; the electromagnetic field, the fluid flow field, the temperature field, the reaction kinetics, and the transport of the reaction products.

Electromagnetic Field

The electromagnetic field is calculated using the technique developed by the authors for RF plasma systems/5,6/. In this technique, Maxwell's equations are expressed in terms of the electromagnetic vector potential, **A**, (defined through the relation $\mathbf{B} = \nabla \times \mathbf{A}$, where **B** is the magnetic flux density) as:

$$\nabla^2 \underline{\underline{A}} = \sigma\mu_0 \frac{\partial \underline{\underline{A}}}{\partial t} \tag{1}$$

where σ is the electrical conductivity of the plasma, and μ_0 is the magnetic permeability in vacuum.

The boundary condition necessary for the solution of Equation (1) may be expressed as an integral over the current distribution in the system as:

$$\underline{A}(\underline{r}) = (\mu_0/4\pi) \left[\int_{coil} \underline{J} / | \underline{r} - \underline{r}_c | dv_c + \int_{plasma} \underline{J} / | \underline{r} - \underline{r}_p | dv_p \right] \quad (2)$$

Here, J is the current density vector, **r** is a position vector, dv is a volume element of integration, and the subscripts c and p refer to the coil and to the plasma respectively. This expression permits the individual specification of the location, and associated current, of each coil turn which gives the model a unique flexibility as a design tool for induction plasma reactors. The integrals in Equation (2) were evaluated using the technique of mutual inductance/7/.

Fluid Flow Field

In order to realistically describe the reaction kinetics in the system it is necessary to account for the possibly turbulent nature of the flow. Although the assumption of laminar flow is usually very good in the plasma fireball, in the cooler regions of the torch, especially near the reactant inlets, turbulence effects may be significant. The equations governing the flow field are the continuity equation and the turbulent Navier-Stokes equation:

Conservation of mass

$$\underline{\nabla} \cdot (\rho \underline{V}) = 0 \quad (3)$$

Conservation of momentum

$$\rho \underline{V} \cdot \nabla \underline{V} = - \underline{\nabla} P + \underline{\nabla} \cdot \underline{\tau} + \underline{F} \quad (4)$$

where ρ is the plasma density, P is the pressure, **V** is the plasma velocity, and τ is the effective (laminar plus turbulent) stress tensor. The electromagnetic field interacts with the velocity field through the Lorentz force, **F** = J x B.

The present model uses the well known K-ϵ two equation model of turbulence/8/ with wall functions/9/ to evaluate the effective shear stress tensor, τ.

The boundary conditions on the velocity field require zero velocity at solid surfaces while the velocity at the gas inlets is calculated from the specified gas flow rates. 'Top hat' distributions of velocity are assumed for the reactant and sheath gas inlets, while the following distribution is assumed for the DC torch exit/4/:

$$V = V_0 \left[1 - \left(\frac{r}{R_{DC}} \right)^2 \right]^{3/2} \quad (5)$$

where R_{DC} is the radius of the exit.

Temperature Field

The temperature field is governed by the principle of conservation of energy which for plasmas may be written as:

$$\rho \, \underline{v} \cdot \underline{\nabla} h = \underline{\nabla} \cdot (k \, \underline{\nabla} T) + S_J - S_R + S_C \tag{6}$$

where k is the effective thermal conductivity, and T and h are the plasma temperature and specific enthalpy respectively. The temperature field is essentially determined by the Joule heating, $S_J = J^2 / \sigma$, the radiation loss, S_R, and the heat required for the chemical reaction S_C.

The gas was assumed to enter at room temperature through the reactant and sheath inlets while the temperature at the DC torch inlet was assumed to have the following form/4/:

$$T = T_{room} + (T_0 - T_{room}) \; [1 - (\frac{r}{R_{DC}})^2]^{0.5} \tag{7}$$

Chemical Reaction Kinetics

Materials synthesis in plasma systems involves the chemical reaction of gases with the formation of solid products. A quantitative description of the reactions in the system requires the solution of the species mass conservation equations together with the appropriate chemical kinetics data.

The mass conservation equation for a general chemical species may be written as follows:

$$\rho \, \underline{V} \cdot \underline{\nabla} C_i = \underline{\nabla} \cdot (D_i \, \underline{\nabla} C_i) + R_i \tag{8}$$

where C_i, D_i, and R_i are the mass concentration, diffusivity, and volumetric chemical rate of formation of species i respectively.

In the present work the thermal decomposition of silicon tetrachloride to silicon is studied since reliable data on the rate constants at plasma temperatures are available for this reaction. Thermal decomposition of silicon tetrachloride proceeds through the successive removal of chlorine atoms according to the following reactions:

$$SiCl_4 = SiCl_2 + Cl_2$$

$$SiCl_2 = Si + Cl_2$$

Vurzel and Polak/10/ found the second reaction, the thermal decomposition of $SiCl_2$, to be the rate controlling step and determined the corresponding chemical rate constant. The volumetric rate of depletion of $SiCl_4$ in kilograms per cubic meter may be expressed as:

$$R_{SiCl_4} = - \rho C_{SiCl_4} \; 5 \times 10^7 \exp(-1.26 \times 10^5/RT) \qquad (9)$$

where R is the universal gas constant and T is the absolute temperature of the plasma.

From the stoichiometry of the reaction the volumetric rate of formation of silicon is given by:

$$R_{Si} = - 0.167 \; R_{SiCl_4} \qquad (10)$$

The reaction kinetics interact with the temperature field through the heat of reaction which totals 652.1 KJ/mole for the two stage reaction described above. In terms of eq. (6), in Watts per cubic meter:

$$S_C = - 23.2 \times 10^6 \; R_{Si} \qquad (11)$$

All calculations are performed for argon at 1 atmosphere and the plasma properties are taken from Liu/11/ and Evans and Tankin/12/. It is assumed that the concentration of other gases is low enough not to affect the properties of argon. To facilitate the mass balance required by Equation (8) it is necessary to assume that the density varies with the mass fraction of the reacting gases, according to the relation:

$$\rho = \rho_{argon} / \; (1 - 0.76 \; C_{SiCl_4}) \qquad (12)$$

The boundary conditions on the concentration of silicon tetrachloride express the requirement of zero diffusion through the solid walls of the reactor and across the center line. In order to obtain an estimate of the silicon recovery at the torch exit it is assumed that the walls act as perfect silicon sinks, and that the concentration of silicon there is zero.

Equations (1) to (12) comprise a general mathematical description of the plasma synthesis of materials. In the present study these equations are solved using the numerical techniques described in detail in references/6,7,13/.

Results and Discussion

Calculations were performed for a hybrid torch similar to the one described by Yoshida, et al., for the production of silicon nitride/4/, sketched in Figure 1. In essence, the hybrid reactor consists of a low power DC torch discharging into a conventional RF induction torch which consists of a quartz tube surrounded by a circular induction coil. The reactor wall is cooled by a sheath gas introduced through an annular opening adjacent to the wall, while the reacting gases are introduced through an annular opening immediately surrounding the DC torch.

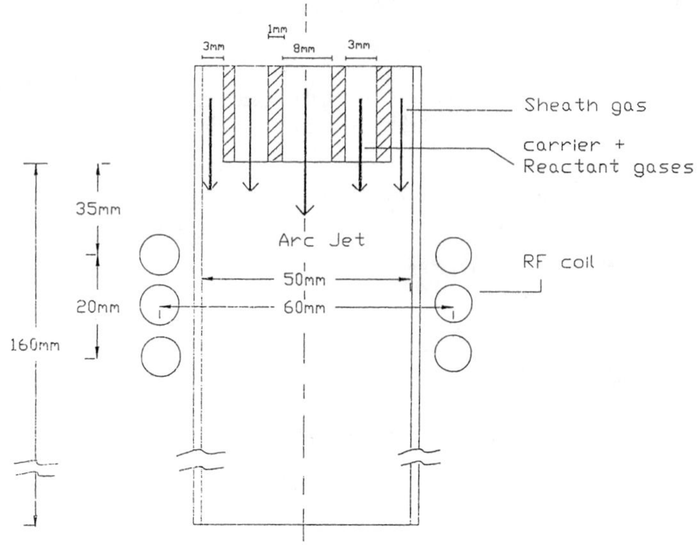

Figure 1. Sketch of the hybrid plasma reactor including dimensions.

In this study the thermal decomposition of silicon tetrachloride in the hybrid system operating at two different RF power levels is investigated. In order to evaluate the performance of the hybrid reactor calculations are also performed with the system operating as a pure DC torch and as a conventional RF torch. The operating conditions for the torch are summarized in Table I.

TABLE I. TORCH PARAMETERS

CARRIER GAS AND REACTANT FLOW RATE	10	liters/min
VOLUME FRACTION OF $SiCl_4$ AT INLET	10%	
SHEATH GAS FLOW RATE	20	liters/min
EXCITATION FREQUENCY	4.0	MHz
DC TORCH EXIT VELOCITY (V_0)	100	m/s
DC TORCH EXIT TEMPERATURE (T_0)	10,000	Kelvin

In interpreting the results of the model emphasis will be placed on defining the role of the fluid flow in convecting reaction products, not only to the torch exit where they are collected, but also to the reactor walls where they may condense and decrease product recovery. Table II summarizes the computed overall conversion rates and silicon recovery, together with the coil currents and torch powers, for the four cases considered.

TABLE II. SUMMARY OF MODEL RESULTS

Reactor	RF Current (A)	RF Power (kW)	DC Power (kW)	% $SiCl_4$ reacted	% Si recovery
DC	0.0	0.0	0.56	58	67
RF	115.0	14.0	0.0	100	30
HYBRID	115.0	16.1	0.56	100	48
HYBRID	70.0	3.56	0.56	100	76

DC Plasma Torch

The predicted flow and temperature fields for the DC torch are shown in Figures 2(a) and 2(b) respectively. The flow behaves as a classical, confined, non-isothermal, turbulent jet. The expansion of the jet is readily apparent, and the high temperature region is restricted to the very center of the reactor. It is interesting to note that the temperature of the DC plasma jet drops rapidly to 1000 K within about 5 cm of the inlet which implies that the effective region in which high temperature reactions can occur is very limited.

Figure 3 shows the mass concentration distribution of the silicon produced. The highest levels of silicon concentration effectively correspond to the reaction zone and, as suggested by the temperature field, the reactions occur in a narrow region around the periphery of the DC flame. This is reflected in the small percentage of silicon tetrachloride decomposed into silicon (see Table II). Furthermore, the computed silicon recovery is only about 67% which suggests that there is significant deposition on the wall. This is due largely to the recirculating flow near the torch exit caused by the confinement of the jet. Clearly, with a shorter, or a wider, reaction chamber much of this deposition could be avoided.

(a)

(b)

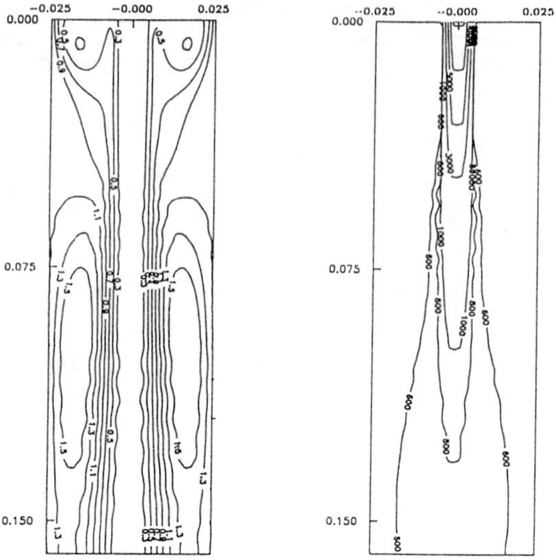

Figure 2. Computed (a) stream lines, and (b) temperature field for DC plasma torch.

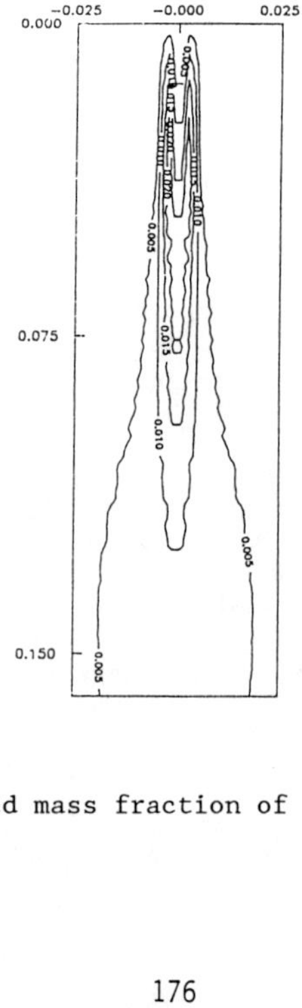

Figure 3. Computed mass fraction of silicon in the DC reactor.

Conventional RF Plasma Torch

Since the small percentage of reactant conversion in the DC torch results from the small plasma volume inherent in such a system, one anticipated advantage of the RF system is the much more extensive plasma volumes produced. Figures 4(a) and 4(b) show the predicted flow and temperature fields respectively for a conventional RF reactor operating at a power of 14.0 kW. The flow field exhibits the characteristically large recirculating loop, caused by the Lorentz force, with a small secondary recirculation downstream of the coil. It is also seen that the plasma region almost completely fills the interior of the reactor.

Figure 5 shows the concentration distribution of silicon. The injected $SiCl_4$ is entrapped by the recirculating flow and is thermally decomposed into silicon mainly at the top of the reactor along the periphery of the fireball. The silicon produced is then transported by the flow towards the wall of the reactor. This is consistent with the observations of Chase/14/ who found that injected particles bounced off the plasma fireball and, subsequently, passed along the reactor wall. Although the reaction is 100% complete, a very large fraction (70%) of the silicon produced condenses on the walls.

The inherent contradiction in the use of conventional RF plasma systems for materials synthesis is that although the electromagnetically driven recirculating flow is very effective in re-distributing the plasma gases to form extended plasma volumes, which result in efficient conversion of the reactants, it also causes entrapment of the reaction products with a consequent decrease in the product recovery. Clearly, a fundamental understanding of the fluid flow and temperature fields in RF systems provides considerable insight into the important problems of reactant conversion and product recovery.

Hybrid Plasma Reactor

The results of the calculations for the DC and the conventional RF reactor suggest that neither of these systems are ideal for materials synthesis. This is the principal reason behind the development of the hybrid plasma reactor which essentially combines the two previous reactors to reduce the effect of the recirculating flow. Such flow modification critically depends upon a balance between the degree of the RF induced recirculation and the velocity of the DC jet. Since these factors, in turn, depend upon the DC and the RF powers, two sets of calculations are performed in which the RF power is varied while maintaining a constant DC power.

The predicted flow field for a hybrid torch operating at the same coil current as the conventional RF torch is shown in Figure 6(a). As expected, the extent of the recirculating flow is considerably reduced, and the gases from the DC torch penetrate through the center of the fireball. Figure 6(b) shows the corresponding temperature field, and it is seen that, except for the very center of the reactor, the temperature and the extent of the fireball is similar to that of the conventional RF plasma. Since the recirculating flow can no longer transport the plasma to the very top of the reactor the upper boundary of the fireball is displaced downwards towards the coil.

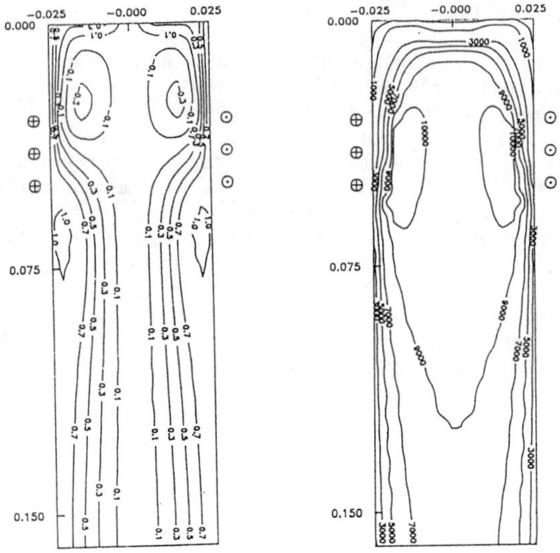

Figure 4. Computed (a) stream lines, and (b) temperature field for the conventional RF plasma torch.

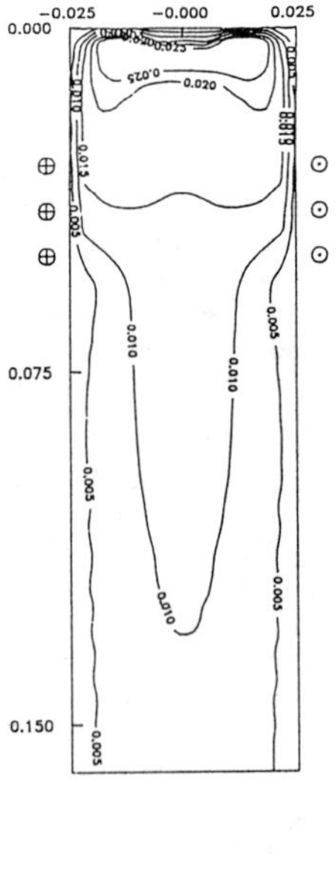

Figure 5. Computed mass fraction of silicon in the Conventional RF reactor.

Figure 7 shows the concentration distribution of silicon within the reactor. Comparison of Figures 5 and 7 shows that the silicon in the hybrid reactor is produced in a region much deeper within the fireball than is possible with the pure RF reactor. As in the conventional RF reactor the reaction is 100% complete. Although some of the silicon is caught up in the back flow the fraction of silicon deposited on the walls drops to 52% compared with 70% for the conventional RF torch. Unlike the RF torch where most of the deposition occurred at the top wall, most of the deposition now appears to occur on the side wall.

The above results suggest that the key element in improving the silicon recovery is to further reduce the back flow velocities. This may be achieved by operating the RF component of the torch at a reduced power. The flow and temperature fields for a hybrid torch operating with a coil current of 70 A are shown in Figures 8(a) and 8(b) respectively. The extent of the recirculating flow and the plasma fireball are significantly reduced with the upper boundary of the fireball being shifted even further towards the coil region. In addition, the fireball is displaced away from the side wall. Despite the decrease in power the general characteristics of the high power and the low power fireballs are very similar. It should be noted that for this coil current the model predicts plasma extinction in the pure RF torch.

The concentration distribution of silicon is shown in Figure 9. It is seen that the concentration immediately adjacent to the walls is lower than in the high power case, and the concentration at the center of the reactor is correspondingly higher. It is important to note that the conversion efficiency is unaffected by the power reduction and that the amount of silicon deposited on the walls is halved, to 24% of the silicon produced.

Conclusions

A computer aided design model for materials synthesis in hybrid plasma reactors has been developed. The model provides a detailed description of the electromagnetic field, the flow and temperature fields, as well as the chemical reaction kinetics in the reactor. The model was used to study the thermal decomposition of silicon tetrachloride into silicon in various reactor designs It was found that the hybrid torch has advantages over both conventional RF and DC reactors for materials synthesis applications. It combines the high reaction rates characteristic of the conventional RF reactor with the high product recovery characteristic of DC torches. In addition, the plasma from the DC torch considerably improves the stability of the RF discharge so that it is possible to operate at much lower power levels than are possible with conventional RF systems.

The result of this study clearly illustrates the importance the CAD approach, not only in determining the appropriate operating conditions for high materials recovery, but also for improving the design of the torch. It is intended that the model described in this paper will provide the theoretical framework for an in-depth understanding of materials synthesis in the hybrid plasma.

Acknowledgments

This work was supported in part by NSF-Alabama EPSCoR Science and Engineering Program, and by GTE Laboratories, Waltham, Massachusetts.

(a) (b)

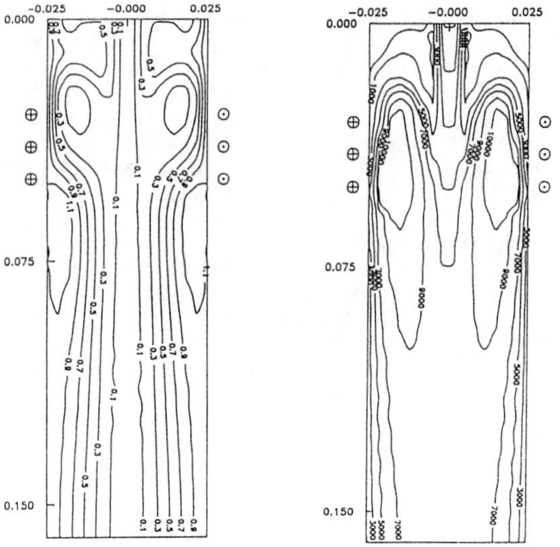

Figure 6. Computed (a) stream lines, and (b) temperature field for the high power Hybrid reactor.

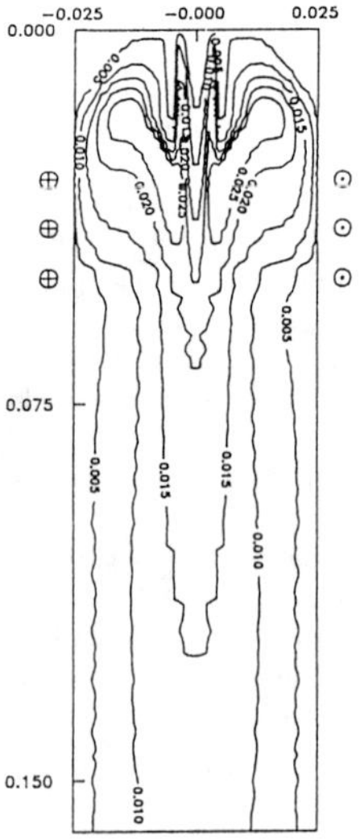

Figure 7. Computed mass fraction of silicon in the high power Hybrid reactor.

(a)

(b)

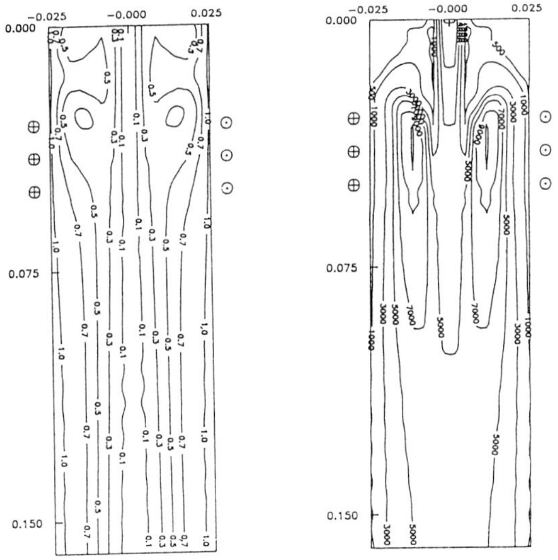

Figure 8. Computed (a) stream lines, and (b) temperature field for the low power Hybrid reactor.

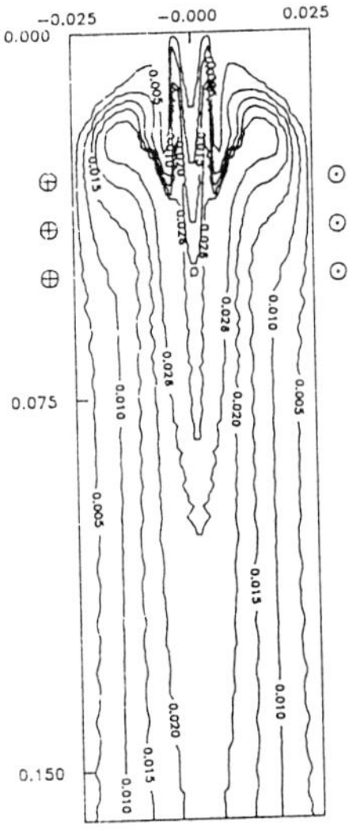

Figure 9. Computed mass fraction of silicon in the low power Hybrid reactor.

REFERENCES:

1. Committee on Plasma Processing of Materials, "Plasma Processing of Materials," (Report NMAB-415, National Materials Advisory Board, 1985).

2. G. J. Vogt, C. Hollabaugh, D. Hull, L. Newkirk, J. Petrovic, "Novel RF Plasma System for the Synthesis of Ultrafine, Ultrapure SiC and SiN3," MRS Symp. Proc. *Plasma Processing and Synthesis of Materials*, vol. 30 (1984), 283-289.

3. S. M. L. Hamblyn, B. G. Reuben, "Use of Radio-Frequency Plasma in Chemical Synthesis," *Advances in Inorganic Chemistry and Radiochemistry*, 17 (1975), 89-114.

4. T. Yoshida, T. Tani, H. Nishimura, K. Akashi, "Characterization of a Hybrid Plasma and Its Application to a Chemical Synthesis," *Journal of Applied Physics*, 54 (1983), 640-646.

5. J. W. McKelliget, N. El-KAddah, "The Effect of Coil Design on Materials Synthesis in an Inductively Coupled Plasma Torch," *Journal of Applied Physics*, 1988 (In Press)

6. J. W. McKelliget, N. El-Kaddah, "Theoretical Prediction of the Effect of Coil Configuration on the Gas Mixing in an Inductively Coupled Plasma," MRS Symp. Proc. *Plasma Processing and Synthesis of Materials*, vol. 98 (1987), 21-27.

7. J. L. Meyer, N. El-Kaddah, J. Szekely, "A New Method for Computing Electromagnetic Force Fields in Induction Furnaces," *IEEE Trans. Magn.*, 23 (1987), 1806-1810.

8. B. E. Launder, D. B. Spalding, *Lectures in Mathematical Models of Turbulence*, (London and New York: Academic Press, 1972).

9. B. E. Launder, D. B. Spalding, "Numerical Computation of Turbulent Flows," *Comp. Meth. in Appl. Mech. Engng.*, 3 (1974), 269-289.

10. F. B. Vurzel, L. S. Polak, *Khim. Vys. Energ.*, 1 (1967), 268.

11. C. Liu, "Numerical Analysis of the Anode Region of High Intensity Arcs" (Ph.D. Thesis, University of Minnesota, Minneapolis, 1977).

12. D. L. Evans, R. S. Tankin,"Measurement of Emission and Absorption of Radiation by an Argon Plasma," *Phys. Fluids*, 10 (1967), 1137-1144.

13. N. El-Kaddah, J. W. McKelliget, " Heat Transfer and Fluid Flow in Plasma Spraying," *Met. Trans. B*, 15B (1984), 59-70.

14. J. D. Chase, "Theoretical and Experimental Investigation of Pressure and Flow in Induction Plasmas," *J. Appl. Phys.*, 42(1979), 4870-4879.

KNOWLEDGE BASED CONTROL OF MATERIALS PROCESSING:

CHALLENGES AND OPPORTUNITIES FOR THE THIRD MILLENNIUM

Diran Apelian and Alex Meystel
College of Engineering
Drexel University
Philadelphia, PA

ABSTRACT

In this paper, an outline is given for a theory of intelligent control (IC) of manufacturing processes (MP) based upon what all of them have in common: all MP's can be represented as distributed networks of sequential/parallel elementary "lumped" models; all elementary models are organized by definite input-output conditions, and by the set of constraints; all elementary models incorporate not only well defined physical knowledge, but also assumed physical knowledge, expert knowledge, expected knowledge to be obtained during the process from sensors, etc. This creates a hierarchy of elementary controllers coordinated by local coordinators of different levels up to a global coordinator. The latter verifies the adequacy of MP by using a holistic image of the process: depictogram. The structure of knowledge based controller is introduced, and the analytical/linguistic tools of control are presented. The concept entails a number of perspectives which are discussed in the conclusions.

Metallurgical Processes for the Year 2000 and Beyond
Edited by H.Y. Sohn and E.S. Geskin
The Minerals, Metals & Materials Society, 1988

1. Introduction

It is becoming increasingly clear to those of us in Engineering and Technology oriented fields that the so called "service industries" do not create wealth. Moreover, it is now accepted that manufacturing is the key to the creation of wealth from a socio-economic point of view. This appreciation of the importance of manufacturing is reflected by the transformations we are seeing in our factories, as well as the changes in both curricula and research activities of our universities. Though we find these changes to be in the right direction, there is yet much more to be done to advance the frontiers of manufacturing and materials processing as we approach the third millenium.

We consider a manufacturing process (MP) an elementary component of a manufacturing system (MS) of any degree of complexity so that $\mathbf{MS}= \bigcup \mathbf{MP_i}$. This means that maximization of productivity and quality of MS is determined by maximization of productivity and quality of all MP involved within the union. Each MP is based upon material processing. Intelligent control of materials processing (ICOMP) based upon AI application has previously been analyzed and advocated in [1]. The need in ICOMP is determined by exponential growth of complexity of contemporary MP, by their rapid dynamic character in many cases, and most importantly, by the lack of knowledge about processes we need to control. Most of the knowledge available should be collected off-line, properly stored and organized, part of the expected control procedures should be precomputed, and control *templates* should be prepared (precomputed sequences of commands for familiar situations). It is important to realize that most of the parameters and variables to be monitored during the process are in fact inaccessible and cannot be measured directly. ICOMP should be capable of judging the control situations based upon indirect information. Thus the capability of *process situation recognition* should be considered an important component of ICOMP.

In addition to the "process situation", the notions are important of *inputs, variables, parameters,* and *outputs*, of the process. Inputs are the physical changes which can affect the process, usually put in a form of *command sequences*. Variables are the physical characteristics of the process which are the part of our model of the process which we can or cannot measure but which are important for predicting the output even if they cannot be measured directly. Parameters are physical characteristics of the objects and devices which are presumed to be relatively unchangeable during the process, or which are not supposed to be changed. Variables are the physical characteristics which are expected to undergo change as reflected in the quantitative or qualitative models of the MP. Outputs consist of the set of parameters which are required as a result of MP. Combination of variables and parameters recorded (or observed) at a particular instance of time will be called the *state* of MP.

General structure of ICOMP is shown in Figure 1. Generally, ICOMP consists of three parts: A-external systems, including specialists, knowledge available, and off-line computer resources, B- on-line computer-controller, and C-technological equipment with processing devices, sensors and various actuators installed. Subsystems A can be considered "Knowledge Base" (KB). KB is always incomplete, contains a lot of knowledge presented in an inadequate, inconvenient, and/or indirect form. Accuracy of this knowledge depend on the source of information and strongly varies. Subsystems B presume on-line dealing with information and producing the planning/control decisions of dealing with the situation. Within computer controller, the current situation must be recognized, put in correspondence with the available knowledge, the contribution of current situation to the desired output should be evaluated, and the "long term" plans, as well as "immediate controls" should be computed. Intuitively it is clear that both "long term" plans as well as "immediate control" solutions are a part of the integral decision considered at different levels of resolution. Multiresolutional control schemes are presently accepted in the area of intelligent control. Subsystems C are actually a hardware incarnation of the control concepts represented within the set of subsystems B.

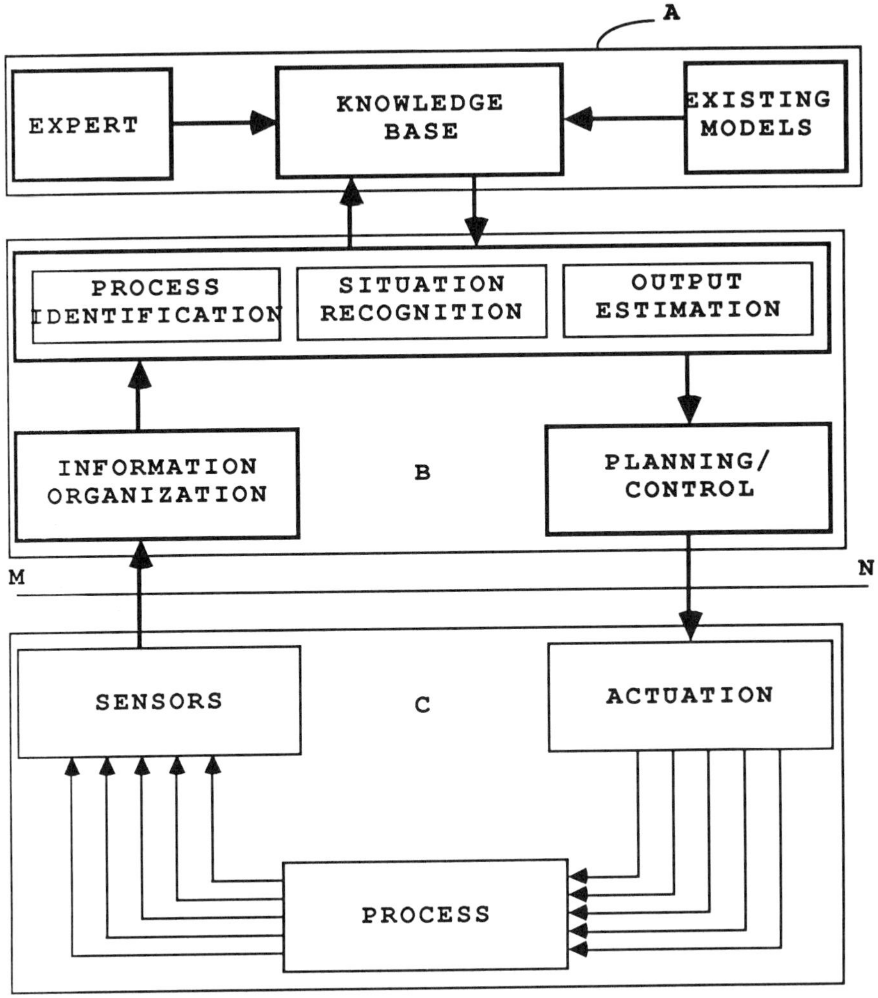

Figure 1: Intelligent processing of materials

The following concepts of intelligent materials processing are taken in account:

-Intelligent control (IC) systems for MP, are being judged upon by the following results of MP:

1) shape (configuration, and its dimensions, with definite accuracy) of the object to be created during MP,
2) microstructure of the critical parts of the object,
3) yield, i.e. ratio between the amount of material properly utilized, and the total losses of the material during the process, and
4) time of MP;

-IC systems for MP, are to be on-line controlled during processing, i.e. on-line sensing is presumed with subsequent on-line incorporation of the results of sensing in the subsystem of control commands generation;

-IC systems for MP, are solving the on-line control problems using substantial amount of knowledge, and control recommendations obtained prior to the on-line control process, thus off-line knowledge collection, organization, and development of the control templates is presumed;

-IC systems for MP, are to be utilized as a source of advanced information about the process, thus the process of learning is considered to be a component of IC system;

-IC systems for MP, are typically systems with delayed result: measurements of sensor output during the MP can give only indirect prediction of the MP final result, and the result of MP can be known only after definite time delay.

Successful implementation of on-line control of materials processing from the total manufacturing point of view involves full integration of design, procurement, control of incoming components, manufacture, assembly, handling, packaging and distribution. The manufacturing chain must and should be considered from a systems point of view rather than in isolated segments of the total process. The analytical method applied in isolation is inadequate when dealing with complex metals processing operations. We will employ a holistic approach to address the issues in material processing as a core of MP.

This approach implies that:

- The whole is different than the sum of its parts, and thus the whole should be controlled independently of its parts (*principle of independence*).
- Each part can be in turn, considered a whole with its decomposition in parts; thus the independence principle is applied to all subsequent decomposition in parts *(principle of multiresolutional representation).*
- The whole determines the nature of its parts *(principle of hierarchical control).*
- The parts are dynamically interrelated, or interdependent *(principle of heterarchical coordination).*

In this paper we propose a methodology for the intelligent control of materials processing in manufacturing which is founded on the principles of knowledge based engineering. This is preceded by a discussion of the the decomposition of manufacturing work-processes as well as a classification of materials processing in manufacturing.

2. Manufacturing Processes: Classification and Trends

A manufacturing process is driven by a package of prescriptions for the set of activities that transform a given combination of raw materials into the unity of final product. These prescriptions are based on a mixture of a) fundamental scientific, engineering, and common sense knowledge, b) empirical domain knowledge (with no analytical models existing), and c) analytically derived and experimentally verified knowledge of processing operations. In fact, the most critical frontier in the science of materials today is this boundary between the ability to actually make things (discovered empirically), and the fundamental understanding of processes with their subsequent representation in a form of consistent models.

It is clear that an accurate description of the process, as well as a good understanding of the system, is a prerequisite for optimizing the manufacturing cycle. There are many ways one can represent a manufacturing system. Figure 2 shows the physical components and functions of a materials fabrication system. This is a description of the physical system showing the various physical components in the workstation associated with the manufacturing system. Other representations are also possible such as cause-and-effect diagrams or parametric models; nevertheless, a prerequisite for on-line control of materials processing is that the process needs to be described, modeled and understood.

Among possible examples of manufacturing processes we will concentrate upon the most complicated ones such as MP of Net Shape manufacturing. This term is used to describe the production of a nearly finished part (with required shape and dimensions) which is produced by the shortest possible (the most efficient) route consistent with service and performance demands. From a historical perspective, foundrymen are the oldest net shape manufacturers. Compared with other metal processing and manufacturing methods such as powder metallurgy, forging, etc., the casting route is the shortest possible route to produce net shaped components. The challenge we face is to produce shaped components highly accurate in their dimensions, with tailored and controlled microstructures, with no substantial loss of material and time.

In an attempt to classify the production routes for net shape castings from the point of view of the incoming melt, one can consider the incoming melt as either being a continuous stream, or a discontinuous stream - i.e., droplets, or an atomized liquid (L). In each case, the initial phase or phases, and the final phase which is in the solidified state (S) are indicated below. The arrow points to the transformation which occurs during the process.

Continuous Stream:

- $(L) \rightarrow S$ — This is the case for the conventional casting route where the melt, a continuous stream of liquid , is poured into the mold cavity and subsequently solidified.

- $(L + S_1) \rightarrow S_2$ — This case represents processes such as rheocasting or squeeze casting where a continuous "stream" of non-newtonian fluid - slurry - is poured into the mold cavity.

Discontinuous Stream:

- $(L) \rightarrow S$ — This case represents the production of powders via atomization. Discontinuous liquid droplets are formed which subsequently solidify to form powder particulates . . . or small castings.

- $L \rightarrow [\,(L) + (L_1 + S_1) + S_2\,] \rightarrow S_3$ — This case represents the recently developed Osprey process where the liquid is atomized and the droplets are allowed to partially solidify and are subsequently consolidated on the substrate/mold.

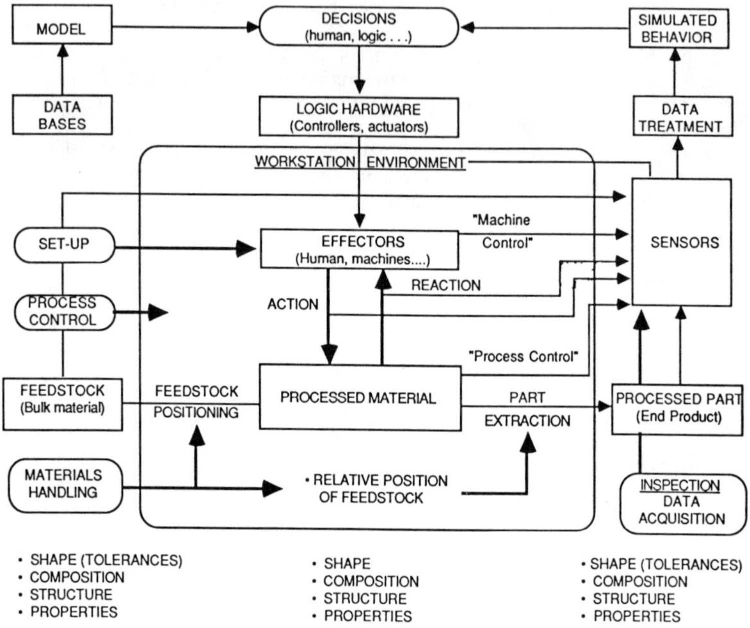

Figure 2: Schematic of the physical components and functions of a materials fabrication system.

During molten metal processing, synthesis can also take place by having the liquid react with another fluid phase - vapor or liquid. For example, in plasma spraying or spray casting (Osprey), the atomizing gas may contain reactive gases. This is shown below:

- $L \rightarrow [\,(L) + (L_1 + S_1) + S_2 + \upsilon\,] \rightarrow [\,(L') + (L_1' + S_1') + S_2'\,] \rightarrow S_3'$

where the primed entities indicate the new phases formed as a result of the reaction with the vapor phase, υ. A similar scheme taking into account the reaction with the vapor phase can be written for each of the other processing routes.

To be complete we should also include those processes where the starting material is in the solid state, such as in deformation processing. Here both shape and microstructural changes occur during processing. This can be represented as

- $S_1 \rightarrow S_2$

where the microstructure of $S_2 \neq S_1$.

3. Decomposition Stages of Manufacturing Work-Processes

Though shape, microstructure, and yield are used as generalized *descriptions of the output* of the work-process in material processing for manufacturing, we can describe the *basic procedures* which take place as: generation, transformation and elimination. *Generation* can be exemplified by the process of melting; *transformation* can be alluded to the microstructural and other state changes which take place during MP, and *elimination* is referred for instance to machining and removal (or yield) issues.

Our purpose is to develop control methodologies for materials processes. To carry out this mission we need to represent the work-process in it's most elementary stages, or decomposition stages. This is done for materials processing in a generic fashion such that the decomposition stages can apply to a broad spectrum of processing schemes. Thus, the basic materials processing stages of decomposition should be able to accommodate a complex process as well as a fairly simple one. In this vein, we propose that every materials processing scheme, assuming a single component system, follows these fundamental sequential/parallel/single/multiple stages:

- Material State preparation (P) stage
- Material State mobilization (M) stage
- Material State transfer (T) stage
- Material State formation (F) stage
- Material State finalizing (O) stage

Moreover, there is a transfer stage between each of the above, such as between the preparation and mobilization stages (P/M); between the mobilization and transformation stages (M/T); between the transformation and formation stages (T/F); and between the formation and finalizing stages (F/O). The overall decomposition stages for a single component "universal" work-process scheme can be given as:

- Material State Preparation Stage (P).
- Material Preparation/Mobilization Transfer Stage (P/M).
- Material State Mobilization Stage (M).
- Material Mobilization/Transfer Stage (M/T).
- Material State Transfer State (T).
- Material Transfer/Formation Transfer Stage (T/F).
- Material State Formation Size (F).
- Material Formation/Finalizing Transfer Stage (F/O).

- Material Stage Finalizing Stage (O).

For illustration purposes, let us apply the above decomposition stages to the Osprey process. In this process, liquid or partially liquid droplets are made to impinge, coalesce and solidify on a substrate, producing preforms of net shape, such as tubulars, disks, billets and/or strip. The spray deposition process can be described briefly as follows. The alloy charge is induction melted in a sealed crucible located on top of the spray chamber. When molten, the alloy exists through a refractory nozzle in the bottom of the crucible at a superheat of about 75°C. In the atomization zone below the crucible, the stream of liquid metal is broken up into a spray of droplets by the atomizing gas. Typically nitrogen or argon is used as the atomizing medium at a pressure in the range 0.6-1.0 MPa. Following atomization, the liquid droplets are cooled by the atomizing gas and accelerated towards a substrate (collector). The gas removed a critical amount of heat from the droplets during flight in the metal spray so that, on impact with the substrate, the droplets flatten and consolidate. More detailed description of the Osprey process are given elsewhere [2,3,6-8].

With this background, the decomposition stages in Osprey processing are:

	Decomposition Stage	*Osprey Process Equivalent Step*
P:	Material Stage Preparation (P)	Melting and Control of Superheat
P/M:	Material Preparation/Mobilization (P/M)	Ejection
M:	Material State Mobilization (M)	Atomization
M/T:	Material Mobilization/Transfer (M/T)	Generation of the droplet pack
T:	Material State Transfer (T)	Motion of the droplet pack
T/F:	Material Transfer/Formation (T/F)	"Landing" of droplets on a substrate
F:	Material State Formation (F)	Formation of the stable surface layer (SL) depending on the substrate motion
F/O:	Material Formation/Finalizing Transfer (F/O)	Hardening of the consecutive SL's
O:	Material State Finalizing (O)	Cooling of the completed object

The letters above give the designation of the particular decomposition stage during processing. For a single component material system which is being processed via Osprey, one can present the processing scheme as shown in Figure 3. When one considers synthesis due to reaction of the melt with a reactive gas during atomization, or impinging the melt stream with another jet of liquid metal ... then we can no longer consider a single component system. Figure 4 shows the complex situation when three components are involved. A similar methodology is followed when there are more than three components involved. The principle of organization for the multiresolutional hierarchy of controllers is illustrated in Figure 5. The elementary stages C11, C12,... are decomposed into elementary actuation controllers AC1, AC2, ... ACm which are being coordinated by the set of coordinator-controllers C11, C12,... . Coordinator-controllers are being coordinated by the meta-coordinators, and so on. The global coordinator Cg is supposed to deal with the generalized representation of the state of MP which is introduced later as *depictogram*..

4. Architecture of the Knowledge-Based Controller

Based on this methodology of the MP decomposition, the following controllers and sensors are defined for the knowledge-based controller.

1. Control of the material state preparation (coordination-controller C11[AC_m,AC_h])
   Melting and Superheat Controller (AC_m)
   Controller of Gradient Heating within Crucible (AC_h)

2. Control of the P/M stage (coordination controller C12[AC_{ep},AC_{ew},AC_{et}])
   Melt Overpressure Gas Controller (AC_{ep})

P: MATERIAL STAGE PREPARATION

P/M: MATERIAL PREPARATION/MATERIAL MOBILIZATION

M: MATERIAL STATE MOBILIZATION

M/T: MATERIAL MOBILIZATION/TRANSFER

T: MATERIAL STATE TRANSFER

T/F: MATERIAL TRANSFER /FORMATION

F: MATERIAL STATE FORMATION

F/O: MATERIAL FORMATION/FINALIZING

O: MATERIAL STATE FINALIZING

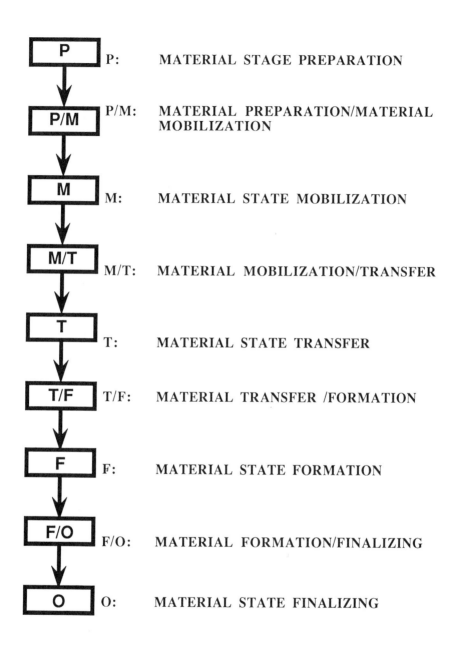

Figure 3: Schematic diagram of the Osprey spray casting process.

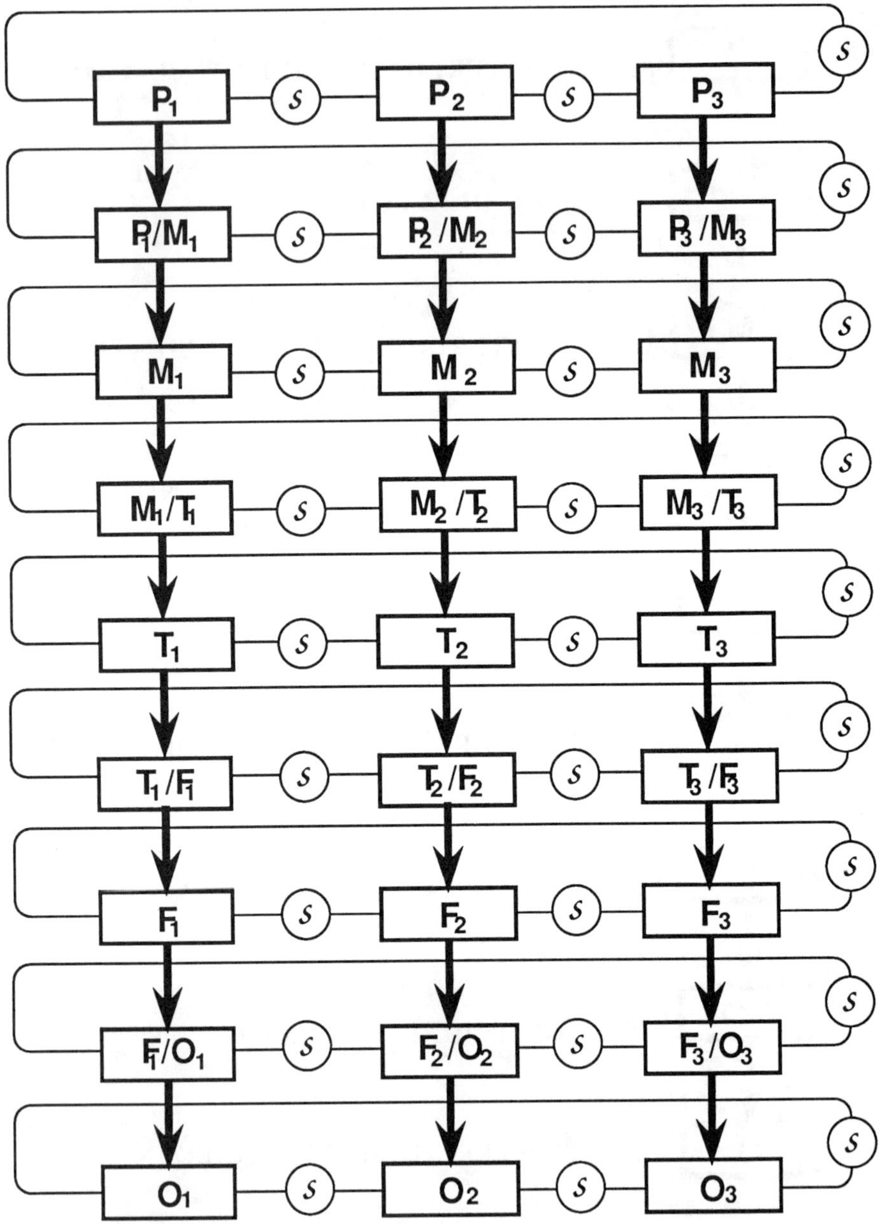

Figure 4: Decomposition stages for the Osprey process for an interdependent multi-component material system. The symbol ***S*** stands for synthesis. All other symbols are the same as in Figure 3.

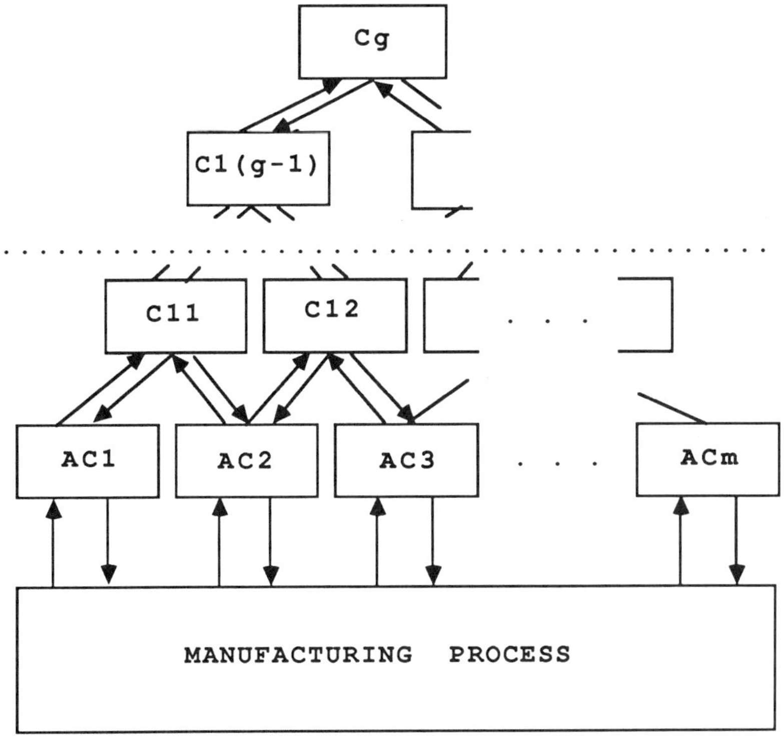

Figure 5: Hierarchical structure of coordination of elementary process (ACi - elementary actuation controllers, Cj - coordinators of the *j*-th level, Cg - global coordinator).

Stopper Rod Removal Controller (AC_{ew})
Induction Heating of Metal Pouring Nozzle (AC_{et})

3. Control of the mobilization (atomization) stage (coordination controller $C13[AC_{as},AC_{ae},AC_{ag},AC_{a1v},AC_{a1t},AC_{a2v},AC_{a2t}]$)
Spiralling of metal stream for more stable atomization (AC_{as})
Number of pouring orifices for more efficient atomization (AC_{ae})
Introduction of hotter atomizing gas for more stable atomization (AC_{ag})
Control of primary gas velocity (AC_{a1v})
Control of primary gas temperature (AC_{a1t})
Control of secondary gas velocity (AC_{a2v})
Control of secondary gas temperature (AC_{a2t})
Changing configuration of atomizing gas nozzles - angle, orifice opening, etc., can be considered as a possible candidate for introduction of elementary actuation controllers.

4. Control of the M/T stage (coordination controller $C14[AC_{tc},AC_{tt},AC_{tdv},AC_{tdt},AC_{tdd},AC_{fd}]$)
Control of cooling gas velocity (AC_{tc})
Control of cooling gas temperature (AC_{tt})
Control of Displacement gas velocity (AC_{tdv})
Control of Displacement gas temperature (AC_{tdt})
Control of Displacement gas direction (AC_{tdd})
Controller for flight distance (AC_{fd})

.　　　　.　　　　.
.　　　　.　　　　.
.　　　　.　　　　.

7. Control of the material formation (coordination controller $C17[AC_{dx},AC_{dy},AC_{dz},AC_{dr}, AC_{dt}]$)
Control of substrate motion in x- direction (AC_{dx})
Control of substrate motion in y- direction (AC_{dy})
Control of substrate motion in z- direction (AC_{dz})
Control of substrate rotation (AC_{dr})
Control of substrate tilt angle (AC_{dt})

.　　　　.　　　　.
.　　　　.　　　　.
.　　　　.　　　　.

Each of the elementary controllers is supposed to entertain one of the available "off-the-shelf" compensatory control strategies. The "plant" to be controlled is modelled as a dymamic system of the utmost 2-nd order. Any substantial nonlinearity can be dealt with by using model-driven and/or self-tuning adaptive algorithm. Since the system is decomposed in substantial degree, one can expect that using PID-controllers at the lowest level can be sufficient enough.

Coordination is performed by negotiating the constraint rules within the Pareto area which is valid for processes of decision making with a limited number of negotiators. Organization of coordinators into meta-coordinators up to the global coordinator is being performed as described above. In a multiactuator system with a multiplicity of stages for MP with synthesis (see Figure 4), coordinating activities at the same level affect each other and should be arbitrated by the negotiators of the meta-level of coordination. This problem can be resolved by forming look-up tables based upon results of the off-line simulation. In a system with learning, more attractive sophisticated methods can be recommended. Feedback integration (via generalization) can be recommended up to the level of general coordinator.

5. Analysis of the Knowledge-based Controller

Intelligent Control at a Level of Resolution

After decomposition is completed, and manufacturing process (MP) is represented as a set of parallel-sequential stages, the following analysis can be done. (We will call this particular case: control with decentralized and distributed observer, or DDO). Let us first introduce an analytical representation of the DDO control system for our distributed plant at a definite resolution level. (Analysis for the control processes at other resolution levels are performed similarly). Each **k**-th stage of MP (**k=1,2,...,s**) can be considered a subsystem with the set of **n** states, and **n** outputs (**j=1,2,...,n**) represented by matrices **x,y** and a set of **m** controls (**i=1,2,...,m**). This subsystem can be represented by equations (1,2)

$$x_{kj}'=A_k\, x_{kj} + B_{k\ i}\, u_{k\ i}; \tag{1}$$
$$y_{kj} = C_{k\ j} x_k; \tag{2}$$

(This representation coincide with the commonly accepted analytical representation of the dynamic system with lumped parameters).

The DDO specifics of our system can be reflected in the fact that states of any of our lumped subsystems are outputs of the preceding subsystem

$$x_{kj}= y_{(k-1)j} \tag{3}$$

whereas the outputs of this subsystem are the states of the subsequent system

$$y_{kj}= x_{(k+1)j}. \tag{4}$$

On the other hand, each k-th stage (k=1,2,...,s) can be partitioned into a set of **p** parallel subsystems pertained to the particular stage. This creates a partition upon the sets of **m** states and outputs into **p** groups, and the equation of the subsystem is being transformed into outputs

$$\Sigma\ x_{k\,r}'= \Sigma\ A_k\ x_{k\,r} + \Sigma\ B_{k\ r}\ u_{k\ r},; \tag{5}$$

$$\Sigma\ y_{k\,r} = \Sigma\ C_{k\ r} x_{k\,r},\quad r=1,2,...,p. \tag{6}$$

It is desired to design a family of $\mathbf{p} \times \mathbf{s}$ dynamic decentralized controllers which can be introduced either in a canonical way as a set of loops

$$\{u=Kx+Ly\}, \tag{7}$$

or the knowledge based programmed, independent (off-line computed) part of the control signal can be added which gives

$$\{u=Kx+Ly+v\}, \tag{8}$$

Let us denote the change of the j-th state during the time from t to t+dt. Then the following equation can be written for the energy balance in each of the lumped subsystems of MP

$$\Delta e = \sum e_j \uparrow \Delta t - \sum e_j \downarrow \Delta t \qquad (9)$$

where Δe is the total increment of energy of a particular subsystem, $e_j \uparrow$ is the increment of the kinetic energy associated with the change in state x_{kj}; $e_j \downarrow$ is the increment of the potential energy associated with the change of the same state. Divide the equation (9) by Δt, then at $\Delta t \to 0$ we will have

$$de/dt = \sum e_j \uparrow - \sum e_j \downarrow , \qquad (10)$$

which is a system of differential equations which satisfy conditions of existence and uniqueness at continuous $e(t)$.

Clearly (10) can be considered an analog of the cost-function of the neural network [9]. This implies that a neural network as a computational structure for a controller with subsystems subordinated to a set of the energy balance conditions, would be a natural engineering solution. After the controllers for the stages of the distributed decentralized MP are designed, the coordinators are to be designed which can be considered simply as devices for maintaining consistency of constraints. In the case of the conflict, the set of constraints should be reassigned which lead to total reassignment of constraints for all lower level controllers involved.

Knowledge Bases for the Intelligent Controller

Assignment of constraints as well as generation of the programmed off-line computed input component **v** (see eq.8), is performed by the knowledge bases of coordination level. Knowledge Bases incorporate rules which are related to the control recommendations depending on the boundary conditions, where change of the control assignment, or restructuring of the controller are definitely required.The set of constraints for the problem can be formulated as follows:

minimize $f(x_1, ..., x_n)$ (11)

subject to $g(x_1,x_2) \leq 0$ (12)

subject to $g(x_1,x_3) \leq 0$ (13)

. . .

subject to $g(x_{n-1},x_n) \leq 0$ (14)

The form for constraints is accepted as follows

constraint g_i (is) active-inactive, i=1,...,n. (15)

The following taxonomy of rules can be foreseen given the prior practice of using the rule based controllers [10]:

Rules of the states existence:

If the problem has solution, **then** $A \leq x_j \leq B$ (**A, B,** x_j-are matrices);

Rules of constraints existence:

If $A \leq x_j \leq B$ (**A, B,** x_j-are matrices) **then** g_j is inactive, i=1,2,...,v; j=1, 2,...,u;

Rules of compatibility of single constraints:
If g_i is active, **then** g_j is inactive, i=1,2,...,v; j=1, 2,...,u;
Rules of compatibility among all possible combinations of constraints:
If $g_i \wedge g_j \wedge ... \wedge g_z$, $i \neq j \neq ... \neq z$, are active, **then** g_l is inactive, i=1,2,...,v; j=1, 2,...,u; (all l-tuples should be combined of constraints, l=1,...,z).

Generators of all {u} including all {v} should be coordinated which is done by rule-based checkers of consistency. When the number of coordinators is too large, coordinators of the upper level (meta-coordinators) are to be included to the structure of intelligent controller

Nonhomogeneous representation in the particular case considered above (mixture of analytical and linguistic representations), does not create too much of a problem because all lower level set of controllers is based upon analytical representation whereas the coordination level is designed as a linguistic controller. The discrepancies are resolved by the translation procedures in the interface. Design is becoming more sophisticated if non-homogeneity holds at a level (e.g. within the set of lower level controllers). In this case another system of knowledge representation is considered: so called *D-structure* (see [11]).

D-structure for Knowledge Integration. Depictogram.

Two interrelated problems are to be addressed specifically: how can the non-homogeneous knowledge be organized, and how can this multiplicity of individual (though coordinated) feedbacks contribute to the unity of the process result. In other words, it would be desirable to find a unified feedback dealing with the united "**y**" vector of the overall MP, and taking care of the DDO as a unity. In this subsection we are going to introduce *D-structure* and *depictogram* as methods of integrated knowledge representation with capability of *pictorial (graphical) representation of generalized morphological portrait* of the process knowledge (or a particular sub-domain of the process knowledge). Visual perception of the generalized morphological portrait of the domain knowledge presumes existence of the internal structure which can be conveniently revealed using methods of computer graphics. The following theoretical premises are formulated for the introduction of depictogram and exercising algorithms for its development.

D-structure is introduced for dealing with knowledge for the case of mostly nonhomogeneous representation of information about the process, system, and the controller. Total knowledge of the process we will name *context of the process* knowledge (COP). COP is understood to be interpretable within *universe of discourse* knowledge (UOD). Certainly,

$$\text{UOD} \supset \text{COP} \tag{16}$$

where the sign of inclusion means that the process knowledge is *a part* of the general scientific knowledge, i.e. inclusion presumes partition to be performed. Another meaning which is implicit at least in the concrete example of inclusion (16) is that statements of COP defined within COP are interpretable within UOD, and interpretability presumes establishing an explanatory paradigm, or additional statement of representation within broader vocabularies V, i.e. we assume that

$$V_{UOD} \supset V_{COP}. \tag{17}$$

Vocabulary is assumed to be a set of labels (set E) for entities, and set of labels (set R) for the relationships among the entities V(E, R); E= $\{e_i\}$, i=1,...,M; R=$\{r_{ij}\}$, j=1,...,N; M is total number of labels (words), N-is a total number of relationships among them (rules). Entities must be connected to each other

$$e_i \in E, \quad e_j \in E, E \supset V \quad r_{ij} \in R, R \supset V \qquad (18)$$
$$\text{such that } e_i \Rightarrow e_j = r_{ij} \text{ ,or } e_j \Leftarrow e_i = r_{ji} .$$

The set of all elementary entity-relational statements in the form constitute the set of all rules. The validity of a rule as a truth statement is limited by the time of validity for a particular set of states.

Entity is defined as a subset of world for which the label is assigned. Label assignment is done on the basis of *unity* observed. The property of unity is declared for the subsets which can be characterized by the spatial *adjacency* of the *similar* parts of the would-to-be entity, or in other words, by the spatial *uniformity* of the zone of the set (uniformity by the property of consideration).These terms: unity, adjacency, similarity, uniformity are properly defined within the theory of D-structure [11], here we will rely upon intuitive interpretation of these terms as related to the subject of consideration.

Similarity is provided by actual existence of many "connectivities", "symmetries", "convexities" and other characteristics typical for "gestalt", all of them in the large vs in the small. Similarity property is also called *classification property* since the algorithms of classification are obtained to arrange the information in the recommended way, and also because we are talking about *similarity classes.*

The difference between *similarity classes* and *equivalence classes* can be explained by defining similarity as the equivalence at a given level of *resolution* . Uniformity can be properly formulated via similarity classes for the objects (sets, mappings, manifolds, varieties, etc.), relations (morphisms) among the objects, functions among the categories, etc.

Resolution of a level is selected based upon the law of parsimony: the number of labels, or words (entities) in the vocabulary (domain) of consideration should be substantially smaller than the number of parts unity of which is being used to form entities. This is just an another form of the well known Ockham's razor: entities are not to be multiplied beyond necessity.

When a definite level of resolution is assigned, the least distinguishable increment is determined of quantitative change in the value of property of consideration. When the increment in the value of property of consideration for two adjacent parts of the space is lower than the level of distinguishability, these two adjacent parts can be unified together. When the resolution level is low the number of parts that can be distinguished from each other is small. Thus, the number of entities recorded is small.

The laws of classification employed within D-structure

One can see that unity implies the partition of the space in components which are becoming partition of the future unity. This means that in fact, statement (17) must be rewritten as follows

$$V_{UOD} \supset \cup V_{COP,i} . i=1,...,k, \text{ number of contexts of interest.} \qquad (19)$$

In turn, the following inclusions can be expected as a result of subsequent partitions

$$V_{COP} \supset \cup V_{PR,i}, \ i=1,...,k, \text{ number of problems of interest.} \qquad (20)$$

$$V_{PR} \supset \cup V_{SYST,i}, \ i=1,...,k, \text{ number of systems of interest.} \qquad (21)$$

$$V_{SYST} \supset \bigcup V_{SUBS.j,i},\ \text{j-number of decomposition of the system into subsystems, i.e. level number, } i=1,...,k, \text{ number of subsystems of interest at the j-th level.} \quad (22)$$

$$\vdots \qquad \vdots \qquad \vdots$$

$$V_{SUBS.j,i} \supset \bigcup V_{OBJ,i}.\ i=1,...,k, \text{ number of objects of interest at the lowest level of system partitioning.} \quad (23)$$

Since each subsequent partitioning is performed at a different resolution level, the levels can be characterized by the increase of precision top down. The validity of statements (17) through (23) and as well as the validity of definitions for all their components, are supposed to hold through the time of process utilization, and this time T is different for all of the levels ($T_{syst} > ... > T_{subs.j} >... > T_{obj.j}$). After partitioning we receive subsystems of rules with smaller scope of attention and with more diversified language (see [12]). In the area of control, it is customary to consider vector spaces as mathematical form of representing sets of rules.

Classification property is a basis of the vector space under consideration (VSUC).So, the problem starts with selection of VSUC for consideration.The overall state space of the system to be controlled is understood as union of VSUCs pertained to different kind of energy transformation: mechanical motion, thermal energy dissipation and/or accumulation, light, acoustic, and other processes.Different VSUCs intersect, usually, by incorporating spatial and temporal dimensions as a part of the space basis. Thus

$$UOD=\bigcup VSUC_i\ ;\ \text{and } VSUC_i \cap VSUC_j \neq \varnothing \quad (24)$$

After the preliminary operations of information organization are completed, all knowledge of the system is being organized into similarity classes. At the low resolution levels we have small number of similarity classes, and we can encompass a substantial area of the space of interest. When the process of partitioning is in progress, the number of entities to be considered is growing which leads to progressing clustering of the entities of the lower resolution levels into entities of the higher resolution levels which results in the partitioning hierarchies, or syntactic graphs which are used sometimes as a means of systems and states representation. Unfortunately, the syntactic graphs do not contain any comprehensive information concerning the laws of adjacency at a level of resolution. Another deficiency of these graphs is their inconvenience for storing and manipulating information in full. (In the control system this leads to the need in narrowing the "scope of attention" [12]).

In our effort not to lose the adjacency information we decided to concentrate on the morphology of boundaries among the entities. It has been demonstrated that the boundaries can be economically represented in the terms of the limited language of the archetypes of boundaries [13]. (The total number of words in this language do not exceed 15-20). Instead of dealing with information about entities we prefer to deal with information about the boundaries (see the example with rule-base: most of the rules are stated about dealing with boundaries). The amount of information required for this is substantially less than in the case of representing the entity as a multiplicity of elementary representations for each of elementary objects with similar properties, and have this representation for all of them!

The archetypes of boundaries correspond to those existing in the human system of visual representation which suggests using transformation of the boundary information into domain of pictorial representation. Methods of computer graphics can be successfully utilized in this case.

6. Conclusions

The new unified approach of analysis of the manufacturing (material) processes is focused on the needs of control system design. It is tempting to develop a theory conducive for efficient methods of design and manufacturing in the area of intelligent control. Scientific theories in the area of materials processing have been traditionally oriented toward greater accord with the form and techniques of models accepted in the areas of fundamental sciences such as physics. Among the multiplicity of possible models (and representations) we have selected a model (and a representation) that should enable us to design and produce machines for automated manufacturing processes using modular unified decisions. We are trying to provide consistency with the fundamental physical representations while making decomposition of the process in stages. Subsequently, the models should be unified (analytical as well as linguistic) , and allow for using standard and/or development advanced computational models.

The following issues seem to be challenging in the are of Knowledge-based ICOMP:

- Methods of coordination are not given sufficient attention in the literature on intelligent control. The only existing results are related to the negotiations within the Pareto area. This approach seems to be straightforward for a round of decision making with a limited number of negotiators. However, in a multiactuator system with a multiplicity of stages for MP with synthesis, coordinating activities at the same level affect each other and should be arbitrated by the negotiators of the meta-level of coordination. This problem is not solved at the present time. We would expect that a practical solution is based on forming look-up tables with results of the off-line simulation.
- It is clear from the structure of hierarchical control and coordination (see Figure 5) that problem solving at the coordination levels (bottom-up) requires feedback integration (generalization) up to the level of general coordinator which monitors the depictogram of the overall MP. Certainly, there is no general theory of developing partial and/or complete depictograms of the MP. Development of such a theory can be considered promising, and there are multiple expected applications for such a theory.
- One of the most challenging problems of knowledge-based ICOMP, is a problem of knowledge acquisition by learning. Since generalization is required anyway due to the required system of coordination activities, the system of conceptual learning may appear as a byproduct of ICOMP development.
- Development of the knowledge based ICOMP employing the modular principle, will require knowledge-bases organized in corresponding fashion: knowledge should be decomposed and "chunked" as to support the process decomposition utilized in the controller.

Having met these challenges, as we approach the third millenium, we will be in a position to control and manipulate material processes/manufacturing systems to ***ensure*** quality of the component . This global challenge is our ultimate goal.

REFERENCES

1. B.G. Kushner, R.A. Geesey, P.A. Parrish, and S.G. Wax, "A Knowledge Acquisition Tool for the Intelligent Processing of Materials", in Artificial Intelligence Applications in Materials Science, published by TMS, Warrendale, PA , pp. 195-204.

2. Apelian, D., Gillen, G. and Leatham, A., "Processing of Structural Metals by Rapid Solidification", eds. Froes, F.H. and Savage, S.J., Amer. Soc. for Metals International, p. 107 (1987).

3. Leatham, A.G., Brooks, R.G. and Yaman, M., Modern Developments in Powder Metallurgy, eds., Aqua, E.N., and Whitman, C.I., Metal Powder Industries Federation, Princeton, N.J., 15, p. 157 (1985).

4. Fiedler, H.C., Sawyer, T.F. and Kopp, R.W., "Spray Forming - An Evaluation Using IN718", General Electric Technical Information Series, 86CRD113, May (1986).

5. Leatham, A.G., Ogilvie, A.J.W. and Chesney, P.F., Modern Developments in P/M, eds.: P.U. Gummeson and D.A. Gustafson, Metal Powder Industries Federation, Princeton, N.J., 18-21, In Press (1988).

6. Gillen, A.G., Mathur, P.C., Apelian, D. and Lawley, A., Progress in Powder Metallurgy, eds., Carlson, E.A., and Gaines, G., Metal Powder Industries Federation, Princeton, N.J., 42, p. 753 (1986).

7. Mathur, P., Apelian, D. and Lawley, A., Acta Metall., In Press (1988).

8. Apelian, D., Lawley, A. Mathur, P.C. and Luo, X.J., Modern Developments in P/M, eds.:P.U. Gummeson and D.A.Gustafson, Metal Powder Industries Federation, Princeton, NJ, 18-21, In Press (1988).

9. J. Hopfield, "Neurons with graded response have collective computational properties like those of two-state neurons", *Proc. Nat'l. Academy of Sciences USA* , Biophysics, Vol. 81, p.p. 3088-3092, May 1984

10. C. Isik, A. Meystel, "Pilot Level of a Hierarchical Controller for an Unmanned Mobile Robot", *IEEE Journal of Robotics and Automation* , Vol. 4, No. 3, June 1988

11. A. Meystel, "Intelligent Control in Robotics", *J.of Robotic Systems* , v. 5, No. 4,

12. A. Meystel, Theoretical Foundations of Planning and Navigation for Autonomous Robots", *Int'l J. of Intelligent Systems* , v. 2, No. 2, 1987

13. R. Thom, Structural Stability and Morphogenesis: An Outline of a General Theory of Models, W.A. Benjamin publ., Reading, MA 1975

WATERJET CUTTING - EMERGING METAL SHAPING TECHNOLOGY

E.S. Geskin, W.L. Chen and A. Vora

Laboratory of Waterjet Cutting
New Jersey Institute of Technology
Newark, New Jersey 07102

Abstract

This paper is concerned with the application of an abrasive waterjet for metal shaping. The waterjet cutting technology uses water compressed to the pressure up to 55,000 psia which is accelerated in a cylindric nozzle. Abrasive particles are then supplied into the water stream and accelerated by the water. Cutting occurs via the interaction between the water-particles stream and a workpiece that causes material erosion at the impingement zone. The mechanism of this interaction enables shaping of practically any material with a minimal effect on the metallurgy of subsurface. The motion of the jet penetrating through a material body is guided by a robotic arm. A waterjet cutter is becoming one of the principal tools for shaping titanium, super-alloys and other hard-to-machine metals and composites. The shaping, resulting from the waterjet cutter, is very similar to laser beam cutting. However, unlike laser processing, there is no damage to the subsurface structure.

The presented work discusses the experimental study of cutting of titanium plates. The results of these experiments demonstrate the potential of the abrasive waterjet as a shaping tool.

Metallurgical Processes for the Year 2000 and Beyond
Edited by H.Y. Sohn and E.S. Geskin
The Minerals, Metals & Materials Society, 1988

Introduction

One of the main avenues in the development of metal shaping is the use of high energy beams as cutting tools (1,2). There are a number of advanced beam oriented material removal technologies, such as laser, electron beam, plasma and abrasive waterjet (AWJ) cutting. The advantages of AWJ cutting are well understood and documented (3–14). This technology has the potential of becoming the principal tool for netshaping and netsurfacing of parts. However, the use of AWJ is currently limited to cutting of hard-to-machine materials such as titanium, superalloys, glass, composites, etc. The cost of AWJ cutting is much higher and the productivity is much lower than those of the thermal cutting. At the same time abrasive waterjet can be used for shaping any existing technological material from fabrics, plastics, aluminum to glass, composites, and diamonds. The material losses during AWJ shaping are minimal. The roughness and waveness of the surfaces generated during AWJ cutting are substantially less than those of the thermal cutting. Due to the size of the jet cross-section total force exerted by the jet on a workpiece does not exceed 30–50 newtons. This terminates the need in the use of complex fixture for material holding. Low applied forces and low surface temperature prevent the damage of metal subsurface in the course of cutting. The small size of the impingement area will determine the limited width of the kerf. Due to the size of kerf and unidirectional character of the water jet the formation of practically any metal contour is possible.

The objective of the presented paper is to discuss the formation of AWJ and the surface generation by the use of these jets. The obtained information enables us to assess the possible application of AWJ cutting in metallurgy.

Mechanism of AWJ Cutting

Waterjet cutting consists of accelerating water in a circular nozzle and aiming the nozzle at a target point. As the jet flow hits the impingement zone, the velocity of the water radically drops. This results in the creation of a high pressure zone on the workpiece surface. If the pressure is sufficient, the interface moves and cutting occurs. If the stresses created in the deformation zone are less than the deformation threshold, cutting is impossible.

Adding abrasive particles to the waterjet changes the cutting mechanism. The velocity of a particle drops in the high pressure zone, however the particles still reach the surface of a workpiece at a comparatively high velocity. In the course of collision, the velocity of a partcicle changes dramatically. The water-hammer effect at the collision site results in the development of pressure exceeding the static pressure of water in the impingement zone and the subsequent removal of a small portion of the workpiece material. The high number of collisions results in the erosion of the workpiece surface.

The mechanism of cutting depends on the ductility of a workpiece (15). Finnie suggested (16,17) that the erosion of ductile material is due to the motion of abrasive particles through the workpiece. The scooping action of the particles results in material removal along the particle trajectory.

According to Sheldon and Finnie (18) and Lawn and Swain (19), penetration of an abrasive particle into brittle material results in the development of lateral cracks and material separation.

Apparatus and Experimental Details

A 5-axes, gantry-type Robotic Waterjet Cutting Cell, manufactured by Ingersoll-Rand, along with the Allen Bradly series 8200 robot controller was used for this study. The hydraulic system included water softener, a booster pump and a 40 h.p. oil driven intensifier. The water pressure, after the intensifier, was maintained at the level of 45-47 ksi. Cutting (Fig. 1) was carried out with the conventional Ingersoll-Rand nozzle set (Fig. 2), consisting of a sapphire cylindrical nozzle, mixing chamber and a carbide tube. The abrasives were pneumatically conveyed into the nozzle body and entered from the side port, via a hose from the abrasive hopper attached to the gantry. The abrasive flow rate was controlled by the voltage applied to the vibratory feeder. The voltage is determined by the set point on the controller of the abrasive flow rate. The performed experimental study involved the examination of the characteristics of the water-particles, jet and jet-workpiece interaction. It also included straight cutting and shaping of titanium and stainless steel and other materials.

Figure 1. Cutting of a composite plate by the use of Ingersoll-Rand Robotic Waterjet Cutting Cell.

The study of the jet included jet photography by the use of a flash light with the time of the exposure of 0.5 μsec. The dynamics of the jet was studied by the use of high speed filming which has been taken by high speed camera (10,000 frames per second) with Cu–vaper laser beam as a source of the light. The forces developed at the jet–workpiece interaction are measured by the Kistler pieoelectric transducer (20).

Jet–workpiece interaction was studied by impinging of a AWJ on surfaces of different materials (stainless steel, aluminum, magnesium) and subsequent examination of the damage of the material surface by optical and scanning electron microscopy.

The study of the cutting results was carried out by cutting of titanium plates at different operational conditions and subsequent measurement of kerf and surface characteristics.

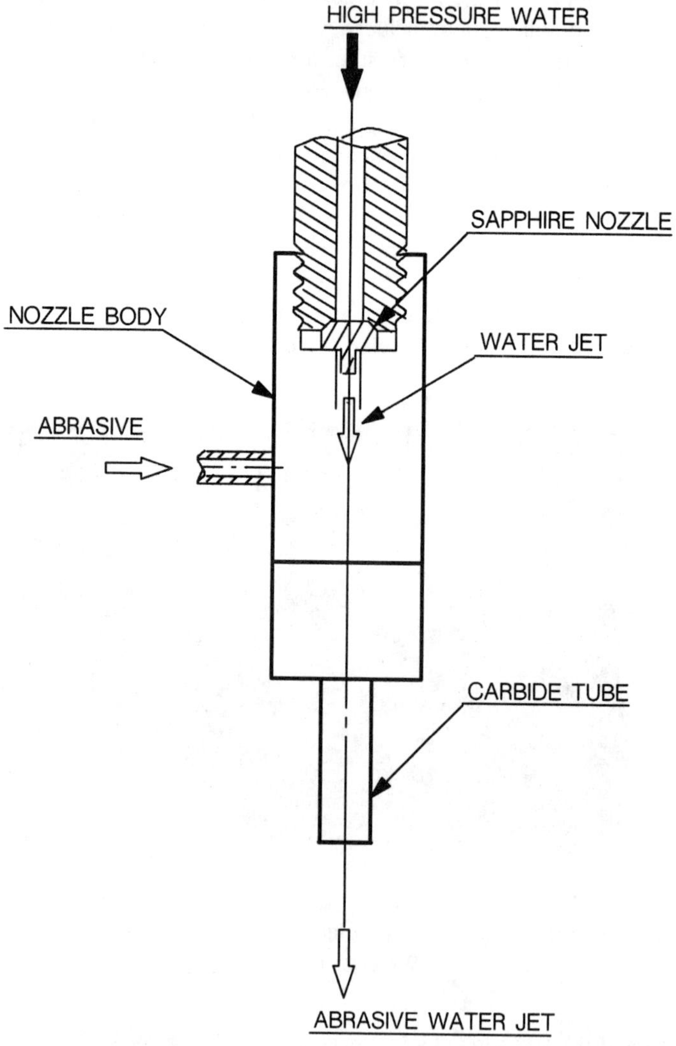

Figure 2. Schematic of nozzle head for the formation of AWJ.

The roughness of the generated surfaces is measured by a stylus type surface analyzer. The top and bottom kerf width and waviness are measured with a Toolmakers microscope, equipped with cartisian x–y digimatic heads. The taper was approximated by the formula

$$\text{Tangent Theta} = ((Us - Bs)/2)/h$$

where, *Us*, *Bs*, and *h* are as shown in Fig. 3.

A characteristic feature of the surface, generated by a high energy beam, is the formation of striation as the cutting depth increases. As the beam progresses into the material, the smoothness of cut, along the beam travel, is reduced. Because of this, two different techniques were used to describe surface topography. The surface roughness was measured at vertical distance of 3 mm from the top surface of the sample (Fig. 4), provided that it is less than 400 micro inches to safeguard the stylus. The stroke of the stylus was 15 mm. The measurement is carried out at three different locations and the mean value of this measurement is used. To measure the striation (waviness), the toolmakers microscope is employed at the bottom side of sample and is estimated by the direct measurement of the distance between the peak and the valley (Fig. 8). The kerf geometry is characterized by measuring the *Us* and *Ub* on the obtained samples (21).

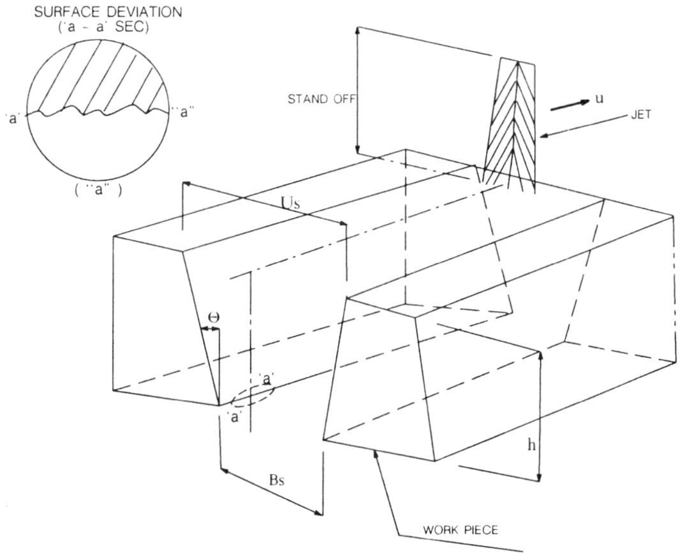

Figure 3. Schematic of material removal in the course of AWJ cutting.

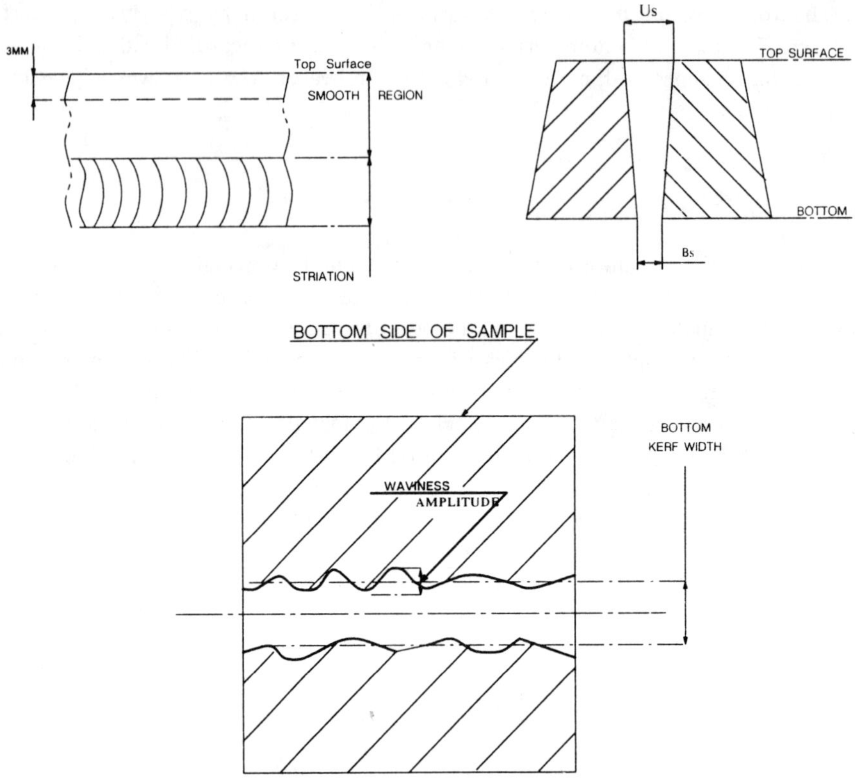

Figure 4. Schematic of measurements of roughness, kerf width and waviness.

Results and Discussion

Examination of flush lights photographics and high speed films show that the core of the jet is surrounded by a mist and is a body with longitudinal and transvertional pulsations. Particles are are distributed evenly across the jet crossection.

The force exerted by the jet on a workpiece depends on the diameter of a saphire nozzle, carbide tube, abrasive flow rate and particles size. The most substantially the force is effected by the diameter of the saphire, which at the fixed water pressure, determine water flow rate (Fig. 5) (20).

As is demonstrated by the SEM picture of the surface (Fig. 6), the material surface is attacked by clusters of particles which create surface cavities. The material removal is carried by the superposition of the clusters of created cavities.

Results of cutting are determined by the combination of process conditions, such as diameter of the sapphire nozzle and the carbide tube, nozzle traverse rate, size and flow rate of abrasives. Relationship between the rate and depth of cutting is depicted in

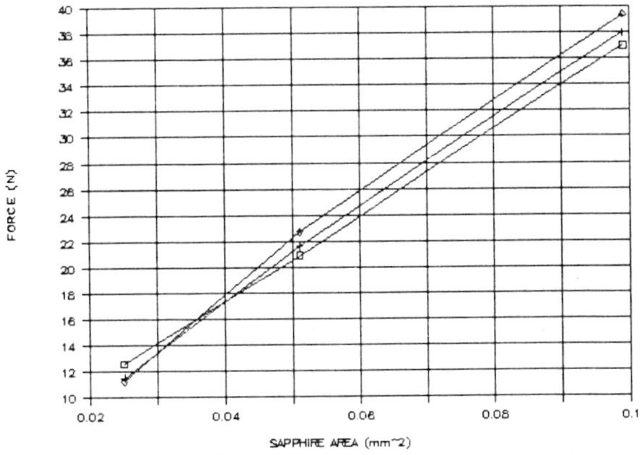

Figure 5. Effect of the nozzle area on the force developed in the impingement zone.
Diameter of carbide = 0.030'
Particle size = 60 ÷ 100 mesh
Particles flow rate: □ = 0, + = 25 lb/min, ◇ = 1 lm/min.
Distance between the nozzle and workpiece surface = 0.5"

Figure 7. The effect of process conditions on the upper kerf width is shown in Figure 8. Similar charts can be constructed for other results of cutting. The correlation between operational conditions and surface geometry enables us to create a database for the process control that will assure the desired output in the course of metal shaping.

The feasibility of the shaping of the hard to machine materials by the use of AWJ is demonstrated by a part shown in Figure 9 (22).

Possible Use of AWJ at Metallurgical Plants

Currently, the only known practical use of AWJ at a metallurgical plant involves metal deburring after casting or thermal cutting. However, many other more important applications can be considered. The most promising use of AWJ is the cutting of expensive metals such as titanium to reduce material losses. Another immediate application involves the cutting of products having thin walls or complex geometry to reduce the cost of cutting facilities and increase the range of possible shapes.

AWJ also can be integrated with a caster or a rolling mill of a micromill to combine metal production and shaping.

Concluding Remarks

It can be expected that in the near future AWJ will remain to be a technology for shaping hard-to-machine materials. However, such expected development as improvement of the cutting effectiveness and increase of the production of new materials (superalloys, composites) will make the AWJ conventional metallurgical facilities competitive with thermal and mechanical devices for cutting and shaping.

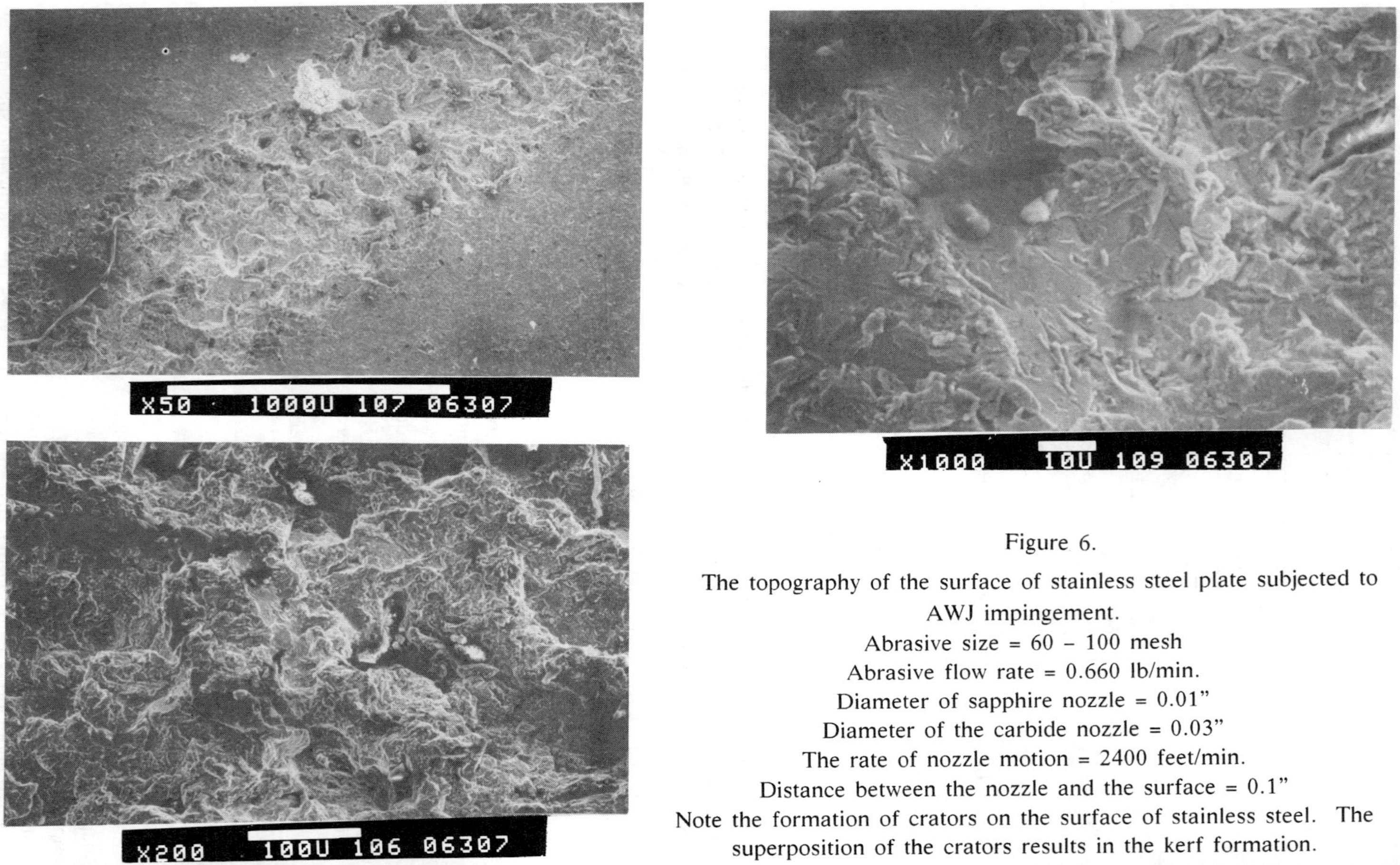

Figure 6.

The topography of the surface of stainless steel plate subjected to AWJ impingement.

Abrasive size = 60 – 100 mesh

Abrasive flow rate = 0.660 lb/min.

Diameter of sapphire nozzle = 0.01"

Diameter of the carbide nozzle = 0.03"

The rate of nozzle motion = 2400 feet/min.

Distance between the nozzle and the surface = 0.1"

Note the formation of crators on the surface of stainless steel. The superposition of the crators results in the kerf formation.

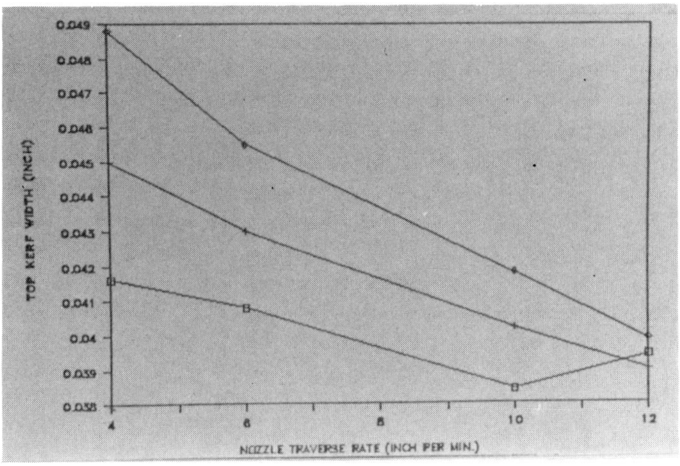

Figure 7. Effects of nozzle traverse rate on the top kerf width.
Diameter of the saphire nozzle = 0.01"
Diameter of the carbide tube = 0.03"
Abrasive flow rate (lb/min): □ = 0.735, + = 0.907, ◇ = 1.032
Abrasive size = 120 mesh
Thickness of the titanium 1/4"

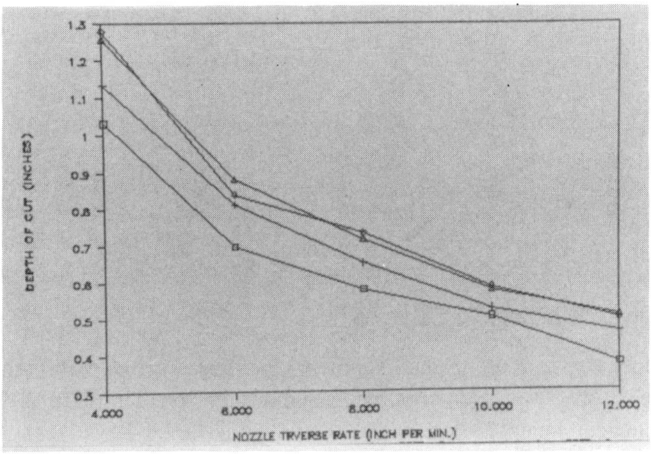

Figure 8. Effect of nozzle traverse rate the depth of cutting
Material: Titanium Grade 2
Diameter of the saphire nozzle = 0.01"
Diameter of the carbide tube = 0.03"
Feed rate of abrasive: □ = 0.474 lb/min., + = 0.660 lb/min.
◇ = 0.915 lb/min., Δ = 0.927 lb/min.
Abrasive size: Barton mines 80 mesh

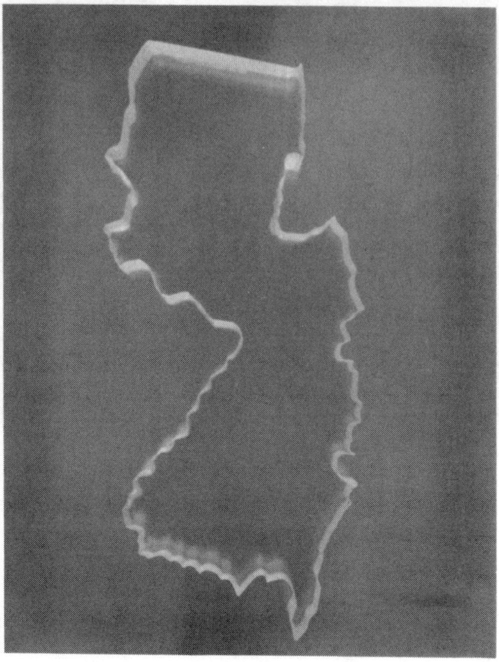

Figure 9. Glass parts produced by AWJ.

References

1. W. Johnson, Impact Strength of Materials, (Crane, Russak, NY, 1972), 336–337.

2. B.V. Amstead, Manufacturing Processes, Seventh Edition, (Wiley), 690–701.

3. P.H. Donnan, "Jet Cutting Development for Special Industrial Applications," (Proceedings of Seventh International Symposium on Jet Cutting Technology, Ottawa, Canada: June 1984), 287–278.

4. J.A. Norwood, et.al., "New Adaptions and Applications for Waterknife Cutting," Ibid, 369–386.

5. "Abrasive Water-Jet Cuts Metals without Heat," Tooling and Production, (May 1985), 64–65.

6. M. Hashish, "Turning with Abrasive Waterjets – A Preliminary Investigation," Advances in Non-Traditional Machining, ed. K.P. Rajkar, (ASME, Anaheim, CA: Dec. 1986), 79–101.

7. K. Przklenk, "Abrasive Flow Machining - A Process for Surface Finishing and Deburing of Workpieces with a Complicated Shape by Means of Abrasive Laden Medium," Ibid, 101–111.

8. M. Hashish, "Application of Abrasive-Waterjets to Metal Cutting," Non-traditional Machining Conference Proceedings, (ASM, Cincinnati, OH: 1985) 1–12.

9. M. Hashish, "Cutting with Abrasive-Waterjets," Mechanical Engineering, (March 1984), 60–62.

10. M. Hashish, "Abrasive-Waterjet Cutting Studies," (Proceedings of 11th Conference on Production Research and Technology, Carnegie-Mellon University, Pittsburgh, PA: May 12–23, 1984), 101–111.

11. J.W. Twigs, "High Pressure Water Jetting Technique," Corrosion Prevention and Control, (April, 1982), 19–23.

12. R.B. Aronson, "Abrasive Water Jets Cut Almost Anything," Machine Design, 21, (1985), 114–117.

13. "Water Jets Work Harder," Engineering, (June 1983), 457–461.

14. H. Kafenkamp, et.al., "Precise Cutting of High Performance Thermoplastics," (Ref. 3), 353–368.

15. A. Engel, Impact Wear of Materials, (Elsevier, 1976), 100–128.

16. I. Finnie, "The Mechanism of Erosion of Ductile Metals," (Proceedings of the 3rd National Congress of Applied Mechanics, ASME, 1958), 527–532.

17. M. Hasith, "Modeling of Abrasive Waterjet Cutting," (Proceedings of VII International Symposium on Jet Cutting Technology, June 1984), 249–265.

18. G.L. Sheldon and I. Finnie, "The Mechanism of Material Removal in the Erosive Cutting of Brittle Materials," Journal of Engineering for Industry, Transaction of ASME, 88 (1966), 393–400.

19. B.R. Lawn and M.V. Swain, "Microfracture Beneath Point Indentation in Brittle Solids," Journal of Material Science, 10 (1975), 113–116.

20. H.Y. Li. Investigation of Forces Developed in the Course of the WaterJet-Workpiece Interaction. (MS thesis, New Jersey Institute of Technology, 1988), 60.

21. A. Vora. Investigation of the Kerf and the Surface Generated in the Course of Cutting Titanium with Abrasive Waterjets. (MS thesis, New Jersey Institute of Technology, 1988).

22. W.T. Lee, Investigation of the Technology for Glass Shaping by the use of Abrasive Waterjet. (MS thesis, New Jersey Institute of Technology, 1988).

TECHNOLOGY NECESSARY TO DEVELOP A TRULY AUTOMATED MANUFACTURING FACILITY

R.C. Progelhof
Chairman, Department of Mechanical Engineering
and
W.F. Ranson
Tamper Professor of Mechanical Engineering
Directory, Center of Industrial Research

University of South Carolina
Columbia, South Carolina 29208

Metallurgical Processes for the Year 2000 and Beyond
Edited by H.Y. Sohn and E.S. Geskin
The Minerals, Metals & Materials Society, 1988

I. THE PROBLEM: MANUFACTURING PRODUCTIVITY

The 1985 President's Commission [1] on Industrial Competitiveness found compelling evidence of the declining economic position of the United States in world markets. The annual rate of increase in the United States productivity is approximately one-seventh that of our major trading partners over the last decade; profit levels in manufacturing investments are below those of the mid-1960's; an increasingly negative trade balance since 1971; and a sharp decline in the United States share of world trade in manufactured goods for more than two decades. A particularly dramatic example of the shifting economic balance is the fact that Japanese productivity growth has been five times greater than that of this country since 1960. In face, Japan's productivity now exceeds that of the United States in such critical industries as steel, transportation equipment, electronics and electrical machinery.

American employees have experienced the competitive consequences of this nation's lagging industrial performance. Between 1980 and 1984 alone, the population of non-farm employees engaged in manufacturing dropped from 22.4% to 20.6%. This reduction in manufacturing must be arrested.

Commissions Recommendations

The comprehensive set of recommendations offered by the President's Commission recognized the nation's "key competitive advantages", technology and human resources, and calls for their strengthening. Particular emphasis is placed on manufacturing technology and the fact that this country does not translate research into new technologies rapidly and fully.

> *"Perhaps the most glaring deficiency in America's technological capabilities has been our failure to devote enough attention to manufacturing or 'process' technology. It does us little good to design state-of-the-art products, if within a short time our foreign competitors can manufacture them more cheaply. The United States has failed to apply its own technologies to manufacturing. Robotics, automation, and statistical quality control were all first developed in the United States, but in recent years, they have been more effectively applied elsewhere".*

Moreover, the Commission recognizes that the use of new manufacturing technologies cannot be limited to high-tech industries.

> *"Mature industries can and should make better use of advanced technologies as part of their own renewal processes. There need be no distinction between high-technology and mature industries--only between industries that have taken advantage of technological advances and those that have not" (1).*

Just what are these new manufacturing technologies and what are their implications for productivity? The computerization of manufacturing very likely represents a revolution comparable in its potential effects to that of the Industrial Revolution. The assembly line made possible the mass production of identical parts at relatively low cost; efficiency was achieved by specializing the factory to product a single product. The system was ideal for the manufacture of items having a long product cycle and requiring heavy capital investment. In a rapidly changing technological marketplace, however, greater flexibility is required than the conventionally organized mass or batch production plant affords.

Recently, improved manufacturing competitiveness has been equated with Computer Integrated Manufacturing (CIM). By appropriate integration of the information and materials processing systems that control the manufacturing process, CIM system makes possible production on demand and rapid product innovation. A truly integrated system also brings efficiencies through significant decreases in product and process design time. But is this really CIM or just an artifact of the true CIM solution to increased manufacturing productivity.

II. COMPUTER INTEGRATED MANUFACTURING

Physically what is CIM and how will CIM improve manufacturing productivity? Firstly, let us look at how other individuals view CIM. In a recent commentary, Beardsley [2], viewed CIM as a strategy:

> *"CIM links people and existing technology to previously independent activities to create a total manufacturing system".*

On the other hand, Chad Frost [3] defined CIM in terms of people and not equipment:

> *"CIM is a people issue, not a technical issue"*

Joel Schnur [4] relates CIM to a dream that if we properly harness advanced technology we will kindle the rebirth of our withered factories. He ask:

> *"But why does it (CIM) elude many of us, while other exuberantly proclaim total success".*

Whereas, most vendors of computer systems view CIM as a trendy marketing caption for hardware and software sales.

Irrespective of from what viewpoint you analyize CIM, the computer is the factor that connects the various aspects of the CIM System. These quotes on CIM are very short sighted and truly do not address the overall concept of increased manufacturing productivity. If a product is being manufactured in a plant with obsolete equipment, it is ludicrous to think that by linking the present antiquated pieces of equipment together by a computer system will make this facility profitable. The computer system will give better process control and scheduling which will result in a better manufacturing process but not necessarily result in the manufacture and sale of a profitable product.

What is needed is a global view of CIM. What is it and how does it relate to production, people and productivity. In essence, we see CIM as a philosophy. It is the marshalling of all aspects of a manufacturing process; materials, labor and capital; in most efficient manner. The computer is only the communications link between the various components of the total system. To characterize or give meaning to CIM it is necessary to state with a definition and then explore its meaning. To us CIM is defined [4] as:

> **The integration of all forms of advanced technology to efficiently and effectively produce goods or services.**

Note that this definition includes all forms of technologies, both present and future, and the word computer is not present. The computer is only one form of advanced technology. We also state that CIM is a philosophy, therefore it must have a meaning and the meaning will be different for different applications.

Our discussions will be limited to a manufacturing environment. The physical components of a CIM system for a manufacturing environment are shown in Figure 1. These are:

Design	Manufacturing Processes
Accounting & Shipping	Automated Materials Handling
Production Scheduling Assembly	Integrated Sensor Metrology

As you would expect the computer network or system is the integrating factor between the various components. A considerable amount of effort [6] has gone into communications standardization; MAP, TOP and OSI. The computer technology to design and analyize a part has matured [7,8] to a relatively high level and the process standardization between systems is occurring [9]. In the same manner, manufacturers find that through parts standardization [10,11] significantly reduces inventory, tooling, and down time. Automation is performing many hazardous or manotous applications [12] as well as transporting material and product within the manufacturing facility [13]. Automation has led to better material scheduling, JIT [14], and the ability to simulate present and future processes [15] once the characteristics of the components are determined. Linkages have been developed between different entities, for example passing design file to manufacturing [16,17] to develop machine tapes.

CIM CONCEPTS

The basic concept of a CIM system is based on two tenants:

1. That all the manufacturing processes used in the system be if a continuous nature. Ideally the rates of production from various machines are direct multiples such that each machine can be operated at its mean full capacity.

2. Information by which decisions are made must flow freely through the system on a real time basis.

Process Technology

The first tenant relates to process technology. For a CIM system to achieve maximum effectiveness each component of the manufacturing process must operate in a continuous manner. Batch operations result in higher than desired WIP.

The distinction between a batch operation and a discrete operation is not clearly discernible. However, the longer the discrete operation the further the process varies from optimum. L.C. Thurow [18] indicates one of the major weakness occurring in the United States is the lack of attention to process technology. Two areas that directly impact on CIM are materials removal and surface modification.

Process Control

To attain maximum part quality requires careful control of each stage of the manufacturing process. Thus, as the raw material is being converted into a product, the geometry of the material must be measured in real time and verified with design specifications. In most instance, the exception is most of the chemical and plastic industry, the parts must be removed from the production line and measured. This approach may result in a portion of the WIP being out of specifications. The percentage depends upon the rate of production and the time frame for measurement. What is need is real time inspection.

Real Time Sensors

At the present time, research and development in computer aided design and automation of machine parts has been developed far beyond the capability of inspection and adaptive control during the manufacturing process. Thus, CIM will require that a sensor based system for automation must be included in the manufacture process of machined parts. Integration of vision technology in the automation will require that this process be included from initial concept which will ultimately influence the design. The ability to manufacture a part certainly influences design, thus sensors as an integral function in the automation of manufactured parts will provide the vision necessary to complete CIM. Figure 1 illustrates symbolically the integration of the elements in a completely automated manufacturing process. At the present time, the broken link occurs between the computer network and the integration of sensor based process. Sensors and their purpose as defined in this paper are to be integrated in the same manner as Computer Aided Design. For example, one of the primary goals of CAD is to produce a solid model complete with tolerances as a complete visualization of the design. A compatible goal of the manufacturing process should be to produce a part complete with measured data with a visualization of the part complete with dimensional tolerances. The sensors integrated with the manufacturing process should be capable of a solid representation of the part as manufactured. Figure 2 illustrates that design, manufacture and sensor control should form an integral part of automation. However, the quality control and measurement capability are not complete in the manufacturing process.

III. MACHINE VISION METROLOGY

The importance of machine vision metrology in manufacturing cannot be overemphasized. For example, at the present time no convenient method exists which can provide automated process control data to a CAD system. However, recent developments in machine vision technology have demonstrated the importance of this technology to produce three dimensional maps during a manufacturing process. Only visual sensors appear to hold the potential for solutions to this problem, however, the CIM solution has seen only limited success compared to other application of vision to manufacturing.

Cost of computers and vision system hardware has decreased in recent years, however, the skill level of engineers and scientists remain the limiting factor. The level of understanding is improving but the growth rate is expected to support no more than 20% per year increases in installation. The United States is the most dominant supplier of machine vision hardware and forecasts indicates that this trend will continue. Also, machine vision in software industries further favors the United States competition with other countries.

Research programs in critical areas of vision research have been established by NSF for robot vision system capable of locating, identifying and inspection objects on a 1 μ scale. As a goal for vision research on the microscale, robotic workstations functionally equivalent to a human operation must be developed. In addition to the robot microscopic hardware, new algorithms of image analysis are currently under development for specific applications which identify and test submicron features in semi-conductor chips.

Another area of established machine vision research which has been identified as important is in manufacturing systems for recognition, inspection or parts and assemblies. Although this area is probably more mature than other machine vision application in manufacturing, improved algorithms for real time inspection in manufacturing systems are needed. Continued research goals in computer vision for assembly and inspection of intelligent manufacturing systems. As part of the long range plans, parallel processing and VLSI implementations are to be integrated into machine vision for inspection of parts and assemblies.

Machine vision for industrial applications in inspection are forecast to constitute approximately 90% of all visual inspection by the mid 1990's. If this projection is to become a reality, then application specific centers must be created and areas of specialty must be developed. The parallels of engineering education are evident in that many disciplines utilize the basic laws of physics as applications to specific problem areas. Although vision technology is generic, solutions will require the specialist as we currently require in the engineering disciplines.

A machine vision system integrated into a manufacturing cell is envisioned to consist of three subsystems. The first subsystem is a CAD design capable of generating and displaying three diminsional solid surfaces. The second subsystem is either a machining or turning center in which tool paths have been generated by the CAD system. The third subsystem is the machine vision process control system. Vision systems represent full field data of an area of an object which can be recorded at one time. Advantages of full field data recording are recording of local anomolies as well as large scale sharpe features to be recorded with the same precision as paint measuring machines. Full field measurements such as vision are much faster than point measuring machines, however, the large amount of data requires the use of sufficient computation power. With the new personal computers, this does not appear to be a limitation and small computer systems may be integrated in a factory work cell. Data from the third system may be compared with the original CAD data in order to complete the cycle. We believe that the vision technology has developed to the level where integration in the factory environment may be readily accomplished. In addition to vision as a method of statistical process control, design of inspection systems may be readily incorporated in the CAD design systems because lighting influences visual inspection is the same manner as shaded solids in three diminsional CAD systems. We, therefore, believe that machine vision represents the best method of integrating SPC in the complete CIM concept.

REFERENCES

1. "Global Competition: The New Reality", President's Commission on Industrial Competitiveness" U.S. Government Printing Office. January 1985.
2. C.W. Beardsley, "CIM As A Strategy", Mechanical Engineering 110, 7 (1988) 8.
3. C.E. Witt, "AMS 88 Offers New Solutions to CIM Complexities", Materials Handling Engineering, 43, 6 (1988) 58.
4. J.A. Schnur, "Our Quest for Manufacturing Excellence", CIM Review, 4 2 (1988) p. 3.
5. R. Progelhof, "Manufacturing Considerations to Become and Remain Competitive in a Global Market: Computer Integrated Manufacturing - What Is It", Fluid Controls Institute Meeting, Dallas, Texas, October 27-29, 1985.
6. G.F. Schwind, MAP/TOP/Q.OSI: New acronyms in world data interchange, Materials Handling Engineering, 43, 8 (1988) 76.
7. N.E. Ryan, "CAD/CAM Integration Yields Quality, Control", Manufacturing Engineering, 2 (1987) p. 43.
8. R.L. Martin, "CAD/CAM-An Even Fuller Menu Ahead", Manufacturing Engineering, 12 (1987) p. 44.
9. G.B. Latamore, "CAD/CAM's $800 Million Winners", High Technology Business, 1 (1988) p. 41.
10. G. Sessler, "Standardization: Fast Becoming a Top Management Priority", Siemens Review, 55, 2 (1988) p. 16.
11. R. Speir, "Maximizing Design Efforts Through Source Standardization", Appliance Manufacturer, 6 (1988) p. 27.
12. T. Zambetis-Jones, "Implementing Robotics in Your Plant", 6 (1988) p. 48.
13. L. Gould, "Could An AGVS Work For You", Modern Materials Handling, 43, 9 (1988) p. 74.
14. I.P. Krepchin, "Here's How 3 Companies Practice JIT", 43, 9 (1988) p. 69.
15. I.P. Krepchin, "We Simulate All Projects", 43, 9 (1988) p. 83.
16. R. Harvey, "CAD/CAM Drives Deep Into The Factory Floor", Iron Age, 6 (1986) p. 23.
17. J.B. Pond, "NC Program Probability is Here, Now", Iron Age, 6 (1986) p. 51.
18. L.C. Thurow, "A Weakness in Process Technology", Science, 238 (1987) p. 1659.

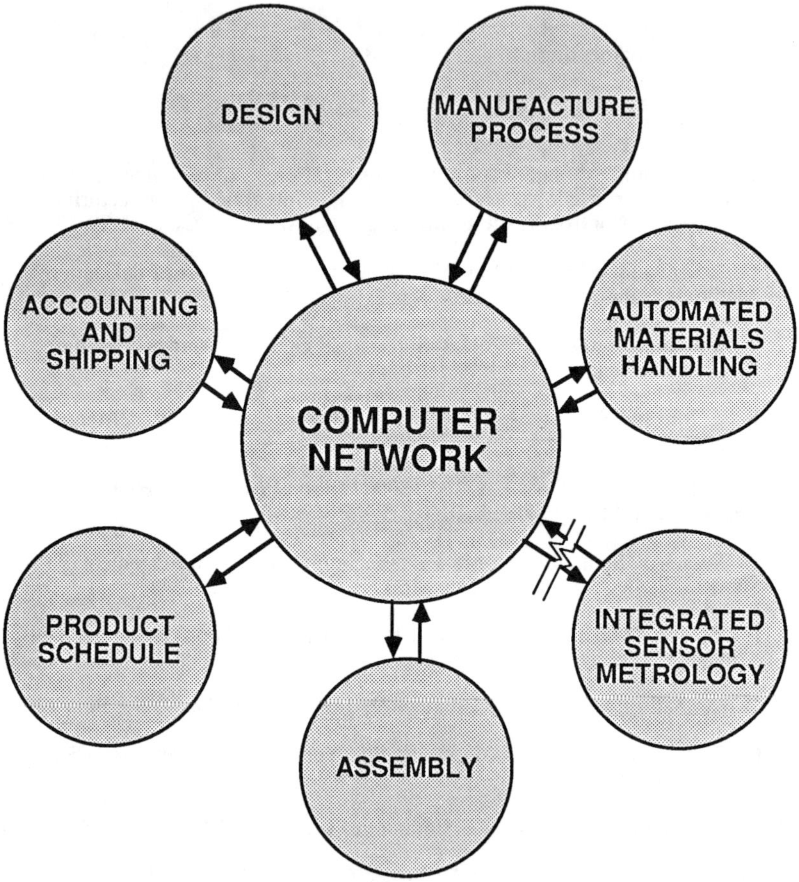

Figure 1. Computer Integrated Manufacturing System

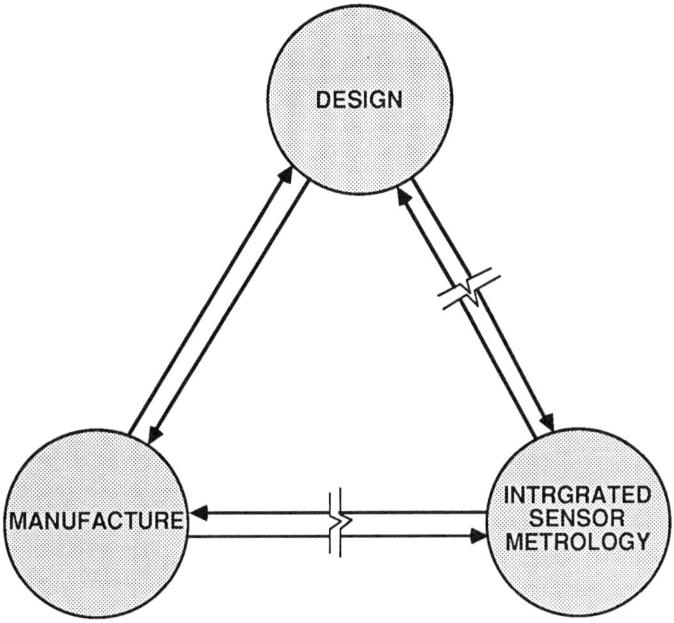

Figure 2. Integration of Sensor Technology in a Manufacturing System.

TECHNOLOGICAL IMPACT OF CUSTOMER REQUIREMENTS IN ALUMINUM METAL WORKING PROCESSES

R. E. Fanning

Fabricating Metallurgy Division
ALCOA Laboratories
Alcoa Center, Pa.

Introduction

The Aluminum industry is one of many materials industries which strive to make their material the preferred choice for large volume products. In some applications one material has a natural advantage over others, but more frequently several materials compete with one another, each providing one or more advantages and some trade-offs. Customers may design their product with several objectives such as minimum system cost, maximum customer appeal or maximum manufacturing flexibility, choosing materials which best combine cost and performance. In such markets it may become necessary for a material supplier to enhance the attractive features of the particular material to outweigh its disadvantages.

It is difficult, if not impossible, to predict how various materials will evolve. Technological breakthrough, government regulations, material availability or sociological issues may redefine the balance between competing materials. In spite of some uncertainty regarding the competitive position of competing materials, there is little doubt that unrelenting pursuit of product quality improvement and cost reduction are essential to future viability of the aluminum industry. It is the intent of this paper to identify some areas where technological improvement would be beneficial. Thus this paper is a collection of briefly described problems that will hopefully stimulate the thoughts and research required for knowledge based solutions.

Rolling Process

A particular example is the beverage can industry where aluminum, steel, glass and plastics vie for a share of the market. The high cost of aluminum relative to steel or glass

Metallurgical Processes for the Year 2000 and Beyond
Edited by H.Y. Sohn and E.S. Geskin
The Minerals, Metals & Materials Society, 1988

requires minimizing the amount of metal per can. The high intrinsic value of aluminum has led to extensive recycling of cans although net metal cost is still higher than that of the competing materials. Intensive engineering of the beverage can led to improved design and better understanding of material requirements. These improvements permitted substantial reduction in sheet thickness over the past decade. The downward trend in metal thickness for beverage can sheet is shown in Figure 1. Reduction in thickness for both body and

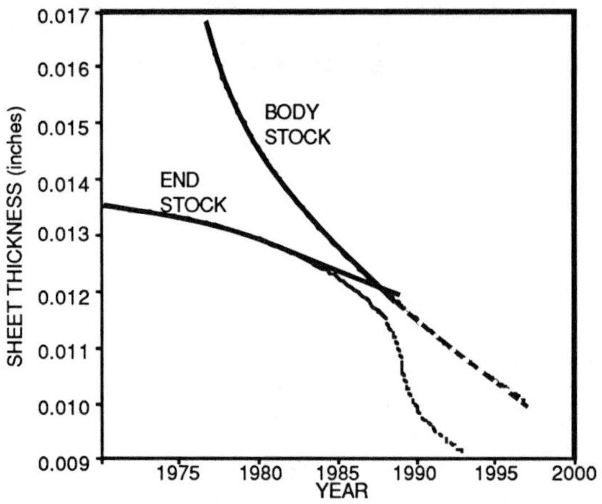

Figure 1 - Beverage Can Sheet Thickness

end stock alloys are shown, although that for body stock has been more extensive. These curves are a smoothed representation of data from historical files of customer thickness specifications. A range of thickness was produced at any time, but the downward trend is the parameter of importance. During the early years of the can business customers were allowed to specify thickness in increments of 0.0005 inches. By 1986 this allowed increment had been reduced to 0.0001 inches. The changes in nominal thickness combined with increased flexibility of specification has led to production of many more product specifications. The variations in specified thickness introduced new scheduling problems since strip width also varied. As sheet thickness was reduced and rolling mill performance improved, the tolerance in sheet thickness was also reduced. The downward trend in thickness tolerance is shown in Figure 2. More specifications produced to tighter tolerances made this seemingly simple business much more diverse. These increasingly stringent quality requirements led ALCOA to build a new rolling mill in their Tennessee Operations. This mill, commissioned in 1987, provides world class hardware and control with fully continuous operation. The process knowledge designed into this mill and its automation strategy was viewed as essential to the operation of a modern multi-stand mill producing can stock in the future.

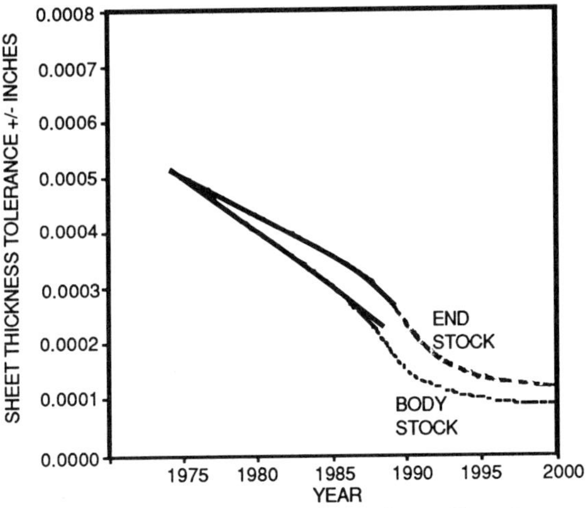

Figure 2 - Beverage Can Sheet Thickness Tolerances

View of the future

The preceding history from the can sheet business provides the context for thoughts concerning the future. It is unlikely that the decrease in thickness or the reduction in thickness tolerance will stop in the near term. Tolerance of +-.0001 inches is very nearly a reality. It is likely that the signal to noise ratio of the thickness measuring device will provide a practical limit on thickness tolerance. Figure 1 contains a projection that body stock sheet thickness will approach 0.010 inches in the 1990's. At the same time end stock thickness is projected to reach 0.009 inches. Although this is contingent on can design changes, experience indicates that changes are likely. The end stock dimensions are probably near to a practical limit considering buckle strength and the necessity of forming the easy-open device. As thickness is reduced material properties become more critical. The stochastic nature of material behavior within an ingot will necessitate statistical analysis of material properties to ensure that product variability is within acceptable limits. Statistical process control or other process control strategies will be required to maintain key process parameters at required values. Different can designs used by various customers will likely require a different balance of sheet thickness and material properties leading to additional process diversity. The high cost of inventory will lead to "just in time" processing. The rolling process will need to be highly flexible with extremely tight control at the desired operating points. Competitive pressures and customer requirements will demand technological excellence and training of people will be required for effective use of technology.

Discussion of required technology may be grouped into categories of quality and process efficiency. Although these groupings are somewhat arbitrary they serve to focus on general areas of product attributes and manufacturing technology.

Quality

In the past material properties were the result of processing according to standard practices defined by experience to provide an acceptable range of product variability. As we look to the future, one can imagine that each piece of metal will require sufficient information at each production location to characterize the input state. As the metal is processed the information base will be updated to the current state. In this way the final product requirements can be assured. It would be valuable to have sensors which could measure texture or other material parameters directly, but in the absence of such sensors other data must be used to define approximate measures. Such use of process data makes the quality of data and the analysis critical. Material science research must become much more quantitative and predictive. Models of material deformation mechanisms should guide development of structure and properties both for the development of deformation practices and for the monitoring of structure evolution. An example of this is the constitutive equation development of Lalli and Sample (1). In this work internal variables combine history and current deformation to obtain flow stress. Such work needs to be extended to incorporate other material property parameters such as texture, and be made applicable to the variety of commercial alloys and variations that make up the can sheet market. This understanding of the response of materials to deformation and temperature will lead to precise definition of the processing path the material must experience. Temperature control during processing will become critical, and for this control to be possible, adequate sensors for measurement of temperature will be required. Temperature measurement of aluminum poses many problems in a production environment as non-contact sensors are often confused by emissivity variations, and contacting devices are subject to a variety of mechanical problems. Model predictions and thermal sensor output could perhaps be combined into a "most probably correct" value for use as feedback to thermal control, thereby overcoming problems of emissivity variation in non-contact temperature measurement. Research is needed in the area of resolution of conflicting data from redundant sensors to support this "most probably correct" concept.

Dimensional quality issues focus on achieving the desired nominal thickness and maintaining acceptable tolerances around this nominal. In the past a mill operator was an integral part of the control. It is unlikely that an operator could cope with the number and rate of process interactions necessary to achieve acceptable product in the future. Thus automatic control will be necessary. The subject of automatic control is widely developed in the literature. No attempt is made here to be inclusive, but rather to suggest work which includes process understanding explicitly. Bryant and Spooner (2) presented a control strategy incorporating process models for steel. DesRochers et al (3) studied control with self learning process models. Kimura et al (4) developed a model based multi-variable control for tandem mill rolling. These investigators have dealt only with the objective of thickness control. Thermal and constitutive behavior models which enable prediction of properties must be combined with the more conventional control to provide total automation of quality. With conventional single input, single output control algorithms each quality parameter would have its own control. A problem arises when dimensional control and temperature control cannot be attained at the same operating point. A multiple objective control algorithm controlling dimension and temperature concurrently would attempt to optimize the operating point to satisfy both requirements. The resultant operating parameters become a compromise of one or both control objectives. Research is needed to provide efficient resolution of this sort of problem. This multi-objective control falls under the classification of state space control or modern control. While there exists a large body of literature on the subject, the author is unaware of any commercial implementation on a rolling mill. More research on control algorithms and the integration of process models into realizable control applications is needed as this more comprehensive control will surely become advantageous in the future.

Process efficiency

Process automation, as described above, can provide repeatable manufacture to given references. In producing many different specifications, it is important that the first metal processed is produced to specification. Process design will likely be required to permit set up and control of a new specification reliably. Both on line and off line design tools are required. On line models, particularly those used as elements of feedback or feedforward control strategies, require high accuracy. In addition, execution time must be much less than the control algorithm cycle time. Thus the models must be simple in structure and in most cases adaption or learning from the process data is required to maintain requisite accuracy. For example, installation of a new work roll in a rolling mill may result in off specification product initially. The new roll has a very different thermal history than the roll that was replaced and the frictional characteristics are not known with certainty. Thus the rolling load and consequent strip flatness may be estimated incorrectly and the mill set up to the wrong references. As subsequent coils are rolled, adaption may be used to tune the models to obtain acceptable setup. However, first coil performance is an important part of process consistency.

For processes in which change of specification is frequent, it is important that design tools be both easy to use and comprehensive. Simple models may suffice for many of the process related decisions, but more complete models such as those based on finite element or boundary integral methods should also be a part of the design package. There are a large number of finite element codes supported by commercial software firms. There are also a number of codes developed in the academic community that may be integrated more readily into a design software package. Thompson (5) has developed an Eulerian analysis of the rolling process using a very simple friction model. More recently Bruce et al (6) have used an elastic plastic formulation to study interface stress in rolling. This work presents evidence that the traditional coefficient of friction model of interface behavior is inappropriate. Relative speed of the roll and strip is shown to be an important factor in defining interface friction. The interaction of surface roughness on the strip and roll in conjunction with applied lubricant on interface friction is not yet known in the point to point detail required by FEM analysis. The impact of this interface between roll and strip on heat transfer is also not known. The mechanisms of interface behavior must be more completely described before the full power of FEM analysis can be brought to bear on production problems. A comprehensive model of interface behavior is also needed to develop simple models suitable for on-line setup and design. It will be a challenge to incorporate such models into user friendly packages that give answers to process design issues in a reasonable time.

The preceding discussion of design models dealt with the process at a given production center and a particular process step. Issues concerning "just in time" operation will require integration of individual process steps into shop floor simulation type models (7-9). These simulation models provide optimization of material flow with respect to specified objectives and constraints. Typically the simulation is based on production time data collected in a plant using a variety of statistically based assumptions about the magnitude and variability of time requirements for each process operation. Processing rules are incorporated in the scheduling algorithms associated with the simulation. The simulations often do not reflect any processing history effects on the metal being processed. When used to schedule production it is possible to attempt processing of specifications which cannot run due to thermal or other process constraints. Research is needed to incorporate process history into the simulation rules to alter scheduling policies to fit a diverse product mix. Research will likely be necessary to transform process knowledge contained in the models for setup into constraints or other rules suitable for incorporation into simulation and optimization programs. This interdisciplinary research should involve people from both process and operations research fields. Very few, if any, cooperative programs in this area are active in the academic community today.

Equipment must be maintained to ensure effective operation. When a piece of equipment is not heavily utilized it is less difficult to schedule maintenance time. As utilization increases preventative maintenance becomes more difficult to schedule and frequently unanticipated failures occur. For a fully continuous mill it becomes advantageous to develop predictive maintenance technology. Pieces of this technology, for example vibration monitoring of gearboxes, are well known. The process control computers monitor a large number of sensors to operate the mill and many predictive features are incorporated to avoid unscheduled stoppage. However integration of the various technologies that would comprise an effective predictive maintenance system has not been developed. Development of this sort of technology as a collection of individual elements is not likely to succeed. Research using a systems approach to problem definition should be invoked to provide a sound basis for subsequent application in industry. Many different tools and skills will be required including statistics, sensors, models and benchmarking. Success of such a system will depend to a great extent on the ease of use. Artificial intelligence or an extensive menu driven user interface will be required. Here also training of the people who must use the technology will be critical.

Computing requirements

The precise thickness control requirements of rolling dictate precise and frequent adjustment of the gap between work rolls. (Figure 3) The actual position control of the screws or

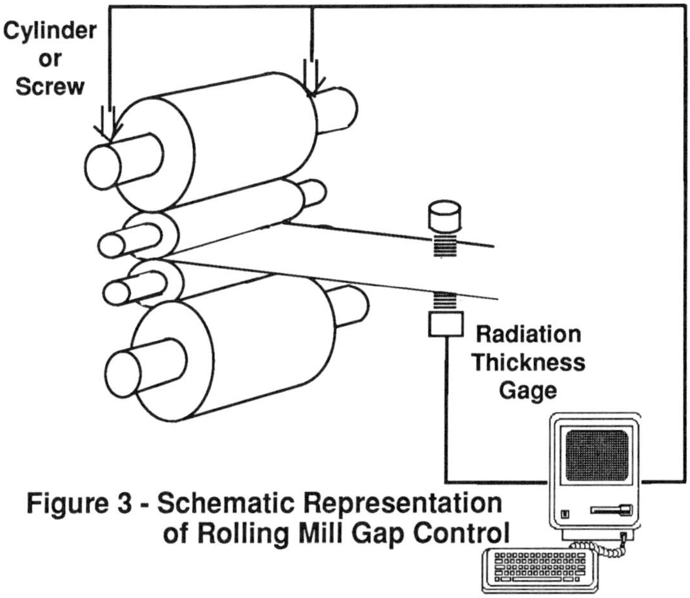

Figure 3 - Schematic Representation of Rolling Mill Gap Control

cylinders which act on the rolls is usually performed by a dedicated micro-computer. The reference position for the gap positioning cylinders is a result of process control output. Motor speed references, roll bending jack references and any cooling spray references are also a result of control output. (Figure 4) The volume of data required for control and the

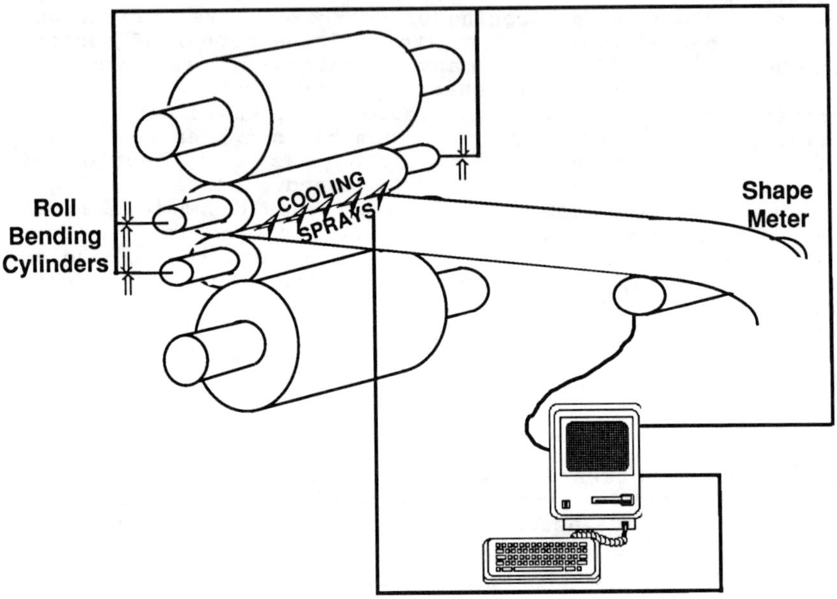

Figure 4 - Schematic Representation of Rolling Mill Shape Control

rate at which this data must be processed and transmitted provides a challenge to computer system design. Each control algorithm needs process and actuator feedback. High speed communication links are available which pass data at 30-50 megabits per second. Normally only a limited amount of data can be transmitted at the highest rate, so the system needs to be developed in a hierarchical form that defines which elements of data are needed at each station and level in the system.

A fully distributed micro computer system is feasible in rolling mill control and commercial systems are available. As the control strategies become more computationally intensive or utilize more process data the control algorithms may not all reside in one computer. Some choices are possible. One can choose to purchase a super mini computer retaining the sequencing of control action in one machine, or a system of "token" passing can be developed to indicate when control action is to be made. In the case of the super mini computer the control is limited by the ability of the computer to process all control algorithms and data handling. For a token based system the increase in data and data processing due to placing information concerning status into the control algorithms may outweigh the inherent computational speed advantage of multiple processors. Another option, which has not been adequately researched, is to develop control algorithms which will operate asynchronously. In this option a control action may be initiated without knowledge that another pre-requisite action in the control sequence is not complete. This usually precludes tuning such a system of control to the highest level of performance due to stability problems. If control algorithms, stability criteria and tuning criteria could be developed for asynchronous control the full power of multi processor distributed computing could be realized in the area of digital process control.

Forging Process

Another deformation process undergoing change due to customer requirements is the forging business. The typical forging often contained extra metal, machining allowances, that were machined away to obtain the desired part. Again, the relatively high metal cost has driven forging design nearer to final part dimensions. Precision forgings have evolved from early 1950's as developments were made in forging process understanding. The precision forging concept allows the designer to specify thinner ribs, smaller radii, thinner webs and other geometrical constraints that would otherwise entail excess metal in the forging. By reducing the amount of metal that must be machined away to obtain the final part, the total cost of manufacture can be reduced. For any particular part there will be an order size which makes precision forging economically attractive. Initial tooling cost and tryout or any extra operations must be balanced against the benefit of reduced machining and increased metal utilization. Figure 5

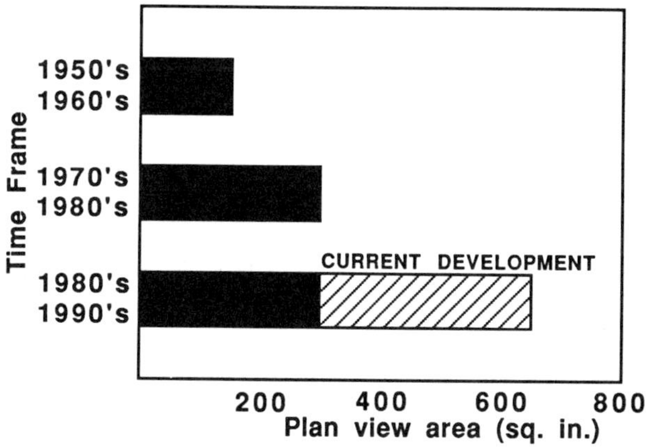

Figure 5 - History of Precision Aluminum Alloy Forging Development

shows the evolution in size of precision forged parts in aluminum (10). The increases in plan view area are related to improvements in tools and in improved die heating capabilities. As the plan view area increases, the unit pressure available on a given press decreases. Thus factors which control forging pressure must be understood and controlled. Much work on modeling of the forging process is currently underway and the work on constitutive modeling described above provides essential support. However, forgings are quite complex three dimensional objects and most of model development is either two dimensional or simple three dimensional.

View of the future

The demand for precision forgings will increase. Customers will ask for larger plan area and will also pressure to reduce the number of parts per order required for economic viability. It is unlikely that modeling capability and increased computing power will permit complete process design for each part. Simpler design software will be required. Rule based design will still be the predominant methodology. User interactions via CAD or artificial intelligence programming will make such packages easier to use allowing more complete design for each customer specified part. Computer integration into die manufacture will increase and feedback from production experience will become available. Research in methodology to incorporate feedback into design software will be needed to reduce initial design error. Less trial and error and costly tool modifications will be necessary to obtain a desired part specification. This will lead to lower cost and reduced flow time for manufacture. These factors should reduce the rate of encroachment of alternative materials into the market. Process efficiency will be an important element in the viability of aluminum forgings. This implies efficiency in use of equipment, metal and time. Quality has always been a strength of this business, however, consistency of dimension and material properties will become increasingly important. Development of technology to address these issues will be critical for this business.

Quality

Consistency in quality has a slightly different context in forging than in rolling. Material properties may vary from point to point in a forging due to variation in deformation and thermal history. It is however, desirable to have consistency at a given point from forging to forging. Metal deformation is sensitive to thermal boundary conditions and to friction between tool and workpiece. To achieve consistency from the first to last part of an order requires control over these interfaces or compensation via process control. Process control is quite different than in rolling in that direct feedback of the critical parameters is virtually impossible. Controlled rate of die closure in forging is a known technology, however implementation is limited in many instances by available press hardware. Also, inadequate process knowledge may not provide quantitative speed references for control purposes. Average strain rate control is not adequate for complex forgings. Lack of temperature feedback and a thermal control strategy are other limitations to speed control. Inferential control integrating model predictions of thermal gradients and other process data could be developed along with improved die heating to minimize thermal variations. Although modeling of heat transfer has developed to a high degree of capability and sophistication, the current ability to characterize interfacial friction forces and heat transfer characteristics limits use on actual production problems. It is difficult to develop such information in the point to point detail required by finite element method analysis as one cannot simply run the problem in reverse to infer interface behavior from forging pressure or workpiece temperature. Other methods must be developed to provide necessary data. It appears that this is a much neglected area of research probably because it is difficult and there has been limited need to date for such data. Simpler analysis methods produce only average values of interface parameters and can be confused by different surface gradients yielding similar averages. Statistical control methodology probably offers a significant opportunity to develop operating practices which produce consistency when forging parts with large numbers of pieces per order. For the small lot size parts there appears to be little alternative to some sort of knowledge based on rules, models and experience. Artificial intelligence would seem to be an excellent user interface to such information. Considerable research will be necessary to provide knowledge based systems that deal with apparently conflicting rules and data. Training of people who deal with forging will also be essential for effective implementation and utilization of this technology.

Process efficiency

Efficient utilization of metal is crucial. Considering the time consuming operations a that a typical forging passes through, failure to meet specifications is costly. Many of the problems that contribute to failures are not highly technical in nature. Each metal handling operation provides opportunity for damage. Extra pieces scheduled to allow for normal rejections become excess in process inventory with extra material moving and storing operations. Excess delay time between operations increases handling requirements as metal must be stored while waiting and then retrieved when needed. Efficient scheduling of equipment and metal movement could be beneficial in reducing such handling and the associated risk of physical damage. Simulation models, as discussed for rolling, will become an essential tool for development of scheduling algorithms and strategies which permit reduction of flow time and excess material handling. The work force must be considered an important process element in material handling. Training of the people who work with the metal in process will be an essential tool in reducing the risk of handling damage, particularly for products whose small volume makes fully automated handling uneconomical.

Press setup time is an important element of process efficiency for small lot sizes. Rapid die exchange has received much attention in the literature (11). Reduction in press idle time is an obvious benefit, however the use of externally aligned die sets with standardized press fasteners will aid in obtaining the tighter tolerances demanded. Attention to detail in all aspects of manufacture is an extension of the rapid die exchange philosophy that will improve both efficiency and quality.

Computing requirements

The computationally intensive nature of many of the problems facing forging dictate use of super mini computers and, in some cases, large main frame scientific computers. The design aspects are built into the tools and operating procedures in an off line mode making the issue of execution time less a problem. This design activity, even though performed off line, is required for every part a customer orders. Ease of use and efficient interface to CAD/CAM equipment is essential. As customers provide more of their part specifications in numerical form the design process can be increasingly automated. Training for the designer becomes essential to permit the routine operations to be performed by the computer. Every design has some commercial compromise and the designer should be freed to concentrate on these issues.

Press control is less complicated than for rolling due to fewer sensors and slower rates of operation. High speed input, output and data transmission are not required, hence the micro computer becomes a viable choice. Data collection and analysis may require the processing power of a mini computer to facilitate interaction with higher level business computers. The choice of computer hardware in forging would seem to be driven mostly by issues such as ease of programming and maintenance, availability of software, networking capabilities for shop floor scheduling and cost.

Summary

Many similarities exist in the technological requirements for processes as diverse as rolling and forging. Indeed, it is safe to assume that many of the issues are common to most metal forming operations even though the specific technologies are somewhat different. Tightening of manufacturing tolerances, the need for process consistency, trends toward smaller lot sizes and product differentiation for specific customers are examples of these generic concerns. Models of thermal and deformation processes will become necessary tools in most industries. Additional research in deformation modeling, constitutive equations and interface characterization will be essential for continued progress in manufacturing capability. Analysis of process data will be essential for obtaining process consistency. Traditional statistical control concepts and time series analysis will provide identification of special cause and time varying factors in production processes. An area of research which has not received sufficient attention is analysis of large volumes of data from production processes. Such analysis requires that graphics software be combined with data base management software to readily select and examine parts of data sets which may contain millions of data points (12). In addition, some form of statistical analysis must be performed. Research is needed in the area of regression analysis of potentially covariant data. These improved data analysis tools should then be integrated into a complete analysis environment which includes both process modeling and statistical analysis.

The temptation to solve all problems with new technology development must be resisted. The process modeling and control technologies of the 1960's and 1970's are not yet adequately implemented in production. In many cases theoretical understanding will never rival the complexity of the actual process. However, theoretical concepts can guide development of process knowledge through knowledge intensive examination of process data. This qualitative or semi-quantitative knowledge, in turn,should be used to develop improved processing parameters, control algorithms or operating procedures to obtain technological benefit long before theory is complete. It is important to understand that the quest for knowledge is never complete, hence the incorporation of technology into production processes is not a one time event. Process improvement, whether based on external technological discoveries, or developed from observation of the process itself, must also be continuous.

User training and simple, yet comprehensive user interfaces to technology are also issues common to most metalworking processes. Little progress toward solution of the problems of manufacture will be made until comprehensive analytical tools become easy to use on a regular basis. People will make the difference, and people are not inherently more intelligent than in the past. People can, however, become much more effective if technology is made available in an environment which is easy to use, efficient in operation and comprehensive enough to deal with problems as they arise in production.

Acknowledgements

The author would like to acknowledge the helpful comments and data of Mssrs. D. Roeber, P. Schilling and L. Lalli. The author also extends a particular acknowledgement to Ms. Veronica Gulick for support with the graphics in this paper.

References

1. V. M. Sample and L. A. Lalli: "Effects of Thermomechanical History on Hardness of Aluminum", Materials Science and Technology, January 1987.

2. G. F. Bryant: Automation of Tandem Mills, Iron and Steel Institute, London, 1973.

3. R. Ramachandran, M. Clifford and A. A. Desrochers: "Optimal Control of a Single Stand Rolling Mill", Robotics and Automation Laboratory Report, Rensselaer Polytechnic Institute, 1984.

4. I. Hoshino, Y. Maekawa, T. Fujimoto, H. Kimura and H. Kimura: "Observer-Based Multivariable Control of the Aluminium Cold Tandem Mill", Sumitomo Light Met. Tech. Rep., July 1987.

5. E. G. Thompson and H. M. Berman: Steady-State Analysis of Elasto-Viscoplastic Flow During Rolling, Numerical Analysis of Forming Processes, John Wiley & Sons Ltd., 1984.

6. R. W. Bruce, M. E. Karabin, S. Panchanadeeswaran, M. L. Devenpeck, C. Y. Lu, T. Sheppard: "Experimental and Analytical Comparison of Interface and Internal Variables in Cold Rolling", To be presented at the winter ASME conference 1988.

7. A. M. Law and W. D. Kelton: Simulation Modeling and Analysis, McGraw-Hill Book Company, 1982.

8. C. D. Pegden: Introduction to SIMAN with Version 3.0 Enhancements, Systems Modeling Corporation, July 1985.

9. A. A. Pritsker and C. D. Pegden: Introduction to Simulation and SLAM, Halsted Press Book, John Wiley & Sons, 1979.

10. G. W. Kuhlman: "The Precision Forging Revolution", Presented at Manufacturing Technology Advisory Group Conference, Mtag 86, New Orleans, Louisiana, November 1986.

11. S. Shingo: A Revolution in Manufacturing: The SMED System, Productivity Press, Stamford, Connecticut, 1985.

12. P. L. Love and M. Simaan: "A Knowledge-Based Approach For Detection and Diagnosis of Out-Of-Control Events In Manufacturing Processes", Proceedings IEEE Conference on Intelligent Control, August 1988.

Non-Ferrous Processing

CONTROL OF POTENTIAL DIFFERENCE IN CONTINUOUS SMELTING SYSTEM

Akira Yazawa

Research Institute of Mineral Dressing and Metallurgy
(SENKEN), Tohoku University
Katahira, Sendai, 980 Japan

Abstract

To realize highly efficient process in extractive metallurgy, continuous operation nearly in equilibrium is preferable, but most metallurgical processes consist of plural steps with different chemical potentials. Traditional two step process of sintering followed by reduction is continuous in each step, but is heterogeneous reaction system far from equilibrium, and the chemical potentials are varied considerably along the distance the charge travels in a reactor. This route will become old-fasioned in next century due to the various inherent problems caused especially by sintering process. To establish reasonable new smelting processes, the important tasks of the metallurgists are how to introduce feed and reacting gas in the reactor, and also how to overcome the difference in chemical potential in the reactor system. Based on the ways to solve theses tasks, recent proposals of new smelting process for copper and lead are classified and discussed with the help of sulfur-oxygen potential diagram. On such potential diagram, each equilibrium step in continuous smelting is expressed at a fixed point, and two reactors with different chemical potentials must be connected by continuous molten flow through launder or partition wall. In the latter case, the reactor is apparently single, but some technical problems must be overcome. While, in a batch smelting like converter, potential variation is successively realized along a path on chemical potential diagram with the time elapsed. To establish immediate equilibrium, continuous steady flow of reactants into well mixed reactor is indispensable. In modern smelting process, the ore is mostly supplied in the form of powder which have huge surface area and fluid character, and consequently injection process must be very attractive. Several examples are discussed for both of copper and lead smelting, and promising future technologies are suggested.

Metallurgical Processes for the Year 2000 and Beyond
Edited by H.Y. Sohn and E.S. Geskin
The Minerals, Metals & Materials Society, 1988

Introduction

In extractive metallurgy, most processes consist of plural steps with different chemical potentials such as matte smelting followed by converting, oxidation followed by reduction in lead or zinc smelting, or reduction followed by oxidation in steel making. To realize highly efficient process, continuous operation nearly in equilibrium in a compact reactor is preferred, but the tasks of metallurgists are how to introduce feed and reacting gas in the reactor, and also how to overcome the difference in chemical potential in the reactor system. Recent proposals of new smelting process always include the challenge for these tasks.

Although there are many interesting proposals in the field of ferrous metallurgy, in the present paper, the proposals for modern copper smelting processes are reviewed, and recent trends of lead smelting are discussed with the help of chemical potential diagrams.

Process of Sintering Followed by Shaft Furnace Reduction

Traditional process of sintering followed by shaft furnace reduction has been widely applied for smelting of copper, nickel, lead, zinc, tin, etc., and if iron making is also taken into account, it is even now most popular process in extractive metallurgy. Undoubtedly, it is a remainder of the process where lump ore was smelted directly, but it must be re-examined if it is suitable or not for treating powder form concentrate.

Although both of sintering and reduction are continuous operations and various improvements have been accumulated, these are essentially heterogeneous reaction system involving lumpy solid, far from equilibrium, and the chemical potentials are varied considerably along the distance the charge travels in each reactor. Agglomeration is an important purpose of sintering to be amenable in shaft furnace, but decreases greatly in reacting surface area of the concentrate. Generally speaking, sintering machine itself is too old metallurgical apparatus: low efficiency non-equilibrium reactor, not convenient for heat and mass transfer, low thermal efficiency, huge amount of exhaust gas, various associated environmental problems, etc. According to the author's opinion, sintering, and also followed shaft furnace smelting, will be better to be substituted by modern high efficient process, even for the iron making in next century. In such modern process, huge reacting surface area of concentrate must contribute to rapid reaction to establish equilibrium immediately in the well mixed reactor, and thus, the chemical potential in the homogeneous products must be a fixed value. The sintering process was already abandoned for copper concentrate, and much more modern processes are operated as described in next.

Evaluation of Modern Copper Smelting Processes

The usefulness of sulfur-oxygen potential diagram for the explanation and evaluation of copper smelting was described previously by the present author [1], but will be reviewed briefly. Although various new copper smelting processes have been proposed during the past two decades, these may be classified as shown in Fig.1 including traditional ones [2], and reasonably explained with the help of potential diagram. In Fig.2, a sulfur-oxygen potential diagram for Cu-Fe-S-O-SiO_2 system at 1200°C is reproduced [1].

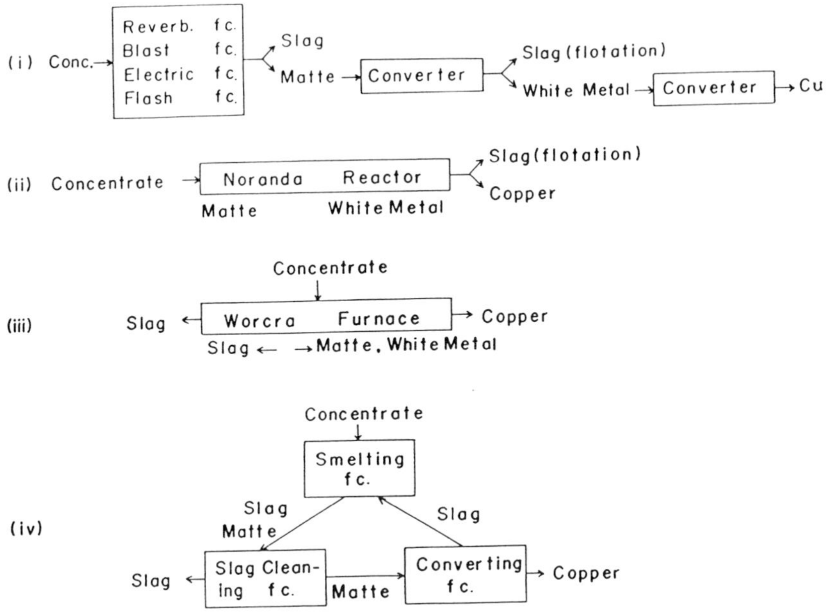

Figure 1 - Classification of various copper smelting processes.
(i) Matte smelting followed by batchwise converting.
(ii) Concurrent direct smelting with a single furnace.
(iii) Countercurrent direct smelting with a single furnace.
(iv) Continuous smelting with plural furnaces.

(i) Matte smelting followed by batchwise converting

The traditional smelting processes are shown by a simple flow sheet in Fig.1(i) in which matte smelting is followed by a converting process consisting of two stages, slag blow and blister blow. In Fig. 2, this conventional oxidation smelting follows the path ABCD. The matte smelting is carried out at region A with a small variation in oxygen potential with increasing matte grade. In the first stage of converting, slag blow, corresponding to between A and B, white metal B is produced without serious solid magnetite separation suggested by MM'. After eliminating the slag, the white metal B is rapidly oxidized to point C in the beginning of the second stage of converting with a drastic variation in oxygen potential. Successive conversion from the white metal to blister is realized at point C, and so called the finishing stage of blister blow corresponds to between C and D. Modern flash furnace smelting process is also classified in this category, but inherent batchwise character in converting process causes trouble in both of smelting efficiency and environmental disruption.

(ii) Concurrent continuous process with a single furnace.

As shown in Fig.1(ii), the Noranda copper mode process is most well known one in this category. In this process, the lateral position in the lengthening converter type reactor corresponds to ABC in Fig.2, and due to the concurrent flow, whole produced slag is in equilibrium with white metal and blister copper at point C. Thus, as suggested from Fig.2, the slag must be troubled by high contents of copper oxide and magnetite. The

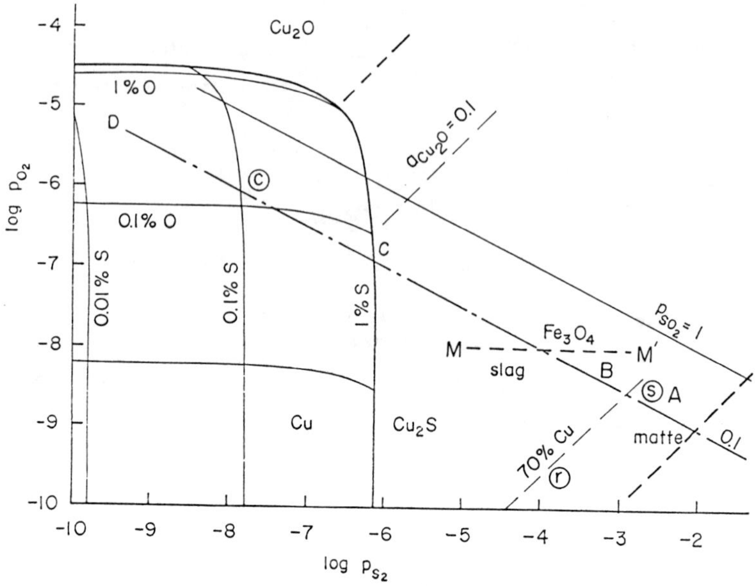

Figure 2 - Sulfur-oxygen potential diagram for Cu-Fe-S-O-SiO_2 system illustrating copper smelting processes at 1200°C.

direct production of blister copper in flash smelting furnace is based on the same principle. These processes will be difficult to develop as a standard method for the usual chalcopyrite type concentrate, but may have possibility just for the low iron concentrate. It is clear from Fig.2 that the troubles associated with the process of this category will be solved greatly if the product is white metal at point B, and this is the case of today's Noranda matte mode process.

(iii) Countercurrent continuous process with a single furnace.

To avoid high copper loss and magnetite trouble in slag, there are some proposals in which matte and slag are in countercurrent flow whereby blister copper and slag are tapped out at opposite ends of the furnace. The Worcra Process shown in Fig.1(iii) is historically famous, in which matte is sent to the direction of higher oxygen potential, A→B→C in Fig. 2, and slag flow is into direction of lower oxygen potential, C→B→A. Such a countercurrent process is theoretically attractive as shown in Fig.2, but to establish three different conditions, blister making, matte smelting and slag cleaning, in a single reactor various technical difficulties must be overcome. To avoid back mixing, very long or U-type furnace, or partition wall was tested, but high efficiency compact furnace is not yet realized.

(iv) Continuous process with plural furnaces

To overcome the problems caused by potential difference in plural steps, continuous operation of plural reactors is a natural solution which is adopted often in hydrometallurgy. Mitsubishi continuous copper smelting process illustrated in Fig.1(iv) is based on this simple principle, but a different type process from conventional copper smelting has been developed. This process consist of three furnaces of matte smelting, blister making(converting) and slag cleaning, and the chemical potentials of each furnace are fixed at the points of Ⓢ, Ⓒ, and Ⓡ, respectively. The bulk of the melts reserved in each furnace can digest immediately small stream of the melt from the previous furnace to reach equilibrium. It should be noticed in Fig.2, in the batchwise operation like conventional converter, the potentials vary gradually along some path such as B→C→D, but in the continuous process realized nearly in equilibrium in a well-mixed reactor, the chemical potentials must be fixed at a single point, and thus, potential gaps are observed between different reactors.

Recent Examples of Potential Control in Copper Smelting

Injection technique

It is observed both in practice and laboratory that molten liquid-liquid equilibria are realized rapidly, but those between gas and melt are not necessarily so. Only when very good contacts between gas and melt are provided, as when dispersed gas bubbles are passed through the melt, the observations in practice come close to the equilibrium assessments [3]. From this point of view, well-mixed reactors are mostly used in modern smelting processes. Flash furnace is one example, but injection will be very promising as today's and future technology. As a proved example to introduce both gaseous and solid reactant materials into the furnace, the injection technique is adopted in Mitsubishi Process, which makes it possible to realize well-mixed equilibrium reactor at a fixed oxygen potential.

Oxygen potential in shaft of flash furnace

The oxygen potential in a commercial copper flash furnace was measured by use of oxygen cell in Toyo Smelter[4]. Fig.3 shows the shaft oxygen potential normalized at 1523K plotted against falling distance from the shaft roof. The measurements were carried out by use of conventional(open marks) and newly improved(closed marks) concentrate burners. As shown in

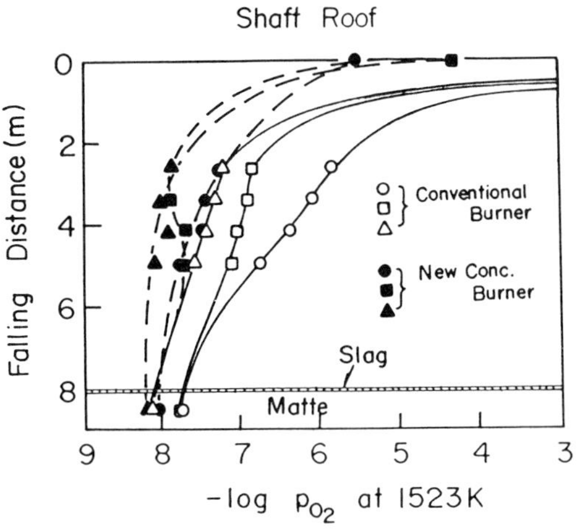

Figure 3 - Variation of log P_{O_2} of the falling melt normalized at 1523 K in flash furnace shaft [4].

the figure, the variations in oxygen potential are considerable in the case of conventional concentrate burner, but not so significant when the improved burner was used to obtain complete mixing. When the falling distance from the improved concentrate burner is more than 2 m, the closed mark oxygen potentials are not so far from those in matte which correspond to equilibrium value approximately. The results in Fig.3 suggest not only important significance of complete mixing, but also the possibility of reduction of the hight of the shaft.

Coal utilization as alternative fuel in flash furnace

Recently, coal has been successfully used in flash furnace as alternative fuel, but an unpredicted effect on the smelting reaction has been noticed. Because the ignition of coal powder is much slower than oil, an appreciable fraction of coal does not burn in the shaft, but reach the hearth to give the reducing reaction for the matte and slag. As the results, furnace operation becomes smooth because of the decreases in magnetite trouble and copper loss in slag [5,6]. In the conventional flash smelting, the reducing condition is given in the forehearth, but the results of coal utilization suggest the possibility of operation without forehearth. Furthermore, combining the improved concentrate burner to realize well-mixed shaft, a single reactor of flash furnace may be available for two step oxidation-reduction smelting. It should be noted that the matte and slag have lower oxygen potential than gaseous phase, but the equilibrium is never attained between gas and melt in the flash hearth which is a settler and never well-mixed.

Classification and Evaluation of Lead Smelting Processes

Various proposals for lead smelting may be classified and evaluated also with the help of a oxygen-sulfur potential diagram following the case of copper described above. Because the traditional standard lead smelting is sintering followed by reduction which was abandoned for copper long time ago, only the category (i) in Fig.1 must be changed. A recalculated version of sulfur-oxygen potential diagram for Pb-S-O system at 1200°C is reproduced in Fig.4 [7].

(i) Traditional sintering-reduction process

In traditional sintering process, concentrate C shown in Fig.4 is oxidized through path CDO to eliminate sulfur. The obtained sinter O is reduced usually with coke in a blast furnace to point R to produce lead bullion. During oxidation, both metallic and oxidic lead are produced, and also as suggested from Fig.2, solid magnetite must be created, but these products make sinter combined with flux by use of updraft sintering machine. It should be noted that the sinter is non-equilibrium heterogeneous product containing substantial level of sulfur. To obtain good sinter, sulfur content in the feed for sintering machine must be kept at around 6 per cent, and thus, considerable amounts of flux, return sinter and slag are mixed with lead concentrate. This essentially results in decrease in lead grade in the sinter, decrease in the capacity of sintering machine, increase in coke consumption in blast furnace, increase in the amounts of slag and exahaust gas which cause environmental problems. In conventional lead smelter, unreasonably huge amount of gas, nearly ten times in comparison with gas amount sent to acid plant, must be treated for environmental protection. Although the usual roasting process is hardly applied for lead concentrate, sintering is also not suitable, and thus, only hopeful way will be oxidation desulfurization in molten state.

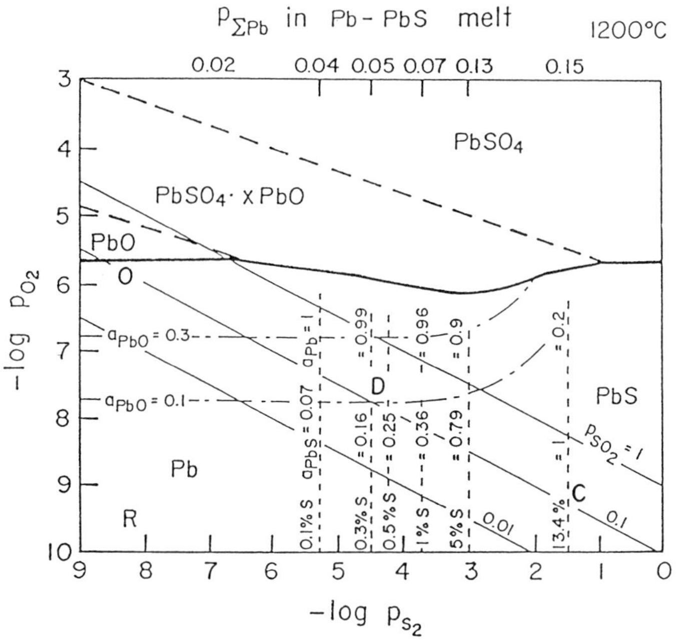

Figure 4 - Sulfur-oxygen potential diagram for Pb-S-O system illustrating lead smelting processes at $1200^{o}C$.

(ii) Concurrent direct production of lead with a single furnace.

As suggested from Fig.4, a simple oxidation with air or oxygen results in the gradual conversion from PbS at point C into Pb labelled D, where the sulfur content in bullion is 0.5 pct or so. This direct smelting is hardly realized by non-equilibrium reactor such as sintering machine, but seems to be hopeful if the smelting potentials are fixed at the region D with well mixed equilibrium reactor. The direct lead production proposed by Outokumpu or Noranda was based on this principle, and many proposals for new lead smelting process considered this possibility at least in their initial stage. However, as decscribed in the next paragragh in detail, due to the high lead losses both in slag and dust, this direct route will have the possibility only for the extremely high grade concentrate.

(iii) Countercurrent continuous process with a single furnace

Following the preliminary proposal by Worcra Process, countercurrent process with a single reactor is realized as QSL process. Originally, the region D in Fig.4 was considered as the oxidation step, but finally high PbO slag produced by oxidation at region O is followed by reduction into region R. To establish two different oxidation potential regions in a single reactor, the furnace is rather long, and the partition wall is necessary to prevent backmixing. Kivcet Process also consists of two steps, over-oxidation is followed by reduction, but oxidation is realized by flash reaction, and immersed partition wall is used to establish two regions with higher and lower oxygen potentials in a single furnace.

(iv) Continuous process with two furnaces

If two steps of oxidation and reduction are indispensable for usual lead smelting, and it seems to be true as described next, we need not insist on one single reactor where the sophisticated techniques are required to prevent the back-mixing and to realize continuous operation. There are some proposals of batchwise two step process like TBRC, but fully continuous two furnace process is not yet realized except the pilot test of Kosaka Lead Smelting Process (KLS) where two oxidation and reduction reactors are connected with launder and each potential is fixed by injection technique. A simplified flow sheet of KLS Process is shown in Fig.5. At a first sight, basic principle is similar to Mitsubishi Process for copper, but Kosaka successfully applied simple injection technique for bulk media of silicate slags involving endothermic reaction [8].

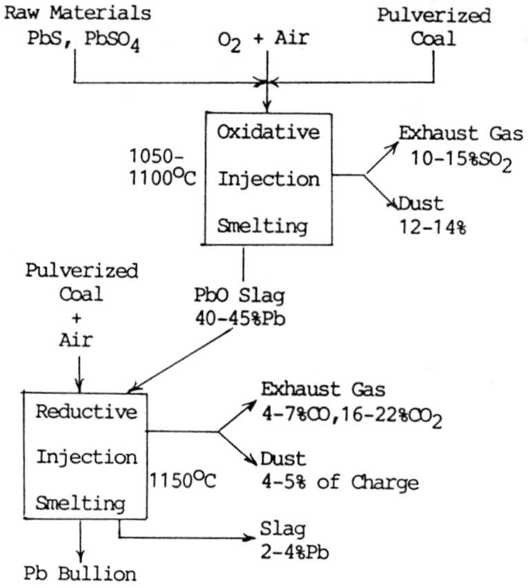

Figure 5 - Simplified flow sheet of Kosaka Lead Smelting Process.

Possibility of Direct Production of Lead Bullion

As described already, the lead bullion seems to be obtained by simple oxidation of lead concentrate if the potentials are kept at region D in Fig.4, but the figure suggests that an activity of PbO of 0.1 to 0.2 is inevitable at 1200°C. Because the activity coefficient of PbO in silicate slag is ten times lower than that of $CuO_{0.5}$, the lead loss in slag must be considerable. Moreover, the effective total pressure, $p_{\Sigma Pb}$, defined as the sum of vapor pressures of all lead species, is nearly 0.07 atm, suggesting that considerable lead loss as dust is also inevitable. In order to decrease vaporization loss, low temperature, low sulfur bullion, and use of oxygen to decrease gas volume will be convenient, but these conditions all tend to increase the activity of PbO.

However, new direct production of lead was proposed by the authors by use of ferrite slag to which the dissolution of lead oxide is only one tenth in comparison with silicate slag [7,9]. Extensive stoichiometric-thermodynamic calculations were carried out to clarify the lead distribution when the direct production of lead is carried out for high grade concentrate of 75 or 66 % Pb. Some results are illustrated in Fig.6 [7] where the distribution of lead at 1200°C is plotted against sulfur content in bullion. It is clear that when 75 % Pb concentrate is used, the recovery of lead in bullion may reach 70 to 80 per cent, but if the concentrate is 66 % Pb and silicate slag is formed(dashed lines), the recovery of lead is less than 50 per cent. Because the grade of common concentrate of lead is less than 60 % Pb, an efficient direct smelting of lead may be quite difficult. Thus, as the standard lead smelting process, two steps of oxidation and reduction must be indispensable.

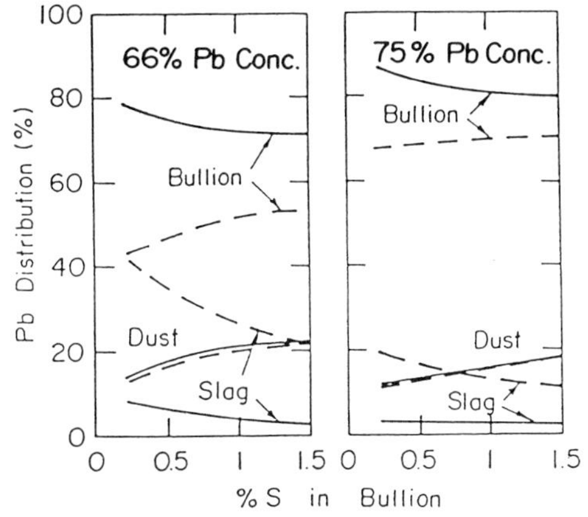

Figure 6 - Distribution of lead in direct smelting of high grade concentrate as a function of sulfur content in bullion. (solid line : ferrite slag, dashed line : silicate slag).

Conclusions --- Future Trends of Smelting Processes

The metallurgists have been liable to have the illusion for both copper and lead metallurgy that the most reasonable extraction process must be direct production of metal by single oxidation process. However, it should be emphasized that the direct production of metal by one step oxidation is possible if we are satisfied with low primary recovery of metal, but is hardly realized as efficient process. Many proposals of new smelting process started based on this illusion, and insist in use of a single reactor, but to realize practically amenable process, the combination of plural steps with different oxygen potentials is necessary. As the simplest solution, well-mixed continuous reactors having each fixed potential may be connected by launder. Installation of partition wall to form plural reaction sectors in a single furnace will be another solution, but there remain some problems to be solved such as complete prevention of back mixing for both the gas and molten phases, or maintenance of the partition wall itself, etc.

From the standpoints of production efficiency and environmental aspects, batchwise process and the reaction between lump ore and gas may be not convenient, and a well-mixed continuous smelting reactor is preferred to establish immediately nearly equilibrium condition with designated oxygen potential. To fulfill these conditions, the method of introduction of the feed and reacting gas into the furnace is very important. Recently, the feeds are almost all in powder form which is convenient for neumatic transportation and has huge surface area. The dispersed introduction of powder feed must be very effective not only for mass and heat transfers, but also to give good mixing to establish equilibrium rapidly, and flash and injection are typical examples to realize these conditions.

Undoubtedly, flash smelting is convenient in the case of exothermic reaction, and may be kept for copper smelting, but tends to cause a considerable mechanical dust, and is not necessarily suitable for endothermic reaction. On the contrary, the injection process is available for both exo- and endo-thermic reaction with evolution of dust caused only by volatile matter. The feeds are first introduced into heavy lower media such as metal or matte to give good mixing condition, and floating separation of produced slag is much faster than settling of matte grain in slag layer [10].

Although some demerits like refractory damage are pointed out, the injection technique is still young, and worthwhile to cultivate as promising future technology applicable for both ferrous and nonferrous industries to realize compact fixed potential reactor. Various trials have been proposed for direct smelting and reduction of iron ore [11,12], and an injection fuming process for zinc was also tested in Japan [13]. In the next century, the process of sintering followed by blast furnace will become obsolete, and will be gradually replaced by combined injection reactors. Combined with extensive oxygen use, the exhaust gas volume from new smelting system will be much lower than the conventional case.

References

1. A. Yazawa, "Thermodynamic Consideration of Copper Smelting," Can. Met. Quart., 13(1974), 443-453.

2. A. Yazawa, "Trends in Modern Copper Smelting Processes," Erzmetall, 30 (1977), 511-517.

3. K. Itagaki and A.Yazawa, "Thermodynamic Evaluation of Distribution Behavior of Arsenic, Antimony and Bismuth in Copper Smelting," Advances in Sulfide Smelting, ed. H. Y. Sohn, D. B. George and A. D. Zunkel, (Warrendale, PA: The Metalllurgical Society, 1983), 119-142.

4. N. Kemori, Y. Shibata, and M. Tomono, "Measurements of Oxygen Pressure in a Copper Flash Smelting Furnace by an Emf Method," Metallurgical Transactions, 17B(1986), 111-117.

5. S.Okada, M.Miyake, A. Hara, and M. Uekawa, "Recent Improvement at Tamano Smelter," Advances in Sulfide Smelting, ed. H. Y. Sohn, D. B. George, and A. D. Zunkel, (Warrendale, PA: The Metallurgical Society, 1983), 855-874.

6. A. Yazawa, T.Okura, and J.Hino, "Chemistry of Coal Utilization in Flash Smelting," (Paper presented at The AusIMM Annual Conference, Newcastle, NSW, May, 1987)

7. A.Yazawa and K.Itagaki, "Novel Process for Lead Smelting by Use of Ferrite Slag," Metallurgical Review of MMIJ, 1 (1984), no.1, 105-117.

8. N. Wakamatsu, Y. Maeda and S. Suzuki,"KLS Process Development for Lead Smelting," (The paper presented at this International Symposium on Metallurgical Processes for the Year 2000 and Beyond, Las Vegas, March, 1989).

9. K. Utsunomiya, K. Itagaki and A. Yazawa, "Basic Study on Direct Smelting of Lead by Injecting Concentrate into Ferrite Slag," Metallurgical Review of MMIJ, 4(1987), no.1, 24-39.

10. A. Yazawa, "Extractive Metallurgical Chemistry with Special Reference to Copper Smelting," (The paper presented at the 28th Congress of IUPAC, Vancouver, Aug.1981).

11. M. Tokuda and S. Kobayashi, "Process Fundamentals of New Ironmaking Processes," 7th Process Technology Conference Proceedings : New Ironmaking and Steelmaking Processes, (Warrendale, PA : Iron and Steel Society, 1988), 3-11.

12. R. B. Smith and M. J. Corbett, "Coal-Based Ironmaking," ibid, 147-178.

13. S.Goto,M.Nishikawa, and M. Fujikawa, "Semi-Pilot Plant Test for Injection Smelting of Zinc Calcine", (The paper presented at this International Symposium on Metallurgical Processes for the Year 2000 and Beyond, Las Vegas, March, 1989).

MITSUBISHI PROCESS - PROSPECTS TO THE FUTURE

AND

ADAPTABILITY TO VARYING CONDITIONS

T. Shibasaki, K. Kanamori, S. Kamio

Mitsubishi Metal Corporation
Naoshima Smelter and Refinery
4049-1 Naoshima-cho, Kagawa-gun,
Kagawa-ken 761-31, Japan

Abstract

Mitsubishi Process has been proven through fifteen years operation to be highly productive, cost efficient and easy to abate environmental requirements. However, in order the process to remain competitive from now on, it should also be adaptable to a variety of varying conditions. Various test works have been conducted from such a stand point of view and some of the new concepts have been implemented to the actual operation.

The paper discusses the adaptability of the process to the change of concentrate grade with respect to copper and also impurities, acceptability of the secondary materials and the potential to increase unit capacity.

Metallurgical Processes for the Year 2000 and Beyond
Edited by H.Y. Sohn and E.S. Geskin
The Minerals, Metals & Materials Society, 1988

Introduction

The first commercial plant of Mitsubishi Process with 48,000 mt copper annual capacity started its operation at Naoshima Smelter of Mitsubishi Metal Corporation in 1974[1], and the second plant with 65,000 st annual capacity started in 1981, at Kidd Smelter, Timmins, Ontario, now Kidd Creek Division of Falconbridge Limited[2]. These plants have increased their capacities to 90,000 mtpa respectively by adding oxygen and acid plant capacities while maintaining the original furnace design unchanged. Thereby both smelters have established more cost efficient operations.[3],[4]

Mitsubishi Process applies so-called multi-furnace continuous system. Figure 1 shows conceptual flowsheet. Backgrounds that required and justified the development of such system are as follows:

(1) The continuous process was considered to be more productive and cost efficient by itself than the conventional processes that incorporate batchwise converting operation.

(2) It was evaluated through theoretical analyses that the multi-furnace system is more efficient with respect to impurities removal and copper recovery than the single-stage direct smelting and converting process.

(3) Furnaces are connected with launders and ladle transportation of sulfur containing molten products inherent with conventional processes can be avoided and the SO_2 gas emission is minimized in limited areas such as melt outlets and launders. Thus it is much easier to abate pollution regulations.

(4) Offgas evolution becomes stable by continuation. The gas volume becomes smaller and SO_2 strength higher because of smaller fuel consumptsion. Thus gas train systems such as boilers, cottrells, and acid plant(s) can be designed much compact.

Such expected advantages of the multi-furnace continuous process had been almost entirely realized by the time when the first commercial plant was constructed. New technologies that have been incorporated to Mitsubishi Process include followings:

Intensive and quick smelting of the raw materials by injection through lances into molten bath.
Oxygen enriched top blow lancing with simple consumable lances.
Smelting operation at high matte grade, e.g., 65 - 68 %, while maintaining slag loss of copper at economically allowable level of 0.5 - 0.6 %.
Direct converting of high grade matte to blister copper with special lime slag formation.
Introduction of fully automatic sampling and analysis systems.
Development of effective process control system.

The productivity of Mitsubishi Process has been improved to almost twice of the original design. However, in order the process to remain competitive from now on, there are at least two important factors to be considered. The first is cost reduction mainly by increasing unit capacity either by increasing the throughput with existing plant or by designing a larger unit. Application of higher oxygen enrichment for lance blast combined with larger diameter lances is one of the most effective ways to realize such requirement. The second is to have flexibility to be adaptable for a variety of raw materials. Itemes of probable variations are as follows:

Concentrate grade with respect to copper.
Impurity content in concentrate.
Treatment of secondary materials.

Treatment of inplant reverts, such as reject anodes and anode scraps.

Naoshima is a custom smelter and thence adjustment to variation in the above mentioned items has always been required. During past fourteen years experience with Mitsubishi Process, various efforts have been done regarding improvement of such adjustability of the process. As a result Mitsubishi Process has become more flexible than the conventional process to the change of raw materials.

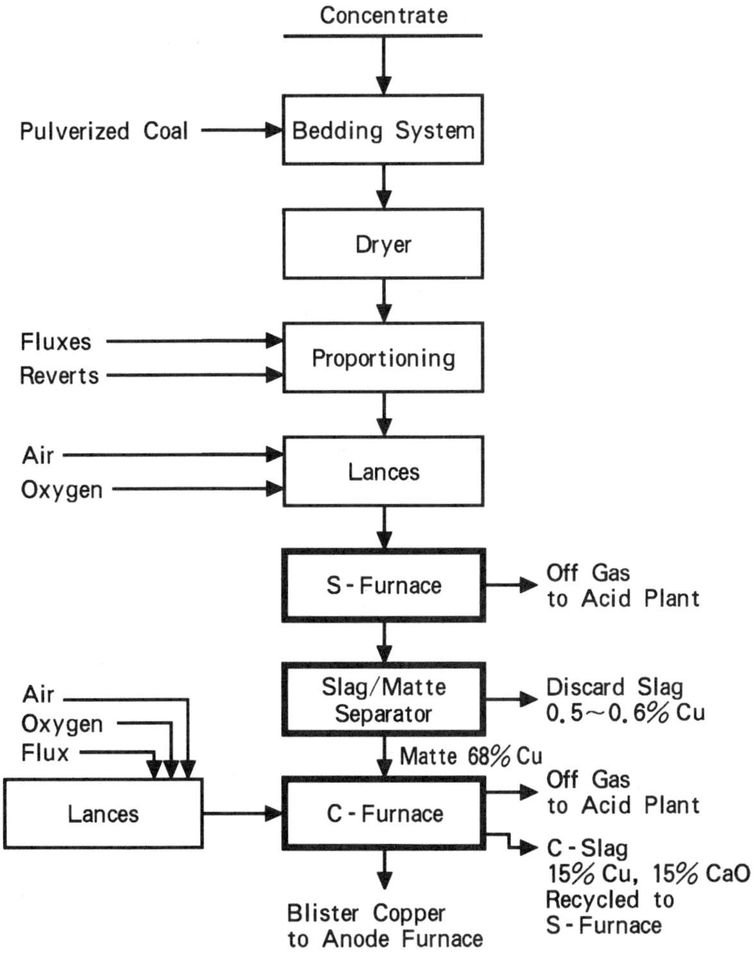

Figure 1 - Conceptual flowsheet of Mitsubishi Process.

Flexibility to the Change of Concentrate Grade

Background

Chalcopyrite type minerals are dominant in a majority of copper mines in the world and concentrate grades are in a relatively narrow range of 25 to 30 % Cu. However, production of high grade concentrates has been increased recently, due to development of new mines where high copper minerals such as covelline, chalcocite and bornite are dominant. Concentrate grade ranges 35 - 43 % in these cases.

The size of freight is another factor to influence the smelter operation. Foreign custom concentrates are now transported by ocean liners with 5,000 - 20,000 mt capacity. The largest lot is equivalent to almost two weeks' furnace feed for a standard size smelter with 1,500 mtpd concentrate charge. As a result, the smelter would require very large amount of concentrate stock, if they want to maintain relatively constant copper grade for the furnace feed although they receive various types of concetrates including regular grade of 25 to 30 % Cu and high grade of >40 % Cu.

So it is quite a serious requirement to treat furnace feed of varying copper grade depending on the constitution of stocked concentrates. In such occasion however, major dificulties will arise for treatment of higher grade feed as will be discussed below.

Unbalance of Heat Requirement

One difficulty is unbalance of heat requirements at S and C furnace. Figures 2 and 3 show fuel requirment at S furnace and lance blowing rate at C furnace of existing Naoshima plant respectively expressed as a function of matte grade. Line I corresponds to standard grade and line II to high grade concentrate. Feed rate for each case is same 40 mtph.

For the standard 68 % matte grade, fuel requirement at S furnace increases from 800 l/hr for standard grade concentrate, point A, to 1,340 l/hr for high grade concentrate, point B, as shown in Figure 2. Similar change is observed in the lance blow rate at C furnace, i.e., from 13,000 Nm^3/hr at 28 % O_2 to 18,000 Nm^3/hr at 24.5 % O_2, see Figure 3. Treatment of high grade concentrate at regular matte grade will thus results in about 70 % increase in fuel requirement at S furnace and about 40 % increase in lance blast at C furnace. Such operating conditions cannot fit with the design of existing smelting facilities, so it is obliged to reduce feed rate during the period of high grade feed treatment. Such unbalance will, however, be quite easily solved by increasing matte grade to white metal grade, e.g., 75 - 76 %, as shown by point C in Figures 2 and 3. Fuel requirement at S furnace and lance blow at C furnace are almost equal to those at the standard operating conditions even the same feed rate is kept. Besides copper output will be increased due to the increased copper content in high grade concentrates.

Slag Loss of Copper in Relation with Magnetite Balance

Another difficulty of high grade concentrate treatment at standard matte grade is increase of slag loss due to inversion of magnetite balance. Low slag loss, 0.6 % Cu at 65 % high grade matte, is one of the most advantageous features of Mitsubishi Process. Recent improvements of operation have enabled to increase matte grade to 68 - 69 %, thereby heat of reactions are utilized to almost the maximum degree. However, slag loss of copper in such high matte grade smelting has a close relation with the magnetite balance in smelting stage.

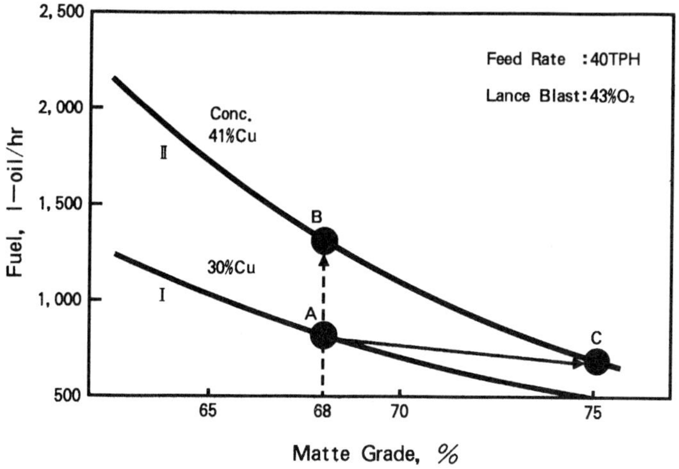

Figure 2 - Fuel requirement at S furnace as a function of matte grade.

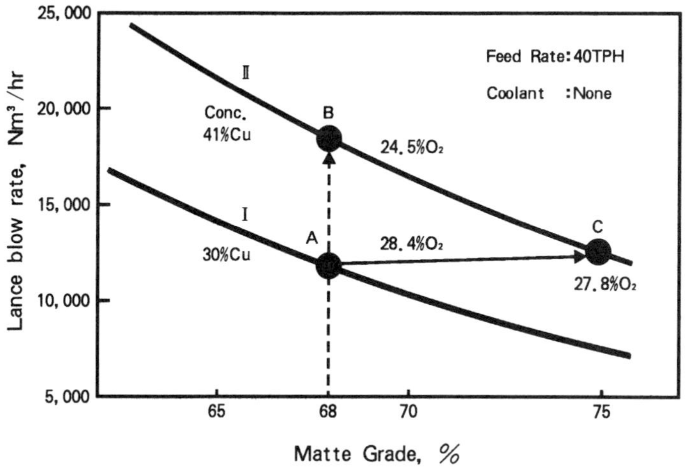

Figure 3 - Lance blow rate at C furnace as a function of matte grade.

Table I. Magnetite Balance at S Furnace

	Case I			Case II			Case III		
Concentrate	Standard, 30% Cu			High Grade, 41% Cu					
Matte grade	Standard, 68% Cu			Standard, 68% Cu			High Grade, 75% Cu		
	Mass kg	Fe_3O_4 %	Fe_3O_4 kg	Mass kg	Fe_3O_4 %	Fe_3O_4 kg	Mass kg	Fe_3O_4 %	Fe_3O_4 kg
<Input>									
Concentrate	100	1.0	1.0	100	1.0	1.0	100	1.0	1.0
C-Slag	9	60.0	5.4	12	60.0	7.2	4	60.0	2.4
Total			6.4			8.2			3.4
<Output>									
S-Slag	67	10.0	6.7	49	10.0	4.9	49	15.0	7.4
Matte	47	1.5	0.7	64	1.5	1.0	56	0.5	0.3
Total			7.4			5.9			7.7
Fe_3O_4 to be reduced			0			2.3			0

Table I shows magnetite balances in various mode of operations. Case I shows one in a standard mode, e.g., smelting standard grade concentrate at standard matte grade, 68 % Cu. Input of magnetite is smaller than output in this case, which means that magnetite is formed in the smelting stage. Equilibrium related with magnetite reaction is generally expressed with equations (1) and (2). Here, partial pressure of SO_2 is considered to be determined by the main smelting reaction (3), and is estimated in the order of 0.2 atm at usual operting condition.

$$3Fe_3O_4 + FeS = 10FeO + SO_2 \tag{1}$$

$$K_1 = \frac{a_{FeO}^{10} \cdot P_{SO_2}}{a_{Fe_3O_4}^{3} \cdot a_{FeS}} \tag{2}$$

$$FeS + 3/2O_2 = FeO + SO_2 \tag{3}$$

Case II shows magnetite balance in smelting of high grade concentrate at standard matte grade. In this case, magnetite input is larger than output, mainly due to increase of C-slag formation and recycle. Thus, partial pressure of SO_2 is considered to be at least 1 atm, because the reaction (1) proceeds toward right direction and foam of SO_2 is generated.

Such equilibrium conditions are calculated from equation (2) and shown in Figure 4. It shows that the activity of magnetite is much larger in case II than in case I. It is well known that the slag loss of copper is closely related with the magnetite concentration in slag and this figure suggests that the inversion of magnetite balance in S furnace will cause serious increase of copper loss even if matte grade is unchanged. Such an interpretation has been confirmed by increasing return C-slag feed rate in the actual operation almost twice of what it was at normal operation. The slag loss was increased to 0.8 to 1.0 % during such period from 0.5 - 0.6 % at standard condition.

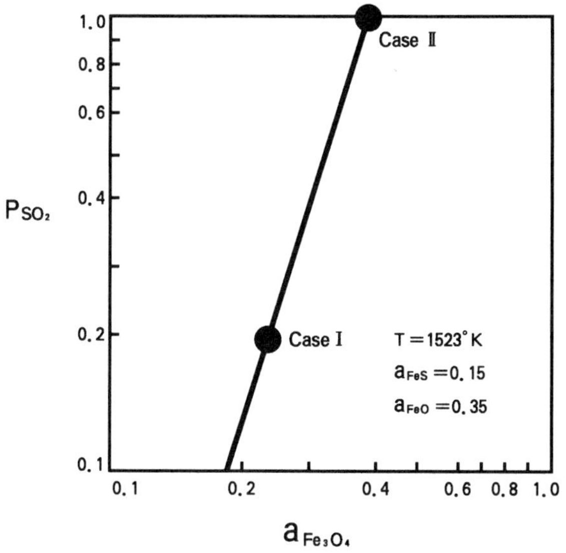

Figure 4 - Relation between P_{SO_2} and $a_{Fe_3O_4}$

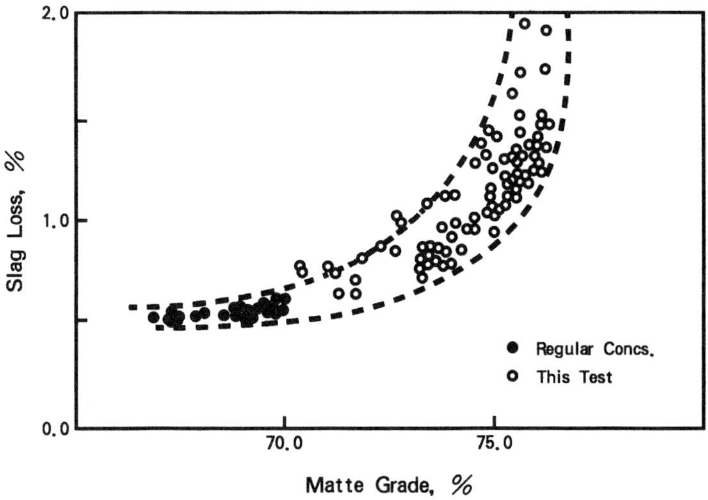

Figure 5 - Relation between slag loss and matte grade.

Inversion of magnetite balance also can be solved by increasing matte grade as shown in Table I, case III. Of course it is not possible to get low copper discard slag at matte grade as high as 75 %. So far as clean slag cannot be obtained and slag cleaning remain a subject to be solved in either cace, it is obvious that the high grade matte smelting gives much benefit as discussed in the previous section. It reduces fuel requirements and gives larger output of blister copper.

Test Operation of High Grade Matte Smelting

High grade matte operation was tested for a week in the actual plant. Table II shows the results together with data for standard mode of operation. Although it was obliged to limit the feed rate at about 35 mtph during the test due to limited anode furnace capacity, it was proved by the test that the heat balance at S and C furnaces were very close to what they were in the standard mode of operation. Moreover, the copper production was increased to more than 10,000 mtpm base from standard capacity of 8,000 mtpm.

Table II Test Results of High Grade Matte Smelting

Mode of Operation		High Grade Matte Mode	Standard Matte Mode
Concentrate grade,	% Cu	41	30
Matte grade,	% Cu	76	68
S-Furnace			
Feed rate,	mtph	34.6	40
Slag,	mtph	18.1	27
Matte,	mtph	19.1	19
Lance air,	Nm^3/h	10,300	14,500
Tonnage oxygen,	Nm^3/h	5,600	7,900
Oxygen concentration,	%	42	42
Pulverized coal,	kg/h	1,350	1,350
C-Furnace			
Blister copper,	mtph	14.2	12.1
C-Slag,	mtph	0.8	3.5
Lance air,	Nm^3/h	7,800	10,200
Tonnage oxygen,	Nm^3/h	1,350	1,500
Oxygen concentration,	%	29.9	28.6
Monthly Capacity			
Concentrate feed,	mtpm	24,600	27,700
Blister copper,	mtpm	10,700	8,300

Slag loss in the high matte grade range is plotted against matte grade and shown in Figure 5. It was high up to 1.5 % as expected at 75 - 76 % matte grade, but was 0.7 - 0.8 % at 70 - 72 % matte grade range. As slag fall is much smaller with >40 % Cu high grade concentrate, distribution of copper to slag is not so large in the latter case as to justify further slag cleaning. See Table III. Considering this fact, feed control strategy at Naoshima has been changed recently. There are two smelter lines at Naoshima, e.g., Mitsubishi Process and the conventional reverberatory furnace line. Grade of concentrate fed to Mitsubishi Process is now occasionally increased to 35 to 40 % Cu, when there are large stock of high grade concentrates. They are smelted at 70 - 72 % Cu semi-high grade matte mode, thereby slag loss is kept

0.7 - 0.8 % at most and slag is discarded without further treatment. Follwing advantages are expected:

- Full capacity feed rate can be maintained irrespective of the type of concentrate.
- It can be avoided to have a large stock of concentrates that would be required if it was intended to keep the grade of furnace feed constant at any time.
- Fuel consumption is minimized.
- By increasing anode furnace capacity, the copper production from the existing furnaces could be increased to more than 10,000 mtpm.

Table III. Analysis of Slag Loss in Semi-High Grade Matte Mode Operation

Mode of Operation		Semi-High Grade Matte Mode	Standard Matte Mode
Concentrate,	Cu %	41	30
Matte grade,	Cu %	72	68
Slag fall,	% of conc	49	67
Slag loss,	% Cu in slag	0.8	0.6
	kg Cu/t conc	3.9	4.0
Recovery of copper*,	%	99.0	98.7

* Only slag loss is counted.

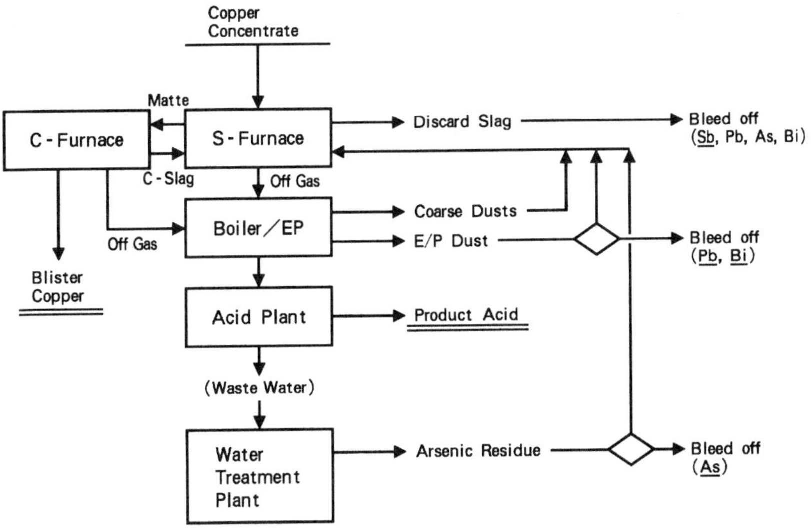

Figure 6 - Behaviour of impurities in Mitsubishi Process.

Flexibility to Impurity Load

Such impurities as arsenic, antimony, lead and bismuth are known to cause problem in the refinery operation at high concentrations in anodes. Their behaviour in the smelting stage as well as in the refining stage is not common and thus the counter measure to meet increased impurity burden is individual to respective element.

Figure 6 shows schematically the behaviour of impurities in Mitsubishi Process. Elements shown at each product name with underlines represent main bleed off of them. Generally they are absorbed in S furnace slag and discarded while their concentrations are low. But vapour pressures of above mentioned impurities are generally high. They are volatilized during smelting and concentrated in precipitator dusts or waste water treatment residue. So, such intermediate products are occasionally bled off for effective impurity removal.

Table IV shows distribution of impurities to anode in Mitsubishi and the conventional process at Naoshima. Impurity removal is made mostly at the smelting stage in Mitsubishi Process, and at the converting stage in conventional process. The PS converter can be operated at extreme conditions, overblowing or high temperature converting for example, when impurity load is very high. However, Mitsubishi Process has an advantage for medium dirty concentrates especially for those containing arsenic and antimony.

Table IV. Distribution of Impurities to Anode

	Distribution to Anode, %				Remarks
	Pb	As	Sb	Bi	
Mitsubishi Process	40 15	8 4	18 15	38 15	All dusts are recycled. EP dust is bled off.
Conventional Reverb/PS Converter	8	9	28	14	Converter EP dust is bled off.

Volatilization of antimony in the smelting stage is not so high, and in the refinery, solubility of antimony ions in the electrolyte is low. So, it is one of the difficult to handle element. Distribution ratio of antimony between slag and matte increases very steeply with matte grade increase as shown in Figure 7. Mitsubishi Process is operated at high matte grade and thus antimony elimination is much better with big margin than in the conventional processes. The PS converter looks better from the figure. However, the conveter slag is either recycled to the smelting stage or milled, then large part of antimony in it is recovered to matte or to copper concentrate.

Regarding arsenic, the solubility of arsenic ions in the electrolyte is very high. Moreover vapour pressure of arsenic is very high, so it is easy to eliminate it in the smelting stage. Thus the increased burden at the waste water treatment plant is the matter for this element. Figure 8 shows arsenic that reports to waste water treatment plant as a function of arsenic content in the smelter feed. It is high in Mitsubishi Process irrespective of concentration, but is a function of concentration in the feed at the reverberatory furnace. From this fact it is deducted that there is an optimum proportioning of arsenic burden between the two smelter lines that minimizes the arsenic burden at waste water plant, and such an optimum point varies depending on the total arsenic burden at the entire Naoshima Smelter.

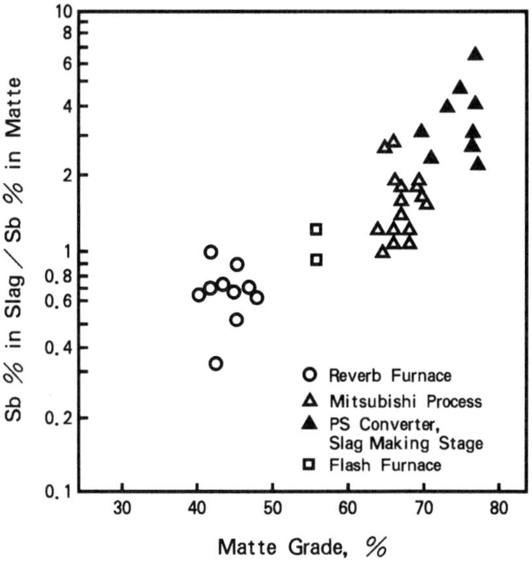

Figure 7 - Relation between slag/matte distribution of antimony and matte grade.

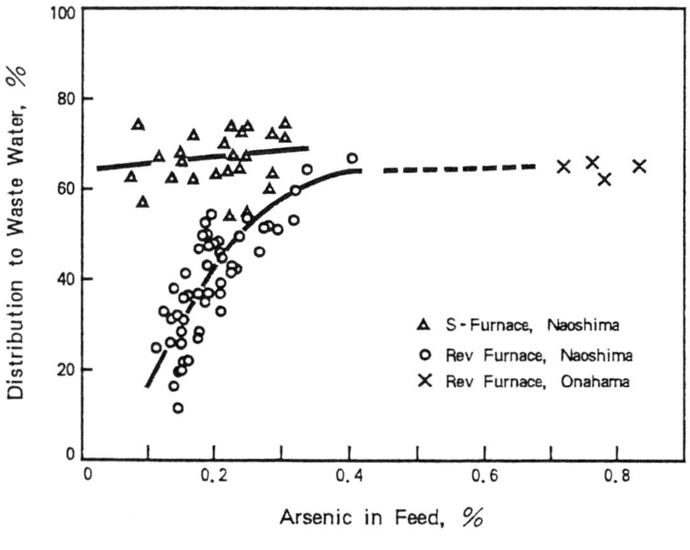

Figure 8 - Relation between distribution of arsenic to waste water treatment plant and arsenic concentration in feed.

The optimization of impurity removal is a subject of another paper by the authers and discussed in detail in it.(5) Just the summary is shown below:

(1) Antimony removal is better at higher matte grade. So, high antimony materials and intermediate products such as slag concentrate and secondary liberater cathode are treated in Mitsubishi Process. By this measure antimony in anodes from Mitsubishi Process has become higher but the total sum amount of antimony in anodes from both smelter lines has been reduced by about 20 %.

(2) Volatility of arsenic is very high, so that there is less concern with respect to anode quality. Thus arsenical materials are proportioned to the two smelter lines so as to minimize the burden at the waste water treatment plant.

Treatment of Secondary Materials

Secondary copper materials are important source of raw materials for even the primary copper smelter depending on the location of plant and availability of such materials. Inplant reverts are similar to them in shape and physical properties, and the smelter should treat them anyway. They are classified into following groups according to the shape and copper grade:

(1) Sludges and pulverized copper bearing secondary materials, such as cement copper, copper cake from zinc plant, various types of copper bearing residues generated in the copper consumer and other area.
(2) Low grade scraps, such as scraps of electronics parts, copper printed electronics circuit board and others.
(3) Regular copper scraps, metallic in nature.
(4) Large shaped but crushable inplant reverts, such as boiler clinker, used copper containing bricks, and solidified chunks of slag and matt
(5) Metallic inplant reverts, such as reject anodes, reject molds for anode casting, anode scraps, and impure cathode from liberator cells.

During the earlier stage of commercialization of Mitsubishi Process, i was considered that the process was not adequate for treatment of such miscellaneous materials and there was no need to treat them either, because the reverberatory furnace line was operating in parallel with the continuous plant and the PS converters were especially suitable for treatment of them For a stand-alone continous plant however, situation is quite different Even at Naoshima it is more convenient if secondary materials can be fed to either of the two smelter lines. On this reason various test works hav performed and now they are actually treated or it was demonstrated that the can be treated in the continuous plant as well in a manner as shown in Figur 9.

Secondary materials of type-1 are treated just by blending with copper concentrates. Crushable reverts of type-4 can also be treated in a same manner after being crushed. Materials of type 2 often include or are mixture of the materials of following properties:

Generally copper grade is low.
Containing high levels of impurities such as Pb, Ni, Al and Zn.
Containing large amounts of flammable materials, such as PVC, rubber, bakelite and FRP.
Shapes and mechanical properties are not simple, that it is not possib to pulverize with one simple crushing unit to the degree that they ca be fed through lances.

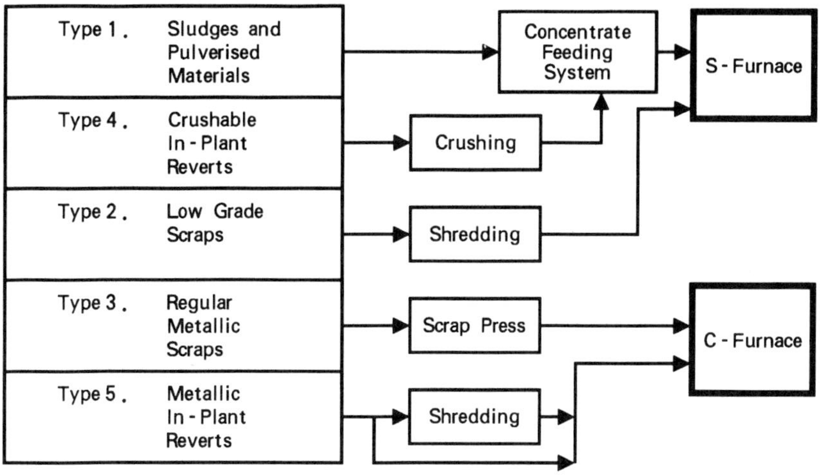

Figure 9 - Treatment of secondary materials in Mitsubishi Process.

Flammable materials are very common components in these materials and they cause emission of black smoke, burn-out of converter primary hood and over burden at the boilers, when fed to the converter in large batches. Classification and calcination are often required. Mitsibishi Furnaces do not have such a large mouth as a converter but they can receive these materials if they are shredded to less than about 50 mm. Considering the low grade of copper, S furnace is suitable for treatment of them. Calorific value of contained plastics will compensate the heat required for smelting of them. From such considerations, 10 mtpd test feeding facility was constructed in early 1987. Shredded low grade scraps with flammable materials are now fed steadily at 100 - 300 mtpm rate with this equipment. There is no change observed in slag loss, fuel requirement or other conditions.

For treatment of metallic materials of type 3 and 5, the PS converter is most suitable. Judging from copper grade of them, they can be treated in C furnace of Mitsubishi Process too. For the actual treatment there are three important factors to be considered. The first factor is heat balance at C furnace. With the larger throughput or higher oxygen enrichment of lance blast, this subject can be solved. The second is handling of miscellaneous and large shaped metallic scraps. However, most part of the metallic scraps and reverts can be fed to C furnace provided apparent shape is made uniform either by shredding or by pressing into cubes. At Kidd, anode scraps were shredded to 25 kg pieces and fed to C furnace at 7 mtph to test and confirm such idea. Any metallurgical limitation was not found. At Naoshima, purchased scraps are pressed to 400 kg cubes and fed to PS converters normally. For the test purpose, they were fed to C furnace at 2 cubes per hour through a hole at the side wall of offgas uptake. Although test was not continued long due to mechanical limitations unfortunately, such scrap feeding did not interfered the normal converting operation. The third factor of consideration is impurity contamination. In Mitsubishi Process, impurity is removed mostly during smelting stage and elimination is not sufficient at converting stage as previously mentioned. So some selections will be required to receive larger quantities of such secondary materials at C furnace.

As a conclusion, it was proved or demonstrated through a series of test works that Mitsubishi Furnaces can receive various types of secondary materials, although there are some limitations with impurity burden at C furnace.

Improvements of Productivity

For improvement of productivity at a given size of smelting facilities, there are following possible limitations:

- # Kinetics of related metallurgical reactions.
- # Physical fitness of the existing plant with the increased amount of materials and products.
- # Thermal balance or thermally autogenous point.

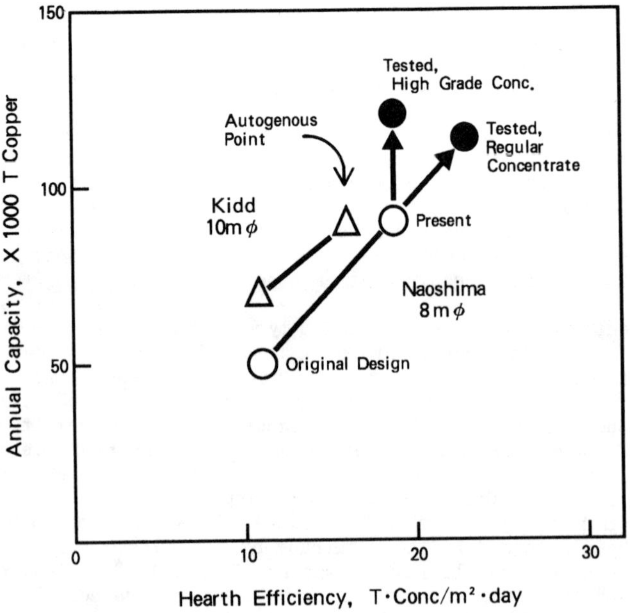

Figure 10 - Improvement of productivity at Naoshima and Kidd Smelter.

The metallurgical reactions related to Mitsubishi Process mainly proceed in narrow space just beneath lances and complete almost instantaneously. So it is considered that the rates of reactions are very fast and do not limit the improvement of productivity. Items that belong to the second factor include, (a) various limitations in such mechanical capacities as dryer and conveyer, (b) total oxygen supplying capabilities through lances, (c) off gas handling capability, and (d) product handling capacities. The first and fourth items could be problems only when improvements are intended with the existing plant and even in such a case they can be solved rather easily. For the second and third items, they can be solved just by adding capacity to the tonnage oxygen plant. Figure 10 shows productivity increases at Naoshima and Kidd that have been achieved according to such a line.

The third limiting factor is thermal autogenousness of the smelting operation at S furnace. They apply 10 mϕ S furnace at Kidd and it is already at the autogenous point. However at Naoshima, operation is not autogenous yet, and still have enough room to increase production just by increasing oxygen enrichment and feed rate. Such difference comes from differences in calorific value of concentrates, amount of dust recycle and number of jackets applied to the furnace structure.

The expected productivity at 10 mϕ S furnace will then be calculated for model cases below. Table V shows basic design or operatational parameters of the larger Mitsubishi Furnace, almost equivalent to that at Kidd except for the type of concentrates.

Tabele V. Basic Design Parameters for Model Calculation

Design parameters		S Furnace	C Furnace
Furnace size,	mϕ	10	8
Heat loss,	10^6cal/hr	4,400	5,000
Number of lances,		8	6
Size of lance,	in ϕ	4	4
Total lance blow,	Nm^3/Hr	30,000	22,000
Oxygen enrichment,	% O_2	50	35

Type of concentrate	Type 1	Type 2	Type 3
Cu %	30.0	35.0	40.0
Fe %	24.7	22.8	20.8
S %	30.1	28.8	27.5
Matte grade, % Cu	67	71	75
Feed rate, mtph	70	75	80

Figures 11 and 12 show operating conditions at S and C furnace respectively at varying grade of concentrate. Operating conditions for respective type of concentrate are determined by adjusting matte grade and concentrate feed rate so that the lance blow rates at both furnaces do not exceed the conditions cited in Table V. Such pairs of matte grade and feed rate are summarized and shown in the same table.

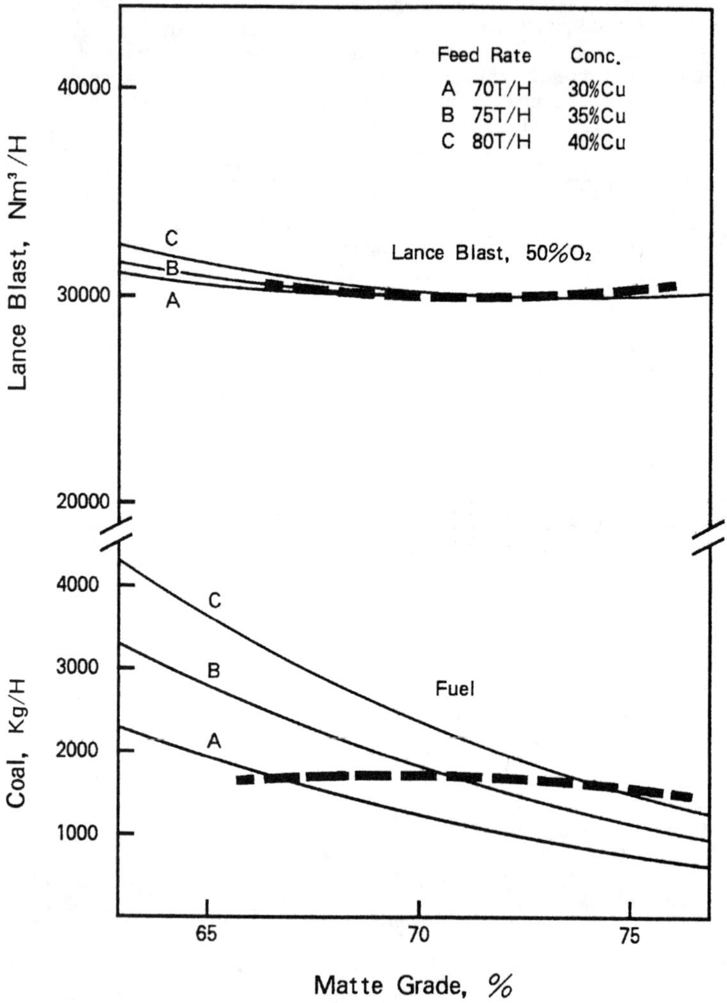

Figure 11 - Operating conditions of 10 mφ S furnace at varying concentrate grade.

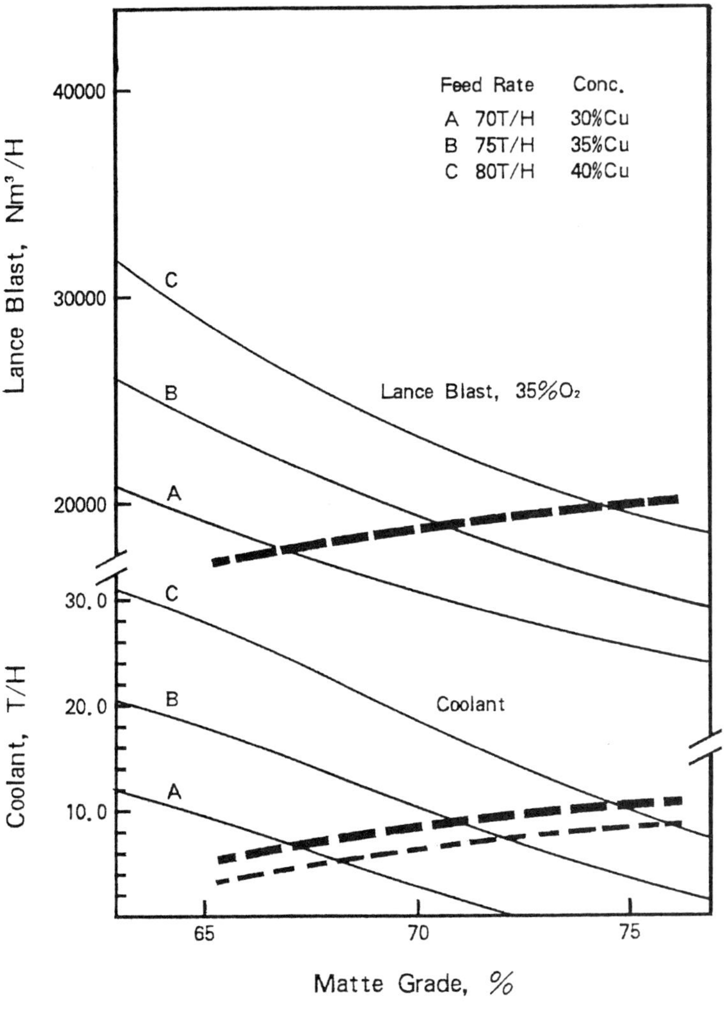

Figure 12 - Operating conditions of 8 mø C furnace at varying concentrate grade.

Lance blow rates and fuel consumptions for each cases are plotted against matte grade on Figure 11. None of the operationg conditions is thermally autogenous. Pulverized coal is used as fuel which is fed together with concentrates through lances. Lance blow includes what is required for combustion of coal.

Lance blow rate and coolant requirement, equivalent to metallic copper, at C furnace are shown in Figure 12. The degree of oxygen enrichment is set at 35 % which can be reduced depending on availability of coolant to the extent that lance blow rate does not exceed maximum of 22,000 Nm^3/Hr. A thin dotted line shows the sum of calculated amount of anode scraps and other metallic reverts. Coolant requirements are larger than generation of such inplant scraps. Followings are available to cope with the situation:

- Use purchased scraps.
- Recycle whole anodes as coolants.
- Recycle C-slag as coolants.
- Design C furnace larger.
- Reduce oxygen concentration of lance blow.

The choice depends on availability of purchased scraps and should be subjected to analyses from economical and technical stand points of view.

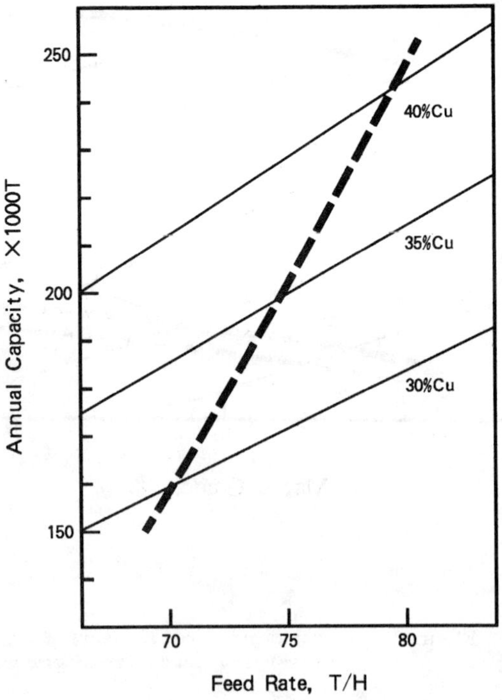

Figure 13 - Annual copper production capacity as a function of feed rate and concentrate grade.

Assuming 340 operating days in a year and 95 % availability during operation, annual capacities excluding scrap treatment are calculated for respective cases and shown in Figure 13. Annual capacity is 160,000 mtpa with 30 % Cu concentrate, 200,000 mtpa with 35 % Cu, and 240,000 mtpa with 40 % Cu. As was discussed in the second chapter, the identical smelter line can be operated at either of the three, or at any intermediate condition among them depending on the grade of currently available concentrate.

Summary

Mitsubishi Furnaces at Naoshima and Kidd Smelter have improved their productivity to almost twice of the original design by adding oxygen and acid plant capacities while maintaining the original furnaces essentially unchanged. It was also demonstrated that Mitsubishi Process has flexibilities against the variation in grade of concentrate and increased impurity burden, and it has acceptability of various types of secondary materials and inplant reverts.

Especially it was proved by plant tests that Mitsubishi Process can treat, if required, high grade concentrate at higher matte grade and at the same to or larger concentrate feed rate than the normal operation. Such concept has been partially incorporated to the actual operation at Naoshima as semi-high grade matte operation and thereby fuel consumption has been saved, the large stock of concentrates has been avoided and copper production has been increased by about 25% during such operation period.

Based on such experiences, it was demonstrated by model calculations that the larger unit can be operated at 160,000 to 240,000 mt copper annual capacity depending on the grade of concentrate.

References

1. T. Nagano and T. Suzuki, "Commercial Operation of Mitsubishi Continuous Copper Smelting and Converting Process","Extractive Metallurgy of Copper", ed. J. C. Yannopoulos and J. C. Agarwal (New York, NY: The Met. Soc. of AIME, 1976), pp.439-457.

2. R. M. Sweetin, C. J. Newman, and A. G. Storey, "The Kidd Smelter - Start-up and Early Operation","Advance in Sulfide Smelting", ed. H.Y. Sohn etal. (New York, NY: The Met. Soc. of AIME, 1983), pp.789 - 815.

3. M. Goto, S. Kawakita, N. Kikumoto and O. Iida, "High Intensity Operation at Naoshima Smelter", J. Metals, Sep. 1986, pp.43 - 46.

4. C. J. Newman, A. G. Storey and K. Molnar, "Expansion of the Kidd Creek Copper Smelter" (Paper Presented at the 115th Annual Meeting of AIME, New Orleans, Louisiana, March 2 - 6, 1986).

5. T. Shibasaki, S. Kamio and O. Iida, "Optimization of Impurity Elimination at Mitsubishi's Naoshima Smelter and Refinery" (Paper to be Presented at the 118th Annual Meeting of AIME, Las Vegas, Nevada, February 1989).

TECHNOLOGIES FOR LOW COST RETROFITTING OF THE ER&S COPPER SMELTER

Peter G. Cooper

ER&S Company Limited
Port Kembla NSW 2505 Australia

Abstract

During 1987, the feasibility of modernising the ER&S smelter at Port Kembla, NSW, Australia was studied. The selection of appropriate technologies to carry the smelter into the 21st century considered issues such as capital cost, existing infrastructure, feed mix, capacity, product quality, flexibility and potential for further development. Both hydrometallurgical and pyrometallurgical process schemes were considered. Some novel combinations were assessed but ultimately rejected. Final technology selections included proven smelting and anode furnace processes, pyrometallurgical slag cleaning and modifications to existing equipment. This process scheme balanced risk against development potential within the constraints imposed by the existing plant.

Metallurgical Processes for the Year 2000 and Beyond
Edited by H.Y. Sohn and E.S. Geskin
The Minerals, Metals & Materials Society, 1988

Background

The Electrolytic Refining and Smelting Company of Australia (ER&S) has operated a custom copper smelter at Port Kembla since 1908. Existing technology includes a sinter plant/blast furnace, three small Peirce Smith converters, two reverberatory anode furnaces and a conventional tankhouse. These process units were all installed prior to 1960 and have proven to be flexible enough to process the wide variety of domestic feed materials purchased by ER&S.

However, by the early 1980s, threats to domestic concentrate supplies highlighted the need for major changes to reduce operating costs to world competitive levels. Various studies in the ensuing years failed to identify an economically attractive solution. Furthermore, pressures were increasing to improve the environmental performance of the smelter.

During 1986 the options for redevelopment were again considered and by the end of 1987 technologies for the retrofit had been selected and a full feasibility study was completed.

Existing Operations

Most aspects of the existing operations have been reported previously (1, 2, 3). The major process units all require modification for either economic or environmental reasons.

Sinter Plant/Blast Furnace

Sinter Plant capacity is only sufficient to treat 60% of the concentrate feed to the smelter. The remainder is fed directly to the blast furnace. Resultant poor permeability has restricted copper production from blast furnace feed to less than 20 000 t/yr over the last five years. Development of cold briquetting techniques to improve permeability has been successful (4) but even with a totally lump feed charge, the blast furnace can only treat 90 000 t/yr of concentrate. Smelting capacity is therefore the limit to production for the site and its removal is the key to a reduction in overall operating costs. Furthermore, poor environmental performance of the sinter plant and high maintenance costs provide a substantial incentive to replace this technology.

Converters

Three 3 m diameter Peirce Smith converters are used to treat low grade matte (50% Cu) from the blast furnace and substantial quantities of low grade scrap. The converter hooding is in a poor state of repair and fugitive emissions create the major environmental problem for ER&S (Fig. 1). Containment of process gases from the converters is mandatory to achieve a high standard of environmental performance.

Anode Furnaces

All blister is cast and charged with baled scrap to one of two reverberatory anode furnaces. Working capital associated with blister stocks and the cost of natural gas for melting provide a major incentive for replacement of these furnaces with new technology.

Figure 1 - Existing Peirce Smith Converter

Gas Treatment

Gases from the sinter plant and blast furnace are combined and passed to a shaker baghouse. Converter primary and secondary gases are combined and also filtered in the baghouse before discharge to atmosphere via a 200 m concrete stack. The baghouse and stack are in good condition and a major asset to the redevelopment of the site.

Tankhouse and Other Plants

The existing tankhouse is in need of some refurbishment and mechanisation to reduce labour costs. A number of minor changes to other plants are required to increase recoveries and reduce operating costs.

Objectives of the Retrofit

The primary objectives of the retrofit are to:

- Establish ER&S as a cost competitive smelter in world terms.
- Substantially improve environmental performance.
- Minimise the capital requirement.
- Maintain flexibility as a custom smelter.
- Generate sufficient return to justify the capital.

These primary objectives can be translated into a number of technology selection objectives which ultimately reflect in the project economics (5).

Capital Cost

To minimise the capital requirement and maximise project returns it is essential to make maximum use of existing equipment. Specifically, replacement of the smelting technology needs to be achieved within the limitations of the existing converter aisle layout and maximum use should be made of the existing tankhouse, gas cleaning equipment and materials handling facilities.

Raw Materials

ER&S's business has been built on the custom treatment of a wide range of copper containing materials. In this respect, ER&S is very different from other smelters in that it currently sources copper equally from primary and secondary raw materials. In the future, ER&S will continue to treat about 20 000 t/yr of copper from secondary sources with the remainder from concentrates. Furthermore, ER&S will continue to treat low grade and 'dirty' raw materials.

To maintain ER&S's competitive position into the 21st century, the new technologies must be capable of treating raw materials which vary widely in size and analysis. The design raw material mix shown in Table I illustrates the range of materials likely to be treated by the smelter in the future.

Table I: Future Raw Materials Mix

Raw Material			Tonnes/yr
Concentrates:	conc. 1	30% Cu	107 000
	conc. 2	26% Cu	23 100
	conc. 3	25% Cu	24 000
	conc. 4	24% Cu	20 800
	conc. 5	23% Cu	22 100
	conc. 6	21% Cu	28 600
	conc. 7	12% Cu	3 000
Residues:	Residue 1	48% Cu	4 500
	Residue 2	40% Cu	1 000
Scrap:	Non-Sulphide Residues		4 000
	Domestic High Grade		10 000
	Alloy		5 300
	Irony		2 750
	Covered Wire		1 333
	Leady Covered Cables		2 000
	Bare Wire		1 000

Environmental Performance

ER&S treats materials which contain minor quantities of lead. Fugitive emissions of process gases containing lead have resulted in high levels of lead in ambient air (Fig. 2). Following the retrofit, ER&S must be capable of meeting the National Health and Medical Research Council (NHMRC) goal of 1.5 ug/Nm^3 lead.

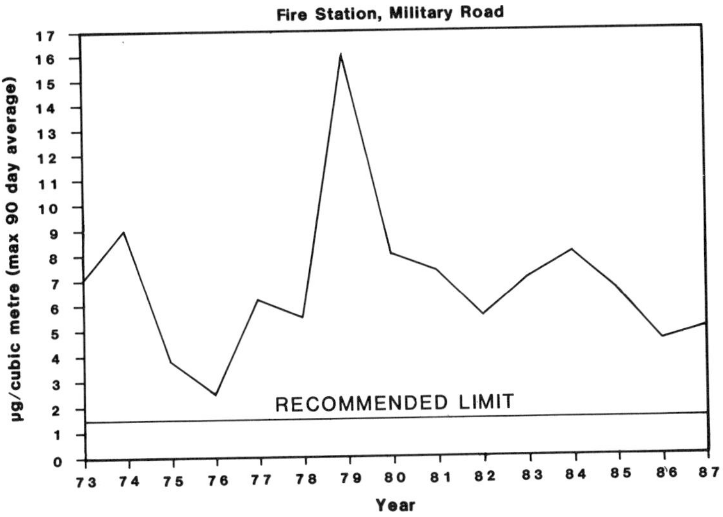

Figure 2: Historical Lead in Air Levels

Currently process gases containing sulphur dioxide are dispersed via the 200 m stack. This occasionally creates complaints when weather conditions cause the plume to reach ground level in residential areas. A corporate objective is to minimise any deleterious impact of ER&S operations on the local community. Thus, with an increase in sulphur input to the site, a large fraction of the input sulphur should be "fixed." The new smelting technology must therefore produce process gases suitable for conversion to sulphuric acid which is saleable to an adjacent fertilizer manufacturer.

Operating Cost

To achieve a cost competitive position and generate sufficient project returns to justify the capital expenditure, the direct operating cost of the new technologies must be substantially improved over current levels. The major avenues for operating cost reduction in decreasing importance are:

- Capacity
- Labour requirement
- Energy requirement

Capacity Selection

A wide range of capacity options was considered. These were ultimately condensed to three alternatives:-

- 60 000 t/year cathode
- 80 000 t/year cathode
- 120 000 t/year cathode

The major issues in capacity selection were capital cost per unit of production, concentrate supplies, metal marketing and operating cost. Improvements in operating cost gained through higher capacity were found to be partially offset by reduced revenues from metal marketing and reduced treatment charges for the incremental concentrate requirement. The major issue which remained was capital cost per unit of production. As the capacity of the smelter retrofit is increased, less of the existing equipment can be used and thus capital cost increases due to the dual effect of increases in process capacity and the number of processes which have to be modified.

Capital cost estimates were prepared using the same smelting technology at each capacity level. Table II shows the escalation of capital cost in each plant area. The ultimate capacity selection of 80 000 t/year cathode reflects the particular circumstances of the ER&S retrofit and consequent optimum use of existing facilities.

Table II - Capital cost for alternative Capacities

Capacity (t/yr Cathode)	60 000	80 000	120 000
CAPITAL COST (A$M)			
Raw Materials Handling	4	7	21
Smelting Unit	13	14	18
Acid Plant	32	37	76
Slag Cleaning	3	9	13
Converters	6	7	19
Anode Furnaces	10	10	10
Tankhouse	11	11	11
Other	19	20	27
TOTAL	98	115	195
Capital Cost $A/annual tonne	1 630	1 440	1 630

Selection of Smelting Technology

Criteria

Concentrates. Operating as a custom smelter in Australia, ER&S requires flexibility in both capacity and metallurgy. For 80 000 t/yr total cathode, 60 000 t/yr must be sourced from concentrate. To achieve this, the smelter must treat between 200 000 and 250 000 t/yr of concentrate depending on grade. However, under some conditions the smelter may have to operate at half this capacity. Thus the 'concentrate' criterion specified a design capacity of 240 000 t/year concentrate with 50% turndown.

Secondaries. The remaining 20 000 t/yr cathode is sourced from secondaries. These include up to 10 000 t/yr slags, drosses and low grade scraps. Whilst it is not mandatory that these materials be treated directly in the smelting unit, it is nonetheless desirable for maximum smelter flexibility. This criterion is therefore significant in the selection of smelting technology. At the very least these secondaries must be consumed at some point in the smelting circuit.

Retrofit. To minimise capital cost the smelting unit needs to be integrated with the existing converter aisle and smelter layout. Satisfying this criterion was considered to be essential in technology selection.

Energy. Operation close to autogenous smelting is desirable to minimise energy consumption.

Offgas Strength. Sulphur dioxide strength from the smelting unit must be suitable for sulphuric acid manufacture.

Proven Performance. With business success dependent on the smelting unit, proven commercial performance is essential.

Environmental. Apart from the requirement for high sulphur recovery, overall control of fugitive emissions must be acceptable in the workplace and capable of achieving the NHMRC targets.

Sulphur Recovery. For economic reasons, the converter gases will not be treated in the sulphuric acid plant. Thus operation at high matte grades with concomitant high sulphur recovery to acid is desirable.

Table III - Assessment of Smelting Technologies

	CRITERIA							
	CONCENTRATE	SECONDARIES	RETROFIT	ENERGY	OFFGAS STRENGTH	PROVEN	ENVIRON-MENTAL	SULPHUR RECOVERY
Blast Furnace (existing)	NO	NO	YES	NO	NO	YES	NO	NO
Reverberatory Furnace	YES	NO	NO	NO	NO	YES	NO	NO
Electric Furnace	YES	YES	NO	NO	NO	YES	YES	NO
Noranda Reactor	YES	YES	YES	YES	YES	YES	YES	YES
Flash Furnace	YES	NO	NO	YES	YES	YES	YES	YES
Mitsubishi	YES	NO	NO	YES	YES	YES	YES	YES
Kivcet	YES	NO	NO	YES	YES	YES	YES	YES
TBRC (Kaldo)	NO	YES	YES	YES	YES	YES	YES	NO
Flame Cyclone Reactor	YES	NO	YES	YES	YES	YES	YES	YES
Sirosmelt	NO	YES	YES	YES	YES	NO	YES	YES
Smelt in Melt	YES	YES	NO	YES	YES	NO	YES	YES
Hydrometallurgical	YES	NO	NO	NO	-	NO	YES	YES

Assessment

Table III indicates the acceptability of each process against these criteria. In the case of the Flame Cyclone Reactor (FCR), the technology is considered proven because the pilot plant is close to a commercial size and has been operated for long periods. The hydrometallurgical processes were not considered adequate for secondaries because of the different leaching conditions necessary for sulphide, oxide and metallic feeds.

Because the TBRC, FCR and Noranda Process satisfied the retrofit and proven criteria, preliminary capital cost estimates were made for all three processes. However, for the overall retrofit of each technology the capital requirements were indistinguishable within the error bands of the estimates.

The Noranda Process met all criteria and its capital requirement in the ER&S circumstances was equal to or better than the other alternatives. This process was therefore selected as the most appropriate smelting technology for ER&S.

Alternative Overall Process Schemes

Using the Noranda Process as the core smelting technology, both hydrometallurgical and pyrometallurgical routes for high grade matte processing were considered.

Hydrometallurgical

The combination of Noranda smelting technology and the BHAS process (6) for hydrometallurgical treatment of high grade matte to produce electrowon copper is illustrated by the flowsheet of Fig. 3. This scheme was attractive for a number of technical reasons:-

- All sulphur was converted to either sulphuric acid or saleable elemental sulphur.
- All residues could be recycled to produce environmentally acceptable discard slag and saleable fume.
- The hydrometallurgical flowsheet was simplified through treatment of a single uniform feed (high grade matte).
- The Noranda Reactor would act as a feed 'conditioner' removing most impurities before hydrometallurgical processing .
- Materials handling could be automated and labour minimised.
- Expansion would be facilitated through the modular nature of the hydrometallurgical section.

However the scheme had three serious limitations:-

- It was technically and commercially unproven.
- Very little use could be made of existing facilities.

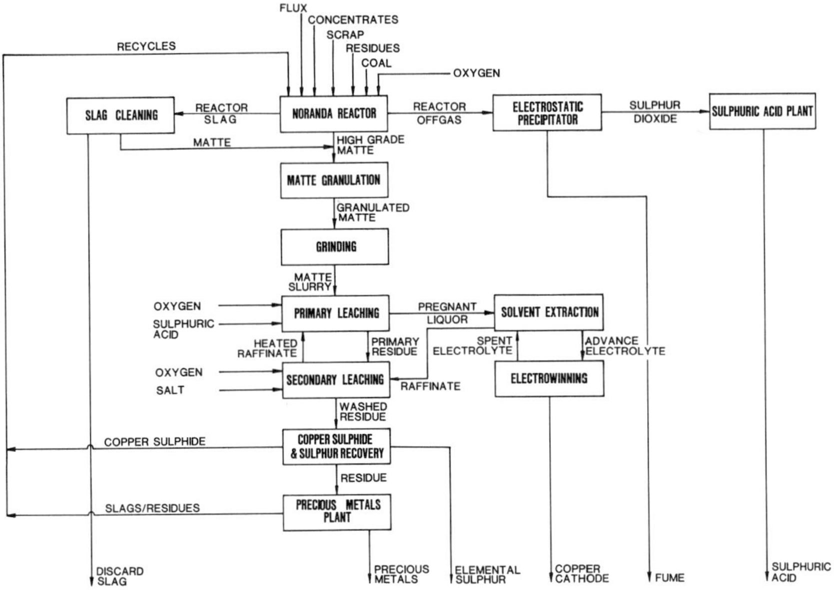

Figure 3 - Hydrometallurgical Scheme

All scrap had to be treated via the Noranda Reactor which required significant expense in equipment to reduce scrap to - 75 mm. The alternative was a change in scrap purchasing policy and a reduction in revenues from scrap.

Pyrometallurgical

The flowsheet of Fig. 4 illustrates the retrofit of the Noranda process using existing ER&S infrastructure. This scheme allowed maximum use of existing facilities and was commercially proven.

Broad capital and operating cost estimates for both hydrometallurgical and pyrometallurgical options were prepared (Table IV). The benefits of retrofitting which apply to the pyrometallurgical option are the capital cost savings associated with using existing equipment and the lower energy and supplies costs. The hydrometallurgical option was rejected because technical risk was considered too high, operating costs were exposed to increases in electricity cost, and the option became more capital intensive at high capacities.

Slag Cleaning

In 1975, ER&S installed a pilot plant to test the technical feasibility of cleaning converter slag using a submerged combustion reactor (7). Because Noranda Reactor slag is similar to converter slag, this option was considered for reactor slag cleaning. Preliminary capital and operating cost estimates using this technology were prepared in 1986. These indicated that submerged combustion would have considerable economic advantages over electric furnace slag cleaning.

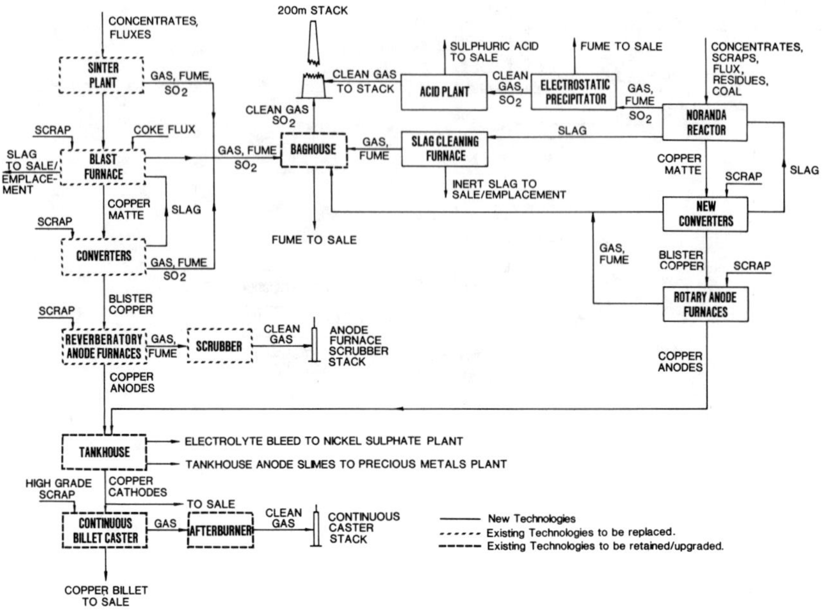

Figure 4: Pyrometallurgical Scheme

Table IV
Comparison of Hydrometallurgical and Pyrometallurgical Matte Treatment

	Capital Cost $A/annual t cathode	
	Hydrometallurgical	Pyrometallurgical
Noranda Reactor and Ancillaries	260	260
Slag Cleaning	120	120
Acid Plant	460	460
Matte Treatment	700	600
Total	1 540	1 440
	Operating Cost A¢/lb cathode	
Labour	6.1	8.3
Energy	10.9	6.2
Other (including overheads)	6.3	5.3
Total Direct	23.3	19.8

Consequently, in 1987 a programme of testwork was commissioned using a one tonne reactor to determine the feasibility of applying this technology to Noranda Reactor slags. Results indicated that with further development submerged combustion could be used to reduce copper levels in discard slag below 1%. An added benefit was that zinc and lead values could be recovered as a fume suitable for direct charging to a zinc-lead blast furnace.

To apply submerged combustion technology in the ER&S case would have required a 40 to 50 tonne vessel. This level of scale-up and the lack of time available to develop the process was considered to create too great a commercial risk and this option was rejected in favour of electric furnace slag cleaning.

Flotation was not considered as an option because an environmentally acceptable disposal scheme for tailings could not be found.

Selected Technologies

Final process selection consisted of:-

Smelting Unit	Noranda Reactor
Slag Cleaning	Electric Furnace
Smelter Gas Handling	Sulphuric Acid Plant
Matte Converting	Two larger Peirce Smith Converters
Scrap Melting and Fire Refining	Rotary Anode Furnaces
Converter Gas Handling	Existing Baghouse and 200m stack
Electrolytic Refining	Automation and refurbishment of existing tankhouse
Precious Metals Plant	Refurbishment and replacement of existing technology

The installed equipment cost for the retrofit has been estimated to be approximately $115 M. The cost has been contained by making judicious use of existing equipment whilst achieving significant improvements throughout the whole complex.

Development Potential

Located adjacent to a major fabricator, the Company will be in a position to expand production in line with anticipated expansion in exports of fabricated products. Similarly, the installation of an acid plant will allow an adjacent fertilizer plant to be recommissioned. This plant has a large installed capacity and the availability of low cost metallurgical acid will provide an opportunity to improve competitiveness and expand markets.

The results of the slag cleaning testwork (8) show that if zinc or lead values are present in the slag, they could be fumed by installation of a submerged combustion furnace operating in conjunction with the electric furnace. Development of this option will depend on the feed materials available for treatment in the 1990's and beyond.

Improvements in the recovery of gold, silver, platinum and palladium will allow the smelter to take advantage of opportunities in the treatment of primary gold containing materials as well as secondary materials including electronic scraps, spent catalysts and residues.

Retrofitting of the smelter and refinery complex will therefore provide a sound technical and commercial base for expansion of the ER&S business into the next century.

Acknowledgement

The contribution of ER&S technical staff to the selection of technologies is gratefully acknowledged. The author wishes to thank ER&S for permission to publish this paper.

References

1. P.D. Wand, "Copper Smelting at Electrolytic Refining and Smelting Company of Australia Limited, Port Kembla, NSW "Mining and Metallurgical Practices in Australasia", ed J.T. Woodcock (Parkville, Vic, Aust. 1980: The Australian Institute of Mining and Metallurgy), 335 - 340

2. C.P. Dixon, "Copper Refining at Electrolytic Refining and Smelting Company of Australia Limited, Port Kembla, NSW "Mining and Metallurgical Practices in Australasia", ed J.T. Woodcock (Parkville, Vic, Aust. 1980: The Australian Institute of Mining and Metallurgy), 353 - 355

3. A. Christlo, "Developments in Electrolytic Refining of Copper" (Paper presented at the Aus IMM Conference, Illawarra, May 1976), 169 - 173

4. David Lane et al, "Developments in Cold Briquetting of Mixed Copper Bearing Fines to Increase Blast Furnace Smelting Rates", (paper presented at 27th CIM Annual Conference of Metallurgists, Montreal, August 28-31, 1988)

5. M.A. Smales and P.G. Cooper, "Economic Influences on a Base Metals Project", (paper presented at the Aus IMM 1988 Annual Conference, Sydney, Australia 11 - 15 July 1988), 207 - 210

6. R. Lal and J.H. McNicol, "The BHAS Copper Leach Plant", (paper presented at the 116th AIME Annual Meeting, Denver, Colorado, 1987)

7. D.S. Conochie et al, "One Tonne Trials of Copper Converter Slag Cleaning using Submerged Combustion" (Report VI 6/6, CSIRO Division of Chemical Engineering, November 1975)

8. G.E. Binks, "Copper Smelting Slag Cleaning by Submerged Combustion, Stage 2 Trial Results", (Technical Report No. TR 88/037, CRA Advanced Technical Development, 15th July 1988)

THE DEVELOPMENT TRENDS OF THE OUTOKUMPU FLASH SMELTING PROCESS FOR THE YEAR 2000

P. Hanniala, J. Sulanto

Outokumpu Oy
Engineering Division
P.O. Box 86
02210 Espoo, Finland

Abstract

The requirement for the reduction of sulphur dioxide emissions from copper smelters is becoming more and more important. Therefore, the most important question in smelting technology will be how to produce even, high strength SO_2 gas in smelter and how to combine smelting and converting. The best solution in certain cases is to eliminate converters totally.

Outokumpu is already responding to this challenge. As a result of extensive research work, Outokumpu Oy has developed the Flash Smelting Process for direct smelting of blister copper from concentrates in one step.

The newest process development is the Kennecott-Outokumpu Flash Converting process, where solidified copper matte is converted into blister copper. The production of low volume high SO_2 content gases, separation of smelting and converting operations to allow increased on-line availability, and simplified process control are the major benefits of the process which translates to reduced smelter capital and operating costs.

Also, it has been studied the use of the same Outokumpu furnace for smelting and converting alternately when the production capacity is low.

In this paper, the developments and improvements of the Outokumpu Process are described from the operation and investment cost standpoints.

KLS PROCESS DEVELOPMENT FOR LEAD SMELTING

N.Wakamatsu, Y.Maeda and S.Suzuki
Kosaka Smelter, Dowa Mining Co.,Ltd.
Kosaka, Akita, 017-02, Japan

Abstract

A new pyrometallugical technology, named KLS process, has been developed by Dowa Mining at its Kosaka Smelter to smelt various kinds of lead raw materials.

The process consists of clearly separated two reaction steps, smelting and reduction. For this purpose, respective stationary furnaces are installed which are connected by a slag launder.
Dry raw materials, together with pulverized coal and oxygen-enriched air, are injected through a lance pipe into the molten slag in the first furnace. It is kept under a high oxygen potential to remove sulfur completely. Resultant PbO rich slag overflows to the second furnace where pulverized coal and insufficient air are also injected through a lance pipe to reduce PbO to lead bullion and to compensate for thermal deficit. The entire operation is carried out continuously. The process development started back in 1981. In late 1986, a pilot plant with a capacity of 3.5t/H was constructed.
Already nearly 20,000 tons of raw materials, mostly lead sulfate mixtures have been smelted successfully.

Metallurgical Processes for the Year 2000 and Beyond
Edited by H.Y. Sohn and E.S. Geskin
The Minerals, Metals & Materials Society, 1988

Introduction

For a long time, main raw materials for lead smelting at Kosaka smelter have been such secondary incombustibles as lead sulfate and lead-barium sulfate. The former is leach residue of the copper smelter flue dust and the latter originates from Iijima zinc refinery. While the most of lead concentrates produced locally has been smelted by an ISF of Hachinohe smelter in which Dowa Mining has a share.
The surplus lead concentrates were processed at Kosaka by a hydrometallurgical method to convert PbS to $PbSO_4$ as pretreatment, and followed by simultaneous smelting together with the above secondary materials (1). Approximate output has been 1,500t/M of lead bullion and slightly more than 2,000t/M of refined lead by adding lead bullion returned back from Hachinohe smelter (2).

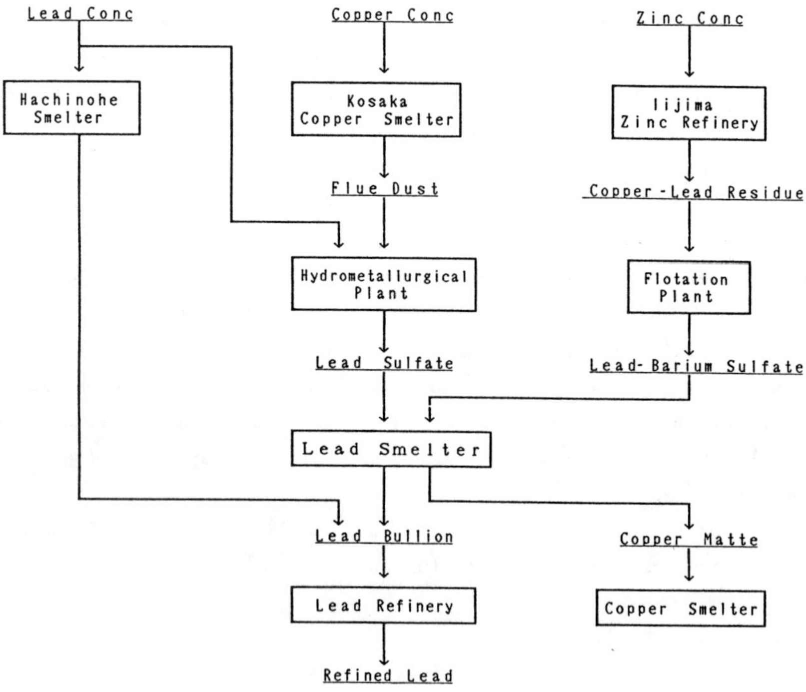

Figure 1-Basic flow sheet for lead production at Kosaka

To smelt these incombustibles raw materials, so called iron-cementation process has been applied in an electric furnace with 12 MVA transformer.
This process is commonly used in small plants, however it has several disadvantages e.g. high consumption of metallic iron and use of expensive energy, low lead recovery due to the inevitable metal loss into the by-produced matte and generation of waste gases containing weak SO_2 gas.

As an alternative, conventional sintering-blast smelting process was once examined. However, besides the low economical feasibility if the production is limited in only small scale, there was a doubt whether the sintering of such incombustible raw materials was practical or not. As a matter of fact, at the beginning of 1980's, there were already several new emerging processes for lead smelting e.g. KIVCET, QSL, Flash Smelting, SIROSMELT etc. But none of them wasn't yet commercially used and their common basic subject was seemed to treat chiefly combustible lead concentrates. Under these circumstances, in 1981 we started research and development for lead smelting.

Theoretical Considerations

To develop a new smelting process, many aspects should be considered, above all thermodynamics and its inherent energy cost are of the most importance. Discussion will be focussed on these points.

Thermodynamics

Table I shows typical assays of the lead raw materials to be treated at Kosaka. About 95% of lead in the lead sulfate, lead-barium sulfate and zinc leach residue exist in a form of sulfate, while the remainder of lead is silicate compound and the most of Zn and Fe are oxides. It is another feature of the raw materials that Cu content is rather high.

Reaction of the iron-cementation process are expressed in the following equations:

$$PbSO_4 + 2C = PbS + 2CO_2 \quad (1)$$
$$PbS + Fe = Pb + FeS \quad (2)$$

Table I. Typical Assays of Raw Materials

(%)

	Pb	Cu	Zn	Fe	S	SiO2	BaO
Pb sulfate	38.5	2.5	2.9	8.5	10.6	4.5	1.5
Pb-Ba sufate	41.0	1.5	1.4	1.2	17.4	6.0	17.0
Zn leach residue	26.2	0.1	10.2	17.9	8.1	8.1	

The formation of FeS means the by-production of matte in which metallic lead and lead sulfide dissolve mercilessly. A cutback of the matte production will de realized by elevating the oxygen potential to expel sulfur from the melt as SO_2 gas, however it's liable to result in the greater consumption of metallic iron due to the preferential iron oxidation as shown in Figure 2. It is a real dilemma of this process.

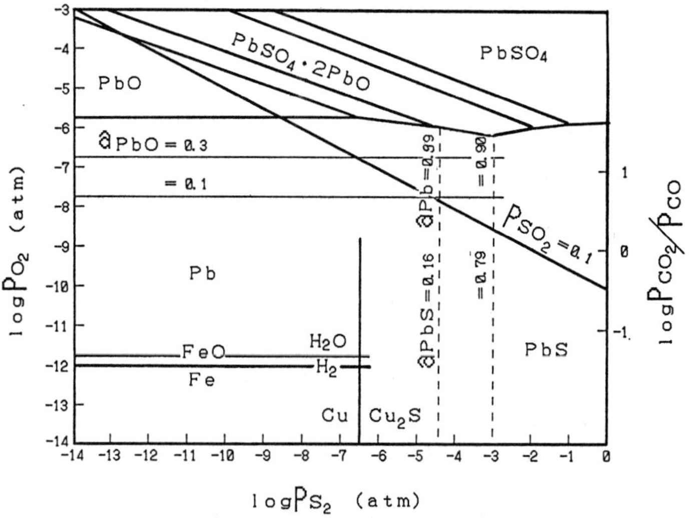

Figure 2. Sulfur-oxygen potential diagram for lead smelting at 1200 °C

In accordance with the strengthening demand to develop a new smelting technology in place of the sinter-blast furnace process, many authors have already discussed in their papers the best ways to smelt lead concentrates (3) (4). It is quite natural that there are several basic consensus among them, namely ;

(a) Two reaction steps, desulfurization smelting and reduction, are indispensable, expect for extremely high grade lead concentrates.

(b) It may be preferable to maintain the first reaction under the slightly higher oxygen potentioal than the center of metallic lead appearance zone, in order to reduce flue dust generation due to PbS evaporation. The higher reaction temperature brings the wider potentiality of direct metallic lead production, though it increases dust generation.

(c) The reduction of PbO rich slag dosen't necessitate so stronge reduction atmosphere as zinc or iron oxides reduction, and the evaporation of lead compound is not so serious comparing with smelting reaction.

(d) PbO rich slag produced at the first reaction should be transferred to the following step in molten state, not in solid like sintered ore, from the viewpoints of the total energy consumption and reaction kinetics.

(e) Furnace waste gas volume should be minimized in order to reduce the flue dust.

The above ideas should be applicable when lead sulfate is processed, even though there is a big difference in the first reaction itself.

$$PbS + 3/2\ O_2 = PbO + SO_2 \qquad (3)$$

$$PbSO_4 = PbO + SO_2 + 1/2\ O_2 \qquad (4)$$

Equation(3) is an exothermic reaction, but on the contrary the latter equation(4) requires a lot of energy to be put in ,while liberating oxygen. The utilization of this oxygen must be important to reduce the formation of corrosive SO_3 gas and compensate the energy defficit to some extent by adding fuel or lead concentrate.
There must be any distinguished difference in PbO reduction between lead concentrates and lead sulfate.

$$PbO + C = Pb + CO \qquad (5)$$

As a whole by a clear separation of the reaction into two steps, lead sulfate must be treated without metallic iron, resulting in the better lead recovery and the stronger SO_2 gas emission.

Energy and Reagent

From the above thermodynamic considerations, electricity or fuel oxygen-enriched air will be selected as energy source. Electricity is preferable the ratio of waste gas volume/effective calorific value, however it may be difficult to control the oxygen potential at a rather high level like the smelting reaction and it is the most expensive energy in our locality. Recently the price of pulverized coal, per effective calorific value, has been slightly lower than that of oil, however it may be reversed in the future. Thus it is better that a new process can use both of them depending on circumstances time to time. Concerning the oxygen content in the blast air, it is predicted that the most advantageous point lies between 40-60% O_2 in our case, taking account of the flue dust recycling. Anyway a combination of pulverized coal and approx.50% oxygen-enriched air is the best selection at this moment.

As an absorbent for oxygen liberated from the reaction(4), lead concentrates must be the best selection, of course pulverized coal can be a substitute. To reduce PbO to metallic lead, it is predicted from Figure 2 that not only carbon but also hydrogen gas must be fully effective. Therefore as a reduction reagent, coal or oil is usable equally to cokes. To promote the endothermic reaction and to compensate the furnace heat loss, additional energy input is required. For this purpose, a wide variety of energy sources could be chosen.

Preliminary Experiments

In 1981 we made a start of preliminary experiments by using crucibles to confirm the reactions(4) and (5). As a matter of course the crucible experiments proved the basic reactions to be valid in place of iron-cementation process.

Next, small electric furnaces were used for the experiments, because we thought that we might be able to save investment if the existing 12 MVA electric furnace should be still usable to a new process after minor modification. The smelting operation was continued more than 100 hours in 240 kVA electric furnace at a charge rate of 100kg/H. After filling the molten PbO rich slag in the furnace, several reduction tests were also carried out by floating guranular cokes or coal on the surface of molten slag.

As a result, it was found that the electric furnace process wasn't practical, because controlling of the oxygen potential during the smelting reaction was very difficult due to a huge consumption of carbon electrodes and due to the sooner evaporation of PbS in spite of the thorough blending and pelletizing of the lead sulfate bearing materials, and because the reduction rate was uncontrollable.

Basing on the experience, we then chose a much faster and more homogeneous reaction mechanism, namely bath smelting by means of injection. Flash smelting seemed to be unfavorable, because the smelting of the lead sulfate ore isn't autogeneouse and so it requires a very high reaction shaft which must cause various trouble.

Injection smelting was tested for years in the small electric furnace. Basically the raw materials, after being dried and blending with pulverized coal, were injected together with oxygen-enriched air into the molten slag through a lance pipe at an approximate feed rate of 100kg/H. While by injecting similarly pulverized coal and insufficient air to the PbO rich slag, reduction tests were continued. Of course we encountered many such problems as lance pipe melting, plugging etc., however it was found that the injection smelting had many advantages and could be used even for the incombustible raw materials like sulfate from both technical and economical viewpoints.

Concept of KLS Process

In 1985 the preliminary experiments had finished and thereafter we made a conclusion that a new process should be developed to smelt mainly lead sulfate containing raw materials at Kosaka. It was also determined that the new process was named KLS process and should have the following basic concept;

(a) Two furnaces, smelting and reduction, should be provided independently, but being connected each other with a slag launder for PbO rich slag.
(b) The reactions should be carried out in molten phase but not in gas phase
(c) Lance injection should be applied for both furnaces
(d) The cheapest energy should be used, nowadays it is the combination of pulverized coal and oxygen-enriched air.

(e) The operation should be a continuous mode.

Pilot Plant

A pilot plant for KLS process was constructed in 1986. It has a design capacity of 3.5 t/H of charging rate, including recycling dusts and pulverized coal, which would allow a continuous slag overflow from the smelting furnace to reduction furnace without any care. A simplified flow chart and specifications of the main equipment are shown in Figure 3 and Table II respectively.

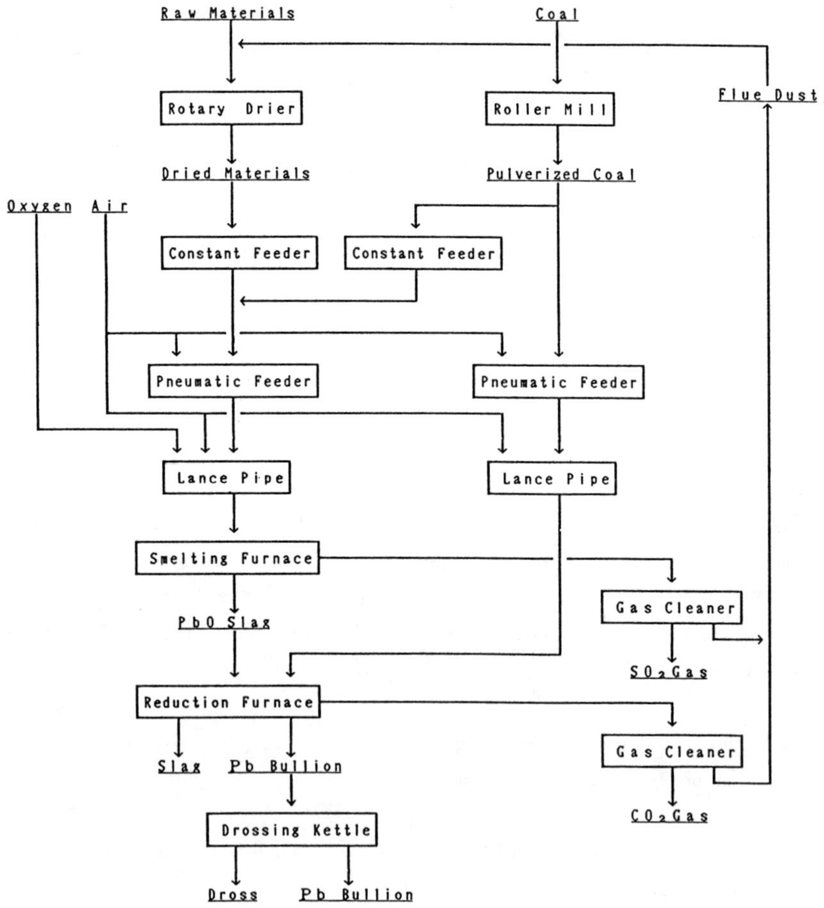

Figure 3 Flow Sheet of KLS Pilot Plant

Table II. Specification of KLS Pilot Plant

Equipment	Unit	Specification
Rotary Drier	1	Inner paddle equipped type Capacity : 3.2 t/H Moisture : Inlet 5.0 % Outlet 0.5 %
Roller Mill	1	Vertical type Capacity : 1.0 t/H Moisture : Inlet 10.0 % Outlet 1.0 % Particle Size : Inlet 10-20mm Outlet -200 mesh 90 %
Constant Feeder (for dried materials)	1	Screw converyor with hopper scale Holding cap. : 1,000 kg
Constant Feeder (for pulverized coal)	1	Disk feeder with hopper scale Holding cap. : 120 kg
Pneumatic Feeder (for smelting furnace)	1	Double hoppers with constant discharger Discharge cap. : 3.5 t/H
Smelting Furnace	1	Stationary, vertical cylinder Bricks : $MgO-Cr_2O_3$ Inner Dimension : 2.0mϕ × 2.0 mH
Gas Cleaner (for smelting furnace)	1	Dry gas cooler and cleaner, consisting of a settling chamber, jacket coolers, cyclones and an electrostatic precipitater.
Pneumatic Feeder (for reduction furnace)	1	Double hoppers with constant discharger Discharge cap. : 300 kg/H
Reduction Furnace	1	Stationary, vertical cylinder Bricks : $MgO-Cr_2O_3$ Inner Dimension : 2.0mϕ × 2.0 mH
Gas Cleaner (for reduction furnace)	1	Wet gas cleaner, consisting of a settling chamber, a jet scrubber, a TCA, a packed tower and a flare stack.
Drossing Kettle	1	Holding cap. : 15 ton

Pilot Plant Operation

The KLS pilot plant commenced its operation in October 1986. To accelerate the development, the operation has been carried out incessantly by three shifts. Four men were distributed to a shift.

After blending, the raw materials were dried to about 0.5-0.8% moisture content and then stocked in a silo. To avoid plugging trouble, lumpy materials were screened out before the drying.

Coal used for the operation had about 7,500 kcal/kg of calorific value and 20% of volatile matter. Particle size of the pulverized coal was normally controlled at 60-80% passing 200 mesh. The capacity of the roller mill was so big that the pulverized coal was produced intermittently.

Since the mixing of the pulverized coal into the raw materials was very important and so it was accurately controlled by constant feed weighers. Coal/raw materials ratio and oxygen/raw materials ratio were chiefly determined by the composition of raw materials and then slightly adjusted according to the resultant oxygen potential, the slag temperature and the charge rate. These operation was aided by a computer.

After preheating the smelting furnace by an auxiliary oil burner the injection was started. Charging materials was pneumatically conveyed by means of compressed air of small volume to the top of the lance pipe where secondary air and 91% oxygen were also blown in. Oxygen concentration in the blast air was maintained typically around 45-50% dependent on the oxygen gas availability. The higher oxygenenriched, up to 70% O_2, was tried for a week without any trouble. The lance pipe was inserted through a hole that was located at the center of furnace roof. Gas velocity in the lance pipe was normally kept at 120-150m/sec. At such gas velocities, it was found that the distance between the tip of lance pipe and the top of settled slag phase, if it was less than 1m, didn't affect the reaction efficiency so much. The slag temperature was controlled between 1,100-1,130 $^{\circ}$C to allow contineous slag overflow to the succeeding reduction furnace. Lead content in the slag reached to 43-47%. As an example, typical operation condition for the smelting furnace are summerized in Table III.

Table III. Operating Condition for Smelting Furnace

Raw materials	Blending ratio	Lead sulfate	37 %
		Lead-barium sulfate	30
		Zinc leach residue	15
		Recycled dust	18
		Total	100 %
Raw materials chagrge rate			3.20 t/H
Oxygen % in injection air			49 %
Slag temperature targeted			1,110℃
Pulverized coal addition			0.28 t/H
Injection air volume			950 Nm³

In the reduction furnace, pulverized coal and insufficient volume of air were injected to slag phase, very similarly to the smelting furnace. However in this case, since the charged coal didn't penetrate the slag phase so deeply as the smelting furnace due to the smaller solid/gas ratio in weight, and therfore the higher injection velocity and the closer setting of the lance pipe were necessary to get a maximum reaction efficiency. It was aimed that the slag temperature should be kept at 1,130-1,160 $^{\circ}C$ and that Pb content in the discard slag be about 2% which was equivalent to roughly 1.8% lead loss. These aimes could be easily accomplished by controlling of coal and/or air supplies dependently or independently.

Normally 200-220 kg/H of pulverized coal was injected to the reduction furnace. About a half of it was simply burnt by the injected air as fuel. From the reduction furnace, lead bullion and slag were tapped out discontineously, because their volumes were too small to discharge constantly.

Both in the smelting and reduction furnace, slag always foamed up 20-40 cm high during the injection. However there was no experience in melting or plugging of the lance pipes being made of stainless steel.

The drossing kettle was later added, where molten lead was agitated and cooled down to 400-450 $^{\circ}C$ to float dross and then it was cast into ingot. The drossing operation was not yet skilled well and so entrainment of metallic lead to the dross was considerable.

The furnace waste gas were cooled and cleaned in respective facilities. The flue dust collected were recycled to the blending yard for raw materials. At the outlet of the electrostatic precipitator, the smelting furnace off-gas still contained 10-15% SO_2 gas, while the reduction furnace off-gas consisted of 5-7% CO and 14-17% CO_2 gases.

Operation Result

Metal Distribution

During the one and a half years operation, the smelted raw materials totaled to 20,000 tons. In Tables IV and V, chemical assay of the products and metal distribution are shown respectively.

Table IV. Chemical Assay of Feed and Products

(%)

	Pb	Cu	Zn	Fe	S	SiO_2	BaO	As	Sb	Bi
Feed Materials	40.8	2.8	3.5	6.4	12.3	4.3	2.5	0.69	0.73	0.40
S-Furnace Slag	46.2	3.8	5.2	10.4	0.2	7.7	4.2	0.71	1.30	0.23
S-Furnace Dust	57.8	0.6	1.8	0.7	9.3	1.2	0.1	1.31	0.36	1.38
Lead Bullion	97.9	0.1	0.0	0.0	0.0	–	–	0.01	0.98	0.74
Dross	54.3	27.5	0.1	0.2	0.4	–	–	3.50	5.97	0.29
R-Furnace Slag	3.2	0.3	5.0	28.3	0.1	21.5	13.0	0.04	0.14	0.03
R-Furnace Dust	51.9	0.1	22.4	0.5	1.8	1.4	0.0	0.93	0.31	0.24

Table V. Typical Metal Distribution

(%)

		Pb	Cu	Zn	Fe	S	SiO_2	BaO	As	Sb	Bi
Input (including coal)		100	100	100	100	100	100	100	100	100	100
Output	Lead Bullion	57	1	0	0	0	–	–	1	28	43
	Dross	13	96	0	0	1	–	–	51	66	7
	S-Furnace Dust	18	3	17	2	9	2	–	29	7	46
	R-Furnace Dust	7	0	40	0	1	2	0	8	3	4
	R-Furnace Slag	2	2	33	96	0	93	98	2	4	2
	Waste Gases	–	–	–	–	89	–	–	–	–	–
	Unknown	3	Δ 2	10	2	–	3	2	9	Δ 8	Δ 2

Lead and copper losses to the discard slag were low enough if the high slag fall was taken into account. There was no trace of metallic lead droplet in the slag even though it was tapped out during vigorous injection. The higher direct lead recovery is expected to be attained by acquiring of the drossing skill and by increasing of the oxygen-enrichement not only for smelting but also for reduction.

As the development progressed, flue dust generation during the smelting became lower and settled down at about 13-15% of the charged raw materiales and 5% in the course of the reduction.

Energy Consumption

It was one of our object to utilize the reduction furnace

off-gas as fuel. But it was vain, since CO gas content in it remained only at several percents. If electricity or highly oxygen-enriched air should be input, instead of normal air, CO gas content must go up and our object should be well achieved. However in this case, another economical examination is essential.

When a full size plant gose into operation, it is forecast that 10-20% or more energy will be saved comparing the pilot plant results.

Table VI. Energy Consumption List

Fuel Oil	For drying of raw materils	147 ℓ/t-H_2O
	For production of pulverized coal	192 ℓ/t-H_2O
	For furnaces preheating	6 ℓ/t-raw materials
Pulverized Coal	For smelting	89 kg/t-raw materials
	For reduction	85 kg/t-raw materials
91 % Oxygen		120Nm^3/t-raw materials
Electricity		126 kWH/t-raw materials

Note: The above figures were obtained during the operation being specified in Table III.

Problems

The most serious problem we encountered was the plugging phenomena at the small up-take ducts where splashed slag and precipitated flue dust made hard accretion. It took several hours to clean out them at an interval once a few days. Interruption of the operation due to this reason affected straightly such test results as energy requirements and direct lead recovery efficiency. However this problem is expected to become smaller, if adequate design is made on the experience or if the plant capacity becomes big enough to allow some dust accretion.

Next problem was bricks wear. Both the smelting and reduction furnaces were lined with direct-bonded $MgO-Cr_2O_3$ bricks, and these lining were checked by measure at regular intervals. The side wall bricks of the reduction furnace had been worn out after only a half year operation. Therefore it was relined with fused $MgO-Cr_2O_3$ clinker semi-rebonded bricks. Judging from the one year use afterwords, the life of the relined bricks are expected more than two years. Comparatively the bricks of the smelting furnace were much durable, however repairing was necessary after 1.5 years use at side and bottom parts.

Conclusion

Though there are still minor problems to be solved,however it was fundamentally proved that KLS process must be one of possible process to smelt incombustible lead raw materials. It may be worth-while noting that the alternative use of lead concentrate in place of coal as an absorbent of excess oxygen liberated in the smelting step didn't affect the operations and results.

Acknowledgments

The authers wish to thank the managements of Dowa Mining Co.,Ltd. for permission to this paper. We are also grateful to Dr. Yazawa for his informative advice.

References

1. K.Ueda,K.Kobayashi and K.Yamaguchi,"Hydrometallugical Pretreatment of Lead Concentrates for Electric Smelting and Electrolytic Refining," SME-AIME FALL MEETING, Honolulu,Sept.1982,Paper No.82-374
2. J.Minoura and Y.Maeda,"Current Operation at Kosaka Smelter and Refinery," Metallurgical Review of MMIJ, Vol.1,No.2,Sept.1984
3. A.G.Matyas and P.J.Mackey,"Metallurgy of the Direct Smelting of Lead," Journal of Metals, 28(1976) Nov.10-15
4. A.Yazawa and T.Ookura,"Direct Smelting of Lead," The Bulletin of the Research Institute of Mineral Dressing and Metallurgy, Tohoku University, Vol.34,No.2 Dec. 1978

SEMI-PILOT PLANT TEST FOR INJECTION SMELTING OF ZINC CALCINE

S. Goto, M. Nishikawa and M. Fujikawa

Committee of Joint Research
Organization of Injection Metallurgy
3-6, 1-chome, Uchisaiwaicho
Chiyodaku, Tokyo, JAPAN

Abstract

The semi-pilot scale plant for injection smelting of zinc calcine was operated under the committee of Joint Research Organization of Injection Metallurgy. The capacity of the plant is 10 ton metallic zinc per day. The plant was designed by the results of basic and bench scale tests of the capacity of 1 ton metallic zinc per day which was published at "Zinc '85". The equipments of the plants, operational conditions, results of the operation such as recovery of zinc in injection furnace and condenser, combustion rate of powdered coke and consumption of energy in this process are presented in this paper. The results of the study of the feasibility of the commercial plant of 100,000 ton zinc per year by the injection smelting are also discussed in the paper.

Metallurgical Processes for the Year 2000 and Beyond
Edited by H.Y. Sohn and E.S. Geskin
The Minerals, Metals & Materials Society, 1988

Introduction

The research committee of new smelting process of zinc by injection was established in 1981 under the finantial support of Mining Promoting Foundation of Japan. Six zinc producing companies joined in this organization. The price of zinc was very low at that time and the energy cost , especially, cost of electric power in Japan was extremely high. Invention of a new economical process for zinc extraction had been stronglydesired.
The author proposed a injection smelting of zinc as a economical process in energy saving after a calculation of thermodynamical equilibrium and heat balance(1)(2).

The research committee has been studying the injection process since 1981. The small basic experiment by using a crucible furnace was carried out from October 1981 to March 1983 and the reaction of reduction of calcine by injection smelting was ascertained (2). Vaporizatipon of zinc and behaviours of impurities were examined. After the basic experiment, the bench scale test with the production capacity of zinc of 1 ton per day was carried on from November 1983 to October 1984.
The bench scale tests were carried out to obtain the proper design of feeding devices of calcines and coke into a furnace and also to obtain fundamental data for enlargement of this smelting process(3,4). The recovery of metallic zinc in this small equipments was about 50% and main loss of zinc was formation of dross in the condenser.

Semi-pilot plant of 10 ton zinc per day for injection process was constructed from August 1984 to October 1985 and the plant was operated from October 1985 and the tests were finished on September 1987. This plant was modified several times after the test operation. The average one operating period at a constant conditon was 30 hours and total 84 tests were performed. The best operational condition in this plant was obtained after many trials and the feasibility study by this process for commercial plant of 100,000 ton zinc per year was performed.

Semi-pilot plant

The schematic flow diagram of the plant is shown in Fig.1 and the main equipments are shown in Table 1.
The main equipments of this plant are explained.

Milling device for fine coke

The equipmet which was used for milling coke breeze are shown in Fig.2. Coke breeze from a outer plant is stocked in coke bins and sent to a small holding bin attached to a mill plant. The coke is fed into the mill through a center shoot by a chain feeder. The roller mill of Ube Roshe type LM8 was used. Coke is fed in a center of the mill and crushed between rollers and the table. During crushing, coke is dried by hot air at about 250°C

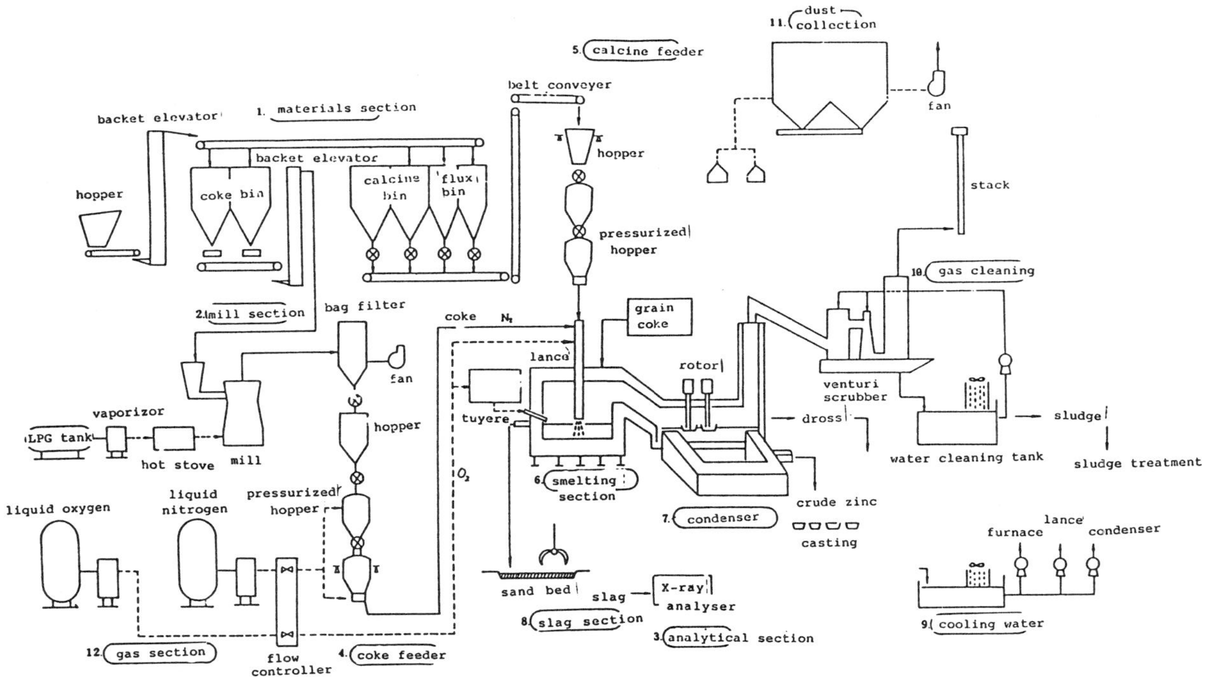

Fig. 1 Equipment flow of the semi-pilot plant

Table 1 Main equipments of semi-pilot plant

section	equipments	specifications
1.Materials	Hopper	10 m^3
	Coke bin	15 m^3 x2
	Calcine bin	15 m^3 x2
	Flux bin	2 m^3 x2
2. Mill	Roller mill	Roshe type 1 ton/hr mill 30 Kw separator 3.7 Kw fan 100 m^3/min,570 mmAq 18.5 Kw
	Fine coke bin	20 m^3
	Hot air generator	3,000 Nm^3/hr, temperature 250 °C
3. Coke feeder	Constant feeder	capacity 300-1,200 Kg/hr reservoir tank 1.0 m^3 blow tank 1.4 m^3
	Air compressor	2.8m^3/min x8.5Kgf/cm^2,22Kw (with dehumidifier)
	Grain coke feeder	100 Kg-1,000 Kg/hr hopper 13 m^3 kiln 300 mm dia.x 1,000 mmL pan conveyer 10 m/min,0.4 Kw
4. Calcine feeder	Constant feeder	capacity 200-2,000 Kg/hr service tank 0.5 m^3 measure tank 1.0 m^3
	Air compressor	5.8 m^3x9.9Kgf/cm^2,37 KW (with dehumidifier)
5. Smelting furnace	Furnace	slag holding amount 15 ton inner dimension 3,000 mm dia x 3,300 mm H refractory Chrome-magnesia brick
	Lances	coke lance 3,960 mm L calcine lance 2,970 mm L lance of coke and calcine 4,200 mm L oxygen tuyere 1,350 mm L x LPG-O_2 burner 3,880 mm L
6. Condenser	Condenser	lead amount 46 ton
	Rotor	x 4 (total 130 Kw)
	Lead pump	250 l/min x 2 mH, 5.5 Kw 400 l/min x 3 mH, 11 Kw
	Casting mould	1 ton x 5
	LPG heating equip.	LPG burner

7. Slag treatment equip.	
Sand bed	4,000mmW x 8,350mmL x 500mmH
Crab hoist crane	2.8 ton
8. Cooling water	
Pump for furnace	1.34 m^3/min x 28 mH, 11 Kw
Pump for lance	0.67 m^3/min x 59 mH, 11 Kw
Circulation equip.	pump 2 m^3/min x 15 mH,11 KW pits 6 m^3, 20 m^3 cooling tower 120 m^3/hr, fan 3.7 Kw x 2
9. Gas cleaning equip.	
Cleaning tower	gas velocity 26.5 m/sec, pressure drop 130 mm Aq.
Venturi scrubber	gas velocity 78 m/sec pressure drop 850 mm Aq.
Spray tower	500 mm dia.
Mist eliminator	1,500 Nm^3/hr
Overflow tank	1.5 m^3
Overflow pump	0.7 m^3/min x 40mH, 11 Kw
Cooling tower	30 m^3/hr
Spray pump	0.75 m^3 x 40.6 mH, 15 Kw
Dust thickner	
Primary thickner	4 m^3
Slurry pump	0.3m^3 x20mH(60g/l),3.7 Kw
Secondary thickner	15 m^3
spigot pump	0.3m^3 x15mH, 2.2 Kw
Exhaust fan	30Nm^3/hr x1,800mmAq, 30 Kw
Flare stack	300 mm dia x 10mH
10. Dust collector	
Material transport system	
bag filter	120 m^2
exhaust fan	250 m^3/min x 270mmAq, 22Kw
Smelting area	
bag filter	320 m^2
exhaust fan	225 m^3/min x 320mmAq, 22Kw
11. Gas supply system	
Oxygen	1,000 Nm^3/hr
Holding tanks	16.7 m^3 , 12.3 m^3 holding capacity 20,780 Nm^3
Nitrogen	150 Nm^3
Holding tank	5 m^3, holding capacity 2,533 Nm^3
LPG	500 Kg/hr
holding tank	35.3 m^3, 15 ton
12. Analytical equipment	
X-ray fluorecent equipment	

The rotating separator is used for separating fine coke from larger one by controlling a rotaing speed. Fine coke is transported with hot air and caught by a bag filter and contained in a hopper tank. To mill coke is generally difficult and the design of roller and table of the mill was modified after the several pilot tests.One of the test results of milling coke is shown in Table 2.

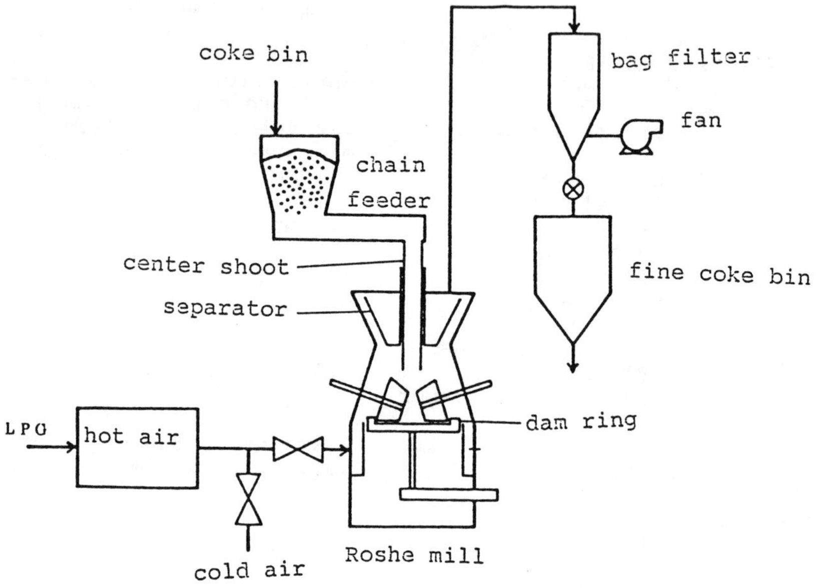

Fig 2 Mill equipment

Table 2 Results of Milling test of coke

Feed coke of the mill		
Brand	Miike coke	Miike coke
Moisture	<17 %	< 8%
size	< 7 mm	< 5 mm
Milling condition		
Mill electric current	100-110 A	90-100 A
Rotaing speed of separator	1,250 rpm	1,250 rpm
Pressure drop in the mill	300 mm Aq	300 mm Aq
Result of the test		
fineness(distribution % of -200 mesh)	81 %	85 %
moisture	2.9 %	0.6 %
Capacity Kg/hr	955	952

Constant coke feeder

Constant amounts of fine coke must be fed in the smelting furnace, so the apparatus as shown in Fig 3 was used.The apparatus was designed and constructed by Nippon Steel Corp. and is composed of three holding tanks, that is, a fine coke bin, a reservoir tank and a blow tank. Total weight of the reservoir and the blow tank with fine coke is measured and recorded continuously by a pair of load cells. The fine coke bin is connected to the reservoir tank by a flexible pipe. Fine coke in the blow tank is fluidized and pressurized by nitrogen and fed to the lance by air through a flexible pipe. The feeding amount of fine coke is controlled by controlling valve under the blow tank and the pressure in the blow tank. Fine coke in the reservoir tank is fed into the blow tank at a constant interval. The coke in the reservoir tank is also fluidized to avoid sticking when coke is discharged to the blow tank.

The operation of this feeding system is controlled automatically and a example of fluctuation of amount of injection of coke is shown in Fig. 4. The maximum fluctuation from the average feeding amount is 2.81 % and this value is less than the setting value of 3.0 %.

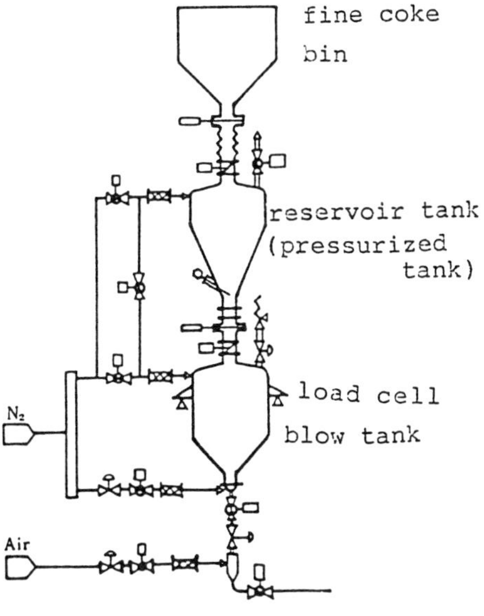

Fig 3 Coke feeder

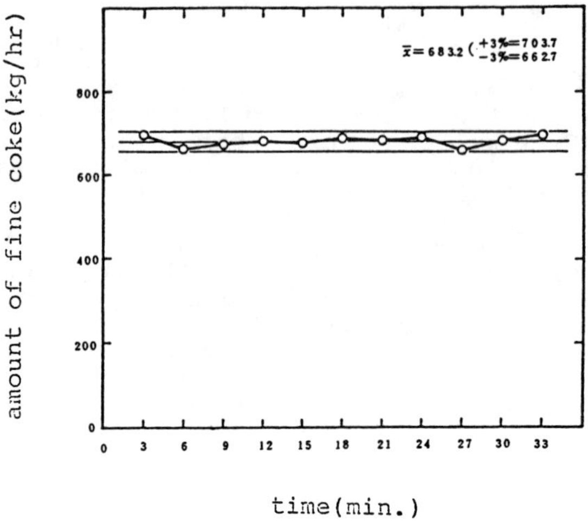

Fig.4 Change of a injected amount of fine coke

Pipes and valves was protected from abrasion of coke by using ceramic materials and an anti-abrasion flexible pipe with ceramic powder is used successfully.

Feeding system of calcine ,flux and returned materials

Calcine and other materials are sent to the hopper through belt conveyer. A constant amount of materials are fed to the survice tank by a rotary valve. After closing the rotary valve, the survice tank is pressurized . The valve between the survice tank and the measuring tank is opened and the materials are fed into the measuring tank . A neccessry amount of the materials in the meassuring tank are sent to the lance with air through a flexible pipe. The idea of this feedidng system is shown in Fig. 5.

The lances

Several types of the lance for injection of fine coke, oxygen and raw materials into the smelting furnace were tested. At the beginning of the test, fine coke and materials were fed from the separate lances. The fine coke and oxygen are fed from the same lance and the raw materials are fed from the other lance. Two kinds of a lance for fine coke and oxygen as shown in Fig. 6 were tested.

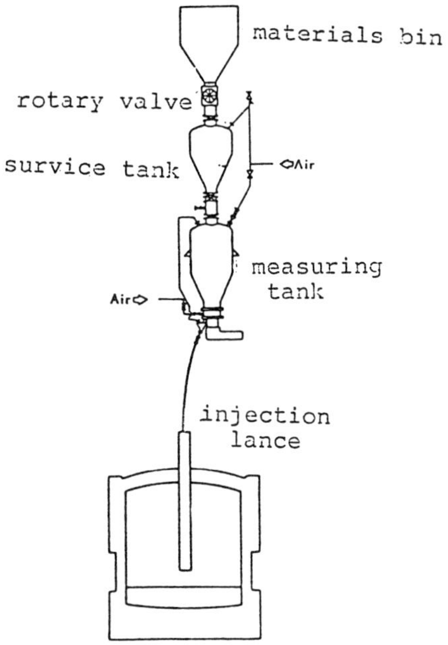

Fig. 5 Supply system of raw materials

Lance with single hole: The fine coke and carrier gas are transported into a center pipe of the lance, and oxygen is fed in the outer pipe . Two kinds of the above feeding materials are mixed near the head of the lance and injected through a single injection hole with a diameter of 30 mm.

Lance with multi holes:In case of the lance with multi holes, coke and carrier gas are injected through a center hole of the lance and oxygen is fed from the three outer holes . Some amount of oxygen is divided to the center pipe to mix with the coke .

In both lances a ceramic pipe is inserted at the head of the lances to prevent abrasion by fine coke.

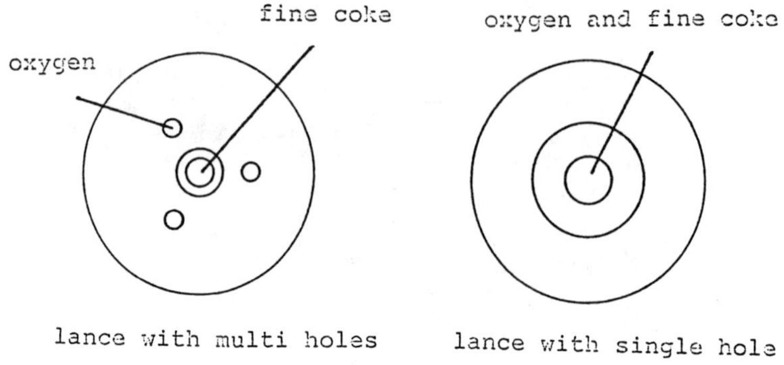

Fig 6 Injection lance of fine coke

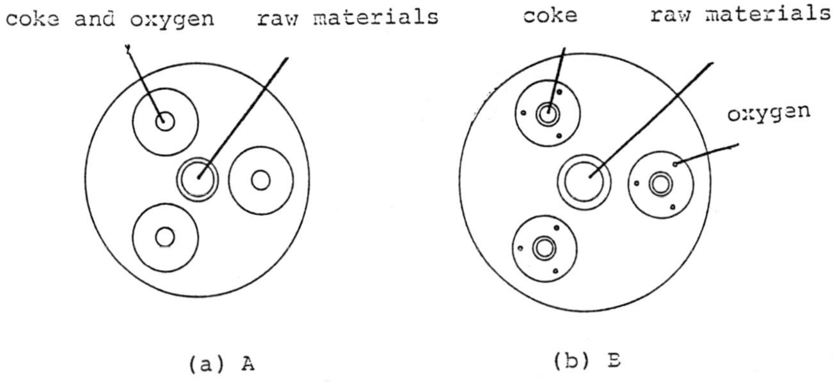

Fig 7 Simultaneous injection lance of fine coke and raw materials

Two types of the simultaneous injection lance were also tested. Fine coke, oxygen and raw materials are injected into the smelting furnace through one lance. The schematic figures of the head of the lance are shown in Fig. 7 and the dimensions are listed in Table 3.

Type A : The same type of this lance is used for L-D converter of steel making. The main body of the lance is steel pipe and the head part is oxygen-free copper. The center pipe is used for injection of raw materials with carrier gas. The fine coke is fed to the top of the lance and divided to three outer pipes and mixed with oxygen near the head of the lance. The head of the lance for coke blow is covered by ceramic tube.

Table 3 Size of the simultaneous injection lance

	Type A	Type B
Diameter of injection pipe for coke and oxygen	15 mm x 3	-
Diameter of injection pipe for coke	-	14 mm x 3
Diameter of injection pipe for oxygen		7 mm x 9
Diameter of injection pipe for materials	27.6 mm	30 mm
Oxygen injection speed m/s*	350	230
Outer diameter	190.7 mm	216.3 mm
Total length	4,200 mm	4,220 mm

* : Total injection amount of oxygen is 670 Nm^3/hr

Type B : The main body of the lance is stainless steel(SUS 316) and the head is copper also. As shown in Fig.7 the raw materials are injected through the center pipe and three sets of lance with multi holes for coke and oxygen are provided around the center injection hole. Fine coke and oxygen are injected through separate holes as shown as in Fig. 6. The head of the lance for coke is covered with ceramics also. Oxygen injection speed in Type B is less than that of type A .

Construction of the simultaneous injection lances are more complex and more expensive than the separate types. However,raw materials are injected into the slag surrounded by coke-oxygen flows , so reactions between charged materials by this simultaneous lance proceed better than the separate injection type and also carry-over of the raw materials into the condenser becomes less than the separate type. At the final tests , high speed of injection was recognized to be better for recovery of zinc, so the lance was modified to be provided to get higher injection speed. The results of this plants explained in the latter section is obtained by using this simultaneous injection lance of type A.

Grain coke feeder

The exit gas from the furnace to the condenser must be provided to high value of CO/CO_2 ratio for high recovery of zinc in the condenser. Coke must be combusted at high CO level with unsufficient amount of oxygen and some amount of unburnt coke was caught in the condenser and this unburnt fine coke was recognizd as one of the main sourses of zinc loss to dross.

Less amount of fine coke can be burned easily by sufficient amount of injected oxygen from the injection lance, and the reducing atmosphere and necessary heat for keeping the furnace temperature were obtained by combustion of the grain coke layer . The grain coke is combusted by 4 tuyeres which are set near the slag surface .

The grain coke is charged into the furnace after preheated in a rotary kiln as shown schematically in Fig.8. The coke is preheated to 500-600°C and charged through a pan conveyer . The slag in the furnace is melted by this coke layer and combustion heat and also recovery of zinc in the condenser becomes high by less amount of flying unburnt coke , dust of raw materials and splash of molten slag.

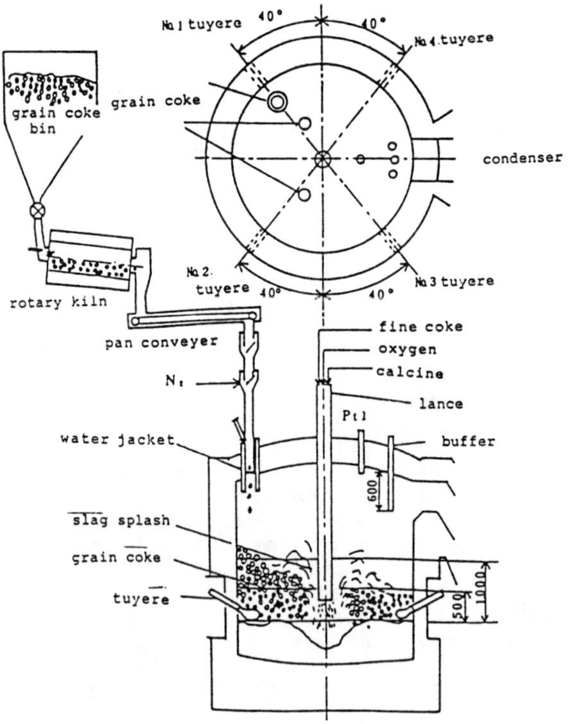

Fig. 8 Grain coke feeder

Smelting furnace

The smelting furnace of the bench scale test contained 1.5 ton slag and the inner diametewr of the furnace was 1,000 mm and the height was 2,000 mm. The production of zinc was 1 ton per day. The plan of this semi-pilot plant is to produce 15 ton metallic zinc per day, that is, 10 times greater than the bench scale test. The amount of slag which is kept in the furnace was planned to be 15 tons. The thickness of the molten slag layer in the furnace was planned to be 700 mm, so the cross sectional area of the furnace was calculated about 9 times greater than the bench scale furnace,that is, the diameter of the furnace becomes 3,000 mm instead of 1,000 mm of the bench scale furnace. The height of the furnace from the surface of slag layer was 1,400 mm at the bench scale furnace and it was designed to be 2,100 mm , and the inner height from the hearth to the roof was about 2,800 mm.

Chrome-magnesia brick was used at the hearth, roof and walls. The thickness of the brick was normally 350 mm. The brick of the wall at the slag level was attacked severely by the molten slag ,so the thinner brick was used and spray water was used for cooling .

LPG-oxygen burner

At the start of the operation LPG burner was used for heating the furnace and to melt the slag. The length is 3,880 mm and the outer diameter is 101.6 mm. The capacity of the combustion of LPG burner is 50 Nm^3/hr and is enough to heat the furnace up to 1,300 ^oC.

Condenser of zinc vapor

The condenser for zinc is the same one as the ISP type condenser.The width of the condenser is 1,350 mm , the length is 5,000 mm and the height is 220 mm. The retention time of the gas in the condenser is about 2.2 sec. Four rotors for splashing lead were used. The detail of the dimension and operational conditions are restricted to be reported here.

Equipments for environmental protection

Gas from the condenser is washed by venturi scrubber, and the dust which includes zinc is recovered from the cleaning tank. Gas after venturi scrubber is burnt by a flare stack. Dust which comes from materials handling systems is collected in bag filter and the plant is kept clean.

The operation and the results

Operations

The flow of the operation are schematically shown in Fig.1. Powder coke is so crushed by Roshe type mill in an atmosphere of hot air that more than 80 % of the fine coke is obtained to be -

200 mesh. A constant amount of this fine coke is supplied by the coke feeder which was specially designed for this purpose as shown in Fig. 3 and is carried to the lance with air.

Oxygen gas is supplied through a flow controller and is mixed with a fine coke near the head of the lance and injected into the furnace. Calcine, flux and returned dross are mixed at a desired ratio and this mixture is stocked in the hopper. A constant amount of this materials is fed through the constant feeder to the lance with air.

The mixed raw materials are injected into the furnace and fine coke with oxygen are also injected into the furnace by the lance. The single simultaneous injection lance was used after testing several kinds of lances.

Grain coke is also charged in the furnace after preheating and is reacted with the oxygen which is injected from tuyeres.

Slag is tapped intermittently on a sand bed after detecting the thickness of molten slag layer.

The exit gas from the furnace which includes zinc vapor is introduced in the condenser of ISF type and the gas is rapidly cooled by lead splash and zinc vapor is absorbed to molten lead. Lead which absorbs zinc is taken off by a lead pump and cooled to separate zinc. Some amount of zinc in the gas is not absorbed in the molten lead . Some zinc goes to dross and some zinc escapes to the exit gas without absorption.

Exit gas from the condenser is washed by venturi scrubber and zinc is recovered as sludge. The gas after cleaning is not recovered in this plant and is burnt by a flare stack .

One example of the typical compositions of calcine and coke which were used in this test plant are shown in Table 4 and 5.

Table 4 Chemical analysis of calcine(%)

Zn	Pb	Cd	S	Cu	Ag g/t	Fe	SiO_2	CaO
60.6	1.4	0.11	0.4	0.42	57	10.1	2.2	0.5

Al_2O_3	MgO	MnO	C	Bi	AS	Sb	Sn	Cl
0.3	0.3	0.7	-	0.011	0.082	0.012	0.010	<0.01

Table 5 Composition of coke

F.C	VM	ASH	H_2O	Kcal/Kg
81.6	2.3	14.0	2.2	6596

Composition of ash

SiO_2	Al_2O_3	Fe_2O_3	CaO	MgO
47.8	21.1	6.8	3.6	1.6

One of the conditions of the operation and the results are shown in Table 6. In this operations the simultaneous injection lance of type A was used.

Table 6 Operational conditions and results

Conditions for operations	Run No. 81-1	Run No.81-2
Lance	calcine and coke fed from one lance	
center hole for coke and calcine	20 mm dia.	
outer holes for oxygen blow	13 mm dia x 3	
Height of lance from slag surface	440 mm	420 mm
Amount of fine coke ,Kg/hr	523	520
Amount of grain coke,Kg/hr	99	96
Lance oxygen , Nm^3/hr	586	580
Lance oxygen/fine coke, Nm^3/ton	1,120	1,120
Total oxygen/total coke, Nm^3/ton	940	940
Calcine and returned dross , Kg/hr	1,163	1,185
Lead tempreture at pump sump , ^oC	610	610
Results of the operations		
CO_2/CO ratio in exit gas	0.90	0.88
Zn content in slag ,wt%	10.6	9.6
Formation of dross, %*	16.3	14.3
Recovery of zinc metal,%**	70.4	73.9
Coke consumption ton/ton metallic zinc	1.50	1.39

One example of the compositions of metallic zinc from the condenser are shown in Table 7. Many impurities are included and refining process is necessary for shipment.

The composition of slag is also shown in Table 8.

Table 7 Compositions of metallic zinc(wt%)

Zn	Pb	Cd	Cu	Ag g/t	Fe	Bi	As	Sb	Sn
97.5	1.68	0.18	0.13	72	0.045	<0.001	0.067	0.006	0.011

Table 8 Compositions of slag(wt%)

Zn	Pb	Cd	S	Cu	Ag g/t	Fe	SiO_2	CaO	Al_2O_3
7.95	0.57	0.0019	0.38	0.32	8	25.9	20.0	8.8	6.5

MgO	MnO	Bi	As	Sb	Sn
6.6	1.7	0.007	0.014	0.009	0.011

Discussions

Many tests were performed by this semi-pilot plant. The operational conditions and the equipments of the plant were changed several times after the tests. Low recovery of metallic zinc less than 60 % was obtained at the initial tests. The reasons of this low recovery of zinc were discussed as follows:

1) Fine coke is not sufficiently burnt. Solid fine particles such as fine coke which is not burnt, and fly ash , and carry-overed materials and slag in the condenser become a nucleus of oxidation of zinc vapor , and zinc absorption in lead is disturbed.

2) The temperature of the exit gas from the smelting furnace decreases at the down-comer of the condenser before contacting with splash of lead. The zinc vapor in the gas is thought to be easily oxidized because of its high zinc content and high CO_2.

The next countermeasure was taken to overcome a low recovery of zinc after analyzing the above reasons.

1) To diminish carry-over dust into the condenser

1-1. To obtain high combustion rate of fine coke, next countemeasures were taken.

The grain coke is fed into the smelting furnace to make a grain coke layer, and oxygen or enriched air with oxygen is injected into this coke layer through four tuyeres .

1-2. To diminish carry-over dust, the simultaneous lance with high speed injection was used.

1-3. Buffer bars were equipped at the exit side of the gas in the smelting furnace to catch the dust.

1-4. Slag level was lowered to diminish the slag splash going into the condenser. .

2) The cross-sectional area of the down-comer was narrowed to hasten the gas speed.

These modifications were applied to the plant and operations, and more than 70 % of recovery of zinc was obtained as shown in Table 6.

The results of the bench scale tests, and this semi-pilot plant tests and the operational data of ISP at the small blast furnace (5) are compared in table 9.

The recovery of zinc of the bench scale test is 60% ,but that of the semi-pilot test is 75% . The recovery of zinc of ISF of the capacity of 30 ton zinc /day is 79%. The recovery of zinc will be expected to become higher when a plant is enlarged to a commercial scale.

Table 9 Comparison of the data

	bench scale test	semi-pilot plant	ISP (1958)
raw materials			
calcine	100	100	100(sinter)
dross	30	30	16.8
Kg/h	87	1,185	3,330
Coke Kg/hr	73	616	1,790
(grain coke)	-	(96)	
Oxygen Nm^3/hr	72	569	-
Zn/coke ratio	0.61	0.95	0.89
zinc production Kg/hr	45	443	1,270
zinc % in slag	10.0	9.6	7.9
vaporization ratio of zinc % *	94.3	95.2	93.9
zinc recovery %**	60.4	73.9	79.1
coke consumption ton/ton zinc	2.65	1.39	1.41

*: (total zinc in raw materials - zinc in slag)/(total zinc in raw materials) x 100 %

**: (zinc in crude metal)/(total zinc in raw materials)x 100%

Feasibility study

The production cost of a commercial plant was calculated from the data of the semi-pilot plant tests. The fundamental specifications for feasibility study are listed in Table 10.

Table 10 The specifications of the commercial plant for feasibility study

Plant capacity	100,000 ton of special high grade zinc per year
Raw materials	
zinc concentrates	imported from abroad
coke	imported from abroad
Plant location	domestic and seaside
Days of operation	330 days (smelting)
Index of the operation	
Vaporization ratio of zinc	96 %
Recovery of zinc (primary)	85 %
Coke consumption	1.20 ton/ton vaporized zinc

Index of the operation which was used in this feasibility study is estimated from the data of the semi-pilot plant. The operation data of the semi-pilot plant and the index values are compared in Table 11.

Table 11 Index of operation

	Semi-pilot plant	Commercial plant
vaporization of zinc %	95	96
Zn/carbon ratio	0.95	0.87
condensation efficiency %	80	88.5
zinc recovery(primary) %	75	85
coke consumption, ton		
per ton vaporized zinc	1.10	1.20
per ton of crude zinc	1.35	1.34

The relation between production capacity and condensation efficiency are shown in Fig. 9. The condensation efficiency of the bench scale tests and semi-pilot tests is 63% and 80% respectively. In case of ISP, the condensation efficiency of 30 ton /day is 84%,but that of 240 ton/day is 92 %. The condensation efficiency becomes better by enlargement of the plant, so a primary zinc recovery was assumed to be 85 % in this index for feasibility study.

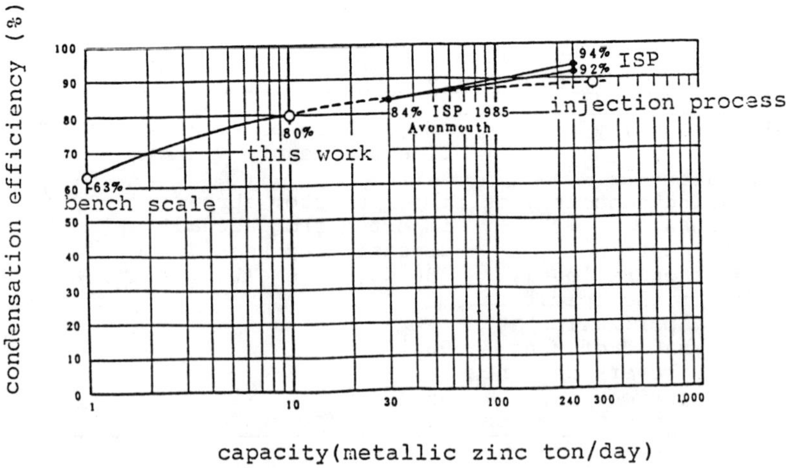

Fig.9 Relation between production capacity and condensation efficiency of zinc

Construction cost

Construction cost of 100,000 ton plant is shown in Table 12.

Table 12 construction cost of the plant

	million yen
Roasting plant	3,274
Sulfuric acid plant	3,381
Smelting plant	4,455
Refining plant	2,448
Electric power plant	3,578
Oxygen plant	2,456
Utility	4,312
Infrastructure	3,796
Green belt	286
Reserve	414
Engineering fee	1,400
Total cost	29,800

Total production cost

Production cost of the plant is compared with hydometallurgical plant of the same capacity in Table 13 and 14.

Table 13 Comparison between injection and hydrometallurgical plant

	Injection plant	hydrometallurgical plant
area of the site	170,000 m^2	2,250,000 m^2
employee	181	300
construction cost (million yen)	29,800	38,000

Table14 Production cost of the plant indicated by index

	Injection plant	Hydrometallurgical plant
Direct cost		
energy cost	14.0	21.0
labour cost	4.0	7.0
repair	6.3	2.8
fixed expense	0.7	3.5
others	2.5	14.9
subtotal	27.5	49.2
Indirect cost	42.3	50.8
Total processing fee	70.0	100.0

Energy consumption and also energy cost for injection process and for hydrometallurgical plant are calculated and shown in Fig.10 .It is concluded that injection process is more profitable in Japan when the same capacity of new plant is constructed.

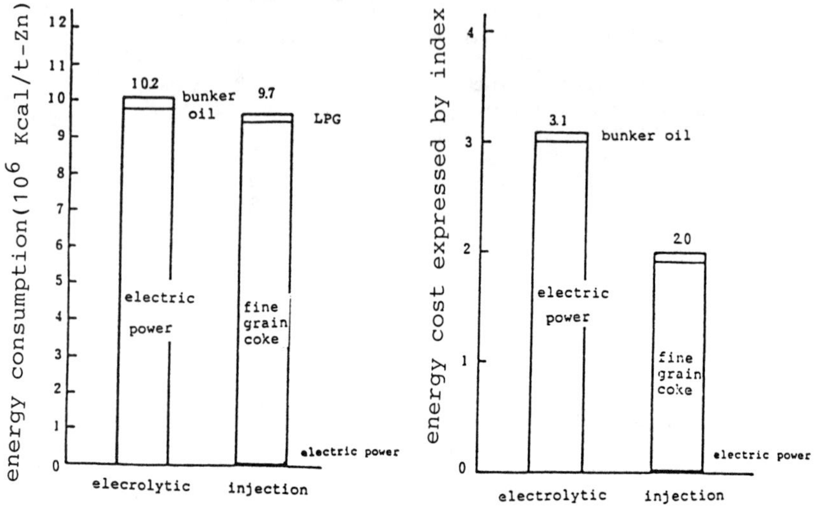

Fig.10 Comparison of energy consumption and energy cost between electrolytic process and injection process

Conclusion

The semi-pilot plant operation for new injection smelting of zinc was performed and about 75% of the primary recovery of zinc was otained in these small scale tests. The condensation efficiency of zinc in the injection smelting is assumed to be lower than that of ISP, because of higher dust formation and higher content of zinc and higher CO_2 in exit gas of injection smelting. But condensation efficiency will be expected to be higher by a basic research works. When this process will be operated in commercial scale, the primary recovery will be assumed to be 85 % and coke consumption per ton of crude zinc will be 1.34 ton. Crude zinc is refined by using the exit gas after condensation of zinc. The feasibility study showed that this process is more economical than electrolytic process at a location of high cost of electric power. Coke consumption of this process is higher than ISP,but cost of coke breeze which is a main energy source is lower than lump coke. The total production cost of this process is expected to be lower than that of ISP in a commercial scale.

Acknowledgment

I wish to thank the members of the Committee of Joint Research Organization of Injection Metallurgy for permitting the publication of this paper and also thank to their finantial supports of Dowa Min. Co., Mitsubishi Metal Corp.,Mitsui Min. Smelt. Co.,Sumitomo Metal Min.Co.,Nippon Min. Co., and Toho Zinc Co.

References

1. S. Goto, "Thermodynamic Consideration for New Smelting Process of Zinc Calcine," J. of Min. Met. Inst. Japan , 97(1981),107-111

2. S. Goto, "Thermodynamic Ccnsideration and Basic Test for New Smelting Process of Zinc Calcine,"AIME,TMS paper Selection, 1984

3. S. Goto ,"Experimental Work of Injection Smelting for New Smelting Process of Zinc Calcine," Zinc '85,(1985), 841-853

4. S. Goto , "Method of Smelting Zinc by Injection Smelting," US Patent No.4,514,221(1985), Europe Pat. No. 0117325, Australia Pat. No. 558,715 (1987), Korea Appl.No. 83-707(1983)

5. S.W.K.Morgan and S.E.Woods, "Avonmouth Zinc Blast Furnace Demonstrates Its Versatility,"Eng. Min. Journal, Sep.(1958)

FUNDAMENTAL STUDIES ON NICHEL METALLURGY

IN CHINA

H.Jiang S.Yang and B.Guo

Department of Chemistry, Central South
University of Technology, Changsha
China

Abstract

The main processes for nickel metallurgy in China consist of electric arc furnace smelting to produce low grade nickel matte and converter to high grade nickel matte, grinding and flotation to separate nickel sulfide from copper sulfide and direct electrolysis of nickel matte. Systematic studies on the properties of electric furnace slag, smelting process for nickel matte, anodic process of nickel matte, cathodic process of nickel electrolysis have been done in our labs. In addition, an approach to a new hydrometallurgy process for nickel matte has been done. The feature of the new process is one stage acidic leaching of high grade nickel matte. The leaching solution could be purified by deposition way or by solvent extraction way. There are two leaching rate-determing steps, chemistry reaction-determing one at temperatures below 60°C and diffusion-determing one at temperatures higher than 60°C. The acidic leaching fraction reacted α varies with leaching time t according to the equation $1-(1-\alpha)^{1/3}=kt$. The experimental data also show that the leaching reaction rate varies with concentrations of H^+, Cl^-, and it is also effected by copper sulfide and Cu^{++}. The Cu^{++} ion can retard the leaching reaction due to forming a copper sulfide layer on the matte surface and the equation $1-2/3\,\alpha-(1-\alpha)^{2/3}=kt$ can be used for the case. However the copper sulfide added into nickel matte can facilitate the leaching process. The research work for nickel metallurgy done by others in China has also been mentioned in this paper.

Metallurgical Processes for the Year 2000 and Beyond
Edited by H.Y. Sohn and E.S. Geskin
The Minerals, Metals & Materials Society, 1988

Studies On Pyrometallurgy of Nickel

Large scale production of nickel in China started in 1960s.The main processes consist of electric arc furnace smelting to produce low grade nickel matte and converter to high grade nickel matte,grinding and floatation to seperate copper sulfide from nickel sulfide,and direct electrolysis of nickel matte.

To reduce nickel loss in slag a reseaech group organized by Jinchuan Company and Central South Institute of Mining and Metallurgy (now the Central South University of Technology) measured the viscosity,surface tention,density,conductance,melting point and structure of the slag(1).It has been found that the physico-chemical properties of the slag depend on the components SiO_2,MgO,and FeO,but FeO in the slag is too high.And the statistic analysis of data from industry gave an equation to correlate the content of Ni in slag with the content of Ni in matte and the composition of slag.Again it was pointed out that the content of FeO in slag effected the loss of Ni in slag very much (2). The ratio (FeO+MgO) to SiO_2 should be less than 1.13 and the content of Fe_3O_4 should be as low as possible,but the reaction

$$FeS + 3\,Fe_3O_4 = 10\,FeO + SO_2 \tag{1}$$

depends on the kinetic condition of nucleation of SO_2.If there is no nucleation center for SO_2,for exaple in homogeneous phase,one can not expect the reaction to reduce Fe_3O_4 (3).

X-ray transmission photogrtaphy is a very powerful method to measure some properties of materials at high temperature, combining the x-ray transmission photography with video can measure the static properties and can watch the kinetic processes too (4-6), for example,watching the matte forming from different roasting processes, one can see that there are more nickel matte pearls attached to gas bubbles if the roasting process produced too much Fe_3O_4 for not enough carbon added.

High grade nickel matte from convertor consists of Ni (48-50%wt), Cu (25-30%), S (10-22%),Co(0.2-06%) and Fe (1-3%). There are NiS_2, Cu_2S and alloy (Ni-Cu-Fe) in it. If one can preserve the content of S but decrease the content of Fe to less than 1% in the high grade nickel matte then the leaching process followed would be much easier. A study has shown that FeO could be extracted to slag and the S could be preserved in matte during the convertering process(7).

Studies on Hydrometallurgy of Nickel

After grinding and floatation, the copper sulfide (copper concentrates) and nickel sulfide (nickel concentrates) are sent to copper system and nickel system respectively. The nickel concentrates are then casted to anode for direct electrolysis.

On electrolysis of nickel matte

The anodic behavior of nickel matte (from nickel concentrates) is rather complicated, much work has been done in china (8-12). The maijor reactions may be represented by

$$Ni_3S_2 = Ni_{3-x}S + x\,Ni^{2+} + 2xe \tag{2}$$

$$Ni_{3-x}S_2 = (3-x)\,Ni^{2+} + 2S + 2(3-x)e \tag{3}$$

$$2S + 3H_2O = S_2O_3^{2-} + 6H^{+} + 4e \tag{4}$$

$$S_2O_3^{2-} + 5H_2O = SO_4^{2-} + 10H^{+} + 8e \tag{5}$$

Where $Ni_{3-x}S_2$ is a series of nickel deficient interme-diates (9,10).In sulfate solution, the intermediates are NiS, Ni_3S_4 and NiS_2, but in chloride solution, NiS may be the only intermediate (10). $S_2O_3^{2-}$ as an intermediate was defected by ring-disk technique(9). The reaction(2) and (3) may be controlled by the diffusion in solid and jointly by discharge(9,10). Before above reactions the adsorption-desorption reactions may be presented as

$$Ni_3S_2 + H^+ = 2Ni^{2+} + NiS + HS^- + 2e \qquad (6)$$

$$NiS + 4H_2O = Ni^{2+} + HSO_4^- + 7H^+ + 8e \qquad (7)$$

The Cl^- has bilateral actions, it can depress the active dissolution and has some activation action too, but its mechanism is not known yet except that it shows adsorption on nickel sulfides (10,11). The complexes of Ni^{2+} with SO_4^{2-} or Cl^- are very weak, the overall stability constants for $NiCl^+$, $NiCl_2$, $NiSO_4$ and $Ni(SO_4)_2^{2-}$ are 0.30, 0.25, 0.24 and 0.20 respectively(13).

The possible mechanism of electrode reduction of Ni^{2+} at nickel electrode is

$$Ni(H_2O)_6^{2+} = Ni(H_2O)_{6-x}^{2+} + x H O \qquad (8)$$

$$Ni(H_2O)_{6-x}^{2+} + e = Ni^+ + (6-x)H_2O \qquad (9)$$

$$Ni^+ + e = Ni \qquad (10)$$

Reaction (9) may be the rate-determing step(14).

On leaching of high grade nickel matte

An alternative way to the grinding and floatation is the separative leaching processes which usually consist of two atmospheric leaching processes and one oxygen pressure leaching process with autoclaves. The autoclaves are difficult to get. Therefore some new atmospheric leaching processes have been tried by us and positive results have been obtained.

$$Ni_3S_2 + 2H^+ + 1/2\,O_2 = 2NiS + Ni^{2+} + H_2O \qquad (11)$$

$$Ni_3S_2 + 2H^+ = Ni^{2+} + 2NiS + H_2\uparrow \qquad (12)$$

and

$$Ni_3S_2 + 6H^+ = 3Ni^{2+} + 2H_2S\uparrow + H_2\uparrow \qquad (13)$$

There are two type of NiS, α-NiS and β-NiS. The later is difficult to dissolve, if it is formed.

If reaction (13) runs fast enough, then the formation of β-NiS is avoidable. The Gibbs free energy ΔG^o of reaction(13) and the equilibrium constant k in Table 1 were calculated according to the thermodynamic data(8), from which we can see that the reaction(13) is very irreversible to the right side.

Table I Values of ΔG^o and k for reaction (13)

temperature, c	25	50	75	100
$-\Delta G^o$(kJ/mol)	47.056	47.621	48.235	48.965

For the purpose of kinetic studies synthetic Ni_3S_2 was made and the high grade nickelmatte from smelter was also used to verify the experimental results of synthetic samples.The synthetic one was made by heating weighed nickle and sulfur powders in a quartz tube under vaccum. Some samples were treated by water hardening.

Table II Compositions of samples

samples	compositions				
	Ni	S	Cu	Fe	Co
before hardening	73.12	26.78	0.02	0.01	0.01
after hardening	73.16	26.43	0.01	0.02	0.01
high grade nickle matte	62.01	22.52	14.78	0.98	0.76

X-ray shows that only those peaks of Ni_3S_2 are in the chart for synthetic one.And the high grade nickel matte consists of Ni-Cu-Fe alloy,Ni_3S_2 and Cu_2S. The leaching expriments carried out in a three opening flask with 2 mole sufuric acid and 2 mole hydro-chloride acid.the agitator was a mechanic one with rotating speed 1020 rpm,they were put in a thermostat.The particle size of the samples(diameter d_o) was 22.64 except otherwise mentioned. The results show in Tables III to VIII and in Fig. 1 to 6. The data in Table III can be illustrated in Fig. 1, i.e.a linear relationship can also be represented by equation (14)

$$1-(1-\alpha)^{1/3} = kt \qquad (14)$$

where α is the leaching fraction reacted, t is the leaching time and k is the overall rate constant.

From the data in Table IV,log k - 1/T plot could be done(Figure 2),the apparent activation energy could be got.In the range of 40 - 60°C it is 55 kJ/mol and in 60 - 90°C is about 10 kJ/mol.It seems that the leaching process at higher temperatue (>60) is controlled by the slow diffusion of bubbles of $H_2$and H_2S through the product layer due to adsorption of them at the suface of the sulfide.

Effect of particle size on leaching rate as shown in Table VI and Fig. 3.

The leaching processes for all these particle size follow the equation (14) very well.The overall rate constant k values were obtained from the slope of eqution (14).The linear of k to $1/d_o$is also an evidence that the leaching processes follow the model of shrinking unreacted core.

The influence of Cl^- again shows two sides on the leaching processes some what like that on the elctrode of nickle matte.(Refer to Fig. 4).

The influence of copper ion on leaching is rather different from it on the elctrode processes,because a layer of Cu_2S formed on the nickel sulfide it retards the leaching proccess.And hydrogen ion or chloride ion can not effect that much.(refer to Table VII).At this time the leaching processes coincide with equation

$$1 - 2/3\,\alpha-(1-\alpha)^{2/3} = k\,t \qquad (15)$$

however if to mix Cu_2S and nickle sulfide powders at high tempe-rature and hardening them then the leaching rate is much faster than the pure nickle sulfide since microgalvanic cell action can accelerate corrosion current.

Fig.5 illustrates the linear relation in equation (15).It implies the insoluble products Cu_2S inhibited leaching reaction,but Table VIII shows Cu_2S mixed with ni-ckel sulfide can accelerate the leaching processes.

Table III. Influence of temperature on leaching rate with time

Temperature (° C)	Time (Min) 1	2	5	10	15	20	25	30	40	50
40	5.57	7.07	10.29	14.53	17.33	17.85				
45	7.35	11.06	16.49	20.07	22.45	23.00				
50	8.37	12.13	19.83	26.62	32.29	33.00				
55	9.23	13.20	23.81	30.41	37.50	44.90	52.70	58.24		
60	10.85	13.90	25.24	32.65		48.10		62.13		
65	11.65	14.30	26.10	33.40	43.15	50.69	57.69	63.73		
70	12.73	15.35	27.35	35.50		52.86		68.11	80.07	84.95
75	13.67	16.93	29.93	36.75	44.39	54.16	62.16	71.63		
80	15.17	18.20	30.35	39.40		58.25		75.59	82.53	87.31
85	16.97	20.37	32.22	47.71	56.23	64.23	70.23	77.15		
90	19.15	23.40	35.49	50.26		66.60		79.34	84.63	90.47

Table IV Value of overall rate constant k at different temperature

temperature	40	45	50	55	60	65	70	75	80	85	90
k·1000	2.432	3.610	5.145	7.341	7.995	8.371	8.880	9.108	9.605	9.692	11.160

Table V Variation of values of overall rate constant k with H^+ concentration

H^+(mol)	3	4	5	6
k 1000	3.45	5.85	6.82	8.55

Table VI Effect of particle diameter (d) on leaching fraction reacted with time (t)

d(μm)	t(min)							
	1	2	5	10	15	20	25	30
22.62	9.23	13.20	23.81	30.41	37.50	44.90	52.70	58.24
44.10	7.00	9.65	11.67	16.95	20.87	23.10	27.23	30.15
54.95	5.82	7.95	9.82	14.02	17.08	20.16	24.20	29.73
66.68	3.92	5.05	6.42	9.70	13.18	16.46	20.32	23.30

Table VII Effect of copper ion on leaching fraction(%)

Cu^{++} (g/l)	H^+ (mol)	Cl^- (mol)	t (min)						
			15	30	45	60	75	90	105
5	3	2	13.15	18.62	23.23	27.62	30.23	32.09	33.31
10	3	0	13.51	19.47	24.16	28.01	31.09	33.31	34.10
10	1	2	11.61	17.08	23.23	26.18	29.55	32.63	

Table VIII Overall rate constants (90°C) of samples prepared differently

samples	air cooling Ni_3S_2	water hardening Ni_3S_2	water hardening $Ni_3S_2+Cu_2S$(25%wt.)
k·1000	6.80	8.16	10.27

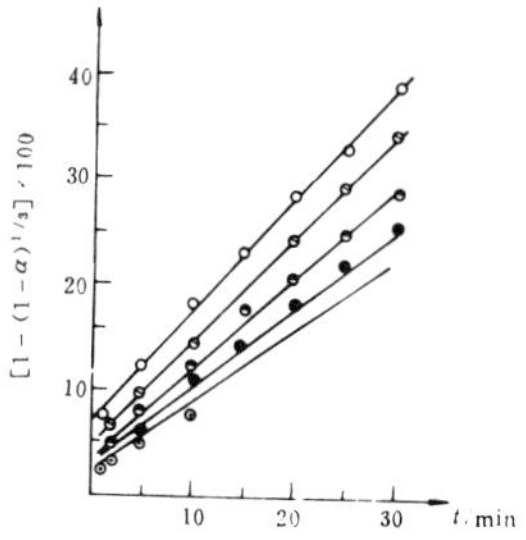

Fig.1 Linear relationship of $(1-(1-\alpha)^{1/3})$ with time at •45, •55, •65, •75 and ∘85°C

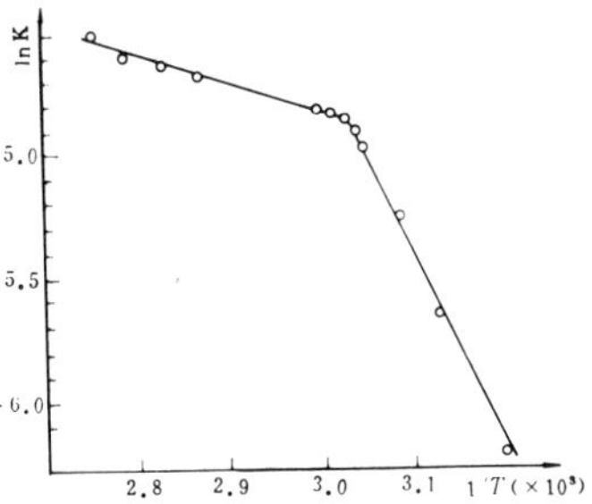

Fig.2 Variation of Ln k with 1/T

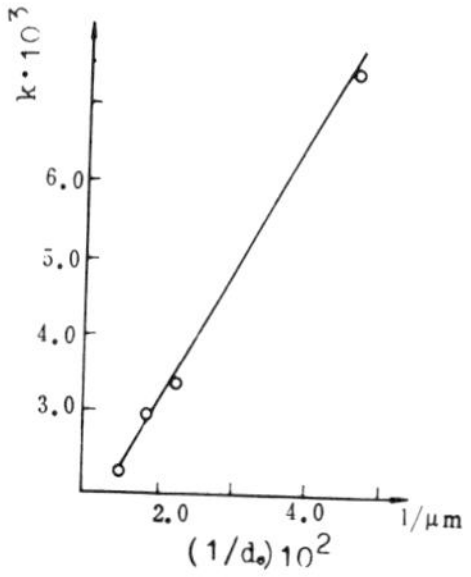

Fig.3 Effect of particle diameter (d_0) of nickel sulfide on leaching rate

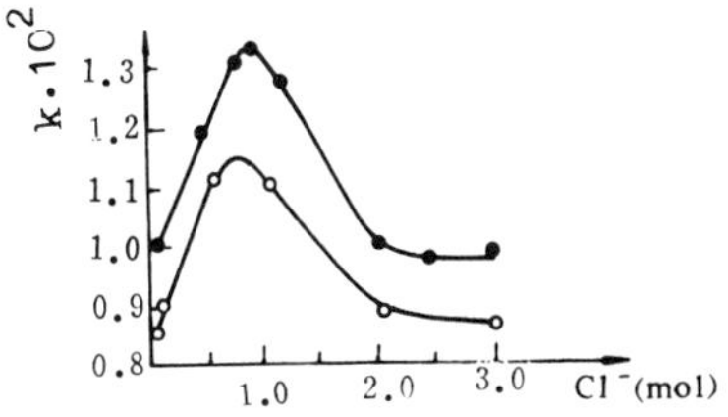

Fig. 4 Influence of Cl^- on leaching rate

During leaching process the solid to liquid ratio (s/l) was very important.s/l from 1/125 to 1/10 the time to reach leaching rate 98% changed from 1 hour to 2 hours (refer to Fig. 6).And there would be more residue left if the agitator worked not so well.The acidity can effect the leaching rate at lower temperaure (60°C and below).The reaction order of hydrogen ion is about 1 (refer to Table V) At higher temperature the acidity did not play a big role.The maijor influence factors to the leaching processes are preperation of samples and the diffusion condition,if temperature is about 90°C or above.Anyhow there were 1 or 2% residues left,sometimes.

The experiments on high grade nickel matte verified the results of synthetic samples

Exepriments On Leaching Residue. An electric leaching method was tried.A graphite cylinder was used as anode and copper as cathode,s/l from 1/5 to 1/10,sulfuric and hydro-chloride acid were used.The current was 2 - 4 amples/ 100 grams of residues. 98 - 99 % of nickel in the residues was leached out and copper sulfide remained in precipitates.

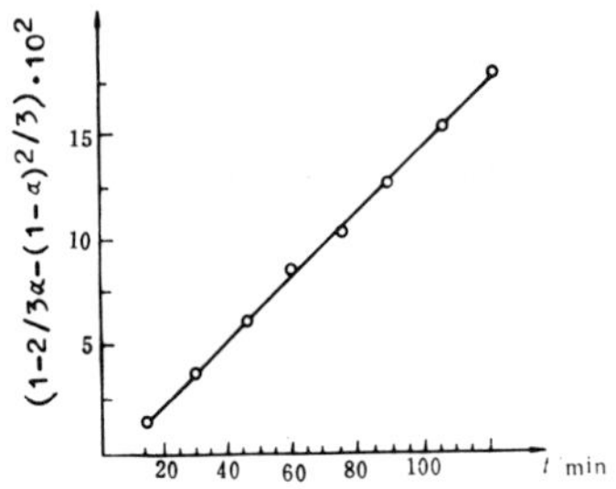

Fig. 5—Linear relationship of $(1-2/3\alpha-(1-\alpha)^{2/3})$ with leaching time t (refer to equation 15)

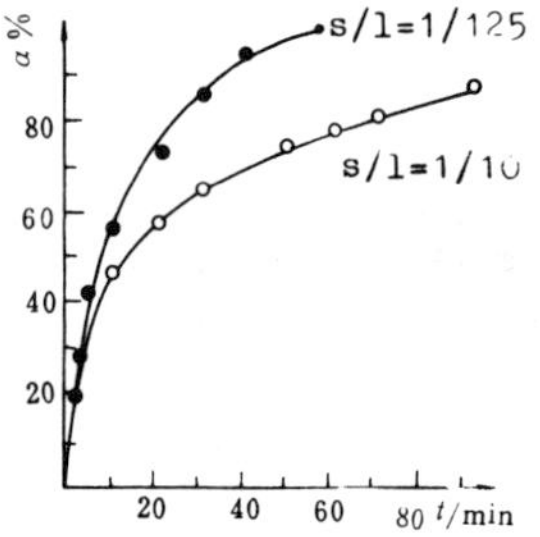

Fig. 6—Effect of s/l ratio on the change of leaching fraction α with time t

Extraction

After nickel and cobalt geting into solution,separation of Ni from Co and purification are very important.To get rid of Fe in solution solvent extraction has been developed in China(15),besides the normal oxidizing and precipitation way. In acidic solution,DEHPA (P 204 in China) and P 538,($(C_nH_{2n+1})_2CHOP(O)(OH)_2$ could be used to extract Fe,to improve extract forward and back processes TRPO, $(C_nH_{2n+1})_3PO$ should be used.To separate Co from Ni in acid solution N 235,$(C_nH_{2n+1})_3)N$ and 7407, $R_3N\!-\!\bigcirc\!-\!CH_2Cl$ have been used here.If Cl^- is more than 100 g/l,Co can be extracted to the ratio Co/Ni large than 10000.After extraction the aqueous solution could be used to electrowin Ni.If more sulfate than chloride in solution MEHPA(P 507 in China)is a good extractant,but the pH value be adjusted to 4-5. P 507 has been used in industry for many years.A new extractant 5709,R′(RO)P(O)(OH) has better extraction efficiency for Co, Cu, Fe, Pb and Zn but Ni remains in solution.The 5709 has been tried in a plant for solvent extraction in nickel metallurgy.To get rid of organic residues in a more economic way than activated charcoal used is not solved yet .

Prospects of Nickel For The Year 2000 In China

Today the maijor mineral deposits of nickel is at Jinchuan.In north-west part of China there is brilliant prospects for other deposits of nickel.However the difficult separative ores and low grade ores should have more work done to meet the needs for nickel and copper.A research group in our university has made progress in oxygen pressue leaching to treat low grade nickel concentrates(16).The needs for nickel in China would be up to 80,000 - 90,000 tons in the year 2000. The flash furnace smelting will be the maijor part for pyrometallurgy,and the present fluid bed roaster and electric arc furnace would also be improved .

References

1. D.Mo,S.Yang and X.Mei."Measurement of The Physico-Chemical Properties of Electric Furnace Slag of Jinchuan Non-Ferrous Metal Company," Jounal of Central South Institute of Mining and Metallurgy 3 (1980) 76-87.

2. D.Mo, K.Huang and J.Li, "A Statistic Analysis on The Content of Ni in Slag". ibid, Supplement 2 (1982) 81-86.

3. K.Huang and M.Sun, "A Study on Reaction of Fe_3O_4 in Slag with FeS in Matt ibid Supplement 2 (1982) 92-98.

4. J.Li, K.Huang and X.Chen, "The Determination of Interfacial Tension Between Nickell Matte and Slag". ibid, 3(1983) 104-111.

5. K.Huang, J.Li, M.Sun, D.Mo and X.Chen. "Investigation of X-ray Transmission Experimental Technik and Installation Used for The Research of Metallurgical Melts At High Temperature" Proceedings of the 5th Symposium on The Physical Chemistry of Process Metallurgy, Xian, China Vol.1.1984 406-412.

6. Q.Liu, K.Huang, D.Ye and X.Chen, "A Study on Surface Tension and Density of Fe-Ni-S system by X-ray Transmission Sessile Method" Journal of Centra South Institute of Mining and Metallurgy, 3(1986) 87-94.

7. B. Li, C. Ji, G. Ye and C.Cui, "A Study on The Converting Process of Jinchuan Ni-Cu Matte and Its Deep Deironization with Preseving Sulfur" Proceedings of the 5th Symposium on the Physical Chemistry of Process Metallurgy, Xian, China. Vol. 1. 1984, 192-201.

8. X.Yang et al "Handbook of Thermodynamic Data for Agueous Solution At High Temperature" (Beijing. Metallurgical Industry Press 1983) 249,436

9. H.Jiang, Y.Shu and Z.Zhang, "Electrochemical Behavior of Nickel Matte" Jonrnal of Central South Institute of Mining and Metallurgy, 4 (1986) 100-109.

10. F.Tu and H.Jiang, "Electrochemistry Kinetics of Nickel Sulfide in Chlori Solution" (Paper Presented at the National Physico-Chemistry Symposium or Hodromelallurgy, Kunming, China, 1987)

11. Z.Fong, "Electrochemistry Study on Anodic Dissolution of Nickel Sulfide" (Ph. D. thesis, Institute of Chemical Metallurgy Academia Sinica,Beijing, (1985)

12. Y. Zhou, et al, "Electrochemical Behavior of Heazle-Woodite Anodes at High Temperature, Journal of Central South Institute of Mining and Metallurgy, 3(1987) 275-280.

13. R.Zhu, Z.Gong and H.Jiang, "A Study of The Chloride and Sulfate Complexes of Nickel" ibid, 3(1983) 119-125.

14. Z.Zeng and A.Yin, "Mechanism of Nickel Electrowinning From Aqueous Sulfa Solution" Nonferrous Metals 40. No.2(1988) 62-65.

15. S.Yu and J.Chen, "Studies on Acidic Organic Phosphorate-Fe (III) Phase Transfer," I and II (Papers Presented at the National Physico-Chemistry Symposium on Hydrometallurgy, Kunming, China 1987)

16. S.Yue, private communication with authers, Metallurgical Research Institute of Central South University of Technology,Changsha, China.

SEPARATION OF VARIOUS ELEMENTS IN

CRUDE METAL OR ALLOY DURING VACUUM DISTILLATION

Dai Yongnian

Research Institute of Vacuum Metallurgy
Kunming Institute of Technology
Kunming, Yunnan, China

ABSTRACT

The feasibility to vacuum distillation of a crude metal or an alloy was discussed. The equilibrium relationship between vapor phase and liquid phase in A-B system ----- $P_A/P_B = \beta_A a^*/b$ has been derived. The separation coefficient $\beta_A = r_A P_A/(r_B P_B)$ is a judgment criterion. By calculation P_A/P_B, the equilibrium composition diagram of vapor phase and liquid phase was drawn. By use of the equation derived by M. Olette, the relationship among the volatilization amounts of various impurities in a crude metal (y_i) was further solved. $1-(X/100)= C =(1-Yi/100)^{1/\alpha i}$. From y_i value obtained in practice, $\bar{\alpha}i$, βi and then can be calculated. The correct volatilization order of various impurities was obtained. These results can be applied to determine the practical conditions (temperture, vacuum degree, type of distillation furnace, times of distillation and control of product composition) in distillation of a crude metal or an alloy.

Metallurgical Processes for the Year 2000 and Beyond
Edited by H.Y. Sohn and E.S. Geskin
The Minerals, Metals & Materials Society, 1988

INTRODUCTION

Refining and separation of various impurities from nonferrous crude metals by vacuum distillation have many the advantages of high metal recovery, impurities recovered in metallic state, flowsheet simplified, good enviromental protection, low operation costs and incomplex equipment. So in recent years this method has been adopted widely.

Vacuum distillation has also been studied and used in separation of various elements from non-ferrous alloys and the area of its application is being extended rapidly.

At this situation, some problems have been put up: What kind of impurities can be removed from crude metals? What kind of alloys can be separated by vacuum distillation? In present paper, attemps have been made to answer these problems and combine with practical results to analyse some valuable data obtained from testwork on vacuum distillation of crude tin and Pb-Sn alloy.

EVALUATION OF DISTILLABILITY FOR SEPARATION OF VARIOUS ELEMENTS FROM CRUDE METALS AND ALLOYS

Evaluation by boiling points or vapor pressures of pure substances. Usually the boiling points of various elements in crude metals and alloys at pure state or their vapor pressures at the same temperature (p_i^o) are compared. The boiling points of some elements, namely the temperatures at vapor pressure of 1 atmosphere are shown in Table 1.

Table 1 Boiling points of some metal elements

Element	Hg	As	Cd	Zn	Ti	Bi	Sb	Pb	In	Ag	Sn
Boiling point °C	357	603	765	907	1460	1564	1678	1740	2073	2200	2623

As shown in Table 1, the difference between boiling points of lead and Zinc is 833°C, so it is determined that Pb-Zn alloy or crude lead containing Zinc can be separated one another by vacuum distillation due to great difference in their boiling points. For the same reason, the similar substances such as Sn-Pb alloy or crude tin containing with lead, cadmium in Zn-Cd alloy or crude Zinc and so on can also be separated by vacuum distillation.

On the other hand, the P_A^o and P_B^o of element A and B at the same temperature can be calculated, if $P_A^o \neq P_B^o$, it is considered that element A and B can be separated by vacuum distillation.

In this evaluation, the difference of action forces between atoms of elements in alloys is not put into consideration, so, for some materials the regurility discribed above can not be adopted and sometimes an impractical result occurs. Alloys of Pb-Bi, Sn-As-Pb and so on belong to such materials. For example, at the part of rich lead (crude lead,$>$90% Pb) in Pb-Sb alloy, lead is selectively vatillized during distillation, and at the part of rich antimony (crude antimony,$>$90% Sb), antimony is preferentially vatillized into gas phase. When crude tin containing arsenic and lead,, lead is selectively vatillized into gas phase rather than arsenic.

Evaluation by practical vapor pressure. When evaluation, not only the values of P_A^o and P_B^o must be considered, but also much attention should be paid to activity a_i of each component (or activity coefficient r_i) and their contents of percentage (a^* stands for A, b for B)or mole fractions N_A and N_B, that is, the practical vapor pressures P_A and P_B produced after the solution of A and B is formed. Since

$$P_i = r_i \; N_i \; P_i^o$$

Here, the change range of Ni is very large, it may be up to several orders of magnitude and that of r_i is of the same extent.

For example

Pb-Zn system (650°C)	Cu-Zn system (500°C)
End in rich lead r_{zn}^o=7.94	End in rich copper r_{zn}^o=0.014
End in rich Zinc r_{Pb}^o=34.6	End in rich Zinc r_{cu}^o=0.018

The difference between two r_{Zn}^o in above binary systems is two orders of magnitude.

The change of r_i and Ni has great inflecence on Pi, so, for evaluating separation of elements, it is more perfect by use of Pi than by use of Pi°.

Evaluation by separation coefficient

Because of the molecular weight of each substance is different at the same practical vapor pressure, it is also different mass in gas. This can be shown by a simple example; at condition of 0°C and 1 atmosphere, the mass of hydrogen in 22.4L is 2gm and that of CO_2 is 44gm, the later is 22 times as heavy as the former.

Thus, the content of one substance i in gas is expressed by vapor density ρ_i

$$\rho_i = Mi\ Pi/(RT) \tag{2.1}$$

Where ρ_i -the mass of gas containing substance i in volume of unit. It is related to molecular weight Mi, practical vapor pressure Pi and the temperature of gas T.

R-gas constant.

Substituting (1) into (2.1)

$$\rho_i = \gamma_i\ Ni\ Mi\ P_i^o/(RT) \tag{2.2}$$

comparing vapor densities of A and B in gas of A-B system, That is,

$$\frac{\rho_A}{\rho_B} = \frac{\gamma_A\ N_A\ M_A\ P_A^o}{\gamma_B\ N_B\ M_B\ P_B^o} \tag{3.1}$$

then changing mole fraction N_A and N_B into weight of percentage a^* and b, equation (3.1) becomes

$$\frac{N_A}{N_B} = \frac{a^*}{b}\frac{M_B}{M_A} \tag{4}$$

If there are the same molecular weights of A and B in both gas and liquid phases, then

let
$$\frac{\rho_A}{\rho_B} = \frac{a^*}{b} \frac{P_A^o}{P_B} \tag{3.2}$$

$$\beta_A = \gamma_A\ P_A^o/(\gamma_B\ P_B^o) \tag{5.1}$$

equation (3.2) turns into

$$\frac{\rho_A}{\rho_B} = \beta_A\ a^*/b \tag{3.3}$$

Where ρ_A/ρ_B is the ratio of contents of two components A and B in gas phase, a^*/b is the ratio of contents in liguid phase and equation (3.3) is the relationship of contents between two components in gas and liquid phases. The distribution of component A in two phases depends on β_A. There are three cases:

1. When $\beta_A > 1$, then $\rho_A/\rho_B > a^*/b$, during distillation A is concentrated in gas phase and B in liquid phase;

2. When $\beta_A < 1$, then $\rho_A/\rho_B < a^*/b$, B is concentrated in gas phase and A in liquid phase. It is opposite to above;

In cases 1 and 2, A and B can be concentrated in gas phase or liquid phase respectively.

3. When $\beta_A = 1$, then $\rho_A/\rho_B = a^*/b$, it means the contents of A and B in gas phase is equal to that in liquid phase, so that A and B can not be separated by distillation. Thus, β_A is a key value for evaluating possibility of separating A and B by distillation, it is called "separation coefficient".

As show in equation (3.3), β_A represents the difference in contents of A and B between two phases by times. The farer is the from 1, the larger is the value of difference and the effect of separation of A and B by distillation is better.

Evaluation of alloys. Taking Pb-Sn alloy as an example the diagram of β_{Pb} - N_{Pb} at temperatures of 1000° and 1100° is shown in Fig 1[5].

β_{Pb} Changes within 10^3 - 10^4 with the change of alloy compositions and temperatures.

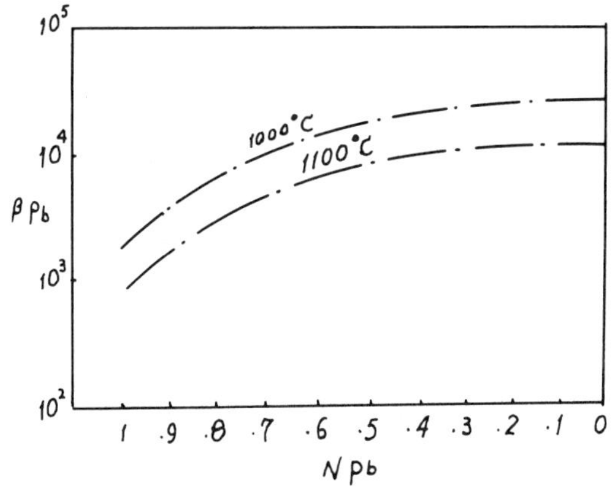

Fig. 1 The β_{Pb} value of Pb-Sn system in gas and liquid phases at balance, at temperatures of 1000^0 and 1100^0C.

As shown in Fig. 1, the ratio of contents of lead and tin in gas and liquid phases is up to thousands to ten thousands. It is determined that lead and tin can be concentrated and purified in gas phase and liquid phase respectivly by vacuum dietillation. The effect of separation for one time distillation is better.

Evaluation of crude metals. Suppose element B is base metal of a crude metal and element A is an impurity. Indeed it is a dilute solution formed by a large amount of B and a small amount of A. Here

$$r_B = 1$$

$$r_A^o = \text{consant}$$

then equation (5.1) is

(5.2)

it is suitable for impurity A in crude metal.

For example of crude tin, by application of data from reference 1, γ_i° of binary system sn-i including tin and various impurities and then β_i° is calculated. They are listed in Table 2

Table 2. β_i° value of sn-i system containing small amount of i at temperature of 1000°C

i	Cu	Ag	In	Bi	Sb	Pb	As
Pi/Psn	$9.3*10^{-2}$	$6.2*10$	$2.3*10^{2}$	$1.4*10^{4}$	$1.42*10^{5}$	$3.3*10^{4}$	1.29*1
γ_i°	0.317	0.187	1.241	1.356	0.411	2.195	
β_i°	$2.948*10^{-2}$	$1.16*10$	$3.85*10^{2}$	$1.898*10^{4}$	$3.84*10^{4}$	$7.23*10^{4}$	

The data fome Table 2 indicated that during vacuum distillation for crude tin, copper remains in residual solution due to $\beta_{cu}^\circ < 1$. Ag, In, Bi, Sb and Pb are impurities wich $\beta_i^\circ > 1$, they can be volatilized and separated from tin. Then according to the value of β_i° of elements from small to large in sequence, the correspondant volatilities of these elements direct form difficult to easy.

RELATIONSHIP BETWEEN DISTILLIZED MATERIALS AND COMPOSITIONS OF ALLOYS

Relationship between contents of constituent in gas and liquid phases is very important for vacuum distillation, it is a necessary datum in distillation process and can be used as an indicator for separation efficiency.

The content of A in gas is

$$100\,\rho_A / (\rho_A + \rho_B) = A\%_{gas}$$

namely

$$100 / [1 + \rho_B / \rho_A] = A\%_{gas} \quad (6)$$

As stated above, when ρ_A / ρ_B is obtained, then at different a^* and b, $A\%_{gas}$ is obtained from equation (6), a diagram of relationship between a and A%gas can be figured, namely a diagram of A%liquid-A%gas. From study of Pb-Sn alloy system by vacuum distillation a equilibrium diagram of relationship between contents of constituent in gas and liquid phases at temperature of 1000°C and 1100°C is produces (expressed by Sn%). As shown in Fig. 2, the diagram can directly show the quantitive relation in tin contents of great difference between gas and liquid phases. If raw material (liquid phase) is only a common tin solder containing about 62%Pb, when distillation lead is preferentially volatized. Tin contained in volatilized lead is only about 0.01%. When at higher distillation temperature of 1100 C tin content is higher about 0.02%. When tin content in liquid phase is 90%, the tin content in volatilized lead is up to 0.1-0.2%.

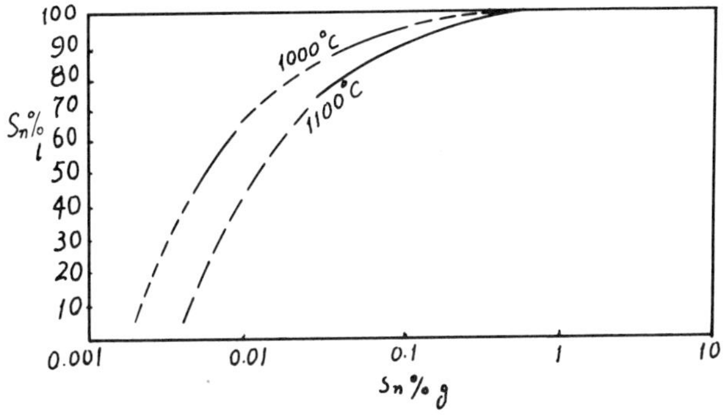

Fig. 2. Equilibrium Relationship between contents of constituent of Pb-Sn system in gas and liquid phases

Equation (5.2) and (6) can also be used for equilibrium composition of various impurities in crude tin between gas and liquid phases. It is shown in Fig. 3.

There are curves of distribution with quantitive values of Pb.Bi. In and Ag when crude tin is processed by vacuum distillation shown in Fig. 3.

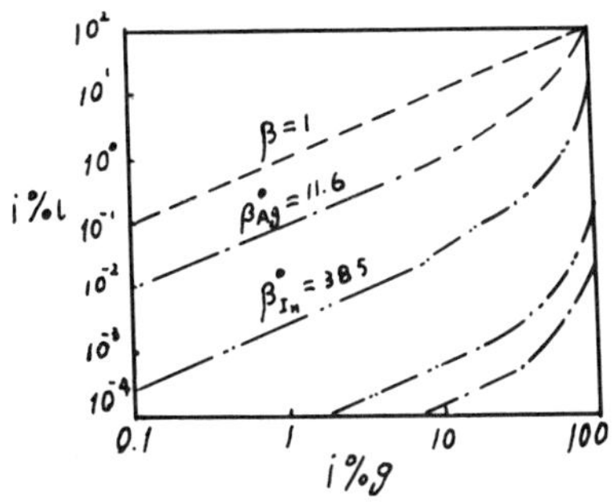

Fig. 3. Equilibrium composition of Sn-i system in gas and liquid phases at temperature of 1000°C

Undoubtly, this diagram has a great important sense in vacuum distillation for crude metals and alloys, when studying crude metals and alloys of Pb-Ag, Pb-Sn and so on by vacuum distillation, it plays a directive role.

THE RELATIONSHIP BETWEEN VOLATITIZING RATES OF VARIOUS IMPURITIES IN CRUDE METAL

A crude metal with B as basic element contains various impurities of 1, 2, 3...... and each element has its own separation coeffecient β_1° , β_2° ...... and activity coefficient γ_1° , γ_2° ... . Distillation coefficients α_1 , α_2 ...... are introduced through application of Langmuir equation by M. Olette, and the relationship between volatilization rates of B and impurity i represented by x and y is shows as follows

$$Yi = 100 - 100 \ (1-x/100) \tag{7.1}$$

$$\alpha_1 = \gamma_1^\circ \ P_1^\circ \ \sqrt{M_B} \ / \ (\gamma_B^\circ \ P_B^\circ \ \sqrt{M_1} \ ) \tag{8.1}$$

since suppose α_1 is a constant at integral to establish equation (7.1), so, equation (7.1) can only be used in this condition. Impurities in crude metals belong to this classification. For crude tin, there is

$$\alpha_i = \gamma_i^\circ \ P_i^\circ \ \sqrt{M_{Sn}} \ / \ P_{Sn}^\circ \ \sqrt{M_i} \tag{8.2}$$

after considering equation (5.2)

$$\alpha_i = \beta_i^\circ \ \sqrt{M_{Sn}} / \sqrt{M_i} \tag{8.3}$$

x gained from equation (7.1)

$$x = 100 \ [1 - \sqrt[\alpha_1]{1 - Yi/100} \ ] \tag{7.2}$$

for various impurities Pb, Bi...... there is

$$x = 100 \ [1 - \sqrt[\alpha_{pb}]{1 - Y \ /100}]$$

$$= 100 \left[1 - \sqrt[\alpha_{Bi}]{1 - Y /100} \right]$$

$$= \ldots\ldots$$

then obtain

$$1-x/100 = (1 - Y /100)^{1/\alpha_{pb}} = (1 - Y /100)^{1/\alpha_{Bi}} = \ldots = c \quad (7.3)$$

for one imputity can obtain

$$\log (1 - Yi/100) = \alpha_i \log c \qquad (7.4)$$

Equation (7.3) shows the relationship between volatilization rates of various impurites in crude metals and α_i and Yi are important values in equation (7.3).. volatilization rates of various impurities are obtained through testwork, then α_i and γ_i° can be calculated.

The content of impurity is obtained when crude tin is treated by test of vacuum distillation at 1200°C, 1.33 pa, and 2-15 minutes, let γ_{pb} = 1.2, by use of equation above α_{Bi} and its average value $\bar{\alpha}_{Bi}$ of impurity Bi are calculated and shown in Table 3.

Table 3. α_{Bi} and $\bar{\alpha}_{Bi}$ abtained from volatilization rates of lead and bismuth in crude tin

volatilization time(min)	2	3	5	7	10	15
Y_{pb}	20.1	76.55	93.16	78.4	99.77	99.95
Y_{Bi}	61.21	70.91	91.92	96.77	99.84	99.84
α_{Bi}	2.036*10^3	2.24*10^3	2.43*10^3	2.158*10^3	2.75*10^3	2.202*10^3
$\bar{\alpha}_{Bi}$			2.303*10^3			

put $\bar{\alpha}_{Bi}$ into equation (7.4), the calculated volatilization rate of bismuth Y_{Bi} is very close to the rate obtained from experiment

Table 4 $Y_{Bi,exp}$ and $Y_{Bi,cal}$

valatilgation time(min)	2	3	5	7	10	15
$Y_{Bi,exp.}$	61.21	70.91	91.92	96.77	99.84	99.84
$Y_{Bi,cal}$	65.72	72.31	90.66	97.42	99.53	99.88

After $\bar{\alpha}_i$ of one impurity in crude tin is gained, then β_i° and γ_i° can be produced, as shown in Table 5.

Table 5 γ_i°, β_i° and $\bar{\alpha}_i$ of impurity i in crude tin (1200°C)

i	Pb	Bi	As	Sb
p_i° / p_{Sn}°	$2.85*10^3$	$3.894*10^4$	$8536*10^7$	$1.778*10^4$
$(M_{Sn}/M_i)^{1/2}$	0.757	0.7536	1.028	0.57
$\bar{\alpha}_i$	$2.60*10^3$	$2.303*10^3$	$1.295*10^3$	$2.501*10^2$
γ_i°	1.2	$7.8*10^{-2}$	$1.475*10^{-5}$	$2.468*10^{-2}$
β_i°	$3.419*10^3$	$3.05*10^3$	$1.295*10^3$	$4.38*10^2$

Note: according to 5, let M_{As} = 1.5 * 74.82 is substance in gas state, bismuth is a single atom, antimony is a molecule of three atoms.

The volatilization rates of impuritues including arsenic and antimony in crude tin calculated form data listed is Table 5 is very close to that obtained form experiments.

It is shown in Table 6.

Table 6 $\gamma_{As,exp.}$, $\gamma_{Sb,exp.}$ and $\gamma_{As,cal}$. $\gamma_{Sb,cal}$

Distillation time (min)	2	3	5	7	10	15
$\gamma_{As,exp.}$	47.2	55.49	71.69	80.25	85.06	98.35
$\gamma_{As,cal.}$	45.24	51.43	73.71	87.25	95.14	97.73
$\gamma_{Sb,exp.}$	9.49	8.57(?)	22.96	28.89	46.66	58.52
$\gamma_{Sb,cal.}$	10.08	13.02	22.73	32.85	44.24	51.86

Note: (?) the data from literature 3 may be inaccurate Among the activity coefficients of arsenic, antimony and bismuth obtained from table 5, γ_{As}° has not yet been studied, while γ_{Sb}° and γ_{Bi}° are very different from the data obtained literature 1. It is shown in Table 7

Table 7 the comparison of γ_i° for As, Sb and Bi in crude tin

i	Sb	As	Bi
γ_i° 1200 C (*)	$2.468*10^{-2}$	$1.475*10^{-5}$	$7.8*10^{-2}$
γ_i°	$4.11*10^{-1}$(677 C)		1.356(326 C)

* - present paper

The reasons for causing this difference are: the first is the data listed in reference 1 belong to pure binary system with no influence of other impurities, but the calculated data in present paper are produced by use of practical data as calculating basis, the system for experiment contains several impurities which interreact one another, so later data are more practical, the second is the influence in determination method and its accuracy and so on in testwork.

According to calculated values β_i° in present paper and arranging them from large to small, the sequence of corresponding volatilization of various impurities is as follows.

Pb, Bi, As, Sb, In, Ag

According to the boiling point of pure substances only, the sequence is

As, Bi, Sb, Pb, In, Ag

There is a great difference between two sequences and the former is closer to practice.

CONCLUSION

1. Evaluation of a meterial containing metals can whether or not be separated or purified by vacuum distillation has been investigated. For impurity i in crude metal B, equation $\beta_i^\circ = \gamma_i^\circ \cdot P_i^\circ / P_B^\circ$ is used. For alloy A-B, it is suitable for using equation $\beta_A = \gamma_A P_A^\circ / (\gamma_B P_B^\circ)$. Only if β_i° or β_A is larger than 1, i or A can be volatilized and concentrated in gas phase, if β_i° or β_A is less then 1, i or A remains in distilled residual alloy, if β_i° or β_A is equal to 1, it means i and A can not be separated by distillation.

This method for evaluation is more satisfactory than that by use of Pi° or Pi.

2. The relationship between volatilization rates of various impurities in crude metals is expressed by equation $\log(1-Yi/100) = \alpha_i \log c$. It plays a great important role in practice. There is a difference between calculated results by data obtained from pure binary system and crude metal in distillation practice. The reason is the former containing no associated impurities inter-reacting one another. It is more accurate by use of corrected γ_i° and β_i° in the later.

3. γ_i°, β_i°, $\bar{\alpha}_i$ and the equilibrium composition diagram in gas

and liquid phases obtained in present paper have great importance in practice. It is a supplement to basic data and can be used to predict quality of product, structure of equipment and scheme of distillation.

REFERENCES

1. A.N. Nesmeyanov, Vapor Pressure of the Chemical Elements, (1963)
2. R.Hultgren et al. Selected Values of the Thermodymic Properties of Binary Alloys. (1973)
3. S.C.Pearce Lead-Zinc-Tin (1980). 754-768
4. M.Olette, physical Chemistry of Process Metallurgy, (1961), pary 2. 1065; Proceedings 4th International Conference on vacuum Metallurgy Section 1, (New York. London. Interscience publishers 1973). 29
5. Dai Yongnian. He Aiping. "Vacuum Distillation at Lead-Tin Alloy". (1987)

1988. 3.

THE NEW CONCEPT IN ZINC ELECTROWINNING OPERATION

K.Kaneko
General Manager
H.Ohba
Superintendent of Tank House
T.Kimura
Assistant Surerintendent of Maintenance Section
Akita Zinc Refinery, Mitsubishi Metal Corporation
Akita, Japan

Abstract

Much effort has been made to reduce the energy requirement in zinc electrowinning. In order to keep energy consumption low, either to make cell voltage low or current efficiency high is necessary. Decrease of cell voltage practically depends on how to reduce ohmic resistance in the electrolyte and the anode crust. To realize these, new concept was proposed and the pilot plant has been operated. The space between a cathode and an anode is halved by utilizing the frame insulator. With these frames the anodes and cathodes compose of a unit of electrodes, and this unit is handled by newly designed electrode handling machine (E.H.M.). The process includes lifting up electrodes, expanding the space between a cathode and an anode, cleaning them, stripping the deposited zinc and returning the unit to the cell. As a result, cell voltage is kept as low as calculated and predictable energy reduction has been achieved.

Metallurgical Processes for the Year 2000 and Beyond
Edited by H.Y. Sohn and E.S. Geskin
The Minerals, Metals & Materials Society, 1988

Introduction

In the recent trend of a rapid appreciation of the Japanese yen, Akita Zinc Refinery of Mitsubishi Metal Corporation has been confronted by problems of relatively high operating costs. In order to compete with other overseas zinc refineries, it is necessary to reduce the cost of zinc production, especially the cost of electric power which now accounts for up to 35 % of the entire cost of production. Three fourth of the electric power used at Akita Zinc Refinery is for the electrolysis. Therefore it is extremely important to reduce the cost of electrolysis.

There are two possible measures for solving the problem; one is to use more nighttime electricity taking the advantage of its inexpensive power price, and the other is to reduce electric power consumption.

As a part of a rationalization plan carried out in 1986, Akita Zinc Refinery increased the number of operating cells. As a result, when nighttime electricity is used, the silicone rectifier has been able to operate at full capacity. Moreover, by new schemes of "increase of area for electrodeposition by raising the liquid level of electrolyte" and "increase of the number of electrodes in each cell", current density could be lowered and as a result, the electric power consumption has been improved. In order to raise nighttime ratio even higher, installation of an additional rectifier and cells would be required.

On the other hand, reducing cell voltage or improving current efficiency may contribute to reducing electric power consumption. Since current efficiency has already reached nearly 90 %, a big leap in its improvement cannot be expected. Contrarily, the current cell voltage is about 3.4 V (at 500 A/m^2) compared with theoretical value of 2.28 V. Therefore, there is a possibility of voltage reduction by "reducing the distance between electrodes" and "reducing ohmic resistance of anode crust".

Under these circumstances, the following plans was set forth; transferring the silicone rectifier which was installed at Hosokura Zinc Refinery to Akita Zinc Refinery, and a production increase by increasing the number of cells.

A fundamental plan to adopt a new electrowinning method for zinc has been devised using the newly installed cell, which is completely different from the conventional concept.[1] According to the plan, by means of reducing the distance between electrodes and frequently removing anode crust, cell voltage will be reduced. In that way, a major improvement in electric power consumption can be achieved. After a year and a half of field testings, satisfactory results has been obtained. A report on the details of the tests and results for the new electrowinning method for zinc are as follows.

Electric Power Consumption

The two major factors that determine electric power consumption are cell voltage and current efficiency, which are formularized as follows ;[2,3]

$$W = E / (1.22 \times \eta_1 \times \eta_2) \times 10^7 \text{ -----------(1)}$$

W: electric power consumption (AC kWH/t - Cathode Zinc)
E: cell voltage (V)
η_1: current efficiency (%)
η_2: current converting efficiency of silicone rectifier (%)
1.22: electrochemical equivalent for zinc per 1 AH (g•Zn)

As it is clear from the formula, by minimizing E and maximizing η_1 the value of W can be reduced.

The value of cell voltage E can be obtained from the following formula;

$$E = 2.67 + \beta i \text{ --------------(2)}$$

2.67: apparent decomposition voltage (V)
β: total coefficient for voltage drop (Ω cm^2)

$$\beta = \beta_1 + \beta_2 + \beta_3$$
$$\beta_1 = \ell / k$$

ℓ: inter electrode distance (face to face) (cm)
k: specific electric conductivity of electrolyte ($\Omega^{-1}cm^{-1}$)
β_2: coefficient for anode crust
β_3: coefficient for other cause (bus bar switch)
i: current density (A/cm^2)

Specific electric conductivity k of electrolyte at Akita Zinc Refinery can be formularized in the following empirical formula.

$$k = 0.376 + 0.0064(t - 35) - 0.00242(Zn - 55) - 0.0065(Mg - 8) + 0.00142(H_2SO_4 - 140) \text{ --------(3)}$$

t: temperature of electrolyte (^{0}C)
Zn: Zn concentration in electrolyte (g/1)
Mg: Mg concentration in electrolyte (g/1)
H_2SO_4: sulfuric acid concentration in electrolyte (g/1)

Current efficiency at Akita Zinc Refinery can be formularized in the following empirical formula.

$$\eta_1 = 76.67 + 921 i - 13{,}700 i^2 \text{ -----(4)}$$

Table-1 shows normal operational conditions of the zinc electrowinning process at Akita Zinc Refinery.

Composition of Electrolyte in the Cell		
Zn :	60 (g/l)	
H_2SO_4 :	180 (g/l)	
Mg :	8 (g/l)	
Temperature	39 (^{0}C)	
Max. Operation Current Density		
i :	0.05 (A/cm^2)	
Inter Electrode Distance (face to face)		
ℓ :	3.05 (cm)	
Total Cell Voltage		
E :	3.42 (V)	(at 0.05 A/cm^2)
Current Converting Efficiency of Silicone Rectifier		
η_2 :	96.0 (%)	

Table-1 Normal Operational Conditions of the Zinc Electrowinning Process at Akita Zinc Refinery

In order to obtain the total coefficient for voltage drop β under the above conditions, the following formulas can be obtained.

From the formula (3); $k = 0.4463$ (Ω^{-1}cm^{-1})

and by using $\ell = 3.05$; $\beta_1 = \ell / k = 6.83$ (Ω cm^2)

Also by substituting the values of E = 3.42 (V) and i = 0.05 (A/cm^2) in the formula (2); $\beta = 15.0$ (Ω cm^2)

can be obtained.

Based on current operation at Akita Zinc Refinery, the anodes are usually cleaned in 26 day cycles during which period the coefficient increase has been measured to increase 0.1 (Ω cm^2) per day. Therefore the mean coefficient during the period of 26 days is 0.1 × 26 / 2 . Moreover, although the anodes are cleaned mechanically every 26 days, a small amount of residue still remains after cleaning. Compared to completely clean anodes without any crusts, the coefficient of the mechanically cleaned anode is higher by 0.6 (Ω cm^2). So, coefficient β_2 caused by anode crust can be calculated as follows;

$$\beta_2 = 0.6\ (\Omega\ cm^2) + 0.1\ (\Omega\ cm^2/day) \times (26/2)\ (day) = 1.9\ (\Omega\ cm^2)$$

As a result, β_3 can be determined as 6.27 (Ω cm^2).

$$\beta_1 = 6.83\ (\Omega\ cm^2) \qquad \beta_2 = 1.9\ (\Omega\ cm^2) \qquad \beta_3 = 6.27\ (\Omega\ cm^2)$$

$$\beta = \beta_1 + \beta_2 + \beta_3 = 15.0\ (\Omega\ cm^2)$$

Electric power consumption W obtained from the formula (1) is 3,301 kWH/t-cathode which is the electric power consumption of Akita Zinc Refinery at maximum current density.

The New Concept of Electrowinning of Zinc

Reducing the distance between electrodes is one of the useful measures to reduce cell voltage in an attempt to reduce electric power consumption.

However, there has been electric short-circuit problems when trying to reduce the distance considerably between electrodes due to the lack of proper devices to keep the correct distance. As a result, a considerable reduction of electric power consumption has not been realized. In order to keep the correct distance, an edge insulator, called "the frame", was attached to the anode. A sketch of the frame is shown in Fig-1. By using the frame, the distance between a cathode and an anode could be kept exactly. Fig-2 shows a cathode and an anode with a frame. Fig-3 shows their state when being submerged in a cell.

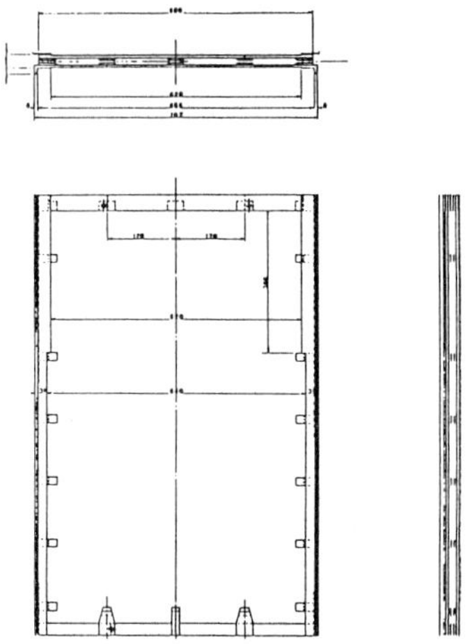

Figure-1 A Sketch of the Frame

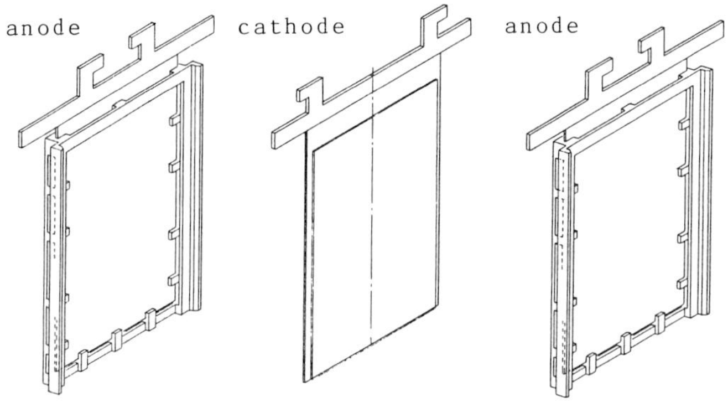

Figure-2 An Electrode Installed with Frame

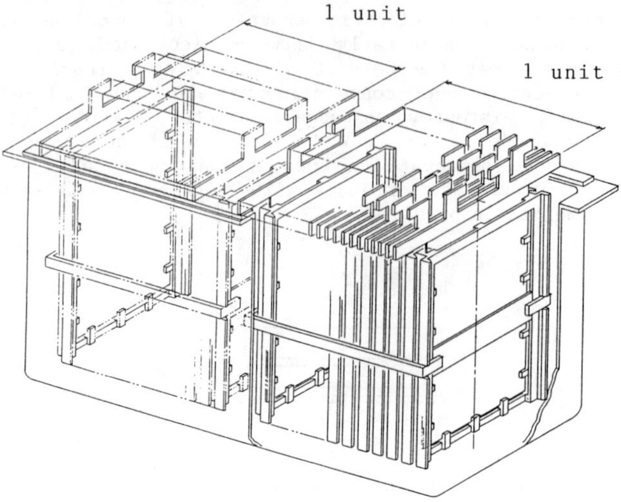

Figure-3 An Electrode Unit Being Submerged in a Cell

Another effective method of reducing the electric power consumption is reducing resistance caused by anode crust. In normal electrolysis for zinc, cathodes and anodes are treated separately.

As shown in Fig-3, an entire electrode unit is pulled out to be treated; at the time of stripping zinc from cathode, the anode is also cleaned.

The positive effects of introducing these two ideas can be calculated as follows.

In the formula for β_1 the distance between electrodes ℓ should be as small as possible. However, there are certain limitations in the hardware, mainly the thickness of the head bar and the hook for handling electrode. As a consequence value of ℓ = 1.45 cm was set up in the experimental stage. Therefore the following result can be obtained.

$$\beta_1 = \ell / k = 3.25 \ (\Omega \ cm^2)$$

When cathodes and anodes are pulled out of cells every 2 days, the anode is cleaned to remove crusts. As a result, the following value can be obtained;

$$\beta_2 = 0 + 0.1 \times (2 / 2) = 0.1 \ (\Omega \ cm^2)$$

Assuming that the value for β_3 is fixed at 6.27 (Ω cm^2), the following figures can be obtained.

$$\beta_1 = 3.25 \ (\Omega \ cm^2)$$
$$\beta_2 = 0.1 \ (\Omega \ cm^2) \qquad \beta = \beta_1 + \beta_2 + \beta_3 = 9.62 \ (\Omega \ cm^2)$$
$$\beta_3 = 6.27 \ (\Omega \ cm^2)$$

Electric power consumption W at the maximum current density of i = 0.05 (A/cm^2) is expected to be 3,041 kWH/t-cathode, a figure which is smaller than the amount needed for normal electrolysis by 260 kWH/t-cathode.

The following problems, in the above mentioned newly developed operation of electrolysis of zinc, should be taken into consideration; 1. loss of current efficiency due to increased incidences of electric short-circuits, 2. higher Pb contamination due to increase of electric short-circuits, and 3. loss of current efficiency due to reduction of circulation of the electrolyte. In order to identify these problems, the following tests was carried out.

Tests and Results on the New Electrowinning Method

1: Tests and Results Using an Experimental Cell

During the period between June and December in 1987, electrolysis tests on the new method had been carried out using an existing cell which was remodeled for the tests. Table-2 shows the electrolytic conditions. Fig-4 shows the results of the tests.

Electrolytic Condition	Zn H_2SO_4 Mg Temp. Cell Overflow	55 - 62 g/l 170 - 180 g/l 7 - 8 g/l 38 - 40 ^{0}C 36 - 40 l/min
Cell Size Material Electrodeposited Area Current Density Electrodeposited Time	1355L × 736W × 1380H mm PVC with steel frame 1.29 m^2 484 A/m^2 Max. 48 Hr	
Distance between center of each cathode Inter electrode distance (face to face) Total coefficient for voltage drop β	41 mm 14.5 mm 9.6 Ω cm^2	
Cathode Stripping Cleaning Polishing Anode Maintenance	48 Hr cycle submerged in a hot bath 48 Hr cycle Wet Method 48 Hr cycle using high pressure water	

Table-2 Electrolytic Conditions of Testing at an Experimental Cell

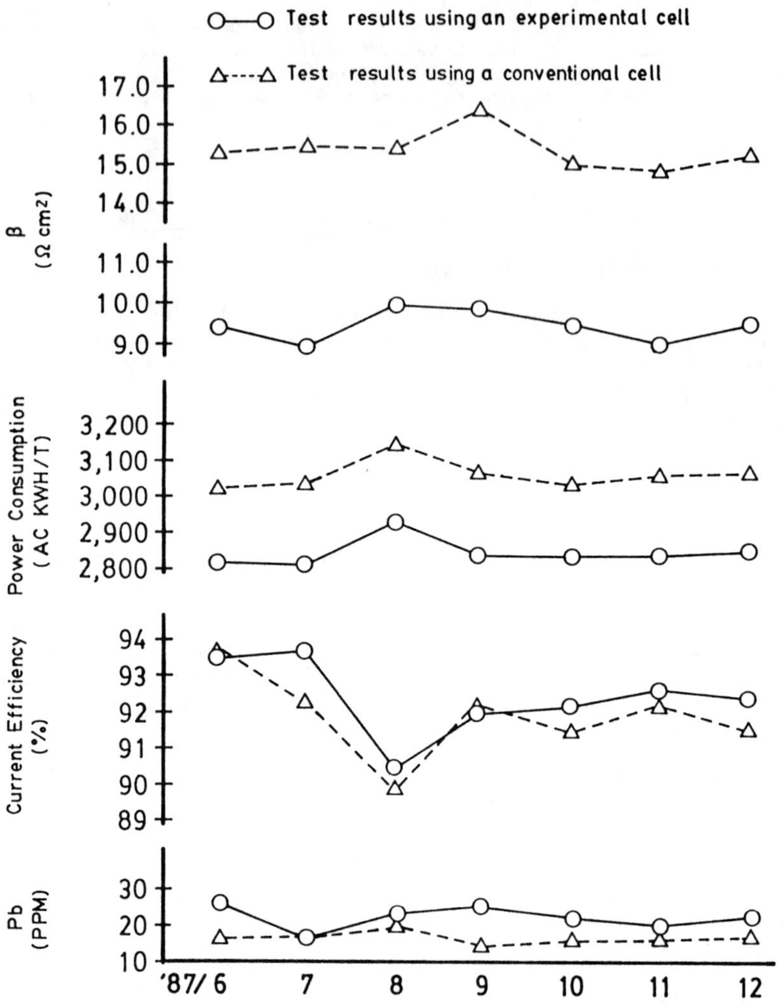

Figure-4 Test Results Using an Experimental Cell

The following are the test results.

(1) Electric power consumption : calculated figure 2,871 kWH/t
actual test result 2,843 kWH/t

(2) Current efficiency : no difference is observed.

(3) Pb contents in cathode : slightly high, although it is well under the JIS limit of 0.003 %

Based on the above, the results of the new process were considered sufficient to apply for full scale operation.

2: Tests and Results in the Pilot Plant

From December 1987 to May 1988, pilot plant tests had been carried out using the cells with the same layout as newly designed tank house. Table-3 shows electrolysis conditions, and Fig-5 shows test results.

Electrolytic Condition	Zn H_2SO_4 Mg Temp. Cell Overflow	53 - 73 g/1 155 - 220 g/1 7 - 8 g/1 38 - 40 ^{0}C 60 - 70 1/min
Cell Size Material Electrodeposited Area Current Density Electrodeposited Time	1465L × 760W × 1500H mm Pre-cast concrete Lining with semi-hard PVC 1.29 m^2 181 - 491 A/m^2 48 Hr	
Number of Electrode (in one cell)	Cathode 30 plates (15 plates/unit×2) Anode 32 plates (16 plates/unit×2)	
Head Bar: Cathode Hook Anode Hook	869L × 63H × 14t mm A 240 Type 316L: 120H × 4t mm 869L × 63H × 14t mm A 240 Type 316L: 121H × 4t mm	
Distance between center of each cathode Inter electrode distance (face to face) Total coefficient for voltage drop β	41 mm 14.5 mm 9.6 Ω cm^2	
Cathode Stripping Cleaning Polishing Anode Maintenance	48 Hr cycle using high pressure water 48 Hr cycle Wet Method 48 Hr cycle using high pressure water	

Table-3 Electrolytic Conditions of Testing in the Pilot Plant

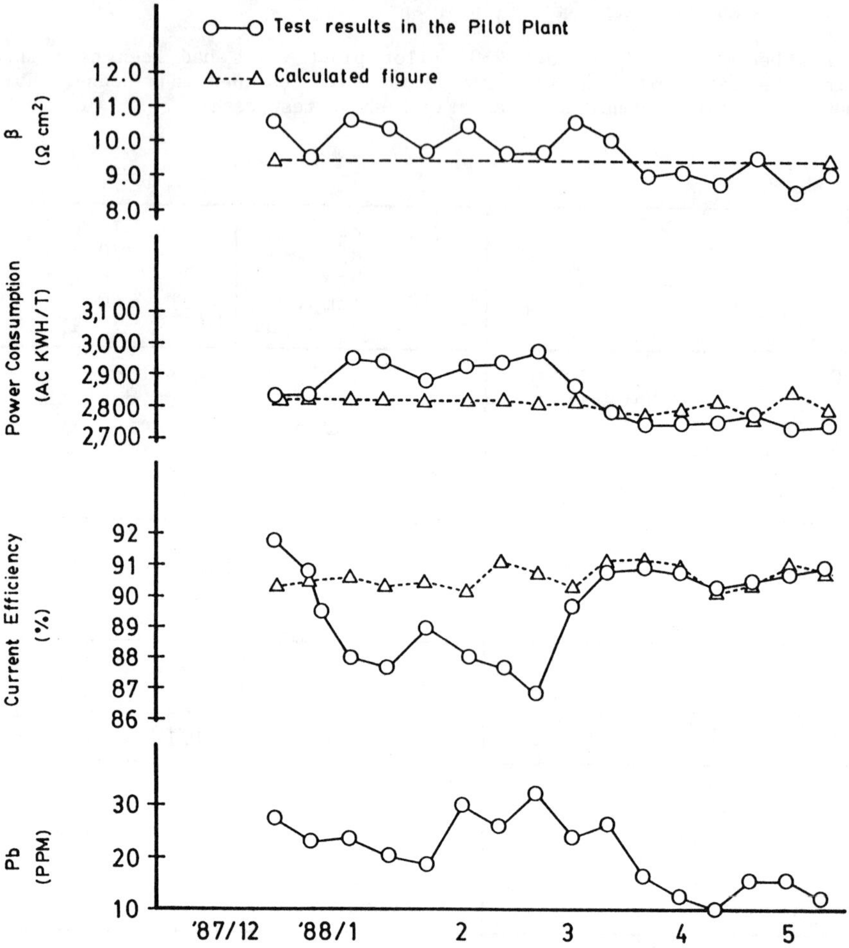

Figure-5 Test Results in the Pilot Plant

Results of the tests carried out during the period between December 10th and 31st were almost the same as the calculated values, but results during the period between January 1st and March 3rd did not show expected figures. During that period, electric power consumption and current efficiency were considerably worse.

Various factors have been analyzed accordingly. As a result, it has been concluded that the cause was the fact that electricity did not flow sufficiently and evenly due to increased contact resistance between the bus bar and electrodes. Table-4 shows the phenomena, causes and countermeasures taken.

	Phenomena	Cause	Countermeasure
(1)	Bus bar got deformed.	Since it was structured so that the thin bus bar of 8 mm thickness was supported at three points, the weight of electrodes caused a deformation of the bus bar.	A liner was placed to cover the entire area underneath the bus bar and reduced the deformation to a minimum.
(2)	The switch section of electrodes did not touch the bus bar.	The clamp needed for the electrode unit was installed so tight that it prevented the electrodes from making contact with the bus bar with their own weight.	The width of the clamp was widened slightly so that the unit could touch the bus bar with its own weight.
(3)	Electro-depositing of zinc on cathodes was insuffi-cient.	Cathode head bar was hand made for temporary use so that the switch section was not good enough for electricity to fully flow through cathodes.	Cathode head bar was replaced with one of the usually used type with a switch section that had the proper specifications.
(4)	Electricity failed to flow fully despite the proper contact conditions.	Bus bar was soiled so that contact resistance increased.	Bus bar was cleaned thoroughly and as a result a sufficient flow of electricity was observed.

Table-4 Phenomena, Causes and Countermeasures When Electric Power Consumption and Current Efficiency Were Worse

By taking the above countermeasures, electric power consumption and current efficiency were improved. As a result, it was confirmed that the new method was effective.

Concept on Fully Automatic Treatment of Electrodes

The silicone rectifiers, which were previously installed at Hosokura Zinc Refinery, are planned to be transferred to Akita Zinc Refinery for the new electrowinning method. Each Galvanic capacity of the rectifier is 9,000 A, and parallel connecting two rectifiers make the maximum two-fold current.

$$9{,}000\ (A) \times 2 = 18{,}000\ (A)$$

In order to maintain current density under 500 A/m^2, even during maximum current, 30 cathodes per cell was determined.

Then, 15 cathodes and 16 anodes are combined to form a unit. Two units are put into each cell. Since the conventional method of treating electrodes requires considerable manpower, a conveyance device called "Electrode Handling Machine" will be used to do the entire sequences of treatment automatically; lifting a unit from a cell, treatment of electrodes and putting the unit into the cell. Fig-6 shows a flowsheet of treatment of unit in the new electrowinning method using E.H.M..

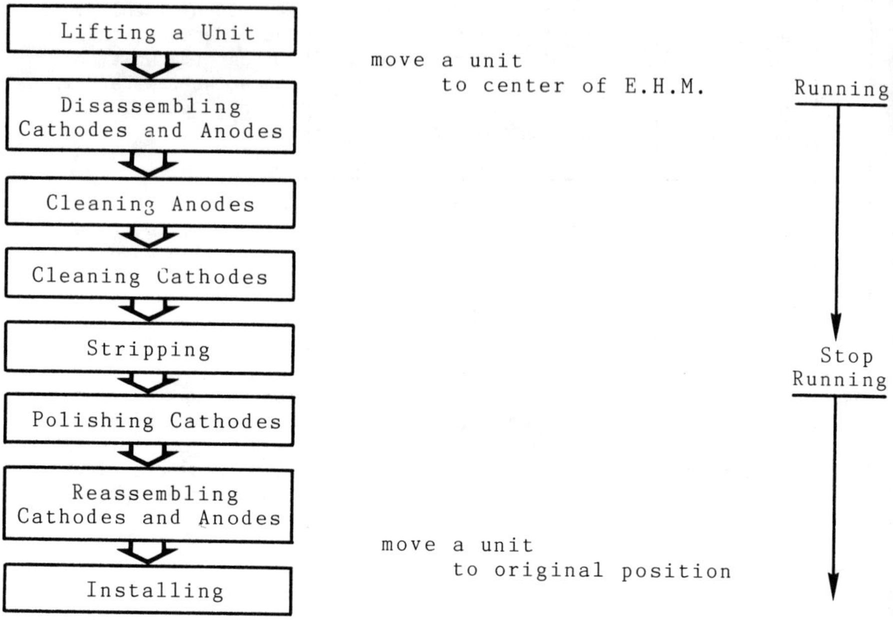

Figure-6 Flowsheet of Treatment of the Unit Using E.H.M.

The following is an outline of the operational pattern of E.H.M..

(1) When E.H.M. is in a programmed position over a cell, it closes the hooks on 15 cathodes and 16 anodes.
(2) Then it pulls out 31 pieces of anode and cathode as a unit.
(3) This unit moves to the center of E.H.M..
(4) After completion of shifting of the unit, E.H.M. travels toward the electrodes treatment device. E.H.M. travels all through the process from (5) to (8).

(5) While traveling, the hand-arm of E.H.M. grabs the clamp bar for fixing electrodes.
(6) The compressive arms of E.H.M. come down to compress the electrodes.
(7) The hand-arm removes the clamp bars and the compressive arms ascend.
(8) The electrodes unit is disassembled.
(9) The electrodes are opened to 320 mm pitch so that anodes can be cleaned as E.H.M. goes through an anode cleaning device.
(10) E.H.M. then travels through a cathode cleaning device to wash the cathodes.
(11) After completion of pre-treatment, cathodes are transferred to a stripping machine. After deposited zinc is stripped off from the cathodes (aluminum sheet) automatically, E.H.M. pulls out the cathodes.
(12) Then the cathodes are polished as E.H.M. goes through a cathode polishing machine.
(13) When all of the electrode treatment is completed, cathodes and anodes are reassembled back into a unit in the center of E.H.M..
(14) Then the hand-arms comes down again rearrange the electrodes back in place using a correction device for the anode-cathode correction rod.
(15) The compressing arms come down again to compress the distance between electrodes.
(16) The clamp bars is put back in place to fix the electrode unit.
(17) E.H.M. travels back to the programmed position to put the unit back in place.
(18) The unit is moved back down into the original cell and then the hooks are reopened.
(19) E.H.M. moves to another unit of same cell or the next cell according the sequential program.

Various tests have been carried out for many of the above points to identify and solve any of the problems.

Summary

The above is an outline of the new electrowinning method for zinc. The test results of the process have proved to be almost satisfactory.
Moreover, applications for 6 patents have been submitted in Japan in relation with this new method. In addition, 2 kinds of overseas patents, which summarize the above 6 items, have been applied for.

Reference

1. G.Freeman, A.Pyatt, et al., "Comparison of Cellhouse Concepts in Electrolytic Zinc Plants", Lead-Zinc-Tin '80, 1980, pp.222-246

2. N. Masuko, K. Musiake, "Feasibility Study on Energy Saving of Zinc Electrowinning", Zinc '85, 1985, pp.337-348

3. M.Takahashi, N.Masuko, - Chemistry of Industrial Electrolytic Processes (Kogyo-Denkai no Kagaku) AGNE, 1979

COMPUTER-AIDED ANALYSIS AND SIMULTION OF HIGH-TEMPERATURE PROCESSES

I. Barin, G. Eriksson, and F. Sauert

KHD Humboldt Wedag AG
5000 Cologne 91
West Germany

Abstract

Computer-aided calculations using data bases and program systems are more and more needed for the rational development and optimization of high-temperature processes. The calculation of mass and energy fluxes between different process steps as well as the determination of the chemical equilibria will be discussed, giving examples from nonferrous metallurgy. The simulation of a waste-material treatment process including high-temperature combustion and gas-cleaning systems will also be demonstrated.

Metallurgical Processes for the Year 2000 and Beyond
Edited by H.Y. Sohn and E.S. Geskin
The Minerals, Metals & Materials Society, 1988

ON-LINE MEASUREMENTS AT ELEVATED TEMPERATURES IN METALLIC SOLUTIONS, MATTES AND MOLTEN SALTS USING SOLID ELECTROLYTES

Derek J. Fray

Department of Materials Science and Metallurgy
University of Cambridge
Pembroke Street, Cambridge, England CB2 3QZ

Abstract

A need exists for fast and accurate analysis of species in various phases at elevated temperatures. Although modern analytical techniques are very efficient, the equipment is usually situated some distance from the industrial operation and time is spent taking the sample and sending it to the analytical facility. Several minutes can easily lapse and, during this time, heat must be supplied in order to keep the system molten and, if the composition is changing rapidly due to a refining process or reaction with the environment, the information may only be of retrospective use. Perhaps more important reasons for the use of on-line sensors are that instantaneous analysis gives better control, process optimisation and the minimisation of energy losses and pollution. To date oxygen has been the only element which is measured widely using solid electrolytes. However, the analysis of many more elements is required and possible electrolytes for other species are discussed. For some elements, there is no suitable conducting electrolyte and, therefore, a different approach is adopted where the electrolyte, in which one species is mobile, comes to equilibrium with another species.

Metallurgical Processes for the Year 2000 and Beyond
Edited by H.Y. Sohn and E.S. Geskin
The Minerals, Metals & Materials Society, 1988

Introduction

There is an ever increasing demand for the development of sensors for on-line measurements in most areas of technology and, in particular, metal processing. This is especially relevant in situations where greater control can be achieved by monitoring the chemical composition of the melt during processing and where a minor constituent in the melt can have a major effect on the properties of the finished alloy. Although modern analytical techniques are fast and reliable, the equipment due to its complexity, is frequently situated away from the process area and valuable time may be spent in sampling and transportation. During this period, heat may have to be supplied to prevent solidification and the refining reactions may be continuing, and, therefore, the analytical information may not be of direct use.

The theme of this paper is to discuss electrochemical sensors for on-line measurements in metallic solutions which are based upon solid-state galvanic cells as these have been shown to offer the possibility of providing an instantaneous measurement using a sample, rugged, reliable and relatively inexpensive piece of equipment. The overall arrangement is shown in Figure 1.

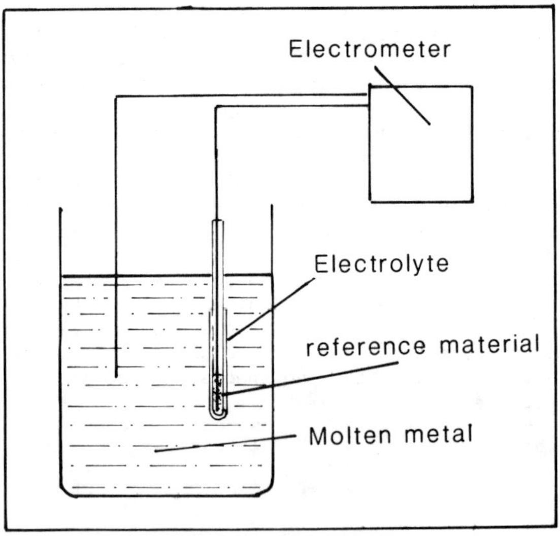

Figure 1 Experimental arrangement for on-line analysis using a sensor based on a solid electrolyte.

There are several modes in which solid electrolytes can be used to monitor the concentration of species in gases, metals or molten salts. The simplest arrangement can be represented by

$$W,\ M_{(ref)}|M^{+}\ \text{Conductor}|M_{(metal)},\ W$$

where W are the leads, $M_{(ref)}$ is the reference material of known activity or concentration, M^{+} conductor is an ionic conductor of M^{+} ions and M is the activity or concentration of M in the metallic solution. The potential across the cell is given by the Nernst equation

$$-ZEF = RT\ln \frac{a_m}{a_{m(ref)}} \qquad (1)$$

where Z is the number of units of charge carried, E is the potential measured by a high impedance voltmeter, F is Faraday's constant, R is the gas constant, T is the temperature, a_m is the activity of M in the metallic solution and $a_{m(ref)}$ is the activity of M in the reference. The activities can be related to the atom fractions through the relationship

$$a_m = \gamma_m X_m \qquad (2)$$

where X_m is the atom fraction and γ_m is the activity coefficient. The key parameters in making a satisfactory measurement are the selection of the electrolyte which must have a reasonably high ionic conductivity, preferably greater than 10^{-5} ohm^{-1} cm^{-1}, good reversibility to M^+ ions at the electrolyte/reference and electrolyte/metal interfaces, low electronic conductivity and the reference, which preferably should be both an electronic and an ionic conductor.

Probably, the best known solid electrolyte is stabilised zirconia ($CaO.ZrO_2$) in which the oxygen ion is mobile and has been used for several decades to monitor the oxygen contents of gases and liquid metals (1,2). However, sensors for other species have been developed and it is perhaps worth considering other electrolytes which might find application, in the future, in systems of metallurgical interest.

Sodium β-alumina, $Na_2O.11Al_2O_3$, is a fast ion conductor of sodium ions and is very easy to prepare from the constituent oxides (3). One of its remarkable properties is that it is possible to ion exchange the sodium for other ions. Originally, it was thought that only monovalent β-aluminas such as Li^+, K^+, Ag^+, Rb^+, Tl^+, NH_4^+, In^+, NO^+, Ga^+, Cu^+ and H_3O^+ could be prepared this way (3). Sensors using Ag^+ (2) and H_3O^+ (3) β-alumina have been reported. However, Dunn and Farrington (6) showed that for β"-alumina, which has a slightly different structure from β-alumina, it is possible to ion exchange divalent ions such as Ca^{2+}, Sr^{2+}, Pb^{2+}, Ba^{2+}, Cd^{2+}, Cu^{2+}, Fe^{2+}, Zn^{2+}, Ni^{2+}, Mn^{2+} and Hg^{2+}. A sensor based on Ca^{2+} β-alumina has been reported (7). More recently, it has been found that it is possible to form trivalent ion β"-aluminas by immersing sodium β"-alumina in the appropriate nitrate or chloride salt at 800-900K. Bi^{3+}, Nd^{3+}, Eu^{3+}, Sm^{3+} and Gd^{3+} can be prepared this way (8) and, although these materials may not have wide metallurgical applications, certain of these compounds have shown promising properties in novel laser phosphor and electro-optic devices.

Whereas β and β"-aluminas have planes in which the ionic species have high mobility, Nasicon electrolytes have tunnels within the structure through which the ions move (9). A typical composition might be $Na_{1+x}Zr_2Si_xP_{3-x}O_{12}$ and like sodium β-alumina it is possible to substitute other ions for the sodium. Lithium (10) and copper (11) compounds have been synthesised for use in sensors and a hydrogen ion conductor has been made by ion exchange for steam electrolysis and fuel cells (12). High temperature proton conducting oxides are also favourable materials for hydrogen extraction and the sensing of hydrogen. One possible material in $SrCeO_3$ which is a p-type conductor in an atmosphere free of hydrogen or water vapour (13). However, when water vapour or hydrogen is introduced into the atmosphere, electronic conductivity decreases and protonic conduction appears.

Low temperature proton conductors have been developed which are based upon hydrates and hydrated acids such as hydrogen uranyl phosphate tetrahydrate (14), tungsten phosphoric acid hydrate or $HMO_3.xH_2O$ where M = Sb, Nb or Ta. Other low temperature polymeric proton conductors are Nafion (per fluorocarbon sulphonic acid) (15) and poly-amps (Poly(2-acrylamido-2-methyl-1-propane sulphonic acid)) (16). As will be shown later in the paper, these materials have applications in hydrogen sensors.

Sensors for Elements whose Ions are not Mobile in Solid Electrolytes

So far this paper has considered sensors in which the species to be measured is mobile in the electrolyte. However, this approach excludes many elements which are not mobile in the electrolyte but by identifying electrolytes which contain the desired species, although this may not be the mobile species, it is possible to measure the non-mobile species. Examples of this approach are the determination of calcium in indium using calcium fluoride where F^- is the mobile ion (17). The mechanism by which the sensor operates is that the F^- ion comes to equilibrium with the calcium in the alloy via the reaction

$$CaF_2 = Ca + F_2 \qquad K = \frac{P_{F_2} a_{Ca}}{a_{CaF_2}} \tag{3}$$

The electrolyte detects the changes in fluorine content which are directly related to the changes in calcium content as the activity of calcium fluoride is constant. In a similar approach sulphur dioxide, carbon dioxide and nitric oxide have been detected by using K_2SO_4, K_2CO_3 and $Ba(NO_3)_2$ as electrolytes (18). Other electrolytes, such as sodium β-alumina and Nasicon have been used with an auxiliary phase of sodium sulphate and sodium carbonate for sulphur dioxide and carbon dioxide sensors (19). In these cases, the sulphur dioxide and carbon dioxide alter the activity of the sodium in the sulphate and carbonate and this is detected by the sodium β-alumina electrolyte.

In order to measure elements like phosphorus, arsenic and antimony in metallic solution it is necessary to take this concept one step further as coating phosphides and arsenides onto the surface of an ionic conductor is not easy and, secondly, hydrolysis of these phases can easily occur. Sodium phosphate and sodium arsenate have been coated onto the surface of sodium β-alumina (20). The equilibrium that is set up on the surface is

$$Na_3PO_4 = 4\underline{O} + \underline{P} + 3\underline{Na} \qquad K = \frac{a_{Na}^3 . a_P . a_O^4}{a_{Na_3PO_4}} \tag{4}$$

Usually, the activity of the phosphate phase is fixed and the activity of the phosphorus is given by the composition of the alloy. The main unknown is the oxygen activity in the alloy and this can either be measured or it can be regarded as constant which is controlled by the metal/metal oxide equilibrium as most metallic baths are exposed to the environment. By altering the phosphorus content, the sodium activity in the phosphate changes and this can be detected by the sodium β-alumina sensor using a ferric oxide-sodium ferrite reference (20).

Measurement of High Concentration of Elements

One of the problems of using a sensor based upon the Nernst expression (equation 1) is that the emf change is the same for 1 ppm to 10 ppm as it is for 10,000 to 100,000 ppm. The devices, therefore, appear to become considerably less sensitive at the higher concentrations. This is a problem which has been faced in the control of lean burn engines, where changes in the air/fuel ratio only have a small effect on the oxygen partial pressure and, via the Nernst expression, this results in only a small change in the emf. In order to increase the sensitivity of the sensor, coulometric devices have been developed which operate by applying a potential across the electrolyte and measuring the current flow (21). The current, which is measured, is directly related to the flux of oxygen ions through the electrolyte and this, in turn, depends upon the rate of arrival of oxygen at the surface of the electrolyte. The current flow is, therefore, directly proportional to the diffusional flux of oxygen and, therefore, linearly dependent on the concentration of oxygen in the gas phase rather than logarithmically in the case of the potentiometric method. This approach could also be used in the case of high concentrations of solutes in metallic solvents, as shown in Figure 2.

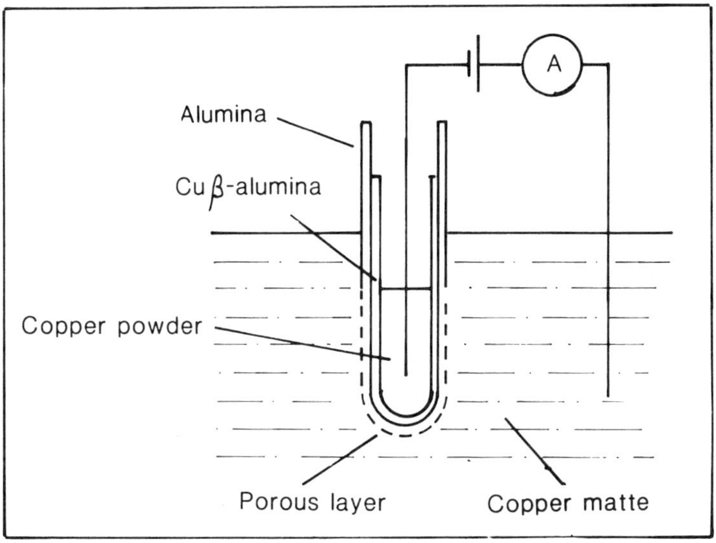

Figure 2 - Arrangement for the determination of copper in copper matte.

Reference Materials

The reference material is required to control the activity or pressure of the mobile species at a precise value on one side of the electrolyte. It, also, needs to come to equilibrium quickly at a given temperature and should be both an ionic and, also, an electronic conductor. For sensors, used in the determination of oxygen, sulphur, hydrogen and phosphorus, it is possible to use a mixture of a metal and its oxide (1,2), sulphide (22), hydride (23) or phosphide (24), respectively. For systems in which the component can be gaseous, i.e. oxygen or hydrogen, it is possible to use the gas or a mixture of gases to fix the equilibrium (2). For more reactive elements, this simple approach is not possible as the pure elements are reactive such as sodium, lithium and calcium and most of the intermetallic compounds are unstable at elevated temperatures and are subject to oxidation. However, it is still possible to establish a reference sodium activity by using a mixture of sodium β-alumina and α-alumina or a mixture of sodium ferrite and iron oxide to fix the sodium oxide activity (25). In order to maintain a constant sodium activity, it is necessary to maintain a constant oxygen partial pressure and this can be achieved by either using a metal/metal oxide mixture or, alternatively, exposing the reference to atmospheric oxygen. Applying this concept in another system, Dubreuil and Pelton (26) have used the equilibria

$$Na_3AlF_6 + 3Na = 6NaF + Al \tag{5}$$

and

$$Li_3AlF_6 + 3Li = 6LiF + Al \tag{6}$$

to give constant sodium and lithium activities.

Having presented the basic techniques of on-line analysis using solid electrolytes, various examples will be given of their application.

Oxygen

Oxygen sensors have been used for two decades for the control of the oxygen content of gases, molten metals, slags and mattes and will not be discussed further except to mention the improvements brought about by magnesia stabilised zirconia (27), which has better thermal shock properties and higher temperature corrosion resistance by oxide melts and fused sulphides (28).

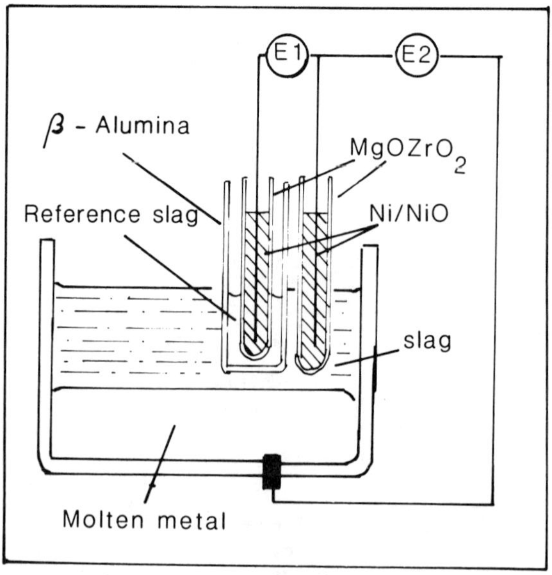

Figure 3 - Basicity Sensor

Yamaguchi and Goto (29) have combined an oxygen conducting electrolyte with sodium β-alumina to measure the activity of sodium oxide in silicate melts using the arrangement shown in Figure 3. The potential E_1 relates to the activity difference between the sodium oxide in the reference slag and the slag. The advantage of this approach is that by combining the sodium β-alumina electrolyte, which is exclusively a sodium ion conductor, the overall device gives the activity of sodium oxide in the slag regardless of the difference in oxygen partial pressure. The potential between the stabilised zirconia and the metal phase (E_2) gives the oxygen pressure at the interface between the slag and the iron. This approach allows the measurement of changes in the soda activity and oxygen pressure during the dephosphorization of pig iron by the addition of sodium carbonate.

$$2\underline{P} + 5Na_2CO_3 = P_2O_5 + 5Na_2O + 5CO \qquad (6)$$

Measurements in Molten Salts

There have been very few measurements of species in molten salts using solid electrolytes but one important system is the electrolyte in the Hall-Heroult cell which is basically alumina dissolved in cryolite Na_3AlF_6. The main parameters are the amount of alumina dissolved in the electrolyte and, also, the NaF/AlF_3 ratio. As the vapour pressure of NaF and AlF_3 are different, the composition of the electrolyte can gradually change with time and this can affect the current efficiency of the cell. In order to carry out measurements in the laboratory, Brisley and Fray (30) measured the sodium activity in molten super purity aluminium which was in equilibrium with various sodium fluoride-aluminium fluoride ratios between 1273 and 1323K. The results showed that the following equilibrium was attained:

$$3NaF_{(\ell)} + Al_{(\ell)} \quad AlF_{3(\ell)} + 3\underline{Na} \qquad K = \frac{a^3_{Na} \cdot a_{AlF}}{a^3_{NaF} \cdot a_{Al}} \qquad (7)$$

and
$$a_{Na} = \frac{K^{1/3} a_{NaF}}{a^{1/3}_{AlF_3}} \qquad (8)$$

It is possible to determine the bath ratio using this approach. The sodium β-alumina electrolyte was either in the form of a tube or a pellet of sodium β-alumina sealed into a silica tube and sheathed with alumina. When the probes were immersed directly into cryolite, the lifetime was only a few seconds, whereas, immersed in the molten aluminium layer in equilibrium with cryolite resulted in lifetimes of more than one hour. Using a lithium-conducting electrolyte (10), this approach could be used to determine the lithium content of the electrolyte bath and, obviously, the lithium and sodium contents of the aluminium pool at the bottom of the Hall-Heroult cell could also be measured.

Direct Measurement of As,Sb and P in Molten Metals

Zinc from the Imperial Smelting Furnace contains several impurities, one of which is arsenic in concentrations of up to 500 ppm. The arsenic is usually removed by reaction with sodium to form a dross which is removed from the surface of the zinc bath. The addition of sodium is not straightforward as it evaporates and oxidises in the environment so that excess sodium is invariably added. However, as soon as the sodium has combined with all the arsenic, the excess reacts with considerable amounts of zinc to increase the amount of dross removed. Accurate control of the sodium addition would eliminate this problem. Broadhurst and Fray (31) used a sodium β-alumina sensor with a reference of sodium ferrite and ferric oxide and found that there was a reproducible relationship between the potential and the sodium content, as is shown in Figure 4.

Rather than determining the sodium content of the metal and correlating with the metalloid content, it would be preferable to measure the metalloid content directly using a solid-state probe. As mentioned earlier, it is possible to perform this measurement by combining the electrolyte with a compound which contains both the mobile species and, also, the species to be measured. To date, most measurements have been performed on the detection of phosphorus in liquid aluminium using a sodium β-alumina probe with a thin layer of sodium phosphate-platinum on the working surface of the electrolyte to facilitate the coupling between phosphorus in the melt and the sodium ion (20).

The phosphorus sensor can be represented by the following galvanic cell:

$$(-)\ \text{S.Steel}\,\Big|\,\begin{matrix}\text{liquid}\\ \text{metal}\end{matrix}\,\Big|\,Na_3PO_4+Pt\,\big|\,\text{Na }\beta\text{-alumina}\,\big|\,\text{reference}\,\big|\,\text{S.Steel}\ (+)$$

using these sensors measurements were carried out on molten Al-Si alloy containing known amounts of phosphorus. The coupling reaction at the working surface can be represented as:

$$3Na_3PO_{4(s)} + 8Al_{(\ell)} \rightleftarrows 9\underline{Na} + 3\underline{P} + 4Al_2O_3 \qquad (9)$$

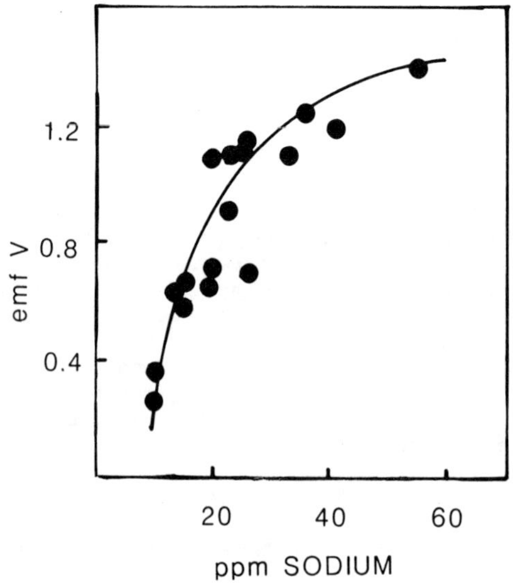

Figure 4 - Relationship between potential and sodium content of zinc.

A plot of emf vs log [ppm P] is shown in Figure 5. Additions of phosphorus were made in the form of P_2O_5 which was reduced by the aluminium. Each addition of P_2O_5 resulted in a change in the emf until a value of -0.260V was achieved when further additions of P_2O_5 gave no additional change in the emf value. This observation is consistent with the attainment of P saturation and the onset of AlP precipitation, as is suggested by the Al-P phase diagram.

Similar measurements were performed in molten tin where the coupling on the electrode interface is provided by the following equilibrium between the dissolved, Na, P and O in molten tin and the solid Na_3PO_4:

$$3\underline{Na} + \underline{P} + 4\underline{O} = Na_3PO_{4(s)} \tag{10}$$

independent measurement with oxygen sensors revealed that the oxygen activity in the melt corresponded to Sn/SnO_2 equilibrium and were unaffected by P additions to the melt. This approach can also be used to determine the phosphorus contents of copper, and the arsenic and antimony contents of non-ferrous metals.

Silicon has been detected in cast iron by Romero et al. (32) using a silicate melt as an electrolyte and, recently, Iwase (33) has adopted the approach outlined in this paper. Stabilised zirconia was used as the electrolyte with a mixture of ZrO_2 and $ZrSiO_4$ as the auxiliary electrode. The equilibrium with the melt is given by

$$ZrO_{2(s)} + \underline{Si} + 2\underline{O} = ZrSiO_4 \qquad K = \frac{a_{ZrSiO_4}}{a_{ZrO_2} \cdot a_O^2 \cdot a_{Si}}$$

The changes in oxygen, detected by the stabilised zirconia, are related to the activities of silicon in the melt.

Determination of Copper in Matte

As was discussed earlier in the paper, by simple immersion or electrochemical exchange, it is possible to substitute other ions for the sodium in the β-alumina structure. One electrolyte which has been studied in detail is copper β-alumina (34) and this has been used to investigate the thermodynamics of the copper-tin system and good agreement data was obtained using other techniques. However, the probe is unlikely to be particularly sensitive when operated in the potentiometric mode due to the high concentration of copper in many systems. If, however, the sensors are operated in the coulometric mode with the rate controlling step being the diffusion of copper to the interface, the sensitivity will increase. This arrangement is shown in Figure 2.

One of the disadvantages of the substituted β-aluminas is that most of them have to be made by exchange for the Na^+ in the structure rather than directly from the constituent oxides. This is a time consuming process and, at the end of the transformation, the solid electrolyte has to be separated from the fused salt. The ionic properties of the $Na_{1+x}ZrSi_xP_{3-x}O_{12}$ (Nasicon) system were investigated by Hong in 1976 (9). It was found that the conductivity of the x = 2 composition had a similar value to that of sodium β-alumina but as the activation energy for Nasicon is higher, the conductivity is greater at higher temperatures, and, secondly, the sintering temperature is 300K lower than that for sodium β-alumina. It, therefore, seemed possible that an equivalent copper compound could be made directly from the constituents. Yao and Fray (11)

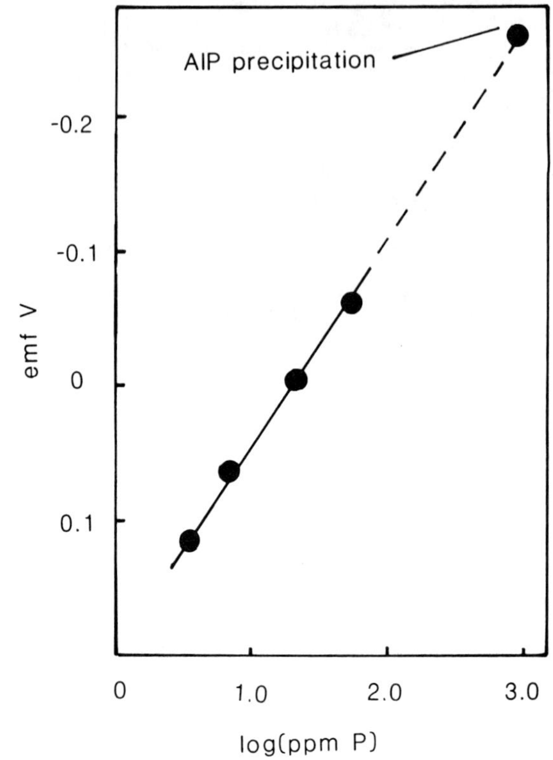

Figure 5 - Plot of emf versus log [ppm P].

reported the preparation, lattice parameters and the ionic conductivity of $CuZr_2(PO_4)_3$ and its use in determining the standard free energy of formation of cuprous sulphide. Sensors based upon this electrolyte might be used to monitor the copper content of mattes.

Sulphur and Sulphur-Containing Gases

The electrochemical determination of sulphur and sulphur dioxide is nowhere as straightforward as the detection of oxygen as the electrolytes are neither as stable nor as conducting. In 1977, Schmalzried proposed several sulphur sensors with the solid electrolytes calcium fluoride and stabilised zirconia but due to the complicated electrode reaction it proved difficult to achieve equilibrium (35). In order to solve this problem Nagata and Goto (36) used calcium sulphide, which is assumed to be a pure Ca^{2+} conductor in the following cell:

$$\text{Powder mixture}|CaS(Ca^{2+}\text{ conductor})|H_2\text{-}H_2S$$

where the powder mixtures, Mo-MoS_2, Cu-Cu_2S, Pb-PbS or Fe-FeS fix the sulphur potentials. It should be noted that the electrolyte responds to the changes in calcium activity which is related to the change in sulphur potential. Using the same electrolyte, Fischer and Janke (37) measured sulphur in liquid copper at 1423K but the measured potential was less than the calculated potential and this was explained by partial electronic conduction in the electrolyte. In an attempt to improve on these measurements, Ono, Oishi and Moryama (38) doped the CaS with Y_2S_3 or ZrS_2 and Egami, Onoye and Nasita (39) have used a sulphur sensor with a CaS_2-1% TiS_2 electrolyte to measure the sulphur content of iron between 0.005 and 0.5 wt.% at 1400^oC. Iwase and Jacob (40) have used two-phase electrolytes of CaS and CaO-ZrO_2 and, also, Na_2S and $Na_2O,11Al_2O_3$. At the working electrodes the following equilibria are set up for the two cells:

$$CaS + \tfrac{1}{2}O_2 = CaO(in\ ZrO_2) + \tfrac{1}{2}S_2 \tag{12}$$

$$Na_2S + \tfrac{1}{2}O_2 = Na_2O\ (in\ Na_2O.11Al_2O_3) + \tfrac{1}{2}S_2 \tag{13}$$

As the activities of CaS, CaO and Na_2S, Na_2O are fixed, the oxygen pressure must be proportional to the sulphur pressure. However, due to its instability, it is impossible to monitor sulphur in oxidising environments with calcium sulphide.

At higher oxygen pressures, the sulphate of the alkali metals are thermodynamically stable and can be used to detect sulphur dioxide as the sulphate can be considered to dissociate into SO_2 and oxygen according to the equation

$$SO_4^{\ 2-} = SO_2 + O_2 + 2e^- \tag{14}$$

and the emf can be expressed by

$$E = \frac{RT}{ZF} \ln \frac{P''_{O_2}}{P'_{O_2}} + \frac{RT}{ZF} \ln \frac{P''_{SO_2}}{P'_{SO_2}} \tag{15}$$

If the oxygen partial pressure is fixed at both electrodes, the potential is given by the ratio of the partial pressures of sulphur dioxide. Other workers have used Nasicon and sodium β-alumina (19) as the electrolyte where it is assumed that sodium sulphate forms on the surface of the electrolyte.

Liu and Worrell (41) have developed a solid state electrochemical sensor for the measurement of sulphur dioxide or sulphur trioxide using a mixed electrolyte of Li_2SO_4-Ag_2SO_4 with a silver reference electrode. The overall cell reaction is

$$SO_{3(g)} + \tfrac{1}{2}O_{2(g)} + 2Ag = Ag_2SO_4 \tag{16}$$

The two-phase electrolyte cell exhibited excellent long term chemical stability over a six-month period over the concentration range of sulphur of 3-10,000 ppm.

Hydrogen

Due to the fact that the vast majority of molten metals came into contact with the environment which, inevitably contains water vapour, hydrogen is introduced into molten metals. In most cases hydrogen does not form compounds with the metal and is relatively innocuous in the molten state. However, on solidification, the solubility of hydrogen generally decreases dramatically with the resulting formation of porosity. In the case of solid steel, hydrogen at high pressure is soluble and is capable of diffusing to grain boundaries and inclusions to generate cracks. It is very important to be able to measure hydrogen both in the molten and the solid states. Unfortunately, proton-conducting solids are unstable at temperatures above about 973K so it is difficult to foresee hydrogen being directly measured in molten steel and copper. Aluminium melts at a lower temperature at which both hydrogen β-alumina (5), made by substitution in sodium β-alumina, and calcium hydride (42) are stable. Williams and McGeehin (5) reported the use of hydrogen β-alumina electrolyte to measure hydrogen in aluminium but it was found that the sensor also responded to the presence of sodium, an impurity which is always present in aluminium. Gee and Fray (42) used calcium hydride, which is a H^- ion conductor, as an electrolyte and a mixture of calcium and calcium hydride as the reference. As the calcium hydride is unstable in the environment the sensor, was sealed with cement at one end and aluminium foil at the working electrode. On immersion in the molten aluminium, the foil dissolved and the sensor responded to the hydrogen content of the aluminium. The lifetime of the sensor was about 20 minutes before the electrolyte degraded due to attack by the aluminium.

Recently, two low temperature hydrogen sensors have been developed based upon hydrogen uranyl phosphate tetrahydrate (43) and Nafion (44) (Figure 6). At low temperatures it is important that the electrode kinetics should be fast and this is achieved by incorporating platinum black into the surface of the working electrode. The same applies at the reference electrode where it was found that WO_3, which intercalates hydrogen, can be used or preferably a redox mixture of Fe^{2+} and Fe^{3+}. The sensors responded virtually instantaneously to changes in hydrogen content of inert gases and gave a Nernstian response down to a few ppm as is shown in Figure 7 (44). If the sensor was placed upon the surface of a steel component containing hydrogen, which could have been introduced by a corrosion reaction, over cathodic protection or electroplating, the sensor responded to the hydrogen content of the component. Although these sensors only operate at temperatures close to room temperature, it is possible by equilibrating an inert gas with the hydrogen content of a molten metal, to measure the hydrogen content at room temperature (Figure 8). This has been achieved for liquid aluminum, copper and liquid steel and the results for

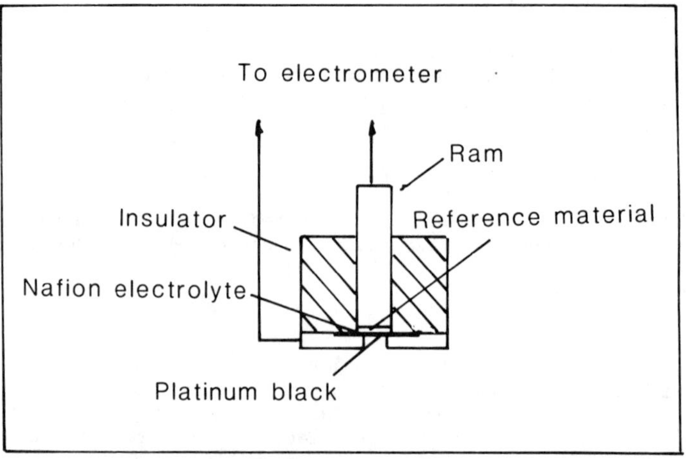

Figure 6 - Hydrogen sensor based on Nafion electrolyte.

aluminum are shown in Figure 9.

There are also possible applications at moderate temperatures 9001200K and in order to meet this requirement, a novel hydrogen sensor based upon strontium chloride (45) electrolyte with either a Ag|AgCl or Ni|$NiCl_2$ reference electrode has been developed. Strontium chloride is a chloride ion conductor and in order to facilitate the coupling between hydrogen gas and the chloride ion, the surface of the electrolyte is covered with a thin layer of SrHCl and $SrCl_2$. SrHCl is a stable compound with a melting point of 1073K and it is easily formed by reacting $SrCl_2$ and SrH_2 in an hydrogen

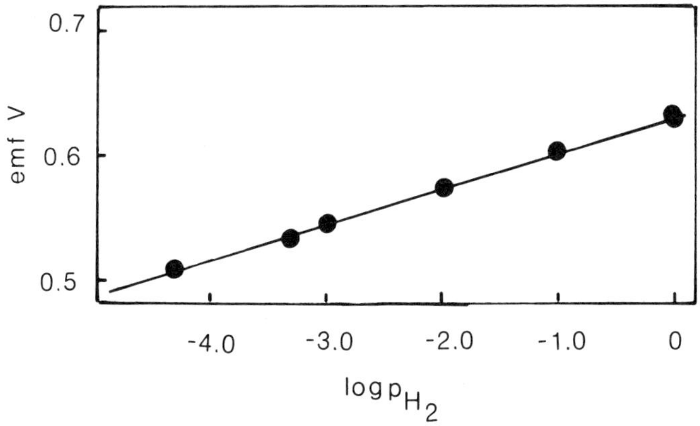

Figure 7 - Nernstian response of Nafion sensor with Fe^{2+}/Fe^{3+} reference.

atmosphere at temperatures of around 900°C. The hydrogen sensor can be represented by the following galvanic cell:

$$- \; H_2|Ar|\begin{matrix}\text{Stainless}\\\text{Steel}\end{matrix}|SrHCl, SrCl_2|SrCl_2|Ag|AgCl|Ag \; +$$

The emf (E) across the cell is determined by the difference in the partial pressures of chlorine at the two electrodes according to the Nernst equation:

$$E = \frac{2.3003RT}{2F}(\log P_{Cl_2(ref)} - \log P_{Cl_2(gas)}) \qquad (17)$$

where $P_{Cl_2(ref)}$ is the known partial pressure of the reference electrode; $P_{Cl_2(gas)}$ is the partial pressure of chlorine on the auxiliary phase as determined by the test gas; R is the gas constant, F is Faraday's constant and T is the temperature. The partial pressure of chlorine on the auxiliary phase is related to the partial pressure of hydrogen in the gas by

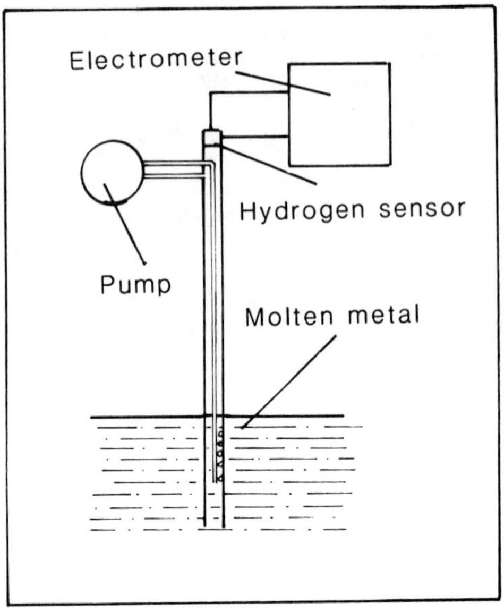

Figure 8 - Arrangement for measuring hydrogen in molten metal by use of an ambient temperature hydrogen sensor.

$$2SrCl_2 + H_2 \rightleftarrows 2SrHCl + Cl_2 \tag{18}$$

and as $SrCl_2$ and SrHCl are at unit activity

$$\log PCl_2 = \log P_{H_2} - \frac{\Delta G^o}{2.303RT} \tag{19}$$

Where ΔG^o is the standard-free energy change for the above chemical reaction. Substituting into equation (17) gives

$$\log P_{H_2} = \log P_{Cl_2(ref)} + \frac{\Delta G^o}{2.303RT} - \frac{2FE}{2.303RT} \tag{20}$$

The results at 673K and 873K are shown in Figure 10.

Discussion and Conclusions

Solid electrolytes have been investigated for nearly a century as ionic conductors were used in 1908 (46) to determine thermodynamic data. Three decades later, Tubandt (47) confirmed that many halides were ionic conductors but most applications were restricted to simple cells operating over narrow temperature ranges due either to the low melting points of the halides or the occurrence of non-stoichiometry at high temperatures. Unfortunately, in many of these salts, both anions and cations were mobile dictating that the electrodes had to be reversible to two different ionic species. Interest in these systems then lapsed until 1957 when Kiukkola

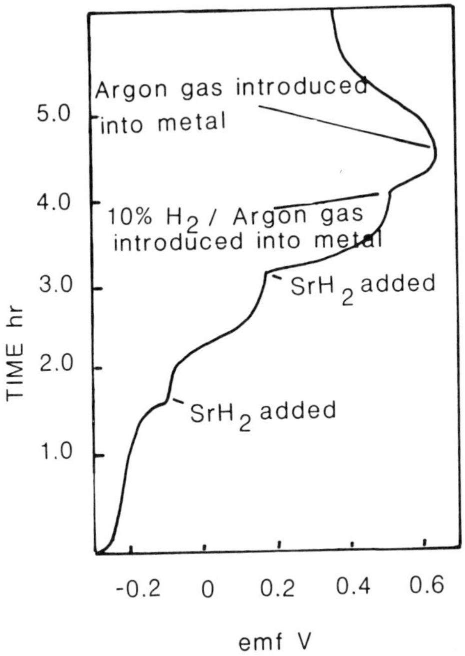

Figure 9 - Measurement of hydrogen in molten aluminum using Nafion sensor.

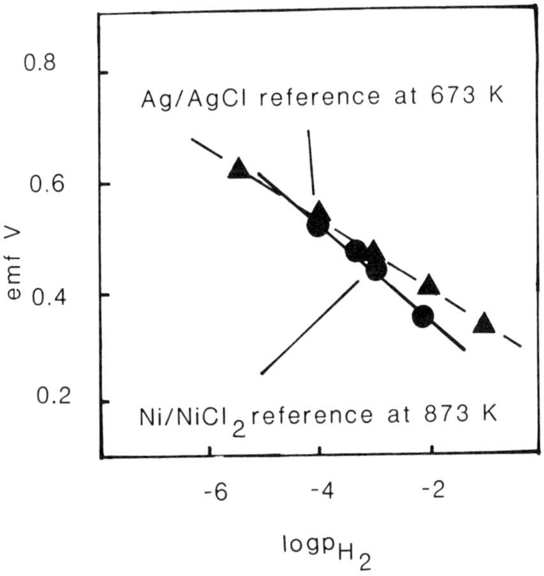

Figure 10 - Results from sensor based on SrHCl,$SrCl_2$.

and Wagner (48-50), Ure and Liddiard (51) restimulated interest in high temperature galvanic cells incorporating electrolyte phases with only one mobile species which, in the case of stabilised zirconia, has found wide application.

However, recently, due mainly to the interest in battery systems, many solid electrolytes which may be considered as fast ion conductors have been developed or discovered. This has extended the range of possible sensors from oxygen to most monovalent elements and some divalent and trivalent elements. However, there were still a wide range of elements of metallurgical interest which formed part of the anion lattice in the conductors and were relatively immobile. In order to be able to measure these species, compounds which contained both the species and the mobile ion were placed between the sensor and the metallic solvent.

Early sensors used simple elements as the reference materials, but now it is possible to use mixture of compounds and even redox mixtures and this has further extended the number of elements which can be measured. Furthermore, solid electrolytes used in the potentiometric mode have always been notoriously insensitive to small changes in composition at high concentrations. However, by operating the sensor in the coulometric mode, the sensitivity at high concentrations can greatly be increased.

From this paper it can be concluded that it is now possible to devise sensors which will measure most species but whether they function adequately depends upon electrode kinetics, exchange currents, and the resistance to attack of the sensor to its environment.

References

1. T.H. Etsell, S. Zador and C.B. Alcock, Metal-Slag-Gas Reactions and Processes, Eds. Z.A. Foroulis and W.W. Smeltzer. (Princeton: The Electrochemical Society, 1975), 834-50.

2. E.T. Turkdogan and R.J. Fruehan, Review of oxygen sensors for use in steelmaking and of deoxidation equilibria. Can. Met. Quart. 11 (1972), 371-82.

3. J.T. Kummer, Progress in Solid State Chemistry, Vol. 7, Ed. H. Reiss and J.O. McCaldin, (Oxford: Pergamon 1972), 141-175.

4. D.J. Fray, Extraction Metallurgy '81. (London: The Institution of Mining and Metallurgy, 1981), 321-330.

5. D.E. Williams and P. McGeehin, Roy. Soc. Chem. Specialist Periodical Reports, Electrochem., 9 (1984), 246-90.

6. B. Dunn and G.C. Farrington, Fast divalent conduction in Ba^{++}, Cd^{++} and Sr^{++} beta alumina, Mat. Res. Bulletin, 15 (1980), 1773-77.

7. J.T. Whiter and D.J. Fray, The preparation and electrical properties of polycrystalline calcium β"-alumina. Solid state Ionics, 17 (1985), 1-6.

8. G.C. Farrington et al., The Lanthanide β"-aluminas, Applied Physics A., 32 (1983), 159-161.

9. H.Y.-P. Hong, Crystal Structures and Crystal Chemistry in the system $Na_{1+x}Zr_2Si_xP_{3-x}O_{12}$, Mat. Res. Bull., 11 (1976), 173-82.

10. P.C. Yao and D.J. Fray, Determination of the lithium content of molten aluminum using a solid electrolyte, Trans. Met. Soc. B, 16B (1985), 41-46.

11. P.C. Yao and D.J. Fray, The preparation and properties of the solid state ionic conductor $CuZr_2(PO_4)_3$. Solid State Ionics, 8 (1983), 35-42.

12. J. Guleus et al., Hydrogen electrolysis using Nasicon solid state protonic conductor. 6th International Conference on Solid State Ionics, Garmisch, FRG 1957, 185-6.

13. H. Iwahara et al., Proton conduction in sintered oxides and its application to steam electrolysis for hydrogen production. Solid State Ionics, 3/4 (1981), 359-63.

14. P.E. Childs et al., Battery and other applications of a new proton conductors: hydrogen uranyl phosphate tetrahydrate, $HUO_2PO_4.4H_2O$, J. Power Sources, 3 (1978), 105-14.

15. V. Choudhury et al., New inorganic proton conductors, Mat. Res. Bulletin, 17 (1982), 917-33.

16. J.P. Randin, Ion Containing Polymers as Semisolid Electrolytes in WO_3-based electrochromic devices. J. Electrochem. Soc., 129 (1982), 1215-20.

17. J. Delcet and J.J. Egan, Thermodynamics of liquid Ce-Ag and Ca-In alloys. J. Less Common Metals, 59 (1978), 229-236.

18. M. Gauthier and A. Chamberlain, Solid state detectors for the potentiometric determination of gaseous oxides. 1. Measurement in Air, J. Electrochem. Soc., 124 (1979), 1579-83.

19. M. Itoh and Z. Kozuka, Reaction of β-alumina solid electrolyte with sulphur oxide (SO_x) (x = 2,3) gas. Trans. Japan, Inst. of Metals, 26 (1985), 17-25.

20. R.V. Kumar and D.J. Fray, Unpublished work.

21. R.E. Hetrick et al., Oxygen sensing by electrochemical pumping, Appl. Phys. Lett, 38 (1981), 390-2.

22. W.L. Worrell, Metal-Slag-Gas Reaction and Processes, Ed. Z.A. Foroulis and W.W. Smeltzer. (Princeton: The Electrochemical Society, 1975), 822-33.

23. R. Gee and D.J. Fray, Instantaneous determination of hydrogen content in molten aluminum and its alloys, Met. Trans. Soc., 9B (1978), 427-430.

24. V.W.A. Fischer and D. Janke, Electrochemische und Elektrische Unter suchungen im system eisen-phosphor-sauerstoff-kalk. Teill. Arch. Eisenhuttw., 37 (1966), 853-62.

25. D.J. Fray, Determination of sodium in molten aluminum and aluminum alloys using a beta alumina probe. Metall. Trans. 8B (1977), 152-6.

26. A.A. Dubreuil and A.D. Pelton, Light Metals '85, Ed. H.O. Bohner, (Warrendale PA: The Metallurgical Society, 1985), 1197-1205.

27. K.S. Goto, M. Sasabe and M. Someno, Change of chemical potential of oxygen in liquid metals and liquid oxide phases. Trans. Met. Soc. 242 (1968), 1757-9.

28. K. Nagata et al., Measurement of oxygen potential and temperature in liquid slag, metal and gas phase of Q-BOP converter by oxygen concentration cell. Tetsu-to-Hagane, 68 (1982), 1271-83.

29. S. Yamaguchi and K.S. Goto. Unpublished work.

30. R.J. Brisley and D.J. Fray, The determination of the thermodynamics of the NaF-AlF_3-Al_2O_3 system with a solid electrolyte cell. Trans. Met. Soc. 15B (1983), 135-9.

31. B. Broadhurst and D.J. Fray. Unpublished work.

32. A. Romero et al., Foundry Processes Their Chemistry and Physics, Ed. S. Katz and C.F. Laudefeld, (New York: Plenum Press, 1988), 219-238.

33. M. Iwase. Rapid determination of silicon activities in hot metal by means of solid state electrochemical sensors equipped with an auxiliary electrode. Scandinavian J. of Met. 17 (1988), 50-56.

34. J.A. Little and D.J. Fray, Determination of the activity of copper in liquid copper-tin alloys by use of copper beta alumina. Trans. IMM 88 (1979), C229-233.

35. H. Schmalzried, The Development of Sulphur ion-conducting electrolytes, Arch. Eisenhutt., 48 (1977), 319-22.

36. K. Nagata and K.S. Goto, On recent development of solid electrolyte sensors. Tetsu-to-Hagane 67(11) (1981), 1899-908.

37. W.A. Fischer and D. Janke, eds., Metallurgische Elektrochemie, Springer-Verlag 1975.

38. K. Ono et al., Measurements on galvanic cells involving solid sulphide electrolytes. Solid State Ionics, 3/4 (1981), 555-58.

39. A. Egami et al., Solid electrolyte for the determination of sulfur in liquid iron. Solid State Ionics, 3/4 (1981), 617-20.

40. M. Iwase and K.T. Jacob, High Temperature Materials Chemistry, Ed. D.D. Cubicciotti and D.L. Hildebrand, (New Jersey: Electrochem. Soc., 1982), 431.

41. Q.C. Lui and W.L. Worrell, Phys. Chem. Extr. Metall. Proc. Int. Symp., Ed. V. Kudryk and Y.R. Rao (Warrendale, PA: The Metallurgical Society, 1985), 387-96.

42. R. Gee and D.J. Fray, Instantaneous determination of hydrogen content in molten aluminum and its alloys. Met. Trans. 9B (1978), 427-30.

43. S.B. Lyon and D.J. Fray, Determination of hydrogen generated in electrochemical processes by use of a solid electrolyte probe. Brit.

Corrosion J. 19 (1984), 23-29.

44. D.R. Morris, R.V. Kumar and D.J. Fray, Development of an electrochemical sensor for hydrogen and deuterium, Submitted to J. Metals.

45. R.V. Kumar and D.J. Fray, Solid state hydrogen sensors based on $SrCl_2$ electrolyte. Submitted to J. Solid State Ionics.

46. M. Katayama, Uber Amalgamkonzentrationskelten, Chemische Kelten und Daneillkelten mit festen Elektrolyten. Z. Phys. Chem. 61 (1908), 566-87.

47. C. Tubandt et al., Handbuch der Experimental Physik (Leipzig: Akademie Verlagges) 12 (Pt 1) (1932) 459.

48. K. Kiukkola and C. Wagner, Galvanic cells for the determination of the standard molar-free energy of formation of metal halides, oxides and sulphides at elevated temperatures. J. Electrochem. Soc. 104 (1957), 308-16.

49. K. Kiukkola and C. Wagner, Measurements on Galvanic cells involving solid electrolytes. J. Electrochem. Soc. 104 (1957), 379-87.

50. R.W. Ure, Ionic conductivity of calcium fluoride crystals. J. Chem. Phys. 26 (1957), 1363-73.

51. A.B. Lidiard in Handbuch der Physik, Ed. S. Flugge (Berlin, Springer-Verlag 1957), 20, 246.

A MAMUFACTURE PROCESS OF NEW TYPE TUNGSTATE AND HYDROGEN REDUETION OF TUNGSTEN OXIDE THEREOF

Cao rong-jiang   Tang Xin-he

General Research Institute for Non-Ferrous Metals
Beijing P.R.China

Abstract

In ordinary manufacture process of tungsten powder, a preferable intermediate product used is ammonium paratungstate (APT). The particle size of tungten powder made from it through $W_{20}O_{58}$ is larger than 1.0 μm. In order to obtain ultrafine tungsten powder with narrow size distribution, a new type of tungstate, hydrazinium tungstate, was studied. Its formula is $5(N_2H_5)_2O \cdot 12WO_3$, with tetragonal crystal structure of a=5.821, c=18.468.

Three methods for preparing hydrazinium tungstate (ART) were set up and ART can be produced by current industrial APT process with little modification of operation and without any change of equipment.

Under certain PH and operation condition of the solution from which ART was produced, an autoreduction of tungsten occured, which produced a lower valency tungstate with blue color. when ART was heated at different temperatures, in N atomsphere, tungsten bronze, tetragonal $WO_{2.90}$ and $WO_{2.72}$ could be obtained respectively. They all have the superiorities, in the production of tungsten powder, of superfine size and narrow size distribution [average size of 0.5 μm or even smaller].

The rate of autoreduction of ART is very fast at elevated temperature. The reduction reaction, which gaves tetragonal $WO_{2.90}$, could be handled smoothly without explosion. To research their behaviors at autoreduction temperatures and the products of autoreduction will be new advances of tungsten chemistry and may also serve tungsten industry with new processes and materials.

Metallurgical Processes for the Year 2000 and Beyond
Edited by H.Y. Sohn and E.S. Geskin
The Minerals, Metals & Materials Society, 1988

Introduction

There are signficant advantages of powder metallurgical technology as compared to casting and forging with regard to material and process efficiency. As to powder production process, such as for titanium and refractory steels, melting and atomizing are usually used. But for tungsten, as its high melting point, there is difficulty to melt it, therefore reduction of tungsten oxides made form APT is used in ordinary process for manufacture of tungsten powder.

From commercial source, the particle size of finest tungsten powder is at the range of 0.7-0.8 μm. But that of ultrafine tungsten powder is much smaller. the suggusted manufacture process of ultrafine tungsten powder is hydrogen reduction of WCl which is still at development stage. there is no stable commercial supply of it.

According to our experience, particle size of a tungsten powders of about 0.3 μm with narrow size distribution range show excellent characterstics for hard metal industry. In this paper , a new product used as raw material for production of above mentioned powder from H_2WO_4 is reported. From the improved qualitis of hard metal for cutting and drilling made from it, we expect this brand tungsten powder can get a broad selling market in our country and abroad.

Experimental

Main products of a caustic pressure leaching of wolframite are H_2WO_4 $W_{20}O_{58}$ APT WO_3 etc. After purification from P, As, Si, Mo, a solution of Na_2WO_4 may be used as raw material for production of tungsten powder. It may be solvent extracted to produce APT, or neutralized with acid and H_2WO_4 is preci pitated from it with a innovated process. In our case, We used $(NH_4)_2WO_4$ solution, tungsten loaded extractant solution as starting materials for experiments, all of them are made from H_2WO_4.

Material:

(1). H_2WO_4 H_2WO_4 was supplied by Dungtai tungsten acid plant. Its typical chemical analysies are shown in table I

Table I. Impurities of H_2WO_4 (ppm)

element	As	Si	Mg	Al	Cr	Fe	Ni	Mo
content	21	58	3	6.5	1	5	≤1	22
element	Sn	Ca	Au	Ti	Co	Sb	Pb	V
content	<1	18	<1	2	<3	<1	<1	<1

(2). solution of $(NH_4)_2WO_4$. Dissolving H_2WO_4 with conc. NH_4OH solution, then filtering to separate nonsoluable solid out, the solution of $(NH_4)_2WO_4$ with concentration of 249.2 g/l WO_3 and 5.8m of free NH_4OH is used as raw material for precipitation of the new product.

(3). Hydrazine commercial hydrazine is a 33.3% N_2H_4 water solution .

(4). N 235 A chinese trade name of tertiary amine with the composition similar to Adogen 368.

(5). Octanol and Kerosene These are used as modifier and diluent for extractant respcetively.

Procedure:

Our new process is only a part of whole innovated process for manufacture of tungsten powder from wolframite concentrate . Other innovated parts are complexation precipitation of H_2WO_4,[1] solvent extration to separate Mo, P and As,[2] decomposition of APT with rotary kilm to produce $WO_{2.90}(T)$ etc. The expriment procedure reported in this papar with the aim to connect it to the main ondinary flow diagram or to the innovated process easily. It focused at the preparetion of a new product to replace APT only.

(1). Added solid H_2WO_4 direct to the hydrazine solution at 20℃ with mixing to prepare $5(N_2H_5)_2O\cdot12\ WO_3$ appeared as solid→solid reaction.

(2). A solution of $(NH_4)_2WO_4$ was prepared from H_2WO_4 with conc. NH_4OH. Its conceutration is 1.015M WO_3. mixed it with hydrazine solution at 80 ℃, 61℃ and 35℃ The effect of temperature on the rates of reaction were measured.

(3). A imitative solution of Na_2WO_4 with a conc. of 221.5 g/l WO_3, acidfied with H_2SO to PH 2.5-3 was used as aquaeus phase for solvent extraction. Extractant was a organic mixture of N-235 , octanol and kerosene. Striping the tungsten loaded organic phase with hydrazine and NH_4OH solution mixture and standing the separated solution, Hydrazine tungstate was precipitated from it.

Results and discussions

(1). Adding solid H_2WO_4 into Hydrazine solution, a hetrogenueus reaction in situ was taken place , the H_2WO_4 gradually transformed to $5(N_2H_5)_2O\cdot12W\ O_3$. If the time was not sufficient there was a mixture of H_2WO_4 and hydrazine tungstete. the mole ratio effect of N_2H_4 to WO_3 on the quality of product is shown in table II

Table II N_2H_4/WO_3 effect on the quality of the solid product

N_2H_4/WO_3	0.65	0.76	0.86	1.09	2.17
Solid product	ART** H_2WO_4	ART H_2WO_4	ART	ART	ART
PH(final)	7.0	7.0	7.2	8.9	9.2

** ART is Abbreviation of Auto reduction tungstate to indicate hydrazine paratungstate.

The transformmation may be represented by the following reactions

$$5H_2WO_4(s)+10N_2H_4(aq)\rightarrow5(N_2H_5)_2WO_4(S) \quad (1)$$

$$5(N_2H_5)\ WO_4(S)+7H_2WO_4(s)\rightarrow5(N_2H_5)_2O\cdot12WO_3(s)+7H_2O \quad (2)$$

or

$$12H_2WO_4(s)+10N_2H_4(aq)\rightarrow5(N_2H_5)_2O\cdot12WO_3(s)+7H_2O \quad (3)$$

(2). At different temperature mixing $(NH_4)_2WO_4$ solution with hydrazine a

* $WO_{2.90}(T)$ is a tentative symbol for tetragonal cryster form of O/W=2.90 in this paper.

precipitation occured . As a function of temperature, the rate of transformation was plotted as Fig 1

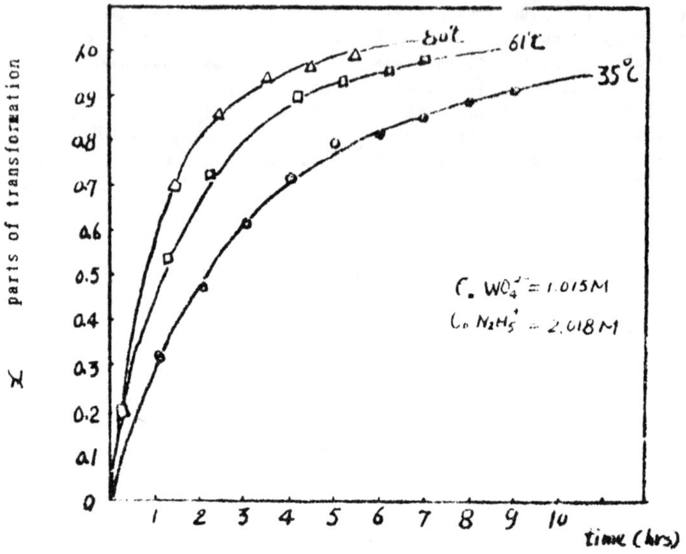

Fig1 The temperature effect on the rate of transformation

From these data ,the apparent active energy was calculated as 20.9 KJ/mol.

(3). strip solution is a mixture solution of hydrazine tungstate and amonium tungstate. If hydrazine only used as stripping solution, there are precipitates in the stripphase making difficulty to the operation of mixer settler. Stripped with mixture of ammonium hydroxide and hydrazine with proper ratio, strip phase on standing 3 minutes or more a solid was precipited therefrom.

From the three above mentioned method all the new product was hydrazine paratungstate . It was proved by chemical analysis, and thermoanalysis; with X-ray diffractometry the crystal sturcuture is hexagonal with a=5.821A c=18.468A Its chemcal formular is $5(N_2H_5)_2O \cdot 12WO_3$.We called it ART already in Table II.

(4). Decomposition Heating ART in nitrogen atmosphere at 500℃, ART is decomposed to tetragonal $WO_{2.90}$.There are two crystal froms of $WO_{2.90}$, one is monoclinic, its molecular formula is written as $W_{20}O_{58}$. Another is tetragonal with the same O/W ratio , but the molecular unit is not determined yet. In this paper , we designate it as $WO_{2.90}$(T). there after.

It is known that from APT under proper condition of H_2 atmosphere and temperature, $WO_{2.90}$(T) also can be made,(3) but it is not pure crystal. Some monoclinic $W_{20}O_{58}$ is present in it. Use of ART as raw material, $WO_{2.90}$(T) is easily made and is crystal pure product without $W_{20}O_{58}$.

(5). Reduction of $WO_{2.90}$(T) to β-W. Reducing of $WO_{2.90}$(t) with dry hydrogen (dew piont -30°C) at different temperature and time durations β-W, α-w, and their mixture were produced as shown in Table III.

Table III The results of Hydrogen reduction of $WO_{2.90}$(T)

Temperature(°C)	time(min)	results
520	70	$WO_{2.90}$(T) 80% β-W 20%
530	60	β-W
540	50	β-W 90% α-W 5% WO_2 5%
540	70	β-W 85% α-W 15%
580	70	αW 70% β-W 30%
600	90	αW 90% β-W 10%
650	60	α-W

For dopped tungsten powder used as raw material to produce lamp filement, β-W is much favored as a middle phase presented (4). From Table III at 520-580°C, $WO_{2.90}$(T) reduced with hydrogen can get β-W in larger amount than that reduced from $W_{20}O_{58}$.

In tungsten industry, Hydrogen reduction of tungsten oxides, blue or yellow are carried out in mutitube furnace or rotary tube furnace. The tungsten powder is always in agglomerater (5). Therefore its particule size must distinguished between that of single particle and agglomerater. If we use ART instead of APT and then $WO_{2.90}$(T) instead of $W_{20}O_{58}$, the powder size of tungsten powder determined with centrifugal disc method is <0.5μm. It is reasonable to attribute the finess of the powder to the ART and more directly to the $WO_{2.90}$(T). The character of $WO_{2.90}$(T) are shown in Table IV.

Table IV Character of $WO_{2.90}$(T)

Surface area (m^2/g)	average size(μ)	size distribution(μ)	density g/cm^3	morphology
30.1	1.4	0.5-4.0	6.2	hexagonal

From the SEM micrograph of α-W in Fig 2, one can see grain sizes of it are about 0.3 μm. These are still of agglomerators. The true particle size measured with X-ray diffraction is about 0.02 μm.

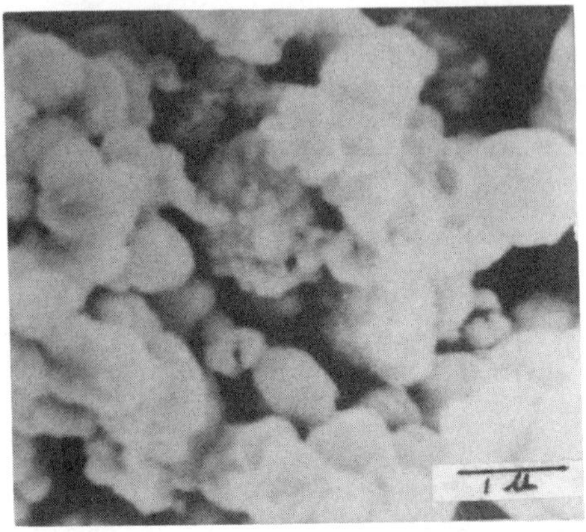

Fig 2 SET Micrograph of α-W hydrogen reduced from $WO_{2.90}$(T)

In recent years, hydrazine is used as reducing agent in separation of uranium and plotonum with a solvent extraction method (6) and in precipitation of copper powder from aqueous solution etc. The autoreducing power of ART in solution appears at that its color changes from colorless to bright blue as acidified with H_2SO_4 to PH 3. Since the autoreducing power of it , or in other words the more unstable property of it , at decomposition temp, the rate of autoreduction is so fast as an explosive reaction. It is reason for the fine particle size of $WO_{2.90}$(T).

Reduction from $WO_{2.90}$(T) to α-W may pass through β-w; at the stage of β-w the morphology of it is same as $WO_{2.90}$(T). There is a further break of particle at the stage of β-W transforming to α-w. As a result of it. The super fine crystals of tungsten are obtained.

From above mentioned , a new process for crystals pure $WO_{2.90}$(T) and superfine tungsten powder can be developed as shown in Fig 3 in a condensed flow diagram.

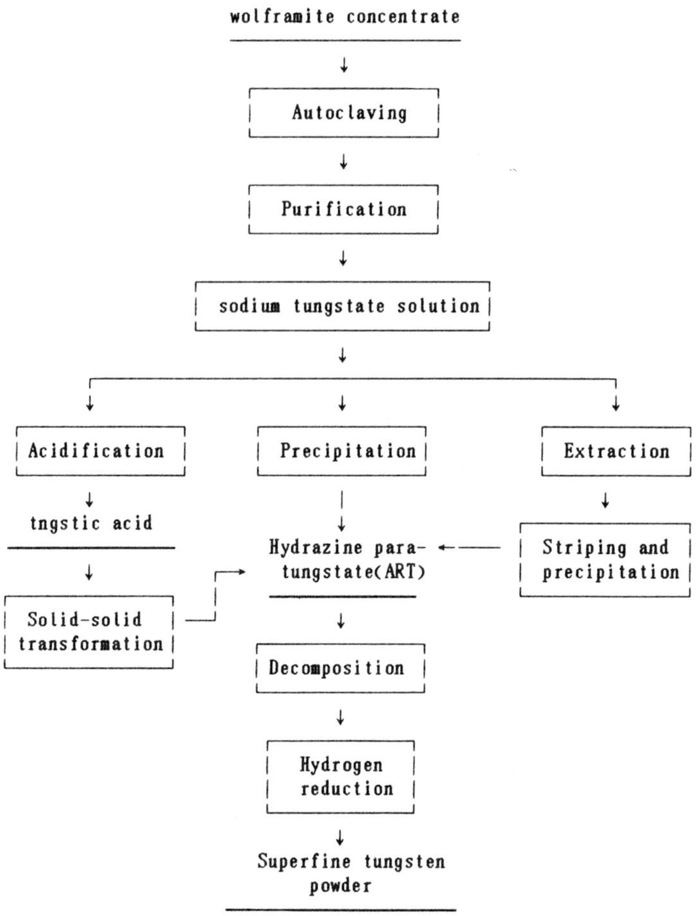

Fig 3 A new process for $WO_{2.90}$(T) and superfine tungsten powder from ART

Conclusions

(1). A new middle product , ART can be prepared from H_2WO_4 (s) ,$(HN_4)_2WO_4$ solution , and tungsten loaded extractant solution.

(2). The ART is autoreducible to lower valancy state in solution and decomposed very rapid in nitrogen atmospher to $WO_{2.90}$(T) with pariticle size average of 1.4 μm. or less at about 450-500℃

(3). The blue oxide $WO_{2.90}$(T) is a best blue oxide among a group of blue oxides i.e. tungsten bronse $W_{20}O_{58}$, $WO_{2.70}$(blue) used as the raw material for production of tungsten powder.

(4). In wet hydrogen atmosphere ,such as $P_{H_2O}/P_{H_2}=2.70\times10^{-2}$, β-W can be obtained in more quantity in the mixture of α-W+β-W from blue oxide $WO_{2.90}$(T) which is made from ART than that from $W_{20}O_{58}$ which is made from APT or from $WO_{2.96}$(T) + $W_{20}O_{58}$ mixture.

(5). Superfine tungsten powder with grain size < 0.5 μm and subgrain size <0.02 μm was obtained by hydrogen reduction of $WO_{2.96}$(T) which was made from ART.

(6). To sum up , A suggested flow diagram for production of superfine tungsten powder from tungsten ore concentrates through ART was presented as Fig 3.

References

1. Jiang Anren et al., Chinese patent 86100321.

2. Su Yuanfu an Lee Daren, "separation of tungsten and phosphorous with two step extraction" (paper presented at 2nd all China tungsten and molybdenum science meeting 1987).

3. L. Barth et al. "A study of the APT and some intermediate products during the W metal powder production " paper presented at plansee seminar RM 11 1985.

4. Tang Xinhe ; "Hydrogen reduction of tungsten oxide" (Doctor thesis partII, General Research Institute for Nonferrons Metals 1987).

5. Tao Zhenji "Morphorlogy; structure and particle size of raw material powder for hard metals" (Report of Zhu Zhou hard metal factory)1983.

6. D. A. Orth; "Purex: Process and equipment perfromance " (Paper presented at ISEC '86 Munchen FRG.

PHYSICO-CHEMICAL FUNDAMENTALS OF MELTING NON-FERROUS CONCENTRATES WITH THE PRODUCTION OF METAL (MATTE) & FERRITE-CALCIUM SLAGS

Dr. A. I. Okunev

Institute of Metals
Ural Dept. of Academy of Science
USSR

Abstract

The investigation results are submitted on separation into layers in the systems Cu (Ni) - Fe - Ca - O - S - (Si), as well as on physico-chemical properties of oxide - ferrite, oxide-sulfide melts (density, surface tention, viscosity, electric conductivity), on kinetics and oxidation macromechanism of iron and sulfur in the oxide-silicate, oxide-ferrite and oxide-sulfide melts, kinetics of reduction of oxide-silicate and oxide-ferrite systems. Conditions are discussed for copper, nickel and zinc recovery from oxide-silicate and oxide-ferrite slag in liquid and solid states. Some promising technique is the technique of future, as is considered.

Metallurgical Processes for the Year 2000 and Beyond
Edited by H.Y. Sohn and E.S. Geskin
The Minerals, Metals & Materials Society, 1988

Biological and Aqueous Processing

FLOTATION OF MICROORGANISMS - IMPLICATIONS IN THE REMOVAL OF METAL IONS FROM AQUEOUS STREAMS

R. W. Smith* Z. Yang* R. A. Wharton, Jr.**

*Department of Chemical and Metallurgical Engineering
University of Nevada, Reno
Reno, Nevada 89557

**Desert Research Institute and
Department of Biological Sciences
University of Nevada, Reno
Reno, Nevada 89557

Abstract

Microorganisms have the ability to remarkably adsorb and/or absorb heavy metal ions. Hence they are being studied widely as a means of removing metal ions from aqueous solutions both in order to clean up toxic streams and to recover valuable metals from these and other streams. It is often difficult to harvest organisms such as bacteria and unicellular algae since they are usually too small to be readily screened out. Froth flotation may prove to be a feasible method to harvest such organisms subsequent to the ad/absortion of the metal ions. The present paper discusses the surface/electrokinetic properties of microorganisms and the modifier-collect conditions under which they should float. The zeta potential of the green alga Chlorella vulgaris was measured in the presence of Al(III) and Pb(II) ions and in the presence of dodecylammonium chloride and sodium dodecylsulfate. Present and past data suggest that the organisms should behave similarly to other finely divided, charged (polar) solids with their behavior sometimes modified by the non-polar surface regions present.

Metallurgical Processes for the Year 2000 and Beyond
Edited by H.Y. Sohn and E.S. Geskin
The Minerals, Metals & Materials Society, 1988

Introduction

The overall field of bioengineering is growing at a remarkably accelerating rate. Hydrometallurgy and mineral processing are included among those disciplines where there has been actual or proposed commercial use of bioorganisms. Further, the actual or proposed use includes bacterial enhancement of leaching of a number of different types of ores (1-5), modification of the flotation or oil agglomeration separation of minerals by bacteria (6,7) and microorganism (algae, bacteria or fungi) concentration of metal ions present in effluent or other mineral processing or hydrometallurgy aqueous streams (8-23).

Considering the latter item, microorganisms are able to remove metal ions from aqueous solutions in a remarkable fashion (8-11). The uptake of the ions can be both by adsorption onto the surface of living or dead cells or by absorption through the cell walls of the living organisms (11-13, 15-19, 21-25). Although the concentration of metal ions may be by as much as 10^5 times or more (8-10) because of the small size of the organisms there may exist a practical problem of how to harvest such small organisms. In some cases flocculation followed by filtration may be possible. Also, there is the possibility of froth or foam flotation concentration of the organisms since it is known that most microorganisms can be readily floated (26-28).

Concentration of Metal Ions by Microorganisms

Electrokinetic studies of both bacteria and unicellular algae have shown that the organisms in the absence of significant concentrations of multivalent metal ions are negatively charged at all but quite acidic pH values. Daniels (29) lists measured isoelectric points (i.e.p.'s) for 48 different bacteria. In the absence of added buffers the i.e.p. values varied from pH 0.5 to pH 3.6. Smith, et.al. (28) found the i.e.p. value for the green alga Chlorella vulgaris to be at about pH 1.5. The origin of the charges on microorganisms arises from the presence of functional groups on their surfaces. These groups are usually those associated with amino acids, fatty substances, cellulose, lecithin and similar structures. Thus, there can be at least carboxylate, phosphate, sulfide, hydroxyl and amino groups present in the cell surfaces (19,23,27-35). Sulfate groups also may in some cases be present. Metal ions can certainly bond with any of the anionic groups and perhaps also with the hydroxyl groups (35). In addition it is known that some living microorganisms can concentrate metal ions within the interior of the cell by favored passage through cell membranes (36). Considering the quantities of functional groups, the overall negative surface charge and the cells' ability sometimes to absorb metal ions, it is thus apparent why such organisms so readily concentrate multivalent metal ions.

Flotation of Microorganisms

Dognon (37,38) in 1941 reported on the flotation of the bacteria Escherichia coli and Mycobacterium tuberculosis var. hominus. Subsequently, Hopper and McCowen (39) showed experimentally that the protozoa Endamoeba hystolytica could be successfully removed from waste water by flotation. Boyles and Lincoln (40) experimentally found that the bacteria Serratia marcescens, Brucella suis and Pasteurella tularensis could be floated and that spores of Bacillus anthracis could be separated from its cells by flotation. Gaudin and co-workers (41-45) were able to float the bacteria Escherchia coli, Bacillus subtilis and Serratia marcescens. Then, Levin, et al. (46) backed by experimental flotation studies on the green alga Chlorella vulgaris suggested froth flotation as a practical means of harvesting algae. Additional studies on the flotation of Chlorella vulgaris were performed by Smith, et al. (47), Funk, et al (48) and Grieves and Wang (49). Also Rubin and co-workers (50-53) studied the microflotation of the bacteria Escherichia coli, Bacillus cereus and Aerobacter aerogenes and of the algae, Chlamydomonas reinhardtii and Chlorella ellipsoidea. The work of Rubin and Lackey (52) is especially interesting in that these researchers studied the effect of coagulation on the microflotation of Bacillus cereus. In addition to these investigations, a number of researchers have reported on the flotation of algae and bacteria in connection with treatment of waste waters or harvesting of algae (54-60).

The various flotation studies have found that some of the organisms are naturally floatable in the absence of a conventional flotation collector or any other additive (37,38,46,61). In some cases a collector was not necessary but flotation could be induced only after addition of some salt. Sometimes only a simple salt such as sodium chloride or sodium dihydrogen phosphate was required (37,38,43,45). Figure 1 from Gaudin, et al. (45) illustrates this phenomenon. Sometimes a small quantity of a flocculating agent, such as an aluminum salt would induce flotation provided sufficient was added to induce flocculation (48,54,55). In other cases a collector aided separation or was required (40,43,45,50,53,62). That some of the organisms were floatable in the absence of additives (collectors or inorganic salts) is not surprizing since the organisms' surfaces contain various fatty-non-polar groups in addition to the various functional groups. In the case of the addition of a simple salt inducing flotation Gaudin (43) suggested that the salt might cause the surface of the organism to become folded due to osmotic shrinkage in the cell. The folded skin then disposes itself in such a manner that the hydrocarbon groups point outward and thereby the surface becomes hydrophobic. However, it also may be that the function of the salt is simply to reduce the zeta potential of the organism through the resulting compression of the double layer surrounding the organism. The cells should then aggregate making them more susceptible to bubble encounter and, hence, flotation. In this regard the work of Funk, et al. (48) is of interest. They showed that rather modest amounts of Fe(III) ionic species effectively promote the flotation of Chlorella vulgaris. These species should coagulate the algae providing for better bubble-algal cell interaction. Figure 2 abstracted from Rubin and Lackey (52) illustrates the effect of pH and Al(III) ionic species on laurylamine flotation of Bacillus cereus. They also found that an anionic surfactant would readily collect such organisms in the presence of aluminum salts at acidic pH values. These authors concluded from their experimental results and observation of the systems that best flotation took place when the bacteria were present in clumps.

Since obviously microorganisms can be floated it should be possible by flotation to harvest (concentrate) small algae or bacteria that have previously concentrated multivalent metal ions. However, to the present time no work has been performed on the flotation or electrokinetic behavior of microorganisms that have previously adsorbed multivalent ions except for aluminum, ferric or ferrous ions. Subsequent experimental work and discussion in this paper includes electrokinetic studies on the alga Chlorella vulgaris in the presence and absence of aluminum and lead salts and an anionic and a cationic collector.

Experimental Procedure

The Chlorella vulgaris used in the experimentation was grown from a bacteria free culture purchased from Carolina Biological Supply Company. For experimentation the organism was cultured in a nutrient solution

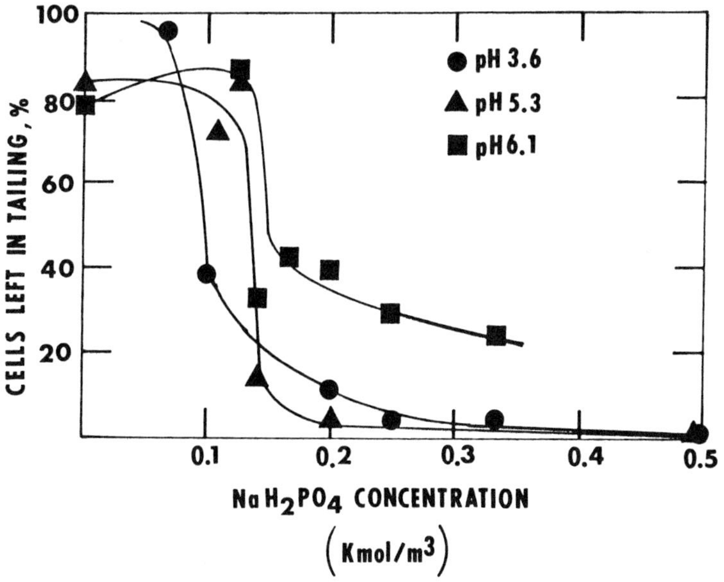

Figure 1. Flotation removal of Escherichia coli cells as a function of NaH_2PO_4 concentration at three different constant pH values. From Gaudin, et.al. (45).

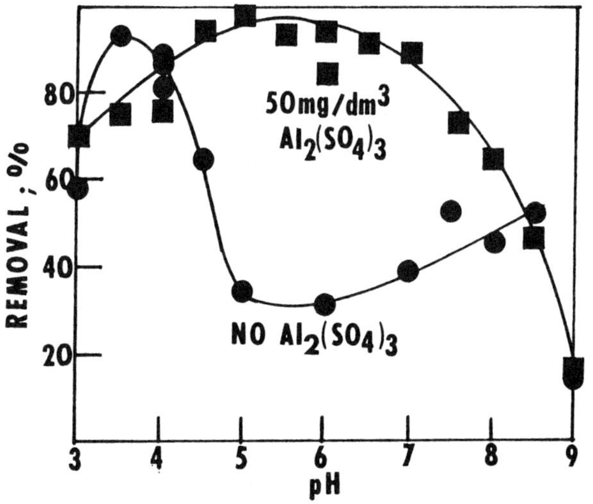

Figure 2. Flotation removal of Bacillus cereus as a function of pH using 20 mg/dm^3 of laurylamine as collector in the absence and presence of $Al_2(SO_4)_3$. From Rubin and Lackey (52).

containing $15 g/dm^3$ agar, $4.36\ g/dm^3$ NaEDTA, $4.36\ g/dm^3$ $FeCl_2$, and trace amounts of Ca^{2+}, Mg^{2+}, Cn^{2+}, Zn^{2+}, Co^{2+}, Mn^{2+}, MoO_4^{2-}, NO_3^- and HPO_4^{2-}. Culturing was done in 250 ml flasks continously shaken in a water shaker bath held at 25°C. Algae used for electrokinetic measurement were taken near the end of the exponential portion of their growth curves (5-7 days). This was done in order to insure the presence of a maximum number of live, active cells. All inorganic chemicals used either in the culture medium or subsequent of algae growth were of reagent grade. The collectors studied, sodium dodecylsulfate and dodecylammonium chloride were purchased and were of high purity.

The electrokinetic measurements were performed using a Komline-Sanderson Zeta Reader in a conventional manner. In the procedure the algae culture was diluted by distilled water addition prior to electrokinetic measurements. The amount of dilution had little effect on the measurements. If metal ions were added to the algae suspension it was done after dilution and a 2 hr. conditioning time allowed to elapse before electrokinetic measurements. The pH was adjusted by HCl or NaOH addition.

Experimental Results

Figure 3 from Smith, et al. (28) illustrates the effect of $Al_2(SO_4)_3$ addition on the zeta potential of Chlorella vulgaris as a function of pH. The curves generated are typical of adsorption and surface precipitation of aluminum hydroxy species onto any solid substrate. Also shown is the electrokinetic curve for the organism in the absence of $Al_2(SO_4)_3$ taken from Smith, et al. (28). The curves should be compared to the flotation results of Figure 2 noting the parallel between zeta potential and flotation as a function of pH when an aluminum salt is present.

Figure 4 indicates the effect of increasing lead acetate, $Pb(Ac)_2$, addition at pH 8.0 ± 0.2 on zeta potential of Chlorella vulgaris. It is apparent that this lead salt can readily reverse the sign of zeta potential of the organism.

Figure 5 shows the effect of $1 \times 10^{-3}\ kmol/m^3$ $Pb(Ac)_2$ addition on the zeta potential of Chlorella vulgaris as a function of pH when 1×10^{-4} $kmol/m^3$ dodecylammonium chloride was present. Also shown are curves for the presence of $1 \times 10^{-4}\ kmol/m^3$ amine in the absence of $Pb(Ac)_2$ and for the absence of both the amine and the lead salt. Note that the addition of amine alone shifts the zeta potential versus pH curve to somewhat more

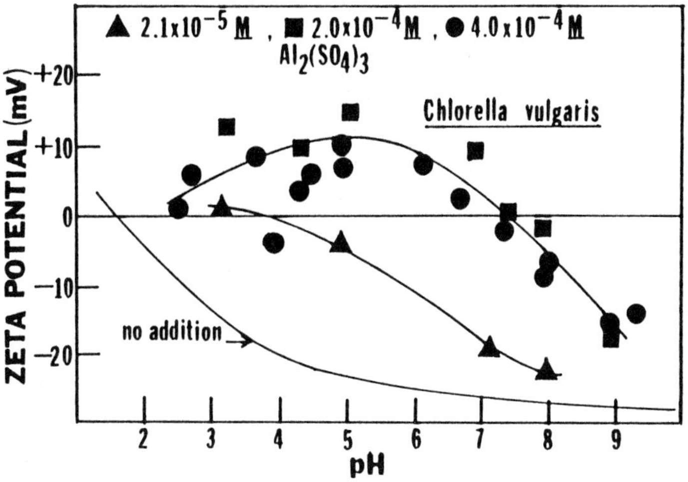

Figure 3. Effect of $Al_2(SO_4)_3$ addition on the zeta potential of Chlorella vulgaris as a function of pH. From Smith et.al. (28).

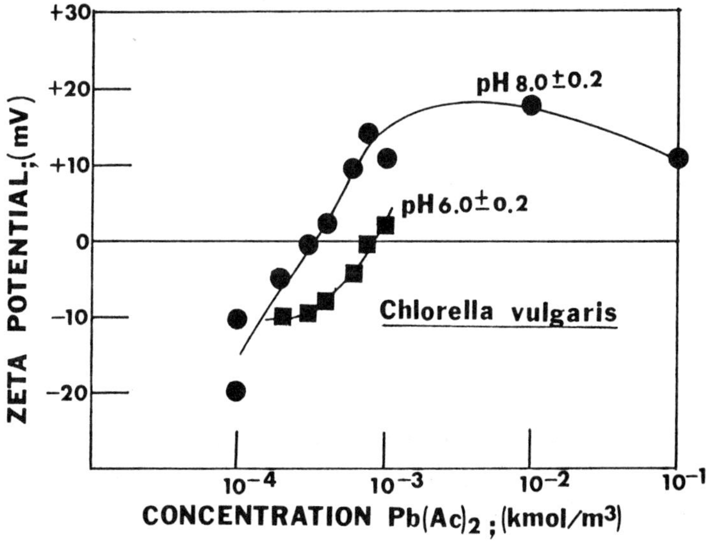

Figure 4. Zeta potential of Chlorella vulgaris as a function of lead acetate concentration at pH 8.0 $\pm$ 0.2 and pH 6.0 $\pm$ 0.2.

positive zeta potential values and that the pH of zero zeta potential is shifted to a more basic pH value. The further addition of $Pb(Ac)_2$ affects the zeta potential value less at pH values below about pH 4, but at higher pH values markedly affects zeta potential. Note the two zero zeta potential values obtained at about pH 6.6 and pH 10.5. The results are similar to those for addition of the aluminum salt and are typical of systems where metal hydroxy species adsorb or surface precipitate onto a solid substrate.

Figure 6 shows similar data to that of Figure 5 except that 1×10^{-4} $kmol/m^3$ sodium dodecylsulfate was present rather than dodecylammonium chloride. At acid pH values the alkylsulfate apparently adsorbs onto the algae resulting in more negative zeta potential values. However, as the pH becomes more basic and the organism's surface becomes more negative the presence of the alkylsulfate fails to increase the negative charge. In fact the zeta potential appears to become more positive. The reason for this latter phenomenon is not known.

When 1×10^{-3} $kmol/m^3$ $Pb(Ac)_2$ is also present the curve generated is almost identical to the similar curve of Figure 5 and the same comments obtain.

Discussion

Many researchers have shown beyond doubt that microorganisms remove from aqueous solutions and remarkably concentrate multivalent metal ions, and other researchers have shown that microorganisms can be readily floated. Rubin and Lackey (52) have also shown that the organisms can still be floated even if a multivalent metal ions (Al[III] ions) are present in the systems and in fact the multivalent ions can enhance flotation.

Preliminary work in our laboratory indicates that even minute quantities of *Chlorella vulgaris* (<0.5 wt.%) can remove lead from an aqueous solution. Table I lists our preliminary data.

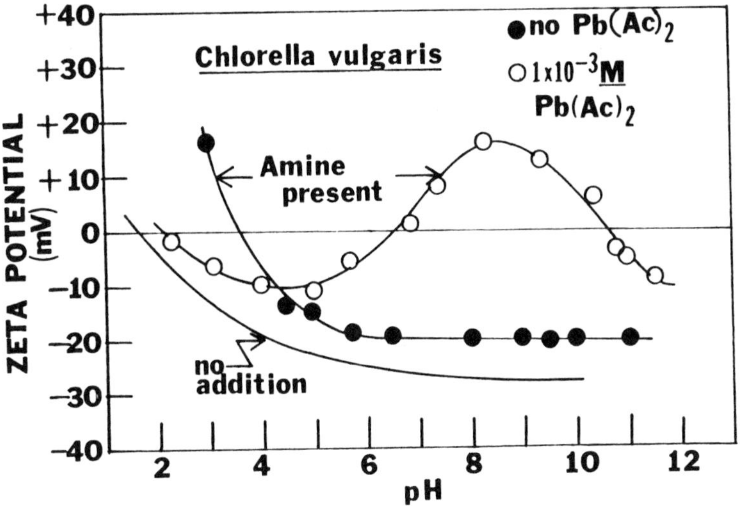

Figure 5. Effect of the presence of 1x10^{-4} kmol/m^3 dodecylamine and 1x10^{-4} kmol/m^3 dodecylamine plus 1x10^{-3} kmol/m^3 lead acetate on the zeta potential of Chlorella vulgaris as a function of pH.

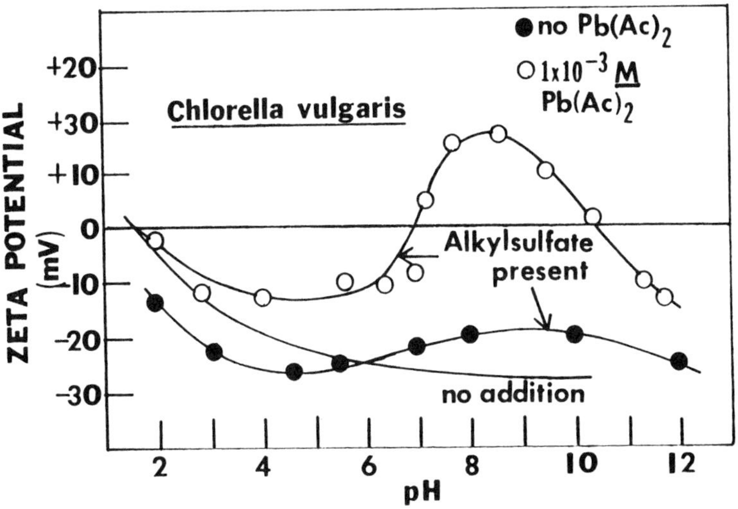

Figure 6. Effect of the presence of 1x10^{-} kmol/m^3 sodium dodecylsulfate and of 1x10^{-4} kmol/m^3 sodium dodecylsulfate plus 1x10^{-3} kmol/m^3 lead acetate on the zeta potential of Chlorella vulgaris as a function of pH.

TABLE 1

Removal of Pb(II) From Solution by Chlorella vulgaris
(pH 8.0 ± 0.1)

Concentration Pb Before Algae Addition ($kmol/m^3$)	Concentration Pb After Algae Addition ($kmol/m^3$)	% Removal
	SAMPLE 1	
2.41×10^{-4}	6.54×10^{-5}	72.9
4.83×10^{-4}	1.86×10^{-4}	61.8
7.24×10^{-4}	2.84×10^{-4}	60.7
9.65×10^{-4}	3.62×10^{-4}	62.5
1.21×10^{-3}	7.09×10^{-4}	41.3
	SAMPLE 2	
2.41×10^{-4}	6.67×10^{-5}	72.4
4.83×10^{-4}	2.03×10^{-4}	58.0
7.24×10^{-4}	3.25×10^{-4}	55.1
9.65×10^{-4}	4.44×10^{-4}	54.0
1.21×10^{-3}	5.90×10^{-4}	51.1

In this experimentation ICP analysis was used for lead determination. The algae were allowed to react with the Pb(II) solution for one day before analysis of the filtered solution. Work by Greene et al. (22) confirms the strong uptake of Pb(II) onto Chlorella vulgaris. Figure 7 from these authors demonstrates this uptake plus the uptake of several other metals ions. Recent electron microscopy work by Gołab (63) has shown that lead strongly adsorbs onto the cell walls of Bacillus mycoides and does not appear to enter the interior of this organism.

Our electrokinetic measurements indicate that Pb(II) functions in a manner very similar to Al(III) and thus, Chlorella vulgaris which has concentrated Pb(II) should be readily floated by the froth or foam flotation processes. It may well be that due to the small and light nature of such organisms a column type flotation device would be most suitable for microorganism flotation.

Judging from Figure 7, it is evident that many other metal ions are concentrated by Chlorella vulgaris and could also be concentrated by

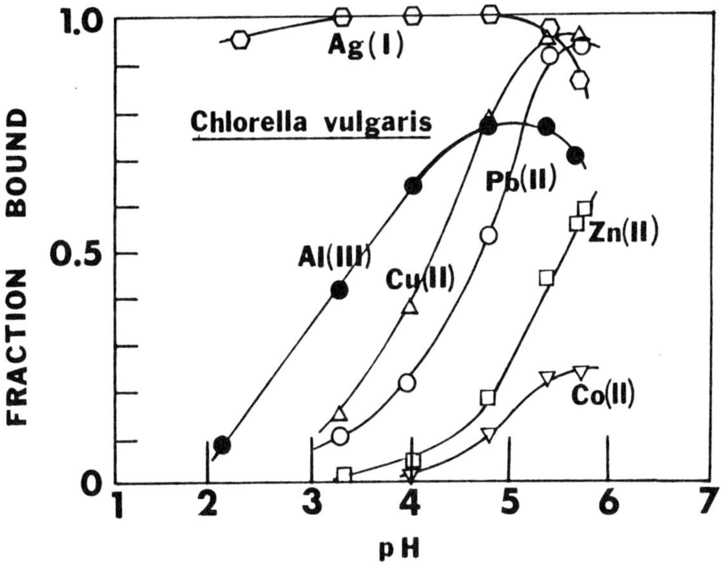

Figure 7. The pH dependence of metal ion sorption by Chlorella vulgaris from solutions containing equimolar concentrations of various metal ions. Lyophilized C. vularis cells were washed and suspended at 5 mg/ml for 2 hr. in solutions containing 0.1 mM in each of the metal ions. From Greene, et.al. (22).

flotation. Also there is no reason to suppose that Chlorella vulgaris is an optimum microorginism ad or absorbent for metal ions. Thus many other microorganisms should work as well or better than this alga .

Overall, then, the prospect of using microorganisms to clean up metal containing aqueous waste streams or to recover valuable metals from such streams or various process streams is indeed intriguing. Obviously, however, much more research is necessary in order to devise economic, optimum schemes for the flotation of microrganisms that have concentrated metal ions.

Aknowledgement

This research was supported by the Department of the Interior's Mineral Institutes Program administered by the Bueau of Mines through the Generic Mineral Technology Center for Mineral Industry Waste Treatment and Recovery under allotment grant number G1125132.

References

1. L.C. Bryner and R. Anderson, "Microorganisms in the Leaching of Sulfide Minerals," Ind. Eng. Chem., 49(1957) 1721-1724.

2. E.E. Malouf and J.D. Prater, "Role of Bacteria in the Alternation of Sulfide Minerals," J. Metals, 14(1961) 353-356.

3. A. Bruynesteyn and R.P. Hackl, "The Biotank Leach Process for the Treatment of Refractory Gold/Silver Concentrates," in: Microbiological Effects on Metallurgical Processes, eds. J.A. Clum and L.A. Haas, The Metallurgical Society, (1986) 121-127.

4. Y. Attia, J. Litchfield and L. Vaaler, "Applications of Biotechnology in the Recovery of Gold," in Microbiological Effects on Metallurgical Processes, eds. J.A. Clum and L.A. Haas, The Metallurgical Society (1986) 111-120.

5. A.B. King and A.W. Dudeney, "Bioleaching of Nepheline," Hydromet., 19 (1987) 69-81.

6. A.G. Kempton, N. Moneib, G.R.L. McCready and C.E. Capes, "Removal of Pyrite from Coal by Conditiong with Thiobacillus ferrooxidans Followed by Oil Agglomeration," Hydromet., 5 (1980) 117-125.

7. M. Ismail and Y.A. Attia, "Biodepression of Pyrite Floatability in Oxidized Coal Cleaning," (Paper presented at 19th Annual Meeting of the Fine Particle Society, Santa Clara, CA, July 19-22, 1988.)

8. P. Denny and R.P. Welsh, "Lead Accumulation in Plankton Blooms from Ullswater, The English Lake District," Environ. Pollut., 18 (1979) 1-9.
9. J.C. Jennett, J.M. Hassett, and J.E. Smith, "Control of Heavy Metals in the Environment Using Algae," in: International Conference, Management and Control of Heavy Metals in the Environment. London (1979) 210-217.

10. S. Aaronson, T. Berner, and Z. Dubinsky, "Microalgae as a Source of Chemicals and Natural Products," in: Algae Biomass, Eds. G. Sheleg and C.J. Soeder, Elsevier (1980) 575-601.

11. D. Khummongkol, G.S. Canterford, and C. Fryer, "Accumulation of Heavy Metals in Unicellular Algae," Biotech. Bioeng., 24 (1982) 2643-2660.

12. G.W. Strandberg, S.E. Shumate II, and J.R. Parrott, "Microbiol Cells as Biosorbents for Heavy Metals: Accumulation of Uranium by Saccharomces cerevisiae and Pseudomonas aeruginosa," Appl. Environ. Microbiol., 41 (1981) 237-245.

13. J.A. Brierley and C.L. Brierley, "Biological Accumulation of Some Heavy Metals - Biotechnological Applications," in: Biomineralzation and Biological Metal Accumulation, eds, P. Westbrolk, and E.W. de Jong, D. Reidel Publ. Co. (1983) 499-509.

14. F.J. Sloan, A.R. Abernathy, J.C. Jennett, and G.V. Goodman, "Removal of Metal Ions from Wastewater by Algae," in: Industrial Waste Conference Proceedings, 38th (1983) 423-429.

15. D.W. Darnall, B. Greene, J.M. Hosea, R.A. McPherson, M. Henzl and M.D. Alexander, "Recovery of Heavy Metals by Immobilized Algae," in: Trace Metal Removal from Aqueous Solution, ed. R. Thompson, Special Publ. No. 61, The Royal Soc. of Chem., London (1986) 1-24.

16. D.W. Darnall, B. Greene, M.T. Henzl, J.M. Hosea, R.A. McPherson, J. Sneddon and M.D. Alexander, "Selective Recovery of Gold and Other Metal Ions from an Algal Biomass," Environ, Sci. Technol., 20, (1986) 206-208.

17. B. Greene, J.M. Hosea, R. McPherson, M. Henzl, M.D. Alexander and D.W. Darnall, "Interaction of Gold (I) and Gold (III) Complexes with Algal Biomass," Environ. Sci. Technol., 20 (1986) 627-632.

18. B. Green, M.T. Henzl, J.M. Hosea and D.W. Darnall, "Elimination of Bicarbonate Interference in the Binding of U(VI) in Mill-Waters to Freeze-dried Chlorella Vulgaris," Biotech. Bioeng., 28 (1986) 764-767.

19. I.C. Hancock, "The Use of Gram-positive Bacteria for the Removal of Metals from Aqueous Solution," in: Trace Metal Removal from Aqueous Solution, ed. R. Thampson, Special Publ. No. 61, The Royal Soc. of Chem., London (1986) 25-43.

20. J.M. Hosea, B. Greene, R. McPherson, M. Henzl, M.D. Alexander and D.W. Darnall, "Accummulation of Elemental Gold on the Alga Chlorella vulgaris," Inorg. Chem. Acta. 123 (1986) 161-165.

21. S.R. Hutchins, M.S. Davidson, J.A. Brierley, and C.L. Brierley, "Microorganisms in Reclamation of Metals," Ann. Rev. Microbiol., 40 (1986) 311-336.

22. B. Greene, R. McPherson and D. Darnall, "Algal Sorbents for Selective Metal Ion Recovery," in: Metals Speciation, Separation and Recovery, eds. J.W. Patterson and R. Passimo, Lewis Publ. Inc. (1987) 315-338.

23. T.J. Beveridge, "Mechanisms of the Binding of Metallic Ions to Bacterial Walls and the Possible Impact on Microbiol Ecology," in: Curr. Perspect. Microb. Ecol. Proc., Int. Symp., 3rd, eds., M.J. Klug and C.A. Reddy, Am. Soc. Microb. (1984) 601-607.

24. T.J. Beveridge and S.F. Koval, "Binding of Metals to Cell Envelopes of Escherichia coli K-12." Appl. Environ. Microbiol. 42 (1981) 325-335.

25. R.H. Crist, K. Oberholser, N. Shank and M. Nguyen, "Nature of Bonding between Metal Ions and Algal Cell Walls," Environ. Sci. Tech. 15 (1981) 1212-1217.

26. P. Somasundaran, "Foam Separation Methods," in: Separation and Purification Methods, Vol. 1, eds, E.S. Perry and C.J. Van Oss, (1973) 117-199.

27. R.W. Smith, "Flotation of Algae Bacteria and Other Microorganisms," in press, to be published in Mineral Processing and Extractive Metallurgy Reviews (1988).

28. R.W. Smith, M.A. Beck and R.A. Wharton, Jr., "Microorganism Flotation," (Paper presented at the Annual Meeting of the Fine Particle Society, Santa Clara, CA, July 22-26, 1988, and being reviewed for publication by the Fine Particle Society).

29. S.L. Daniels, "Separation of Bacteria by Adsorption onto Ion Exchange Resins," Ph.D. Thesis, Univ. of Michigan, Ann Arbor, Dist. Abstr., 29, 1336B (1968), (1967) 466 pp.

30. M.R.J. Salton, The Bacterial Cell Wall, Elsevier Publ. Co., (1964) 293 pp.

31. S. Heptinstall, A.R. Archibald and J. Baddiley, "Teichoic Acids and Membrane Function in Bacteria," Nature, 225 (1970) 519-521.

32. S.L. Daniels, "The Adsorption of Microorganisms onto Solid Surfaces: A Review," Dev. in Indust. Microbiol., Vol. 13 (1972) 211-253.

33. S.L. Daniels, "Mechanisms Involved in Sorption of Microorganisms to Solid Surfaces," in: Adsorption of Microorganisms to Surfaces, eds. G. Britton and K.C. Marshall, John Wiley and Sons (1980) 7-58.

34. R.H. Crist, K. Oberholser, D. Schwartz, J. Marzoff, D. Ryder and D.R. Crist, "Interactions of Metals and Protons with Algae," Environ. Sci. Technol., 22 (1988) 755-760.

35. S. Hunt, "Diversity of Biopolymer Structure and Its Potential for Ion-Bonding Applications," in: Immobilization of Ions by Bio-sorption, eds. H. Eccles and S. Hunt, Ellis Horwood Ltd., Chichester (1986) 15-46.

36. G.M. Gadd, "The Uptake of Heavy Metals by Fungi and Yeasts: The Chemistry and Physiology of the Process and Applications for Biotechnolgy," in: Immobilization of Ions by Bio-sorption, eds. H. Eccles and S. Hunt, Ellis Horwood Ltd., Chichester (1986) 135-147.

37. A. Dognon, "Concentration et Separation des Molecules et des Particules par las Methode des Mousses," Rev. Sci., 79 (1941) 613-619.

38. A. Dognon, "Concentration et Separation par les Mousses," Bull. Soc. Chim., (1941) 249-262.

39. S.H. Hopper and M.C. McCowen, "A Flotation Process for Water Purification," J. Am. Water Works Assoc., (1952) 719-726.

40. W.A. Boyles, and R.E. Lincoln, "Separation and Concentration of Bacterial Spores and Vegetative Cells by Foam Flotation," Applied Microbiol. (1958) 327-334.

41. A.M. Gaudin, A.L. Mular and R. F. O'Connor, "Separation of Microorganisms by Flotation: I Development and Evaluation of Assay Procedures," Applied Microbiol., 8(1960) 84-90.

42. A.M. Gaudin, A.L. Mular and R.F. O'Connor, "Separation of Microorganisms by Flotation: I. Flotation of Spores of Bacillus subtilis var Niger," Applied Microbiol., 8(1960) 91-97.

43. A.M. Gaudin, "Flotation of Microorganisms," Froth Flotation, 50th Anniversary Volume. ed. D.W. Fuerstenau, AIME (1962) 658-667.

44. A.M. Gaudin, N.S. Davis and S.E. Bangs, "Flotation of Escherichia coli with Sodium Inorganic Salts," Biotech. Bioeng., 4(1962) 211-222.

45. A.M. Gaudin, N.S. Davis and S.E. Bangs, "Flotation of Escherichia coli with Some Inorganic Salts," Biotech.Bioeng.,4 (1962) 223-230.

46. G.V. Levin, J.R. Clendenning, A. Gibor and F.D. Bogar, "Harvesting of Algae by Froth Flotation," Applied Microbiol., 10 (1962) 169-175.

47. P.H. Smith, W. H. Funk and D.E Proctor, "Froth Flotation for Harvesting Chlorella Algae," Northwest Science, 42 (1968) 165-171.

48. W.H. Funk, W.J. Sweeney and D.E. Proctor, "Dissolved-Air Flotation for Harvesting Unicellular Algae," Water Sewage Works, 115 (1968) 343-347.

49. R.B. Grieves and S. Wang, "Foam Separation of Escherichia coli with Cationic Surfactant," Biotech. Bioeng., 8 (1966) 323-336.

50. A.J. Rubin, E.A. Cassel, O. Henderson, I.D. Johnson and J.C. Lamb,III, "Microflotation: New Low Gas-Flow Rate Foam Separation Technique for Bacteria and Algae," Biotech. Bioeng., 8(1966) 135-151.

51. A.J. Rubin, "Microflotation: Coagulation and Foam Separation of Aeobacter aerogenes," Biotech. Bioeng., 10(1968) 89-98.

52. A.J. Rubin and S.C. Lackey, "Effect of Coagulation on the Microflotation of Bacillus cereus," J. Am. Water Works Assoc., 60 (1968) 1156-1166.

53. A.J. Rubin, "Removal and Use of Hydrolyzable Metals in Foam Separations," in: Adsorptive Bubble Separation Techniques, ed. R. Lemlich, Academic Press (1972) 199-217.

54. E.A. Cassell, E. Matijevic, F.J. Mangravite, Jr., T.M. Buzzell and S.B. Blabac, "Removal of Colloidal Pollutants by Microflotation," AIChE Journ., 17 (1971) 1486-1492.

55. W.F.R. Bare, N.B. Jones and E.J. Middlebrooks, "Algae Removal Using Dissolved Air Flotation," Journ. Water Pollut. Control Federation, 47 (1975) 153-169.

56. E.J. Middlebrooks, D.B. Porcella, R.A. Gearheart, G.R. Marshall, J.H. Reynolds and W.J. Grenney, "Techniques for Algae Removal from Wastewater Stabilization Ponds," Journ. Water Pollut. Control Federation, 46 (1974) 2676-2695.

57. M.E. Tittlebaum and S. Holtman, "Algae Removal by Induced Air Flotation," Completion Report for Office of Water Research and Technology, U.S. Dept. of the Interior, (1982) 33 pp.

58. J. Arbelaez, B. Koopman and E.P. Lincoln, "Effects of Dissolved Oxygen and Mixing on Algal Autoflotation," Journ. WPCF, 55 (1983) 1075-1079.

59. S.S. Huneycutt, D.A. Wallis and F. Sebba, "A Technique for Harvesting Unicellular Algae Using Colloidal Gas Aphrons," in: Biotech. Bioeng. Symp. No. 13, John Wiley and Sons, (1983) 567-575

60. R. Klute and U. Neis, "Coagulation of Algae and Subsequent Removal by Filtration or Flotation," Water Supply, (1983) 157-162.

61. E.S. Sharpe, A.I. Herman and S.C. Toolan, "Separation of Spores and Parasporal Crystals of Bacillus thuringiensis by Flotation," Journ. of Invertebrate Path., 34 (1979) 315-316.

62. C.A. Hansen and H.B. Gotaas, "Sewage Treatment by Flotation," Sewage Works J., 15 (1943) 212-252.

63. Z. Gołąb, Private communication with R.W. Smith, Technical University of Wrocław, Poland, 13 June 1988.

BIOTECHNOLOGY APPLIED TO RAW MATERIALS PROCESSING

S. K. Kawatra and T. C. Eisele
Department of Metallurgical Engineering
Michigan Technological University
Houghton, MI 49931

Abstract

Recent advances in microbiology have made the application of biotechnology to metallurgical processes possible. Hydrometallurgy stands to gain the most from the use of microorganisms, as they are useful for both dissolution aids and for removing metals from solution. The development of genetic engineering techniques promises to greatly increase the importance of bioprocessing of metals and minerals by the year 2000.

In this paper, the basic effects of microorganisms on metallurgical processes are reviewed, and the applications which are currently either in use or near being used are presented. Representative data collected at Michigan Technological University is also given.

Metallurgical Processes for the Year 2000 and Beyond
Edited by H.Y. Sohn and E.S. Geskin
The Minerals, Metals & Materials Society, 1988

Introduction

Many hydrometallurgical operations exist which are thermodynamically possible, but which are industrially impractical due to slow kinetics, high reagent costs, or the need for a highly corrosive environment. However, recent discoveries in microbiology indicate that in many cases these difficulties can be reduced greatly by the action of bacteria and other microorganisms (1-8).

A major advantage of using living organisms in hydrometallurgy is that once the culture is established the microorganisms produce their own reagents either from the material being processed or from low-cost nutrient supplements. They thus allow extremely complex chemical reactions to be carried out at reasonable cost. The major drawback of using microorganisms is their inability to tolerate extremely high temperatures, excessive concentrations of toxic metals, or very highly acidic, alkaline, or corrosive conditions. However, the complex reactions which bacteria make possible frequently eliminate the need for such extreme conditions. Also, in recent years a number of organisms have been isolated from such exotic environments as hot springs and deep-ocean vents which can tolerate temperatures higher than was previously thought possible for any organism, and can catalyze a number of metallurgically important reactions (8).

Of the numerous potential applications for microbial processing, only a few, notably the dump leaching of copper and uranium, are currently used industrially (8). In their present fairly primitive level of development, many biohydrometallurgical processes are too slow to be economical. However, work is underway at Michigan Technological University and elsewhere to develop methods for speeding up these processes. This will result in biohydrometallurgy being well developed and widely used by the beginning of the next century.

Microorganism Characteristics

Microorganisms can be conveniently classified according to what they "eat", as given in Table 1. The two most familiar types of organisms are the chemoheterotrophs, which extract energy from chemical reactions and construct cellular material from organic compounds, and the photoautotrophs, which extract energy from light and construct cellular

Table 1: Classification of Microorganisms

	Energy Source	Carbon Source	Examples
Lithotrophs	inorganic chemical reactions	CO_2	*Thiobacillus ferrooxidans*, *Sulfolobus* species
Chemoheterotrophs	chemical reactions	organic chemicals	most bacteria, all fungi, and protozoans
Photoautotrophs	sunlight	CO_2	photosynthetic bacteria, algae
Photoheterotrophs	sunlight	organic chemicals	purple and green sulfur bacteria

material from carbon dioxide and water. However, the impact of these organisms on minerals and dissolved metals is typically minimal. The lithotrophs are much more important for hydrometallurgical applications, as these organisms extract their metabolic energy from inorganic chemical reactions, such as the oxidation of metals to more soluble forms. Such organisms are therefore often useful for accelerating oxidative leaching reactions (9).

A major consideration in any type of bioprocessing is that, unlike chemical reagents, the cells can reproduce when conditions are favorable, and die when they are not. It is therefore necessary to more carefully control the conditions in a bioreactor than in a reactor which uses only chemical reagents. In general, a bacterial population in a batch process will behave as shown in Figure 1, assuming that the initial conditions are suitable for growth. For a short period after addition, the microorganisms will not reproduce, as they are adjusting their metabolisms to the prevailing conditions. Following this "lag phase", the organisms will enter the exponential growth phase, where cells are actively reproducing and few or none are dying. Eventually, however, some factor such as limited nutrient supply, toxin accumulation, or a shortage of living space causes the microorganisms to die as fast as they reproduce, and the cell numbers become roughly constant. Ultimately, conditions become such that the culture can no longer maintain its numbers, and wholesale death of cells results. The activity of a culture of cells is greatest when they are in the exponential growth phase, and this condition is therefore preferred for biohydrometallurgical applications. Since high cell numbers are also desirable, the optimum condition is exponential growth with the culture just short of entering the stationary phase. The cell concentration at which this occurs can be increased by reducing the influences of limiting factors, such as by increasing the concentrations of critical nutrients and by increasing the tolerance of the microorganisms to the toxins produced by their activity.

Basic Biometallurgical Processes

There are three general ways in which microbes can be useful in metallurgy: 1) as aids in the dissolution of minerals in water solutions; 2) as agents for altering the surface chemistry of mineral particles, and 3) as selective accumulators of metal ions from dilute solutions. Each of these applications depend on completely different properties of the cells used, and therefore the microorganisms best suited for each differ greatly from one another.

Bioleaching

Mineral leaching is the only type of biometallurgical process which is currently used on an industrial scale. The organisms which have been most intensively used are bacteria of the genus *Thiobacillus*, which get their metabolic energy from the oxidation of sulfur compounds to sulfates. Since sulfates are typically much more soluble than the corresponding sulfides, these bacteria are very useful for sulfide mineral leaching. The two most important of these organisms are *Thiobacillus ferrooxidans* and *Thiobacillus thiooxidans*, which frequently occur in a natural mixed culture which oxidizes sulfides more rapidly than can be done by either organism alone. A number of other sulfur-oxidizing bacteria have also been discovered, and there full capabilities are currently being explored (10,11). However, the *Thiobacillus* cultures will naturally arise in heap and dump leaching operations without the need for specific inoculation, which is a powerful argument in their favor (12).

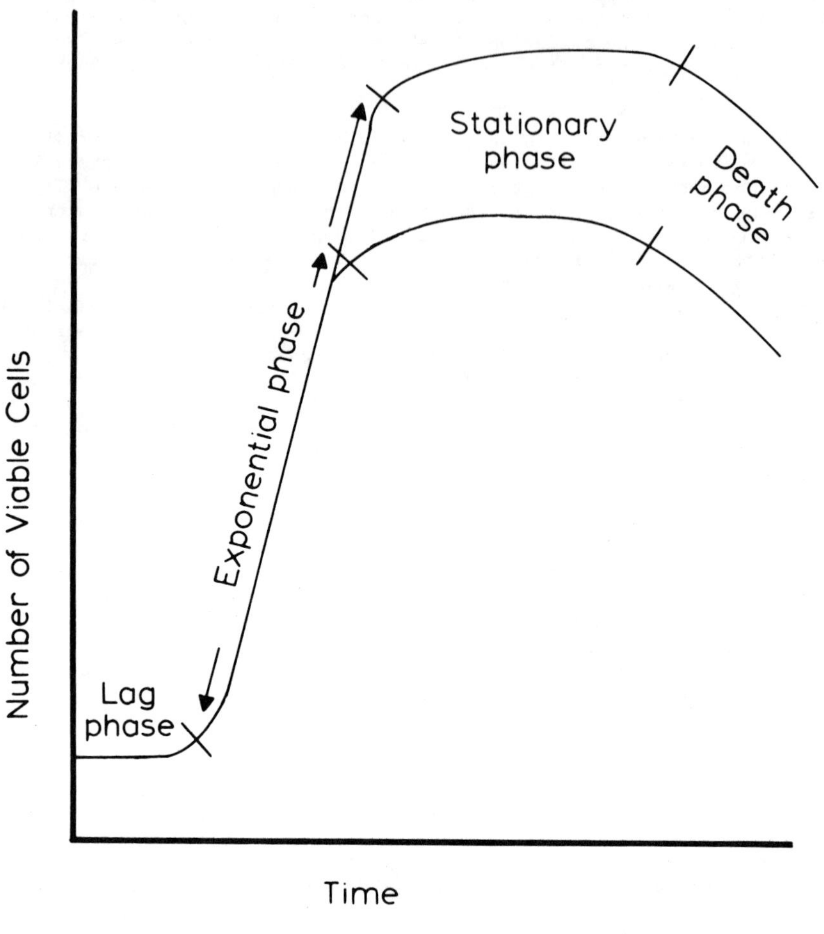

Figure 1. Variation of bacterial numbers over time in a batch culture. The time scale varies from hours for certain chemoheterotrophic bacteria to weeks for lithotrophs such as T. ferrooxidans. The addition of supplemental nutrients will often be able to increase the maximum cell concentration as shown. In a continuous reaction, it is possible to maintain a culture in the expontential growth phase by controlled removal of cells.

The mechanisms through which T. ferrooxidans and T. thiooxidans oxidize sulfide minerals are not fully understood, as many of the biochemical reactions involved require that the bacteria be either contacting or in close proximity to the mineral surface. However, the primary mechanism used by T. ferrooxidans, termed indirect oxidation, can proceed even when the bacteria are completely dissociated from any particular mineral particle, and can therefore be examined fairly easily. In this pathway, the bacteria generate ferric iron from ferrous iron, according to the reaction:

$$4\ Fe^{+2} + O_2 + 4H^+ \xrightarrow{\text{Bacteria}} 4\ Fe^{+3} + 2H_2O$$

Since this is an exothermic reaction, the bacteria can gain metabolic energy from the process. The ferric iron generated can then attack sulfide minerals, pyrite for example:

$$FeS_2 + 14\ Fe^{+3} + 8H_2O \rightarrow 15\ Fe^{+2} + 2SO_4^{-4} + 16H^+$$

with the ferrous iron produced then being reoxidized by the bacteria (13-16). While this reaction will occur even in the absence of bacteria, the regeneration of ferric iron is accelerated several hundred fold by T. ferrooxidans. Since the solubility of ferric iron is very low when the pH is higher than about 4, indirect oxidation can obviously only be important in acid solutions (13). For this reason sulfide-oxidizing bacteria which can use this pathway are specifically adapted to solutions with a pH between 1.5 and 3 and will be killed if the pH exceeds 4.

Although bacteria greatly increase the oxidation rate of sulfide minerals, the reactions are still frequently too slow to be industrially practical. Often the reaction rate could be increased by simply increasing the temperature, but the maximum temperature is limited by the heat tolerance of the organisms used. In the case of the Thiobacilli, this maximum is around 35-40°C. However, numerous examples of bacteria have been isolated from sulfurous hot springs worldwide which live at temperatures from 60-98°C. Some of these bacteria, such as the Sulfolobus genus, oxidize and reduce sulfur and iron compounds (11). Bacteria have also been isolated from undersea hydrothermal vents which grow at high pressure and at temperatures somewhat above 100°C, and may therefore be useful for exotic applications such as in-situ leaching of extremely deep mineral deposits. Unlike the Thiobacillus genus, these high-temperature organisms typically require organic nutrient supplements. Also, as a result of their adaptation to high temperatures the cells are somewhat fragile, and may be torn apart in high-shear applications (11).

Mineral Surface Alteration

Many microorganisms have the ability to attach themselves to solid surfaces, either because the surface supplies some nutrient to the cell or simply to keep the cell from being randomly swept away by flowing water (9). This attachment occurs rapidly, typically in times shorter than 15 minutes, and can be very selective. Lithotrophic bacteria in particular will often specifically attach to those minerals which they can most effectively degrade, and will be attracted to crystalline imperfections where dissolution will occur most rapidly (17). Since the surface chemistry of a bacterial cell is quite different from that of the mineral surface, this would be expected to strongly influence the behavior of the mineral in processes such as froth flotation or oil agglomeration. A typical application would be the use of bacteria to depress pyrite during coal flotation. Since attachment is far more rapid than dissolution, bacterially-aided flotation may very well be preferable to bacterial leaching for certain mineral systems.

Bioaccumulation

It has been found that many microorganisms are capable of specifically adsorbing metal ions from dilute solution (8). This is particularly useful for treatment of mining and processing effluents, with low-cost biomass being used to remove and concentrate heavy metals. Depending on the metal adsorbed, it may be possible to use dead biomass instead of living cells, thus reducing the control problems which result from maintaining live cultures (18).

The microorganisms useful for bioaccumulation can be of any type, and need not be specifically lithotrophic. Often, the accumulation of metals is a side effect of the cell metabolism, with the metals accumulated poisoning the cell. Since the microorganisms are not gaining useful material from the adsorbed ions, and indeed may be killed by them, it is preferable to grow the microbes in a separate vessel, and only add them to the process when they are in the optimum condition for metal adsorption.

Current Applications

A number of promising applications of bacteria to metallurgy have been identified in the laboratory, particularly in the last few years (8). However, very few are currently applied industrially. This limited application is due more to a lack of development than to any insurmountable failing of bacterial processing. The most common problem with bioprocessing, and particularly for leaching processes, is the long leaching time frequently required. The primary thrust of research in bioleaching is therefore to improve its kinetics. Until the dissolution rate can be increased, only leaching processes where speed is not essential, such as dump and in-situ leaching, will be practical.

Copper Leaching

Bioleaching of copper minerals has been carried out since at least the 17th century, although the necessity for bacterial action in acid dump leaching was not discovered until the early 1940's. The bacteria responsible for copper leaching are generally *Thiobacillus* species, particularly *T. ferrooxidans*. Most important copper minerals, including oxides, carbonates, and silicates as well as sulfides, are either naturally soluble in sulfuric acid or become soluble after being indirectly oxidized by bacteria. Bacterial leaching of copper is particularly useful when pyrite is present in the ore, as it can then generate both its own acid and the dissolved iron needed for indirect oxidation. This obviously keeps reagent costs low (19). If insufficient pyrite is present, the necessary iron can be dissolved during cementation of the copper from solution with metallic iron.

Current copper leaching practice is restricted to heap and dump leaching (19). However, in-situ bacterial leaching is also a strong possibility. The major hurdle is the difficulty of uniformly supplying well-oxygenated leaching solutions to the ore body. One method being investigated is to carry out bacterial iron oxidation in reactors on the surface, and then to pump the ferric iron/sulfuric acid solution through the ore body, leaving the bacteria behind, as shown in Figure 2. While this improves the aeration situation, it also eliminates most of the contribution of bacteria which directly oxidize sulfur, and need to be in contact with the mineral surface. In addition to leaching using *Thiobacillus* species, there is potential for using organisms from hot springs and deep-sea hydrothermal vents for in-situ leaching of very deep, high temperature deposits which cannot be economically mined by any other means (8,11).

Coal

Although bioprocessing has not been applied to coal on an industrial scale, a number of possible applications have been examined in the laboratory. The primary thrust has been in coal desulfurization (21-32),

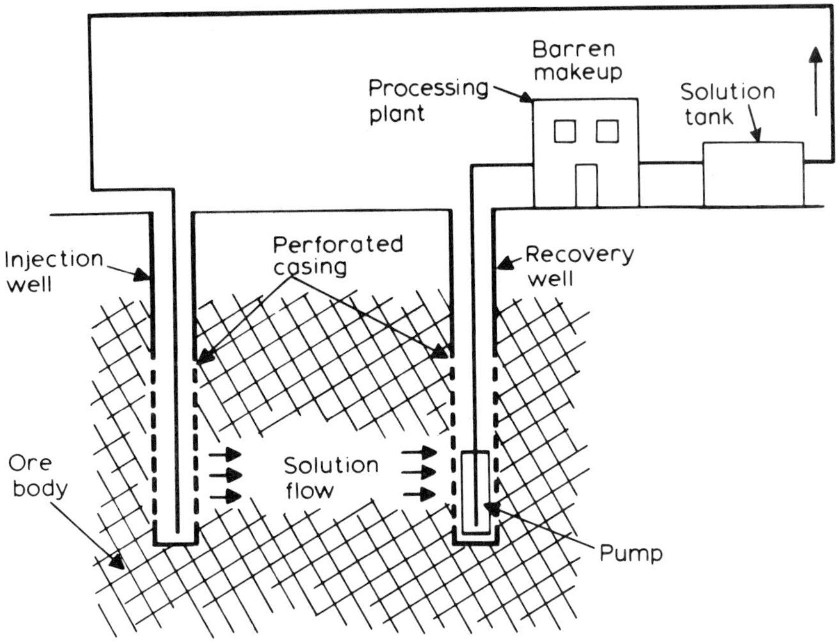

Figure 2. General in-situ leaching arrangement. For leaching of copper by Fe^{+3} in acid solution, the bacterial oxidation of Fe^{+2} to Fe^{+3} can be carried out above-ground. The oxidized solution is then pumped through the permeable ore body. This allows for thorough aeration of the solution during the iron reoxidation, which would be difficult underground.

although some work has also been done recently in biological liquefication (20).

Since a major portion of the sulfur in coal is in the form of pyrite, most research has focused on bacterial dissolution of pyritic sulfur. The reactions are therefore similar to those encountered in copper leaching, and can be carried out by Thiobacillus species. However, heap and dump leaching of coal are not generally suitable for desulfurization, as removal of pyrite is not sufficiently complete. More expensive methods, such as stirred tank reactors, are therefore required, and the one to three week residence times necessary make capital costs prohibitive (30). Thorough oxygenation, such as by the use of pachuca tank reactors, produces a significant increase in leaching rate over simple stirred reactors, as shown in results collected at Michigan Technological University (Figure 3). These results show that for an Ohio coal, pachuca tanks reduce the lag period by about three days, or 25%, and increase the ultimate pyrite removal. However, equipment design has not yet improved leaching performance sufficiently to make this means of coal desulfurization economical.

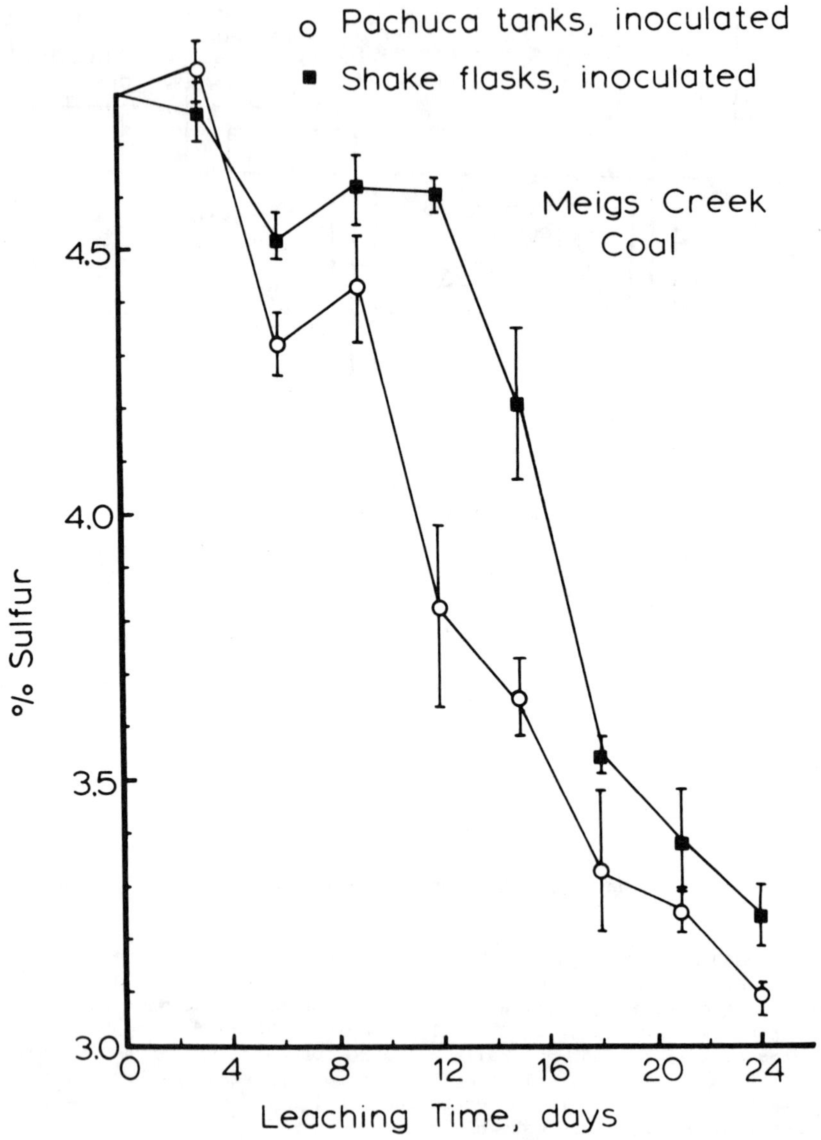

Figure 3. Comparision of the bacterial leaching of pyrite from coal in pachuca tanks and shake flasks. The superior aeration of the pachuca tanks produces a significant increase in the leaching rate, and also results in a greater ultimate removal of pyrite from the coal.

The use of high-temperature bacteria of the Sulfolobus genus is also being investigated for pyrite removal (24) as the increased temperatures produce a corresponding increase in leaching rate. While this decreases the equipment volume needed, the operating cost is increased due to the higher energy consumption. It has been suggested that the leaching operation be installed at the power plant, so that the process can be heated by the waste heat from the generators, thus obtaining the heat for free.

An alternate, much more rapid means of using bacteria to remove pyrite depends on selective attachment of bacteria to the pyrite surface (33-38). The bacteria coating the pyrite particles make the particles strongly hydrophilic, and thus increase the ability of processes such as froth flotation and oil agglomeration to distinguish between coal and pyrite. A number of investigators have used Thiobacillus ferrooxidans for this purpose, with reasonable success. Recent work at Michigan Technological University has shown that many, and perhaps all, other microorganisms are also capable of preventing pyrite flotation, regardless of whether or not they are lithotrophic. Indeed, these other organisms may well be superior to Thiobacillus species for this purpose. Further investigation in this area is currently underway. The major limitation, as with any physical separation, is the need for nearly complete liberation of the pyrite from the coal. This technique will therefore only be practical when the coal is very fine, and will therefore be most beneficial for such processes as column flotation and selective flocculation.

In addition to the sulfur contained in pyrite inclusions, coal contains a substantial amount of organically bound sulfur which is very difficult to remove. There have been some indications, however, that certain bacteria may be able to remove much of this organic sulfur without completely destroying the coal structure. While no organisms have yet shown an unequivocal ability to do this, a number of bacteria are known which can degrade specific sulfur-carbon compounds similar to those found in coal. Further mutation and selection of these bacteria may make possible the low-cost removal of all sulfur from coal (8).

Quite recently, it has been discovered that certain wood-decomposing fungi can partially or completely liquefy some low-rank coals, particularly highly-oxidized lignites (20). The coals are converted into a complex mixture of water-soluble organic liquids, which may be useful in a variety of chemical applications. Initial results are reported to be highly variable, and so the value of this technique is not yet known.

Gold

Probably the most important application of biotechnology to gold recovery will be the treatment of refractory ores to increase gold dissolution during cyanide leaching (39,40). In many such ores, finely divided gold is encapsulated in pyrite particles and cannot be reached by the cyanide solutions unless the ore is very finely ground. An alternative to grinding is to use bacteria to selectively dissolve enough of the pyrite to expose the gold, and then to leach with cyanide in the normal fashion. It is of course necessary to do this in two steps, both because cyanide will kill the bacteria and because pyrite dissolution must be carried out in acid solution while cyanide leaching requires alkaline conditions. The use of bacteria keeps the cost of pyrite dissolution at a reasonable level, and since it is only necessary to dissolve a fraction of the pyrite to expose the gold, the leaching time can be kept to a few days.

Uranium

The dissolution of uranium in acid solution is greatly increased by the presence of *Thiobacillus ferrooxidans*, as the ferric iron produced by this bacteria oxidizes the uranium from UO_2 to the much more soluble UO_2^{-2} (41,42,19). A typical application has been to use acidic mine water containing these bacteria to wash down the stopes and pillars in the mine, and then to extract the dissolved uranium from the water (41). Microbial leaching will also be useful for in-situ mining of uranium, using techniques similar to those being investigated for copper. In addition, uranium concentration by adsorption in biomass is possible (43).

Nickel and Cobalt

Economical dissolution of nickel and cobalt from low-grade laterites requires the addition of complexing agents to increase their solubility. It has been found that the metabolic products which many organisms produce from glucose, such as citric, tartaric, and pyruvic acids, are useful for this purpose (44). Unlike leaching with lithotrophic bacteria, a nutrient sources other than the ore is needed, usually in the form of a low-cost organic additive such as glucose. It may be practical to provide nutrients through the use of photoautotrophic organisms such as algae (45), and thus use sunlight to provide the necessary energy to the process.

The adsorption of nickel from solution by a *Pseudomonas* species has been reported, which may be useful for treating wastewater. The nickel adsorption appears to be a simple exchange for magnesium in the outer membrane. However, nickel tolerance of existing strains is not sufficient for greatly concentrating the nickel (46).

Selenium

Recent work by the Bureau of Mines has been concerned with reducing the amount of selenium in process waste waters and in agricultural runoff. It has been discovered that certain bacteria isolated from brackish-water marshes are capable of reducing soluble selenate ions to selenite and ultimately to insoluble elemental selenium. The organisms which do this are heterotrophic, and therefore require organic nutrients. Approximately 95% of the dissolved selenium is removed by this means, and since the selenium is not adsorbed permanently by the cells they are not killed in the process (47).

Radium

For the adsorption of metals such as radium from wastewater streams, the microorganisms used need not be alive or of a particular species. Work with killed sludge from sewage treatment plants has shown that this material is useful for removing radium from uranium mine runoff, which can then be disposed of properly. The primary difficulty has been that the biomass particles were too small to be easily removed from the runoff stream after loading with radium. This problem has been reduced by the development of a technique for forming biomass pellets which are large enough to be easily removed, and sufficiently porous to be effective adsorbers. Since the material used as adsorbent is itself a waste product, the cost of this technology is quite low (43).

Conclusions

Biological processing of minerals and metals is a rapidly growing, diverse field whose importance will become considerable in the next century. Using naturally occurring microorganisms, a wide variety of techniques are already possible which will reduce the cost of recovering and purifying raw materials. The use of genetic engineering techniques will allow organisms to be precisely tailored to their applications, and thus improve their effectiveness. Also, the utilization of organisms which live in extreme environments will allow biometallurgical processes to be carried out at higher temperatures and under conditions which were once believed to be impossible for living organisms.

The major factor limiting the application of biohydrometallurgy is the short time since most of the applications were first discovered. Implementation of this technology therefore requires a great deal of work in areas such as reactor design and process scaleup.

References

1. D. S. Holmes, "Biotechnology in the Mining and and Metal Processing Industries: Challenges and Opportunities", Minerals and Metallurgical Processing (May, 1988) 49-56.

2. V. I. Lakshmanan, "Industrial Views and Applications: Advantages and Limitations of Biotechnology", Biotechnology and Bioengineering Symposium No. 16, John Wiley & Sons (1986) 351-361.

3. G. J. Olson and R. M. Kelly, "Microbiological Metal Transformations: Biotechnological Applications and Potential", Biotechnology Progress, vol. 2, no. 1 (1986) 1-15.

4. P. L. Wichlacz, "Practical Aspects of Genetic Engineering for the Mining and Mineral Industries", Biotechnology and Bioengineering Symposium No. 16, John Wiley & Sons (1986) 319-326.

5. J. F. Spisak, "Biotechnology and the Extractive Metallurgical Industries: Perspectives for Success", Biotechnology and Bioengineering Symposium No. 16, John Wiley & Sons (1986) 331-341.

6. A. Bruynesteyn, "Biotechnology: Its Potential Impact on the Mining Industry", Biotechnology and Bioengineering Symposium No. 16, John Wiley and Sons (1986) 343-350.

7. M. C. Campbell, H. W. Parsons, A. Jongejan, V. Sanmugasunderam, M. Silver, "Biotechnology for the Mineral Industry", Canadian Metallurgical Quarterly, vol 24, no. 2 (1985) 115-120.

8. B. J. Ralph, "Biotechnology Applied to Raw Minerals Processing", Comprehensive Biotechnology, (Murray Moo-Young, ed.), vol. 4, Pergamon Press, NY (1985) 201-234.

9. T. D., Brock, D. W. Smith and M. T. Madigan, Biology of Microoganisms, Prentice-Hall; Englewood Cliffs, New Jersey (1984).

10. G. Huber, H. Huber, K. O. Stetter, "Isolation and Characterization of New Metal-Mobilizing Bacteria", Biotechnology and Bioengineering Symposium No. 16, John Wiley and Sons (1986) 239-251.

11. R. M. Kelly and J. W. Deming, "Extremely Thermophilic Archaebacteria: Biological and Engineering Considerations", Biotechnology Progress, vol. 4, no. 2 (1988) 47-62.

12. J. R. Yates and D. S. Holmes, "Molecular Probes for the Identification and Quantation of Microorganisms Found in Mines and Mine Tailings", Biotechnology and Bioengineering Symposium No. 16, John Wiley and Sons (1986) 301-309.

13. M. P. Silverman, "Mechanism of Bacterial Pyrite Oxidation", Journal of Bacteriology, vol. 94, no. 4 (1967) 1046-1051.

14. S. B. Yunker and J. M. Radovich, "Enchancement of Growth and Ferrous Iron Oxidation Rates of T. ferrooxidans by Electrochemical Reduction of Ferric Iron", Biotechnology and Bioengineering, vol. 28 (1986) 1867-1875.

15. L. Toro, B. Paponetti, C. Cantalini, "Precipitate formation in the Oxidation of Ferrous Ions in the Presence of Thiobacillus ferrooxidans", Hydrometallurgy, vol. 20 (1988) 1-9.

16. N. LaZaroff, "Sulfate Requirement for Iron Oxidation by Thiobacillus ferrooxidans", J. Bacteriology, vol. 85 (1963) 78-83.

17. G. F. Andrews, "The Selective Adsorption of Thiobacilli to Dislocation Sites on Pyrite Surfaces", Biotechnology and Bioengineering, vol. 31 (1988) 378-381.

18. M. TseZos, "The Selective Extraction of Metals from Solution by Micro-organisms: A Brief Overview", vol. 24 no. 2 (1985) 141-144.

19. L. E. Murr, A. E. Torma, J. A. Brierley (eds.), Metallurgical Applications of Bacterial Leaching and Related Microbiological Phenomena, Academic Press, New York (1978).

20. C. D. Scott, G. W. Strandberg, S. N. Lewis, "Microbial Solubilization of Coal", Biotechnology Progress, vol. 2, no. 3 (1986) 131-139.

21. M. R. Hoffmann, B. C. Faust, F. A. Panda, H. H. Koo, H. M. Tsuchiya, "Kinetics of the Removal of Iron Pyrite from coal by Microbial Catalysis", Applied and Environmental Microbiology, vol. 42, no. 2, (1981) 259-271.

22. J. D. Isbister and E. A. Kobylinski, "Microbial Desulfurization of Coal", Processing and Utilization of High-Sulfur Coals (Y. A. Attia, ed.) Elsevier, New York (1985) 627-641.

23. J. Murphy, E. Riestenberg, R. Mohler, D. Marek, B. Beck, D. Skidmore, "Coal Desulfurization by Microbial Processing", Processing and Utilization of High-Sulfur Coals (Y. A. Attia, ed.) Elsevier, New York (1985) 643-652.

24. F. Kargi and J. M. Robinson, "Microbial Desulfurization of Coal by Thermophilic Microorganism Sulfolobus acidocaldarius", Biotechnology and Bioengineering, vol. 24 (1982) 2115-2121.

25. F. Kargi, "Enhancement of Microbial Removal of Pyritic Sulfur from Coal Using Concentrated Cell Suspension of T. ferrooxidans and an External Carbon Dioxide Supply", Biotechnology and Bioengineering, vol. 24, (1982) 749-752.

26. P. R. Dugan, "Microbiological Desulfurization of Coal and Its Increased Monetary Value", Biotechnology and Bioengineering Symposium No. 16, John Wiley & Sons (1986) 185-203.

27. F. Kargi and J. M. Robinson, "Biological Removal of Pyritic Sulfur from Coal by the Thermophilic Organism Sulfolobus acidocaldarius", Biotechnology and Bioengineering, vol. 27 (1985) 41-49.

28. F. Kargi and J. M. Robinson, "Removal of Organic Sulphur from Bituminous Coal: Use of the Thermophilic Organism Sulfolobus acidocaldarius", Fuel, vol. 65 (1986) 397-399.

29. R.G.L. McCready and M. Zentilli, "Beneficiation of Coal by Bacterial Leaching", Canadian Metallurgical Quarterly, vol. 24, no. 2 (1985) 135-139.

30. C. Rai, "Microbial Desulfurization of Coals in a Slurry Pipeline Reactor Using Thiobacillus ferrooxidans", Biotechnology Progress, vol. 1, no. 3 (1985) 200-204.

31. F. Kargi and J. M. Robinson, "Removal of Sulfur Compounds from Coal by the Thermophilic Organism Sulfolobus acidocaldarius", Applied and Environmental Microbiology, vol. 44, no. 4 (1982) 878-883.

32. S. K. Kawatra, T. C. Eisele and S. Bagley, "Coal Desulfurization by Bacteria", Minerals and Metallurgical Processing, (November 1987) 189-192.

33. C. C. Townsley, A. S. Atkins, A. J. Davis, "Suppression of Pyritic Sulphur During Flotation Tests Using the Bacterium Thiobacillus ferrooxidans", Biotechnology and Bioengineering, vol. 30 (1987) 1-8.

34. A. S. Atkins, E. W. Bridgwood, A. J. Davis, F. D. Pooley, "A Study of the Suppression of Pyritic Sulphur in Coal Froth Flotation by Thiobacillus ferrooxidans", Coal Preparation, vol. 5, (1987) pp. 1-13.

35. M. Elzeky and Y. A. Attia, "Coal Slurries Desulfurization by Flotation Using Thiophilic Bacteria for Pyrite Depression", Coal Preparation, vol. 5 (1987) 15-37.

36. B. J. Butler, A. G. Kempton, R. D. Coleman, C. E. Capes, "The Effect of Particle Size and pH on the Removal of Pyrite from Coal by Conditioning with Bacteria Followed by Oil Agglomeration", Hydrometallurgy, vol. 15 (1986) 325-336.

37. C. E. Capes, A. E. McIlhinney, A. F. Sirianni, I. E. Puddington, "Bacterial Oxidation in Upgrading Pyritic Coals", CIM Bulletin, (November 1973) 88-91.

38. A. G. Kempton, N. Moneib, R.G.L. McCready, C. E. Capes, "Removal of Pyrite from Coal by Conditioning with Thiobacillus ferrooxidans Followed by Oil Agglomeration", Hydrometallurgy, vol. 5 (1980) 117-125.

39. Y. Attia, J. Litchfield, L. Vaaler, "Applications of Biotechnology in the Recovery of Gold", Microbiological Effects on Metallurgical Processes, TMS-AIME (1985) 111-120.

40. A. Bruynesteyn, R. P. Hackl, "The Biotankleach Process for the Treatment of Refractory Gold/Silver Concentrates", Microbiological Effects on Metallurgical Processes, TMS-AIME (1985) 121-127.

41. R.G.L. McCready, D. Wadden, A. Marchbank, "Nutrient Requirements for the In-Place Leaching of Uranium by Thiobacillus ferrooxidans", Hydrometallurgy, vol. 17 (1986) 61-71.

42. L. G. Leduc, G. D. Ferroni, D. Belcourt, "Liquid Scintillation Counting as a Means of Measuring Uranium Leached from Low-Grade Ores by Thiobacillus ferrooxidans", Hydrometallurgy, vol. 18 (1987) 255-263.

43. M. Tsezos, M.H.I. Baird, L. W. Shemilt, "The Use of Immobilized Biomass to Remove and Recover Radium from Elliot Lake Uranium Tailing Streams", Hydrometallurgy, vol. 17 (1987) 357-368.

44. D. I. McKenzie, L. Denys, A. Buchanan, "The Solubilization of Nickel, Cobalt and Iron from Laterites by Means of Organic Chelating Acids at Low pH", International Journal of Mineral Processing, vol. 21 (1987) 275-292.

45. M. J. Hart and J. L. Madgwick, "Utilization of Algae as a Sole Nutrient for Microorganisms Biodegrading Manganese Dioxide", Bull. Proc. Australasian Institute of Mining and Metallurgy, vol. 292, no. 1 (1987) 61-63.

46. A. Bordons and J. Jofre, "Extracellular Adsorption of Nickel by a Strain of Pseudomonas Sp.", Enzyme Microb. Technol., vol. 9 (1987) 709-713.

47. D. M. Larsen, K. R. Gardner, and P. B. Altringer, "Biologically Assisted Control of Selenium in Process Waste Waters", in press (1988).

48. A. D. Agate, "Isolation and Preservation of Bacterial Cultures with Special Reference to Leaching of Indian Copper Ores", Minerals and Metallurgical Processing, (May 1988) 66-68.

49. R. M. Bagdigian and A. S. Myerson, "The Adsorption of Thiobacillus ferrooxidans on Coal Surfaces", Biotechnology and Bioengineering, vol. 28 (1986) 467-479.

50. C.-Y. Chen and D. R. Skidmore, "Attachment of Sulfolobus acidocaldarius cells on Coal Particles", Biotechnology Progress, vol. 4, no. 1 (1988) 25-30.

51. Z. M. Dogan, C. F. Gokcay, and E. Atabey, "Bacterial Leaching of Turkish Pyrite Cinders", Progress in Biohydrometallurgy, Cagliari, Italy, (May 1983) 693-703.

52. H. L. Ehrlich, "Recent Advances in Microbial Leaching of Ores", Minerals and Metallurgical Processing, (May 1988) 57-60.

53. S. E. Follin, M. V. Yates, J. A. Brierley and C. L. Brierley, "Microbial Effects on In-Situ Leaching" Microbiological Effects on Metallurgical Processes, TMS-AIME (1985) 129-144.

54. S. Ford, S. C. Simpson, D. Condliffe, and B. H. Olson, "Distribution of Mercury Resistance Genes Among Resistance Determinants in Contaminated Soils in the U.K.", SME Annual Meeting, Phoenix, Arizona, preprint no. 88-25 (1988).

55. A. M. Johnson, D. G. Carlson, S. T. Baley, D. L. Johnson, "In-Situ Bioleaching Investigations of Michigan Chalcocite Ores", SME-AIME Annual Meeting, Phoenix, Arizona, Preprint No. 88-96 (1988).

56. B. H. Kaye, "Fine Particle Characterization Aspects of Predictions Affecting the Efficiency of Microbiological Mining Techniques", Powder Technology, vol. 50, (1987) 177-191.

57. A. B. King, A.W.L. Dudeney, "Bioleaching of Nepheline", Hydrometallurgy, vol. 19, (1987) 69-81.

58. R. W. Lawrence, A. Vizsolyi and R. J. Vos, "The Silver Catalyzed Bioleach Process for Copper Concentrate", Microbiological Effects on Metallurgical Processes, TMS-AIME, Warrendale, PA (1985) 65-82.

59. K. A. Natarajan, "Electrochemical Aspects of Bioleaching Multisulfide Minerals", Minerals and Metallurgical Processing (May 1988) 61-65.

60. D. E. Rawlings, I-M. Pretorius, and D. R. Woods, "Expression of Thiobacillus ferrooxidans Plasmid Functions and the Development of Genetic Systems for the Thiobacilli" Biotechnology and Bioengineering Symposium No. 16, John Wiley & Sons (1986) 281-287.

61. S. E. Shumate and G. W. Strandberg, "Accumulation of Metals by Microbial Cells", Comprehensive Biotechnology, vol. 4 (Murray Moo-Young, ed.) Pergamon Press (1985) 235-247.

62. M. J. Southwood, "Bacterial Leaching: Some Mineralogical Constraints in the Selection of Low-grade Nickel Ores", Mintek Review, no. 1 (1985) 25-30.

63. A. O. Summers, P. Roy, and M. S. Davidson, "Current Techniques for the Genetic Manipulation of Bacteria and their Application to the study of Sulfur-Based Autotrophy in Thiobacillus", Biotechnology and Bioengineering Symposium No. 16, John Wiley & Sons, (1986) 267-279.

64. M.C.M. Van Loosdrecht, J. Lylema, W. Norde, G. Schraa and A.J.B. Zehnder, "Electrophoretic Mobility and Hydrophobicity as a Measure to Predict the Initial Steps of Bacterial Adhesion", Applied and Environmental Microbiology, vol. 53, no. 8 (1987) 1898-1901.

65. M.C.M. Van Loosdrecht, J. Lyklema, W. Norde, G. Schraa, and A.J.B. Zehnder, "The Role of Bacterial Cell Wall Hydrophobicity in Adhesion", Applied and Environmental Microbiology, vol. 53, no. 8, (1987) 1893-1897.

Acknowledgment

The work conducted was supported by the Department of Interior Mineral Institute program administered by the USBM under Grant G1184126.

BIOACCUMULATION OF METALS FROM SOLUTION: NEW TECHNOLOGY FOR RECOVERY, RECYCLING AND PROCESSING

G.J. Olson, F.E. Brinckman, T.K. Trout and D.Johnsonbaugh

Polymers Division
National Institute of Standards and Technology
Gaithersburg, MD 20899

Abstract

Microbiological metal recovery is an emerging technology likely to play an increasing role in commercial ore leaching, metal removal from process and waste streams, and perhaps ultimately in processing to yield metal products in specified forms or oxidation states. We are studying the potential for using microorganisms for the recovery of elements important to emerging materials technologies and for which domestic supplies are limited and/or low grade. Examples to be discussed include the bioaccumulation of yttrium and gallium. These elements are important in the production of new superconductor and semiconductor materials. However, limited domestic reserves and technologies for their recovery may cause supply problems. Microbiological processes may offer new techniques for recovery of these and other strategic elements.

Metallurgical Processes for the Year 2000 and Beyond
Edited by H.Y. Sohn and E.S. Geskin
The Minerals, Metals & Materials Society, 1988

Introduction

Microorganisms in ore leaching

Microorganisms assist in the commercial leaching of copper and uranium from low grade ores (1,2) and recently have been shown to have immediate commercial potential for the pretreatment of refractory gold ores to increase the gold-cyanide recovery yield (3). Copper bioleaching is the best known example of biotechnology applied to metal recovery. Copper can be recovered by commercial leaching of huge piles or dumps of low grade sulfidic ores and waste materials, which are not otherwise processable. Microbial activity in the leach dump environment oxidizes metal sulfides to soluble metal sulfates:

$$MS + 2O_2 \dashrightarrow MSO_4 \qquad (1)$$

MS denotes a divalent metal sulfide. Iron disulfide minerals (especially pyrite) also are often present in metal ore deposits and are oxidized by bacteria, producing acidic ferric sulfate solutions which are good oxidants:

$$2Fe^{3+} + MS \dashrightarrow M^{2+} + S^{o} + 2Fe^{2+} \qquad (2)$$

Certain species of bacteria are indigenous to exposed, weathering metal sulfide deposits and oxidize metal sulfides as energy sources. For example, acidophilic iron-oxidizing bacteria, most notably _Thiobacillus ferrooxidans_, oxidize Fe^{2+} to Fe^{3+}, obtaining metabolic energy from the reaction. This organism requires oxygen for oxidation of iron and uses carbon dioxide as a source of carbon. Reduced sulfur species are also energy sources for _T. ferrooxidans_ and other thiobacilli which inhabit these environments. The combined activity of the metal sulfide oxidizing bacteria and the products that they generate (e.g., ferric ions and sulfuric acid), solubilizes valuable metals which are recovered from solution by techniques such as electrowinning. The technology which uses microorganisms in dump leaching is rather unsophisticated. Microbial growth in huge dumps of low-grade copper sulfide ores and wastes is merely encouraged by the percolation of water and sometimes by the addition of sulfuric acid to lower the pH to levels suitable for growth of the organisms.

Bioleaching recently has been applied to certain sulfidic gold ores which are refractory toward processing, owing to the presence of metal sulfides (i.e., pyrite and arsenopyrite) which restrict access of leaching agents to the gold. Metal sulfide-oxidizing bacteria can be applied to break down the pyrite and arsenopyrite matrix in these ores to improve the access of leaching agents (cyanide). Hutchins and coworkers (4) employed a continuous stirred tank bioreactor to pretreat refractory gold ore concentrates prior to cyanidation. Microbial pretreatments increased gold recovery from 5.5% (untreated) to 56-91% depending on the strain of microorganism. The thermophilic bacterium _Sulfolobus acidocaldarius_ gave the highest gold recovery.

Bioleaching of refractory gold ores is competitive with roasting or pressure oxidation in terms of product recovery and process economics, but requires longer time periods for treatment (4). Additional research and process optimization efforts are accompanying commercialization efforts.

The bioleaching of refractory precious metal ores and concentrates allows a more sophisticated approach to bioprocessing than can be empolyed in the relatively uncontrolled dump leaching environment. Since the product value of precious metal leaching is very high--hundreds of dollars per ounce for gold compared to cents per ounce of copper--more control can be exercised over processing conditions. For example, continuous stirred tank bioreactors achieve better gas (O_2, CO_2) mass transfer and correspondingly higher pyrite oxidation rates.

Further development of bioleaching technology is aimed at extending the range of metals and ores that may be processed. New organisms, especially thermophilic bacteria, may accelerate the rates of metal extraction. There is a need for more fundamental research to determine the factors which limit the rate of microbial ore leaching, the characteristics of the ore bioleaching environment, the interactions of microorganisms and their effect on ore leaching and finally to develop genetic systems with these bacteria. When a better understanding of the leaching organisms is achieved, the factors (e.g., enzymes) which affect bioleaching rates could be enhanced using genetic manipulations.

Microbial recovery of metal ions from solution

In addition to the dissolution of metals from ores, there is considerable interest in the use of microorganisms for the recovery of metal ions from dilute solution. As with bioleaching, this technology to date has been used in simple, unsophisticated yet effective ways. An example is the use of benthic microbial communities in mine ponds (5) or meandering streams to remove lead, zinc and other heavy metals from mine drainage as is practiced in the lead mining district of Missouri (6). The Homestake (South Dakota) gold mine has developed a more sophisticated process involving the use of rotating biological contactors (RBCs) for the oxidation of cyanide and sorption of heavy metals from mine effluents (7). These RBCs contain films of living bacteria on plates and provide high surface area for biochemical reactions with waste stream chemicals.

The cell walls, capsules, exopolymers and membranes of microorganisms including algae (8,9), bacteria (10-13), and fungi (9,14) contain a wide array of chemical functional groups (e.g., amino, carboxylate, phosphate) which bind metal cations and anions. Such properties of the cell envelopes have led to the development of biomass-based products for use in metal recovery from process or waste streams (15,16). These products compare favorably in cost and effectiveness with conventional technologies such as ion-exchange resins, and there are indications that selectivity of recovery can be achieved by manipulation of desorption conditions (17). In some cases, metals are actively accumulated intracellularly by living cells via carrier mediated transport systems (18). Microbial resistance to heavy metals sometimes involves the sequestration of the metals as innocuous compounds. For example, a cadmium-resistant strain of the bacterium *Citrobacter* binds Cd as an insoluble phosphate (19). A nickel-resistant strain of the alga *Synechococcus* produces an intracellular polymer-like material to immobilize Ni (20). Microorganisms attached to surfaces also accumulate metals (21) and organometallic compounds (22).

Although living cells offer interesting prospects for metal recovery, they are inherently more difficult to work with than non-living microbial biomass-based products for metal recovery. Effluents that are selective for particular metabolic types of organisms may initially offer the best prospects for use of living cells in metal recovery. For example, an effluent containing elevated concentrations of heavy metals will restrict growth of all but metal-resistant species which may posess metal resistance mechanisms resulting in a desirable metal transformation (e.g., volatilization, reduction, accumulation).

In most cases, research in and application of microbial metal recovery has aimed at removing heavy metal toxicants from waste streams. Microbial processes for the accumulation of valuable metals from dilute solution have received less attention. As new technologies emerge which require materials composed of rare elements, the prices and availabilities of such elements may be affected. For example, increasing demand for high purity gallium and yttrium will likely result from growth in semiconductor and superconductor industries. Microbial processes may offer prospects for the recovery of these elements from solution for both extraction and recycling applications.

Bioaccumulation of Gallium and Yttrium

Gallium

Gallium has become an important metal to the electronics industry with uses in semiconductors, light emitting diodes and other applications. As late as 1983 there was no domestic production of Ga, making the U. S. reliant on imports to meet domestic demands (23). With the expected increasing demand for Ga, new technologies are sought for exploiting domestic deposits of this metal, which is usually found in very low concentration in ore material and which may be recovered as a by product of bauxite and zinc processing. The Apex Mine in Utah is currently the only primary domestic producer of gallium (24).

Biotechnology may provide routes for improved recovery of Ga from ores and process streams. Such routes might include bioleaching and bioaccumulation. Gallium sulfide is oxidized by *T. ferrooxidans* to release soluble Ga (25). With regard to the bioaccumulation of Ga, microorganisms by virtue of their cell wall characteristics described above, are expected to bind Ga^{3+} ions in aqueous solution. However, it may also be possible to take advantage of more active, selective microbiological mechanisms for the recovery of gallium. For example, aerobic organisms require iron in metabolism, and produce specific iron-chelating molecules, termed siderophores, to acquire iron from its highly insoluble oxides in water (26). Although siderophores have high specificity for Fe^{3+}, Ga^{3+} has a very similar charge/size ratio and is also be bound by siderophores. Emery (27) found that Ga actually displaced iron from a fungal siderophore. Thus, it may be possible to recover Ga from dilute aqueous solution and from insoluble precipitates using siderophores.

We have begun to study the bioaccumulation of Ga by a siderophore-producing strain of *Pseudomonas fluorescens*. When grown under low-iron (micromolar) concentrations in a chemically defined medium (28), the bacterium produces a visible, fluorescent yellow-green siderophore. Figure 1 (top) shows the production of siderophore (monitored as absorbance at 400 nm) in the growth medium amended with various concentrations of iron (as ferric sulfate). Siderophore production was almost totally repressed at an iron concentration 11.5 μM. Other investigators have shown that production of siderophores is repressed at elevated concentrations of iron (26). This is reasonable on a physiological basis since it makes no sense for cells to expend energy to synthesize siderophores if iron is readily available. Cell growth (monitored at 420 nm) was comparable in all four of the flasks (Figure 1, bottom).

Low levels of Ga inhibited the growth of this organism in low iron medium, suggesting that the siderophore produced by the organism complexed with Ga and, in effect, starved the cells for iron. The addition of iron to the growth medium reversed this inhibition (Table 1)

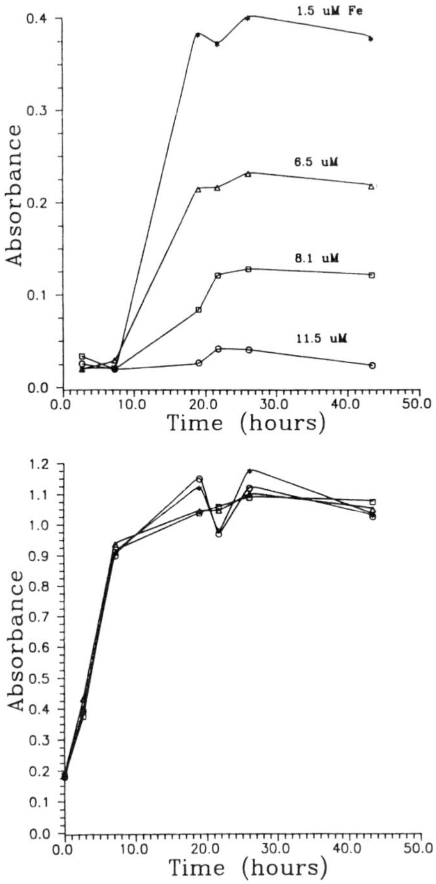

Figure 1. Production of siderophore by Pseudomonas fluorescens (monitored as increase in absorbance at 400 nm) as a function of iron concentration in the growth medium (top). Cell growth as monitored by absorbance at 420 nm is shown at bottom. Cells were grown in defined medium (28) amended with ferric chloride at indicated concentrations and incubated at 28 °C with shaking at 250 rpm. Siderophore was measured after an aliquot of the culture medium has been centrifuged at 12,000 xg for 5 min.

P. fluorescens bound Ga in dilute (4.8 mg/L) solution, accumulating the metal within 13 minutes to 0.23% of cell dry weight as determined by graphite furnace atomic absorption spectrophotometry (GFAA) of acid digested cells.

In addition, qualitative observations indicated that spent culture medium, which fluoresced yellow-green owing to the presence of siderophore, was decolorized on addition of iron or gallium ions to the solution, further suggesting the binding of gallium by the material. We also observed that precipitates of gallium hydroxide were solubilized by spent culture media, suggesting that Ga may be recovered by application of siderophres as leaching agents. Indeed, siderophores can be produced in gram per liter levels by bacteria (26). We are conducting further studies on the specificity of siderophore mediated gallium recovery in the presence of other metals, and the use of immobilized siderophores in columns.

Table I. Growth of P. fluorescens in the Presence of Ga

Ga added (μM)	Cell growth (% of control)
0	100
9	72
27	33
91	2
91 + 91 uM Fe	95

[a]Cells were grown in succinate medium (28) amended with gallium or ferric sulfate in conical flasks at 28°C and shaking at 250 r.p.m. in a gyratory shaker. After 10 hours growth, absorbance at 420 nm was measured.

Yttrium

Yttrium is another example of a comparatively rare element, which occurs in very low concentration in ores and which is increasingly used in new high technology materials applications. These factors will likely result in increasing demand and the need to develop better mining and

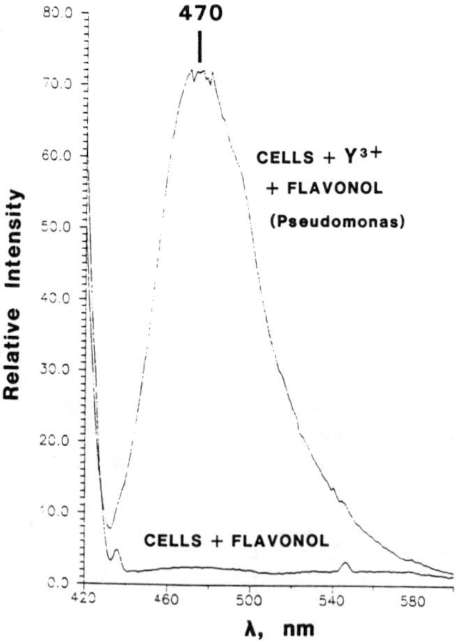

Figure 2. Emission spectra of Pseudomonas fluorescens cells exposed to 20 mg/L of yttrium acetate tetrahydrate in 10 mM PIPES (piperazine bis-2 ethane sulfonic acid)buffer. After 30 min., cells were centrifuged, washed twice in PIPES buffer, resuspended in 1.0 mL of buffer and treated with 100 uL of $5x10^{-3}$ M 3-hydroxyflavone (ethanolic). After 2 min the cells were again washed twice and examined by epifluorescence microscopy at an exciting wavelength of 405 nm. Control cells were not exposed to Y but were carried through the rest of the procedure.

recovery technologies. Yttrium has been used to impart high temperature corrosion resistance to stainless steels, and in laser and microwave applications (29). However, recent development in superconductors usually feature Y-based materials. Thus, demand for this element may grow considerably.

We have found that bacteria also accumulate Y from dilute solution. These studies were conducted using epifluorescence microscopic imaging (EMI) employing a 3-hydroxyflavone, a compound which forms a fluorescent complex with Y, emitting light at 470 nm when excited at 405 nm. Thus, cells which have accumulated Y fluoresce at 470 nm on exposure to flavonol (Figure 2). We have previously used this technique to measure inorganic tin and organotin species in dilute solution (30) and to monitor the bioaccumulation of inorganic tin, organotin and zinc compounds by bacteria (31). The EMI technique has promise for monitoring of cells for metal species accumulation and perhaps also as a survey technique for screening a number of strains for bioaccumulation ability. It may also be possible to use this technique for rapid monitoring of process solutions for metal bioaccumulation.

The flavonol-yttrium complex is also measurable by uv-visible spectrophotometry. At 401 nm, the extinction coefficient of the complex is about 60 times greater than for flavonol alone (Figure 3). P. fluorescens and Escherichia coli rapidly bind Y as shown by Y disappearance from solution in the presence of the organisms (Figure 4) as determined by absorption at 401 nm. We have not yet investigated the selectivity which may be achievable using microorganisms for the recovery of Y in a mixture of metals or rare earth elements.

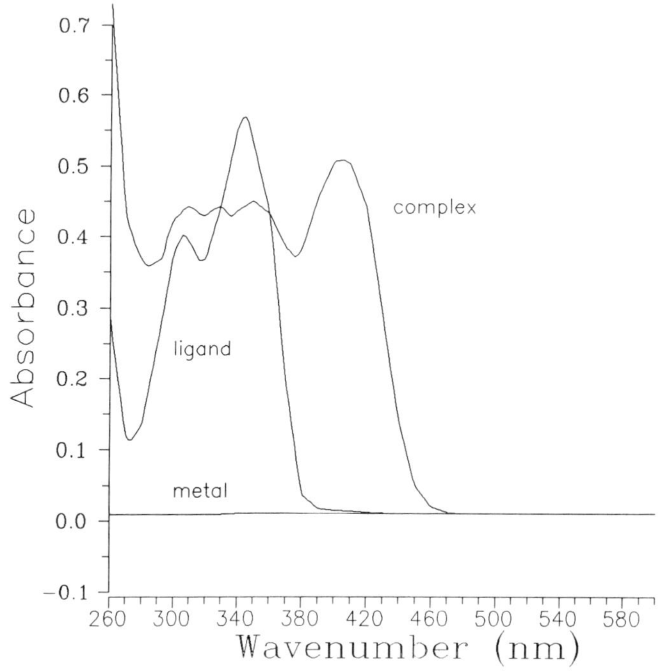

Figure 3. Absorption spectra of yttrium ($5x10^{-3}$ M aqueous, as yttrium acetate tetrahydrate), flavonol ($5x10^{-3}$ M in ethanol) and the yttrium-flavonol complex (0.2 mL each).

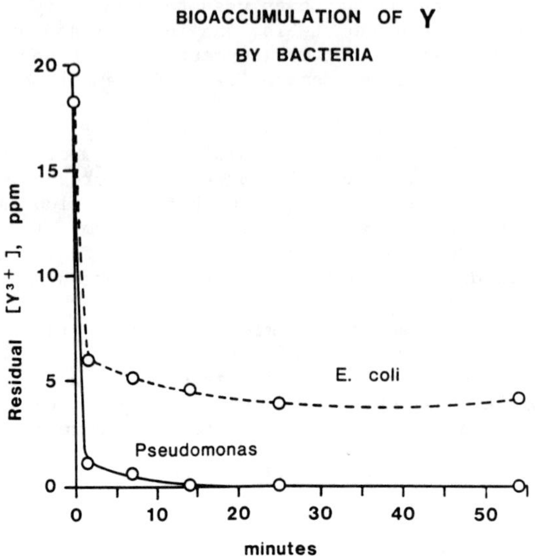

Figure 4. Loss of Y from solution as a function of time in the presence of two different bacterial cultures at a cell density of about $5x10^8$/mL.

Summary and Outlook

The low cost and potentially high specificity of microbial systems for metal recovery make them attractive candidates for future metal recovery applications. However, in contrast to the knowledge of metabolic pathways for carbon metabolism, the full range and mechanisms of microbial reactions with other elements in the periodic table has yet to be fully explored. It may be possible to recover metals in specific oxidation states, particle sizes, or as organometallic forms. For example, magnetotactic bacteria produce sub-micrometer particles of pure magnetite (32). Other organisms produce metabolites which react with precious metal ions in aqueous solution to yield high purity elemental metal particles (33). Better understanding of the mechanisms and rate limiting factors involved in microbial transformations of elements may lead to increasing sophistication in their use for the recovery, recycling and transformation of elements for future needs in materials science.

References

1. A. E. Torma, "Current Standing of Bacterial Heap, Dump and In-Situ Leaching Technology of Copper," Metall, 38 (1984) 1044-1047.

2. D. Wadden and A. Gallant, "The In-Place Leaching of Uranium at Denison Mines," Canadian Metallurgical Quarterly, 24 (1985) 127-134.

3. P. B. Marchant and R. W. Lawrence, "Flowsheet Design, Process Control, and Operating Strategies in the Biooxidation of Refractory Gold Ores," Biominet Proceedings, 3rd, CanMet Special Publication SP 86-9, ed. R. L. McCready (Ottawa, Canada Centre for Mineral and Energy Technology, 1986), 39-51.

4. S. R. Hutchins, J. A. Brierley and C. L. Brierley, "Microbial Pretreatment of Refractory Sulfide and Carbonaceous Ores Improves the Economics of Gold Recovery," Mining Engineering, 40 (1988) 249-254.

5. J. A. Brierley and C. L. Brierley, "Biological Accumulation of Some Heavy Metals--Biotechnological Applications," in Biomineralization and Biological Metal Accumulation, ed. P. Westbroek and E. W. DeJong (Reidel, Dordrecht, The Netherlands, 1983) 499-509.

6. N. L. Gale and B. G. Wixson, "Removal of Heavy Metals from Industrial Effluents by Algae," Developments in Industrial Microbiology 20 (1978) 259-273.

7. J. L. Whitlock, "Biological Degradation and Removal of Cyanides and Metals from Cyanidation Mill and Heap Leach Wastewaters, Residues and Pads" (Paper presented at Workshop on Biotechnology Applied to the Mining, Metal Refining and Mineral Processing Industry, Idaho Falls, Idaho, 26 August 1986).

8. D. W. Darnall et al., "Recovery of Heavy Metals by Immobilized Algae," Trace Metal Removal from Aqueous Solution, Special Publication of the Royal Society of Chemistry (1986) 1-24.

9. N. Kuyucak and B. Volesky, "Biosorbents for Recovery of Metals from Industrial Solutions," Biotechnology Letters, 10 (1988) 137-142.

10. I. C. Hancock, "The Use of Gram-Positive Bacteria for the Removal of Metals from Aqueous Solution," Trace Metal Removal from Aqueous Solution, Special Publication of the Royal Society of Chemistry (1986) 25-43.

11. J. A. Scott, S. J. Palmer and J. Ingham, "Microbial Metal Adsorption Enhancement by Naturally Excreted Polysaccharide Coatings," Immobilization of Ions by Bio-sorption, ed. H. Eccles (Wiley and Sons, New York, 1986) 81-88.

12. T. J. Beveridge and W. S. Fyfe, "Metal Fixation by Bacterial Cell Walls," Canadian Journal of Earth Sciences 22 (1985) 1893-1898.

13. R. M. Sterritt and J. N. Lester, "Heavy Metal Immobilization by Bacterial Extracellular Polymers," Immobilization of Ions by Bio-sorption, ed. H. Eccles (Wiley and Sons, New York, 1986) 121-134.

14. I. S. Ross and C. C. Townsley, "The Uptake of Heavy Metals by Filamentous Fungi," Immobilization of Ions by Bio-Sorption, ed. H. Eccles (Wiley and Sons, New York, 1986) 50-57.

15. B. Greene, R. McPherson and D. Darnall, "Algal Sorbents for Selective Metal Ion Recovery," Metal Speciation, Separation and Recovery, eds. J. W. Patterson and R. Passino (Lewis Publ., Inc. Chelsea, MI, 1987) 315-330.

16. J. A. Brierley, C. L. Brierley and G. M. Goyak, "AMT-Bioclaim, a New Wastewater Treatment and Metal Recovery Technology," Process Metallurgy, Vol. 4, Proceedings of the Sixth International Symposium on Biohydrometallurgy (1986) 291-304.

17. D. W. Darnall et al., "Selective Recovery of Gold and Other Metal Ions from an Algal Biomass," Environmental Science and Technology, 20 (1986) 206-208.

18. S. Silver and J. E. Lusk, "Bacterial Magnesium, Manganese and Zinc Transport," Ion Transport in Prokaryotes (Academic Press, New York, NY, 1987) 165-180.

19. L. E. Macaskie et al., "Cadmium Accumulation by a Citrobacter sp.: the Chemical Nature of the Accumulated Metal Precipitate and its Location on the Bacterial Cells," Journal of General Microbiology, 133 (1987) 539-544.

20. J. M. Wood and H. K. Wang, "Microbial Resistance to Heavy Metals," Environmental Science and Technology, 17 (1983) 582A-590A.

21. G. M. Dunn and A. T. Bull, "Bioaccumulation of Copper by a Defined Community of Activated Sludge Bacteria," European Journal of Applied Microbiology and Biotechnology 17 (1983) 30-34.

22. W. R. Blair et al., "Accumulation and Fate of Tributyltin Species in Microbial Biofilms," Proceedings of the Oceans 88' Conference, (IEEE, Piscataway, NJ 1988) in press.

23. B. Petkof, "Gallium," Mineral Facts and Problems, (U. S. Bureau of Mines, 'Washington, DC, 1985) 291-296.

24. M. P. Wardell and C. F. Davidson, "Acid Leaching Extraction of Ga and Ge," Journal of Metals, 39 (1987) 39-41.

25. A. E. Torma, "Oxidation of Gallium Sulfides by Thiobacillus ferrooxidans," Canadian Journal of Microbiology, 24 (1978) 888-891.

26. J. B. Neilands, "Siderophores," Advances in Inorganic Chemistry, 5 (1983) 137-166.

27. T. Emery, "Iron Metabolism in Humans and Plants," American Scientist, 70 (1982) 626-632.

28. J. M. Meyer and M. A. Abdallah, "The Fluorescent Pigmant of Pseudomonas fluorescens: Biosynthesis, Purification and Physicochemical Properties," Journal of General Microbiology, 107 (1978) 319-328.

29. J. B. Hendrick, "Rare Earth Elements and Yttrium," Mineral Facts and Problems, (U. S. Bureau of Mines, Washington, DC, 1985) 647-664.

30. W. R. Blair et al., Characterization of Organotin Species Using Microbore and Capillary Liquid Chromatographic Techniques with an Epifluorescence Microscope as a Novel Imaging Detector," Journal of Chromatography, 410 (1987) 383-394.

31. F. E. Brinckman et al., "Implications of Molecular Speciation and Topology of Environmental Metals: Uptake Mechanisms and Toxicity of Organotins," Aquatic Toxicology and Hazard Assessment: 10th Volume, ASTM STP 971, ed. W. J. Adams et al. (American Society for Testing and Materials, Philadelphia, 1988) 219-232.

32. R. P. Blakemore, "Magnetotactic Bacteria," Annual Review of Microbiology 36 (1982) 217-238.

33. G. J. Olson et al., "Microbial Metabolites as Agents for Reduction of Metal Compounds to Pure Metals," paper to be presented at Symposium on Biotechnology in Minerals and Metals Processing, SME Annual Meeting, Las Vegas, NV, Feb. 27-Mar 2, 1989, (Society of Mining Engineering, Littleton, CO, in press).

Acknowledgment

The research involving bioaccumulation was supported in part by the Air Force Wright Aeronautical Laboratory, Materials Laboratory, Wright-Patterson Air Force Base, OH.

BIO-LEACHING OF SULFIDE ORES

L. C. Thompson

Gold Fields Mining Corporation

Golden, Colorado 80401

Depletion of known ore reserves and increased demand for precious metals have caused a revolution in biohydrometallurgical processing technology for gold recovery. Known ore deposits and future discoveries are most likely to be lower grade sulfide ores that can be difficult to recover with current leaching technology. Future metal recovery operations will increasingly use bio-oxidation techniques.

Bacteria are an integral part of mineral cycling and mineral transformation in nature. Although biological processes are recognized as part of the mineralization and oxidation reactions in some ore deposits, they have only recently been applied to enhancing metal recovery in hydrometallurgical procedures. This paper explores the traditional metallurgical application of biological leaching and oxidation reactions and reviews the complete range of bacteria available for metal recovery systems. Pre-treatment of ores with bacteria improves metal recoveries in tank leaching systems and shows promise for heap leaching methods.

Metallurgical Processes for the Year 2000 and Beyond
Edited by H.Y. Sohn and E.S. Geskin
The Minerals, Metals & Materials Society, 1988

Introduction

There is no longer any question that bacteria play a natural role in the on-going transformation of some ore deposits (1,2). As applied to sulfide ores, a few strains have been specifically identified that exhibit a natural leaching potential (3). These bacteria can be isolated from waters and soils in old mining districts where they contribute to continuing leaching and pollution problems in acid drainage and heavy metal solubilization. Sulfur and iron bacteria as well as many heterotrophic strains have also been found in ore deposits and sedimentary rocks at depths up to 2000 meters (4). Their presence in both ore deposits and processed ores shows the microbial association with natural leaching and the potential role that bacteria can play in the development of new leaching technology.

The bacteria that are most likely to be found in these specialized environments are the chemolithotrophs, or those bacteria that derive all of their nutrient requirements from inorganic sources. Bacteria included in this category are the genera Thiobacillus, Thiomicrospira, Sulfolobus, Desulfovibrio and Beggiatoa. The cells use iron and sulfur as sources of nutrients for growth, bio-energetics and replication. Any bacteria population mix in these specific environments is in a state of flux where succession populations develop as the environment or the character of the deposit changes. There is evidence that Thiobacillus ferrooxidans is one such bacteria that becomes dominant under aerobic conditions and performs the final stages of sulfide ore transition (5). Other native bacteria operating in a wide range of optimal temperatures, pH, pressure and nutrient media also have the potential for development as bioleaching populations.

Bio-leaching or bio-oxidation processes are essentially metal accumulation/ biomineralization cycles that could be developed as a bioengineered solution for increased precious metal recovery from sulfide ores. The key to the process development is understanding the natural role bacteria can play in ore deposit formation and transformation. Natural processes take place on a geologic time scale and the answer to engineering both time and recovery improvements is found in new bioaugmentation techniques. Bioaugmentation can be best described as the use of microorganisms that have been selected for desirable natural traits and that have then been enhanced in the laboratory. Augmenting a population selectively eliminates competing, non-working bacteria and amplifies the desired characteristics of the working microorganisms. These methods take natural reactions one step further and create a specialized leaching population through selective culturing and randomly induced mutations. The remainder of this review identifies the selection, adaptation and augmentation techniques currently being used or developed in biohydrometallurgy research.

Identification of Ore Transformation Bacteria

Sulfur cycles and oxidation/reduction of metals are the most important reactions to consider in development of bioleaching technology. The wide variety of bacteria that participate in these reactions in nature implies that evolution of specialized bio-oxidation populations should be theoretically possible for most ore-specific leaching problems. It is important to look at both sulfur and iron cycles, as the by-products of one can be either inhibitory or necessary to the other.

The Thiobacilli are unique because they are able to derive all of their energy requirements from the oxidation of inorganic sulfur. These reactions consist of both assimilatory and dissimilatory redox reactions involving sulfides, thiosulfates, elemental sulfur and tetrathionates. The temperature and pH conditions are also significant due to the formation of a variety of rate-limiting or insoluble by-products. The basic sulfur cycle detailed in Figure 1. shows the formation of reaction products and their place in the cycle. Both aerobic and anaerobic bacteria figure in different parts of the sequence. Many equilibrium reactions exist within these transformations and complimentary reactions provide a symmetrical system of synthesis and decomposition.

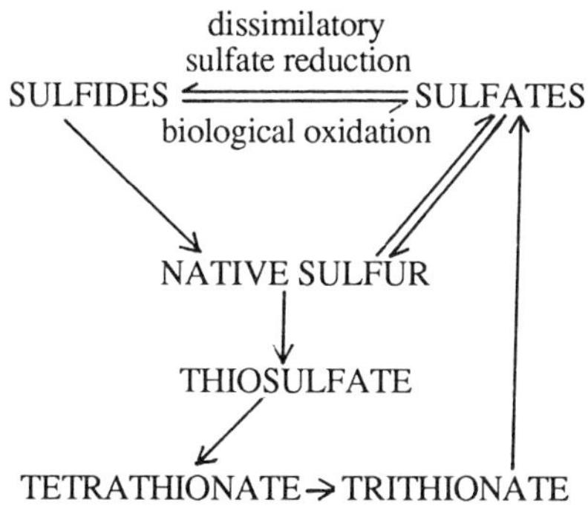

Figure 1. Geochemical and Biological Sulfur Cycles

The sulfur bacteria are defined as any chemotropic microbial population capable of oxidizing inorganic sulfur compounds for the generation of energy. Sulfur bacteria are included in the genera Achromatium, Macromonas, Thiobacterium, Thiospira, Thiovillium, Beggiatoa, Thiothrix, Thioploca, Thiobacillus and Thiomicrospira (6).

The bacterial oxidation of metal sulfides such as pyrite is a complex process which is dependent upon a number of environmental factors such as pH and the presence of oxygen. The most thoroughly studied microorganism regarding sulfide oxidation is Thiobacillus ferrooxidans which grows at an optimal pH range between 2.0 to 3.5 (7). Iron is necessary to many life processes in bacteria as a chelating agent, for nitrogen fixation in some microbes, and as a terminal electron acceptor for cell energy reactions. The processes are a combination of biologically catalyzed reactions and chemical reactions that can potentiate the oxidation of various metal sulfides. The dominant bacterial reactions involving iron are summarized in the following equations.

The oxidation of pyrite by T. ferrooxidans is a pH dependent reaction occurring below pH 4.0. Ferrous sulfate and sulfuric acid are the main reaction products, illustrated by

$$FeS_2 + 3.5\ O_2 + H_2O \longrightarrow FeSO_4 + H_2SO_4 \qquad (1)$$

The system pH drops due to the acid formation and can alter the environment on a microenvironmental scale or, over time, on a macro-environmental scale. The ferrous sulfate that is formed is further oxidized by either Thiobacillus ferrooxidans or T. thiooxidans to form ferric hydroxide or ferric sulfate.

$$FeSO_4 + 0.5\ O_2 + H_2SO_4 \longrightarrow Fe_2(SO_4)_2 + H_2O \qquad (2)$$

At an elevated pH, the ferric iron will be hydrated to form an insoluble ferrous oxide.

$$Fe^{+3} + 3\ H_2O \longrightarrow Fe(OH)_3 + 3\ H^+ \qquad (3)$$

The ferric ion is a strong chemical oxidizing agent which will chemically oxidize pyrite. The ferric ion remains in solution at a pH <3.0 and will continue to catalyze the reaction which yields more ferrous iron.

$$FeS_2 + Fe_2(SO_4)_3 \longrightarrow 3\ FeSO_4 + S^o \qquad (4)$$

The dissimilatory sulfate reducing bacteria have been identified as a source of pyrite in mineral deposit transformation (8). Pyrite is formed in nature from the reduction of sulfates or oxidation of H_2S. The elemental sulfur and the hydrogen sulfide can react with soluble metals to form the insoluble metal sulfides. The microbially mediated iron cycle presented in Figure 2. further summarizes the oxidation and reduction reactions that are catalyzed by bacteria.

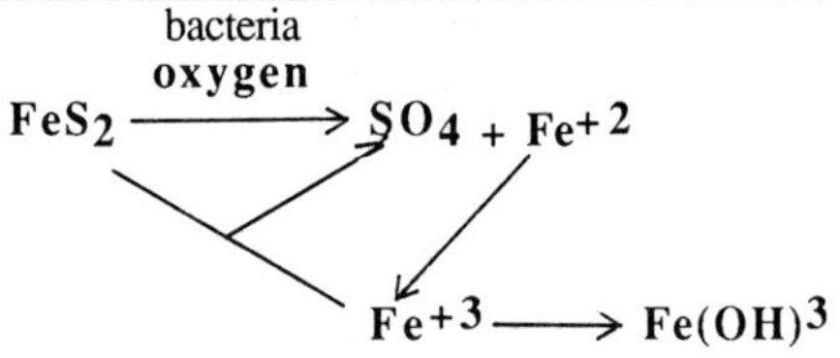

Figure 2. Iron Cycle for Pyrite Oxidation

In addition to T. ferrooxidans and T. thiooxidans, a number of other bacteria have been implicated in the biological decomposition of various mineral sulfides and in the solubilization of metals. The iron bacteria T. ferrooxidans, T. thiooxidans and the Ferrobacilli have proven effective in tank leaching of gold ores and ore concentrates (9,10) and have been used in copper dump leaching (11). Other bacteria that have been found in association with sulfide ores include T. thioparus, T. perometabolis, T. denitrificans, T. neapolitanus, Arthrobacter and Chromatium (12). Growth characteristics are listed in Table I. for select bacteria identified with sulfide transformations. These bacteria show a wide range of growth conditions they can operate in and demonstrate the extensive assortment of bacteria that could be involved in sulfide ore transformation.

Table I. Sulfide Ore Bacteria, Growth Conditions

microorganism	pH	temp., ºC	aerobic	nutrition
Thiobacillus thioparus	4.5-10	10-37	+	autotrophic
T. ferrooxidans	0.5-6.0	15-25	+	"
T. thiooxidans	0.5-6.0	10-37	+	"
T. neapolitanus	3.0-8.5	8-37	+	"
T. denitrificans	4.0-9.5	10-37	+/-	"
T. novellus	5.0-9.2	25-35	+	"
T. intermedius	1.9-7.0	25-35	+	"
T. perometabolis	2.8-6.8	25-35	+	"
Sulfolobus acidocalderius	2.0-5.0	55-85	+	"
Desulfovibrio desulfuricans	5.0-9.0	10-45	-	heterotrophic

Although many of these bacteria are strictly aerobic, there are some microbes such as T. denitrificans that will act as a facultative anaerobe in anoxic conditions. The bacteria are capable of substituting iron, copper or even nitrate for oxygen as a terminal electron acceptor in the energy reactions of the cell. These reactions are defined for only a few microorganisms but raise the question about the role of other potential facultative anaerobes in the transformation of ore deposits.

The main point in identifying these microbial associations with sulfide ores is that bacteria other than the most acidophilic strains probably have some small role to play in ore oxidation and transformation. Conditions in many ore deposits are such that these bacteria apparently exist in only small numbers or less viable populations. The key to using these bacteria in biohydrometallurgical processes is then the adaptation to an ore and augmentation of the population to carry out specific leaching goals. Each ore will present very specific leaching problems but there is evidence that bacteria other than the Thiobacilli can be effective in mineral leaching (13). This broadens the range of conditions for bacterial leaching and shows increased promise for design of fixed-bed reactor bioleaching programs. Even though ore specific leaching problems preclude the use of generic populations, bio-engineered solutions in bacteria population design should be possible for many ore types.

Tank Bioleaching for Sulfide Ores

Tank bioleaching of sulfide ores and ore concentrates has been successfully demonstrated by several companies and is offered as an alternative to other refractory ore treatments such as roasting or pressure oxidation. Tank leaching presents economic and environmental advantages over these traditional processes for ore concentrates or high grade ores (14). With low to medium grade ores, though, the tank bioleaching processes are not as economically attractive. Reagent and energy costs eliminate very refractive, low grade ores from tank bioleaching and process control and recovery problems preclude heap leaching at the present time.

The fundamental principal involved in successful tank bioleaching is that the bioleaching is a pre-treatment step prior to conventional cyanide leaching. Solubilization of gold or silver during the bioleaching is not a goal of the bio-oxidation steps. There are indications that many ores do not need a homogeneous dissolution of the sulfides to effect a successful cyanide leach (15, 16). The tank bioleach process detailed in Figure 3. takes from 24 hours to several days depending on the ore and the pulp densities. The most efficient tank leaches use T. ferrooxidans that have been grown to a working concentration separately from the ore and are then added with nutrients for the bioleach process. The growth requirements of Thiobacillus ferrooxidans are the best defined for all of the Thiobacilli and the bacteria are ubiquitous in processed sulfide ores (17), mine drainage solutions and many transition zone ores.

Tank bioleaching technology can typically improve gold recoveries to 85 to 95% total recovery. Process control of pH and temperature can effect the economy of the process or its workability in some environments. One of the advantages of the tank bioleaching is that it allows for close control of pH and temperatures that would be impossible to control in a heap leach situation.

Bioleaching with an optimized tank design was originally conceived as an improvement to conditions for leaching precious metals. Leaching in natural reactor systems (dump leaching) is an effective process for base metal recovery systems but is not acceptable for precious metal leaching circuits. Tank designs used effectively in different processes include continuously stirred tank reactors or cascade reactor systems for contiuous processing. Both batch and continuous processes are possible with an upper limit of a 10 to 20 tpd capacity (18).

Crush ore
sulfide ores or ore concentrates are crushed to provide a -100 mesh fraction.

↓

Bioleach Ore
Ore is added to the bioleach tank to maintain a pulp density of 15 to 20%. Leach times and bacteria are system-specific. pH and temperature are controlled.

↓

Cyanide Leach
Conventional cyanide leach after bacterial pre-treatment and wash.

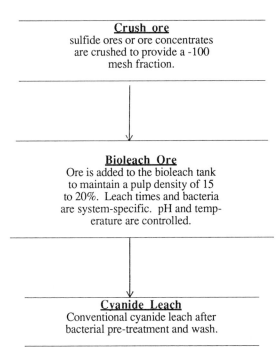

Figure 3. Tank Bioleach Process

Other options for microbial pre-treatment of ores in tank leaching operations include the use of thermophilic microorganisms or elevated pressures to enhance bioleaching rates (19,20,21). While most of the research using thermophiles such as Sulfolobus acidocalderius has been for copper extraction, these microorganisms should have the same potentials for heap or tank bioleaching of precious metals. The advantage of using a balanced population of mesophilic and thermophilic microorganisms is that each population would be variably active depending upon temperatures within a tank leach or a heap leach. The balanced population concept would eliminate the need for the strict temperature control necessary for very specialized leaching populations.

Heap Leaching with Bacteria

The development of a bio-oxidation process for heap leaching of sulfide ores is a combination of bacteria selection/adaptation and process operation design. It is clear that bioleaching or bio-oxidation processes occur naturally within ore deposits and processed ores and that microbial pre-treatment of ores has proven to be effective in tank leaching methods. In any in situ treatment, however, there are additional limiting factors controlling the process design. The problems to be solved in bioleaching design for column or heap leaching include bacteria selection and adaptation, temperature control and pH control. Metal-hydroxide precipitates and

other reaction by-products can form an insoluble coating on a sulfide surface. Formation of any protective reaction products on the sulfide surface can inhibit further bioleaching or subsequent cyanide leaching. Nutrient requirements and possible toxic species in the ore must also be identified in preliminary research.

The first step in the bio-oxidation process design is the isolation and adaptation of the bioleaching population. Bacteria can be isolated directly from the ores or adapted from existing populations. Adaptation to the ore includes growth of the bacteria in chemically defined media and also in ore infusion media. These techniques assure that the leaching population will be both tolerant of possible toxic components in the ore and also able to use the ore as a nutrient source for sulfur and iron. At this development stage, information from flask adaptation and leaching tests can identify metal and reaction by-product solubilization.

The portion of a population native to the ore that will actually participate in the bio-oxidation of the ore can be a very small percentage of the total population. The non-working portion of the population can competetively inhibit the sulfide oxidizing bacteria to the point where the sulfide decomposing strains may be unable to have any significant impact on the bio-oxidation. For this reason, the adaptation/augmentation process must be done for each ore-specific leaching problem. Each specialized population may also require distinctive nutrients which can be identified at this stage of the process research.

A bioaugmentation of the working population is accomplished by selective separation and culturing techniques. The end result of a successful bioaugmentation is a strain of working bacteria that has been selectively enhanced for each ore-specific leaching problem. This working strain can be ideally balanced with bacteria that can either extend the working life of the bio-oxidation bacteria, or be effective under different environmental conditions. The perfectly balanced bio-oxidation strain would contain bacteria that acted as:

1. Primary bioleaching bacteria
2. Secondary strains to remove rate-limiting by-products.
3. Secondary bioleaching strains active at different pH ranges.
4. Secondary bioleaching strains active at elevated temperatures.

This ideal, working population would then be grown separately from the ore in an optimized nutrient media and inoculated on the ore when log phase growth had been reached. This separate growth technique insures that sufficient numbers of working bacteria will be available for the bio-oxidation processes and will not be dependent upon generation of the working population in the ore. Biostimulation techniques where native bacteria are activated by nutrient addition can also be effective in some situations and offers the advantage of eliminating the time-consuming isolation and augmentation methods. The bioaugmentation techniques, however, appear at this time to have the best potential for development of heap leach bio-oxidation technology.

The final problem to deal with in each specialized population devolpment sequence is the tendency of bacteria to mutate and change carefully engineered metabolic characteristics. Bacteria are generally susceptible to enzymatic re-orientation caused by any alteration of their environment. This can mean that the metabolic characteristics that allow the bacteria to perform successful bioleaching can abruptly change when inoculated on the ore or for obscure reasons during the bio-oxidation process. There is no perfect way to control this problem but it can be mitigated by careful preservation of the working strains as back-up. Growth of the bacteria to concentrations significantly greater than those necessary for optimal bioleaching can also dilute the effects of random mutations and changes in the working strains.

The test series for successful bioleaching includes:

1. Tank leaches under carefully controlled conditions to define bacteria growth characteristics and changes, metal solubilization, by-product formation and leaching efficiencies.

2. Column leaches to predict field performance of working bacteria under less controlled conditions with larger sulfide ore fractions.

3. Pilot heap leaches to determine the actual field performance of the bioaugmented bacteria.

Each test stage allows for process adjustment including additions of new bacteria, optimized bioleaching contact times and field re-culturing of large quantities of the working bacteria. It is also possible that selected wash cycles might be necessary to remove bioleaching metabolic by-products or to kill the bacteria at the end of the bio-oxidation cycle. Many of the bacteria that have a potential for sulfide bioleaching can also have a negative impact on cyanide consumption during the cyanide leach cycle.

The research template detailed in Figure 4 summarizes the process developmen scheme for isolation and adaptation of the bio-oxidation bacteria.

Identification/Isolation of Bioleaching Bacteria
1. Isolate bacteria from ore samples.

Adaptation of Bioleaching Population
1. Grow ore isolates in ore infusion media.
2. Stress ore isolates in nutrient and nutrient deficient media.

Bioaugmentation of Bioleaching Bacteria
1. Grow bacteria in chemically defined media.
2. Encourage development of mutant strains.
3. Enhance population through serial sub-culturing techniques.

Tank Bio-contact Tests
1. Identify growth characteristics, metal solubilization and reaction by-product formation.
2. Determine leaching improvement.

Column Tests for Sulfide Bio-oxidation
1. Determine bacteria attachment to ore.
2. Analyze leachate solutions for bacteria, metals and bioreaction products.

Figure 4. Research flowchart for bio-oxidation development.

Conclusions

Bio-extractive metallurgy is a technique that will be necessary to the development of many sulfide ore reserves now and in the future. Bacterial oxidations have proven to be useful in dump recovery operations as well as tank leaching for gold and silver. The techniques offer some economic and environmental advantages over traditional processes but need to be further developed to have an impact on gold recovery in low grade sulfide ores. The key to this development will come through biotechnology methodology using bioaugmentation processes.

References

1. G.I. Karavaiko, S.I. Kuznetsov and A.I. Golonizik, The Bacterial Leaching of Metals from Ores (Stonehouse, Glos., England: Technicopy, Ltd., 1977), 33-37, 61-66.

2. J.B. Davis and D.W. Kirkland, "Bioepigenetic Sulfur Deposits," Economic Geology, 74 (1979) 462-468.

3. L.E. Murr, A.E. Torma and J.A. Brierley, eds., Metallurgical Applications of Bacterial Leaching and Related Microbiological Phenomena (New York, NY: Academic Press, 1978).

4. S.I. Kuznetsov, M.V. Ivanov and N.N. Lyalikova, Introduction to Geological Microbiology (New York, NY: McGraw-Hill Book Co., 1963)

5. T. Fenchel and T.H. Blackburn, Bacteria and Mineral Cycling (New York, NY: Academic Press, 1979(143-146.

6. M.P. Starr, et al, eds., The Prokaryotes, A Handbook on Habitats, Isolation and Identification of Bacteria (New York, NY: Springer-Verlag, 1981).

7. R.E.Buchanan and N.E. Gibbons, eds., Bergey's Manual of Determinative Bacteriology (Baltimore, MD: The Williams and Wilkins Co., 1974).

8. R.F. Commeau, et al., "Chemistry and mineralology of pyrite-enriched sediments at a passive margin sulfide brine seep: abyssal Gulf of Mexico" Earth and Planetary Science Letters 82 (1987) 62-74.

9. H.E. Gibbs, M. Errington and F.D. Pooley, "Economics of Bacterial Leaching" Canadian Metallurgical Quarterly 24(1985) 121-125.

10. P.B. Marchant and R.W. Lawrence "Consideration for the Design, Optimization and Control of Continuous Biological Tank Leaching Operations to Enhance Precious Metals Extraction" (paper presented at the Gold 100 International Conference, Johannesburg, South Africa, 15-19 September 1986).

11. E.E. Malouf and J.D. Prater "Role of Bacteria in the Alteration of Sulfide Minerals" Journal of Metals, May, 1961, 353-356.

12. J.E. Zajic Microbial Biogeochemistry (New York, NY: Academic Press, 1969) 70-71.

13. S.R. Gilbert, C.O. Bounds and R.R. Ice, "Comparative economics of bacterial oxidation and roasting as a pre-treatment step for gold recovery from an auriferous pyrite concentrate" CIM Bulletin, February, 1988. 89-94.

14. N.W. LeRoux, D.S. Wakerley and V.F. Perry, "Leaching of minerals using bacteria other than Thiobacilli" in Metallurgical Applications of Bacterial Leaching and Related Microbiological Phenomena L.E. Murr, et al, eds., (New York, NY: Academic Press, 1978)

15. M.J. Southwood, "The mode of occurrence of gold in pyrite and arsenopyrite and its implications for the release of gold during bacterial leaching" Council for Mineral Technology, Technical Memorandum No. 13255. Johannesburg, South Africa, 31 December 1985. 1-26.

16. M.J. Southwood, "Mineralogical aspects of the bacterial leaching of auriferous sulfide concentrates and a mathematical model for the release of gold" Council for Mineral Technology, Report No. M274. Johannesburg, South Africa. 9 October 1986. 1-12.

17. E.J. Brown, J.M. Forshaug, "Metabolic Properties of Thiobacillus ferrooxidans Isolated from Neutral pH Mine Drainage," Institute of Water Resources, University of Alaska, Fairbanks, AK March, 1983.

18. M. Oertel, "Experience Gained in Operation of an Industrial Scale Bacterial Oxidation Plant at the Fairview Gold Mine since July, 1986" (paper presented at the Randol Gold Forum 88, Scottsdale, AS, 23-23 January 1988).

19. J.A. Brierley, C.L. Brierley "Microbial Leaching of Copper at Ambient and Elevated Temperatures," in Metallurgical Applications of Bacterial Leaching and Related Microbiological Phenomena, L.E. Murr, et al, eds., (New York, NY: Academic Press, 1979).

20. B.W. Madsen and R.D. Groves, "Percolation leaching of a chalcopyrite-bearing ore at ambient and elevated temperatures with bacteria" (U.S. Bureau of Mines) Report of Investigation 8827.

21. M.S. Davidson, A.E. Torma, J.A. Brierley and C.L. Brierley, "Effects of elevated pressures on iron and sulfur oxidizing bacteria" in Biotechnology and Bioengineering Symposium No 11. (New York, NY: John Wiley and Sons, 1981) 603-618.

MODEL FOR BACTERIAL LEACHING OF COPPER ORES

CONTAINING A VARIETY OF SULFIDES

Bradley C. Paul, H. Y. Sohn, and M. K. McCarter

Graduate Student Mining Engineering at University of Utah, Salt Lake, Utah 84112

Prof. of Metallurgy at University of Utah

Chairman and Prof. of Mining Engineering at University of Utah

Abstract

A computer model was constructed for bacterial leaching of the major sulfides found in porphyry deposits, including bornite, chalcocite, chalcopyrite, covellite, cubanite, enargite, native copper, pyrite, and pyrrhotite. The model incorporates equations for the sulfide minerals dispersed in the rock matrix. Leaching occurs by reactions with ferric ion diffusing into the rock fragment. Ferrous ion is regenerated to ferric with oxygen by bacteria at a temperature dependent rate. The model keeps track of the net acid generated, oxygen consumed by the bacterial oxidation of ferrous ion, heat generated, jarosite and hematite precipitated, and copper leached. The model was used to predict the rate of copper recovery, heat generation and oxygen demand for a variety of run-of-mine sized porphyry ores. These results were used in the design of a ventilation system and the economic analysis of a modified in situ extraction method that may be cost competitive with conventional copper production involving open pit mining.

Metallurgical Processes for the Year 2000 and Beyond
Edited by H.Y. Sohn and E.S. Geskin
The Minerals, Metals & Materials Society, 1988

Introduction

A study was undertaken at the University of Utah to evaluate the technological and economic feasibility of mining copper by a method referred to as Flood Drain Leach Cell (FDLC) mining. Cells 60 by 75 meters in plan and 70 meters in height are created throughout the ore body by a modified Vertical Retreat Mining technique that includes the removal of 25% swell material. The swell material is leached in finger dumps on the surface while the cells underground are bulkheaded, flooded, drained, and artificially aerated to promote bacterial ferric sulfate leaching. Solutions are cycled through several cells to reach pregnant grade of 1 gram copper per liter of solution. These solutions are then fed to an underground solvent extraction plant and then a surface electrowinning tankhouse.

FDLC is intended for recovery of sulfide copper from porphyry deposits. Because only 25% of the ore need be moved tremendous savings may be realized in capital, and operating costs, labor dependence, and energy consumption. FDLC mines can double the in situ recovery of copper compared with block caving and do it at a price competitive with conventional surface mining and processing. This improved recovery is made possible by the elimination of channeling, better control of heat build-up, and increased oxygen supply.

A key to both the technical design and economic feasibility of this method is the prediction of leaching rates, and the resulting heat and oxygen consumption loads on the ventilation system. Work began with a review of the geologic characteristics of 200 porphyry copper deposits. This was followed by the development of a mine plan and blast designs that allowed prediction of the size distribution of the resulting material. The next task was to predict the rate of leaching for the ore. Existing simulation models were found inadequate for the task; bornite, enargite, and pyrrhotite are sometimes significant sulfides in porphyry systems, but no model found deals with the bacterial leaching of these minerals. When the leaching of large run-of-mine size ore fragments is simulated, many numerical schemes, particularly implicit schemes used in pre-existing models, become unstable. Additionally, implicit schemes require either linear or linearized kinetic rate equations. None of the models consider that the rate at which bacteria can produce ferric ion might become rate-limiting.

Even if the models could handle the mineralogy and fragment sizes likely in FDLC mining, they provide little information on the leaching environment. The leaching of sulfides in a typical porphyry ore is an exothermic process. Heat must be removed from the system. In FDLC mining forced ventilation is used to cool the rubble and provide oxygen to all parts of the cells. This requires the model to output the rate of heat generation and oxygen consumption.

The chemical reactions involved in leaching a porphyry can both consume and generate acid. Because solution is administered only periodically in a flood, rather than in a constant trickle, it is not possible to constantly add acid to sustain the low pH conditions needed for leaching. Neither is it possible to make up for shortages in acid production by adding large quantities of acid during flooding, since gangue mineral reactions normaly rapidly consume surplus acid below a minimum pH. Since leaching will only proceed at low pH the reactions in the leach cells must produce a surplus of acid. This necessitates keeping track of the net acid balance.

To meet these demands a new computer model, called DLS, was created. The model requires as inputs the temperature of leaching, the types and amounts of sulfide minerals in the ore, the size of the mineral grains, the size distribution of the ore fragments, the porosity, and the maximum rate of ferric generation by bacteria. Output includes the fraction of copper leached, the amount of iron salts precipitated, oxygen demand, heat output, net acid balance, and mineral reaction profiles through the ore fragments.

Theoretical Basis of Model

Diffusion

The DLS computer model assumes that ferric ion generated by bacteria on the surface of the ore diffuses down pores into spherical rock fragments according to the equation,

$$\varepsilon \frac{\partial c}{\partial t} = - R + D_e \left(\frac{\partial^2 c}{\partial r^2} + \frac{2}{r}\frac{\partial c}{\partial r}\right) \qquad [1]$$

where

ε= Porosity
c= Concentration
t= Time
ΣR= Rate of Consumption
D_e Effective Diffusivity
r= radial position

The equation can be simplified by applying the pseudo-steady-state approximation to give (1)

$$\Sigma R = D_e \left(\frac{d^2 c}{dr^2} + \frac{2}{r}\frac{dc}{dr}\right) \qquad [2]$$

The applied boundary conditions are ferric ion concentration equal to bulk solution concentration at the ore solution interface and a zero derivative of ferric concentration with respect to radius at the center of the ore fragment. Numerically, the condition of non-negativity of concentration throughout the fragment is imposed.

Individual Mineral Kinetics

The sum-of-R term in the above equation represents the local rate of ferric ion consumption per unit volume of ore by different mineral species. To obtain a value for the term, leaching rate equations must be known for all the involved mineral species. Satisfactory kinetic expressions have previously been reported for chalcocite, covellite, and pyrite (1). These expressions are used directly in DLS program subroutines. Munoz et al. (2) have developed an expression for the rate of leaching of chalcopyrite through a partially protective sulfur layer. The equation is adapted to the form below and used in a subroutine:

$$\frac{d\alpha}{dt} = \frac{0.0269}{R_p \; \phi} \frac{(1-\alpha)^{1/3}}{(1-[1-\alpha]^{1/3})} \exp\left(\frac{-20{,}000}{R_g \; T}\right) \qquad [3]$$

where

α = Fraction Reacted
R_g = Gas Constant
T = Tempurture
ϕ = Shape Factor
R_p = Radias of Mineral Grain

The rate of ferric consumption by pyrrhotite has been reported by Lowe (3):

$$\frac{d\alpha}{dt} = \frac{0.194}{R_p \; \phi} (1-\alpha)^{2/3} \exp\left(\frac{-7{,}790}{R_g \; T}\right) \qquad [4]$$

Dutrizac and McDonald (4) have reported kinetics for pure specimens of enargite as adapted below:

$$\frac{d\alpha}{dt} = \frac{339}{R_p \; \phi} (1-\alpha)^{2/3} \exp\left(\frac{-13{,}300}{R_g \; T}\right) c^{1/2} \qquad [5]$$

Bornite is a particular problem since it leaches rapidly at first, goes through a slowing trend and forms an intermediate compound which leaches very slowly. For many years researchers could not agree on what the intermediate was since it had kinetics and x-ray patterns characteristic of chalcopyrite, and gave microprobe analysis consistent with Cu_3FeS_2. A recent study by Dutrizac et. al (5) apparently resolved the controversy in favor of a Cu_3FeS_2 intermediate. Bornite kinetics are modeled as a fast diffusion-limited reaction up to 40% copper extraction using a formula developed by Dutrizac (6) and adapted below:

$$\frac{d\alpha}{dt} = \frac{65}{R_p \; \phi} (1-\alpha/.4)^{2/3} \exp\left(\frac{-5{,}200}{R_g \; T}\right) c \qquad [6]$$

for

$c < 6 \times 10^{-5}$ moles/cm
constant at $c = 6 \times 10^{-5}$ rate for higher c

At 40% extraction, kinetics switch to the Munoz equation for reaction across a partially protective sulfur film. The model also incorporates the kinetics for cubanite (7):

$$\frac{d\alpha}{dt} = \frac{20.7}{R_p \; \phi} (1-\alpha)^{2/3} \exp\left(\frac{-11{,}600}{R_g \; T}\right) c^{0.6} \qquad [7]$$

and native copper (8):

$$\frac{d\alpha}{dt} = \frac{751}{R_p \; \phi} (1-\alpha)^{2/3} \exp\left(\frac{-4{,}000}{R_g \; T}\right) c \qquad [8]$$

The former two minerals are more common in massive sulfide deposits than in porphyrys.

Modifications to Spherical Geometry

Although spherical geometry is a necessary assumption in order to allow solution to the equations in a reasonable time, most ore fragments are not spherical. Ore fragments become degraded and more porous as leaching proceeds. This problem has previously been handled by a modification to the diffusivity of ferric ion (1). Such modifications have been called leaching enhancement, or shape factors (1, 9). The leaching enhancement approach of Madsen and Wadsworth (1) is incorporated in the DLS program.

Mineral grains are also often not spherical, and chalcopyrite particularly tends to have a very porous surface (10). The previous approach of adding a shape factor to the rate equations has been used; only in this case the factor is applied to the kinetic equation for the individual mineral grains.

Overall Material and Heat Balances

The DLS program keeps track of and outputs the amounts of reaction products from chemical processes taking place throughout the leached material. The copper leached is the sum of copper removed from each mineral species in each fragment size fraction. The rate of heat generation is the sum of the heats of reaction for all the leached mineral species. The net acid balance is the total of all acid consumed or generated by the leaching of sulfide minerals. It is assumed that any surplus acid produced is consumed by gangue mineral reactions. The oxygen consumption is the total consumed by bacteria to regenerate ferric ion from ferrous. This regeneration is limited either by the total supply of ferrous ion available or the maximum rate at which the bacteria can produce ferric ion. Previous studies give the rate of ferric generation by bacteria as a function of temperature (11). All iron above the solubility limit is assumed to precipitate as either jarosite, if sufficient sulfate has been produced, or as hematite. Thus, bulk ferric concentration either remains at, or tends toward, the equilibrium value.

Numerical Procedure for Solution of Model Equations

The DLS program solves the ferric ion diffusion and reaction equation by finite difference. Ore fragments are divided into shells. Previous studies have indicated that 100 shells will give an essentially grid independent solution (10); however, it was found in this study that large fragments such as those over 0.4 meters in diameter could take as many as 240 shells depending on the mineralogy. Solution to the finite difference equations is obtained by a 4th order Runge-Kutta algorithm (12) coupled with shooting technique. The approach has been used in previous models for oxygen pressure leaching of chalcopyrite (13).

The model allows the user to specify the number of shells and the size of the time increment. Because reactions with several amenable minerals such as chalcocite can go very fast at first and then slow down as the diffusion paths to the mineral become longer, the user is allowed to specify two time steps and the length of leaching time to be simulated before switching to the larger time step.

The results of several model runs were compared with large scale column leach tests (1) to insure that reasonable answers were generated.

Results of Application

Previous modeling efforts have considered the rate limiting effects of mineral kinetics and diffusion for ores that contain only a small number of mineral species. DLS also considers the availability of ferric ion. It was found during the early stages of leaching, especially for fine ores with fast leaching mineralogy, that the supply of ferric ion is rate limiting. Liquid holdup is often as low as 10% of void space (14) and bacteria may only be able to oxidize a few grams of iron per liter of solution in a day (15). Fast reactions can consume more ferric ion than the total soluble iron load. Thus leaching goes through 3 rate limiting regimes, first a ferric supply limited stage, second, a mineral kinetics limited stage, and third, a diffusion limited stage.

While ferric supply limits will produce most of their effects in the first 40 to 100 days, the transition from kinetic limitation to diffusion limitation may take place over a longer period. The trade off between kinetic and diffusion limitation may also be more important in terms of total recovery and the amount of discounting applied to the resulting cash flow. Previous models have considered only a limited number of sulfides in finer size ore fragments. Because DLS considers 7 copper minerals and 2 gangue sulfides and is numerically stable for larger fragments, it provides valuable insights into when and to what extent diffusion dominates the recovery process.

Figures 1a to 1c show the results of leaching fragments of 4 cm, 16 cm, and 32 cm diameter. This ore is similar to that found in the vein systems of Butte Montana and is composed almost totally of fast leaching minerals. The exact characteristics of this and other ores used in illustrations are shown in table 1. Especially for fine ore a sharp slope going through a region of sharp tapering during the first 40 days signals the region of bacterial ferric limitation. Though very subtle and perhaps even beyond the true accuracy of the supporting experiments, it is possible in the numerical simulation to see a region where the fractions leached from different minerals begin to break away from each other slightly. This signals a small kinetic effect, though even here diffusion is dominating. As diffusion control becomes still more pronounced, even the slight separation in the fractions recovered from different minerals begins to disappear.

In this largely diffusion dominated case the fineness of the ore and not the mineralogy will set the rate of recovery. A simpler shrinking core model might be attractive for such an ore except that a shrinking core would usually not include the bacterial rate limiting phase at the start of leaching. The pyrite reaction zone is also rather broad, especially compared to the narrow reaction zone of the copper sulfides. This too is a compromise that must be accepted if a shrinking core is used.

Figures 2a to 4c illustrate a more common case. These ores are similar to the porphyry ores of Bingham Canyon in Utah, the mass sulfide ores of Sudbury in Ontario, or secondary enriched porphyry caps in Arizona. These ores have rapidly leaching minerals that quickly become rate limited by diffusion. Chalcopyrite in all cases reacts slowly creating a reaction zone from the fragment surface to where the remaining ferric is consumed in a narrow reaction zone of fast reacting minerals. Kinetics dominates the leaching of the chalcopyrite portion of the ore while diffusion controls the more rapid minerals. The cubanite in the Sudbury ore is an intermediate case. Its leaching kinetics are slow enough that the kinetic step

Table I

Ore Properties

Ore Type	Grade	Copper Min.	Py:Cu Ratio	Porosity	Min Grain Sz	Frag. Size	Figure Cited
'Butte-Type'	.5% Cu	Bornite 10% Chalcocite20% Covellite 5% Enargite 65%	1.5:1	4%	6 mic.	4cm diam. 16cm diam 32cm diam Simulated Crater Blast	1a 1b 1c 5
'Bingham-Typ'	.5% Cu	Bornite 33% Chalcopyrite 67%	1.5:1	4%	6 mic.	4cm diam 16cm diam 32cm diam Simulated Crater Blast	2a 2b 2c 5
'Sudbury-Type'	.5% Cu	Chalcopyrite 65% Native Copper 2% Cubanite 33%	1.5:1	4%	6 mic.	4cm diam 16cm diam 32cm diam	3a 3b 3c
2ndary Enriched	.6% Cu	Chalcocite25% Covellite 5% Chalcopyrite 70%	1.5:1	4%	6 mic.	4cm diam 16cm diam 32cm diam Simulated Crater Blast	4a 4b 4c 5
'Bornite-Core-Type'	.5%Cu	Bornite 65% Chalcopyrite 35%	.5:1	4%	6 mic.	Simulated Crater Blast	6
'Chalcopyrite-Type'	.4%Cu	Chalcopyrite 100%	2.5:1	4%	6 mic.	Simulated Crater Blast	7

tends to remain pronounced for smaller ore sizes while becoming increasingly eclipsed by the diffusion step in larger fragments. Though not shown in the figures, pyrite is an intermediate mineral like cubanite, and pyrrhotite is an extremely fast leaching mineral. The slow kinetics and broad reaction zone of chalcopyrite is a major violation of the assumptions in simple shrinking core models and limits their usefulness in modeling the leaching of porphyry ores since most contain significant amounts of chalcopyrite. The intermediate behavior of cubanite and pyrite can be problematic for using shrinking core models on finer fractions of ores that contain either of these minerals even if chalcopyrite is not present.

Figures 1a-1c

Copper Recovery from "Butte-Type" Ore

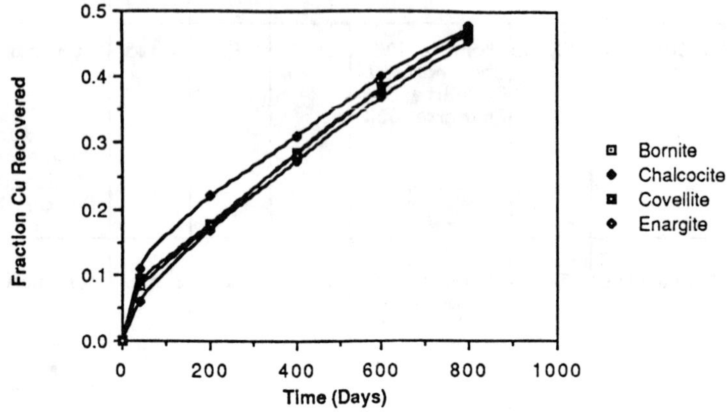

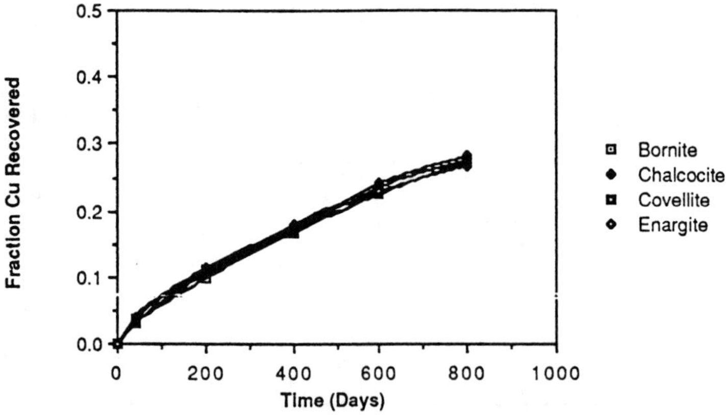

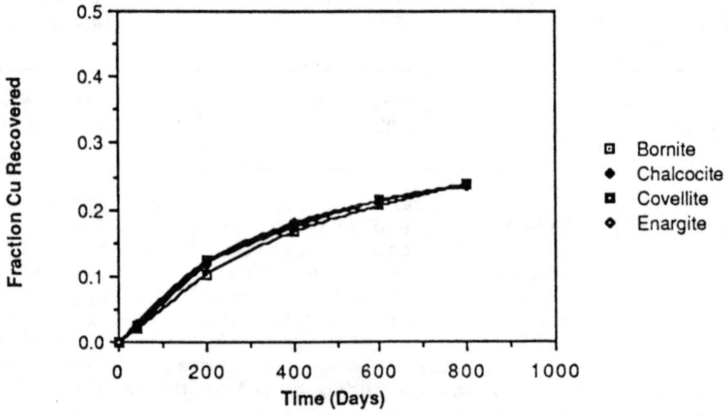

Figures 2a-2c

Copper Recovery from "Bingham-Type" Ore

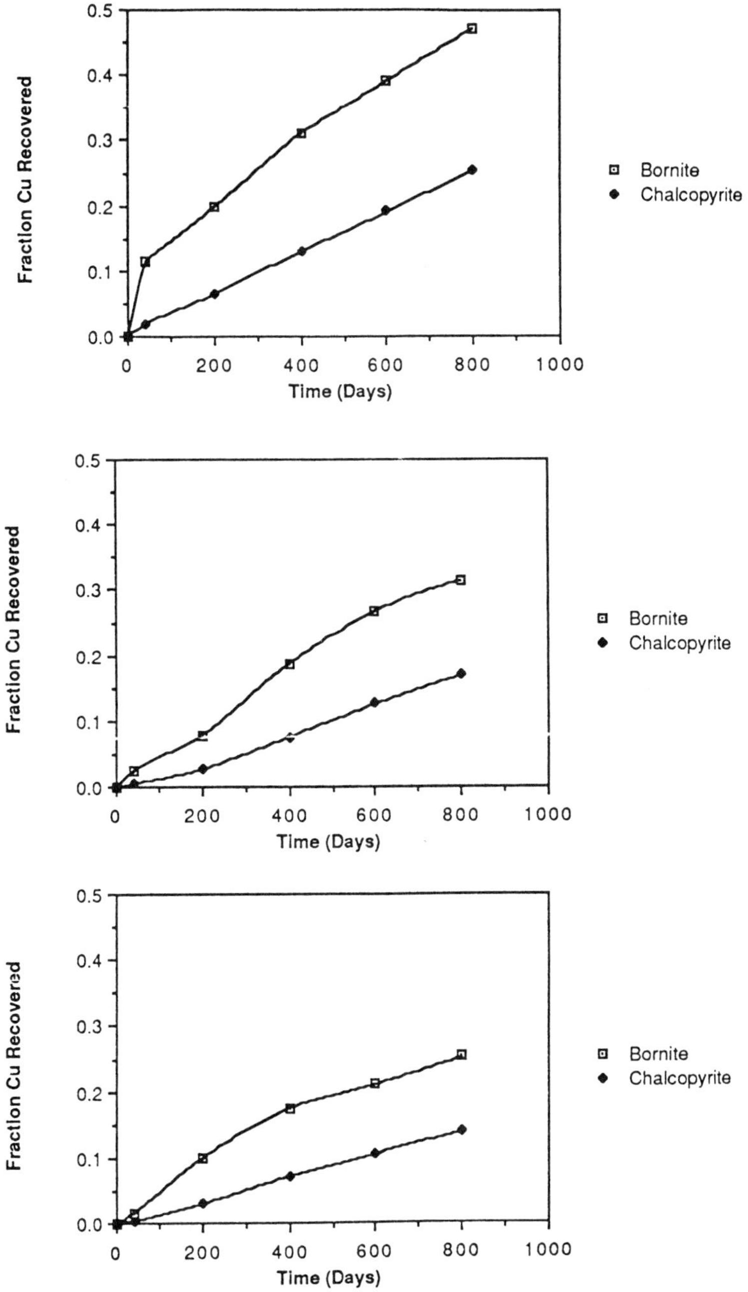

Figures 3a-3c

Copper Recovery from "Sudbury-Type" Ore

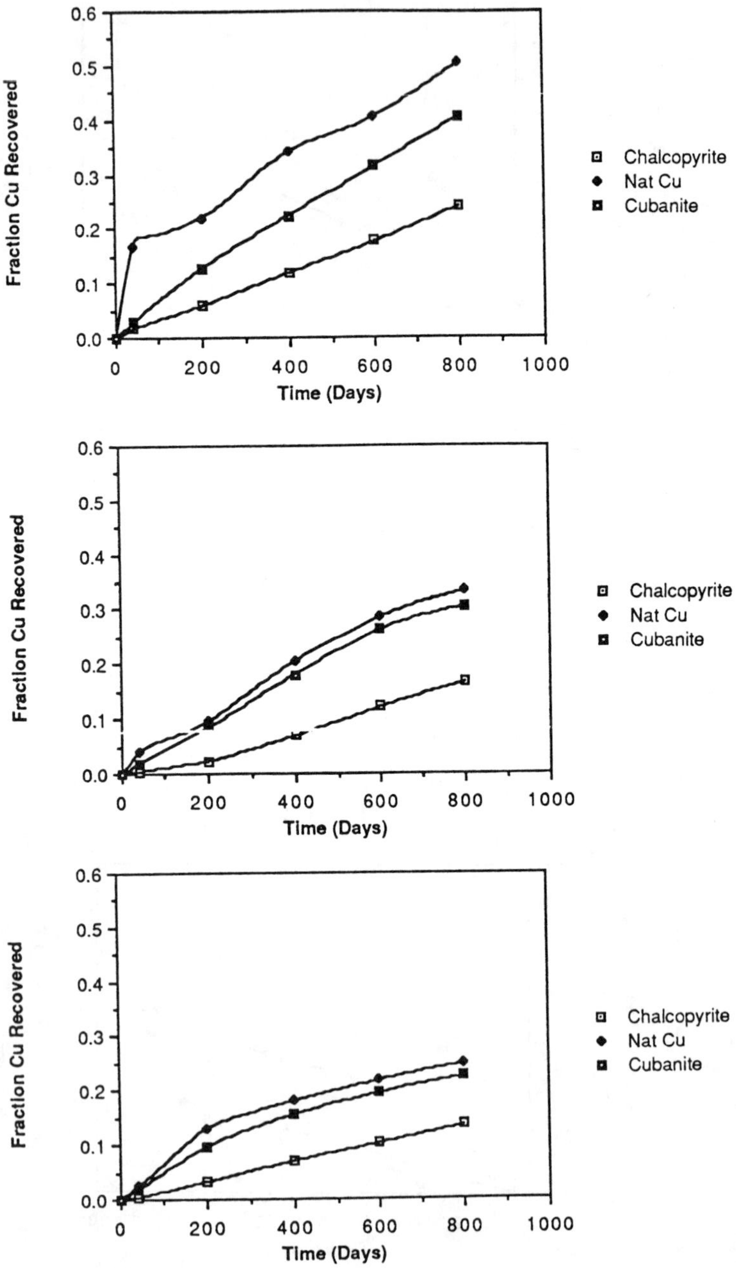

Figures 4a-4c

Copper Recovery from Secondary Enriched Ore

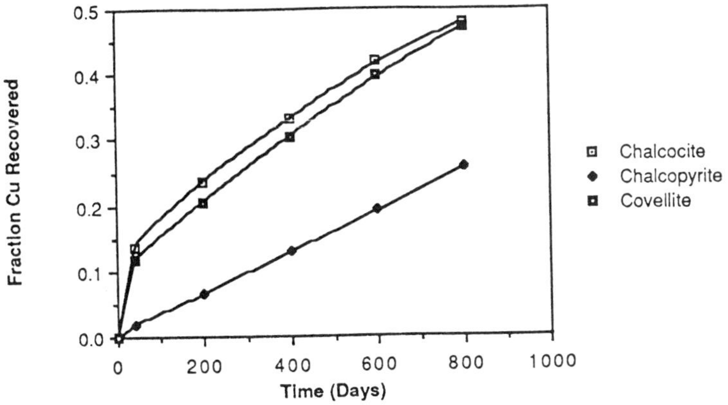

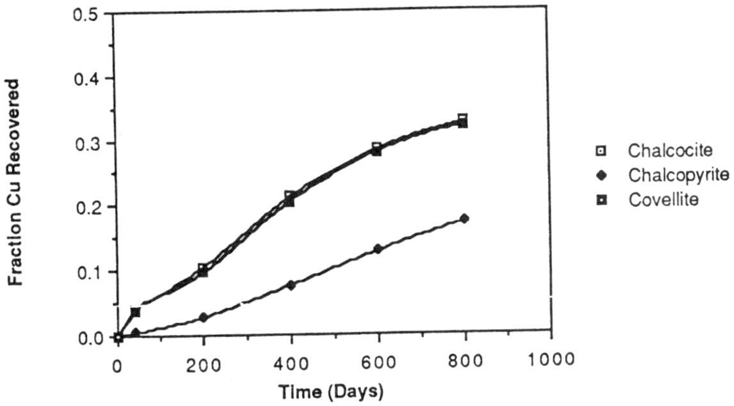

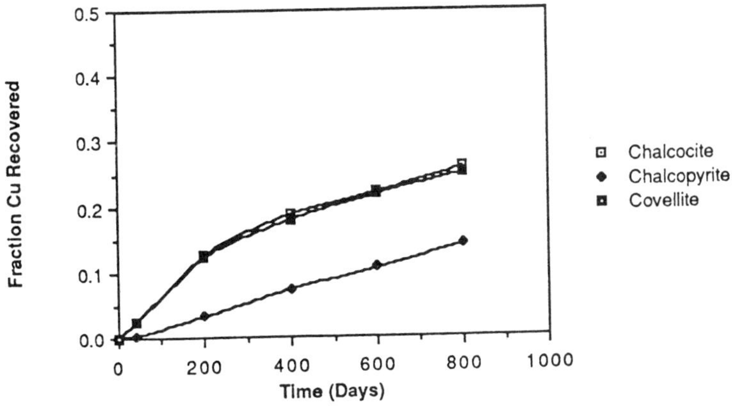

The model was used to obtain the volumetric heat generation and oxygen consumption rate for use in designing the ventilation system for Flood Drain Leach Cell mines. The model used to simulate the ventilation is covered in another paper in preparation and in a dissertation (15). It can be noted from figure 5 that the ratio of moles of oxygen consumed per mole of copper produced is not a constant as has been assumed in previous studies (16). Other studies have simply observed the average range of the oxygen to copper ratio in experiments (17). Because the leaching mineral kinetics of different minerals seldom go in perfect stoichiometric balance, the rate of oxygen consumption is also much higher than simple models would suggest. This may imply that one reason for poor recoveries (20% or less) observed in many dump and in situ leaches is oxygen starvation. Understanding the oxygen demand was important in designing a ventilation system that would overcome the problem and provide high recoveries for Flood Drain Leach Cell mines.

Figure 5

Ratio of Oxygen Consumed to Copper Produced

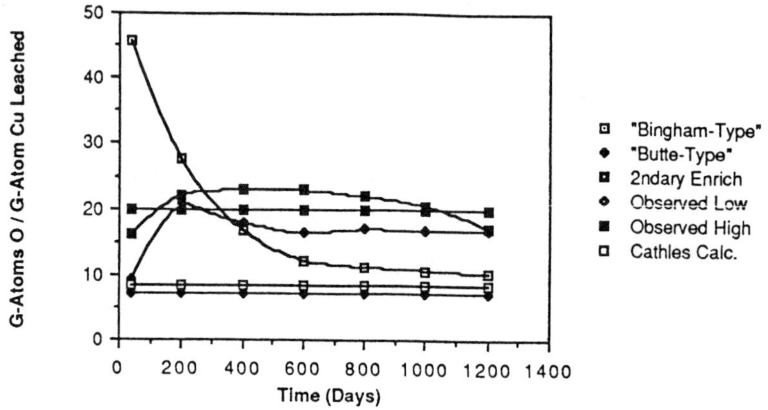

Another effect demonstrated by the model and shown in figures 6 and 7 is that many ores loose the ability to produce surplus acid after a time. This is particularly true of chalcopyrite-pyrite mineral mixes because pyrite leaches faster than chalcopyrite leaving much chalcopyrite in the outer shells. Pyrite becomes confined to deep inner layers. Since pyrite is the primary acid producer in leaching reactions, this results in a loss of ability to produce surplus acid. Because the reactions must produce surplus acid in order to maintain the low pH needed for leaching this probably means that a Flood Drain Leach cycle cannot achieve more than around 60% copper extraction from chalcopyrite based ores.

Economic analysis with the Bureau of Mines Cost Estimating System (18, 19). utilizing copper recoveries predicted by the DLS model show that Flood Drain Leach Cell mines can use underground mining methods to produce copper at from 55 to 91 cents per pound, depending on mineralogy and ore body depth. Prices from about \$0.99 to \$1.60 per pound can provide for opening a new mine with a 15% DCF ROR (15). It is the amenability of the ore mineralogy as predicted by the leaching model that dominates the determination of whether higher or lower prices are needed. Use of the leaching model thus becomes important in identifying deposits that can compete with open pit mines on today's market.

Figure 6

Net Acid Production Rate for "Bornite-Core-Type" Ore

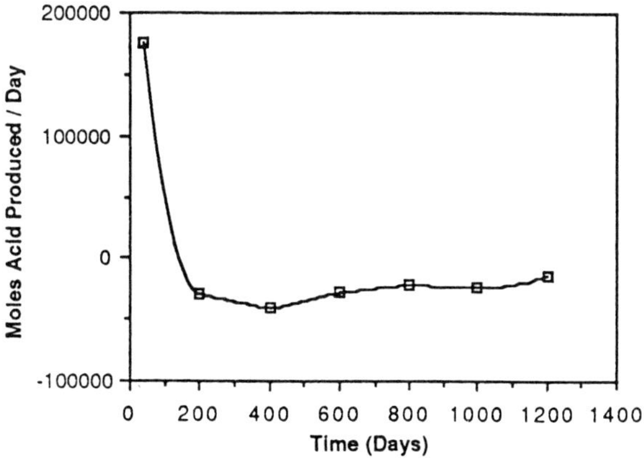

Figure 7

Net Acid Production Rate for "Chalcopyrite-Type" Ore

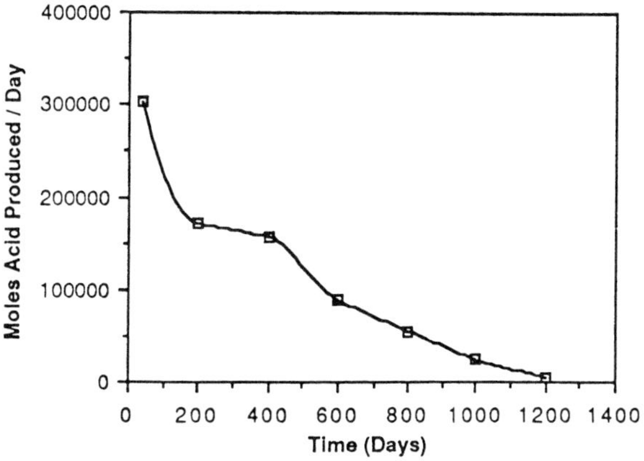

References

(1) B.W. Madsen and M.E. Wadsworth, "A Mixed Kinetics Dump Leaching Model for Ores Containing a Variety of Copper Sulfide Minerals", U.S. Bureau of Mines Report of Investigations 8547, 1981, 44p.

(2) P.B. Munoz, J.D. Miller, and M.E. Wadsworth, "Reaction Mechanism for the Acid Ferric Sulfate Leaching of Chalcopyrite", Metall Trans B, vol.10B, June 1979, pp. 149-158

(3) D.F. Lowe, The Kinetics of the Dissolution Reaction of Copper and Copper-Iron Rich Sulfide Minerals Using Ferric Sulfate Solutions, Ph.D. Thesis, University of Arizona, 1970, 170p.

(4) J.E. Dutrizac, and R.J.C. MacDonald, "The Kinetics of Dissolution of Enargite in Acidified Ferric Sulfate Solutions", Can Metall Quart, vol. 11 #3, 1972, pp. 469-476

(5) J.E. Dutrizac, T.T. Chen, and J.L. Jambor, "Mineralogical Changes Occuring During the Ferric Ion Leaching of Bornite", Metall Trans B, vol. 16B, Dec. 1985, pp. 679-693

(6) J.E. Dutrizac, R.J.C. MacDonald, and T.R. Ingraham, "Kinetics of Dissolution of Bornite in Acidified Ferric Sulfate Solutions", Metall Trans of AIME, vol. 1, Jan. 1970, pp. 225-231

(7) J.E. Dutrizac, R.J.C. MacDonald, T.R. Ingraham, "The Kinetics of Dissolution of Cubanite in Aqueous Acid Ferric Sulfate Solutions", Metall Trans of AIME, vol. 1, Nov. 1970, pp. 3083-3088

(8) E.G. Valdez, A Kinetic Study of the Dissolution of Metallic Copper in Ferric Sulfate Solution, M.S. Thesis, University of Utah, 1968, 36p.

(9) B.W. Madsen, M.E. Wadsworth, and R.D. Groves, "Application of a Mixed Kinetics Model of the Leaching of Low Grade Copper Sulfide Ores", Trans Soc of Mining Engineers of AIME, vol. 258, March 1975, pp. 69-74

(10) J. W. Braithwaite, Simulated Deep Solution Mining of Chalcopyrite and Chalcocite, Ph.D. Thesis, University of Utah, June 1976, 255p.

(11) E.E. Malouf, and J.D. Prater, "Role of Bacteria in the Alteration of Sulfide Minerals", Journal of Metals, vol. 13, May 1961, pp 353-356

(12) D Greenspan, Discrete Numerical Methods in Physics and Engineering, New York, NY, Academic Press, 1974, 312p.

(13) H. K. Lin, and H. Y. Sohn, "Mixed-Control Kinetics of Oxygen Leaching of Chalcopyrite and Pyrite from Porous Primary Ore Fragments", Metall Trans B, 18 B, 1987, pp. 497-504

(14) L.M. Cathles, "Evaluation of an Experiment Involving Large Column Leaching of Low-Grade Copper Sulfide Waste: A Critical Test of a Model of the Waste Leaching Process", Leaching and Recovering Copper from As-Mined Materials, proceedings of the Las Vegas Symposium, New York, NY., AIME, Feb. 26, 1980, pp. 29-47

(15) B. C. Paul, Modified In-Situ Leaching of Copper, Ph. D. Thesis, University of Utah, in preparation

(16) L.M. Cathles, "A Model of the Dump Leaching Process that Incorporates Oxygen Balance, Heat Balance, and Two Dimensional Air Convection", Leaching and Recovering Copper from As-Mined Materials, proceedings of the Las Vegas Symposium, New York, NY., AIME, Feb. 26, 1980, pp. 29-47

(17) M. Adams, J. A. Apps, R. Capuano, D. R. Cole, C. L. Kusik, D. Langmuir, and M. E. Wadsworth, In Situ Leaching and Solution Mining evaluation of the State of the Art, appendix 2, chemistry, Salt Lake, Utah, University of Utah Research Institute, 1983, 346p.

(18) U.S. Bureau of Mines Staff, "Bureau of Mines Cost Estimating Handbook- Surface and Underground Mining", U.S. Bureau of Mines Information Circular 9142, 1987, 631p.

(19) U.S. Bureau of Mines Staff, "Bureau of Mines Cost Estimating Handbook- Mineral Processing", U.S. Bureau of Mines Information Circular 9143, 1987, 566p.

PROCESS CONSIDERATION FOR THE FABRICATION OF

pH MICROELECTRONIC SENSORS

M. MOINPOUR, P. W. CHEUNG, E. LIAO, C. Y. AW, and D.J. BROWN

Microsensor Research Laboratory, Washington Technology Center
Department of Electrical Engineering
University of Washington, Seattle, WA 98195

Abstract

Currently, silicon-based insulated-gate field-effect devices are being widely developed as the basic structural elements in a new generation of chemical microsensors. In this paper, we discuss some of the process considerations related to the fabrication of the Chemically Sensitive Field-Effect Transistor (CHEMFET). This ion-sensitive CHEMFET (or ISFET as it is often called) can be used for pH measurements (i.e., as a pH sensor) when the gate insulator of the electrolyte/insulator interface is made of Si_3N_4 obtained by the LPCVD method. Effects of processing parameters such as the Si_3N_4 deposition rate, gas composition, total flow rate, annealing treatment, etc. on Si/SiO_2, Si/Si_3N_4 and $Si/SiO_2/Si_3N_4$ structures are discussed in relation to the CHEMFET fabrication. Their effects on the electrical characteristics and chemical response of the sensor are also examined. Capacitance-Voltage (C-V) measurements and ESCA analysis have been used to interpret the results. The Si_3N_4/SiO_2 double dialectric layer provides improved ion blocking for the gate insulator in electrolyte solutions with enhanced pH sensitivity over time. The stability and performance of the fabricated CHEMFET sensors were tested by measuring the pH sensitivity in various storage conditions. The SEM analysis was used to evaluate the hydration effect at the ISFET's gate insulator with respect to different storage conditions.

Metallurgical Processes for the Year 2000 and Beyond
Edited by H.Y. Sohn and E.S. Geskin
The Minerals, Metals & Materials Society, 1988

Introduction

There has been a continuous interest in the development and fabrication of various chemical microsensors for chemical and biomedical applications over the last fifteen years. The application of semiconductor (particularly silicon) technology to the area of sensor fabrication has led to an emergence of a new generation of miniaturized, microelectronic, solid-state chemical sensors. There are numerous clinical situations where continuous monitoring of body electrolytes concentrations such as Na^+, K^+, Ca^{+2}, etc. and blood gases (pH, P_{O_2}, and P_{CO_2}) would be essential for the effective diagnosis and management of patients under critical conditions. Solid- state microsensors can be used for direct and continuous *in vivo* measurement of the above parameters in most instances.

A chemical microsensor is broadly defined as a microfabricated device that generates a reproducible electronic signal upon exposure to a chemical stimulus. Probably the best-known chemical microsensor is the Chemically Sensitive Field-Effect Transistor (CHEMFET). First reported by Bergveld in 1970 (1), the CHEMFET is essentially an insulated gate field-effect transistor (IGFET) that has its normal conductive metal gate contact replaced by a chemically sensitive membrane. The CHEMFET can be used in a variety of ways. A sensitive detector for gaseous species can be made if the gate region of the CHEMFET is coated with a thin film of gas-selective membrane (2). In solution the gate region can be coated with an ion-selective membrane; in fact, the ion-sensing CHEMFET (or ISFET as it is often called) with a silicon nitride gate insulator has been fabricated and tested successfully for pH measurements (3).

The solid state process for the fabrication of CHEMFET sensors is very similar to that utilized in standard MOSFET technology due to the similarity of their basic structures. It comprises of diffusion and oxidation steps as well as chemical and physical deposition processes. For pH-sensitive CHEMFETs, Si_3N_4 deposited by LPCVD method, has been widely used as the insulator of the electrolyte/insulator interface. In order to fully characterize the sensor processing and determine the best optimized processing conditions, one has to conduct systematic study of the above-mentioned steps particularly, the gate oxidation (ie, thermal oxidation of Si with O_2) and low pressure chemical vapor deposition (LPCVD) of silicon nitride. The overall performance of the chemical sensor is determined by the properties and characteristics of the gate insulator. Hence, careful characterization of the electrolyte insulator semiconductor (EIS) structures with respect to semiconductor process conditions and their effect on the chemical response of EIS structures become considerably important in CHEMFET fabrication process evaluation. In this paper, the effects of processing parameters such as deposition rate, gas composition, total flow rate, annealing treatment, etc. on Si/SiO_2, Si/Si_3N_4 and $Si/SiO_2/Si_3N_4$ structures are discussed in relation to CHEMFET fabrication and their effects on the electrical characteristics and chemical response of the sensor.

Theory of Operation

The structure of the ion-sensitive field-effect transistor (ISFET) is similar to that of the metal oxide semiconductor field-effect transistor (MOSFET). The construction of the ISFET under consideration differs from that of the conventional MOSFET devices; the metal gate is omitted and the ISFET is rendered pH sensitive by exposing the bare silicon nitride/silicon oxide gate insulator to the sample solution. In order to understand the operational principles of ISFET (note that, in this paper, CHEMFET and ISFET are being used interchangeably to describe the same pH-sensitive chemical microsensor), we will briefly summarize the most important equations and parameters for an n-channel MOS transistor.

The conventional MOSFET has aluminum metal connections to the source, drain and gate insulator (4,5). This is known as a three-terminal device. The equation for the drain current I_D, in the unsaturated region ($V_D < V_G - V_T$) is

$$I_D = \frac{\mu_n C_{ox}}{2}\frac{W}{L}\left[2(V_G - V_T)V_D - V_D^2\right] \quad (1)$$

where I_D is the current flow between source and drain electrodes, μ_n is the electron mobility in the channel, C_{ox} is the oxide capacitance (i.e, the capacitance per unit area between the gate oxide and the silicon substrate), W/L is the channel width–to–length ratio, V_G and V_D are the applied gate-substrate and drain- source voltages, respectively and V_T is the threshold voltage. The threshold voltage is the gate voltage required for the formation of a conducting channel from the source to the drain (i.e, the onset of inversion). Inversion occurs in p-type silicon when electrons begin to outnumber holes in the near surface region, thus forming the n-channel. The threshold voltage can be described as follows:

$$V_T = V_{FB} - \frac{Q_B}{C_{ox}} - 2\phi_F \quad (2)$$

where Q_B is the surface depletion region charge per unit area,ϕ_F is the Fermi potential difference between the doped bulk silicon and intrinsic silicon, and V_{FB} is the flat band voltage, which is given by

$$V_{FB} = \phi_{MS} - \frac{Q_{ss}}{C_o} \quad (3)$$

where ϕ_{MS} is the difference between the metal gate and the silicon work function and Q_{ss} is the fixed surface state charge density per unit area. Q_{ss} is defined in the following manner:

$$Q_{ss} = Q_{IT} + Q_F \quad (4)$$

where Q_{IT} and Q_F are the charge of the interfacial traps and the fixed oxide charge respectively, both per unit area. It should be noted that in the above formulation, the charge contribution due to mobile ions inside the insulator layer has been neglected for simplicity. In the saturated region, where $V_D \gg V_G - V_T$, equations (2), (3) and (4) are still valid, but equation (1) can be replaced by a simpler form,

$$I_D = \frac{\beta}{2}(V_G - V_T)^2 \quad (5)$$

where $\beta = \mu_n C_{ox}\left(\frac{W}{L}\right)$.

In both equations (1) and (5), I_D is a function of the threshold voltage, V_T. It is seen from equations (2), (3) and (4) that, for the MOSFET, the threshold voltage is determined by the property of the semiconductor materials as well as the fabrication process. Summation of equations (2) and (3) yields the following definition of the threshold voltage for the MOSFET:

$$V_T = \phi_{MS} - \frac{Q_{ss}}{C_{ox}} - 2\phi_F - \frac{Q_B}{C_{ox}}. \quad (6)$$

As mentioned before, the structure of the ISFET is essentially that of the metal insulator field-effect transistor (MISFET) without the metal gate electrode. The first ISFET devices were operated without a reference electrode (1,6). However, other investigators (7,8) have shown that proper operation of the ISFET requires a reference electrode to establish the electrolyte potential with respect to the semiconductor substrate. The reference electrode here may be envisioned as the gate electrode of the conventional MOSFET. Hence, the chemically sensitive field-effect transistor (CHEMFET) fabricated and characterized in the authors' laboratory is essentially an extended gate field-effect transistor with the electrochemical potential inserted between the non-metal gate surface of the CHEMFET (i.e, Si_3N_4/SiO_2 double layer) and the reference electrode. The schematic structure of a silicon nitride gate CHEMFET sensitive to pH is shown in Figure 1. This type of CHEMFET which employs only an inorganic insulator film without an ion-selective membrane, and is operated with a reference electrode and consists of a gate structure of electrolyte-insulator-semiconductor (EIS) interfaces, has also been referred to as the electrolyte-insulator-semiconductor field-effect transistor (EISFET) (9).

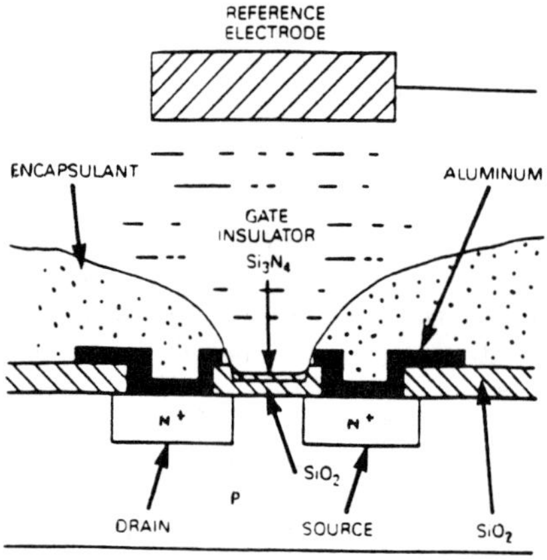

Figure 1 – Structure of silicon nitride gate CHEMFET sensitive to pH.

In order to modify the MOSFET equation (1) to this ISFET configuration, new expressions for V_G an V_T must be established. The response of the ISFET to hydrogen ions in solution may be described by the effect of the electrochemical potential on the threshold voltage. One approach has been to modify V_G, the ground-to-gate metal potential, in equation (1) as discussed by Moss et al (10),

$$V'_G = V_G - E_{REF} + E^\circ + \left(\frac{RT}{nF}\right) \ln a_i \tag{7}$$

where V_G is the applied reference electrode potential, -E_{REF} is the interface potential between the reference electrode and the electrolyte solution. The last two terms in equation (7) constitute the Nernst potential due to an ion activity in the solution generated at the electrolyte/insulator interface. The equation for the drain current of the ISFET then becomes,

$$I_D = \frac{\beta}{2}\left[2\left(V_G + \left(\frac{RT}{nF}\right)\ln a_i - V_T^*\right)V_D - V_D^2\right] \tag{8}$$

and

$$V_T^* = Q_{IS} - 2\phi_F - \frac{Q_{ss}}{C_I} - \frac{Q_B}{C_I} + qE_{REF} - E^\circ. \tag{9}$$

Note that the metal-semiconductor work function difference, ϕ_{MS}, and the oxide capacitance, C_{ox}, have been replaced by an insulator-semiconductor work function difference, ϕ_{IS}, and the insulator capacitance,C_I, respectively. Equation (8) is equivalent to equation (1); an ISFET equivalent gate potential V_G^* is then defined as follows (10),

$$V_G^* = V_G + \left(\frac{RT}{nF}\right)\ln a_i. \tag{10}$$

Thus, comparing equations (1) and (8), it can be seen that the ISFET drain voltage-drain current relationship is analogous to the MOSFET; the main modification lies in the introduction of the Nernst potential term $(RT/nF)\ln(a_i)$ which varies with ion activity, a_i.

More recently, a generalized theory has been developed by Fung et al. (11) in which the chemical response of the ISFET in solution is attributed to changes in the threshold voltage (ΔV_T) due to variation of the interfacial potential at the eletrolyte/gate-insulator interface. Site- binding model with surface association and dissociation of charged species is combined with the MOSFET physics to arrive at this generalized theory for the electrolyte-insulator- semiconductor field-effect transistor (EISFET). The drain current-voltage characteristics of an EISFET in the non- saturation region can be described in the following manner:

$$I_D = \frac{\alpha}{2}\left[2(\phi_0 - \phi_T)V_D - V_D^2\right] \tag{11}$$

and

$$\phi_T = -\left(\frac{Q_{ss}}{C_I}\right) - 2\phi_F - \left(\frac{Q_B}{C_I}\right) \tag{12}$$

where α is a constant and ϕ_0 is defined as the insulator surface potential. The threshold voltage ϕ_T is a quantity determined by the semiconductor substrate, the insulator capacitance, and the fabrication process. Comparison of equation (12) and the threshold voltage of MOSFET, equation (6) yields

$$V_T = \phi_T + \phi_{MS}. \tag{13}$$

Therefore it is possible to estimate the value of ϕ_T by measuring the threshold voltage of a monitoring MOSFET on the same chip. It should be emphasized that the derivation of equation (11) is based on the surface potential ϕ_0 at the insulator.

Both approaches outlined above have demonstrated that as the ionic concentration of the solution varies, the surface charge density at the CHEMFET gate sensing area changes. Electrochemical potential developed at the interface between the electrolyte and the gate insulator modulates the electrical conductance of the physical channel formed between the drain and the source electrodes. Drain current variations as a response to ionic activities or concentrations in solution can be translated to a convenient output voltage.

It can be concluded that the CHEMFET is essentially a field-effect semiconductor device with output electrical characteristics which are sensitive to activities of different ionic species in solution. At the heart of CHEMFET lies the gate sensing area. In order to provide ion selectivity to form a specific ion sensor, various materials can be employed as gate insulator. Si_3N_4 or SiO_2 work well as pH-sensitive layers in the ISFET configuration. Other thin oxide films such as Al_2O_3 (12) and Ta_2O_5 (13) have also been employed.

It is proven that SiO_2, due to its open structure, is not an entirely satisfactory dielectric material. Interest in the Si_3N_4/SiO_2 double insulator gate structure has stemmed largely from the charge instability problems and trapping centers associated with the Si_3N_4/Si interface. Since the Si- SiO_2 structures exhibit no charge instability at room temperature, it is believed that the insertion of the SiO_2 instability problems. pH microelectronic sensors have been successfully fabricated using low pressure chemical vapor deposition (LPCVD) of silicon nitride on top of silicon dioxide as a double gate dielectric(3,14,15). The double dielectric layer provides improved ion blocking for the gate dielectric in solution with enhanced pH sensitivity over time.

Experimental Methods

P-type, boron doped, Cz-grown, 3-inch Si wafers with (100) crystal orientation were used for this investigation. The average thickness of these wafers was about 20 mil, resistivities ranging from 6 to 8 Ω-cm. The insulator/semiconductor structures studied were made according to standard microelectronic technology. Before oxidation, wafers were cleaned in a modified RCA standard cleaning process consisting of immersion in an H_2O_2-H_2SO_4 solution for 10 minutes, a dilute HF solution for 1 minute, followed by an additional 10 minutes in an 80°C HCl solution, with DI water rinses after each step. Thermal oxidation of silicon wafers was carried out in a resistance-heated horizontal furnace. Pure, dry nitrogen flowing at the rate of 1000 cm^3min^{-1}(STP) was passed continuously through the furnace before and after the thermal oxidation. Oxidations were performed at 1000°C in dry O_2 at the flow rate of 1000 cm^3min^{-1}(STP). Oxides were grown to a thickness of about 800 Å.

After oxidation, the wafers were divided in two groups. Ellipsometry was performed on monitor wafers of the first group to allow calculation of oxide thickness and refractive index, n_{ox}. A Gaertner L117 ellipsometer with a He-Ne laser (6328 Å) and 70° angle of incidence was used for these measurements. Ellipsometric data were analyzed by a program developed in the Microsensor Research Laboratory. Following ellipsometry, the monitor wafers were rinsed in DI water and dried. The backside oxides were removed by etching in a semiconductor grade 10:1 buffered HF solution (V:V/40% NH_4F: 49% HF). E-beam evaporation was then employed to evaporate aluminum dots (1.5 mm diameter) on the front side of the wafers. After Al metallization, the monitor wafers were annealed at 400°C for 30 minutes. During postmetal anneal, nitrogen- 6% hydrogen gas mixture was passed continuously through the Al anneal furnace.

High frequency bias temperature capacitance-voltage (BTCV) measurements at

1 MHz were taken to ensure the desired properties of SiO_2 layer and also to qualify the oxidation runs. The applied dc biases were +10 V and -10 V respectively; the temperature of the MOS capacitor was raised to 250°C for 10 minutes. For all runs, the voltage shift of the C-V curves was less than 0.3 V. The C-V characteristics (1 KHz) for the remaining wafers of the first group (i.e, SiO_2/Si structures) were measured on the exposed oxide by means of a mercury probe.

The second group of wafers were directly transferred to a silicon nitride reactor shortly after the thermal oxidation. In addition, fresh, standard-cleaned, unoxidized Si wafers were also included in the nitride runs. A layer of thick silicon nitride was deposited on the top of the SiO_2 layer by the LPCVD technique using silane (SiH_4) in ammonia (NH_3) at approximately 850°C. The schematic diagram of the LPCVD reactor is shown in Figure 2. Depending on different experimental conditions, the thickness of the deposited Si_3N_4 layer varied in the range of 850-1700 Å.

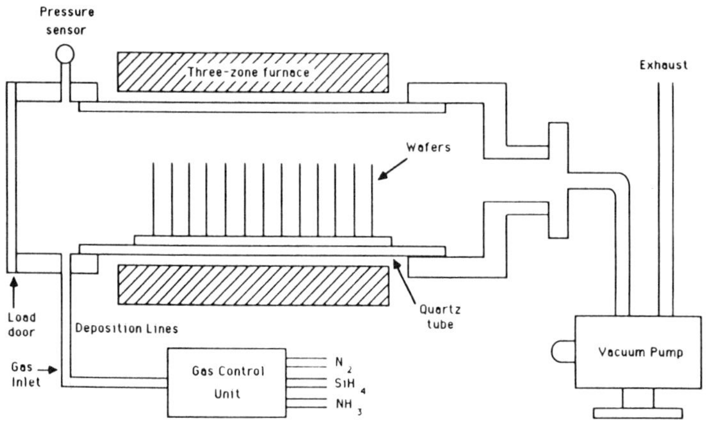

Figure 2 – Schematic diagram of a hot-wall LPCVD system using a three-zone furnace tube.

The resulting Si_3N_4/Si structures (i.e, nitride monitor wafers) were taken for ellipsometry to measure the nitride thickness and the index of refraction n_{nit}. These wafers were then broken into two halves. One half was used to calculate the etch rate of the deposited Si_3N_4 layer. Etch rates in buffered HF solutions (1:10) at 25°C were measured by timing the removal of a layer of known thickness. The other half was used for electron spectroscopy for chemical analysis (ESCA) to determine the composition of the silicon nitride film.

The C-V characteristics (1 KHz) for the remaining wafers of the second group (i.e, Si_3N_4/SiO_2/Si structures) were measured on the exposed nitride with a mercury probe. Also, the capacitance-voltage variations of electrolyte Si_3N_4/SiO_2/Si semiconductor (EIS) structures were examined using the ramp voltage quasistatic method (16). The standard calomel reference electrode with commercial inorganic buffer solutions were used to study the pH sensitivity of the electrodes. The results of these experiments are presented in the next section.

Results and Discussion

I. *Thermal Oxidation:*

Gate oxidation process (i.e, thermal oxidation of Si with O_2) is normally comprised of three steps, namely preoxidation warm-up of silicon wafers in nitrogen, oxidation at 1000°C in dry O_2 at a flow rate of 1000 cm^3min^{-1} (STP) and post-oxidation annealing under N_2 flow. The main parameters of the thermal silicon dioxide growth conditions (e.g., temperature and time of oxidation, and composition and pressure of the oxidizing species) have been substantially investigated and are well documented (17-19). In addition, postoxidation treatments (e.g., high temperature postoxidation anneal in N_2 and low temperature post- metallization anneal in N_2-H_2) aimed at modifying the oxide and interface properties have been extensively studied; it is shown that these treatments can effectively reduce the interfacial traps charges Q_{IT}, and the fixed oxide charges Q_F, thereby improving the electrical properties of MOS devices (18,19). In this study, the focus has been on the preoxidation treatment. One such, routinely done, unintentional preoxidation treatment is the thermal annealing given to wafers during their warm-up inside the oxidation furnace.

The effects of preoxidation anneal time on the voltage responses of both SiO_2/Si and Si_3N_4/SiO_2/Si structures was examined. During the thermal annealing, wafers were maintained in N_2 introduced through the furnace tube. After the annealing treatment, the nitrogen gas was turned off while dry O_2 was simultaneously turned on. Three different anneal times were studied; while the subsequent oxidation, post- oxidation anneal, nitride deposition and postmetallization anneal parameters were all kept constant for these runs. Figures 3 and 4 are plots of the average thickness and the refractive index of the oxide film vs. the preoxidation anneal time, t_A, respectively.

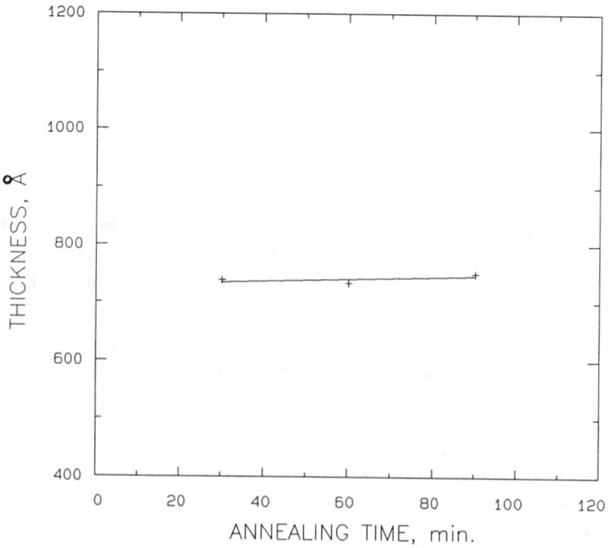

Figure 3 – Silicon dioxide thickness vs. preoxidation annealing in N_2 at 1000°C.

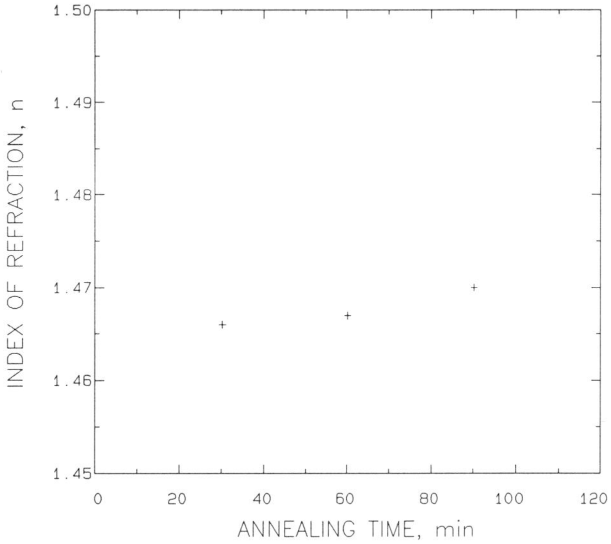

Figure 4 – Refractive index vs. preoxidation annealing time for oxides grown in dry O_2 at 1000°C.

It is seen that they remain independent of t_A. For the $Si_3N_4/SiO_2/Si$ structures, the nitride thickness was approximately 845 Å, and the index of refraction was about 1.992. Both the flat band voltage and the threshold voltage were measured through C-V (1 KHz) analysis using the mercury probe. The results are shown plotted in Figures 5 and 6; related experimental data are listed in Tables I and II. It can be seen from these graphs that increasing the preoxidation anneal time, t_A would cause V_{FB} and V_T to shift to not only less negative but also positive values. For both composite structures, V_T and V_{FB} can be considered linear functions of t_A.

As mentioned above, not much influence of the preannealing time on the final oxide thickness (Figure 3) was observed. This result suggests that even if there have existed mechanisms responsible for an initially enhanced oxidation rate of N_2- treated Si wafers, as proposed by Ruzullo (20), they are being introduced during initial exposure of these wafers to nitrogen. Ruzullo (20) has shown that for oxide thicknesses less than 300 Å, the preoxidation annealing in N_2 for 10 minutes would increase the oxidation rate; at 1100°C the thickness increase due to an increase in N_2 preanneal time from 10 minutes to 30 minutes was negligible.

The growth kinetics of silicon dioxide films with thicknesses greater than 300 Å can be accurately described by a model developed by Deal and Grove (21). Based on the linear parabolic model developed by these authors, for dry oxidation, oxygen diffuses through the existing oxide to the Si/SiO_2 interface where the molecules react with Si to form SiO_2. A thin native oxide layer (20–30 Å) is always present on silicon surface due to atmospheric oxidation even at room temperature. Neglecting the film mass transfer effects, the overall

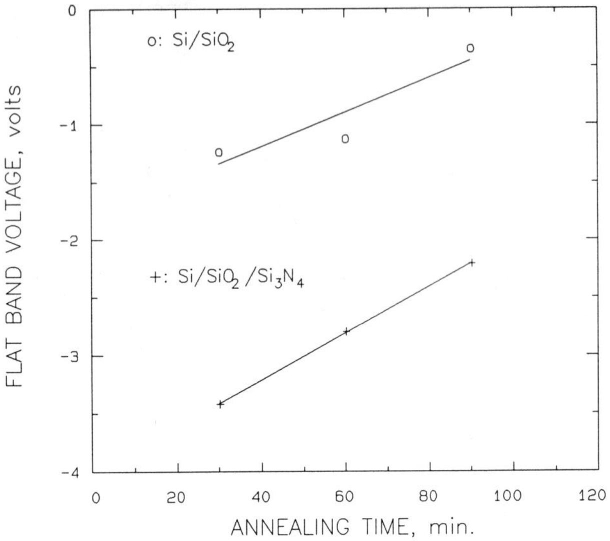

Figure 5 – Effect of preoxidation annealing on flat-band voltage, V_{FB}.

relationship for the oxide thickness, X_o can be written as follows

$$X_0^2 + \frac{2D}{K_s}X_0 = \frac{2DN_0}{n}t \tag{14}$$

where D is the diffusion coefficient for oxidizing species in the oxide in cm^2sec^{-1}, N_0 is the oxygen concentration in the oxide at the oxide surface in molecules/cm^3, n is the number of oxygen molecules that are incorporated into a unit volume of the resulting oxide and K_s is called the rate constant for the chemical reaction at the Si–SiO_2 interface. Equation (14) can be written in the following form,

$$X_0^2 + AX = Bt. \tag{15}$$

Then, for small value of t, the linear region,

$$X_0 = \frac{N_0K_s}{n}t = \left(\frac{B}{A}\right)t \tag{16}$$

and for long oxidation times, the parabolic region,

$$X_0 = \left(\frac{2DN_0}{n}\right)^{\frac{1}{2}} t^{\frac{1}{2}} = B^{\frac{1}{2}}t^{\frac{1}{2}}. \tag{17}$$

The term B/A is called linear rate constant where as the parabolic rate constant is given by B. It follows from equation (16) that the linear rate constant (B/A) should depend on

K_s. It is also known that the reaction rate constant for oxide growth, K_s, should depend on the Si–Si bond breaking mechanism. Therefore, the rate controlling mechanism in the linear region is believed to be the breaking of Si–Si bonds which provides atomic silicon to react with oxygen. Preannealing treatment could potentially effect the process of Si–Si bond breaking and thus changing the oxide growth conditions. In addition, preannealing may reduce the point defect clusters present both in the bulk of Si and at the surface. These clusters can act as surface sites where oxidation induced stacking faults (OISF) can nucleate.

It is believed that OISF are caused by incomplete oxidation at the Si-SiO_2 interface, which results in the formation of Si self-interstitials in this region. The interstitial Si causes OISF formation by nucleation at strain centers (e.g., point defect clusters) in the bulk or at the surface (22). Stacking faults and their accompanied dislocations may act as charge trap centers in the oxide as well as at the Si/SiO_2 interface. It is believed that electrically active stacking faults degrade various MOS device mechanisms. Consequently, preoxidation annealing in N_2 could be responsible for reducing these charged structural defects and thus causing the threshold voltage shifts observed in the SiO_2/Si and the Si_3N_4/SiO_2/Si structures studied. In fact, it has been reported that preoxidation annealing in the N_2/HCl/O_2 mixtures has resulted in annihilation of stacking faults (23).

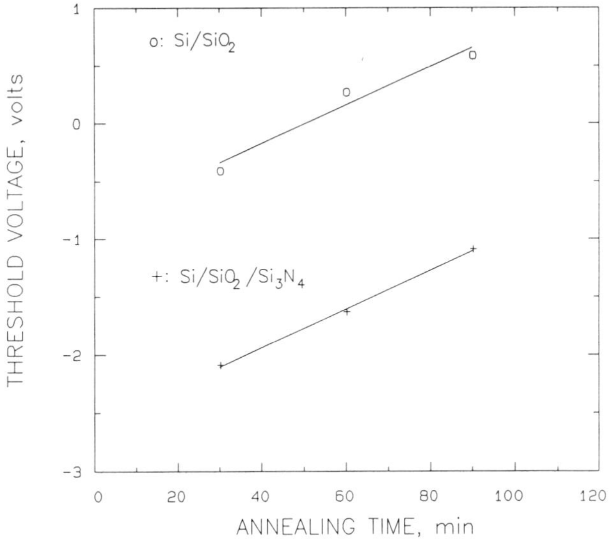

Figure 6 – Effect of preoxidation annealing on the threshold voltage, V_T.

Another possible mechanism responsible for the observed threshold voltage shifts may be related to the nature and properties of Si-SiO_2 interface. Electrical measurements have shown that fixed charges and interfacial traps exist in the oxide within 10-35 Å of the Si/SiO_2 interface (24). These charges and traps exist within a very narrow transition region between Si and SiO_2; and they are originated from local strain and nonstoichiometry at the Si/SiO_2 interface as well as defects such as stacking faults and excess silicon atoms. Thermal annealing treatment in N_2 atmosphere prior to oxidation may cause a formation of Si_xN_y phase on the Si surface as suggested by Ruzullo (20). The presence of

Table I. Experimental Data for the Oxidation of Si/SiO_2 at 1000^oC with Annealing in N_2.

Wafer No.	Anneal time min	V_{FB} volts	V_T volts	Oxide Thickness Å	n_{ox}
Q13	30	-1.23	-0.40	738.6	1.466
Q11	60	-1.12	+0.28	733.5	1.467
Q19	90	-0.34	+0.60	750.4	1.470

native oxide on the wafer surface as well as the possibility of some H_2O and O_2 present in the furnace atmosphere during the N_2 preannealing, could create a situation in which the silicon surface is covered with a random mixture of SiO_x and Si_xN_y phases. These would effect the distribution of Si-Si bond angles and bond energies at the surface resulting in the significant reduction of the concentration of excess silicon and silicon dangling bonds at the Si/SiO_2 interface during subsequent oxidation. Hence, a threshold voltage shift to less negative values in the N_2- treated, preannealed structures. It should be emphasized that Si_xN_y and SiO_x phases would eventually convert to the homogeneous SiO_2 as the oxidation proceeds. This assumption agrees well with the results of oxide's refractive index measurement (Figure 4). If there was an appreciable amount of nitrided oxide left at the end of dry oxidation, the index of refraction would have been in the range of 1.48-1.50 as reported by Ruggles et al. (24).

Table II. Experimental Data for $Si_3N_4/SiO_2/Si$ Structures. The wafers were treated with Preoxidation Annealing in N_2 at 1000^oC.

Wafer No.	Anneal time min	V_{FB} volts	V_T volts
Q14	30	-3.42	-2.09
Q12	60	-2.80	-1.63
Q20	90	-2.21	-1.09

II. *LPCVD of Silicon Nitride:*

The pH-sensitive Si_3N_4 layer for a CHEMFET sensor is obtained by low pressure chemical vapor deposition (LPCVD) technique using pyrolytic reaction between silane (SiH_4) and ammonia (NH_3) in a resistance-heated conventional hot wall reactor. According to Arizumi et al. (25), the nitride deposition reaction is given as follows:

$$3SiH_4 + 4NH_3 \longrightarrow Si_3N_4(s) + 12H_2$$

the standard free energy changes of the deposition reaction in the temperature range of 800°K to 1200°K, calculated from the JANAF thermochemical data (26) is given in Table III. It is seen that in this temperature range, the deposition reaction is thermodynamically feasible.

Table III. Standard Free Energy Changes Associated with Silicon Nitride Deposition Reaction.

$$3SiH_4 + 4NH_3 \text{ ---> } Si_3N_4(s) + 12H_2$$

Temperature K^o	ΔG^o kcal/mol	Equilibrium Const., K
800	-225.1	3.1 x 1061
1000	-245.5	4.5 x 1053
1200	-266.3	3.2 x 1048

The characteristics of the Si_3N_4 film depend on deposition parameters, such as temperature, total pressure and reacting gas flow ratio, $n=NH_3/SiH_4$. In this study, the nitride deposition temperature was fixed at 850°C, various gas flow ratios were employed while keeping the pressure constant at 0.8 torr. The standard run sequence takes about 15 minutes plus the deposition time. After deposition all gases are turned off and the reactor is exhausted to about 0.005 torr pressure after which it is back-filled rapidly with N_2. The wafers are then annealed for 35 minutes in N_2 (atmospheric pressure) at the same temperature. 30 minutes annealing in N_2 prior to the nitride deposition did not have any appreciable effect on deposition rates. The effect of NH_3/SiH_4 ratio on the deposition rate and the nitride thickness are shown in Figures 7 and 8, respectively. Additional experimental data are reported in Table IV. It is seen that lower ratio (i.e., higher SiH_4 concentration in the gas phase) results in faster deposition rate as has been reported in the literature (27,28). The increase in the SiH_4 flow rate and the simultaneous decrease in the NH_3 flow increase the SiH_4 concentration and thus partial pressure. This leads to an increase in the deposition rate as shown in Figure 7. The effect of NH_3/SiH_4 gas ratio on the etch rate of the nitride film can be judged from Figure 9. It is seen that the etch rate has remained constant at approximately 4 Å/min and is independent of the NH_3/SiH_4 ratio.

A series of experiments were performed in which the flow rate of one gaseous species (e.g. NH_3 or SiH_4) was kept constant while varying the flow rate of the other. These results are summarized in Table V. Increasing the SiH_4 flow rate at constant NH_3 flow, that is, decreasing n , results in faster deposition rates. The NH_3 flow rate seems to have negligible effect on the deposition rate. Hence, from Figures 7 and 8 and Tables IV and V, it can be concluded that the growth rate of Si_3N_4 in the SiH_4/NH_3 LPCVD process is determined mainly by the SiH_4 gas flow. This is in good agreement with previous investigations on LPCVD (28) as well as CVD (27) processes of silicon nitride.

The indexes of refraction for deposited nitride layers are given in Tables IV and V. The refractive index of stoichiometric Si_3N_4 is 2.00±0.1 (27). In addition , Makino (29)

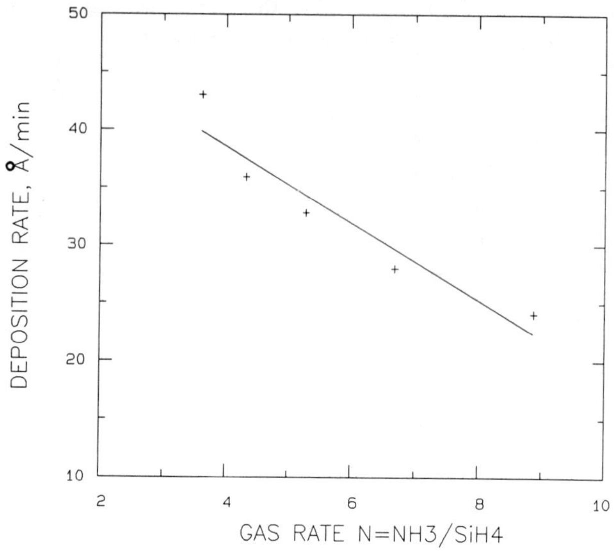

Figure 7 – The deposition rate of LPCVD Si_3N_4 vs. the gas flow ratio NH_3/SiH_4 at 850°C. The total pressure is 0.8 torr.

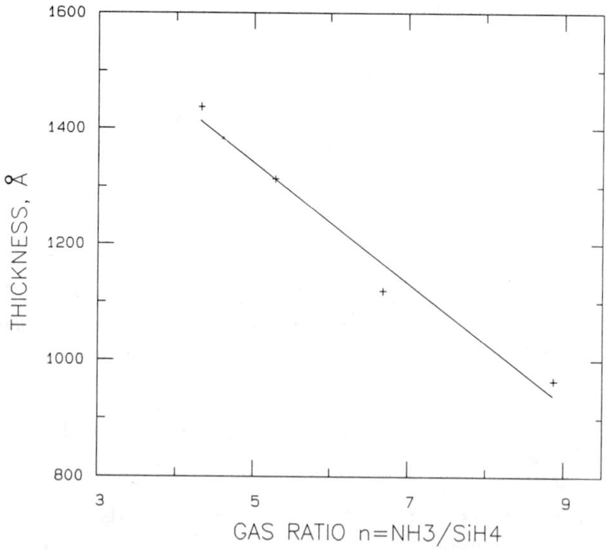

Figure 8 – The nitride layer thickness as a function of n=NH_3/SiH_4.

has suggested that the relationship between the atomic ratio, N/Si, and refractive index is independent of deposition conditions. He has compared his results for SiH_2Cl_2-NH_3 system at 770°C with that of Gyulai et al.(30) for SiH_4-NH_3 system at 850°C and has

Table IV. Experimental Data for LPCVD of Silicon Nitride at 850°C. Total Gas Flow Rate and Total Pressure were kept constant.

Wafer No.	n NH_3/SiH_4	Deposition Rate, Å/min	Nitride Thick, Å	n_{nit}	Etch Rate Å/min
A3	8.86	24.1	965.0	1.999	4.428
A18	6.67	28.0	1119.2	2.012	3.820
A6	5.27	32.8	1312.7	2.031	4.175
A9	4.31	35.9	1437.0	1.991	4.785
A15	3.60	43.0	1720.1	2.003	4.187

proposed a linear relationship between the refractive index and the atomic ratio. Based on his work, stoichiometric Si_3N_4 (N/Si=1.33) would have refractive index in the range of 1.95-2.00. It can be suggested that deposition at lower NH_3/SiH_4 gas ratios (i.e., higher SiH_4 concentration) would result in the formation of Si rich Si_3N_4 films with higher refractive indexes. It should be noted, however, that the amount of Si is not appreciable enough to cause any increase in the etch rate of the Si_3N_4 film as can be seen in Figure 8.

Table V. Experimental Data for LPCVD of Silicon Nitride at 850°C. Total Pressure was in the range of 0.5-0.8 torr.

Wafer No	Flow Rate cm^3min^{-1} (STP) NH_3	SiH_4	n NH_3/SiH_4	Deposition Rate, Å/min	n_{nit}
Q17	60.0	9.0	6.67	21.1	1.992
K14	45.0	11.0	4.10	24.1	2.002
N18	45.0	15.0	3.00	31.1	2.021
N5	30.0	11.0	2.70	24.4	2.002

C-V (1 KHz) measurements on $Si_3N_4/SiO_2/Si$ structures using mercury probe indicates the dependency of flat-band voltage on nitride film properties as shown in Figure 10. For these composite structures, the thickness of thermally grown SiO_2 remained fixed at approximately 760Å. It is seen that Si rich silicon nitride films formed at lower NH_3/SiH_4 gas ratios would result in more negative flat-band voltages. It is believed that instabilities in Si_3N_4/SiO_2 double dielectric structures which is due to the electronic conduction through and trapping in the Si_3N_4 layers, is highly dependent on the exact composition of the nitride film. As mentioned before, there will be some amount of free silicon atoms in the nitride film when it is deposited under ammonia deficient conditions. The work of Doo et al.(31) shows that high SiH_4/NH_3 ratios produce Si-rich films that are more conductive than the stoichiometric Si_3N_4 films produced with low silane ratios. High conductivity nitride films produce large C-V shifts due to the transport of electronic charge through the Si_3N_4 and its subsequent trapping near the SiO_2 interface (32). As we move to low conductivity, stoichiometric nitride, the C-V shift gets smaller. Hence, the shift to move negative flat-band voltages with high silane ratios observed here is probably due to presence of excess silicon atoms with unsatisfied bonds forming positively charged trap centers.

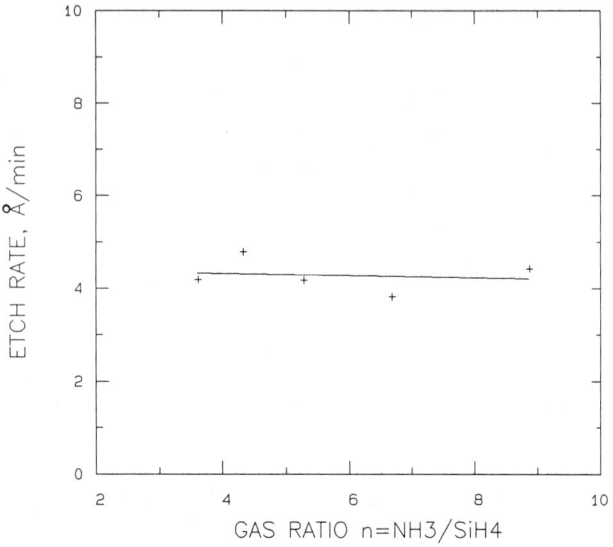

Figure 9 – Effect of NH_3/SiH_4 gas ratio on the etch rate of the nitride film.

The chemical composition of the nitride film prepared here has also been studied using ESCA surface analysis technique. First, the ESCA analysis were performed on as-received nitride samples. After that, the films were treated with a sequence of Xe sputtering steps in the ESCA chamber for various time periods followed by ESCA analysis. The thickness of the top nitride layers removed after 10 minutes of sputtering was in the range of 25-40Å. It should be noted that carbon was found on all samples freshly inserted into the ESCA chamber. After one minute of Xe sputtering, practically all the carbon had desorbed. The results are summarized in Tables VI and VII. In general, results from the ESCA analysis indicate that the surface of the nitride film is rich in oxygen in addition to Si and nitrogen. However, the amount of oxygen decreases rapidly with the depth profile of the film and the nitride film composition approaches the molar ratio of 3 silicon to 4 nitrogen (N/Si=1.33). The oxygen pile-up at the Si_3N_4 surface is probably due to the fact that the layers oxidize slightly when they are unloaded from the LPCVD reactor. This suggests that even at room temperature there could be a slow conversion of silicon nitride.

Janata has argued that Si_3N_4 does not convert to silicon dioxide, but instead, oxygen actually penetrates into silicon nitride and forms a solid solution (Ref. 3, Discussion). He reported the presence of approximately 8% oxygen in silicon nitride films which extends to about 40 Å deep into the nitride. There was no infrared evidence for SiO_2 bond formation.

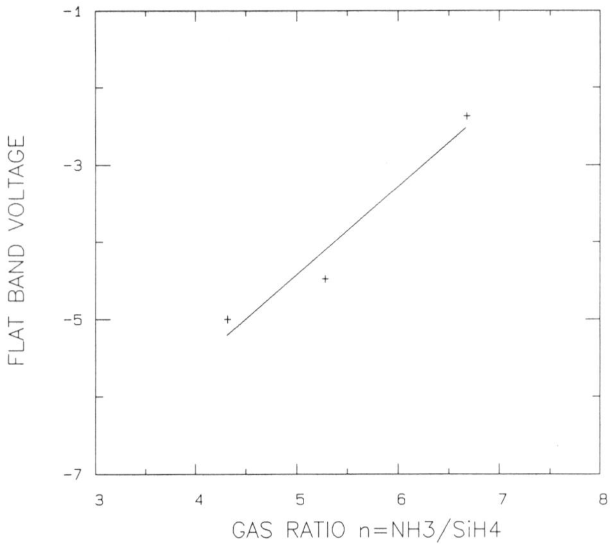

Figure 10 – Effect of deposition conditions on the flat-band voltage. The total gas flow rate and the total pressure are constant.

This observation agrees well with refractive index and etch rate measurements of silicon nitride films as well as the ESCA results. The possibility of oxygen reaction with Si_3N_4 and formation of silicon oxynitride (Si-O_x -N_y) layer at the surface should be treated with caution because such films would have shown refractive indexes in the range of 1.6-1.88 (27,33,34) irrespective of the deposition method employed. We measured the index of refraction of the nitride samples pre and post ESCA analysis; the n_{nit} values remained relatively constant in the range of 1.995 to 2.025. The disturbances of the stoichiometry of Si_3N_4 would give rise to depth dependent etch rate and index of refraction. In addition, studies on structure and composition of oxynitride film deposited by plasma-enchancd CVD technique have shown that in such films the oxygen concentration is fairly constant with depth (33,35).

From Table VI and VII, it can be seen that the percentage of oxygen in the silicon nitride film decreases rapidly as one goes deeper into the surface. A comment should also be made about the role of hydrogen in the silicon nitride films deposited here. Although no quantitative measurement was performed in this study to determine the amount of hydrogen in the Si_3N_4 films, it is believed that hydrogen atoms are introduced in the silicon nitride films as decomposition products from the reactant gases. This has been reported in LPCVD system for both SiH_4/NH_3 (30,36) and SiH_2Cl_2/NH_3 (29) processes. It has been suggested that hydrogen atoms form Si–H and N–H bonds. Silicon oxynitride films would also contain hydrogen in Si–H and N–H bond formation (33,34). The formation of silicon oxynitride layer at the surface of Si_3N_4 film would require a substantial formation of Si–O species in conjunction with the reduction of Si–H bonds in the film bulk. That is, the reduction of hydrogen concentration in the oxynitride film must be due to the increase in the thermodynamically favorable formation of Si–O bonds and a reduction of Si–H bonds in the film. This process would change the hydrogen concentration in both silicon oxynitride and silicon nitride layers specially at the interface. Hydrogen has been

Table VI. ESCA Results for LPCVD Silicon Nitride Films. Total Pressure was in the range of 0.5-0.8.

Wafer No.	$n=NH_3/SiH_4$	X_e time min	Oxygen (Atom%)
Q17	6.67	0	17.91
		1	18.01
		5	16.69
K14	4.1	0	20.96
		1	6.68
		5	5.52
N18	3.0	0	18.84
		1	5.54
		5	5.44
N5	2.7	0	22.84
		1	9.24
		5	5.83

shown to have a strong effect upon the etching properties of deposited Si_3N_4 layers.

Chow et al.(37) prepared nitride layers with hydrogen contents between 4 and 39 a/o and showed that the variation in hydrogen content results in a variation of the etch rate over three orders of magnitude, with the lowest etch rates at the lowest hydrogen concentrations. With this in mind, it is interesting to notice the etch rate of Si_3N_4 films given in Figure 8. It seems reasonable to suggest that hydrogen concentration of these films have not changed despite the fact that they contain various amounts of oxygen. Thus it is concluded that oxygen in nitride surface layers exists in the dissolved state, however, it does not lead to the formation of uniformly distributed single-phase layer of SiO_x–N_y on top of the Si_3N_4 film. Based on experimental results, the amount of dissolved oxygen specially in layers below the first 10–20 Å of the top nitride surface is simply not adequate to transform the nitride to the silicon oxynitride phase.

The experimental results on thermal oxidation of Si and LPCVD of silicon nitride, discussed above, have been incorporated into the fabrication scheme for the CHEMFET sensors.

CHEMFET Sensors

I. *Fabrication*

The solid state process for the fabrication of CHEMFET sensors are similar to those employed in standard MOSFET technology because of the similarity of their basic struc-

Table VII. ESCA Results for LPCVD Silicon Nitride Films. Total Gas Flow Rate and Total Pressure were kept constant.

Wafer No.	$n=NH_3/SiH_4$	X_e Time min	Oxygen (Atom%)
A3	8.86	0	22.42
		1	16.20
		5	6.38
		10	4.87
A18	6.67	0	24.12
		1	10.66
		5	8.60
		10	7.53
A6	5.27	0	25.64
		1	21.14
		5	8.18
		10	5.14
A9	4.31	0	23.17
		1	16.83
		5	5.94
		10	4.80

tures. The most powerful procedure for the fabrication of this device is the general collection of steps known as the planar process. It is based on the use of high resolution photolithography to define selected areas of silicon for the addition of donor and acceptor impurity atoms as well as the other necessary structures such as gate insulators, metal interconnections, etc.. The main thrust of silicon planar processes rests on the premise that thin layers of SiO_2 will adequately "Mask" the diffusion of appropriately selected impurities. Thus, certain desired patterns can be generated onto Si substrate by a sequence of oxidation, photolithographic, and diffusion steps.

The design of the CHEMFET chip is most commonly employed on n-channel metal gate technology. The selection of the n-channel technology was due to the inherently higher transconductance it provides. Transconductance–the rate of change of drain current with respect to change in gate voltage–is directly proportional to the charge carrier mobility in the surface inversion layer. Figure 11 illustrates the major steps involved in the fabrication of a pH sensitive CHEMFET sensor. The starting materials are p-type (100) oriented silicon wafers 3 inches in diameter. The field regions of the device need to be covered with an oxide layer, much thicker than that at the gate regions, to eliminate undesirable parasitic signals. After initial oxidation as shown in step 1 of Figure 11, heavy boron diffusion is employed to form a P^+ guard ring which will enclose the drain region.

It also minimizes leakage between the drain and source region due to surface inversion of the p–type substrate as shown in step 2.

The drain and source regions are then formed by a phosphorus diffusion process. The necessary amounts of phosphorus that forms the source and drain regions of the ISFET sensor is added by solid state diffusion into regions defined by photolithography. The phosphorus dopant is added to the substrate by chemical means in a step generally known as pre-deposition. Following the predeposition step, the phosphorus atoms are driven into the substrate by solid state diffusion (step 3). An approximately 800 Å thick silicon oxide is then formed by thermal oxidation of silicon substrate in dry O_2 at a temperature of 1000°C (step 4). As shown in step 5 of Figure 11, a layer of approximately 850 Å thick silicon nitride is deposited on top of the thermally grown silicon dioxide to form a double layer gate dielectric. This layer of Si_3N_4 is formed by low pressure chemical vapor deposition (LPCVD) using silane in ammonia at 850°C. The final process is contact metallization as shown in step 6 of Figure 11. where windows are subsequently opened up and aluminum film is evaporated and etched to form contacts to the drain and source of the CHEMFET.

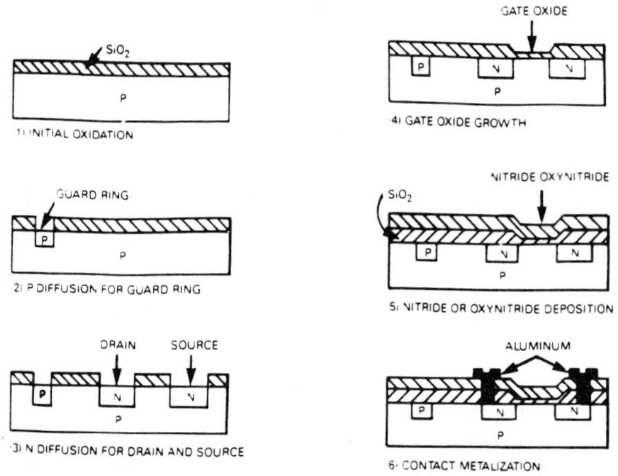

Figure 11 – Illustrations of major steps in the fabrication of CHEMFET sensors.

As mentioned in the previous section, the chemical composition of the nitride film prepared for the CHEMFET sensor has been investigated by ESCA surface analysis technique. Apart from the surface of the nitride film which is rich in oxygen, it can be concluded that the composition of the film prepared by the LPCVD process described here is stoichiometric Si_3N_4. These ISFET devices are thus referred to as Silicon Nitride Gate CHEMFET sensors (Figure 1).

II. *Testing and Performance*

The CHEMFET sensors were mounted on ceramic substrates, wire-bonded and hand encapsulated with an epoxy formulation that is hydrophobic. The pH sensitivity of various CHEMFET sensors was measured using a standard calomel reference electrode with commercial inorganic buffer solutions in a constant temperature water bath. Similar measurements were performed to study the pH sensitivty of the electrolyte-Si_3N_4/SiO_2-semiconductor(EIS) electrodes. The pH response of the EIS structures were in excellent agreement with that of the CHEMFETs reflecting good reproducibility of the fabrication process.

The pH response of the CHEMFET sensors were tested every three days, changes in the FET pH sensitivity, threshold voltage (V_T), and leakage current through the device's encapsulation and packaging were recorded each time. In one series of experiments, it was found that ISFET sensors treated with preoxidation annealing in N_2 during the gate oxidation step demonstrated positive values for the threshold voltage which is the characteristics of enhancement mode device. This feature is desirable from the standpoint of fabrication process because the enhancement mode device does not require a channel stopper to prevent a leakage path between the drain and the source and thus simplifies the overall chip design. Some CHEMFET sensors with no annealing treatment showed slightly negative threshold voltages.

The pH sensitivity of the ISFET sensors is shown in Figure 12; V_{out} is referred to as the sensor's output voltage. It can be noticed that the slope of the sensitivity line is slightly smaller than the theoretical Nernst value (i.e., 59.2 mV at 25°C). No significant variation was noticed in the pH sensitivity of the sensors in dry storage for over 50 days (Figure 12), and the slope was quite time independent. The results of pH sensitivity of the CHEMFET sensors in wet storage are given in Figure 13. These sensors were initially kept in dry storage before testing; they were tested in buffer solutions every three days. After about three weeks of such testing, the surface of these ISFET'S were etched with HF solution to ensure the removal of any contamination and native oxides at the surface. Then, the sensors were subjected to pH testing and in between the tests were stored continuously in pH 7 buffer solution. It is evident from Figure 13 that these ISFETs experience a steady increase in pH sensitivity after the wet storage began. In this case, the pH response approached the theoretical Nernstian potential of 59 mV/pH unit.

The results of the effects of storage conditions on the ISFET sensor's characteristics, plotted in Figures 12 and 13, indicate that there were no measurable shifts in both dry and storage conditions. The measurement of the leakage current for these sensors showed that dry storage did not significantly affect the leakage current. However, an increase in the leakage current up to two orders of magnitude was observed over the two- week period of wet storage. The ISFET sensors were examined under the scanning electron microscopy. The SEM micrographs of the pH-ISFET sensor surface, given in Figures 14 and 15, showed a breakdown of encapsulation on the sensors in the wet storage. This breakdown of the encapsulant is responsible for the significant increase in the leakage current found when the sensors were stored in pH 7 buffer solution (i.e., wet storage). The encapsulation and electrical integrity of the gate and its immediate surroundings are of prime importance in CHEMFET sensor technology. Electrical leakage in this area will usually have an adverse effect on the electrochemical behavior of the transistor and must be prevented. Also, the presence of cracks and pinholes can destroy the masking properties of good nitride films and cause degradation of such ISFET devices.

As mentioned previously, the ISFET sensors, stored in pH 7 buffer solutions, experienced increased pH sensitivity approaching the theoretical Nernstian potential. This behavior could be attributed to the constant hydration of the silicon nitride surface and formation of higher hydroxyl site density at the Si_3N_4 surface. Fung et al.(38,39) have shown that the behavior of the electrolyte-insulator-semiconductor field-effect transistors

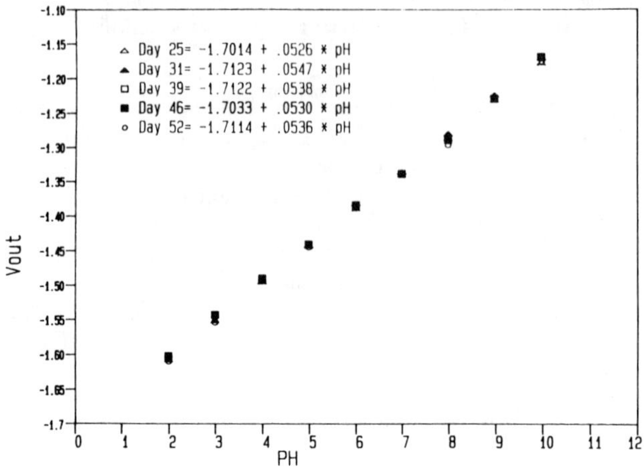

Figure 12 – pH response of CHEMFET sensors in dry storage.

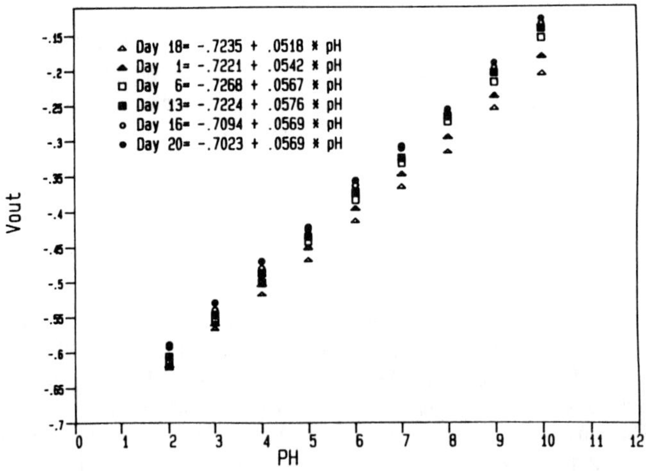

Figure 13 – pH response of CHEMFET sensors in wet storage.

(EISFETs) is primarily attributed to an interfacial potential developed at the insulator-electrolyte interface. The pH sensitivity of the ISFETs is related to a series of surface ionization and complexation events at the Si_3N_4/electrolyte interface as suggested by Cheung and co-workers (11,38) based on the site-binding model (40). The insulator surface potential, ϕ_0 involves numerous parameters such as the number of surface sites, the equilibrium constants of the dissociation and ionization reactions at the surface, and the ion-pair formation related to the ionic strength of the electrolyte. Any of these parame-

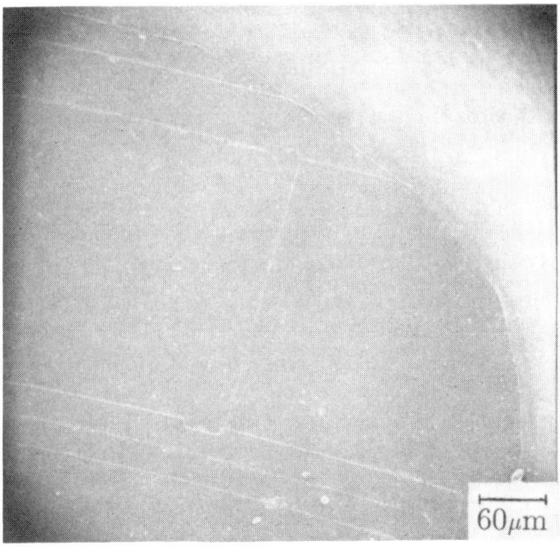

Figure 14 – Scanning electron micrograph of the open gate area of the CHEMFET sensor after treatment in the wet storage, CHEMFET P15-45.

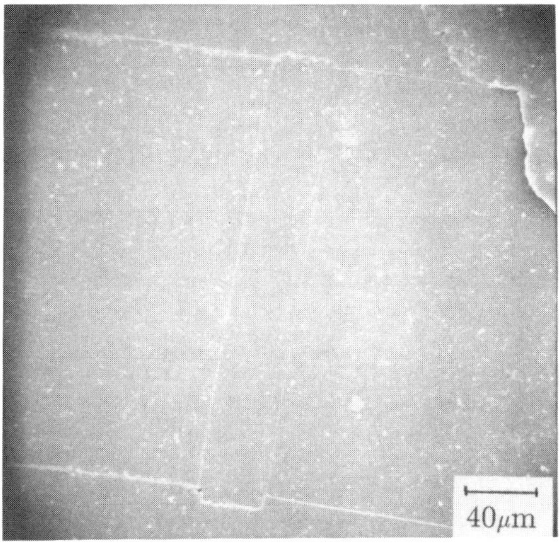

Figure 15 – Scanning electron micrograph of the open gate area of the CHEMFET sensor after treatment in the wet storage, CHEMFET P9-180.

ters may be the controlling factor in determining the surface behavior of the pH sensitive ISFET.

Furthermore, Cohen and Janata (41) have shown that the Si_3N_4 surface conductivity increases dramatically under hydration. Besides, the surface Si_3N_4 diffusion constants of ionic species coming from the electrolyte have to be taken into consideration. It can be concluded then that the constantly hydrated Si_3N_4 layer of the CHEMFETs stored in buffer solution provides a faster pH equilibrium at the silicon nitride/electrolyte interface. This leads to higher hydroxyl site density which is reported to be responsible for the near Nernstian behavior (12,13).

It should be noted that the exact mechanism of interaction between the Si_3N_4 ion sensing layer of CHEMFET sensor and various electrolyte and, in particular, its impact on pH sensitivity is briefly explained here and its detailed treatment would be subject of another publication. More results of the effect of various storage conditions on ISFET performance has been recently reported by the authors (42).

Conclusions

The chemically sensitive field-effect transistor (CHEMFET) is essentially a field-effect semiconductor device with output electrical characteristics which are sensitive to activities of different ionic species in solution. At the heart of the CHEMFET lies the gate sensing area. For pH-sensitive CHEMFETs, Si_3N_4 deposited by LPCVD method has been widely used as the insulator of the electrolyte/insulator interface. Effects of processing parameters such as annealing treatment during the thermal oxidation of Si, and deposition rate, gas composition, and total flow rate during the LPCVD of Si_3N_4 were studied. Chemical and electrical characterization of Si/SiO_2, Si/Si_3N_4, and $Si/SiO_2/Si_3N_4$ structures were carried out using C-V measurements and ESCA analysis.

The results obtained in these experiments were subsequently incorporated into the CHEMFET fabrication process in order to improve the electrical characteristics and the chemical response of the sensor. Two different storage conditions were tested using the sensor at room temperature. Significant differences were observed in the CHEMFET's performance due to the different storage methods. Scanning electron microscopy of the CHEMFETs revealed the degredation of encapsulation on the sensor as a result of continued soaking in the wet storage situation. The CHEMFET sensors experienced a steady increase in pH sensitivity, approaching the Nernstian potential of 59 mV/pH unit, after the wet storage began. This phenomena was attributed to the constant hydration of the silicon nitride surface. The variation of the insulator surface potential, ϕ_o, of the "Site Binding" model may explain the observed behavior.

Acknowledgements

The authors wish to express their gratitude to Mr. John Williams for his help in the Al metallization and Mrs. Cameron Whiting for the pH testing. Funding of this research was provided by the Microsensor Research Program of the Washington Technology Center and a grant from CHEMFET Corp., Bellevue, Washington.

References

1. P. Bergveld, "Development of an Ion-Sensitive Solid State Device for Neurophysiological Measurements," *IEEE Trans Biomed. Engr*, BME-17 (1970), 70-71.

2. I. Lundström and C. Svensson, "Gas-Sensitive Metal Gate Semiconductor Devices," *Solid State Chemical Sensors*, ed. J. Janata and R. J. Huber (Orlando, Fl: Academic Press, 1985), 1-63.

3. P. W. Cheung et al., "Theory, Fabrication, Testing and Chemical Responses of Ion Sensitive Field-Effect-Transistor Devices," *Theory, Design and Biomedical Applications of Solid State Chemical*

Sensor, ed. P. W. Cheung, D. G. Fleming, W. H. Ko, and M. R. Neuman (Boca Raton, Fl: CRC Press, 1978), 91- 117.

4. A. S. Grove, *Physics and Technology of Semiconductor Devices*, (New York, NY: John Wiley & Sons, 1967).
5. S. M. Sze, *Physics and Semiconductor Devices*, (New York, NY: Interscience, Inc., 1969).
6. P. Bergveld, "Development, Operation, and Application of the Ion-Sensitive Field-Effect Transistor as a Tool for Electrophysiology," *IEEE Trans. Biomed. Eng.*, BME-19 (1972), 342-351.
7. S. D. Moss, J. Janata, and C. C. Johnson, "Potassium Ion-Sensitive Field-Effect Transistor," *Anal. Chem.*, 47(1975), 2238-2242.
8. R. G. Kelly, "Microelectronic Approaches to Solid State Ion Selective Electrode," *Electrochimica Acta*, 22(1977), 1-8.
9. C. D. Fung, P. W. Cheung, and W. H. Ko, "Electrolyte-Insulator-Semiconductor Field-Effect Transistor," *IEDM Tech. Dig.*, (1980), 689-692.
10. S. D. Moss, C. C. Johnson, and J. Janata, "Hydrogen, Calcium, and Potassium Ion-Sensing FET Transducers: A Preliminary Report," *IEEE Trans. Biomed. Eng.*, BME-25 (1978),49-54.
11. C. D. Fung, P. W. Cheung and, W. H. Ko, "A Generalized theory of an Electrolyte-Insulator-Semiconductor Field-Effect Transistor," *IEEE Trans. Electron Devices*, ED-33(1) (1986), 8-18.
12. H. Abe, M. Esahi, and T. Matsuo, "ISFET's Using Inorganic Gate Thin Films," *IEEE Trans. Electron Devices*, ED-26(12) (1979), 1939-1944.
13. T. Akiyama et al., "Ion-Sensitive Field-Effect Transistors with Inorganic Gate Oxide for pH sensing," *IEEE Trans. Electron Devices*, ED-29 (12) (1982), 1936-1941.
14. F. Chauvet, A. Amari, and A. Martinez, "Stability of Silicon Nitride/Silicon Dioxide/Silicon Electrodes Used in pH Microelectronic Sensors," *Sensors and Actuators*, 6(1984), 255-267.
15. J. Janata, "Chemically Sensitive Field Effect Transistors," *Solid State Chemical Sensors*, ed. J. Janata and R. J. Huber (Orlando, Fl: Academic Press, 1985), 65-118.
16. M. Kuhn, "A Quasi-Static Technique for MOS C-V and Surface State Measurements," *Solid State Electron.*, 13(1970), 873-885.
17. R. A. Colclaser, *Microelectronics Processing and Device Design*, (New York, NY: John Wiley & Sons, 1980).
18. S. K. Ghandi, *VLSI Fabrication Principles*, (New York, NY: John Wiley & Sons,1983).
19. S. Wolf and R. N. Tauber, *Silicon Processing for the VLSI Era Vol.1-Process Technology*, (Sunset Beach, CA: Lattice Press, 1986).
20. J. Ruzyllo, "Effects of Preoxidation Ambient in Very Thin Thermal Oxide on Silicon," *J. Electrochem. Soc.*, 133(8)(1986), 1677-1681.
21. B. E. Deal and A. S. Grove, "General Relationship for the Thermal Oxidation of Silicon," *J. Appl. Phys.*, 36(1965), 3770-3779.
22. S. M. Hu, "Formation of Stacking Faults and Enhanced Diffusion in the Oxidation of Silicon," *J. Appl. Phys.*, 45(1974), 1567-1573.
23. T. Hattori and T. Suzuki,"Elimination of Stacking Fault Formation in Silicon by Preoxidation Annealing in $N_2/HCl/O_2$ Mixtures," *Appl. Phys. Lett.*, 33(1978), 347-352.
24. G. A. Ruggles and J. R. Monokowski, "An Investigation of Fixed Charge Buildup in Nitrided Oxides," *J. Electrochem Soc.*, 133(4)(1986), 787-793.

25. T. Arizunni, T. Nishinaga, and H. Ogawa, "Thermodynamical Analyses and Experiments for the Preparation of Silicon Nitride," *Japan J. Appl. Phys.*, 7(9)(1968), 1021- 1027.

26. D. R. Stull and H. Prophet, *JANAF Thermochemical Table*, 2nd ed., NSRDS-NBS37 (Washington, D.C.: U.S. Department of Commerce, 1971), 1141 pp.

27. J. T. Milek, *Silicon Nitride for Microelectronic Applications*, Handbook of Electronic Materials, Vol. 3 (New York, NY: IFI/Plenum, 1971), 118 pp.

28. R. S. Rosler, "Low Pressure CVD Production Processes for Poly, Nitride and Oxide," *Solid State Tech.*, (1977), 63-70.

29. T. Makino, "Composition and Structure Control by Source Gas Ratio in LPCVD SiN_x," *J. Electrochem. Soc.*, 130(2)(1985), 450-455.

30. J. Gyulai et al., "Analysis of Silicon Nitride Layers on Silicon by Backscattering and Channeling Effect Measurements," *Appl. Phys. Lett.*, 16(1970), 232-234.

31. V. Y. Doo, D. R. Kerr, and D. R. Nichols, "Property Changes in Pyrolytic Silicon Nitride with Reactant Composition Changes," *J. Electrochem. Soc.*, 115(1)(1968), 61-64.

32. M. H. Woods, "Instabilities in Double Dielectric Structures," *12th Annual Proceedings, IEEE Reliability Physics Symposium*, (1974), 259-266.

33. W. A. P. Claassen et al., "Characterization of Silicon-Oxynitride Films Deposited by Plasma-Enhanced CVD," *J. Electrochem. Soc.*,133(7)(1986), 1458-1464.

34. V. S. Nguyen, W. A. Lanford, and A. L. Rieger, "Variation of Hydrogen Bonding, Depth Profiles, and Spin Density in Plasma-Deposited Silicon Nitride and Oxynitride Film with Deposition Mechanism," *J. Electrochem. Soc.*, 133(5)(1986), 970-975.

35. H. G. Maguire and P. D. Augutus, "The Detection of Silicon-Oxynitride Layers on the Surface of Silicon-Nitride Films by Auger Electron Emission," *J. Electrochem. Soc.*, 119(6)(1972), 791-793.

36. A. H. Van Ommen et al., "Etch Rate Modification of Si_3N_4 Layers by Ion Bombardment and Annealing," *J. Electrochem. Soc.*, 133(10)(1986), 2140-2147.

37. R. Chow et al., "Hydrogen Content of a Variety of Plasma-Deposited Silicon Nitrides," *J. Appl. Phys.*, 53(1982), 5630-5633.

38. C. D. Fung et al., "Investigation of Insulator Materials on the Response of Ion-Sensitive Field-Effect Transistors (ISFET's)," *Extended Abstracts Electrochem. Soc. Meeting*, (1978), 197-199.

39. C. D. Fung, "Characterization and Theory of Electrolyte-Insulator-Semiconductor Field-Effect Transistor," (Ph.D. Thesis, Case Western Reserve University, 1980).

40. D. E. Yates, S. Levine, and T. W. Healy, "Site- Binding Model of the Electrical Double Layer at the Oxide/Water interface," *J. Chem. Faraday Trans. I*, 70(11)(1974), 1807- 1818.

41. M. Cohen and J. Janata, "The Surface Conductivity of Silicon Oxynitride," *Thin Solid Films*, 109(1983), 329-338.

42. E. Liao et al., "Effects of Storage Conditions on ISFET Performance," *Proceedings of IEEE EMBS 10th Annual International Conference, New Orleans, LA, 1988.*

Electrolytic Processing

POTENTIAL FOR FUSED SALT ELECTROLYSIS FOR METAL WINNING AND REFINING

Derek J. Fray

Department of Materials Science and Metallurgy
University of Cambridge
Pembroke Street, Cambridge, England CB2 3QZ

Abstract

In order to extend the application of fused salts in the electrowinning and electrorefining of metals, it is necessary to maximise the current density, minimise the voltage drop and greatly increase the area of electrode surface so as to compete with the space-time yield of pyrometallurgical reactions. The various designs of electrowinning cell are reviewed and it is shown that by the application of a modest centrifugal field it is possible to reduce the anode to cathode interelectrode spacing to 4 mm whilst still maintaining a current efficiency of 85-90%.

The thermodynamic potential for electrorefining is far less than for electrowinning yet fused salt electrorefining is only applied in limited cases. The reasons for this are discussed and the important design parameters evaluated. It is concluded that by using packed bed electrodes, separated by a diaphragm, the possibility of electrorefining with very low energy consumption can be achieved. Carbothermically-produced metals are always cheaper than those produced by electrolysis but suffer from the disadvantage of low purity. It is proposed that fused salt electrorefining may offer an energy efficient way of upgrading the metal.

Metallurgical Processes for the Year 2000 and Beyond
Edited by H.Y. Sohn and E.S. Geskin
The Minerals, Metals & Materials Society, 1988

Introduction

The majority of metals can be deposited from fused salts (1,2) but as commercial production is restricted to the more reactive elements, it is worthwhile considering the advantages and disadvantages of fused salt electrolysis as compared to smelting and aqueous electrolysis. The advantages over aqueous extraction are as follows:

1) the much higher conductivities and diffusivities result in much lower IR losses and higher current densities can be achieved at modest voltages;

2) the elevated temperatures of operation result in larger exchange currents which give rapid kinetics and lower activation polarisation for the electrode reactions;

3) the absence of water as a solvent means that it is not necessary to consider the evolution of hydrogen as a competing cathodic reaction. Furthermore, alkali halide salts, often used as solvents in fused salt electrolytes, have high deposition potentials;

4) molten salts generally are mutually soluble and, therefore, concentration polarisation effects are minimised.

Compared to smelting, the advantages are less apparent but are generally considered to be a higher purity product and, perhaps, intrinsically cleaner processes. However, as few metals are produced or refined using fused salt electrolysis, these advantages are normally outweighed by the disadvantages which include:

1) pyrometallurgical processes are generally more energy efficient as reduction, using carbon, occurs directly rather than having to burn the carbon to generate heat which eventually is converted into electricity via the Carnot cycle;

2) the higher operating temperatures, corrosive nature of the melts, and the tendency to fume, due to the high vapour pressures of molten salts, lead to difficulties with containment and process control;

3) for metals whose melting point exceeds the boiling point of the electrolyte, the metals are deposited in a solid, dendritic form for which, unless special precautions are taken, there may be difficulties in extracting the product;

4) the output of metal per unit volume and the metal production rate (space-time yield) are relatively low.

For many metals, the last point is, perhaps, the most significant and this is a result of the cell design which, in many cases, can be regarded as a two-dimensional reactor with large interelectrode spacing. This can be contrasted with pyrometallurgical reactors where the reactions occur three-dimensionally, throughout the whole reactor, with very short diffusion distances and high surface areas. As well as a poor production rate per unit volume or floor surface area, the net result is a high energy consumption which is usually at least two or three times the theoretical minimum (2,3).

In spite of these disadvantages, a few metals are produced commercially by molten salt electrolysis and these include the refractory metals, alkali metals such as lithium, sodium and potassium and, more significantly, magnesium and aluminum (2-6). It is perhaps worth noting that the demand for many of these metals is expanding rather than contracting.

From this brief introduction, it is apparent that to make molten salt electrolysis processes more attractive, a "three-dimensional" reactor is required with substantially reduced interelectrode spacing which would increase both the space-time yield and the energy efficiency. Some advances have already been made in this area and it is worthwhile to review existing cell designs and to project future improvements.

Cell Design

Basic Cell

The most basic type of cell consists of an anode and cathode simply dipping into a molten salt bath as is shown in Figure 1. In this case, no attempt has been made to minimise the anode to cathode distances and, consequently, high voltage losses are incurred. Some refractory metals are produced in this type of cell (7-10), although a solid dendritic deposit is usually formed and, frequently, a ceramic diaphragm is placed between the electrodes to separate the reaction products. However, the main problems associated with the deposition of refractory metals are choice of electrolyte and maintaining its purity rather than cell design. A slight modification of this design is the cell which is used to produce lithium metal (Figure 2) from lithium chloride-potassium chloride eutectic (6). From the design, it is apparent that with the large anode to cathode distance of several cms, the current efficiency is likely to be high (90%) but at the expense of the energy consumption (40 kWh/kg), compared to the theoretical energy consumption of 10 kWh/kg (12).

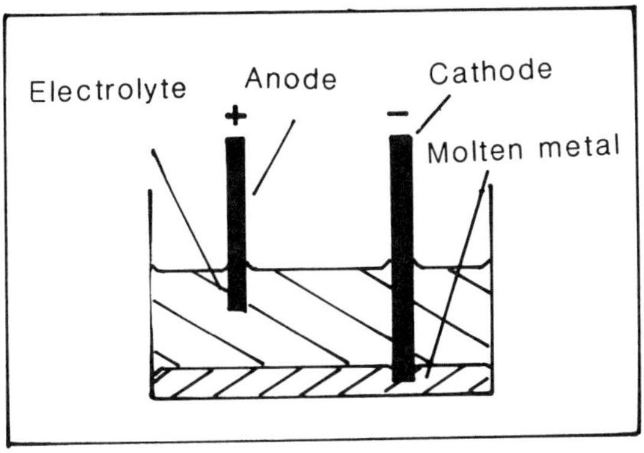

Figure 1 - Basic electrowinning cell.

Diaphragm Cells

A cell which is similar to the lithium cell but incorporates a series of electrodes is the I.G. cell for the production of magnesium in which the feed is dehydrated magnesium chloride (5,13). The cell, shown in Figure 3, is a steel tank, lined with refractory insulation, with the electrode assemblies consisting of graphite anode plates with cast steel cathode plates on either side. Each cathode contains slots which permit the lighter magnesium to collect and, also, allow circulation of the electrolyte. Between the anodes and cathodes are refractory diaphragms which permit separation of the products and, thereby, prevent losses in current efficiency. The overall result of this design is that high current efficiency is attained although the diaphragms and the large anode to cathode spacing result in a high cell voltage of approximately 7 volts, compared with the reversible decomposition potential of 2.52V for magnesium chloride, giving an overall energy consumption of 15-18 kWh kg^{-1} compared to minimum theoretical values of 5.5 kWh kg^{-1}

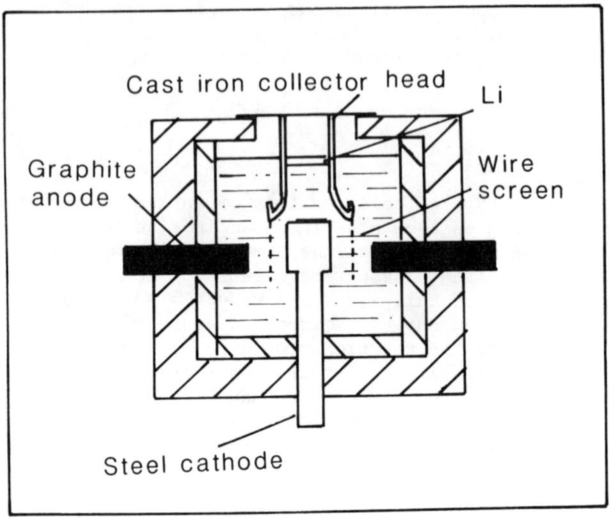

Figure 2 - Cell for the electrowinning of lithium.

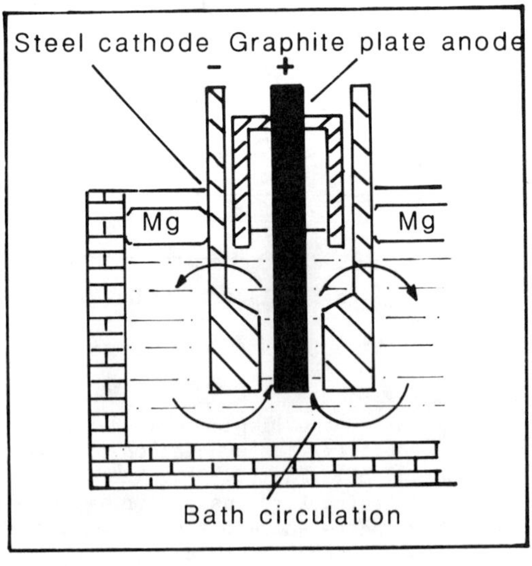

Figure 3 - I.G. cell for the production of magnesium.

Cells with Consumable Anodes

In the U.S., the majority of the magnesium extracted electrolytically is produced in the Dow cell (Figure 4) in which the feed is partially hydrated magnesium chloride and consists of an externally heated steel tank with vertical graphite anodes and horizontal steel cathodes (4). Any water in the feed which is not evaporated is decomposed to give oxygen at the anode with which it reacts resulting in a high rate of anode consumption. Due to the fact that the main anodic reaction is the evolution of chlorine, the reaction of the oxygen with the graphite does not lead to a decrease in the cell potential as in the Hall-Heroult cell for aluminum

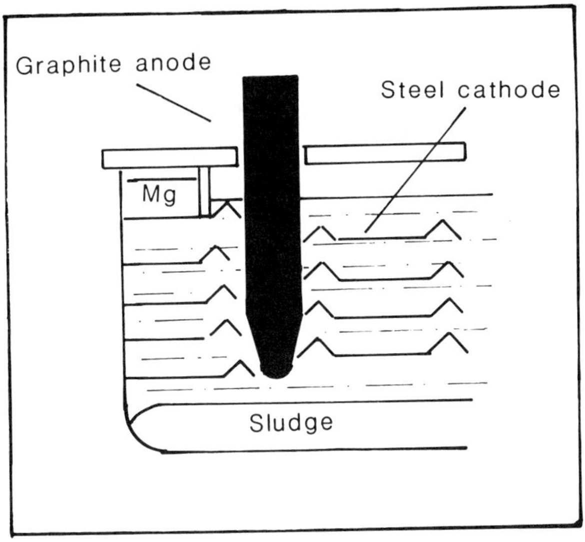

Figure 4 - The Dow cell for magnesium production.

production. In order to prevent the magnesium combining with the anode gas, the lighter magnesium spirals up through the louvered steel spiral cathodes to collect at the surface. As the anode to cathode spacing is only 38 mm, some back reaction occurs resulting in a current efficiency of 80% and an overall energy consumption of 18 kWh kg^{-1}.

However, the cell with consumable anodes which produces the most metal is the Hall-Heroult cell for the extraction of aluminum (3,5,14). A typical cell is shown in Figure 5 and consists of an insulated rectangular steel shell about 9-12m long, 3-4m wide and 1.2m high lined with carbon. The design of the cell is totally different from the magnesium cells due to liquid aluminum being more dense than the electrolyte and the electrolyte, alumina dissolved in cryolite, being far more aggressive than a simple mixture of fused chlorides. As a result the bottom of the cell is covered with the liquid aluminum upon which the aluminum is deposited, while several carbon blocks form the consumable anodes. The frozen electrolyte layer, around the sides of the cell, and the aluminum pool are to prevent the carbon directly coming into contact with the electrolyte as intercalation can occur which causes the carbon to expand. This metal pool gives rise to several problems. Firstly, a large surface area is made available for the back reaction between aluminum and the evolved carbon dioxide, leading to a reduction in current efficiency and, secondly, a resistive carbide layer forms between the aluminum and the carbon. However, the most significant problem is that, due to the intense magnetic fields generated by the high flow of current, the metal pool is unstable with the following effects:

1) vertical static displacement leading to a greater depth of the metal pool in the centre of up to 50 mm;

2) a wave on the surface of the aluminum pool, circling the cell;

3) metal flow with flow rates up to 0.1-0.2m s^{-1}.

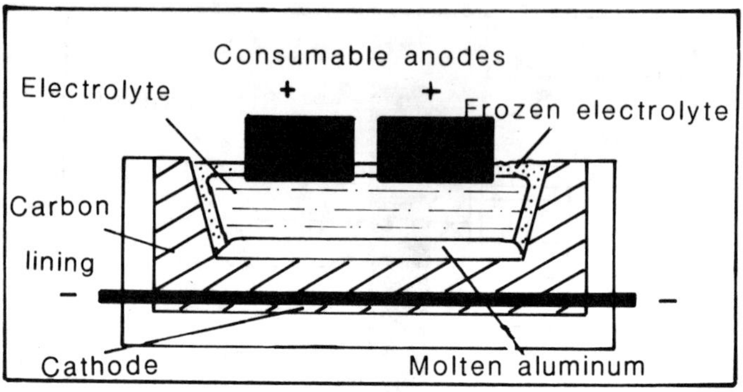

Figure 5 - Hall-Heroult cell for the production of aluminum.

These lead to losses in current efficiency, in spite of a relatively large anode to cathode separation, and erosion of the cell wall. A large electrode gap, 40-45 mm, is required in order to avoid shorting between the anode and cathode due to the wave effect caused by the magnetic fields and to reduce the chance of the aluminum and the carbon dioxide, evolved at the anode, recombining. The net result is high resistance losses and a relatively low overall energy efficiency. The large potential drop, relative to the overall cell potential is shown in Table 1.

Table 1 Contributions to the overall cell potential in an Aluminum Reduction Cell

Electrochemical Reactions	27%
Polarisation effects	17%
IR drop in liquid metal	8%
IR drop in electrolyte	39%
IR drop in anode	7%
IR drop in current leads	2%

Overall, the major problems with the design of this cell are as follows:

1) large heat losses in order to maintain a solidified layer of electrolyte to prevent failure of carbon lining;

2) the deep unstable liquid aluminum pool;

3) the relatively large anode to cathode distance;

4) the consumable anodes.

These are very considerable problems and without a dramatic improvement in materials technology, it is difficult to envisage how these problems can be overcome. However, some success has been achieved in developing inert anode materials (2,3,14) and cathode materials. Ideally, inert anode materials should have the following properties:

1) a low reactivity with oxygen, fluorine and the electrolyte;

2) adequate mechanical strength and resistance to thermal shock;

3) high electronic conductivity;

4) ability to be fabricated into large shapes;

5) relatively low cost.

These criteria are very demanding although ferrites and mixtures of ferrites and copper-nickel have been investigated. Inert anodes would overcome the problem of each of the carbon anodes in a Hall Heroult cell being in a different state of consumption at any one time and, therefore, to ensure even distribution of the current, the anode to cathode distance are varied for each anode. Inert anodes would obviously result in much greater control over the interelectrode spacing. Similar properties are demanded by inert cathode materials in conjunction with the proviso that the material be wetted by molten aluminum in order to eliminate the molten pool of aluminum from the cell (2,3,5,15). The most suitable materials are the borides and carbides of titanium and zirconium. The development of these materials are probably more advanced than that of the inert anode materials. An inert cathode would remove the necessity for a metal pool and, therefore, remove the problem of the instability of the metal pool and allow a much smaller anode-to-cathode distance. Table 2 (5) shows a comparison of the cell voltages in experimental industrial cells and the projected figures for inert electrodes and a reduced interelectrode gap.

Table 2 (5) Cell Voltages for Various Cell Designs

Cell	Standard Hall-Heroult	Inert Anode	Inert Anode and wetted cathode
Interelectrode gap/mm	45	45	19
Voltage/V	4.64	5.19	4.18

The increase in voltage for the inert anode is due to the fact that the fuel cell reaction of the carbon with oxygen does not take place and, therefore, the voltage of the cell increases. If both an inert anode and a wetted cathode are used, there is an overall voltage saving of 10% if the interelectrode spacing can be reduced by 57%, which results in a reduction in electrical energy consumption of approximately 1.5 kWh kg^{-1}, down to 11.7 kWh kg^{-1} for the most efficient industrial cells. However, with a theoretical minimum of 6.2 kWh kg^{-1}, this still leaves considerable possibilities for improvement.

Composite Anodes

Recently, experiments with composite anodes made of alumina and carbon and magnesium oxide and carbon have been carried out in electrowinning cells (16,17). The overall reaction in the alumina case is the same as in the Hall-Heroult process:

$$Al_2O_3 + 3/2C = 2Al + 3/2\ CO_2$$

The main advantage of the composite anode is that the alumina and carbon are fed together in their stoichiometric ratio. A high solubility of alumina in the electrolyte is not necessary and, perhaps, could be regarded as an advantage as this allows lower melting point electrolytes with higher conductivities to be used. Alumina powder is removed from the pot room giving a cleaner operation and the crust breaking procedure to feed alumina to the cells is eliminated. Furthermore, as there is always a supply of alumina, anode effects are no longer observed in the cell. However, these advantages are partially offset by the poor electronic conductivity of the composite anode which can only be overcome by inserting consumable aluminum conductors into the anode.

The incentive for the development of this technology to magnesium extraction is perhaps greater as the preparation of magnesium chloride feed to the cells consumes about 50% of the cost and energy consumption for the production of magnesium (18). Withers and Loutfy (17) showed that the process worked at high efficiency but, again, there were problems with the conductivity of the composite anodes which were overcome by inserting rods of magnesium into the anodes.

It may be possible to electrowin magnesium in a cell where magnesium oxide is actually reacted with carbon and chlorine in the anode section of the cell to form magnesium chloride (8). The U.S. Bureau of Mines reported on electrowinning magnesium into liquid lead and aluminum cathodes from melts containing a slurry of magnesium oxide and carbon particles (19). The evolved chlorine from the anode reacts exothermically with magnesium oxide and carbon to give magnesium chloride:

$$MgO(s) + C(s) + Cl_2(g) = MgCl_2(\ell) + CO(g)$$

$$MgO(s) + \tfrac{1}{2}C(s) + Cl_2(g) = MgCl_2(\ell) + \tfrac{1}{2}O_2(g)$$

The U.S. Bureau of Mines demonstrated that 98-99% of the chlorine reacted giving a gaseous product of 80-90% CO_2, 8-15% CO and a few percent chlorine. Cathode efficiencies were around 90% but a major difficulty was found to be a build up of sludge consisting of magnesium oxide and carbon, on top of the liquid metal cathode.

A similar process has been used by Sintim-Damao et al. (20) where a lithium compound is either dissolved or dispersed in a fluoride or chloride metal and electrolysed. Again, the cathode consists of a solvent metal, heavier than the electrolyte, which collects the lithium at the bottom of the cell. The solvent metal also reduces the activity of the lithium which helps prevent the reaction of the lithium with the oxidic species in the melt. The Nippon Soda Company of Japan (21) have operated a conventional lithium cell with a feed of lithium carbonate and carbon powder. The cell was found to operate at a current efficiency of 87%, producing metal of 98% purity with an energy consumption of 33.3 kWh/kg lithium.

Bipolar Cells

As well as materials problems with existing cells, the other major problem with electrolysis cells is the low space-time yield due to the two dimensional nature of the electrode arrangements. The space-time yield can be increased if a bipolar cell, such as that shown in Figure 6, is used (2,22,23). With this arrangement, electronically conducting plates are inserted in the electrolyte between an anode and cathode so that during electrolysis, the intermediate plates become bipolar, with alternate anodic and cathodic surfaces down the stack. One of the advantages of this arrangement is that there is only one set of electrode contacts for many plates, thereby, reducing resistance losses. However, of greater significance, is the greater surface area per unit volume and hence output which increases the space-time yield. However, this arrangement usually results in a slightly lower current efficiency because of current leakage around the edges of the intermediate plates which can be mitigated against by insulating the edges (24). Overall, a lower energy consumption is possible with a bipolar cell because of the reduced overall resistance per pair of electrodes.

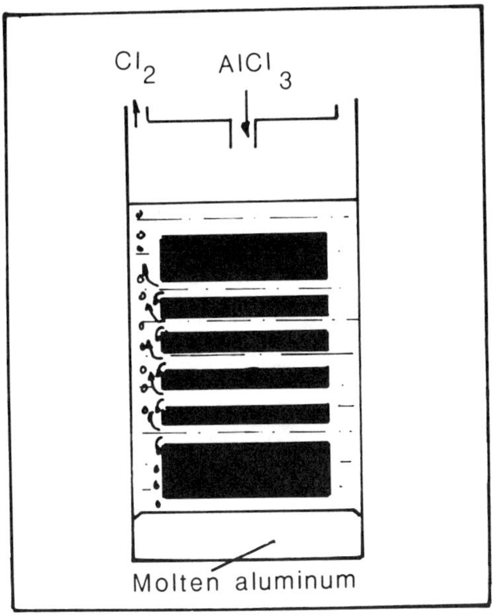

Figure 6 - Schematic diagram of Alcoa's bipolar cell.

The U.S. Bureau of Mines has electrolysed lead chloride in both vertical and horizontal plate bipolar cells (24). Insulation problems and changes in current density as the bath level varied affected the vertical plate cell but the horizontal design was more successful with a current efficiency of 86% and an overall energy consumption of 1.1 kWh kg^{-1} compared with 1.5 kWh kg^{-1} for the comparable monopolar cell, both with a relatively large interelectrode spacing (24-27). A similar cell, shown in Figure 7, was developed for the electrolysis of zinc chloride with ceramic insulation around the edges of the electrodes to prevent current leakage, molten zinc cathodes to prevent the formation of small zinc droplets on the cathode surfaces and thus minimise the chance of back reaction and, finally, the use of the evolved chlorine to pump the electrolyte through the cell. A maximum current efficiency of 70% was achieved with an electrode gap of 12 mm and an energy consumption of only 3.7 kWh kg^{-1} compared with 4.6 kWh kg^{-1} for the equivalent monopolar cell.

Alcan have also developed a bipolar cell to extract magnesium from magnesium chloride which uses the evolved chlorine to improve circulation of the electrolyte (28,29). In this case, the magnesium is lighter than the electrolyte and the electrodes are arranged vertically with a weir system at the top to draw off the metal. Rapid circulation of the electrolyte is achieved by using small interelectrode spacing and a high

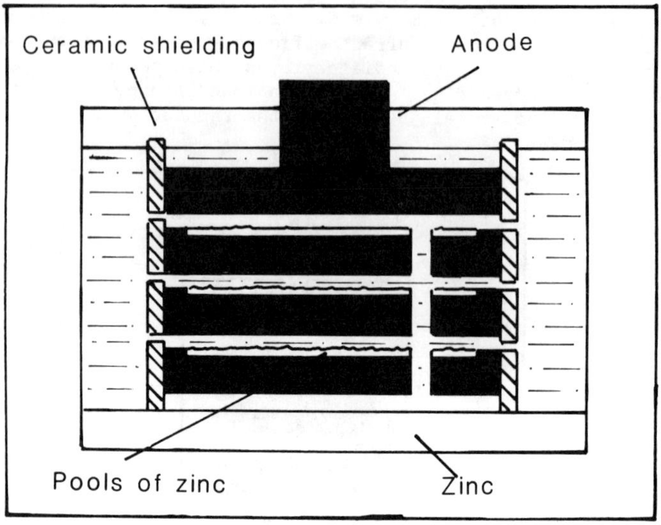

Figure 7 - The bipolar zinc chloride cell with zinc pool cathodes and ceramic shielding.

current density, which results in a high gas lift rate generated by the large amount of chlorine produced. The cell has been run with gaps as small as 8 mm, compared with 50 mm or more in existing commercial cells, and a current efficiency of 70%. However, no overall energy consumption is quoted and it may not be significantly more efficient than existing cell designs, as a result of the large amount of gas between the electrodes, leading to a relatively large resistance.

The most advanced industrial scale project involving a bipolar cell is Alcoa's aluminum chloride cell (22,30), Figure 6, which, again, uses the evolved chlorine to improve circulation in the cell by sweeping the aluminum off the cathode surface and drawing in fresh electrolyte. Coupled with the fact that graphite electrodes are effectively inert in the aluminum chloride based melt, the electrode gap could be reduced to 10 mm, as compared with approximately 50 mm in the case of the conventional Hall- Heroult cell. This resulted in an energy consumption of approximately 9 kWh kg^{-1} which is about 66% of the value achieved in modern Hall-Heroult cells. Unfortunately, although the cell the operated successfully, problems associated with the preparation of the aluminum chloride feed to the cell have led to the curtailment of the project.

Recently, Jarrett (31) has proposed retrofitting bipolar inert electrodes into existing Hall-Heroult cells.

Use of Rotating Electrodes

From the previous discussion, it is apparent that a further reduction in cell voltage can only be achieved if the anode-to-cathode distance can be decreased but, with existing technology, this would be at the expense of

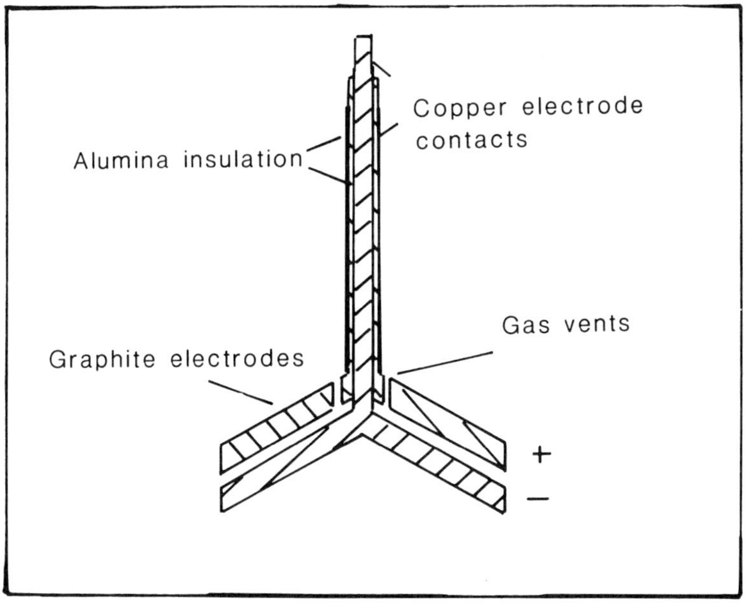

Figure 8 - Schematic diagram of rotating electrode.

increased back reaction of the products and a lower current efficiency. Elimination of the back reaction can only be accomplished if the products of electrolysis can be removed more efficiently. One possible way of meeting this objective is to rotate the electrodes in order to apply a modest centrifugal field to encourage the separation of the products of electrolysis and, thereby, allow the interelectrode gap to be reduced (32). A centrifugal field acts on a mixture in the same manner as a gravitational field but the former can be varied by changes in rotational speed or the dimensions of the equipment with the result that a much better separation can be attained. In the particular case of molten salt electrolysis, there are three phases - liquid metal, molten salt and gas. According to centrifugal theory, the gas will flow inwards, whereas the denser fluids will be thrown outwards with the net result of a more rapid and efficient removal of the electrolytic products from the interelectrode gap improving the current efficiency at small electrode separations. Another possible advantage of the application of centrifugal fields is that the metallic electronically-conducting phase will be ejected beyond the edge of the electrodes reducing the chances of the partial short circuiting the electrodes. This may be particularly important for bipolar electrodes.

In order to evaluate this concept, Copham (33) investigated the effect of rotation of circular electrodes 100 mm in diameter and 10 mm thick on the electrolysis of a zinc chloride-potassium chloride-sodium chloride electrolyte. It was found that, at very small anode to cathode spacings, it was possible to achieve high current efficiencies at relatively modest rotation speeds. By selecting a 4 mm interelectrode spacing, it was found that by having the electrodes arranged as cones, it was possible to increase the current efficiencies even further to close to 90% for a 40° angle (Figure 9). Under these conditions both gravitational and centrifugal forces were being combined.

On the basis of the encouraging results obtained with the laboratory scale experiments, a pilot scale cell was built to test the feasibility of scaling up the rotating conical electrodes. The electrodes were machined from graphite and were 200 mm in diameter, 20 mm thick with a cone angle of 35° to the horizontal. This angle was chosen as the electrodes were of a larger diameter than for the laboratory experiments, as it was expected that the optimum cone angle would be smaller. It was found that the peak current efficiency of 81.7% occurred at 44 r.p.m., whereas further increases in rotation speed resulted in the current efficiency decreasing due to increasing turbulence in the interelectrode space.

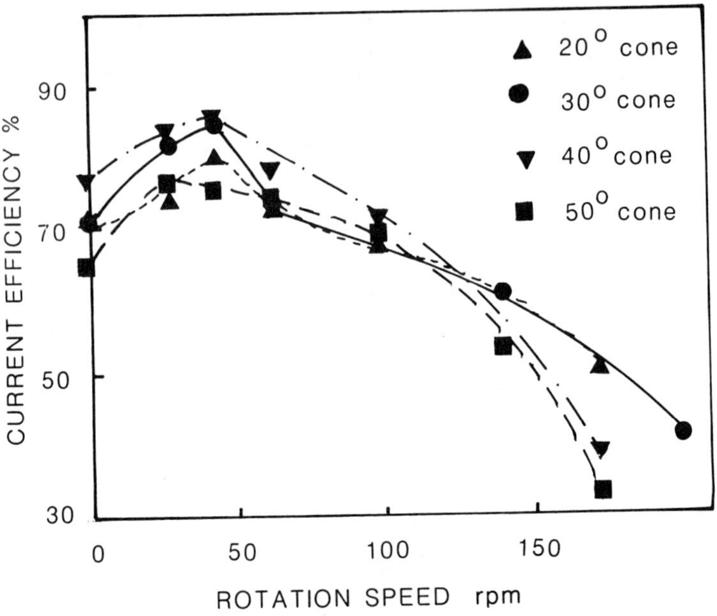

Figure 9 - Variation in current efficiency with rotation speed for different cone angles.

It is felt that the rotation speed is so modest that the energy input to rotate the electrodes is virtually insignificant compared to the energy savings achieved by reducing the resistive losses in the interelectrode gap.

The excellent results from the 200 mm diameter electrodes were particularly encouraging as Jarrett (31) has proposed approximately this size for the electrodes for the Advanced Technology Reduction Cell for aluminum extraction. This work has concentrated on carbon electrodes but obviously the concept would equally apply to titanium boride or steel cathodes and inert anode material, thereby extending the approach to other metals.

Discussion

Fused salt electrowinning cells have undergone considerable evolution since the cells of 100 years ago with the overall energy consumption falling to about 1/3rd. However, with existing materials, further improvements are only likely to be incremental in nature. The Hall-Heroult cell, in particular, is in need of improved materials for refractories which are inert to cryolite, inert anode materials and wetted cathode materials. Some progress has been made in this direction but none of the materials which have been developed are entirely satisfactory. Fortunately, there is considerable effort in materials technology and it is likely that novel and improved materials will be developed at an ever increasing rate. However, even where inert materials are available, for example, in magnesium and lithium electrowinning, the energy efficiency is still quite low with large anode-to-cathode spacing.

The other major disadvantage of fused salt electrowinning cells is the two-dimensional nature of the electrodes which is partially overcome by the bipolar arrangement of electrodes but, even so, the output compared to the equivalent volume of a pyrometallurgical reactor is poor.

Much effort has been devoted to utilising the gas evolved at the anode to pump the electrolyte around the cell and, generally, to raise the mass transfer within the cells. However, more is needed in order to be able to reduce the interelectrode spacing further and it may be necessary to apply other forces, besides gravity, to separate the products of electrolysis. Centrifugal forces are a possibility in that the magnitude of the force can be varied. It has been shown that with simple conical electrodes, the application of a modest centrifugal field permits a considerable reduction in the interelectrode spacing whilst, at the same time, maintaining a high current efficiency. Some of the work was done using a size of electrodes which might find industrial application in the near future. With careful design, it may also be possible to apply the concept to more complicated electrode arrangements, thereby, approaching the output of pyrometallurgical reactors.

Fused Salt Electrorefining

Fused salt electrorefining has considerable promise as a technique for metal refining, with possible advantages over pyrometallurgical and hydrometallurgical techniques. The thermodynamic potential requirement is very low and is given by

$$-ZEF = RT\ln a \qquad (1)$$

where Z is the number of units of charge carried, E is the thermodynamic potential, F is the Faraday, R is the gas constant, T is the temperature and a is the activity of the metal in the anode pool. It is assumed that the activity of the metal at the cathode is unity. In all systems there is going to be more than one element in the anode and the ease by which the elements dissolve is given by the position in the electrochemical series (Table 4, taken from Winter and Strachan (34)), and the relative position depends, to a certain extent, on the electrolyte (35). Elements near the top of the table will ionize more easily than those near the bottom and will, therefore, be taken into solution.

Advantages of electrorefining are that a dross-free product is produced and, generally, less atmospheric pollution is generated than for pyrometallurgical refining processes. As an example, the chlorination of aluminum scrap to remove magnesium as the chloride (36) produces a hygroscopic flux of low or negative commercial value with some unreacted chlorine escaping into the environment. However, most existing fused salt electrorefining cell designs are inefficient, with large anode-cathode distances, resulting in excessive voltage drops and energy losses (2). The two dimensional nature of present-day cell designs leads to lower space-time yields and poor mass transfer compared to pyrometallurgical reactors, which are always three-dimensional in nature. This causes difficulties such as the depletion of the least noble element in the anode metal pool, allowing other elements to be transferred. This depletion may be diminished by enhancing mass transfer in the anode pool, as attempted by Mullins and Leary (37) in a refining cell for plutonium, by stirring the metal with a ceramic impeller. Secondly, the anodic current density may be reduced to balance the diffusional flux in the metal pool but this leads to economic penalties with conventional cells.

Table 4. Standard Electrode Potentials in 0.48 NaCl - 0.52 $CaCl_2$ at 1000K against chlorine reference electrode. Taken from Winter and Strachan (34).

	V
Mg^{2+}/Mg	-2.596
Mn^{2+}/Mn	-2.001
Al^{3+}/Al	-1.838
Zn^{2+}/Zn	-1.582
Fe^{2+}/Fe	-1.272
Pb^{2+}/Pb	-1.112
Sn^{2+}/Sn	-1.041
Cu^{+}/Cu	-1.010
Sb^{3+}/Sb	-0.821

Cell Design

As it is obvious that the advantages of fused salt electrorefining over other metal refining techniques cannot be exploited unless efficient cell designs are used, it is worthwhile outlining the various possible cell designs.

Basic Cell

The basic design, shown in Figure 10, consists simply of liquid metal pools separated by an insulating wall submerged in a layer of fused salt electrolyte. This design is unsatisfactory as it requires a large floor area and a long and irregular anode-cathode path which results in an excessive voltage drop in spite of the high conductivity of the electrolyte.

Three-Layer Cells

The above difficulties are reduced in the three-layer refining cell (2,6) for aluminum (Figure 1) in which the density of the impure aluminum anode is increased by a 30 wt.% copper addition, and the density of the molten cryolite electrolyte is increased by the addition of barium and calcium fluoride additions so that it falls between that of the high purity electrorefined aluminum and the aluminum-copper alloy. However, as the density differences between three layers are small, the interfaces are relatively unstable, the electrolyte layer must be approximately 20 cms thick in order to prevent physical transfer of anode metal into the cathode layer. As a result, the voltage drop in the electrolyte is very large, around 7V at 0.4A cm^{-2}, which can be compared to a thermodynamic potential requirement for transfer of aluminum from Al-30 wt.% Cu to pure aluminum of 7 mV. Tiwari and Sharma (38) have used the three-layer approach to electrolytically remove magnesium from aluminum magnesium alloy.

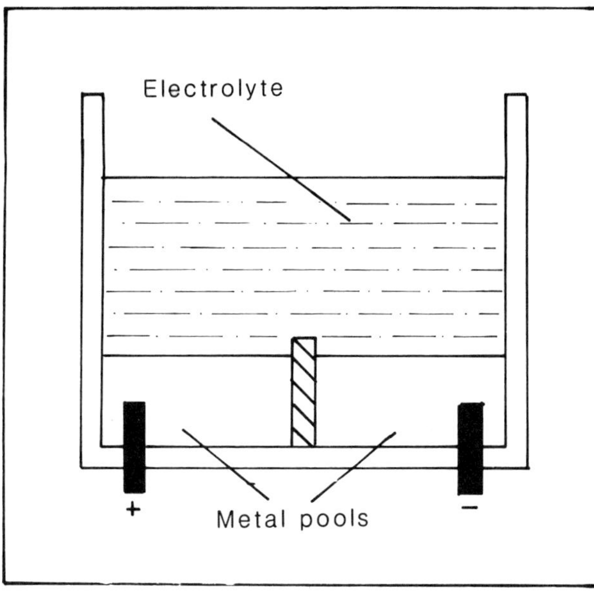

Figure 10 - Simple electrorefining cell.

The anode-cathode distance can be reduced if a porous diaphragm saturated with molten salt were used to separate the liquid metal pools but early work with porous ceramic diaphragms was unsuccessful due to the mechanical weakness of the diaphragm material (39). An elegant solution to this problem was given by Amstein and co-workers (40) in a cell to refine lead-antimony alloys using a $PbCl_2$-KCl-NaCl electrolyte. The lead-antimony anode was supported on a flexible blanket of alumina-silica fibres, saturated with electrolyte, held on alumina rods above the lead cathode pool (Figure 12). In order to prevent metal deposition at the cathode/diaphragm interface, a layer of electrolyte was maintained between the cathode and the blanket by careful control of hydraulic heads within the cell. Although laboratory trials were sufficiently encouraging to construct a pilot plant, the project was ultimately abandoned due to the eventual growth of "liquid metal stringers" through the blanket causing the cell to short circuit. The operating cell voltage was 2.4V with a current density of 2.3A cm^{-2}, a considerable improvement on the three-layer cell for aluminum refining.

A further development of the diaphragm cell was patented by Alcoa (41) for aluminum refining in which the diaphragm was a porous carbon membrane which was permeable to the molten alkali chloride-aluminum chloride electrolyte, but impervious to the molten metal (Figure 13). As the carbon membrane was electronically conducting, it was also used as the anode

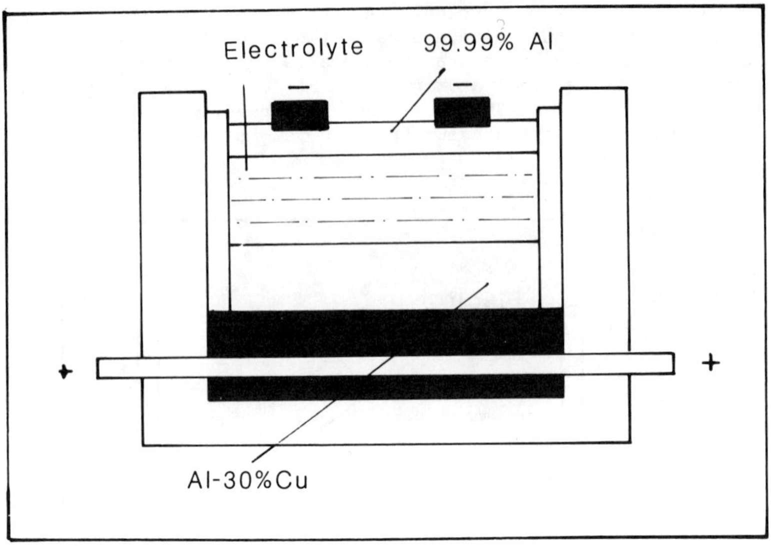

Figure 11 - Three-layer refining cell for aluminum.

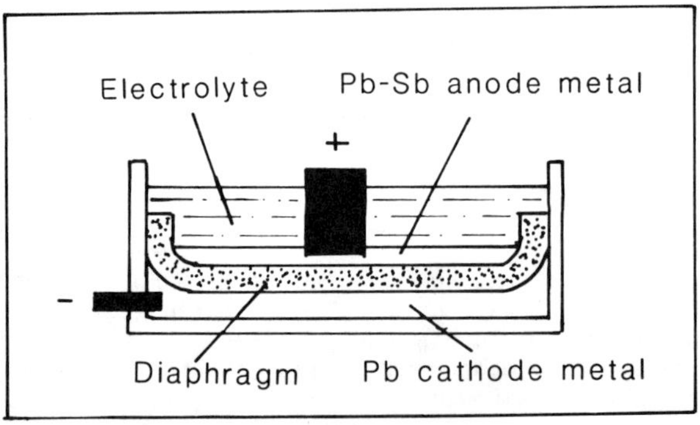

Figure 12 - Diaphragm cell for the electrorefining of lead.

feeder electrode. No chlorine evolution was detected as the potential for this reaction is much larger for the anodic dissolution of aluminum.

Other workers have found that anodic depletion is a major problem in fused salt electrorefining. For example, Geolff and co-workers (42) found in electrorefining manganese from ferro-manganese, the iron content of the electrolyte, from the ferro-manganese anode, was found to increase progressively to such a level that the cathodic deposit of manganese became unacceptably contaminated. The problem was overcome by using a flow-through electrochemical cell in which the anolyte was continuously removed, passed through a purification reactor, and returned to the cathode side of the cell.

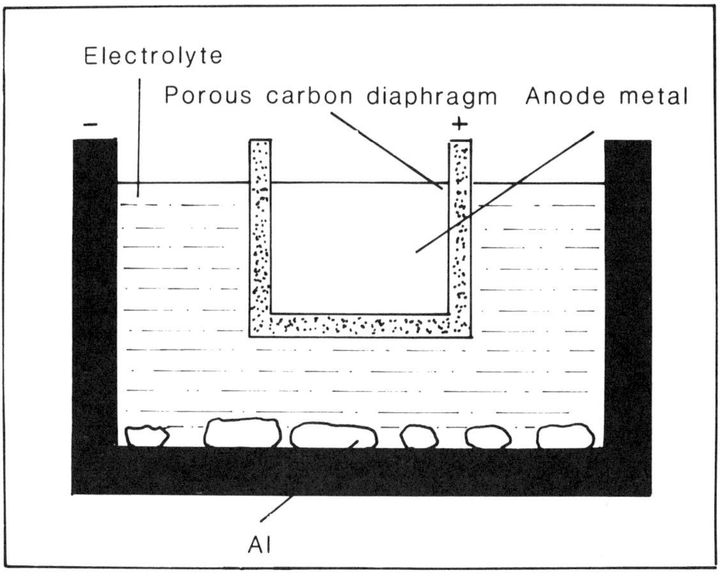

Figure 13 - Diaphragm cell for refining aluminum.

Packed Bed Electrorefining Cell

In order to address the above problems, Cleland and Fray (43,44,45) developed the packed bed electrorefining cell which comprised two packed bed electrodes of conducting particles in a side by side configuration, separated by a ceramic fibre cloth or blanket. The beds are flooded with electrolyte, and liquid metal is allowed to percolate through the beds (Figure 14). The great advantage of this cell compared to the other designs is that although the current density on the diaphragm can be very high, the current density on each individual droplet of metal, as it meanders down the cell, can be very low due to the large surface area of droplets in the bed. This will obviously decrease the likelihood of surface depletion but, perhaps, more significantly the motion of the droplets in the bed ensures that the metal is continually agitated, thereby, increasing mass transfer within the droplet. Furthermore, unlike other designs, the load of metal on the diaphragm will be virtually insignificant as droplets are on the order of a few mms in diameter. Various studies have been performed on this type of cell with voltages as low as 0.205V at 5000 A m^{-2} across the diaphragm for the electrorefining of zinc (46). With careful cell design, it is anticipated that these figures can be improved.

Unlike electrowinning where there are many examples, it is perhaps instructive to consider areas in which fused salt electrorefining could find application.

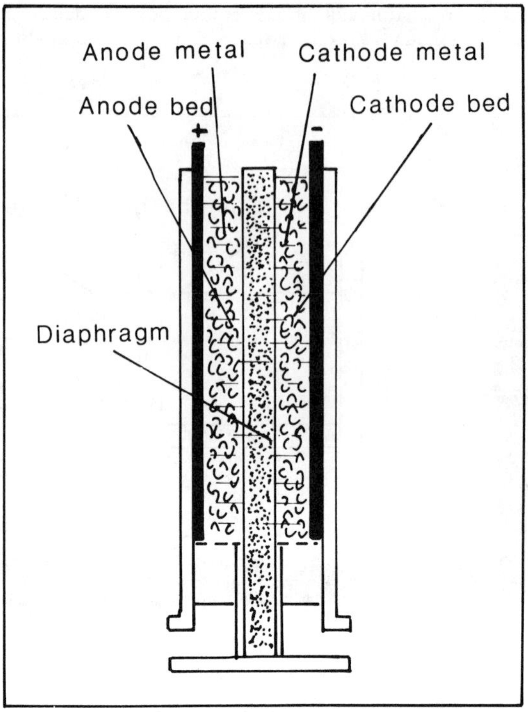

Figure 14 - Schematic diagram of packed bed cell.

Upgrading of 99.85% Al to 99.99% Al

There is a growing market for high purity aluminum in the electronics industry but the three-layer cell is expensive to run making the refining of aluminum an expensive step. Cleland and Fray (44) showed that it was possible to electrorefine aluminum using the packed bed cell filled with an electrolyte containing aluminum chloride. The current densities were much higher and the cell voltage much lower than for the conventional process.

Recycling of Scrap Aluminium

Due to the very high energy input necessary to produce primary aluminum, it is likely that most of the primary production will eventually move to those areas of the world where electrical energy is relatively cheap, namely, Australia, New Zealand, Latin America and Canada. On the assumption that industry will still require high grade aluminum for its products, a greater proportion of aluminum will have to be imported into Europe and North America and, eventually, this aluminum will need to be recycled.

At the present time, the secondary aluminum industry supplies much of the ingot metal for the foundry and die casting industries. Wrought products are not generally provided by the secondary aluminum industry as these products require less impurities than cast products, as common alloying elements such as copper and silicon greatly reduce the ductility of the aluminum. Furthermore, it is virtually impossible to produce secondary metal having the same degree of purity as primary metal as it is difficult, using existing technology, to economically remove metallic impurities, except for magnesium from aluminum; and without dilution of the scrap with primary metal, it is virtually impossible to meet the specification of many casting alloys. It should be noted that those alloys with a high impurity content are sold at a substantial discount compared to primary metal and this, perhaps, gives the incentive for electrorefining, especially electrorefining using the packed bed cell which offers the opportunity for very low energy consumption.

There appears to be two opportunities for the packed bed cell in secondary aluminum. The first is to remove magnesium from aluminum to produce magnesium metal as laboratory results show that electrochemical separation gives the same aluminum product as chlorination but, at the same time, produces magnesium of commercial value (44). The process is also quiet and free from pollution. The second application is to upgrade low grade secondary aluminum to pure aluminum. Electrochemically, there is no problem in separating aluminum from the impurities other than magnesium and, perhaps, manganese. If magnesium and manganese were present, these elements would be removed initially. The major problem is that with an aluminum feed, laden with impurities, any removal of aluminum is likely to cause precipitation of intermetallic compounds. In order to overcome this problem it may prove possible to cool the aluminum, external to the cell, and to filter the aluminum intermetallics using conventional ceramic filters, thereby, overcoming precipitation in the packed bed electrode. This material may find a market in reducing oxides in the thermite process. Overall, electrorefining would give the secondary smelter access to the wrought aluminum market.

Carbothermic Reduction followed by Fused Salt Electrorefining

For the production of magnesium, there are basically two methods. Firstly, in the metallothermic processes, magnesium oxide is reacted with a reducing agent in an electric furnace under vacuum and, secondly, there are the electrolytic processes, described earlier in this paper. The principal drawbacks of the batch thermic operation are the need to operate at greatly reduced pressure, a solid rather than a molten product is obtained, the purity of the product is low and the condensers are relatively inefficient. The main drawback of the electrolytic process is the high cost of feed preparation, which was mentioned earlier. There have been several attempts to use carbothermic reduction but in order to get the reaction

$$MgO + C = Mg_{(v)} + CO \qquad (2)$$

to proceed, the temperature has to be around 2300K (47,48). However, on cooling the gases to condense the magnesium vapour, the reaction reverses producing magnesium oxide and carbon. Anderson and Parlee (49,50) have suggested carry out the reaction in a molten solvent, to reduce the activity of the magnesium, driving the reaction to the right, and then removing the magnesium by distillation. Solvents which have been suggested include tin, bismuth, antimony or mixtures of these metals (52). As the overall reaction is endothermic, there is unlikely to be an excess of heat available, and this would appear to make distillation an expensive process. All the solvent metals are electrochemically more noble than magnesium indicating that fused salt electrorefining might be a suitable route for separation. In this case, the solvent metal, containing the magnesium, is transferred to the electrorefining cell, where the magnesium is transferred to the cathode. It should be emphasised that the reduced activity of magnesium only has a small effect on the cell potential as the potential is related logarithmically to the activity, as is shown in equation (1).

The electrolytic route, as well as giving a pure product, would overcome the problem of the formation of constant boiling mixtures as described by Howell et al. (52). The combination of carbothermic reduction in a metallic solvent followed by electrorefining could be applied to other metals such as aluminum and, perhaps, titanium and zirconium.

Some reactive metals are already prepared carbothermically in solution in another metal and these include the ferroalloys of manganese, chromium, vanadium and titanium. Unfortunately, iron melts at too high a temperature to be of use as a solvent in fused salt electrorefining cells and in order to overcome this problem, Godsell and Fray (53) dissolved the ferroalloy in a metallic solvent. Fortunately, a search of the published phase diagrams shows that there are several noble, relatively low melting point metals which have high solubilities for manganese, chromium, vanadium and titanium. For the specific case of manganese, bismuth shows substantial solubility for manganese in bismuth, whereas iron is virtually insoluble in bismuth. However, for the other elements, there does not appear to be a metallic solvent which will only dissolve the desired element without also taking the iron into solution. Possible solvents for these elements are lead, antimony and tin and, in these cases, the solvent metal would be saturated with iron to prevent dissolution of iron from the ferroalloy. Godsell and Fray examined, in depth, the treatment of ferromanganese and found that bismuth very readily leached manganese from low carbon ferromanganese, but it was found that less satisfactory results were obtained for high carbon ferromanganese. Instead of electrorefining to pure manganese, it was easy to electrorefine the manganese into molten aluminum to produce an aluminum-manganese master alloy.

Electrowinning and Electrorefining

Earlier in this paper (19), it was suggested that it might be possible to carry out carbon chlorination in the anode part of an electrowinning cell. Unfortunately, carbon frequently introduces impurities in the melt leading to an impure cathodic product. One way of alleviating this problem would be to deposit the cathodic metal in a metallic solvent and then refine the metal from the solvent. A design which allows both the electrowinning and electrorefining steps to be carried out in the same cell has been described by Slatin (54) and is shown in Figure 15. The cell is divided into two compartments by an inert non-conducting barrier that extends from above the fused salt level until a seal is made with the bipolar, liquid metal pool in the bottom. On one side of the barrier, metal is electrowon into the metal pool with diffusion transporting the electrowon metal into the electrorefining side of the cell in which the melt is electrorefined and collected on top of the melt. It is apparent that such a system is only possible for metals which are lighter than the electrolyte such as aluminum, magnesium and lithium.

Discussion of Electrorefining

Although fused salt electrorefining is considerably less energy intensive than electrowinning, the only commercial application is in purifying aluminum from the Hall-Heroult cell. The limited application of fused salt electrorefining is due almost entirely to the present day design of the cells which do not take advantage of the high electrical conductivity of fused salts and the very low thermodynamic potential required for electrorefining. Designs of cell have been suggested and evaluated on a laboratory scale which indicate that voltages as low as 0.2V can be achieved

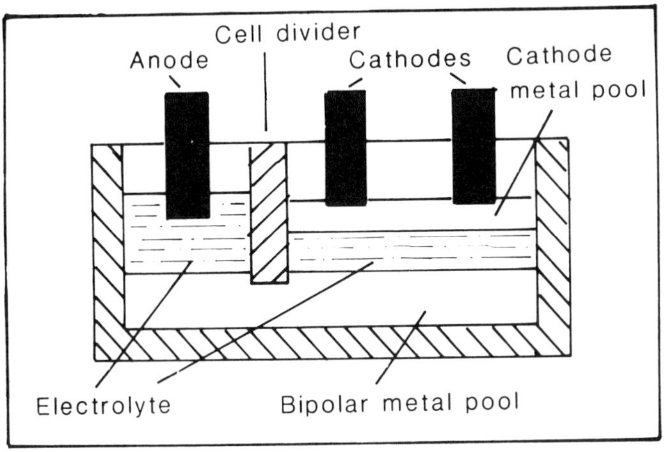

Figure 15 - Diagram of combined electrowinning-electrorefining cell.

at acceptable current densities. It appears that with an energy efficient cell, the possibilities for fused salt electrorefining are quite extensive ranging from treatment of scrap, combination with carbothermic reduction and, lastly, in combination with electrowinning. It is perhaps in combination with carbothermic reduction that the greatest possibilities exist as carbothermic reduction, although energy efficient and rapid, generally gives a product which is impure and needs further treatment. Efficient fused salt electrorefining should give a very pure product at relatively low energy consumption. In addition, unlike purification using air or chlorine, all the metals remain in the metallic state and, therefore, retain their commercial value rather than appearing as a residue.

General Discussion

Fused salt electrowinning and electrorefining face different challenges in the next decade. Fused salt electrowinning is essentially a mature industry and, generally, one can only see incremental improvements in performance, without dramatic changes in cell design. These changes would be brought about by new materials for cell construction, anodes and cathodes, but it is difficult to foresee how the cells can be made competitive with pyrometallurgical reactors without a dramatic change in cell design and a substantial decrease in anode to cathode spacing. To a certain extent this can be achieved by modelling and careful control of the gas, evolved at the anode, to maximise mass transfer without increasing the recombination of the anodic and cathodic products. For a more striking change, one may need to consider other means to remove the products and in this paper it is suggested that centrifugal forces may find application in fused salt electrowinning. Using a modest centrifugal field, substantial decreases in the anode to cathode spacing have been achieved without any loss of current efficiency.

One interesting concept is to carbochlorinate in the anodic part of the cell, thereby, eliminating the high cost of chlorinating oxides, external to the cell. This would also have the effect of reducing the cell potential, but, perhaps, at the cost of a decrease in the purity of the cathodic product. In these cases, it may be necessary to combine electrowinning with electrorefining.

The development of electrorefining has been very much hindered by the lack of suitable cell designs to give energy efficient processing. In this paper, indications of possible design changes are given which make fused salt electrorefining more attractive. This opens up the opportunity for fused salt electrorefining to be applied to the treatment of scrap, combined with carbothermic reduction to give a high grade product and, lastly, combined with electrowinning.

References

1. P.G. Lovering, Molten Salt Technology, (New York, NY: Plenum Press, 1982).

2. D.J. Fray, Energy Considerations in Electrolytic Processes (London: Society of Chemical Industry, 1980), 99-111.

3. K. Grjotheim et al. "Aluminum Electrolysis" (Dusseldorf, W. Germany: Aluminum-Verlag GmbH, 1977).

4. N. Jarrett, "Metal Treatises" Edited by J.K. Tien and J.F. Elliott, (Warrendale, PA: The Metallurgical Society, 1981), 159-169.

5. N. Jarrett, W.B. Frank and R. Keller, "Metal Treatises", (Warrendale, PA: The Metallurgical Society, 1981), 137-157.

6. C.L. Mantell, "Electrochemical Engineering", (New York, N.Y.: McGraw-Hill, 1960).

7. G.W. Mellors and S. Senderoff, Electrodeposition of coherent deposits of refractory metals, I.Niobium, J. Electrochem. Soc., 112, (1967), 266-72.

8. D. Inman and S.H. White, Molten Salt Electrolysis in Metal Production, (London: IMM, 1977), 51-61.

9. O.Q. Leone, H. Knuden and D. Couch, High-purity titanium electrowon from titanium tetrachloride, J. Metals, 19(3) (1967), 18-23.

10. M. Broc, G. Chauvin and H. Coriov, Molten Salt Electrolysis in Metal Production, (London: IMM, 1977), 69-73.

11. M. Nardin, E. Chassaing and G. Lorthioir, Molten Salt Electrolysis in Metal Production (London: IMM, 1977), 36-41.

12. P. Mahi, A.A.J. Smeets, D.J. Fray and J.A. Charles, Lithium-metal of the future, J. Metals, 38(11) (1986), 20-6.

13. A.S. Emley, Principles of Magnesium Technology (Oxford, England: Pergamon 1966).

14. A.R. Burkin, Production of Aluminum and Alumina (Chichester, England: John Wiley,1987).

15. L.G. Boxall, A.V. Cooke and H.W. Hayden, TiB_2 cathode material: application in conventional USS cells. J. Metals 36(11) (1984), 35-39.

16. T.R. Beck, J.C. Withers and R.O. Loutfy, Light Metals 1986, Ed. R.E. Miller (Warrendale, PA: The Metallurgical Society, 1986).

17. J.C. Withers and R.O. Loutfy, Light Metals 1986, Ed. R.E. Miller, (Warrendale, PA: The Metallurgical Society, 1986) 1013-18.

18. M.C. Flemings et al., "An Assessment of the magnesium Primary Production Technology" (U.S. Department of Energy, Final Report, EX-76-A-01-2295, 1981).

19. B. Cartwright, L.R. Mechels and S.F. Ravitz, "Electrolysis of Magnesium into Liquid Cathodes from magnesium oxide-carbon suspensions in Molten Chlorides" (U.S. Bureau of Mines, Report of Investigations No. 3805, 1945).

20. K. Sintim-Damao, S. Srinivasa, N. Keddy and S. McCormack, Electrolytic Production of lithium metal, U.S. Patent 4,455,202 (1984).

21. Nippon Soda Co. Ltd., Lithium by electrolysis, Japanese Patent 59,200,731 (1984).

22. M.B. Dell et al., Electrolytic manufacturers of aluminum, U.S. Patent Appl. No. 3822195 (1974).

23. F. Goodridge, Some recent developments of monopolar and bipolar fluidised bed electrodes, Electrochimica Acta 22 (1977), 929-33.

24. M.M. Wong and F.P. Haver, Molten Salt Electrolysis in Metal Production (London: IMM, 1977), 21-29.

25. F.P. Haver, D.E. Shanks, D.L. Bixey and M.M. Wong, Recovery of zinc from zinc chloride by fused salt electrolysis (U.S. Bureau of Mines R.I. 8133, 1976).

26. D.E. Shanks, F.P. Haver, C.H. Elges and M.M. Wong, Electrowinning Zinc from zinc chloride-alkali metal chloride electrolytes, (U.S. Bureau of Mines R.I. 8343, 1979).

27. S.D. Hill, O.L. Pool and S.A. Smyres, Electrowinning zinc from zinc chloride in monopolar and bipolar fused salt cells (U.S. Bureau of Mines R.I. 8524, 1981).

28. O.G. Sivilotti, Metal production by electrolysis of a molten electrolyte, European Patent Appl. No. 96,990 (1983).

29. O.G. Sivilotti, Metal production by electrolysis of a molten electrolyte, European Patent Appl. No. 101,243 (1984).

30. M.B. Dell, W.E. Haupin and A.S. Russell, Electrolytic Production of Aluminum, U.S. Patent Appl. No. 3893899 (1975).

31. N. Jarrett, United States Extractive Metallurgy - The 80's and beyond, Met. Trans. B., 18B (1987), 289-34.

32. J.A. Charles, D.J. Fray and P.M. Copham, Metal Separation Process, European Patent Appl. No. 87309052.6 (1987).

33. P.M. Copham, Rotating electrodes in molten salt electrowinning (Ph.D. Thesis, University of Cambridge, 1987).

34. D.G. Winter and A.M. Strachan, Advances in Extractive Metallurgy, Ed. M.J. Jones, (London: Institution of Mining and Metallurgy, 1977), 177-184.

35. Yu.K. Delimarskii and B.F. Markov, Electrochemistry of Fused Salts (Washington: Sigma Press, 1961).

36. B.L. Tiwari, Demagging processes for aluminum alloy scrap, J. Metals, 34(7) (1982), 54-58.

37. L.J. Mullins and J.A. Leary, Fused salt electrorefining of plutonium and its alloys by the LAMEX process, Ind. Eng. Chem. Proc. Des. Div. (4) (1965), 394-400.

38. B.L. Tiwari and R.A. Sharma, Electrolytic removal of magnesium from scrap aluminum, J. Metals, 36(7) (1984), 41-43. 41-43.

39. I.G. Pavlenko and A.P. Grinyuk, Electrolytic processing of lead in melts using porous diaphragms, Ukr. Khim. Zh. 29 (1963), 868-73. 868-73.

40. E.H. Amstein, W.D. Davis and C. Hillyer, "Advances in Extractive Metallurgy and Refining", Ed. M. Jones (London: Institution of Mining and Merallurgy 1971) 399-412.

41. K.A. Bowman, Electrolytic purification of metals, U.K. Patent Appl. No. GB203300A (1980).

42. P. Geolff, E. Bruneel, A. Fontana, R. Winand, Proc. Int. Symp. Molten Salt Ferromanganese, Chem. Technol. (1983), 233-37.

43. J.H. Cleland and D.J. Fray, Advances in Extractive Metallurgy" (London: Institution of Mining and Metallurgy 1977), 141-146.

44. J.H. Cleland and D.J. Fray, Packed bed electrorefining of Al, Bi and Zn alloys, Trans. IMM, 88 (1973), 191C-196C.

45. D.J. Fray and J.H. Cleland, Packed Bed Electrorefining and Electrolysis, British Patent No. 1515216 (1978).

46. A.J. Godsell, Fused Salt electrorefining of ferroalloys (Ph.D. Thesis, University of Cambridge 1987).

47. T.A. Dungan, Production of Mg by the carbothermic process at Permanente, Trans AIME, 159 (1944), 308-14.

48. A.M. Cameron et al., "Pyrometallurgy '87" (London: The Institution of Mining and Metallurgy 1987), pp 196-222.

49. R.N. Anderson and N.A.D. Parlee, Carbothermic reduction of an oxide of a reactive metal, U.S. Patent Appl. No. 3,794,482 (1974).

50. R.N. Anderson and N.A.D. Parlee, Carbothermic reduction of refractory metals, J. Vac. Sci. Tech. 13 (1976), 526-29.

51. C.A. Eckert, R.B. Irwin and C.W. Graves, Liquid metal solvent selection: the MgO reduction reaction, Ind. Eng. Chem. Proc. Des. Dev., 23 (1984), 210-17.

52. W.J. Howell, C.A. Edkert and R.N. Anderson, Carbothermic reduction using liquid metal solvents, J. Metals, 40(7) (1988), 21-23.

53. A.J. Godsell and D.J. Fray, Unpublished work.

54. H.L. Slatin, Electrolytic production of aluminum, U.S. Patent Appl. No. 2,919,234 (1959).

THE REDUCTION OF ALUMINA BEYOND THE YEAR 2000
OVERVIEW OF EXISTING AND NEW PROCESSES

A. F. Saavedra, C. J. McMinn, and N. E. Richards

Reynolds Metals Company
Manufacturing Technology Laboratory
Extractive Metallurgy Department
P.O. Box 1200
Sheffield, AL 35660

Abstract

The aluminum industry in general and Reynolds Metals Company in particular have continued to support development work for improvement of the Hall cell design and for alternative processes for the reduction of alumina. The evolution of the Hall cell design will continue into the next century as a result of further development of mathematical models, sophisticated control systems and new materials for cell components. This paper discusses the areas in which the new generation cell advances will differ from the present day design. Additionally, Reynolds Metals has supported work for the development of a carbothermic process to reduce alumina. Theoretical work has indicated that this process could possibly have economic advantage over the Hall process. Several complicated furnace configurations have been proposed to provide the conditions for successfully reducing alumina. The results of small pilot scale operation and most promising furnace design are discussed. For comparison, projected capital investment costs and operation costs are discussed.

Metallurgical Processes for the Year 2000 and Beyond
Edited by H.Y. Sohn and E.S. Geskin
The Minerals, Metals & Materials Society, 1988

Introduction

The evolution of the Hall-Heroult process will continue into the next century. This evolution will be influenced by economics, more restrictive environment and worker protection requirements, and technological development. These factors will also cause changes in the competitive position of alternate processes for aluminum production.

Several aluminum producers have made major investments in alternate processes for aluminum production in the past three decades and there is continued support in some companies for improvements in the Hall-Heroult process and for alternate processes. The more visible development work in reduction technology, not aimed at significant process changes, is in the areas of higher amperage cells with bus bar designs to reduce harmful magnetic effects, more advanced and automated feed, voltage and electrolyte composition control, improved energy efficiency and solid waste disposal. Efforts to bring significant change to the reduction process are being made in development of materials that can function as inert anodes and cathodes. These materials could lead to the development of reduction cells with multiple bipolar electrodes operating on oxide feed and fluoride electrolytes. Development work on new processes was reduced in the last decade by several companies because of economic and technological realities. Reynolds Metals supported work on a carbothermic process to reduce alumina and produce commercial pure aluminum. This work progressed through small pilot plant operation, and resulted in development of furnace designs and operation regimes. This paper reviews how this and other technologies could influence the reduction of alumina beyond the year 2000.

Evolution of Present Reduction Process

It appears certain that the present reduction process will remain dominant into the next century. Evolutionary changes will also certainly continue to be made under the pressure of factors discussed previously. These factors that dictate change will vary in importance and degree of acuteness as demand on energy resources, environmental needs and economic cycles develop.

During the last two decades, energy requirements for the best alumina reduction cells have decreased from 17 kwh/kg to slightly below 13 kwh/kg. This brings the process to an energy efficiency of about 50%; therefore, there appears to be further opportunity for power savings. There has also been significant improvement in plant working conditions and in control of emissions harmful to the environment. The history of the Hall-Heroult process was documented on the occasion of the 1986 Centennial volume sponsored by TMS - Light Metals (1). Articles have also been published recently that give many statistical details on aluminum production history along with discussions on future developments (2, 3).

Reduction cell size has continued to increase throughout the life of the process. At present, 280 kA pots are commercial and have operated at a current efficiency of 95% and a d.c. power consumption of 12.9 kwh/kg (4). Technological developments, including mathematical models that have allowed advances in bus bar design and magnetic stability, and sophisticated control systems led to the utilization of the economics of scale. Already, designs conceived are near 500 kA and this probably will be realized in the next century. Patent literature on bus bar addresses 600 kA designs (5). Factors other than technology may dictate that the larger cells not be used under certain conditions. These conditions are where 200-240 cells of the larger size represent too much capacity or investment, or not enough power is available at a location.

How many reduction plants in developed countries with limited low-cost power will survive into the next century is dependent on economics and in some instances on national policy. Plants that survive will certainly be retrofitted to meet environmental requirements, to increase efficiency, and reduce power usage.

Inert Anodes and Cathodes in Bipolar Cells

Substantial efforts have been made during the past four decades to develop inert electrodes and to develop commercially viable bipolar cells. Advantages that such an electrolytic process offers are very important in meeting manufacturing requirements that may prevail beyond the year 2000. Single units with multiple bipolar electrodes could significantly reduce power requirements, operate as a closed system producing oxygen gas and reduce plant area.

At the 1967 AIME meeting, G. de Varda and A. Vajna de Pava described a 10-year effort by Montecatini Corp. to develop a multi-cell electrolytic furnace with bipolar electrodes (6). They piloted seven-cell units operating at 4-6 kA. Corrosion of the cell lining was a major difficulty. Anode consumption and replacement, with their carbon electrode system, would certainly have been another problem.

Alcoa came much closer to a commercial process with The Alcoa Smelting Process, made public in 1973. Aluminum chloride was electrolyzed permitting carbon to be used as an inert electrode in a bipolar cell where chlorine was produced at the anode. The thermal energy generation was reduced in this operation to below that for the Hall-Heroult cell (7). The cost advantages of the bipolar cell in this Alcoa process were offset by the carbochlorination step added to the reduction process and by difficulties in maintaining efficient cell performance over long periods. The corrosiveness of the chlorides was a significant factor.

The incorporation of inert anodes and stable, wetted cathodes into a bipolar cell operating on alumina feed would appear to improve the potential of the bipolar arrangement. The key to this process appears to reside in the success of the development of materials for the electrodes, and other materials necessary for cell construction.

Inert, wetted cathodes have been under development for about 35 years but have not reached commercialization. The emphasis on the development has been for retrofitting aging reduction cells to combat the rising cost of electric power.

Work on non-consumable or inert anodes became evident in 1966 when a patent was granted to Marincek in Switzerland (8). Work supported by DOE-Alcoa produced a new generation of anodes. Presently, DOE-PNL has modified these materials and anodes will be exposed to pilot scale testing in the very near future.

Carbothermic Reduction

Over the years, efforts have been directed to the development of alternative processes to the Hall-Heroult electrolytic process for the production of aluminum. Among these is the carbothermic reduction process, which can be divided into two major routes: alloy route and pure metal route. Figure 1 shows the different options of both routes.

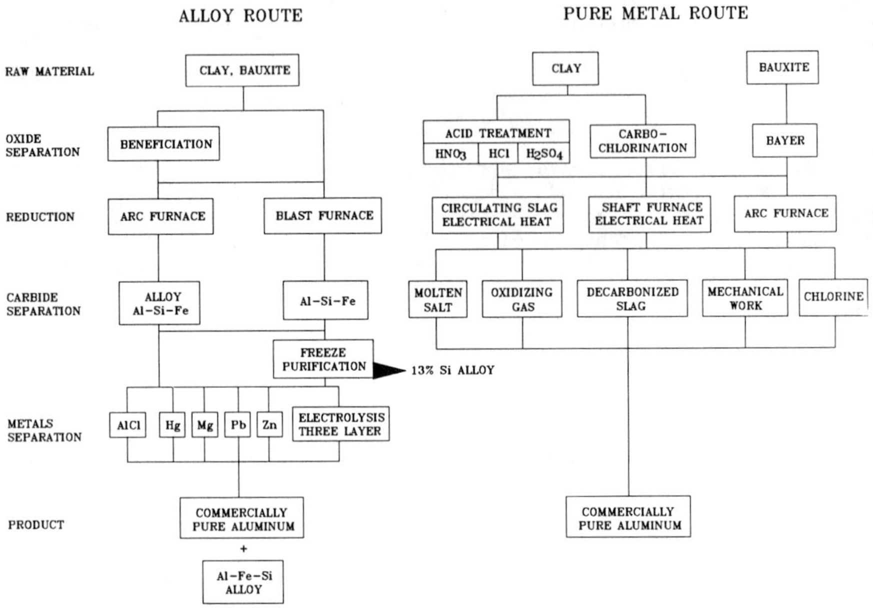

Figure 1. Alternate Routes for Carbothermic Production of Aluminum

The alloy route allows the use of raw materials with or without beneficiation and further purification to produce alloys by reacting the ore with coke or coal. From the raw alloy, pure aluminum metal can be produced along with byproduct alloys. Some of the energy requirements for this route could be supplied by combustion of coal as in the case of shaft furnaces.

The pure metal route requires the use of high grade alumina. A purification process for the raw materials, i.e. Bayer process, is needed for this route. Pure coke is used as the reductant to minimize the introduction of impurities into the final product.

Reynolds Metals Company's interest in carbothermic reduction of alumina started with the work by Schmidt and Robinson on the smelting of bauxite for the production of aluminum-silicon alloys (9, 10).

Carbothermic reduction of alumina has been compared with the carbothermic production of silicon metal in which quartz is reduced by coke in a submerged arc furnace. Quartz mixed with coke and wood chips is charged to the furnace burden in which the reduction reactions take place according to the following reactions (11) as the feed materials in the burden move down and the reaction gases move up:

$$2C(s) + SiO(g) \rightleftarrows SiC(s) + CO(g)$$

$$SiC(s) + 2SiO_2(l) \rightleftarrows 3SiO(g) + CO(g)$$

$$2SiO(g) \rightleftarrows Si(l) + SiO_2(l)$$

$$2SiO_2(l) + 4C(s) \rightleftarrows SiO(g) + SiC(s) + CO(g)$$

$$SiC(s) + 2SiO_2(l) \rightleftarrows 3SiO(g) + CO(g)$$

Silicon is produced at temperatures close to its melting point, at which the solubility of silicon carbide is small and acceptable; the liquids present in the burden, mostly SiO_2, react with the silicon carbide forming gaseous silicon monoxide (SiO). The SiO moves upward in the burden and as it condenses in the cooler regions of the burden it dissociates in elemental silicon and silica. The liquid silicon moves down to the furnace hearth from where it is tapped periodically.

The operating temperatures involved in carbothermic reduction of alumina are nearly three times that of the melting point of the pure metal and the solubility of the carbide in aluminum at these temperatures is approximately 35%. Such a product is useless commercially. It does not flow as a liquid below 2000°C. The burden in an aluminum furnace contains a much higher proportion of liquid slags, which do not necessarily form gaseous products upon further reaction to help maintain a permeable burden as in the silicon furnace.

Several approaches have been taken for the carbothermic reduction of alumina; in most cases the furnace used has been similar to the silicon furnace. We found this type of furnace impractical for the production of aluminum due to the difficulty of maintaining a burden in place and with the necessary permeability to allow the vaporization products to contact the feed materials contained in the burden. The formation of slags and production of oxycarbide and carbide in the burden hindered structural strength of the burden and the power delivery to the hearth of the furnace. The amount of liquid slag on the hearth is greater in the aluminum furnace and the metal produced floats on top of the slag. This presents a problem of electrical shorting between electrodes impeding operation in the submerged mode.

Other Concepts for Carbothermic Production of Aluminum

A comprehensive analysis and practical evaluation of the production of Al, and Al-Si alloys, by three concepts for the direct reduction of aluminum-silicon ores was described by Bruno (12). The three extraction processes were based upon a combustion-heated blast furnace, a combination blast-arc furnace, and a submerged arc furnace. While the process energy requirement for a process producing pure aluminum indicated a preference for the blast furnace route using bauxite and/or clay, the work demonstrated that a low pressure, low iron blast furnace was not technically feasible. This was largely a consequence of the vapor pressure of SiO, Al_2O and Al at the source of the area for reduction ($C+O_2$) and the volume of CO being much greater than stoichiometric effecting transpiration of those compounds out of the reaction zone without either capture or incorporation into useful, partially reduced, intermediate products.

One route to aluminum for which Alcoa did demonstrate potential viability, was to use a submerged arc furnace to produce Al-Si for which two further processes were perceived. The product from clay, alumina, and coke (62% Al-30% Si and some iron and titanium) was treated by fractional crystallization, when intermetallics of Fe-Si-Al-Ti were separated yielding an alloy 12-17% Si, 1% Fe containing about 75% of the initial aluminum.

The second refining step was based upon an electrolytic diaphragm transport cell which was patented by Alcoa (13).

Bruno indicated that for a plant capacity of 353,000 ton/yr, 85% Al, 15% Si alloy (in 1981 dollars), the capital cost would be about $2,300/ton year and operating cost of $1.01 lb/Al-Si alloy with potential for credits of $0.81/lb Al-Si alloy.

We have seen no recent literature on the Kuwahara process (14) based upon the carbothermic reduction of aluminous ores, partial beneficiation by fractional crystallization and extraction of aluminum in a recirculating stream of lead. By cycling the temperature of the Pb-Al alloy, an aluminum-rich product can be obtained (17). Without further energy intensive purification, however, residual concentration of lead would preclude use in many aluminum products.

The feasibility of an aluminum blast furnace was also investigated very extensively at the Isukuba Research Center, Ibaraki, Japan (15, 16). From bench scale experiments in shaft furnaces equipped with tuyeres, and where low aluminum alloy (20% Al, 80% Si) could be produced, it was discovered that in the very zone where the briquettes of bauxite and carbon need to be above 2270°K for reaction, the oxygen partial pressure was enough to reoxidize or volatilize aluminum. The low yields could be slightly improved by coating the charge with powdered coke, but the temperatures down the vertical column of descending charge were too low in both the bench and later pilot scale blast furnaces to initiate the prereduction reactions that need to occur before the final conversion to aluminum (with minimal conversion to Al_2O) occurred.

The inherent problems of sourcing aluminum from bauxite or other aluminous ores are chemically produced heat, localization of the high temperature zone, large volumes of gas from combustion on which SiO, Al_2O, Al must be transported because of their partial pressures, problematic management of the descending burden and down stream processing to pure aluminum, which vitiate strongly against the practical fulfillment of this method of production.

A concept for an extraction scheme for aluminum of commercial quality based on research done at M.I.T. has been evaluated by A. D. Little (18) for the Department of Energy.

Elliot, Frank, and Finn (19), in bench scale studies, showed that alumina could be carbothermically reduced over the solvent metal tin, to produce at about 1800°C, an alloy containing up to 9 wt.% Al. The reaction was rapid at reduced pressure, utilizing the phase diagram of Sn and Al, partial separation of Sn-rich, and Al-rich phases through controlled cooling, rolling, and slitting. The research also demonstrated that the carbothermic reduction of alumina could not be effected in any shaft furnace without electrical energy.

Based on this research, a technical team (18) devised a flow sheet comprising unit processes that, with further development and demonstration, could integrate to a 200,000 tonne Al/yr plant, and for which preliminary economic estimates could be made. The flow sheet consisted of:

- Reduction of alumina in submerged arc furnaces. The submerged arc furnace was to be fed with pelletized Bayer alumina with a carbonaceous reductant at a temperature of about 2000°C and at 1 atmosphere of pressure using tin as a solvent for the aluminum.

- Cooling the aluminum-tin alloy to crystallize the aluminum.

- Separating the aluminum crystals from the tin at a temperature of about 330°C (some 100°C above the melting point of tin). Tin was to be recycled to the electric arc furnace by exchanging heat with the hot tin from the furnace.

- Crystallized aluminum was to be freed of entrained tin.

- Aluminum alloy crystals were then to be melted and the residual tin was to be reduced to 50 ppm through the use of sodium extraction.

Estimates for the capital cost of the carbothermic plant (excluding power plant as an option to recover energy in CO) were $3340/annual tonne Al compared with $3210 for Hall-Heroult. The operating production costs for this process were estimated to be 15% higher than the Hall-Heroult process.

While the net unit energy requirements via the carbothermic route were estimated to be 64% of those by conventional electrolytic, there were reservations expressed about the reliability of the regenerator in which tin from the sodium extraction unit was recycled. In the innovative heat exchanger, enthalpy was extracted from the Sn-Al product tapped from the submerged arc furnace at temperatures close to 2000°C by counter current flow into the recycled tin. This device would need further development and demonstration before we could consider this operable.

The Reynolds Carbothermic Reduction Process

The Reynolds carbothermic reduction process follows a thermodynamically-based model (20, 21, 22) for the pseudobinary Al_2O_3-Al_4C_3 phase diagram system shown in Figure 2. This process assumes that selected chemical reactions go to thermodynamic equilibrium and determine the composition of exit streams at different stages in the process. The reactions involved are:

$$2Al_2O_3(s) + 3C \rightleftarrows Al_4O_4C(s) + 2CO(g) \qquad R1$$

$$2Al_2O_3(l) + 9C \rightleftarrows Al_4C_3(s) + 6CO(g) \qquad R2$$

$$Al_4O_4C(l) + 6C \rightleftarrows Al_4C_3(s) + 4CO(g) \qquad R3$$

$$Al_2O_3(l) + Al_4C_3(s) \rightleftarrows 6Al(l) + 3CO(g) \qquad R4$$

$$Al_2O_3(l) + Al_4C_3(l) \rightleftarrows 6Al(l) + 3CO(g) \qquad R5$$

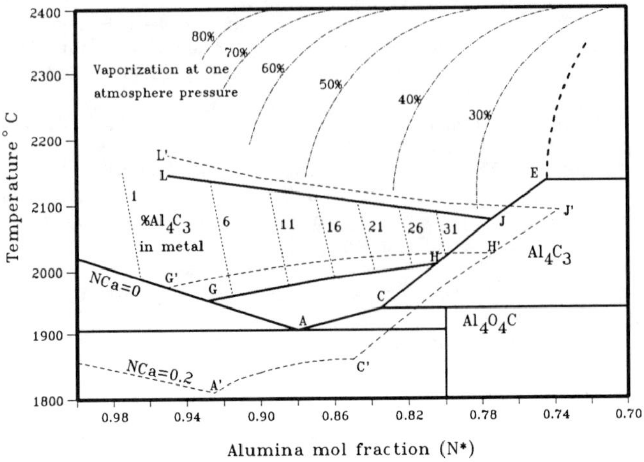

Figure 2. The Al_2O_3-Al_4C_3 Pseduo Binary System. Phase and Reaction Equilibrium and Vaporization At 1 Atmosphere

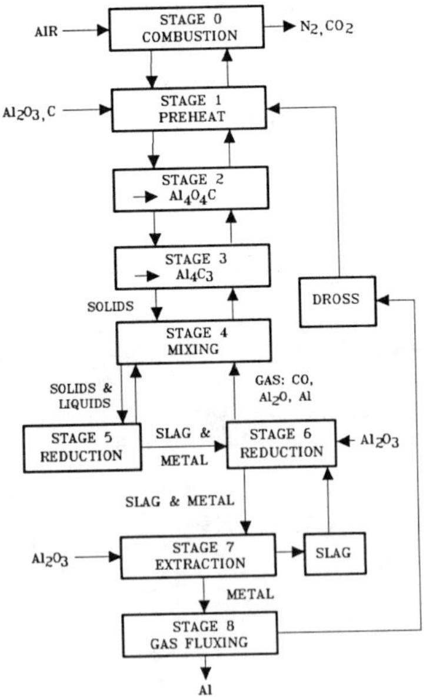

Figure 3. Flow Diagram of the Reynolds Carbothermic Reduction Process

$$Al(l) \rightleftarrows Al(g) \qquad R6$$

$$4Al(g) + Al_2O_3(l) \rightleftarrows 3Al_2O(g) \qquad R7$$

$$Al_2O_3(l) + 2C(s) \rightleftarrows Al_2O(g) + 2CO(g) \qquad R8$$

$$Al_4O_4C(l) + C(s) \rightleftarrows 2Al_2O(g) + 2CO(g) \qquad R9$$

$$Al_2O_3(l) + Al_4C_3(s) \rightleftarrows 3Al_2O(g) + 3C \qquad R10$$

$$4Al(g) + 3C \rightleftarrows Al_4C_3(s) \qquad R11$$

The carbothermic reduction of alumina as practiced by the Reynolds process involves eight stages as shown in Figure 3. Stages 0 to 3 take place in the upper regions of the furnace whereas Stages 4 to 7 take place in the hearth.

Stage 0 is the fume collection stage, Stage 1 is the charge column preheating stage, Stage 2 is the pre-reduction stage in which Al_4O_4C is produced. In Stage 3, further pre-reduction of the feed materials and the exit gases takes place that leads to the formation of Al_4C_3. Al_2O moving up the feed column forms Al_4O_4C in Stages 2 and 3, and Al_2O_3 in Stages 0 and 1.

Stage 4 is a mixing stage in which the composition of the slag is adjusted for the metal production stages by adding alumina or recycling slag. Any remaining carbon in this stage reacts forming Al_4C_3. In Stage 5, the slag and solid Al_4C_3 react yielding carbide-rich aluminum metal according to the reduction decarbonizing reaction (R4). In Stage 6, metal production continues by slag decomposition reaction R5. The last two stages have vaporization losses according to reactions R6 and R7.

Stage 7 is another decarbonizing stage in which alumina is consumed and decarbonization takes place by the extraction mode according to:

$$Al_4C_3(m) + 4Al_2O_3(slag) \rightleftarrows 3Al_4O_4C(slag)$$

In the final stage, the metal is further decarbonized by Tri-Gas fluxing, which yields two fractions: one is the metal product and the other a dross containing aluminum, carbide and slag. The dross is recycled to the first stage.

Experimental Program

Initially, the experimental work was directed to determine the feasibility of producing aluminum metal according to the process stages described above. The feasibility program was followed by laboratory evaluation of the commercial system.

Experimental Program - Feasibility Studies

The initial experimental work was conducted in 50 to 100 kw d.c. electric furnaces. Furnaces were constructed to test the concept of producing a carbide-containing metallic product in one furnace and removing the carbide in another furnace. The furnace that produced $Al-Al_4C_3$ alloy was referred to as the primary furnace. The purification furnace was called the decarbonization furnace.

The configuration of the primary furnace is shown in Figure 4. Its most salient features were: the single top entry electrode (anode); the hearth stubs or bottom electrodes (cathode); the carbon hearth and lining; and, the retaining shelf or ring also made of carbon. The shelf was used to contain the burden and control the admission of feed to the reaction zone. This furnace was operated in week-long campaigns with taps every 24 hours.

The operation was in three phases or stages: slag pool building; metal making; and decarbonization. These phases were repeated for each tap cycle and their duration was determined by the pool size, carbon content of the slag, and carbide content of the metallic product.

The feed materials were formulations developed for use in the different stages. The composition of these formulations ranged from 100% metallurgical alumina, to balancing mixtures of alumina and carbon, to 100% carbon.

The pool building state, which operated in submerged electrode mode, consisted of charging alumina and mixed formulations of carbon and alumina to produce a tetraoxycarbide (Al_4O_4C)-rich melt(sl). When this melt reached a preset volume, the metal making phase was started by feeding carbon-rich feed compositions to reduce further the Al_4O_4C and produce Al-Al_4C_3 alloy(m).

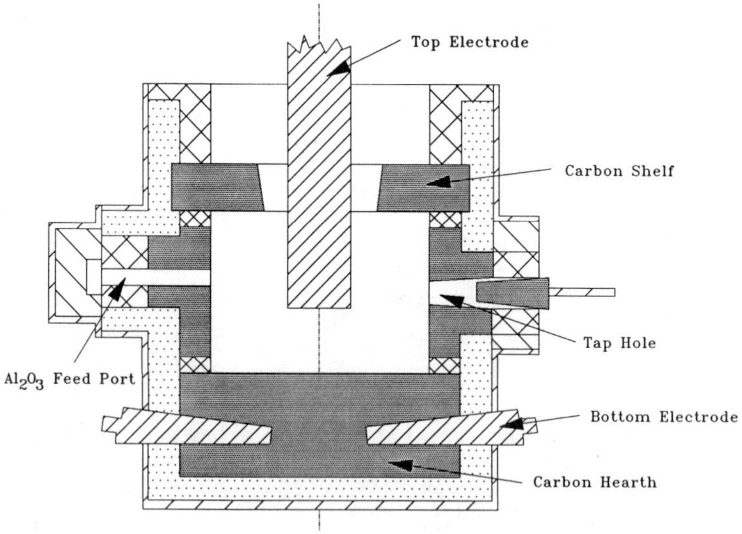

Figure 4. D.C. Primary Furnace Used in Feasibility Studies

At the beginning of the metal phase, carbon-rich feed materials were placed on top of the retainer ring and around the electrode to pre-react as well as to absorb vapors leaving the furnace. This pre-reacted burden gradually entered the reaction zone and new material was placed in the burden as its size decreased.

The presence of a metallic product during the metal making phase lowered the resistance of the pool forcing a switch in the furnace operation to open arc mode to avoid electrical shorting through the supernatant metal layer on the pool.

The decarbonization stage began when the pool contained mostly Al-Al_4C_3 alloy; at this point, feeding of alumina was started to lower the carbide content of the metal. The amount of alumina needed was determined by the charge history and estimates of the feed materials left on the burden. The furnace was tapped at the end of the decarbonization stage to recover the metal produced. The product tapped was mostly Al-Al_4C_3 alloy containing from 66 to 77% Al, 7 to 13% Al_4C_3 and minor amounts of slag.

The primary furnace product was treated in another furnace (decarbonization furnace) that contained an alumina-rich melt to further decarbonize the aluminum-carbide alloy(m). The product from this furnace contained from 1 to 3% Al_4C_3, which was purified by standard Tri-Gas fluxing techniques.

From this experimental work we learned that:

1. Metal with an average carbide content of 9% can be produced in the primary furnace by a three-stage operating sequence.

2. Carbon-rich feed shrinks the molten pool and alumina-rich feed results in high vaporization losses.

3. Feed control and maintenance of a burden is enhanced by the use of a retainer shelf or ring. This ring has to be properly isolated electrically to prevent shorting between electrodes through the furnace interior lining.

4. Proper electrode management results in good control of melt position and composition.

5. There are two decarbonization regimes, which are related to the operating temperature:

   A. Reduction decarbonization, which takes place at temperatures high enough to produce CO according to:

   $$Al_4C_3(m) + Al_2O_3(sl) \rightarrow 6Al(m) + 3CO(g)$$

   B. Extraction decarbonization, which takes place at lower temperatures without CO production according to:

   $$Al_4C_3(m) + 4Al_2O_3(sl) \rightarrow 3Al_4O_4C(sl)$$

6. The metallic product from the decarbonization furnace contained from 1 to 3% Al_4C_3, which can be purified by standard fluxing techniques.

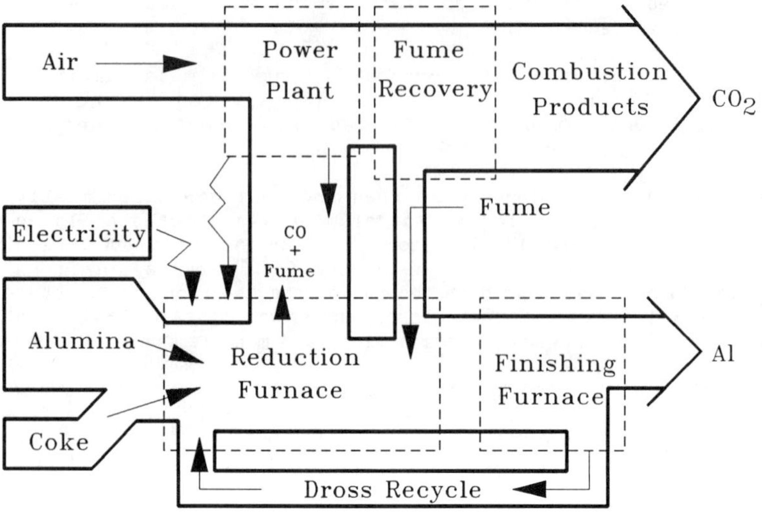

Figure 5. Conceptual Flow Diagram of Commercial Scale Carbothermic Process

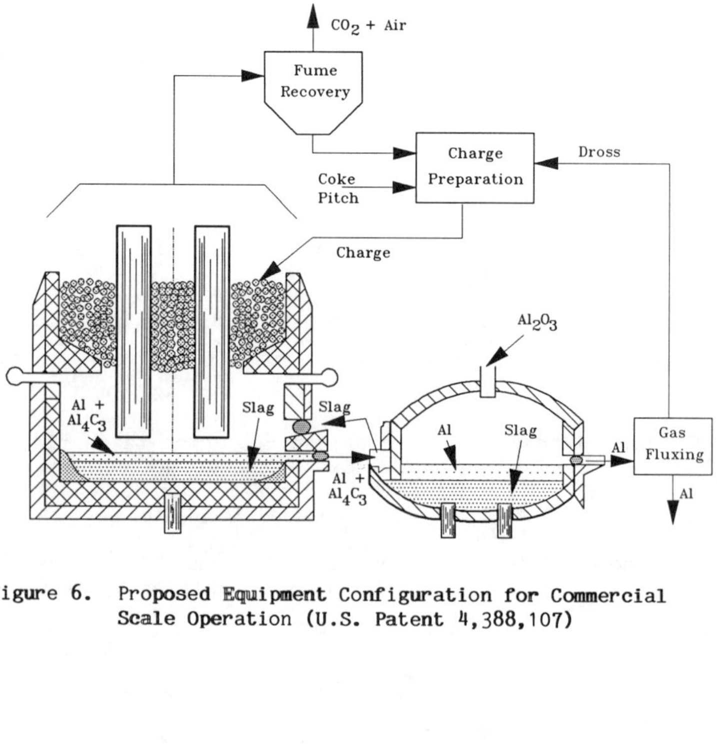

Figure 6. Proposed Equipment Configuration for Commercial Scale Operation (U.S. Patent 4,388,107)

Definition of the Commercial Process

Based on the process steps carried out in the laboratory feasibility work with d.c. furnaces, a commercial process was defined and it is presented in a simplified form in the flowsheet seen in Figure 5. The idea was to stay close to proven electrical metallurgical furnace designs to determine the extent of modifications needed to meet the process demands.

Several furnace configurations were proposed as candidates for commercial use (Patent No. 4,388,107) (20). The configuration in Figure 6 was selected as the one offering the conditions for the process to operate according to the steps outlined in Figure 3. Stages 1 through 6 take place in the Primary or Reduction Furnace. Stage 7 occurs in the Decarbonization Furnace and the slag of this stage is transferred in the molten state to the hearth of the Primary Furnace.

There were a number of issues related to power delivery and charge column management which needed to be resolved. The study was undertaken in a 200 kw furnace, but at this scale it was considered impractical to use two furnaces and transfer molten slag. The procedure was changed to facilitate the use of solid slag and alumina to the hearth of the furnace instead of molten slag.

Experimental Program - Commercial System

The furnace design used initially in this phase of the program was a direct result of the experience with the single electrode d.c. furnace. The new furnace was single phase a.c. powered, and accommodated two top entry electrodes to simulate a commercial type furnace. The power supply was rated at 200 kw with multiple operating voltage configurations.

The major obstacle in the development work with the a.c. furnace was to overcome the electrical shorting between electrodes through the charge materials used in the burden and the structural materials used to control charge feeding.

It was necessary to prevent the flow of electrical current between electrodes through undesirable paths to maintain power to the molten pool. Some of the feed formulations used in the burden became conductive as their temperature increased. At the temperatures involved in this process, refractory materials commercially available could not survive to use as electrical insulators.

Through the evolution of different furnace configurations, we were successful in overcoming these problems. By effectively isolating the electrodes by structural means and strategic management of the burden charge, it has been possible to operate the a.c. furnace for several days without electrical or structural failure. At this point, we were able to observe the impact of the electrode operating mode without perturbations from sporadic electrical shorting.

Figure 7 depicts the a.c. twin electrode furnace at the time of the latest experiment conducted. In the most productive operating campaign, the furnace was operated for a period of 10 days during which several taps were conducted and metal was recovered consistently. After reaching operating temperatures, the tap cycles averaged 13 hours.

Figure 7. A.C. Primary Furnace Used in Commercial Simulation Tests

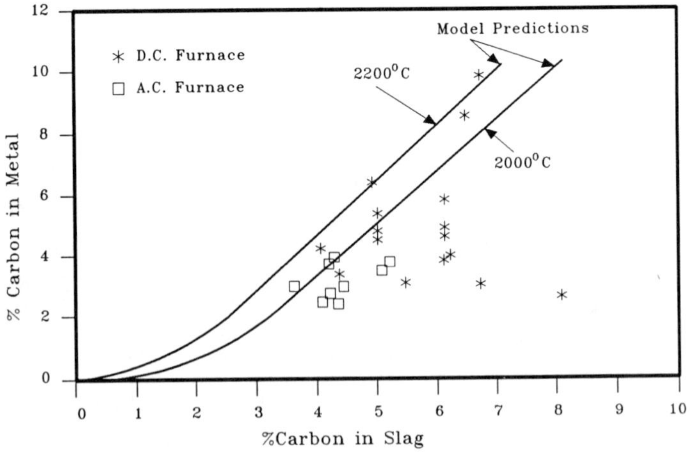

Figure 8. Comparison of Model and Experimental Results for Carbon Distribution Between Metal and Slag Phases

As in the early experimental work, the metal tapped from the primary furnace contained from 7 to 13% Al_4C_3; however, the aluminum content was generally higher in the a.c. furnace product. The aluminum content ranged from 70 to 88%, averaging 80% for all the taps conducted.

The results of model simulation studies were compared with the results of actual operations, indicating that our test results correlated very favorably with the model predictions, as shown in Figure 8. The carbide content of the metallic product tapped was lower than the model predicted in most of the taps.

The importance of the electrode operating mode was another point learned to achieve the reaction rates needed and to avoid high vaporization losses that result from sustained operation in the open arc mode.

Economic Considerations

The direction of aluminum production taken beyond the year 2000 will depend on factors such as environmental regulations, advances in materials technology, and domestic market conditions for the different aluminum producing countries.

Cochran, in his paper, "Alternate Smelting Processes for Aluminum," (7) identified the cost sectors in which there is potential for cost reduction in the present day Bayer/Hall-Heroult process and in alternate processes.

Our experience with carbothermic reduction of alumina has indicated that the process is viable in conventional electrical furnaces with modifications to deliver power efficiently and accommodate the charge columns accordingly. It appears that a carbothermic plant might require up to 30% less capital investment than an electrolytic facility of comparable capacity. Process cost comparison for electrolytic processes with the Reynolds carbothermic process were conducted for a plant with an aluminum capacity of 260,000 tons/year, Table I. The projections were based on process modeling results for the carbothermic process and available data in the open literature for the other processes.

There appears to be a cost advantage for the carbothermic process as practiced by Reynolds. These costs are for a U.S. based operation; there could be cost advantages for other processes in countries with lower energy, labor, and raw material costs. Environmental concerns could become one of the most important factors in shaping the future of the primary aluminum industry and ultimately determine which technology remains or becomes most advantageous to meet the aluminum needs of a particular country.

TABLE I. Process Cost Comparisons

Cost Element	CENTS/LB ALUMINUM Typical Electrolytic	Advanced Electrolytic	Alcoa Chloride	Reynolds Carbothermic
Power (kwhr/lb)	22.0 (7.55)	17.8 (6.1)	13.7 (4.7)	12.2 (4.2)
Carbon	10.4	10.4	10.8	10.0
Alumina	19.0	19.0	19.0	19.0
Chemicals	3.0	2.0	2.3	0.1
Labor	5.5	3.5	5.5	1.5
Depreciation	0.4	3.1	2.8	2.3
Other Costs	9.1	9.1	9.1	9.1
TOTAL	69.4	64.9	63.2	54.2

Outlook For Primary Aluminum Production

Most of the primary aluminum production at the turn of the century will come from 1960-1980 era technology Hall-Heroult cells. A small fraction of the production will be from larger, more efficient Hall-Heroult cells that are now being developed. The evolution of the process will continue through the first decades of the next century with emphasis on automation and decreasing the impact on the environment. Most of the new primary capacity will continue to be built in areas of the world with ample supplies of low cost power.

Emphasis on recycling and reclamation will play an increasingly important role in the supply of aluminum, particularly in developed countries. Processes to sort and purify this recycled metal will become more important and will supplant an increasing volume of primary capacity.

Environmental and energy concerns will cause continuing interest in inert anode and cathode development and their incorporation into bipolar cells. Successful advances in materials could lead to commercial installations during the first two decades of the next century.

Other alternative processes have a potential for capital investment savings. However, the technical problems remaining in their development and the plentiful energy and raw material resources of the present day favors the Hall-Heroult process well beyond the year 2000.

At this time, it is not foreseen that any carbothermic process will be in commercial operation in the early 2000's, unless world-wide events, such as another energy crisis, put pressure on industrialized nations to seek alternative processes. Under these circumstances, it is not likely that commercial scale facilities will be ready before the year 2010 since substantial engineering and development work remains today to achieve a fully commercial scale carbothermic process for aluminum production.

References

1. Hall-Heroult Centennial - First Century of Aluminum Process Technology, Edited by W. S. Peterson and R. E. Miller, The Metallurgical Society, Inc., Warrendale, PA, 1986.

2. Production of Aluminum and Alumina, Edited by A. R. Burkin, John Wiley & Sons, New York, 1987.

3. The Aluminum Statistical Review, Aluminum Association, Inc., Washington, D.C., 1982.

4. B. Langon and P. Varin, "Aluminum Pechiney 280 kA Pots," Light Metals 1986, Edited by R. E. Miller, The Metallurgical Society, Inc., Warrendale, PA, 1986, pp. 343-347.

5. J. Chaffy, and B. Langon, "Circuit for the Electrical Connection of Rows of Electrolysis Cells for the Production of Aluminum at Very High Current," U.S. Patent 4,696,730, Sept. 29, 1987

6. G. de Varda and A. Vajna de Pava, "Primary Aluminum Production in a Multicell Electrolytic Furnace," 96th AIME Annual Meeting, Los Angeles, Feb. 19-23, 1967.

7. C. N. Cochran, "Alternate Smelting Processes for Aluminum," Light Metals 1987, Edited by R. D. Zabreznik, The Metallurgical Society, Inc., Warrendale, PA, 1987, pp. 429-443.

8. B. Marincek, "Apparatus for the Electrolysis of Molten Electrolytes," U.K. Patent 1,152,124, Assigned to Swiss Aluminum Ltd., May 14, 1969.

9. W. Schmidt, "Thermal Reduction," U.S. Patent 3,257,199, June 21, 1966.

10. G. C. Robinson, et al., "Centrifugal Separation," U.S. Patent 3,374,059, March 10, 1968.

11. S. Selmer-Olsen, "The Silicon Metal Process, (A Thermodynamical Study)," Electrical Furnace Proceedings, Vol. 39, The Iron and Steel Society of AIME, Warrendale, PA, 1981, pp. 310-318.

12. M. J. Bruno, "Overview of Alcoa Direct Reduction Process Technology," Light Metals 1984, Edited by J. P. McGeer, The Metallurgical Society of AIME, Warrendale, PA, 1984, pp. 1571-1612.

13. S. K. Das, C N. Cochran, R. Milita, W. Hill, R. Mazgaga, "Aluminum Purification," U.S. Patent 4,115,215, Sept. 19, 1978.

14. K. Kuwahara,"Method of Carbothermically Producing Aluminum," U.S. Patent 4,394,167, July 19, 1983.

15. M. Dokiya, M. Fujishige, T. Kameyama, H. Yokokawa, S. Ujiie, K. Fukuda, and A. Motoe, "Aluminum Blast Furnace," Light Metals 1983, Edited by E. M. Adkins, The Metallurgical Society of AIME, Warrendale, PA, 1982, pp. 651-670.

16. H. Yokokawa, M. Fujishige, S. Ujiie, and M. Dokiya, "Phase Relations Associated with the Aluminum Blast Furnace: Aluminum Oxycarbide Melts and Al-C-x(x=Fe, Si) Liquid Alloys," Metallurgical Transactions, Vol. 18B, No. 2, June 1987, pp. 433-444.

17. E. W. Dewing, "Direct Reduction Process for Production of Aluminum," U.S. Patent 3,836,357, Sept. 17. 1974.

18. "Techno-Economic Assessment of a Carbothermic Alumina Reduction Process," A. D. Little, Inc., ADL Ref. #61532, July 1988.

19. J. F. Elliott and C. W. Finn, "Analysis of Solvent Carbothermic Smelting of Alumina," 117th TMS Annual Meeting, Phoenix, Jan. 25-28, 1988.

20. R. M. Kibby, "Minimum Energy Process for Carbothermic Reduction of Alumina," U.S. Patent 4,388,107, June 14, 1988.

21. L. M. Ruch, A. F. Saavedra, R. M. Kibby, "Carbon Partition Between Carbothermically Reduced Al and Al_2O_3-Al_4C_3 Melts," Light Metals 1984, Edited by J. P. McGeer, The Metallurgical Society of AIME, Warrendale, PA, 1984, pp. 589-599.

22. R. M. Kibby, A. F. Saavedra, "Mathematical Models for the Study of Carbothermic Reduction of Alumina," Mathematical Modelling of Materials Processing Operations, Edited by J. Szekely, L. B. Hales, H. Henein, N. Jarrett, K. Rajamani, and I. Samarsekera, The Metallurgical Society, Inc., Warrendale, PA, 1987, pp. 627-642.

FUTURE TECHNOLOGICAL DEVELOPMENTS FOR ALUMINIUM SM

by

Kai Grjotheim, Department of Chemistry, University of Oslo, Norway,

and

Barry Welch, Department of Chemical & Materials Engineering, University of Auckland, New Zealand

Abstract

The capital intensive nature of the present process for aluminium smelting, the low productivity per unit reactor, and pressure to reduce the electrical energy demand have motivated the search for an alternative process. Optional routes include carbo-thermal reduction of alumina (using an alloy phase immediate), chlorination followed by electrolysis of aluminium chloride, and electrolytic decomposition of alumina using inert electrodes. Materials performance and reactor design constraints have been a major limitation for alternatives. However parallel research has led to considerable advances in the process efficiencies and scale of both the Bayer process and the Hall-Heroult cells.

Future development of alternatives is dependent on breakthroughs in Materials Science such as the development of stable conducting anodes and stable wettable cathode materials. However if successful, it is probable that some spin-offs will be retrofitted to existing cells. Therefore the basic Hall-Heroult technology will continue as the dominant process for at least the next half century.

Metallurgical Processes for the Year 2000 and Beyond
Edited by H.Y. Sohn and E.S. Geskin
The Minerals, Metals & Materials Society, 1988

Introduction

The present Bayer/Hall-Heroult process for production of aluminium is just over 100 years old. Frequently, during the century of its technological development questions have been raised as to whether it is the ideal route for metal production. The questioning was particularly intense in the period between the mid 1950s and the mid 1060s when the limitations of the Hall-Heroult electrolytic stage were evident. Then the cells were only 30 to 35% efficient with respect to the utilisation of the hydro electricity (requiring 16 - 18 kWh/kg, thus making it strongly dependent on cheap electrical energy), while the size of cells being installed were 80 to 130 kA. Hence the process was capital intensive, the productivity per reactor volume low, and the process was labour intensive (requiring about 15 man hours per tonne of aluminium). In addition, the use of pitches for Soderberg anodes and the fluoride electrolyte were creating problems through emission. At that time material science had paid little attention to the manufacture and use of carbonaceous materials or the development of alternatives that were corrosion resistant to cryolite or liquid aluminium. Hence cell lives were short - typically 600 to 1000 days - thus adding to operating costs. Thus, the grass over the fence, as foreseen in 1960s vintage crystal balls, appeared much greener for alternative processes.

In the last quarter century many millions of dollars have been expended investigating alternative processes as well as on parallel R & D to upgrade the existing Hall-Heroult technology by solving problems and improve process economics. Therefore many of the arguments in favour of alternative processes are different to those of the 1960s. However the experience gained in all the R & D in the last quarter of a century enables one to objectively assess the likely trends for the future.

The Production Cycle

Virtually all applications of aluminium are dependent on the product being a high quality metal (>99.8%) with the absence of specific contaminants. The true starting point can however be varied as a diverse range of mineralogical oxides exist. All the process routes tested are governed by the overall thermodynamic cycle, illustrated in Figure 1.

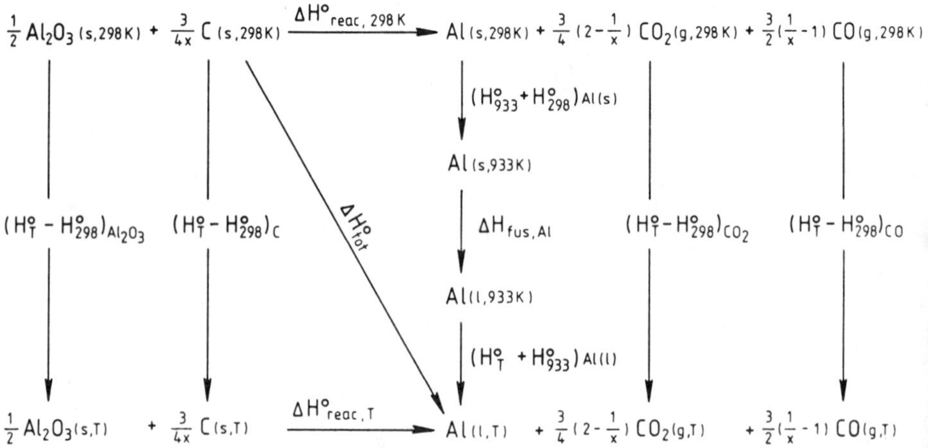

Figure 1 - The Thermodynamic Cycle for Production of Aluminium

The starting point is always an impure oxide mineral, the finishing point the pure metal, and during the transformation the oxide ions of the alumina are usually combined with carbon to form carbon oxides. Processes can differ as to when the purification stage is introduced, and how the associated mineralogical impurities are handled. It is most convenient to consider these options by reference against key features of the present process.

Bayer Purification

By using bauxite, which is an abundant alumina-rich mineral, separation of the alumina from the other oxides can be effected at a high level of purity. The resulting product, however, represents approximately 40% of the cost for metal production, and subsequently imposes purity restrictions on carbon used as electrodes in the cells and other raw materials used. Obviously one processing alternative is to eliminate the Bayer process and consider purification of the metal at the final stage.

Carbon

The thermodynamic cycle involves carbon and for the traditional process it must be of high purity. Therefore petroleum or coal tar pitch coke is used. Because of their high quality, the production of the consumable electrodes is performed carefully and contributes significantly (10 - 20%) to the metal production cost. If the purification were carried out with the metal, potentially carbons from coal or other cheap sources could be used.

Capital and labour

Based on the 1960s technology level, the productivity per unit reactor volume was less than 10% of the comparable productivity of a blast furnace reactor. The cell design was relatively complex and involved anode changing for carbon regeneration. Only a limited amount of metal could be removed from each cell daily (typically 0.6 - 0.9 tonnes). Obviously it was desirable to increase the productivity per unit volume and lower the man hours. This could be addressed by either increasing the size of cells (as will be demonstrated and has been successfully achieved) or going to a totally different design or process.

Process temperature and reaction balance

The present cells have a maximum temperature of approximately 1000°C and the by-product is predominantly carbon dioxide although based on the thermodynamic cycle presented in Figure 1, we can generate a mixture of carbon monoxide and carbon dioxide. The gas balance depends on variations to the proportions of reaction according to equations (1) and (3):

$$2Al_2O_3 + 3C \rightarrow 3CO_2 + 4Al \quad (1)$$

$$\text{where } \Delta G°_T = 1679.9 - 0.433T \quad (2)$$

$$\text{and } 2Al_2O_3 + 6C \rightarrow 6CO + 4Al \quad (3)$$

$$\text{where } \Delta G°_T = 2529 - 0.682T \quad (4)$$

and T is in K, and $\Delta G°_T$ in kJ.

For reaction (1) to dominate, provision of energy (electrically) is always required, while for reaction (3) to proceed without additional electrical energy, temperatures in excess of 2000°C are required. Thus, adjusting process temperature can adjust the proportion of electrical energy that is being added. It results in changing the proportions of carbon monoxide and carbon dioxide generated and, hence, the amount of carbon required. Simply the Hall-Heroult cell maximises the amount of CO_2 at a given temperature by using electrical energy. Other process routes can be devised varying these proportions as is discussed below.

Electrical energy

In the 1960s its utilisation was only approximately 30% energy efficient. If the source of the electrical energy is from combustion of coal, and this inefficiency is coupled with the 33% efficiency for the power generation, the use of electrical energy then compares extremely unfavourably with conventional blast furnaces processes. For example, an iron blast furnace is approximately 80% energy efficient. In the 1960s, most of the electrical energy was generated from hydro sources, but today there is an increasing proportion generated thermally. However, in the intervening time, the energy efficiency of the present smelting cells has been lifted to between 45 and 50% for the latest technology and much of the capacity expansion has taken place in regions where abundant hydro (or byproduct) electricity is generated. Despite these impressive achievements and trends, reducing electrical energy demand is clearly a processing option. This can be achieved by using other aluminium compounds as the intermediate. For example, converting the oxide through carbo chlorination to aluminium chloride - a process that can be carried out thermally - and then electrowinning the aluminium from the chloride. Such a route does not alter the overall cycle, but readily enables proportions of reactions and electrical energy demands to be changed.

Technological change

From the various comments made, it is obvious that a diverse range of approaches are theoretically possible. One can focus attention on incremental changes to the existing process, (making each stage more efficient), or alternatively develop a new process, addressing one or more of the options outlined. All routes have been tried.

Advances in Hall-Heroult Technology

As demonstrated in Table I, there has been a dramatic change in cell design and performance since 1960. Several factors have contributed to these advances, most of which revolve around a better understanding of the chemical and electrochemical processes, and consequential improvements to engineering design, frequently through modelling. Major areas of achievement have included:

- Improvements to process control.
- Engineering design changes for alumina conveying and feeding.
- Improvements to materials for cell construction.
- Optimisation of anode manufacture.
- Mathematical modelling for magnetic and thermal design.
- Refinements to operating procedures and electrolytes used.

The accessibility of computers has also obviously played an important part.

TABLE I: CHANGES IN PREBAKE CELL TECHNOLOGY

Era/Period	1955-1965	1980's	Achievable
Cell Type	CWPB & SWPB	PFPB	PFPB
Cell Size (kA)	80-140	175-280	>280
Current Efficiency (%)	85-89	93-95	95
Electrical Energy (DC.kWh/kg)	15.2-17	12.9-13.2	<12.3
Man Hours/Tonne Al.	10-15	3.5-5.8	<2.5
Net Carbon (kg/kg Al.)	0.46-0.49	0.40-0.43	<0.40
Age Cell Life (months)	26-30	65-85	>85

CWPB = centre worked (fed) prebake anode cell
SWPB = side fed prebake anode cell PFPB = point fed prebake anode cell

Process control

The only signals available for process control are cell voltage and line current and since these interact and fluctuate, fast data acquisition and intelligent interpretation has been necessary for sensible process control. It was theoretically shown in 1965 (1) that, for fixed current density and inter-electrode distance, the cell voltage passed through a minimum at an alumina concentration between 3 and 5%. The absolute voltage and concentration of the minimum depended on other cell conditions. This relationship and the rate of change in a voltage with alumina has subsequently been used as the basis for process control. Rather than use voltage, it is usually converted to a pseudo resistance (2, 3) so that fluctuations in current are compensated for.

Figure 2 compares the shape of two resistance v's alumina concentration curves (for different inter-electrode distances) with the real behaviour data in another cell design. It demonstrates how the absolute value of resistance, and rate of change in resistance can be used to optimise the inter-electrode distance and alumina concentration once the actual characteristics are known.

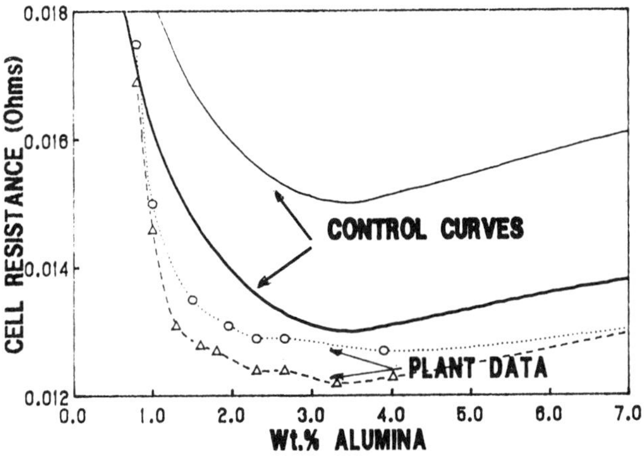

Figure 2 - Typical smelter cell resistance versus alumina profiles at different inter-electrode spacings.

Alumina feeders

Successful implementation of "resistance tracking" control strategy necessitated development of a feeder that would add alumina in small, but repeatable, amounts. Over the last 20 years several point feeder designs have been developed (4, 5), one of which is illustrated in Figure 3. This feeding unit has a pneumatically operated crust breaker that pierces a hole in the crust to give direct access to a measured volume of alumina that is discharged from the feeder section. Today these feeders can add as little as 0.5 kg of alumina per action and operate at relatively short intervals, either on a pre-programmed basis or on resistance demand.

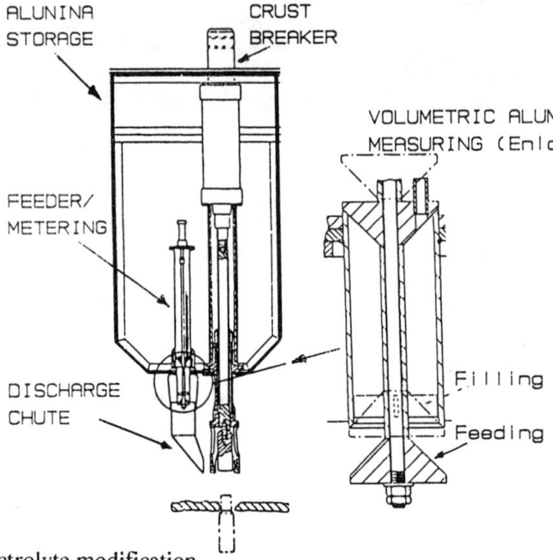

Figure 3 - A modern point feeder for alumina additions (ref. 2.)

Electrolyte modification

With the ability to control alumina concentrations in a narrow band, knowledge of the physico-chemical properties of the electrolyte can be applied so that the Faradaic efficiencies are maximised. The electrolyte is usually modified by addition of or aluminium fluoride plus calcium fluoride and sometimes lithium fluoride to lower the melting point and metal solubility (6, 7). Unfortunately, in all cases the saturation solubility of alumina is also reduced. Use of high aluminium fluoride concentrations is finding greater acceptability in modern cells, but this has only become possible through the development of efficient fume scrubbing systems that recycle the sodium tetrafluoraluminate vapours and hydrogen fluoride emitted from the cell (8). Although the capital cost of emission scrubbing systems is high, operating cost are readily compensated for, through the recovered fluoride value.

Modelling

Increasing line current from the approximately 100 kA level to values approaching 300 kA has only been possible through astute compensation for the magnetic forces that otherwise hydrodynamically distort the metal pad and generate abnormal flow patterns (9). Mathematical modelling has played an important role here, typical ensuring the variation the height of the metal does not exceed 2 cm across most of the cell, whilst lowering the metal velocities to below 10 cm/sec. Mathematical modelling has also played an important role in improving design for optimum heat balance control and mass transfer and distribution of the dissolved alumina (10).

Costs and Productivity per cell

The improved process control, modified bath chemistry and upgraded operating strategies have been rerofittable to existing installed capacity. Therefore smelters have been able to operate within a narrower band of variance. Consequently many smelters, have increased their operating line current with the appropriate increase in cell productivity. The changes in Faradaic efficiency and cell productivity for two smelters which have introduced incremental changes is illustrated in Figure 4.

While the capital costs per annual tonne have been reduced by increasing cell productivity, another major contribution has been the dramatic (approximately three-fold) increase in cell life (11, 12). This has been as a consequence of greater attention being given to the material science of carbons, and selection of refractories used to construct the cathodes, coupled with the understanding of the mechanisms of cathode degradation and energy dissipation (13, 14). It is probable further advances will be made in this area, since it is only in the present decade that a true understanding has been gained for the concurrent importance of heat balance and heat transfer within cells (10, 15). Consequently, today's cathodes are complex in construction, but it is likely that 100 months will become the new standard for life expectancy because of the consolidation in knowledge of material science that is resulting.

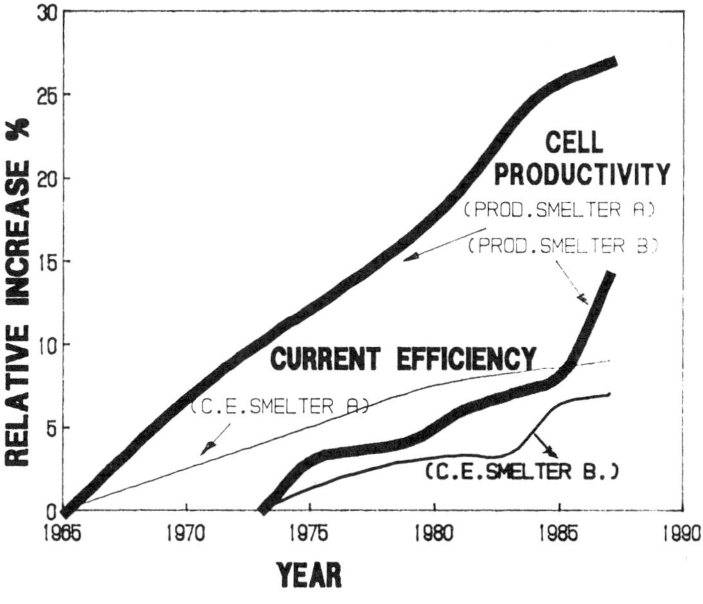

Figure 4 - Typical improvements in performance for existing smelters.

Future Variations to Hall-Heroult Technology

The achievable performance levels cited in Table I demonstrate that the cell size is large and energy efficiencies are approaching 50%. Increasing the cell size further is one of three areas of development currently disclosed (16). Without introducing any additional knowledge, but simply using advanced modelling techniques, construction of these new cells is feasible with limited innovation. These new (400 to 500 kA) cells are likely to lead to performance levels at least comparable to the best technology today. They have the obvious advantage that labour costs are lowered and productivity per unit reactor cost and volume is increased. Large cells, however, suffer from the major step in incremental capacity of the metal - a full potline of 400,000 ampere cells would increase the world capacity by more than 5%.

Two other areas for potential further advancement are the introduction of drained cathodes and the use of non-consumable anodes. The driving force for the drained cathode concept is to give greater design flexibility by eliminating magnetically-induced movement of the metal pad. High current efficiencies and shorter inter-electrode distances (and hence lower voltages) would result. Titanium diboride, and various composites containing it, have shown the greatest promise as material for construction of a wettable cathode. However, the twenty years of research and development on these materials attests to the difficult materials problem with which any process for the production of aluminium is faced. Not only is the metal reactive, but the electrolyte is also extremely corrosive. Present indications are that the use of composites for drained cathodes is on the horizon (17, 18), but full testing, design, and evaluation will take at least another decade before widespread application if the materials prove to be successful.

Contrasting, the development of a non-consumable anode is lagging further behind (19). While high temperature conducting oxide electrode materials can be formulated, the fabricated shapes generally have high resistance (and therefore are energy inefficient). Service life has been disappointing as they soon become deactivated through corrosion or migration of ions. In any case, as discussed below, if the non-consumable anodes are developed, the only resemblance cells would then have to the existing Hall-Heroult technology is the use of a common electrolyte.

Carbothermic Reduction Options

As given in more detail elsewhere (20-22), direct carbothermic processing attacks two of the largest cost factors, namely energy and capital cost, providing the design concept is similar to efficient iron making furnaces. The thermodynamic data given in equations 2 and 4 indicate extremely high temperatures will be required, with the probable need for supplementary electrical heating. Reaction between alumina and carbon is much more complex (22) than given by these reactions and begins near 1950°C. Initially a liquid aluminium oxycarbide solution is formed (20) according to the following simplified reaction:

$$2Al_2O_3 + 9C \rightarrow Al_4C_3 \text{ (in } Al_2O_3\text{)} + 6CO \qquad (5)$$

As the temperature is increased above 2050°C, reaction between the aluminium carbide and alumina (of the oxycarbide) will take place in the absence of carbon, forming a new liquid phase of aluminium containing aluminium carbide and simultaneously evolving carbon monoxide.

$$Al_4C_3 + Al_2O_3 \rightarrow 6Al \text{ (in } Al_4C_3\text{)} + 3CO \qquad (6)$$

Liquid phases are crucial to operation, but the formation of two separate liquids necessitates a very delicate balance in temperature versus reaction and conversion. Also, cooling the aluminium rich aluminium carbide phase necessitates a fluxing agent for effective coalescence. The process is hindered by vapourisation of aluminium and aluminium sub oxide (Al_2O) as the carbon monoxide is simultaneously evolved. These back react, not only decreasing the yields dramatically, but also generating operating problems by forming bridging refractory layers in other zones.

Most impurities present would be reduced preferentially to the alumina because of their less negative Gibbs energy of formation.

Carbothermic reduction of unrefined ores to form an alloy with subsequent aluminium purification is an alternative. Thermodynamically reductions would be marginally easier and yields fractionally higher (because of alloying). Practically the process still suffers from similar problems that carbothermal reduction of alumina has. Reaction yields have always been low, bridging problems from reflux metal and sub oxide vapours hamper operations. High temperatures are still needed and electrical heating is still necessary to achieve the temperatures. This approach also has difficult subsequent metal purification stages as well as producing an imbalance of alloy products.

Based on these factors, carbothermal reductions are not currently foreseen as being a competitive alternative for several decades - if at all.

Carbothermal plus Electrochemical Options

As noted in the preceding discussion, various aluminium compounds can be produced from alumina via carbothermal reaction and these products are used as intermediates to subsequently produce aluminium electrolytically. These processing schemes always involve a certain proportion of electrochemical energy and can be represented by the following two general equations:

$$Al_2O_3 + 3C + 2X \rightarrow 2AlX + 3CO \quad (7)$$

$$\text{or } 2Al_2O_3 + 3C + 4X \rightarrow 4AlX + 3CO_2 \quad (7a)$$

where X represents the appropriate amount of chlorine, nitrogen, sulphur or similar element. The aluminium-containing species AlX is then electrochemically decomposed according to the following reaction:

$$AlX \rightarrow Al + X \quad (8)$$

Obviously the latter reaction needs an appropriate ionically conducting solvent for electrolysis to take place. Generally, the higher the temperature that is required to produce the compound AlX, the lower will be the amount of electrochemical energy theoretically required for the electrolysis reaction but a greater amount of carbon will be needed. For example, aluminium chloride and carbon dioxide are readily produced at temperatures above 600°C (thermodynamic plus kinetic considerations) and the electrolysis of aluminium chloride subsequently requires about two volts. Contrasting, the nitrides and sulphide formation requires temperatures well above 1000°C for production, the by-product being pure CO, but subsequently less than 1 volt is theoretically required for electrolysis.

Practically, it has proven difficult to form aluminium nitride or aluminium sulphide according to appropriate reaction similar to Equation (7). In the case of the nitride, reactants and products are solid and therefore rates become controlled by diffusion. Contrasting, aluminium chloride formation can be successfully carried out and at temperatures above 600°C with the formation of carbon dioxide as the by-product, as described by Equation (9), thus reducing the amount of carbon required:

$$Al_2O_3 + 3/2C + 3Cl_2 \rightarrow 2AlCl_3 + 3/2CO_2 \qquad (9)$$

This reaction is aided by volatilisation of the aluminium chloride. Production of aluminium via equation (9), and the following electrolytic decomposition reaction

$$2AlCl_3 \rightarrow 2Al + 3Cl_2 \qquad (10)$$

formed the basis of what became known as the Alcoa Smelting Process. This process showed considerable potential at the bench scale level and was tested at a substantial pilot plant scale. Although it has proven to be uneconomic - particularly because of difficulties in producing aluminium chloride of the purity required while conforming to stringent environmental standards - it has introduced some interesting processing design concepts.

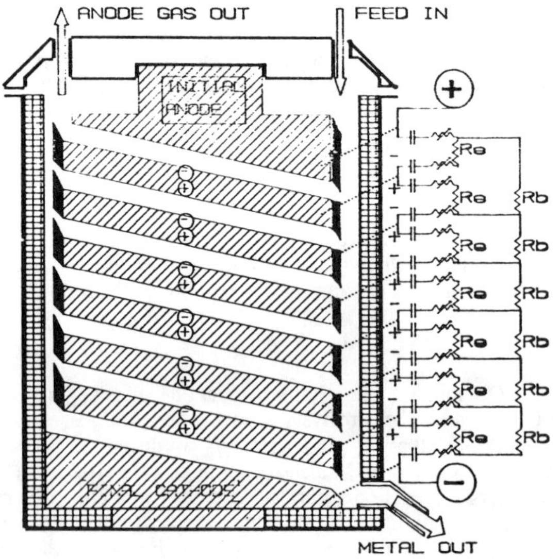

Figure 5 - Schematic of a bipolar cell and the equivalent circuit showing current by-pass.

One of the most attractive features of the Alcoa Smelting Process was the potential for using a bipolar cell design for the electrolysis. A bipolar cell design (illustrated in Figure 5) is appropriate whenever the metallic product is liquid, the anode is not consumed, and the anode product is a separate phase from the liquid metal. In the case of the Alcoa Smelting Process, carefully machined graphite slabs were used as the bipolar cells. As illustrated in Figure 5, a number of these slabs (6 illustrated) are placed between the normal anode and cathode of the cell. Each face of these slabs act as electrodes of electrolytic cells in series. Thus the capital cost for the increasing number of electrode pairs is minimal and the productivity per unit volume can be increased substantially. Because the electrodes are non-consumable, they can be designed for optimum gas release (hence being inclined at a small angle, as well as being grooved) and positioned at the optimum inter-electrode spacing at all times.

Practically, a bipolar cell must have a ready supply of fresh reactants between each electrode plate, as well as allowing ready removal of the gaseous anode product (chlorine, which is recycled) and the liquid metal. However, the flow of electrolyte around the ends of the electrode plates permits a bypass current path, thus lowering the Faradaic efficiency below that expected. The equivalent circuit is also presented in Figure 5. A good design ensures the electrical resistance of the electrolyte (R_e) is much less than the bypass resistance around the ends of the electrode plate (R_b). This goal can be aided by use of appropriate end-plate for aiding the flow of electrolyte while increasing resistance. Practically, current efficiencies seldom exceed 70% in bipolar cells. Despite this, the Alcoa Smelting Process was claimed to be able to produce metal while using less than 12 kilowatts of electricity per kilogram of metal. Although carbon efficiencies were similar to those of the conventional Hall-Heroult cell in the chlorination stage, the saving in capital costs achieved by the cell design are countered by the extra processing stages for producing the high purity chloride.

An interesting variation that has been proposed has been to carry out the reaction according to Equations (9) and (10) simultaneously in an electrolytic cell (23). This necessiates making a composite anode of alumina and carbon in the appropriate stoichiometric proportions and using an aluminium chloride based electrolyte. Practically, the resistivity of the composite anode and its structural integrity have hindered this development. Also, this approach does not have the advantage of non-consumable anodes, while additional costs are incurred in producing the composite anode.

Total Electrolytic Approach

This is going from the extreme of total carbothermal reduction through the proportionate chemical-electrochemical systems back to a totally electrochemical process that does not use carbon. It is feasible if inherent design limitations and capital costs can be overcome. The specific process considered has been to electrolyse alumina dissolved in cryolite (essentially the conventional electrolyte) but performing the electrolysis in a bipolar cell using a non-consumable anode and a wettable cathode. Schematically, the overall process is as illustrated in Figure 6, with electrolysis proceeding according to the reaction:

$$2Al_2O_3 \rightarrow 4Al + 3O_2 \qquad (11)$$

It is seen here that environmentally acceptable oxygen is formed, eliminating the need for carbon.

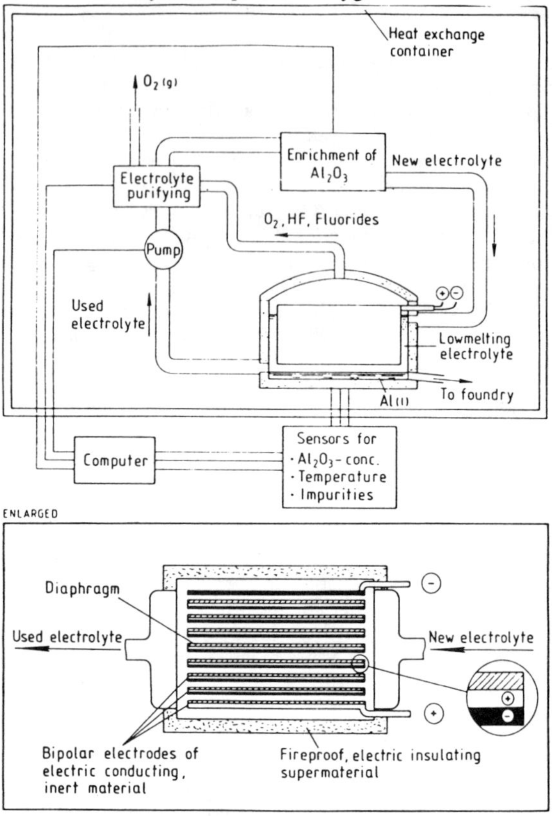

Figure 6 - Idealized arrangement for electrolysis of alumina using a bipolar cell.
Upper: Overall arrangement for optimum operation.
Lower: Concepts of an efficient bipolar cell design.

The evolution of oxygen is at the expense of increasing the cell voltage by approximately 1 volt. Contrasting however, if the simplicity of bipolar cell designs can be utilised, voltage savings can be made by eliminating ohmic effects associated with ancillary circuitry. Optimistically it may be hypothesised that the oxygen over-voltage will be significantly lower than that for evolving carbon dioxide.

Hitherto, this approach has been limited by the failure to make suitable electrode materials for both the non-consumable anode and wettable cathodes. While the titanium diboride based cathode materials are being advanced further, much more research is required for the non-consumable oxide anode. Even when this is achieved successfully, there are still materials jointing problems in order to make the bipolar plates of the two structurally different materials.

It is obvious the most attractive feature of this cell would be to reduce labour costs and increase productivity - it would not have a significant effect on the electrical energy demands for metal production. It is also obvious that success is some time away.

Summary of Process Options and the Future

No alternative process has been demonstrated that approaches economic competitiveness with the present Hall-Heroult process - especially since the latter still shows potential for substantial improvements. Electrical energy is required in all options (including carbothermal processing to heat and achieve the very high temperatures involved) and the various alternatives either involve more processing stages or are much less efficient with respect to the chemical reactions.

Contrasting, steady advancements are continuing to be made in the conventional Hall-Heroult process and future design options appear to offer viable ways of reducing the present limitations. While the energy efficiency will continue to be less than conventional carbothermal processing, it is a process that is suitable for locations where surplus hydro-capacity potentially exists. It is interesting to note that in the last two decades most of the new smelters have been built in area where there is the abundance of hydro electricity but it has no other alternative use. For example, Venezuela, Brazil, Indonesia are recent growth areas. Thus the inefficiencies associated with converting coal to electricity are not applicable.

The most likely development in the short term is better cathode materials which will enable the drained concept. This would be retrofittable to existing cells, making them more economic and competitive, thus further distancing the economic gains that need to be achieved by alternative processes.

In developing new process options in the future, it is obvious that formation of gaseous product streams (as experienced in carbothermal reductions) is undesirable because of back reaction. It is also obvious that any new technology involving electrolytic processes should utilise the bipolar concept so that high current densities, high productivities and low capital costs can be achieved.

With all the obstacles that still need to be overcome, and a time frame for development and research, it is clear that we will be seeing Hall-Heroult based cells being constructed for some time to come. Material science breakthroughs are required if either minor or major variations from this concept are to be achieved. In any case, much of the new developments will utilise the electrochemical and physicochemical knowledge already gained for the electrolyte and electrolysis of alumina dissolved in cryolite. We will however, see bigger cells that are more automated and meet even higher standards environmentally.

Acknowledgements

The authors wish to record their appreciation to many people in industry for providing technical material used here and also as background information. We also wish to record our appreciation to the Royal Norwegian Council for Scientific and Industrial Research (NTNF) for financial support that enabled the preparation of the material for this manuscript. Assistance from Comalco Research Centre is also greatly appreciated.

REFERENCES

1. B.J. Welch, "A voltage analysis under conditions of varying alumina concentration", pp.1-19, in Aus. IM&M Proceedings No. 214 (1965).

2. P. Bonny, J-L Gerphagnon, G. Laboure, M. Keinborg, P. Homsi & B. Langon, U.S. patent No. 4,431,491, Feb. 1984.

3. T. Moen, J. Aalbu & P. Borg "Adaptive control of alumina reduction cells with point feeders", pp.459-469 in AIME Light Metals (1985).

4. S. Casdas. U.S. Patent No. 4,435,255 March 6, (1984).

5. R.L. Lowe, U.S. Patent 3,681,229 Aug 1 (1972).

6. G.L. Bullard, & D.O. Przybycien, "D.T.A. determinations of bath liquidus temperatures : Effect of LiF", pp.437-444 in AIME Light Metals (1986).

7. R. Odegard, A. Sterten & J. Thonstad, "The Solubility of Aluminium in Cryolitic Melts", pp.389-398 in AIME Light Metals (1987).

8. E. Keul, "Emission Control Systems" Proc. N.H.T. Aluminium Smelter Technology Course (eds. H. Φye. & D. Bratland), Trondheim Norway (1987).

9. R. Huglen, "Influence of Magnetic Fields", pp.25-62 in "Understanding the Hall-Heroult Process for production of Aluminium" (eds. K. Grjotheim & H. Kvande), Al-Verlag publ. Dusseldorf, (1986).

10. M.P. Taylor, B.J. Welch & M. O'Sullivan in Chemeca 83 - Proc 11th Aust. Chem. Eng. Conf. Brisbane (1983) pp.493-500.

11. K. Etzel, F. Brandmair, P. Aeschbach & H. Friedli, "Gluing of Cathode and Anodes - A proven technology", pp.885-896 in AIME Light Metals (1983).

12. B.J. Welch "Aluminium Reduction Technology - some Specific Advances", J. Metals - Nov. (1988).

13. K. Grjotheim & B.J. Welch, "Aluminium Smelting Technology - a pure and applied approach", ch.5 pp.119-153, (Al-Verlag Dusseldorf) 1988.

14. C. Krohn, M. Sorlie & H.A. Φye, "Penetration of Sodium and bath constituents into cathode carbon materials used in Industrial cells", p.311-324 in AIME Light Metals (1982).

15. M.P. Taylor & B.J. Welch, "Bath/Freeze heat transfer coefficients: experimental determination and industrial applications", p.781-792, AIME Light Metals 1985.

16. R. Pawlek "Primary Aluminium Smelters v. Producers of the World" (Al.-Verlag Dusseldorf.) 1987.

17. L.G. Boxall & A.B. Cooke, "Use of TiB_2 Cathode Material Application and benefits in Conventional VSS cells', pp.573-588 in AIME Light Metals (1984).

18. K.W. Tucker, J.T. Gee, J.R. Shaner, L.A. Joo, A.T. Tabereaux, D.V. Stewart, & N.E. Richards, ''Stable TiB_2 - Graphite Cathodes for Aluminium production'', pp.345-349 in AIME Light Metals 1987.

19. A.D. McLeod, J-M Lihrmann, J.S. Haggerty, D.R. Sadoway, ''Selection and Testing of Inert Anode Materials for Hall Cells'' pp.357-366 in AIME Light Metals (1987).

20. C.N. Cochran ''Alternative Smelting processes for Aluminium'' pp.429-443 in AIME Light Metals (1987).

21. K. Grjotheim & J.B. See ''The Hall-Heroult Process and Alternative Processes for the Manufacture of Aluminium'' Mineral Sci & Eng 11 (2), 80-98, (1979).

22. K. Grjotheim, H. Kvande, K. Motzfeldt & A. Schei ''Carbo Thermal Production of Aluminium '' (Al.-Verlag Dusseldorf 1989).

23. T.R. Beck J.C. Withers & R.O. Loutfy ''Composite-anode Aluminium Reduction Technology''. pp.261-266 AIME Light Metals (1986).

LUNAR PRODUCTION OF ALUMINUM, SILICON AND OXYGEN

Rudolf Keller

EMEC Consultants
R.D. 3, Roundtop Road
Export, PA 15632

Abstract

EMEC Consultants is developing a process to extract oxygen, aluminum and silicon from lunar anorthositic soil. The process will provide oxygen in space for propulsion, aluminum as a structural material, and possibly silicon for the construction of solar cells.

The evolution of the process concept in the course of our present NASA-sponsored SBIR effort is discussed. A molten salt electrolysis producing aluminum and oxygen is the central step; inert anode technology is to be employed. Silicon will be produced by reducing the anorthite, $CaAl_2Si_2O_8$, with aluminum. In a steady-state operation, all components of the ore added have to be removed from the process, and the removal of calcium is a major topic of our experimental studies.

Metallurgical Processes for the Year 2000 and Beyond
Edited by H.Y. Sohn and E.S. Geskin
The Minerals, Metals & Materials Society, 1988

Utilization of Lunar Resources

Utilization of lunar resources may play an essential role in the exploration of the solar system and the conquest of space. High costs are involved in overcoming Earth's gravity to supply missions in space. Even at a low projected cost of $ 300 for transporting a pound of payload into Earth orbit -- today's costs are many times higher -- it appears attractive to use resources in space to produce needed materials. Processing of lunar resources on the lunar surface is an obvious possibility.

Oxygen is a product of major interest. It is required for support of human life in space. Even greater quantities may be needed for propulsion, to conduct further missions into space. To satisfy this demand for oxygen through lunar processing requires a quite different solution than to supply oxygen on Earth, where oxygen gas is abundant and easy to extract from air or water. On the lunar surface, solid oxide materials have to be used. Hydrogen reduction of ilmenite and subsequent electrolysis of the resulting water is being studied [1]. Other suggestions include fluorination of lunar soil [2], reduction with lithium and electrolytic decomposition of the resulting lithium oxide [3], and aqueous HF leaching [4].

In our process, oxygen is produced electrochemically from lunar soil constituents dissolved in a molten salt mixture. The process is designed to produce other valuable materials in addition to oxygen: aluminum as a structural material or as a propellant [5] [6], silicon for potential use in solar cells, and calcium either as a good electric conductor or to alloy with aluminum to produce structural metals, or calcium oxide for the production of lunar concrete.

EMEC Consultants is developing a process concept to produce aluminum, silicon and oxygen from lunar soil. Production of 1000 t O_2 per year and, correspondingly, 500 t Al and 500 t Si annually is envisioned. A process scheme is being developed, details of the process steps investigated, and a lunar installation conceptualized.

Anorthosite as Raw Material

Ample supplies of suitable raw materials for our process are present in the lunar highlands. Table I lists the composition of some lunar soil samples recovered on Apollo missions [7]. They are remarkably low in sodium but contain, for processing purposes, relatively much iron.

Using a 1 g sample of lunar soil, EXPORTech Company, Inc. (under subcontract to EMEC Consultants and with the support of Prof. L. A. Taylor, University of Tennessee) conducted a magnetic separation that produced essentially pure anorthite in the diamagnetic final cut that amounted to about 10 percent of the total sample weight.

Anorthite, $CaAl_2Si_2O_8$, is the nominal feed material of our process. Its elemental composition is given in Table II. With magnetic separation experiments indicating that anorthositic soil can be beneficiated with relative ease, we can expect that suitable process feed materials will be available.

Table I. Composition of Lunar Highland Soil Samples # 67710, 67460 & 64420, and Mare Soil Sample # 71500

Soil Sample	# 67710	# 67460	# 64420	# 71500
Si	21.0 %	20.9 %	21.0 %	18.79 %
SiO_2	44.9 %	44.8 %	44.9 %	40.27 %
Al	15.8 %	15.3 %	14.6 %	5.96 %
Al_2O_3	29.9 %	29.0 %	27.6 %	11.26 %
Ca	11.8 %	12.0 %	11.3 %	7.84 %
CaO	16.5 %	16.9 %	15.8 %	10.97 %
Fe	2.34 %	3.38 %	3.92 %	13.68 %
FeO	3.01 %	4.35 %	5.03 %	17.61 %
Ti	0.16 %	0.22 %	0.33 %	5.77 %
TiO_2	0.26 %	0.37 %	0.55 %	9.63 %
Mn	0.47 %	0.05 %	0.05 %	0.19 %
MnO	0.6 %	0.07 %	0.06 %	0.25 %
Mg	2.37 %	2.53 %	3.24 %	5.81 %
MgO	3.86 %	4.20 %	5.35 %	9.62 %
Na	0.55 %	0.33 %	0.29 %	0.24 %
Na_2O	0.73 %	0.44 %	0.39 %	0.32 %
K	0.09 %	0.05 %	0.08 %	0.06 %
K_2O	0.11 %	0.06 %	0.10 %	0.07 %
O	45.4 %	45.2 %	45.1 %	41.7 %

(wt%)

Table II. Elemental Composition of Anorthite

$CaAl_2Si_2O_8$	20.2 wt%	Si
	19.4 wt%	Al
	14.4 wt%	Ca
	46.0 wt%	O

The Original Process Concept

Figure 1 represents the process as originally conceived [8]. It is similar to one suggested by Jarrett, Das and Haupin [9] in which aluminum and silicon were to be co-deposited electrochemically, along with iron and titanium, in a bipolar cell with a fluoride electrolyte. Instead of the co-deposition, our process, however, employs two sequential reductions. Silicon is reduced in a first major process step, the reduction produced by aluminum rather than electrochemically. In a second major process step, aluminum oxide is decomposed electrochemically to aluminum metal and oxygen. This step supplies both the aluminum required for the reduction of the silicon in the first process step as well as aluminum product. The main purpose of the third major process step is to avoid the accumulation in the salt melt of calcium contained in the process feed. Calcium is removed as the cathodic product of a calcium oxide electrolysis.

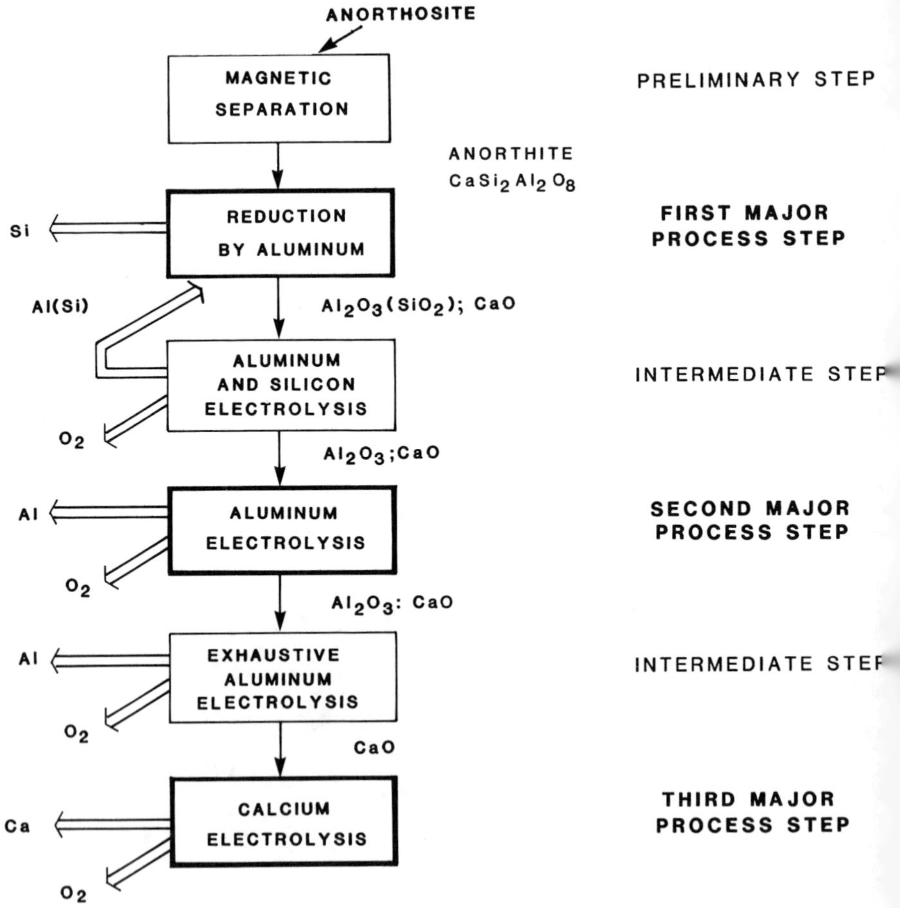

Figure 1. Original Process Concept to Produce Silicon, Aluminum and Oxygen from Lunar Ore [8]

According to this original concept, anorthite is dissolved in a cryolite-like molten salt. Aluminum is then added to reduce most of the silicon component of the anorthite, according to:

$$3\ SiO_2 + 4\ Al \longrightarrow 3\ Si + 2\ Al_2O_3 \qquad (1)$$

The metallic silicon is removed as product, e.g., by filtration. In the reduction, aluminum oxide forms and remains in the electrolyte, along with the aluminum oxide component of the anorthite feed. After completion of this reaction, the electrolyte is moved to the next process step.

According to the original concept, the silicon dioxide is not reduced quantitatively in the first major process step, to avoid complications caused by the presence of residual aluminum on the separation of the silicon produced. On the other hand, all silicon dioxide has to be removed from the electrolyte to produce pure aluminum in the second major process step. It was suggested to accomplish this by an auxiliary intermediate process step in which a mixture of silicon and aluminum were to be produced by electrolysis. The product is used in the first process step as a reductant.

The electrolyte melt containing dissolved aluminum oxide is transferred to the electrolysis cells for the second major process step. These cells are equipped with oxygen-producing anodes. Such inert anodes are presently being developed for application in an advanced Hall-Heroult process [10] [11]. Aluminum and oxygen are obtained according to

at the cathode $\quad Al^{3+} + 3\ e^- \longrightarrow Al \qquad (2)$

at the anode $\quad 2\ O^{2-} \longrightarrow O_2 + 4\ e^- \qquad (3)$

overall reaction $\quad 2\ Al_2O_3 \longrightarrow 4\ Al + 3\ O_2 \qquad (4)$

Pure aluminum and oxygen gas are recovered as products.

To obtain calcium by electrolysis, it appeared necessary to remove all aluminum from the electrolyte in a well-controlled additional process step. After this auxiliary step, the fluoride-based electrolyte containing calcium oxide is subjected to electrolysis in the third major process step, to form calcium and oxygen.

With this original process concept as a starting point, experimental studies of process details were initiated early in 1986. In the following, modifications of the process concept based on the results of our investigations are discussed.

Evolution of the Process Concept

Reduction of Silicon

When we dissolved quartz or terrestrial anorthosite in cryolite melt and added some aluminum, dendritic silicon particles were observed in the salt melt after solidification. Upon addition of more aluminum, "silicon nuggets" could be recovered. When we sectioned such products, we discovered that substantial amounts of aluminum were still present, although the product outwardly looked more like silicon than aluminum. Such a product, including a polished cross section, is pictured in Figure 2.

These reduction products essentially consisted of silicon particles in an aluminum-silicon matrix of eutectic composition. We decided to modify our process concept to produce a hypereutectic Si-Al alloy in the first major process step rather than pure silicon. According to the phase diagram [12]

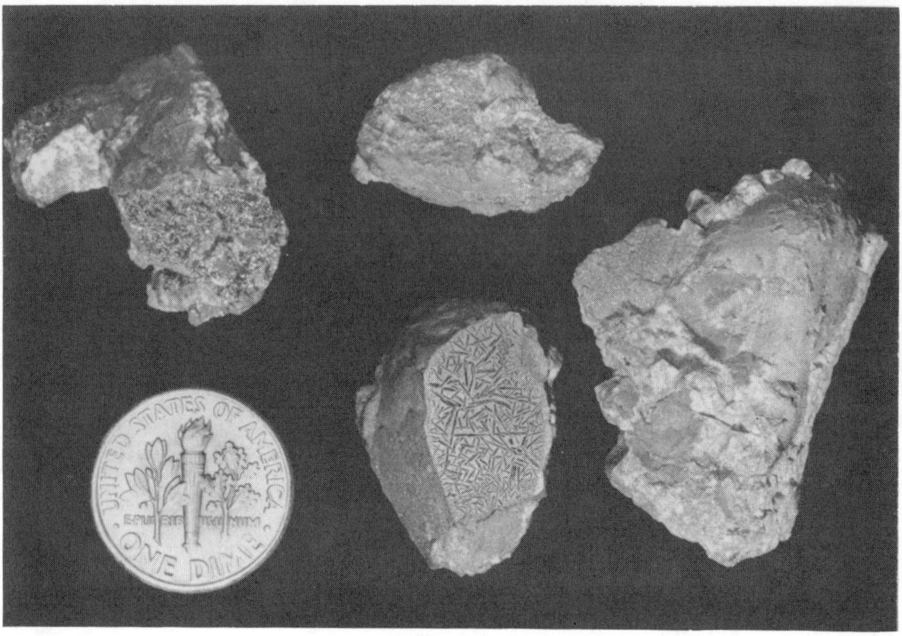

Figure 2. Experimental Si-Al Products

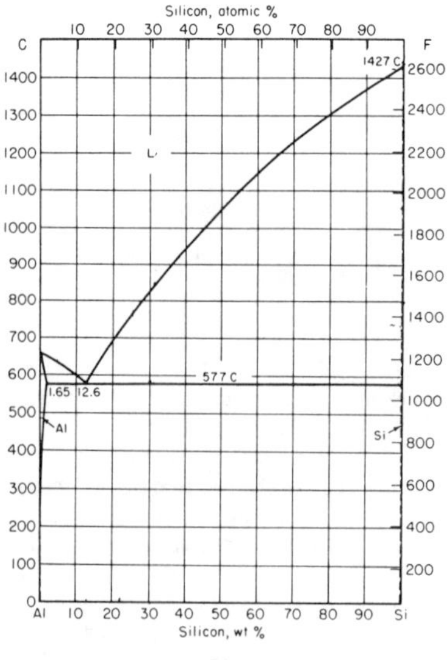

Figure 3. Silicon-Aluminum Phase Diagram [12]

represented in Figure 3, about 50 % silicon is soluble in aluminum at 1050 °C. The eutectic composition is 12.6 % Si - 87.4 % Al at 577 °C. After production of a hypereutectic Si-Al alloy at elevated temperature, the product can be cooled to near-eutectic temperature, whereby essentially pure silicon crystallizes and can subsequently be recovered. This procedure is actually a part of a process patented by Alcoa [13] to produce high-purity silicon.

In our present studies, we attempt to separate the silicon from the aluminum (or rather from the Al-Si alloy of near-eutectic composition) by filtration and draining. Only partial separation has been achieved; improvements through centrifuging or pressing could be sought. The degree of separation, however, appears to be sufficient for the basic process concept if a pure silicon product is not required. Actually, we are considering the possibility of using Si-Al products with high silicon contents as structural materials.

In our experiments, we found the molten metal in the reduction step most manageable if amounts of aluminum reactant and anorthite additions were tailored to yield fully liquid aluminum-silicon mixtures, without formation of solid silicon. At 1050 °C, a silicon content of 50 % should not be exceeded. Note that the reduction of the silicon oxide component is now carried to completion , and the intermediate additional reduction step of the original process concept is not necessary.

We continue to find that silicon formed by the reduction of SiO_2 dissolved in molten salt at the surface of liquid aluminum does not fully alloy into the molten aluminum and we are studying this effect quantitatively. Typically 30 - 40 percent of the reduced silicon is found in the electrolyte, suspended as solid particles. This needs to be considered when the electrolyte is transferred for aluminum electrolysis, as suspended silicon will be reoxidized and rereduced into the cathodic deposit. Special steps have to be taken if one wants to produce a pure aluminum product.

Aluminum Electrolysis

In this second major process step, an electrolysis, aluminum is produced at the cathode and oxygen at the anode. The electrolysis needs to be conducted with high energy efficiency, adequate stability of electrodes and other cell parts, and low losses of electrolyte components. At the present time, we are not conducting any experimental effort in this area, since critical aspects of this technology are being addressed in other ongoing developments [10][11].

Electrolysis conditions, however, most likely will deviate from optimum conditions of terrestrial aluminum electrolysis. The selection of the electrolyte composition has to accommodate other process steps, particularly the removal of calcium. It may be difficult to maintain high levels of oxide concentrations, as have been found beneficial to the life of inert anodes [10]. As our work progresses, such issues will have to be addressed.

Removal of Calcium

In a process operating at steady state, calcium oxide must be removed from the system at the rate it is being added. We think that electrolytic reduction may be the most promising approach, although removal of CaO would be a viable alternative.

Calcium fluoride is one of the thermodynamically most stable compounds, along with lithium fluoride. This is indicated by the data represented in

Figure 4 [14]. We envision electrolyzing CaO added to a CaF_2-LiF electrolyte. Moderate temperatures have to be employed because metal solubility in the electrolyte is excessive at elevated temperatures; molten calcium and calcium fluoride are actually miscible in all proportions [15].

Reducing calcium into an aluminum cathode may be necessary, as in this manner the activity of calcium can be reduced substantially [16]. The calcium subsequently is separated from the aluminum, e.g., by vacuum distillation.

In our experiments, we found moderate co-reduction of lithium. Lithium must also be separated, to be returned to the process to avoid losses that would otherwise have to be made up from terrestrial supplies.

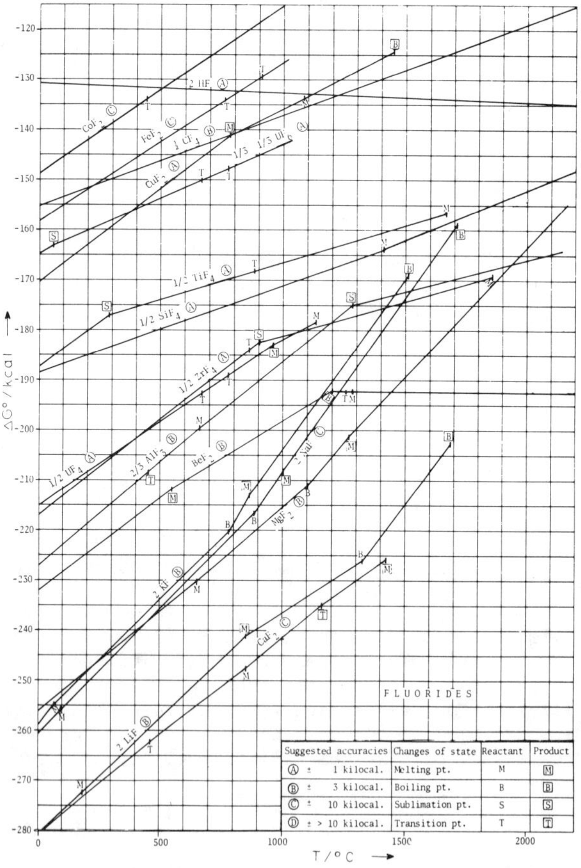

Figure 4. Thermodynamic Data Representing the Stability of Various Metal Fluorides [14]

Our Present Concept

As our laboratory investigations have progressed, our process concept changed; it probably will be further modified in the future. Our present version is displayed in Figure 5.

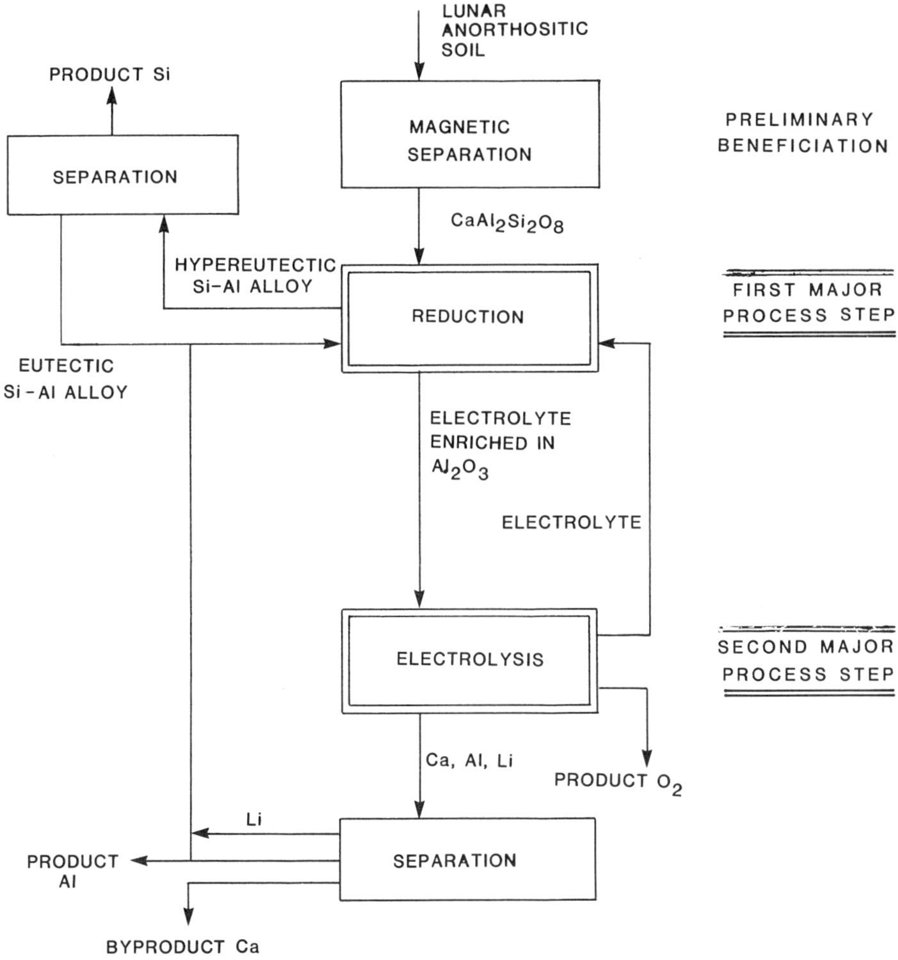

Figure 5. Modified Process Concept

Anorthositic lunar soil is beneficiated by magnetic separation. Anorthite then is fed to the first of two major process steps.

In the first major process step, anorthite is dissolved in a fluoride salt melt and the silicon reduced by metallic aluminum. The reduction is conducted at about 1050 °C to produce a hypereutectic silicon-aluminum alloy, with a silicon content of approximately 45 %. This alloy is separated from the salt and cooled to about 650 °C. Silicon precipitates on cooling. The remaining liquid metal of near-eutectic composition is drained off and returned to the reactor of the first major process step. Additional reactant for the reduction is provided by the second major process step.

By the addition of anorthite and the reduction reaction

$$3\ CaAl_2Si_2O_8 + 8\ Al \longrightarrow 3\ CaO + 7\ Al_2O_3 + 6\ Si \qquad (5)$$

the electrolyte of the first reaction step becomes enriched in aluminum oxide and calcium oxide. This electrolyte is transferred to an electrolysis cell. Oxygen and a metal product are then produced electrochemically in the second major process step.

Oxygen gas is collected as a product. It can be purified by adsorption of contaminants on cell feed material. In this manner, fluoride components evaporating from the bath are recovered.

The electrolysis is conducted to yield a cathodic aluminum product with appropriate amounts of calcium. The electrolyte is probably a CaF_2-LiF mixture, to dissolve aluminum oxide and calcium oxide. This composition permits conducting the electrolysis at relatively low temperatures and high current efficiencies. Aluminum is reduced preferentially, but calcium is also deposited when the aluminum concentration in the electrolyte becomes sufficiently low. Some lithium is expected to be co-reduced.

Calcium and lithium are extracted from the product in an auxiliary separation step. Vacuum distillation or an electrochemical refining technique may be employed for this separation. Pure calcium is recovered as a product and lithium is transferred to the first major process step to reduce anorthite.

Unconventional Technology

The production of metals and oxygen from unconventional raw materials and under unusual conditions requires innovative solutions to technical problems. Special problems result mainly from: (1) the necessity to produce oxygen from solid oxides; (2) the employment of a process feed consisting of a mixture of oxides; (3) the need to strictly minimize terrestrial make-up supplies; and (4) the reduced gravity on the lunar surface.

The following process elements require novel technical solutions:

(1) Sufficiently stable oxygen-producing anodes. Such anodes are being developed for terrestrial use [10] [11]. The flexibility in adjusting process parameters to maximize anode performance, however, is more limited. On the other hand, solutions not feasible for terrestrial use because of intolerable metal contamination or excessive material cost may be possible.

(2) Transport of liquid metal and molten salt between reactor vessels of different process steps present engineering and material problems.

(3) It may be necessary to stabilize liquid-liquid systems physically, e.g. by providing surfaces that are readily wetted by molten metal.

(4) Unconventional separation techniques such as high-temperature centrifuging may be necessary.

We believe that the challenge to produce oxygen and metals in the lunar environment can be met by combining elements of proven technology with innovative solutions to arrive at a viable process.

Acknowledgements

The described development is conducted under Small Business Innovation Research Contract NAS 9-17811, administered by the NASA Lyndon B. Johnson Space Center, Houston, Texas. We appreciate the support by NASA and, in particular, the interest expressed by Dr. D. S. McKay, Technical Manager.

The contributions of D. L. Anthony, C. N. Cochran, W. C. Cochran, W. E. Haupin, and K. T. Larimer to EMEC Consultants' effort are gratefully acknowledged.

References

[1] M. A. Gibson and C. W. Knudsen, "Lunar Oxygen Production from Ilmenite," Paper No. LBS-88-056, Second Symposium on Lunar Bases & Space Activities of the 21st Century, April 1988, Houston TX.

[2] D. M. Burt, "Lunar Mining of Oxygen Using Fluorine," Paper No. LBS-88-072, Second Symposium on Lunar Bases & Space Activities of the 21st Century, April 1988, Houston TX.

[3] K. W. Semkov and A. F. Sammells, "The Indirect Electrochemical Refining of Lunar Ores," J. Electrochem. Society 134, 2088 (1987).

[4] R. D. Waldron, "Total Separation and Refinement of Lunar Soils by the HF Acid Leach Process," Space Manufacturing 5, Proceedings of the 7th Princeton/AIAA/SSI Conference, p. 132-149, 1985.

[5] H. H. Koelle, "The Influence of Lunar Propellant Production on the Cost-Effectiveness of Cis-Lunar Transportation Systems," Paper No. LBS-88-113, Second Symposium on Lunar Bases & Space Activities of the 21st Century, April 1988, Houston TX.

[6] W. N. Agosto and J. H. Wickman, "Lunar Fuels Derived from Aqueous Processing of Lunar Materials," Paper No. LBS-88-7090, Second Symposium on Lunar Bases & Space Activities of the 21st Century, April 1988, Houston TX.

[7] R. V. Morris, R. Score, C. Dardano, and G. Heiken, "Handbook of Lunar Soils," NASA Lyndon B. Johnson Space Center, Planetary Materials Branch, Publication 67, July 1983.

[8] EMEC Consultants, "Dry Extraction of Silicon and Aluminum from Lunar Ores," SBIR Phase I Contract NAS 9-17575, Final Report, July 1986.

[9] N. Jarrett, S. K. Das and W. E. Haupin, "Extraction of Oxygen and Metals from Lunar Ores," Space Solar Power Review, Vol.1, p. 281-287, 1980.

[10] J. D. Weyand, D. H. DeYoung, S. P. Ray, G. P. Tarcy, and F. W. Baker, "Inert Anodes for Aluminum Smelting," Final Report, Contract DE-FC07-80CS40158, February 1986.

[11] Battelle Pacific Northwest Laboratories, "Inert Electrode Program," Topical Reports, Contract DE-AC06-76RL01830, 1988.

[12] K. R. VanHorn, "Aluminum. Vol. 1, Properties, Physical Metallurgy and Phase Diagrams," American Society for Metals, 1967.

[13] R. K. Dawless, "Silicon Purification Process," U. S. Patents No. 4,246,249 (1981), 4,256,717 (1981), 4,312,847 (1982).

[14] T. Rosenqvist, "Thermochemical Data for Metallurgists," Tapir Verlag, 1970.

[15] B. D. Lichter and M. A. Bredig, "Solid and Liquid Phase Miscibility of Calcium Metal and Calcium Fluoride," J. Electrochem. Society, 112, 506 (1956).

[16] K. T. Jacob, S. Srikanth and Y. Waseda, "Activities, Concentration Fluctuations and Complexing in Liquid Ca-Al Alloys," Trans. Japan Institute of Metals, 29 (1), 50-59 (1988).

THE CATHODE PROCESS OF ALUMINUM

CHLORIDE ELECTROLYSIS

Y.J.Zhang

Central South Univ. Technology, Hunan, P.R.C.

and

R.Tunold

Norwegian Inst. Technology 7034 Trondheim NTH, Norway

Abstract

The cathodic kinetics of aluminum deposition from aluminum chloride-alkali chloride melts was studied by use of electro-chemical transient techniques at 700 °C. Aluminum deposited as an insoluble product on a glassy carbon electrode and as a partly soluble product on a tungsten electrode. The deposition was found to be a simple diffusion controlled process involving the exchange of three electrons. The diffusion coefficient of aluminum chloride was calculated - 4.5×10^{-5} $cm^2 \cdot S^{-1}$ in $AlCl_3$ -NaCl-LiCl melts and 2.6×10^{-5} $cm^2 \cdot S^{-1}$ in $AlCl_3$ -NaCl-KCl melts. For $AlCl_3$ -NaCl-CsCl system the diffusion coefficient exhibited some uncertainty due to nucleation. The determined rate constant of aluminum deposition reaction was on the order of 10^{-1} -10^{-2} $cm \cdot S^{-1}$. Constant potential electrolysis was carried out. TiB_2 was well wetted by aluminum. The species of melts penetrated into graphite and glassy carbon but not into tungsten. The diffusion coefficient of species in glassy carbon was evaluated.

Metallurgical Processes for the Year 2000 and Beyond
Edited by H.Y. Sohn and E.S. Geskin
The Minerals, Metals & Materials Society, 1988

Introduction

Aluminum chloride electrolysis is attractive for its power saving of 30%. Different process to circumvent the conventional chlorization to obtain aluminum chloride economically, such as directly chlorizating mixture of Al_2O_3 and C in a cell with ʌ-shaped laminated bipolar electrodes by rising chlorine (1), or by use of a composite anode (2), will promote the electrolysis technology to be realized industrially. With the appearance of the Alcoa process periodically there are studies on the kinetics of aluminum deposition. Some discrepancies seem to exist on the mechanism of aluminum deposition. Some studies (3,4,5) showed that the deposition of aluminum was hindered by diffusion process in the melts. M.Gabco et al. (6) claimed a preceding dissociation of a complex being the rate controlled step. This paper presents our results on the mechanism of aluminum deposition from the study by use of transient techniques and some results from constant potential electrolysis.

Mechanism of aluminum deposition

Experiments were carried out at 700 °C in equimolar NaCl-KCl, NaCl-LiCl and NaCl-CsCl melts with $AlCl_3$ concentration $2\cdot10^{-5}$-$5\cdot10^{-4}$ mol·cm^{-3}. The cathode was glassy carbon and tungsten rods. As reference electrode aluminum was used. The salt and melts used were purified by strict procedures. Measurements were carried out with a PAR Model 173 potentiostat/galvanostat with an IR compensation unit and a PAR Model 175 Universal programmer. Slow responses were measured by an X-Y recorder whereas fast responses were recorded and stored on a Nicolet Explorer IIIA oscilloscope.

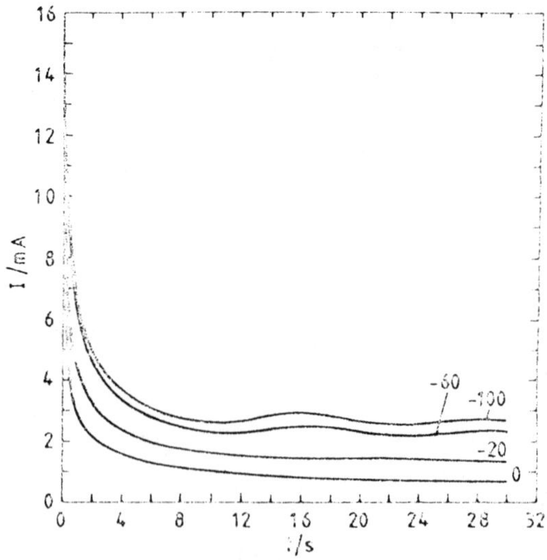

Fig.1. Chronoamperomogram at a glassy carbon electrode. (Figures adhered to curves are applied potentials, mV

C_{AlCl_3} = 2.13×10^{-4} mol·cm^{-3}

Cyclic voltammetry, chronoamperometry and chronopotentiometry were used for measurements. The cyclic voltammograms showed that aluminum deposits as an insoluble product on the glassy carbon electrode, but as a partly soluble product on the tungsten electrode.

For $AlCl_3$-NaCl-LiCl and $AlCl_3$-NaCl-KCl system the peak potential remained constant when the scanning rate was lower than 1000 $mV \cdot S^{-1}$. The plots of the cathodic peak current VS. the square root of the sweep rate exhibited straight line passing through the origin. In chronoamperometric and chronopotentiometric measurements the diffusion limiting current VS$\cdot t^{-\frac{1}{2}}$ plots and the $\tau^{0.5}$ VS$\cdot$ i^{-1} plots were also straight line passing through the origin. The facts above indicated the nature of diffusion control of aluminum deposition.

For $AlCl_3$-NaCl-CsCl system, voltammograms obtained seemed not to be very different from the corresponding diagrmas of two other melts, and the Ip VS. $v^{1/2}$ plots were straight line showing the diffusion control nature. But there were some abnomalities. The Ip VS. $v^{1/2}$ plots were not through the origin. There were maxima in current-time transients repeatedly as shown in Fig. 1. The Id - $t^{-1/2}$ plot

relating to the transients before the first maximum started with a straight line but not passing through the origin, and departed from the line with

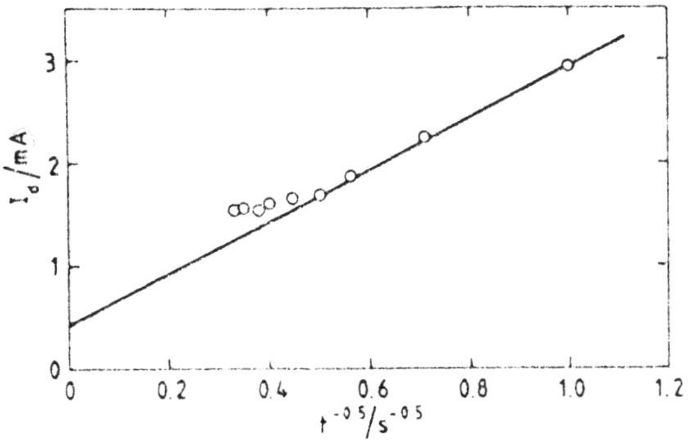

Fig. 2.An I_d - $t^{-1/2}$ plot at a glassy carbon electrode C_{AlCl_3} - 6.14 x 10^{-5}mol$\cdot$cm^{-3}

time increasing. (Fig.2). In chronopotentiometric transient there was an over shoot at the beginning (Fig.3), although the $\tau^{0.5}$ VS$\cdot$ i^{-1} was straight lines through the origin. All these phenomena pointed out that a nucleation process played a part for the overall process in $AlCl_3$-NaCl-CsCl system. In the present study no sign of a preceding chemical reaction was found. Shaking the electrode at chronopotentiometric measurement obviously showed the effect of convecton on the current even for $AlCl_3$ -NaCl-CsCl system (Fig.4), which confirms the diffusion control of the process.

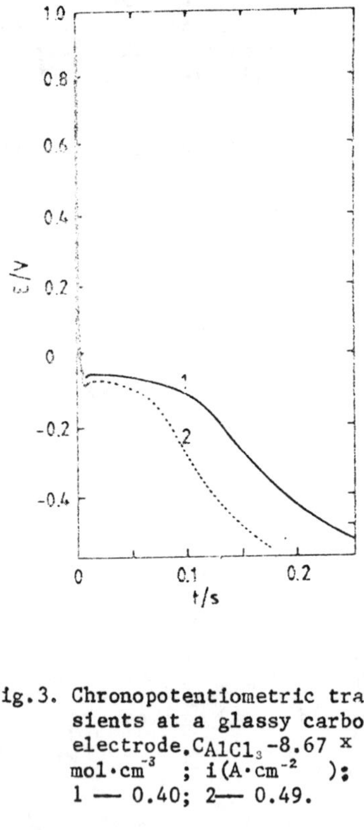

Fig.3. Chronopotentiometric transients at a glassy carbon electrode. C_{AlCl_3} -8.67 x 10^{-5} $mol\cdot cm^{-3}$; i($A\cdot cm^{-2}$); 1 — 0.40; 2— 0.49.

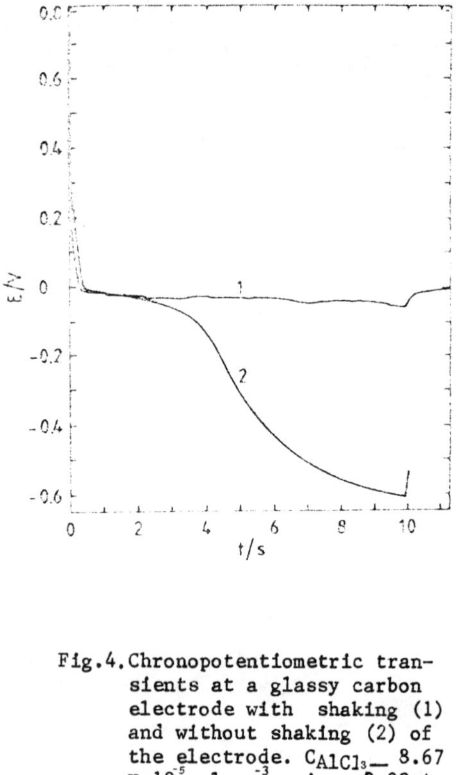

Fig.4. Chronopotentiometric transients at a glassy carbon electrode with shaking (1) and without shaking (2) of the electrode. C_{AlCl_3}— 8.67 x $10^{-5}mol\cdot cm^{-3}$; i — C.09 $A\cdot cm^{-2}$.

The electron number involved in the deposition reaction at glassy carbon electrode was derived from the equations:

$$E^c_p - E^c_{p/2} = -0.77\ RT/nF \quad (1)$$

$$E = E^o + (RT/nF)\ln[(Id-I)/Id] \quad (2)$$

and

$$E = E^o + (RT/nF)\ \ln[(\tau^{\frac{1}{2}} - t^{\frac{1}{2}})/\tau^{\frac{1}{2}}] \quad (3)$$

The number is close to 3. The deposition of aluminum can be considered as a simple rapid reaction involving the exchange of three electrons:

$$AlCl_4^- + 3e = Al + 4Cl^- \quad (4)$$

The diffusion coefficient was calculated from

$$i_p = 6.35x10^6\ c^{\circ}\ n^{\frac{3}{2}}\ D^{\frac{1}{2}}\ \ v^{\frac{1}{2}}\ \ /T^{\frac{1}{2}} \quad (5)$$

$$Id = -nFAD^{\frac{1}{2}}\ \ C^O/(\ \pi^{\frac{1}{2}}\ t^{\frac{1}{2}}\ ) \quad (6)$$

and

$$i\,\tau^{\frac{1}{2}} = \frac{nFc^{\circ}(\pi D)^{\frac{1}{2}}}{2} \quad (7)$$

The value is about $4.5x10^{-5}$ cm^2s^{-1} for $AlCl_3$-NaCl-LiCl, and $2.6x10^{-5}$ cm^2s^{-1} for $AlCl_3$-NaCl-KCl melts. For $AlCl_3$-NaCl-CsCl system the result from different methods differs significantly from $0.4x10^{-5}$ to $4.2x10^{-5}$ $cm^2{\cdot}S^{-1}$ due to the effects of nucleation.

The Reaction Rate of Aluminum Deposition

The rate of aluminum deposition reaction was determined with $AlCl_3$-NaCl-KCl melts by chronoamperometric measurement.

After a potential pulse being applied, for very short times the reaction is kinetically controlled. The current varied with time (7) according to the equation;

$$i = nFAK_fC_0^b\left[1 - \frac{2K_ft^{\frac{1}{2}}}{\pi^{\frac{1}{2}}D_0^{\frac{1}{2}}}\right] \quad (8)$$

when $t \rightarrow 0$, the current approaches a definite value:

$$i_{t\rightarrow 0} = nFAK_fC_0^b = nFAC_0^bK^o\exp[-\frac{\alpha nF}{RT}(E-E^{\circ})] \quad (9)$$

where K_f and K^O are rate constant at potential E and E^O (equilibrium potential) Due to the aluminum reference electrode being used, E^O=0. Hence

$$\log i_{t\rightarrow 0} = \log nFAC_0^bK^{\circ} - \frac{\alpha nFE}{2.303RT} \quad (10)$$

Equation (10) shows $\log i_{t\rightarrow o}$ varies linealy with E. From the intersect of $\log i_{t\rightarrow o}$ VS. E plot K^O can be obtained. Fig.5 is the diagram of $\log i_{t\rightarrow o}$

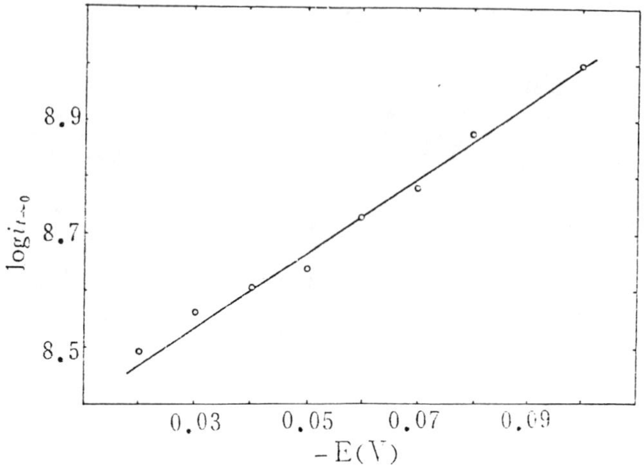

Fig.5. $\log i_{t\to o}$-E plot
$C_{AlCl_3} = 2.7\times10^{-5}$ mol·cm^{-3}

VS.E, the maximum time for obtaining $i_{t\to o}$ by extraporating was 200 µs . Fig.5 gives

$$\log i_{t\to o} = 0.84 - 5.95\ E \quad (11)$$

From equations (10) and (11), $K^{o}=8.8\times10^{-1}$ cm·S^{-1} is obtained. This value is somewhat different from the value from voltammetric measurement (0.2cm·S^{-1}) and from chronocoulometric measurement (1x10^{-2} cm·S^{-1}) (8). The discrepancy might be caused by some simplified assumption in calculation equations and by the difficulty of controlling the experimental conditions at high temperatures such as inadequate IR compensation in voltammetric measurement, incomplete elimination of the effects of residue current in chronocoulometric measurement and the extrapolating time being not short enough in chronoamperometric method. Considering the experiment being performed at high temperature, the result is rather satisfactory. All the results indicate that the reaction of aluminum deposition is a rapid process.

The penetration of species into the cathode

Electrolysis was conducted in 4 wt% $AlCl_3$+42 wt% NaCl+54 wt% KCl and 5 wt% $AlCl_3$ + 40 wt% NaCl +55 wt% LiCl melts at constant potential-10mV (VS. aluminum reference electrode). After electrolysis the cathode was withdrawn and cut along penetrating direction to subject to electron microprobe x-ray analysis. Results showed that aluminum, sodium, potassium and chloride ion did not penetrate into the tungsten electrode, but penetrated into glassy carbon for 454 µm after 90 mins. Concentration calculation showed, the mole ratio of Al+Na+K to Cl in glassy carbon was more than unity, which indicates that the penetrating species includes both metals and ions. The diffusion coefficient was evaluated according to the semi-infinite conditions by the equation:

$$C = Co\,[1 - \mathrm{erf}\,(x/2\sqrt{Dt})] \qquad (12)$$

Co being the concentration of the species in the melts. The D value was found on the order of 10^{-8} $cm^2 \cdot S^{-1}$ for Al, Na and K.

In above mentioned $AlCl_3$-NaCl-KCl melts graphite swelled and broke seriously soon after commencement of the electrolysis perhaps due to the effect of potassium. In $AlCl_3$ -NaCl-LiCl melts graphite remained perfect without spalling through the electrolysis. The species penetrated into the graphite mainly through pores as showed in Fig.6 (a).

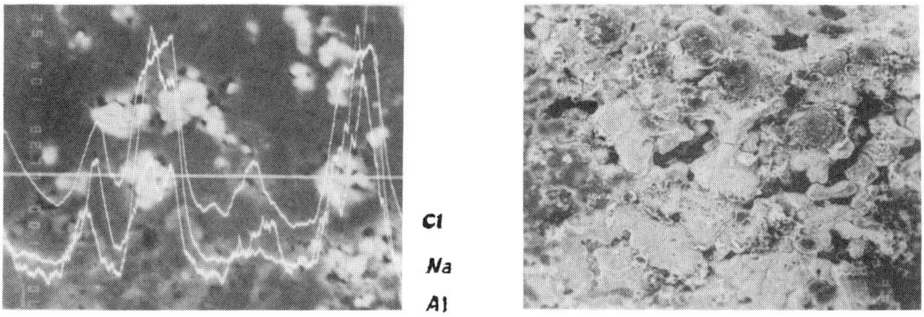

(a) (b)

Fig.6. The electron probe micrograph of (a) graphite cathode after electrolysis (BEI, 450 x) and (b) the deposit state of aluminum on surface of graphite with TiB_2 coating (BEI, 40 x)

The deposition state of aluminum on the cathode surface

Due to the disolution of chlorine into the melts and its recombination with deposited aluminum. It is obvious that the dispersion of aluminum would seriously impair the current yield. Electrolysis with different material as the cathode was carried out at constant potential by passing through small quantity of current. On graphite and prebaked carbon (anthracite) block aluminum deposited as discrete droplets sparsely distributing on the surface. On the graphite with TiB_2 coating (1-2 mm) aluminum densely distributed on the surface(Fig. 6 (b), aluminum oxidized to some extent during washing from the electrolyte with alcohol). Aluminum wets the substrate well and adhers to it firmly. It could imagine that the boundary between aluminum pieces would overlap and they would join together into a compact layer at longer electrolysis or larger current.

Conclusions

Aluminum deposition from $AlCl_3$ -alkali chloride melts is a simple rapid reaction with the exchange of three electrons:

$$AlCl_4^{-1} + 3e = Al + 4Cl^-$$

In the $AlCl_3$ -NaCl-C_sCl system the nucleation plays a part in the overall process. Species of melts penetrate into glassy carbon and graphite electrode but not into tungsten electrode. Among different substrates aluminum shows best wetability to TiB_2

References

(1) Zhao Wuwei, "Cone Bipolar Electrolysis", Nonferrous Metals, 3 (1981),71-77 (Chinese).

(2) T.R.Beck, J.C.Withers, and R.O.Loutfy, "Composite-Anode Aluminum Reduction Technology", Light Metals , 1986, 261-266.

(3) V.I.Sal'nikov, V.P.Butorov, B.V.Mel' nikov, V.A.Lebedev and I.F.Nichkov, "Diffusion coefficients of aluminum ions in an equimolar mixture of potassium and sodium chlorides." Elektrokhimiya,8(1974),1199-1201 (Russ)

(4) S.L.Gol'dshtein, Z.P.Raspopin and V.A.Fedorov, "Kinetics of the discharge of potassium (1+) sodium (1+) and aluminum (3+) ions on liquid aluminum." Elektrokhimya, 12(1977), 1791-1795 (Russ).

(5) R.Φdegård, A.Bjørgum, A.Sterten, J.Thonstad and R.Tunold,"Kinetics of aluminium deposition from aluminium chloride-alkali chloride melts." Electrochimica, 27(1982), 1595-1598.

(6) M.Gabco, P.Fellner and Z.Lubyova, "Chronopotentiometric Study of aluminium deposition from MCl-AlCl melts (M = Na,K and C_s ", Electrochimica, 29 (1984), 397-401.

(7) D.D.MacDonald, Transient Techniques in Electrochemistry (Plenum Press. New York and London, 1977), 78

(8) R.Tunold, Y.J.Zhang and R.ødegard, "kinetics of the deposition of aluminium from chloride melts," Proceedings of the 5th international symposium on molten salts. 86-1, 408-423.

ELECTROMETALLURGY OF SILICON

Gopalakrishna M. Rao

Department of the Air Force
The Frank J. Seiler Research Laboratory
United States Air Force Academy
Colorado Springs, Colorado 80840–6528

Abstract

Several industrial and government laboratories are of the opinion that solar–cell technology will grow into a billion–dollar industry by the year 2000, especially if oil prices rise again. Of the variety of materials that have been investigated as the major component of photovoltaic systems, silicon appears to have an outstanding advantage. The cost factor is the primary importance in photovoltaic development, since the technology to make silicon cells of acceptable efficiency and life time has existed since the early stage of the space program. Electrometallurgy is one of the low–cost processes for the production of silicon, since it allows direct, single–stage production from some abundant, inexpensive source materials such as silica and fluorosilicates. Among several systems, two are given particular attention, using inorganic baths with SiO_2 or K_2SiF_6 as the source of silicon respectively. They appear capable of development to commercial production. Conceptual designs of a commercial–scale cell for silicon production at temperatures above its melting point, and of a pilot plant for plating silicon from an all–fluoride bath are presented.

Metallurgical Processes for the Year 2000 and Beyond
Edited by H.Y. Sohn and E.S. Geskin
The Minerals, Metals & Materials Society, 1988

Energy Prospects

The major changes in the world energy since the first energy crisis in 1970 have impacted on our community. When oil was $40 a barrel in the mid–1970s, we became obsessed with the energy conservation and a search for alternate energies. Even though the crude oil fell below $15 a barrel during the last several years, our energy security fears continue to haunt us. This time it is not the scarcity, but the abundance of foreign oil and the decline in the domestic production.

The prospects of energy depends on the demand and supply of energy. These, in turn, depend on future energy prices, which are linked to economic factors, political events, and unforeseen circumstances. Although no one can predict the future with certainty, three energy consumption patterns used by the Ford Foundation are as shown in Fig. 1.

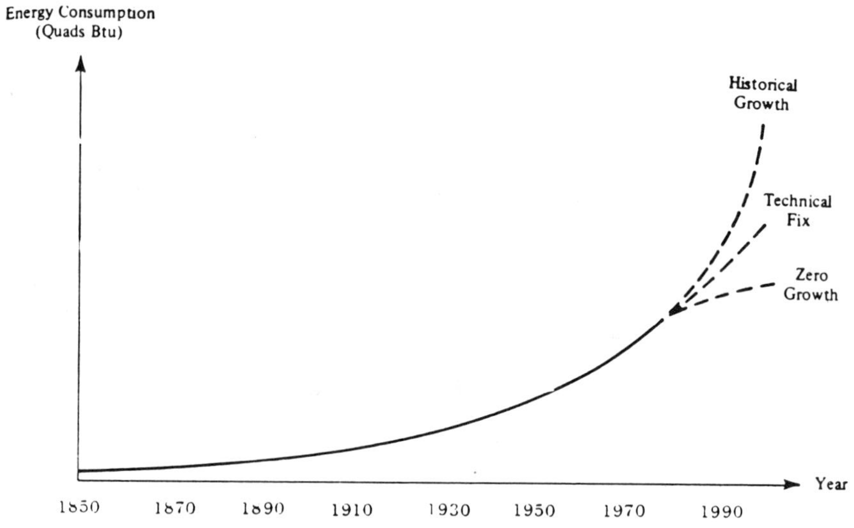

Figure 1 – Patterns of energy consumption.

The first forecast is based on a continuation of a historical exponential growth rate of energy consumption. The second forecast is based on a concerted national effort to use energy more efficiently by introducing energy–conserving technology (technological fix). The final forecast is a tapering off of the rate of increase of energy consumption until zero growth is reached in about the year 2000.

According to a recent report, Energy 2000 (1), the consumption of all forms of energy should increase by about 2% by now and the year 2000 except the oil. Then, the most likely scenario would be technological fix pattern. However, sustained current low oil prices, combined with higher rates of world economic growth than assumed 3% (1) and current low priority to improve the efficiency of energy in industry, building and transportation (2), would certainly favor the historical growth pattern.

The United States is currently importing about 38% of its total oil supply and the figure threatens to pass 50% by the mid–1990s. Furthermore, identified United States reserves are estimated to last only 12 years at current production levels. Coal, in abundant supply over 100 years, is currently accepted as an alternative energy of the future. However, the numerous environmental problems involved in the mining, transportation, and burning of coal will impose stringent standard on its utilization. While nuclear power, once regarded as the "ultimate" energy solution, is presumably besieged with a bewildering array of problems that considerably endangers its future prospects. Growing doubts over nuclear safety have recently been reinforced by the accident at the Three Mile Island plant in the United States and the Soviet catastrophe at Chernobyl. Radioactive waste disposal is proving to be much more difficult than was indicated by earlier, optimistic pronouncements.

It is very important to note that reserves of our principal energy resources, the fossil fuel (oil, coal and natural gas) and uranium are finite and essentially non–renewable. By far, the world's most important inexhaustible source, however, is solar energy, which can be harnessed both in the direct form of sunlight or in more indirect forms as the energy stored in the wind, plants, and water impounded in elevated reservoirs. Photovoltaic or solar cells, which convert sunlight directly to electricity, belongs to the former group. The most significant barrier to widespread utilization is the high cost of the fabrication of these devices. However, manufacturing cost reduction similar to those already achieved in related electronic industries could make solar cell economical for an expanding range of applications; a detailed study concerning the economic issues and the economic feasibility was given by Walton and Warren, Jr. (3). It is not clear how much of our future energy needs will be supplied by solar energy. A speculative guess would be about 15% to 33% of the total primary energy use by the year 2000 depending on the commitment made both to improving energy efficiency and the deploying solar technologies. Interestingly, several industrial and government laboratories are of the opinion that solar–cell technology will grow into a billion–dollar industry by the year 2000, especially if oil prices rise again.

Silicon as the Solar Material

Of the variety of materials that have been investigated as the major component of photovoltaic systems, silicon appears to have the outstanding advantages. Although its bandgap and absorbing properties are not ideal, its availability, technological development, low toxicity, and relative ease of fabrication make it the leading contender to satisfy the needs of the solar cell industry in the next decades. Currently, commercially available silicon photovoltaic cells operate at approximately 10–16% sunlight to electricity conversion efficiency. Although, the theoretical efficiency limit is 25% for a single silicon semiconductor device operating at room temperature, the most efficient silicon device yet produced, the point–contact photovoltaic cell, a system of concentrating solar collector lens in conjunction with photovoltaic cell, has achieved an unprecedented 28.2% efficiency. Moreover, an efficiency of 31% was recorded for a new photovoltaic multijunction solar cell which consisted of stacked layers of silicon and gallium arsenide.

It is important to note that semiconductor grade silicon is too expensive if solar cell costs are to be reduced to the level where they are acceptable for domestic or utility applications. Elwell and Feigelson (4) gave an account on the various competing processes for the production of low cost silicon. One of the promising processes is electrometallurgy since it allows direct, single–stage production of silicon from some inexpensive source materials.

Source Materials

Silicon is the Earth's second most abundant element. It occurs primarily in the form of silicate rocks but the most important source mineral for silicon production is its oxide. SiO_2, which is the main constituent of sand. Silica occurs extensively as quartzite, a rocky crystalline variety with typically less than 1% of metallic impurities (Al, Fe, Ti) and often occurring in very high purity. These materials have been used in processes aimed at producing "solar grade" silicon (SGSi) with a purity substantially higher than that of the metallurgical grade (MGSi). The relatively pure lump sources cost around $0.50kg^{-1} while the less pure Illinois material costs around $0.10kg^{-1} in bulk.

Another important source of silicon is fluorosilicic acid which is a by–product of phosphate fertilizer production. Silicates occur together with phosphates in the rocks from which the fertilizer is extracted, and the extraction process yields fluorosilicic acid of 98–99% purity and at low cost. The silicon costs less than $1 kg^{-1}. The acid may be neutralized to form the sodium or potassium salt:

$$H_2SiF_6 + 2KOH \rightleftharpoons K_2SiF_6 + 2H_2O$$

or the dilute acid is reacted with NaF to give hydrofluoric acid as a by–product:

$$H_2SiF_6 + 2NaF \rightleftharpoons Na_2SiF_6 + 2HF$$

Another potential source of silicon which has not been exploited commercially is rice hulls. It is the silica particles in the rice hulls which gives them their stiffness, and United States production alone could yield 100,000 tons of silicon annually if an extraction process could be developed. The hulls can be bought for about $0.20kg^{-1} and the metallic impurities are comparable with those in a high purity mineral silica. Carbon is, of course, a major constituent and the C:Si ratio in rice hulls is about 4:1.

Processes for the Year 2000 and Beyond

Shortly after the invention of the first battery, the Volta piles, electrolysis was used to produce light metals and it is not surprising that attempts to electrodeposit silicon date back to the mid–19th century. The first attempt on record is that of St. Claire DeVille, who claimed that silicon was produced by electrolysing an impure melt of $NaAlCl_4$, but his material did not oxidize at white heat (for example, see reference (4)). Progress in silicon electrodeposition and the possible applications of electrolytic silicon have been reviewed by a number of authors. Elwell (5) and Fulop and Taylor (6) surveyed the whole field of electrodeposition of semiconductors. Elwell and Feigelson (4) discussed the possible application of electrolysis to produce SGSi, while Rao and Elwell (7,8) considered the prospects for economic electrowinning of silicon for more general applications and for a possible role in industry for the electrolytic production of silicon. A general review of silicon electrowinning and refining was given by Monnier (9) with particular emphasis on the SiO_2/cryolite system. Among several systems, silica–based electrolytes and fluorosilicate–based electrolytes appear capable of development to commercial production.

Silica–based Electrolyte

Since SiO_2 melts at temperatures above 1700°C, a binary or ternary melt composition is required for deposition at about 1450°C and the simplest choice appears to be an alkali– or alkaline earth–silicate system. Electrolysis of molten silicates may produce silicon, silicon monoxide or the metal as the cathodic product. The respective reactions may be represented schematically by the equations:

$$3MSiO_3 \rightleftharpoons 3MO + Si + 2SiO_2 + O_2$$

$$4MSiO_3 \rightleftharpoons 4\ MO + 2SiO + 2SiO_2 + O_2$$

$$2MSiO_3 \rightleftharpoons 2M + 2SiO_2 + O_2$$

Thermodynamic data is in general not available for calculations of the likely reactions in systems of interest. Poris and Huggins (10) made some theoretical calculations. They found that Ca and Mg silicates appeared particularly favorable for Si deposition; Li silicate less promising; while K and Na silicates should yield the alkali metal on electrolysis. It was confirmed experimentally that potassium was liberated on electrolysis of K_2O/SiO_2 melts at 1450°C. Electrolysis of Li silicate melts gave silicon, but not at temperatures above its melting point. Mg and Ca silicates did not yield silicon, and the best results were obtained using the BaO/SiO_2 eutectic (11). BaF_2 was added to facilitate the reaction between $BaCO_3$ and SiO_2 to form the melt, and to lower the melt viscosity. Silicon was also deposited by electrolysis of baths in the $SrO/SiO_2/SrF_2$ system.

The BaO/SiO_2 system has a single eutectic of composition 53% SiO_2/47% BaO by weight and melting at 1370°C. About 15% of BaF_2 was added to this composition and purified melts were electrolysed at about 1450°C in graphite crucible and using graphite electrodes. Applied potentials were normally in the range 1–8 V with currents of 0.1–2.0 A on an immersed area of about 2 cm^2. Silicon has a lower density than the bath, and the molten Si deposits floated around the cathode and gathered into roughly spherical drops, in a horizontal layer near the surface. The largest lump obtained by cooling a melt and dissolving the solidified silicate weighed over 1.6g and contained over 1/3 of the total silicon deposited from a 125g bath. The Faradaic efficiency of deposition was typically 20%, although values as high as 40% were observed. The efficiency tended to decrease with applied potential difference, possibly because of increasing liberation of SiO. This correlation was not firmly established, however, and the SiO could have been evolved because of a slow reaction between the electrodeposited silicon and the silicate bath. Clearly an efficiency in the range 20–40% would not be acceptable for a commercial process but the main limitation could be associated with the small size of the apparatus used. A reverse reaction between anodic and cathodic products is to be expected in a small system, and is known to be significant in Hall–Heroult systems when the anode–cathode separation is less than about 2 cm.

The purity of Si electrowon from the barium silicate melts was about 99.98% by weight, the main impurities being 60 p.p.m. of Ti and 20 p.p.m. each of Al and Fe. Silicon of this purity is close to the quality required to produce 10% efficient solar cells by a single stage of directional solidification. The major impurities originate in the SiO_2 and can be reduced by purification of the silica in HCl gas at about 800°C, or by the use of a purer starting material. The SiO_2 used in this investigation was Illinois Mineral material of 99.5–99.8% purity.

Since electrolytically produced silicon is of lower density than the bath, a modified magnesium (12) or Down's cell for sodium (13) might be suitable. A possible design for a commercial cell for silicon electro–deposition is given in Fig. 2.

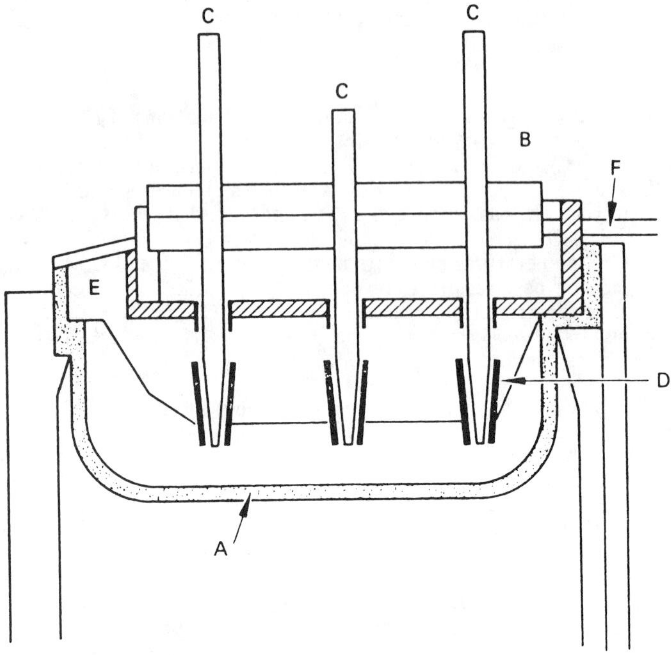

Figure 2 – Schematic of a commercial scale cell for depositing silicon above its melting point, adapted from (12). The steel container (A) is fitted with a ceramic cover (B) through which pass the graphite anodes (C). The silicon is deposited on the cathodes (D) and rises into a collection area (E). Gaseous anodic products are withdrawn through a vent (F).

The cell is of steel construction with the cathode supports welded to the tub–like container that holds the melt. These are fitted with refractory covers, which act as ports for the graphite anodes, metal suction system and solute feed. The anodes are suspended in a manner that permits them to be adjusted independently as the graphite is consumed during electrolysis, to maintain the proper spacing and centering with respect to the cathode. The outside area of the steel container is surrounded by a refractory chamber with arrangements for gas or electric heating. This allows flexibility in adjusting to various electrical loads, especially as the cells are restarted after a complete shutdown. Since a similar system, oxide/cryolite, is used in aluminum electrolysis the cell construction materials used in aluminum cells are generally an appropriate choice. There is, however, the requirement that any materials used must be stable and inert up to 1500°C. Without a protective coat on the steel cathode support, the initial product deposited at the cathode is likely to be iron silicide. If this silicide formation ceased on continued electrolysis, steel would be an inexpensive choice for the silicon production. Otherwise, a suitable coating material on the steel would certainly be required. This coat would also have the role of minimizing the iron contamination of the silicon. Our laboratory experiments (11) suggest graphite as a suitable coating material and inner liner for the steel container.

Alternatively TiB_2, TiC or TiB_2–TiC mixtures, which have been successfully tested in aluminum pilot scale cells (14,15), could be tried as possible coating materials. Although consumable graphite anodes are the initial choice, the inert anodes in development for aluminum cells (15) should also be tested in this system. The silicon produced at the cathode rises into the collection chamber in front of the cell. Molten silicon can be periodically pumped out and cast into ingots. Alternatively, the silicon could be pulled directly from the collection chamber as relatively pure boules because of the additional purification inherent in controlled solidification. Although not shown in Fig. 2, it is desirable to use a diaphragm either to separate the anode and cathode or to act as a hood surrounding the upper region of the anode but immersed 2–3 cm into the melt. The purpose of this arrangement is to minimize the back reaction. To maintain the heat balance in the cell and to control the mass balance in the bath, the cells should be fed semicontinuously through a screw–type feeder, not shown in Fig. 2.

An inert gas atmosphere is required to protect the graphite and to prevent oxidation of the floating silicon. This requirement would add to construction and operating costs but the argon would be recirculated after passing through scrubbers to remove fluoride traces.

The cost of power is an important factor affecting the commercial viability of a silicon plant. Commercial plants could be designed in which most or all of the heat required to maintain the bath at about 1450°C would originate from Joule heating by the deposition current. In a plant with a current 10^5 A at 6 V, a Joule component of 20% would provide 150 kW of "waste" heat which should be ample to compensate for heat losses from a well–insulated container. Plants used for aluminum production normally require active (i.e. forced) cooling. In general, the cost of silicon produced by this method should be comparable with that of aluminum from the Hall–Heroult process.

Fluorosilicate–based Electrolytes

Fluorosilicates as source materials have been investigated intensively but their continued availability as an inexpensive source is dependent on the fertilizer industry. The commercial production of silicon using a fluorosilicate–fluoride system is analogous to the molten salt process for titanium. The latter process was tested successfully on a pilot plant scale (16,17) but did not proceed to full–scale development due to market conditions rather than to any technical problems, at least for the Dow–Howmet cell. It is therefore clear that silicon produced from an analogous system would be attractive to industry only if the product had some special application. Film deposits have shown promise for direct application in low–cost solar cell fabrication and as a corrosion protection coating (7,18), especially at elevated temperatures. Also, electrolytically produced silicon of 4N(99.99%) purity or better (as layer, dendrite, sponge or powder) has the potential to replace MGSi, which needs further purification, as a charge for directional solidification or distillation as a volatile species (e.g. trichlorosilane or tribromosilane) to produce electronic grade silicon.

The deposition of silicon films using K_2SiF_6 began at the Stanford University Center for Materials Research in the early seventies. Although several fluoride solvent systems were studied, only LiF–KF and LiF–KF–NaF melts gave acceptable quality deposits. In early work, cohen (19) showed that single crystal epitaxial layers could be electrodeposited from solutions of K_2SiF_6 in a LiF–KF eutectic and that continuous films could be produced by electrorefining using a dissolving silicon anode. Later work shifted to electrowinning of silicon using an inexpensive graphite anode. The ternary LiF–KF–NaF eutectic or the binary LiF–KF eutectic at 750°C were used as solvents and had the advantage of a high solubility for K_2SiF_6. In the earlier experiments (18,20) silver was chosen as cathode material because of easy nucleation of silicon. Relatively inexpensive graphite substrates, including low grade porous material, were used in later experiments (21,22). Since a low K_2SiF_6 concentration normally resulted in a non–uniform powdery or dendritic deposit on top of a thin, coherent layer about 2 μm thick (20), the K_2SiF_6 concentration was maintained between 4–20 m/o to grow thicker, continuous silicon deposits. The silicon was electrodeposited at a constant current (10–25 $mAcm^{-2}$) or at a constant potential (–0.74 ± 0.04 V against Pt) for 2–4 days. Well–adherent, coherent and continuous films up to 3–4 mm in thickness were prepared in this study. Growth was columnar with a normal grain size up to 250 μm. The current efficiency for silicon deposition was as high as 80%. The purity of electrodeposited silicon was normally 4N, but the level of impurities in the best samples was less than 10 p.p.m. Undoped samples were normally *n*–type with resistivity up to 3 Ωcm, carrier mobility 100 $cm^2V^{-1}s^{-1}$ and carrier concentration 10^{17} cm^{-3} (23).

Other recently published studies (24–27) on this system have added additional complementary information. Olson and Carleton (24) used a silicon–copper anode to simulate the electrorefining of the metallurgical grade silicon, while Sharma and Mukerjee (26) demonstrated the semi–continuous production of 99.99% pure silicon powder from impure (2.5% impurity) MGSi. Bouteillon *et al.* (25,27) showed that improvements in deposit morphology and the purity of the silicon (impurity levels less than 1 p.p.m.), both during electrowinning and electrorefining, could be achieved by pulsed electrolysis, which is currently practiced in commercial copper electrolytic cells.

The quality, purity and electronic properties of electrolytically produced silicon from the fluorosilicate–fluoride system show promise for commercial applications. However, all the experiments to date have been done on a small laboratory scale. Future study should include pilot plant investigations aimed at solving electrochemical engineering problems. A pilot scale design is shown in Fig. 3, which is adapted from the pilot scale cell for titanium electrowinning developed by NL Industries (17). Both Si and Ti processes involve a molten salt medium and require

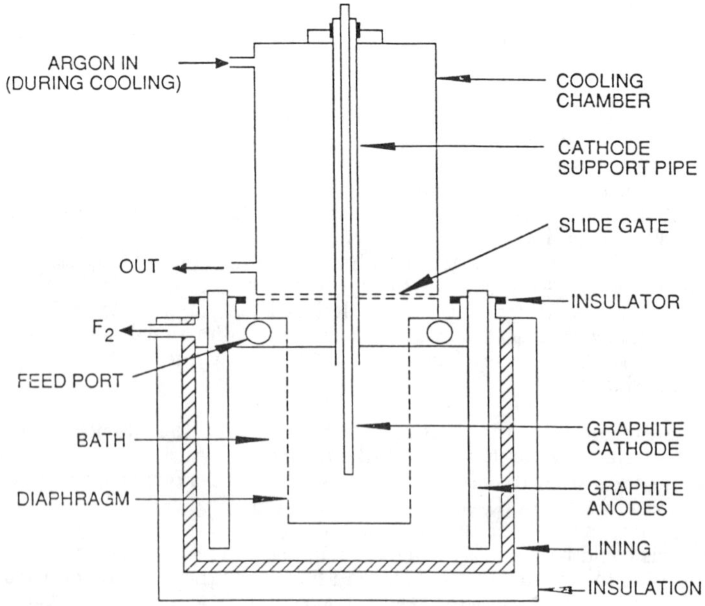

Figure 3 – Design for a pilot cell for electrodeposition of silicon from a fluorosilicate–fluoride bath (adapted from (17)).

four electrons for the reduction from M^{4+} to M, although the actual steps are more complex (27,28). The capital cost, the use of refractory metals and other working conditions are also comparable. The initial choice of electrode materials would be graphite. If it is economically possible, graphite– or silver–coated steel could replace graphite cathodes. The inert electrodes discussed in the preceding section and the new diaphragm materials must be tested thoroughly prior to use even in pilot production cells.

Conclusions

It is obvious that the current low oil prices and the abundance of foreign oil supply, and other fossil fuels may not last forever. Depending on the demand and supply, environmental restrictions, and safety constraints, the cost of energy production will eventually increase. The long term solution, therefore, is to make a commitment now to develop technologies based on the energy from renewable sources such as solar energy to avoid the long gasoline lines and energy shortage by the year 2000. An important requirement of the solar cell development has been to produce a material of acceptable quality and cost.

The availability, low toxicity and high degree of technological development make silicon the most likely material to be used in terrestrial solar cells. Silicon dioxide, SiO_2, occurs in nature as pure deposits which can be mined inexpensively. Although carbothermic reduction is well established as a practical method of producing silicon of metallurgical grade, electrometallurgy offers a viable alternative of producing silicon of significantly higher purity at a cost comparable with that of aluminum. The relative absence of carbon in electrodeposited silicon should offer advantages for some high–grade metallurgical applications, and the electrodeposited silicon should be an attractive starting material for solar applications or as a feed material for the production of semiconductor or detector grade silicon.

The rate of production of silicon as a solid may be too low for commercial viability, and materials such as Na, Mg and Al in the same row of the periodic table are all commercially produced at temperatures above their respective melting points. Designs have been presented for a plant for production of silicon as a liquid, by electrolysis of a barium silicate–fluoride bath. Its rather high melting point of 1412^{0}C is clearly a disadvantage but should not be an impossible hurdle and is, for example, much lower than that of iron. The yield of silicon needs to be improved over that obtained in small–scale laboratory studies but an improvement is to be expected on scale–up. This challenge and that of developing a method for efficient collection of the electrolytically–produced silicon are reasonable and should not require a very high investment.

The alternative of using an all–fluoride bath, with K_2SiF_6 as the source of silicon, may also find a commercial role. The purity and morphology of the deposits from laboratory scale experiments show promise for the direct fabrication of solar cells. The most promising application of this technology is in producing hard, chemically stable coatings on metal surfaces. Also, silicon dendrites, sponge and powder of 4N purity could replace MGSi as the charge to produce higher purity silicon by directional solidification.

Acknowledgements

This work was done while the author held a National Research Council – Frank J. Seiler Research Laboratory Guest Scientist position.

References

1. J. Guilmot et al., Energy 2000 (Cambridge University, Cambridge, **1987**).

2. R. Seltzer, "Energy Efficiency Efforts Losing Steam," Chemical and Engineering News, (**1988**, April 4) 6.

3. A.L. Walton, and E.H. Warren, Jr., The Solar Alternative – An Economic Perspective (Prentice – Hall, Inc, NJ, **1982**).

4. D. Elwell, and R.S. Feigelson, "Electrodeposition of Solar Silicon," Solar Energy Materials, 6 (**1982**) 123–145.

5. D. Elwell, "Electrocrystallization of Semiconducting Materials from Molten Salt and Organic Solutions," J. Crystal Growth, 52 (**1981**) 741–752.

6. G.F. Fulop and R.M. Taylor, "Electrodeposition of Semiconductors," Ann. Rev. Mater. Sci, 15 (**1985**) 197–210.

7. G.M. Rao, and D. Elwell, "Electrolytic Production of Silicon," Light Metals, ed. E.M. Adkins (Warrendale, PA: The Metallurgical Society, **1983**) 1107–1116.

8. D. Elwell, and G.M. Rao, "Electrolytic Production of Silicon," J. Appl. Electrochem, 18 (**1988**) 15–22.

9. R. Monnier, "L'Obtention et le Raffinage du Silicium par Voie Electrochimique," Chimica, 37 (**1983**, April) 109–124.

10. J.A. Poris, and R.A. Huggins, unpublished results, Stanford University, **1977**.

11. R.C. DeMattei, D. Elwell, and R.S. Feigelson, "Electrodeposition of Silicon at Temperature above its Melting Point," J. Electrochem. Soc., 128 (**1981**) 1712–1714.

12. L.F. Lockwood et al., "Magnesium and Magnesium Alloys," Kirk – Othmer: Encyclopedia of Chemical Technology, ed. W.H. Gross, vol. 14 (Wiley, NY, **1981**) 579.

13. D. Pletcher, Industrial Electrochemistry (Chapman and Hall, NY, **1982**) 122.

14. R.C. Darwood, "Energy Consumption of Aluminium Cells Containing Solid Wetted Cathodes," J. Appl. Electrochem., 13 (**1983**) 569–575.

15. K. Grjotheim et al., Aluminium Electrolysis – Fundamentals of the Hall–Heroult Process (Aluminium – Verlag, Dusseldorf, **1982**).

16. Y. Ito, and S. Yoshizawa, Advances in Molten Salt Chemistry, eds. G. Mamantov and J. Braunstein, vol. 4 (Plenum, NY, **1981**) 391.

17. D.G. Lovering, and D.F. Williams, Molten Salt Technology, ed. D.G. Lovering (Plenum, NY, **1982**) 91.

18. G.M. Rao, D. Elwell, and R.S. Feigelson, "Electrocoating of Silicon and its Dependence on the Time of Electrolysis," Surface Technology, 13 (**1981**) 331–337.

19. U. Cohen, "Some Prospective Applications of Silicon Electrodeposition from Molten Fluorides to Solar Cell Fabrication," J. Electron. Mater., 6 (**1977**) 607–643.

20. G.M. Rao, D. Elwell, and R.S. Feigelson, "Electrowinning of Silicon from K_2SiF_6 – Molten Fluoride Systems," J. Electrochem. Soc., 127 (**1980**) 1940–1944.

21. G.M. Rao, D. Elwell, and R.S. Feigelson, "Electrodeposition of Silicon onto Graphite," J. Electrochem. Soc., 128 (**1981**) 1708–1711.

22. G.M. Rao, D. Elwell, and R.S. Feigelson, "The Morphology of Silicon Electrodeposits on Graphite Substrates," J. Electrochem. Soc., 130 (**1983**) 1021–1025.

23. G.M. Rao, D. Elwell, and R.S. Feigelson, "Characterization of Electrodeposited Silicon on Graphite," Solar Energy Materials, 7 (**1982**) 15–21.

24. J.M. Olson, and K.L. Carleton, "A Semipermeable Anode for Silicon Electrorefining," J. Electrochem. Soc., 128 (**1981**) 2698–2699.

25. R. Boen, and J. Bouteillon, "The Electrodeposition of Silicon in Fluoride Melts," J. Appl. Electrochem., 13 (**1983**) 277–288.

26. I.G. Sharma, and T.K. Mukherjee, "A Study on Purification of Metallurgical Grade Silicon by Molten Salt Electrorefining," Metall. Trans., 17B (**1986**, June) 395–397.

27. J. De Lepinary et al., "Electroplating of Silicon and Titanium in Molten Fluoride Media," J. Appl. Electrochem., 17 (**1987**) 294–302.

28. D. Elwell, and G.M. Rao, "Mechanism of Electrodeposition of Silicon from K_2SiF_6 – Flinak," Electrochimica Acta, 27 (**1982**) 673–676.

MATHEMATIC MODELS OF LOST CURRENTS IN BIPOLAL CELLS.

Zhao Guangwen, Duan Shuzhen, Tian Qiuzhan & Wu Tan

Department of Physical Chemistry
University of Science and Technology Beijing
Beijing, China.

Abstract

A thorough study of lost currents in bipolar cells has been made for the development of the great potential of bipolar cells in energy saving. A proper equivalent circuit for bipolar cells has been found, and mathematic methods for computing lost currents at all kinds of polarization have been developed.

At linear polarization, the lost currents can be computed by solving either a tridiagonal matrix or an ordinary differential equation of second order. This kind of values can be used to evaluate the minimal lost currents at given conditions.

At other kinds of polarization, the lost currents can be computed by the same methods as above if the necessary experiments of measuring the cell voltages of every electrolytic compartment have been made. This kind of values can be used to design bipolar cells at practical conditions.

Extensive experiments have been made to evaluate the effects of electrode polarization, emf of galvanic cell for the formation of electrolyzed substance,gaseous products, geometric dimensions of electrolytic cells,special resistance of electrolyte used, and the total and ordinal number of electrolytic compartments on lost currents. It is the first time to make such a systematic research by experiments in the world. Experiments showed that theoretic models were in good agreement with measured data. Experiments also showed that D.C. energy consumption of 0.45 kwh/kg of Pb for lead electrolysis from molten chlorides could be achieved in a cell with 7 bipolar electrodes.

Metallurgical Processes for the Year 2000 and Beyond
Edited by H.Y. Sohn and E.S. Geskin
The Minerals, Metals & Materials Society, 1988

Introduction

It is very promising to apply bipolar electrolytic cells to electrolysis of aluminum, magnesium, lead, zinc and perchlorinates for their great potential of energy saving. for example, bipolar cells had been used in the production of magnesium and saved more than 30% of energy, being one of the three main techniques of energy saving in the production of titanium sponge (1).

It is known that the energy efficiency is the product of voltage efficiency and current efficiency. For molten salt electrolysis, the voltage efficiency is much higher than that of monopolar cells, but their current efficiency is always lower than that of monopolar cells. The bipolar cells therefore must be designed correctly in order to keep the lost currents in them in a suitable level and then to keep high level of energy efficiency. A thorough study on the lost currents in bipolar cells is one of the keys for developing the great potential of bipolar cells in energy saving.

The problems to be solved

The lost currents in bipolar cells were first described using a equivalent circuit of three kinds of resistance by Wilson (2), and later by Rousar (3,4), and similar problem in high voltage battery was studied by Ksenzhek (5) and Onishchuk (6). there were two main problems which they had not solved. one was the proper equivalent circuit of bipolar cells, the other mathematic methods for solving the problems. Thiele (7) had made some progress in mathematic methods and set up a model of the lost currents in bipolar cells. Unfortunately, Thiele's model was based on Wilson's equivalent circuit in which electrolytic cells were simplified as three kinds of constant pure resistance. This simplification suggested that the equivalent resistance of a electrochemical reaction did not vary with the unequal electrolytic currents flowing through different compartments. This simplification is valid only in linear polarization, suggesting that Thiele's model was in fact made on the basis of linear polarization, and could not be used in practical calculation because any electrolytic process would not be carried out at linear polarization which is close to equilibrium. In other kinds of polarization, electrolytic cells which are in steady state can also be simplified as resistances, but in that case, the equivalent resistance of electrolytic reaction varies with the unequal electrolytic currents flowing through different compartments, and the relationship between the equivalent resistance of electrolytic reaction and the currents is different in different kinds of polarization. thus, the kinds of polarization must be taken into account and more effective mathematic methods must be found if one wants to get a model which can be used in practical calculation. Besides, systematic experimental research must be made because such a research has never been carried out before due to the lack of knowledge about lost currents as to people did not know how well the models were once they were set up.

The equivalent circuit of bipolar cells

A proper equivalent circuit is the basis of the correct description of bipolar cells. Wilson's equivalent circuit is too simple to take the effects of electrode polarization into account. In order to consider these important effects, the electromotive forces at given polarized conditions, E_k, is included in our equivalent circuit except the three kinds of resistance of Wilson's circuit as shown in Fig. 1.

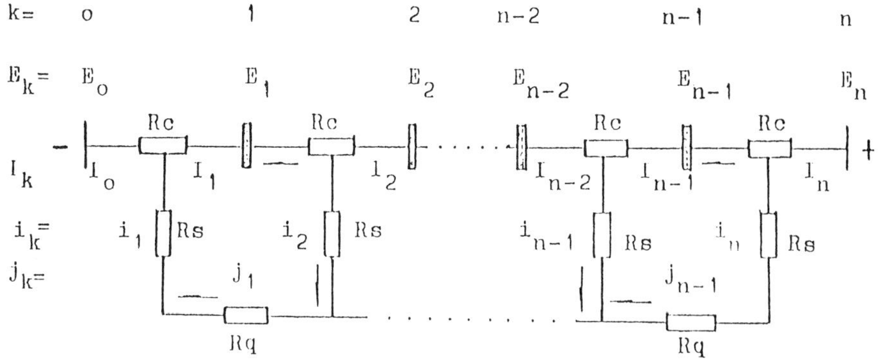

Fig.1. The equivalent circuit of bipolar electrolytic cells

where

n -- The total number of electrolytic compartments in a bipolar cell

k -- The ordinal number of electrods

E_k -- The emf of the k-th electrolytic compartment at given polarized conditions

I_k -- The electrolytic current flowing through the k-th electrode

i_k -- The lost currents flowing through the k-th Rs

j_k -- The lost currents flowing through the k-th Rq

Rc -- The resistance of the electrolyte in each electrolytic compartment

Rs -- the resistance of electrolyte in by-pass, e.g., in channels for the Mg-collecting or slag-discharging of each electrolytic compartment

Rq -- The resistance of electrolyte in by-pass, e.g., in every part of the Mg-collecting compartment between every two sets of channels for Mg-collecting or slag-discharging.

This equivalent circuit gives a solid fundation for correct description of the bipolar cells. Since E_k is included in the equivalent circuit, the thermodynamic and kinetic features of a electrochemical reaction can be taken into account, therefore , more correct and accurate results can be expected.

The lost currents at different kinds of polarization

According to Kirchhoff's law and Fig. 1, the following equations can be given

$$E_k + 1/2 I_k Rc + 1/2 I_k Rc + i_k Rs - j_k Rq - i_{k+1} Rs = 0 \quad (1)$$

$$i_k = j_{k-1} - j_k \quad (2)$$

making use of the boundary conditions $j_0 = j_n = 0$, we can obtain:

$$I_k = I_0 - j_k \quad (3)$$

In the equation (1)

$$E_k = E_e + \eta_k \quad (4)$$

where η_k is the sum of cathodic and anodic polarization in the k-th electrolytic compartment and E_e is the emf of the galvanic cell of formation of electrolyzed substance. At different kinds of polarization there will be different kinds of relationship between η_k and I_k. therefore, equation (1) and then j_k will be given in different forms.

The lost currents at linear polarization

If the cathode and anode are all in linear polarization, the following equation can be given

$$\eta_k = (RT/nFi_0)(I_k/S_k) = Rp(I_k/S_k) \quad (5)$$

where Rp -- The equivalent resistance of the electrolytic reaction, which is independent of I_k.

S_k -- The area of the apparent surface of the electrods

Combined equations from (1) to (5) and rearranged, the following equation can be given

$$Rsj_{k-1} - (2Rs + R + Rq)j_k + Rsj_{k+1} = -U_0 \quad (6)$$

$$\text{where } R = Rp + Rc \quad (7)$$

$$U_0 = E_e + I_0R \quad (8)$$

where U_0 is an imagined cell voltage under linear polarization and applied current I_0 and can be calculated from the emf of the reaction involved, the exchange current density i_0 and the special resistance of the electrolyte used.

For a bipolar cell with n electrolytic compartments, (n-1) equations can be obtained from equation (6), forming a tridiagonal matrix which can be easily solved by computer to get the values of j_k, i.e., the distribution of the lost currents among the electrolytic compartments. A set of data computed are shown in Table 1. This method of computation is refered to TM for short.

Table 1. A set of computed data of j_k and i_k

n=10   Rs=800   Rq=18   R=8 U_0=3.2

k	j_k (in A)		i_k (in A)
	TM	DE	
1	.01396	.01399	-.02798
2	.02437	.02442	-.02087
3	.03158	.03164	-.01444
4	.03581	.03588	-.00848
5	.03720	.03728	-.00280
6	.03581	.03588	.00280
7	.03158	.03164	.00848
8	.02437	.02442	.01444
9	.01396	.01399	.02087
10			.02798

The only assumption we have so far made is that the electrolytic cells are in steady state, the rest follows the defination of electrochemistry, and the direct method for solving the tridiagonal matrix gives the exact solution of the matrix. thus, this method can be used as a criterion for judging whether other methods are good or not.

Solving the tridiagonal matrix by computer to get the values of j_k is very easy, but it can not give the functional expresions of j_k, while the differential equation method can do so. Thus equation (6) can be rewrited by

$$(j_{k-1} - 2j_k + j_{k+1}) - bj_k = -U_0/Rs \tag{9}$$

where $b = (R+Rq)/Rs$ (10)

b and U_0/Rs are positive constants when the temperature, composition of the electrolyte used and dimemsions of the cells have been given. Suppose that k is a continuous variable, equation (9) can be approximated by the following ordinary differential equation of second order:

$$d^2j_k/dk^2 - bj_k = -U_0/Rs \tag{11}$$

with boundary conditions

$$j_0 = j_n = 0 \tag{12}$$

The general solution of equation (11) is given by

$$j_k = C_1\exp(b^{1/2}k) + C_2\exp(-b^{1/2}k) + U_0/(Rq+R) \tag{13}$$

substituting the boundary conditions into equation (13), the solution fitted with boundary conditions can be given by

$$j_k=[U_0/(Rq+R)]\{1-\cosh[(k-n/2)b^{1/2}]/\cosh[(n/2)b^{1/2}]\} \tag{14}$$

a set data of j_k computed by equation (14) is also shown in Table 1. This method is refered to DE for short.

It is obvious from Table 1 that the results from TM and DE were in good agreement. The differences between them increased very slowly with decreasing values of n and with increasing values of b. But the results from these two methods were still approximated even b is far away from the practical range. For example, the difference between the mean values of a set of j_k from the two methods was less than 1% even at n=12, b=5. Thus, equation (14) derived from the differential equation can be used to analyze the regularities of the lost currents in bipolar cells.

It can be seen from equation (14) that j_k depended on n, k, thermodynamic (including in U_0) and kinetic (including in U_0 and R) factors, geometric factors of the cells (including in Rs, Rq, and R) and special resistance of the electrolyte used. j_k will reach maximum when k=n/2 if n is a even number.

From equations (2) and (14), i_k can be given by

$$i_k=\left(\frac{U_0}{R+Rq}\right)\{2\sinh[(k-(n+1)/2)b^{1/2}]\sinh(b^{1/2}/2)/\cosh[nb^{1/2}/2]\} \tag{15}$$

A set of computed values of i_k is also shown in Table 1. i_k is negative when k < (n+1)/2, meaning that i_k is opposite in direction to that of Fig. 1. Under such condition, the potential difference of each Rs (see Fig.1) are controlled by the potential of wired cathode of the bipolar cells, the potential at upper end of each Rs is lower than that at the lower end and that makes i_k negative. The farther the distance from the wired cathode, the less the potential difference on the Rs, and so the i_k. The $i_k=0$ when $k=(n+1)/2$ if n is a add number. When $k > (n+1)/2$, the situation is similar but controlled by the wired anode of the bipolar cells, and i_k are positive, the bipolar electrods are polarized oppositely, leading to corrosion of the bipolar electrods.

The lost currents at strong polarization

If the cathode and anode of each electrolytic compartment of bipolar cells or one of them are at strong polarization, there will be a nonlinear relationship between η_k and I_k. another method for simplifying equation (1) must be used then.

The first three terms of equation (1) is the cell voltage of the k-th electrolytic compartment, U_k, i.e.,

$$U_k = E_k + 1/2 I_k Rc + 1/2 I_k Rc \qquad (16)$$

from equations (1), (2) and (16), the following equation can be given

$$Rs j_{k-1} - (2Rs + Rq) j_k + Rs j_{k+1} = -U_k \qquad (17)$$

where U_k should be a variable depended on I_k or k, but its values can be easily measured accurately so that U_k can be treated as given values if they have been measured before computation. Thus, equation (17) can be solved by using a tridiagonal matrix as above or can be approximated by a ordinary differential equation of second order if the necessary experiments for measuring the cell voltages of each electrolytic compartment of the bipolar cell have been made. The approximate differential equation can be given by

$$d^2 j_k/dk^2 - (Rq/Rs) j_k = -U_k/Rs \qquad (18)$$

The solution of equation (18) is given by

$$j_k=(U_k/Rq)\{1-\cosh[(k-n/2)(Rq/Rs)^{1/2}]/\cosh[(n/2)(Rq/Rs)^{1/2}]\} \quad (19)$$

except for the measured U_k, no additional assumption has been made in the case of strong polarization, thus, equations (17) and (19) should be as reliable as equatios (6) and (14). The thermodynamic and kinetic factors of the electrolytic reaction are concentrated in U_k/Rq, and the effects of n, k and geometric factors of the bipolar cells can be found in the rest of equation (19).

The lost currents at concentration polarization

Eqution (19) can be applied to concentration polarization because the effects of polarization on lost currents should be reflected in the experimentally measured values of U_k. eqution (19) will be applied to this case.

The lost currents when gaseous products form

The effects of formation of gaseous products on polarization of electrods and on apparent special resistance of electrolyte used, and then on the lost currents will be reflected in U_k and Rc. The correction of Rc had been studied (8), therefore, eqution (19) will be applied to this situation.

Experiments

Although the lost currents in bipolar cells have been being concerned for a long time, systematic research by experiments has never been made because knowledge was not enough for people to do so. Rousar (3) had measured the cell voltages of bipolar cells, but some of their measued values deviated from the calculated ones greatly. Thiele's study (7) on the lost currents in the bipolar cells was considered a good one, but their model did not show all-sidely the effects of the main factors on the lost currents and did not make any experiments. Making some additional assumptions, they made some calculation by using some data from a factory and made some discussion, such as how to enlarge a bipolar cell, and how to arrange the inlet and outlet of the electrolyte reasonablely, and so on. Making use of the proper equivalent circuit and effective mathematic methods, we have found the quantitative relations between lost currents and main factors and made systematic experimental research on the lost currents under the powerful theorical guidance. After several times of improvement, all requirements in the accuracy of fabrication and assembly of bipolar cells, in the changable ranges of controlled conditions, in the analogy to the practical problems, etc., were achieved in our experimental cells, and good results were obtained. The time in which the design of bipolar cells was improved by experience completely in order to decrease the lost currents has been over.

The bipolar cells for experiments were made of organic glass. Errors in dimensions of cells were less than 0.5%. The electrolyte used was prepared by analytical reagents and deionized water, and contained 430 g/l of $ZnSO_4 \cdot 7H_2O$, 30 g/l of $Al_2(SO_4)_3 \cdot 18H_2O$,50 g/l of $KAl(SO_4)_2 \cdot 12H_2O$,50 g/l of $Na_2SO_4 \cdot 10H_2O$ Mild steel plates were used as cathods and subjected to chemical polishing before use. Zinc or graphite plates were used as anodes to change the thermodynamic and kinetic conditions of electrochemical reactions. The geometric factors (Rs and Rq) could be changed greatly. The special resistance of electrolyte would change with temperature, composition, and PH of solution during experiments and could not be controlled at the exactly same value in every experiment. Thus, the special resistance was simultaneously measured while every main experiment was making. the level of electrolyte in each compartment was controlled by a general overflow.

Experiments were arranged according to an orthogonal array. Five factors with two levels each were selected as shown in Table 2.

In order to find out the effects of the interactions among the arrangred factors, a orthogonal array which includes fifteen factors with two levels each was chosen. Every experiment was repeated once more for better accuracy.

Table 2. The arranged factors and their levels

Factors		Level 1	Level 2
Total number of compartments	n	12	8
By-pass resistance	Rs	44.6r	14.9r
By-pass resistance	Rq	.337r	.931r
Applied current	I_0	1.5 A/dm^2	1.0 A/dm^2
Materials for anodes	an	Zinc	Graphite

A regulated D.C. power supply Model 1718 was used for electrolysis, working at constant-current mode. Currents and cell voltages were measured by 4 and 1/2 digital multimeters. Currents were constant within 1 mA. Before and after electrodeposition, samples were washed by water, immered in absolute alcohol twice, and then treated in vacuum for 15 minutes in order to weigh them accurately. Measured current efficiencies of every compartment were computed by the increments of weight of the samples.

Experimental results and their variance analysis

The measured and computed mean values of lost current efficiencies of every experiment are listed in Table 3.

Variance analysis of the experimental data based on orthogonal array L $16(2^{15})$ was carried out by computer as shown in Table 4.

It can be seen from Table 4 that the effects of all arranged factors (n, Rs, Rq, I_0, and an) on the lost currents were of high significance and the experimental error was negligible compared with the main effects of the arranged factors and most of their interactions. The order of the effects of arranged factors was found to be an > Rs > n > Rq > I_0. Most of the interactions among the arranged factors were also of high significance. It is worth noticing that the effects for two of the interactions, Rs*an, and n*an, were much greater than that of the factors Rq and I_0. According to the values of mean squares (see Table 4), the effects could be divided into three levels. The first level included the following factors: an, Rs,

n. The second Rs*an, n*an, Rq. The third included the rest except $n*I_0$, $Rs*I_0$, and $Rq*I_0$.

Table 3. The mean values of lost current efficiencies j (%)

$$j = [\sum_{k=1}^{n-1} (j_k)/(n-1)*I_0]*100\%$$

No.	Measured j (%)		Computed j (%)			
	1st	2nd	TM		DE	
			1st	2nd	1st	2nd
1	3.809	4.011	3.799	3.855	3.881	3.857
2	15.73	16.05	14.70	14.69	14.71	14.70
3	10.37	11.45	9.359	10.27	9.372	10.28
4	3.394	3.514	3.323	3.437	3.327	3.441
5	26.30	26.52	27.32	28.20	27.36	28.25
6	9.946	9.914	8.974	9.233	8.988	9.246
7	7.169	7.151	6.666	7.032	6.684	7.025
8	23.26	23.79	22.10	22.91	22.16	22.98
9	5.455	4.860	5.778	5.723	5.782	5.727
10	1.565	1.588	2.020	1.956	2.021	1.957
11	1.718	1.851	1.792	1.778	1.795	1.780
12	6.258	6.445	7.029	6.993	7.040	7.004
13	5.108	4.900	4.990	5.077	4.999	5.087
14	17.75	17.95	17.66	18.22	17.69	18.25
15	12.98	12.96	12.89	12.14	12.94	12.19
16	4.341	4.050	4.222	4.348	4.238	4.364

Table 4. Variance analysis of experimental data

Sources of Variance	Sum of Squares	Degrees of Freedom	Mean Squares	F	Signif.
n	266.42	1	266.42	1072.61	**
Rs	418.67	1	418.67	1685.60	**
n*Rs	7.34	1	7.34	29.56	**
Rq	29.18	1	29.18	117.48	**
n*Rq	5.73	1	5.73	23.07	**
Rs*Rq	6.60	1	6.60	26.58	**
I_0*an	6.30	1	6.30	23.60	**
I_0	11.04	1	11.04	44.47	**
n*I_0	0.27	1	0.27	1.10	
Rs*I_0	0.56	1	0.56	2.24	
Rq*an	7.95	1	7.95	31.99	**
Rq*I_0	0.20	1	0.20	0.79	
Rs*an	91.53	1	91.53	368.51	**
n*an	64.39	1	64.39	259.25	**
an	844.79	1	844.79	3401.16	**
Error	3.97	16	0.25		
Sum	1759.35	31			

Discussion

Reliability of the models

Measured conditions were changed as wide as posiple to cover the range involved in the models. For example, when anodic materials were changed over from zinc to graphite, the emf of galvanic cell for the formation of the electrolyzed substance changed about 2 volts, anodic polarization turned from rather small for anodic dissolution of zinc into extremely large one for the deposition of oxygen, acompanying without or with formation of gaseous product, resp. The other experimental conditions, such as, the resistances of electrolyte in the by-pass channels, Rs and Rq, the total number of electrolytic compartments, n, and applied current, I_0, were also changed over a wide range. As a result of such a change in experimental conditions, the mean values of lost current efficiencies for every experiment varied in a wide range in which some of the higher lost current efficiencies were far beyond the acceptable range in conventional practice (see Table 3). At such conditions, experimental error was very small compared with the main effects of the factors, showing that the measured data were reliable. On the other hand, the measured values of the lost currents were in good agreement with the corresponding computed ones (see Table 3) and the differences between the values of lost currents computed by TM and by DE were very small, which were smaller than experimental error by an order of magnitude. All these showed that the two methods of computation were reliable and actually identical, and the model we had set up could applied to every kind of polarization.

The important effects of the emf and polarization

The effects of emf and polarization on the lost currents have been pointed out above according to theoretic analysis, here this kind of effects will be discussed by using experimental data. It was found that the most significant factor was materials for anods, its main effects was much larger than that of the next significant factor, and its interactions with Rs and n was very powerful (see Table 4). This was a glaring example to show the important effects of the thermodynamic and kinetic factors of the electrolytic reaction expected by equation (19). It is now sure that any model, such as Thiele's model, in which emf and polarization had not been taken into account properly, can not be considered to be practical.

Applicability of this model to molten salt electrolysis

The current density used in molten salt electrolysis is usually higher than that used in aqueous solution electrolysis by an order of magnitude. Thus, the potential of the application of bipolar cells to molten salt electrolysis is much larger than that to aqueous solution electrolysis. The question we concern very much is if our models can be applied to molten salt electrolysis. Let us remind equation (19) in which the only term connected to the electrolyte systems is U_k/Rq. It is known that $U_k = emf + \eta_c + \eta_a + I_k Rc$. In this study, the anodic polarization for the electrodeposition of oxygen is extremely large, say, 0.8 volts, so that the anodic polarization for the electrodeposition of chlorine or oxygen from molten salts is certainly less than that of this study. The cathodic polarization and $I_k Rc$ drop in molten salts are usually less than or at the most equal to that of this study. In other words, all kinds of polarization for molten salt electrolysis will be within the ranges of polarization of this study. It is now sure that our models should be applied to molten salt electrolysis. But the error of measurements and effects of gas formation are usually larger in molten salts, the design of bipolar cells for molten salt electrolysis will be more difficult.

In spite of these diffculties, the applicability of our models to molten salts is still out of question. For example, we used the idea obtained from this study to design a bipolar cell with 7 bipolar electrods for molten salt electrolysis of lead chloride, and obtained very good results . the D.C. electricity required was o.45 Kwh/Kg of lead, which is much better than that of both bipolar (9) and monopolar (10,11) cells ran by Reno Research Certer, U.S. Bureau of Mines. It should be pointed out that the results of ref.(9) were obtained in a cell with only two bipolar electrods, if seven bipolar electrods had been used as we did, the energy required would have been much higher. The results of refs. (10, 11) were reached after more than ten year's effort and several improvements had been made. All these showed the applicability of our models to molten salt electrolysis, the potential of bipolar cells in energy saving, and advanced performance of bipolar cells designed by us.

Prospect

As the main problem has been solved thoroughly, the application of bipolar cells to the production of metals by electrolysis from molten chlorides will be developed considerabely in 5-10 years. We believe that the energy required will be 450 Kwh/ton of lead, 2000 Kwh/ton of zinc, 8000-9000 Kwh/ton of aluminum, 9000-10000 Kwh/ton of magnesium in bipolar cells with about 10 bipolar electrodes.

References

1. Hiroshi Ishizuka, Eur. Pat., 54, 527 (1982)

2. R. Wilson, Demineralization by Electrodialysis (London: Butterworths, 1960), 265.

3. I. Rousar, & V. Cezner, "Experimental Determination and Calculation of Parasitic Currents in Bipolar Electrolyzers with Application to Chlorate Electrolyzer", J. Electrochem. Soc., 121 (1974), 648-651.

4. I. Rousar, "Calculation of Current density Distribution and Terminal Voltage for Bipolar Electrolyzers; Application to Chlorate Cells", J. Electrochem. Soc., 116 (1969),676-683.

5. O. S. Ksenzhek, "The Lost Currents in Highvoltage Battery with Commen Collector" (in Rus.), Elektrokhimiya, 7 (1971), 353-357.

6. V. A. Onishchuk, "Optimization of Electric Parameter of Multivoltage Battery" (in Rus.), Elektrokhimiya, 8 (1972),698-702.

7. W. Thiele, et al., "Beitrag zur Berechnung und Minimierung von Verluststromem an Bipolaren Elektrolysezellen", Electrochimica Acta, 26 (1981) 1005-1010.

8. R. E. Delarue, & C. W. Tobias, "On the Conductivity of Dispersions," J. Electrochem. Soc., 106 (1959), 827-833.

9. M. M. Wong, and F. P. Haver, "Fused-salt Electrolysis for production of lead and zinc metals," Molten salt Eelectrolysis in Metal Production, Int. Symp., (London: IMM, 1977), 21.

10. J. E. Murphy, et al.,"Molten-salt Electrolysis of Lead Chloride in a 3,000-Ampere Cell with An Improved Electrode Design", Proc. -Electrochem. Soc., 1986, 86-1 (Molten Salts),460.

11. J. E. Murphy, et al., "Electrode Assembly for Molten Metal Production from Molten Electrolytes", U.S. Pat., 4,707,239, (1987).

Ironmaking and Steelmaking

OXYGEN STEELMAKING IN THE FUTURE

Robert D. Pehlke
The University of Michigan
Materials Science and Engineering
2300 Hayward Street
2158 Dow Building
Ann Arbor, Michigan 48109-2136

Abstract

The evolution of oxygen steelmaking is reviewed and new operating processes are described. The developments which could impact this technology are summarized. Staged and continuous steelmaking are considered in the light of new technologies. The future opportunities for oxygen steelmaking processes are overviewed.

Prepared for the International Symposium for the Year 2000 and Beyond, AIME Annual Meeting, February, 1989, Las Vegas, Nevada.

Metallurgical Processes for the Year 2000 and Beyond
Edited by H.Y. Sohn and E.S. Geskin
The Minerals, Metals & Materials Society, 1988

Introduction

The use of oxygen in steelmaking was originally brought on the scene by the development of pneumatic steelmaking. Sir Henry Bessemer in England originally conceptualized the use of gaseous oxygen in steelmaking, and then he and William Kelly in Eddyville, Kentucky independently developed the original pneumatic process which involved blowing air through a bath of molten pig iron to oxidize the carbon and silicon and produce steel. This process and many following processes, including those which would use pure oxygen gas, have been proposed.

Beginning with the first commercial oxygen steelmaking vessels at Linz, Austria in 1952, the world steelmaking industry has adopted this style of processing. In general, the process has involved the use of steel scrap in amounts from 20 to 35% of the metallic charge with molten pig iron produced in blast furnaces making up the remainder. The impurities in the charge are oxidized through the use of oxygen gas injected into the steelmaking vessel.

However, currently being adopted into commerical practice and on the horizon are a number of advances in this process which include the use of iron ores as a source of iron and oxygen, the use of an all scrap charge, incorporation of pre-reduced iron, the use of auxillary fuels to change the balance of cold and liquid iron-base charge materials and numerous refining techniques.

The purpose of this paper is to review briefly a number of these processes as examples of advances in oxygen steelmaking, and to reflect on developments which could advance the state of this steelmaking art. In addition, the concept of a system approach to oxygen steelmaking in the sense of process stages or continuous steelmaking are assessed in terms of these new technologies.

State-Of-The-Art and Evolving Processes

A detailed review of oxygen steelmaking has been presented earlier (1). Current world oxygen steelmaking capacity has been reported by Stone and Michaelis (2). The present trend in basic oxygen steelmaking capacity has been level over the past few years, and is at 538,000,000 metric tons at present. Industrialized countries have closed a number of oxygen steelmaking plants without replacing them, whereas lesser developed countries are increasing capacity and have plans for future additions. In particular, in the Soviet Union which has relied heavily on open hearth steel production, a number of oxygen steelmaking plants are currently planned or under construction to replace present open hearth capacity.

Oxygen steelmaking and related processes are challenged by electric arc furnace steelmaking in most industrialized countries where relatively inexpensive steel scrap sources are available. The future of this process will involve more effective ways to utilize scrap and hot metal (liquid iron units) from the production of smaller units than the traditional blast furnace, and to incorporate other iron-bearing charge materials.

Steelmaking was influenced dramatically when the versatility of the open hearth which could utilize charge materials ranging from almost 100% hot metal to 100% scrap was replaced by the basic oxygen furnace. The BOF is normally limited in the use of hot metal to no less than 70 or 75% which means that the availability of low cost scrap could not be utilized to as great an extent as with open hearth steelmaking (3). Furthermore, the leveling off of steel consumption provided a basis for the growth of the electric arc furnace with its capability of utilizing low cost charge materials. At the same time, direct reduced iron (DRI) offers an opportunity for substitution of high quality iron units where low cost scrap may not be available and on a scale well below that for the most efficient blast furnace/BOF complex (3).

The oxygen steelmaking process at present generally utilizes blast furnace hot metal, scrap and fluxes, principally lime to remove sulfur and phosphorous from the metal. There are three configurations of oxygen steelmaking in current use as illustrated in Figure 1: top blowing, bottom blowing and combined or mixed blowing (4). Combined top and bottom blowing has become an effective and widely adopted technique in oxygen steelmaking.

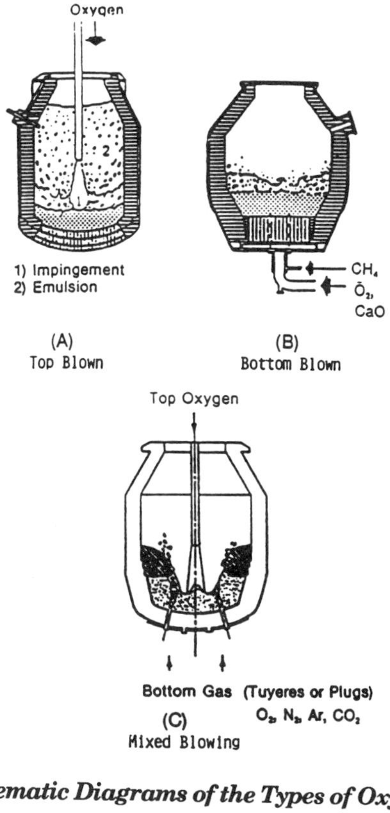

Figure 1- Schematic Diagrams of the Types of Oxygen Steelmaking

Recently IRSID and ARBED have advanced the LBE steelmaking process which combines oxygen lancing in the BOF furnace with bottom injection through porous plugs (5-8). This process, shown schematically in Figure 2, is intended to utilize the advantages of top and bottom blowing, in that it combines early oxidizing slag formation and higher scrap to metal ratios of top blowing with improved metal slag interactions by bottom stirring (5).

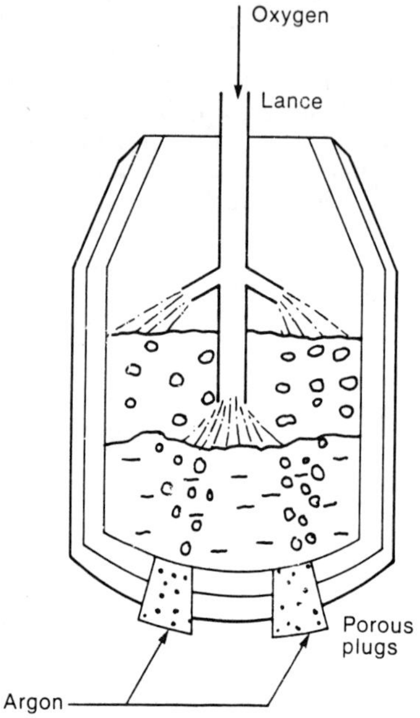

Figure 2 - Schematic Diagram of the LBE Process

The adoption of these technologies has been made possible through the use of secondary refining or ladle metallurgy. In particular, the opportunity to inject oxygen and add refining elements under controlled conditions has made possible a wide variation in the oxygen injection steelmaking process, allowing this process to become a base steel generator with concern for refining reactions, but under conditions where those refining reactions are not limiting to the production capability of the process.

Bottom blowing has been accomplished both through tuyures and porous plugs. The development of the bottom blown process was based on the OBM concept which re-created interest in injection of fuels and fluxs into the iron bath providing energy which could be used to melt additional solid charge materials as in the KMS process illustrated in Figure 3 (9). This technology has led to the innovation of many other injection processes including a recent development at Daido Steel Company of a new scrap melting process using the reaction heat from powdered carbon. The concept of this reactor is shown in Figure 4 and the reactor vessel is illustrated in Figure 5 (10).

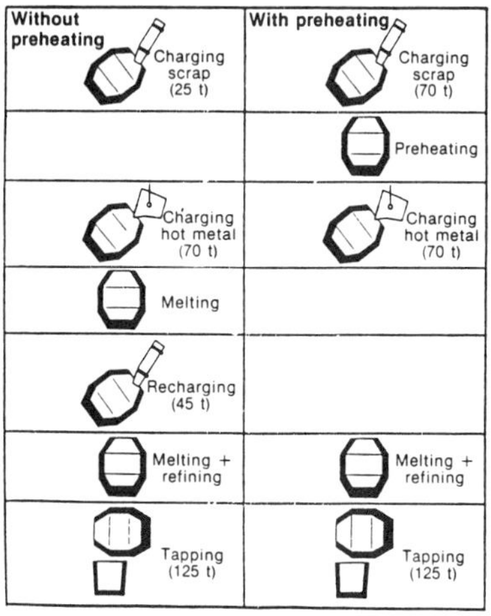

Figure 3 - Comparison of the K(M)S Process Sequence with and without Preheating

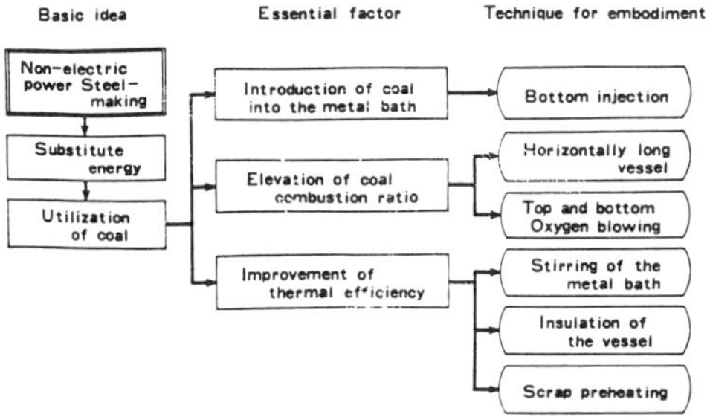

Figure 4 - Conception of the Development of the Reactor Process

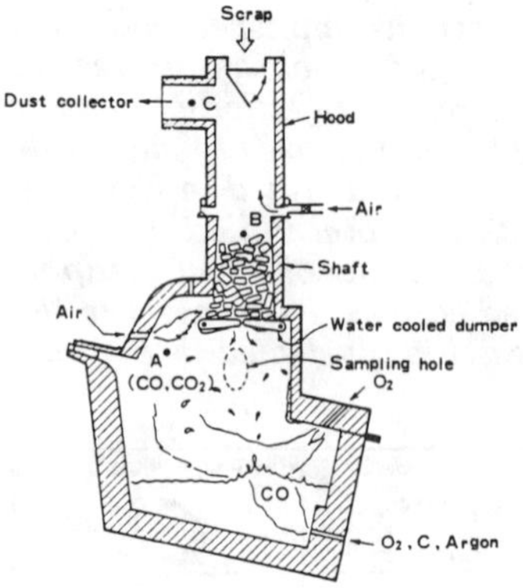

Figure 5 - Schematic Illustration of Reactor

This opportunity to utilize a wide variety of charge materials has led to the development of a proposal by the American Iron and Steel Institute for direct steelmaking in an oxygen blown vessel in which coal and iron would be added to the top of the vessel or injected into the metal and the off gases would be post-combusted to utilize their energy values (11). A schematic diagram of the direct steelmaking system is presented in Figure 6 (11).

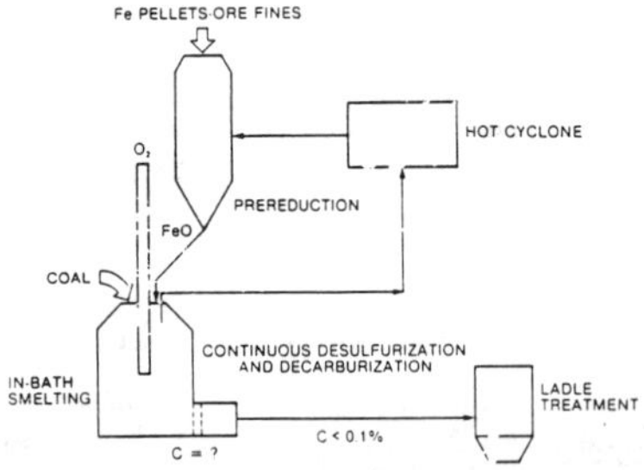

Figure 6 - Schematic Diagram of the Proposed Direct Steelmaking Route

The Future

Earlier predictions suggested that "lasers, electro-optics and infrasonics are an exotic breed of technologies that will assume prime importance for basic production and quality assurance..... Steelmakers will continue the quest after improved productivity through comprehensive computerization, streamlined systems architecture and versatile new thermal and chemical sensors......Advances in digital technology and optimization of the man-machine interface will carry the age-old dream of continuous steelmaking closer" (12).

In a number of ways this early prediction of rapidly advancing technologies and their impact on steelmaking was to become much further extended in time from our present perspective. However, indeed a number of those technologies that were cited are much closer today and have a high probability of implementation on a broad scale within the next two decades.

The citing of laser technology was directed toward high temperature processing and product quality assurance, and at present, there is an intense activity in this area. However, in fact, the laser is now implemented on a daily basis to monitor refractory contours of steelmaking furnaces which provides a basis for optimizing production capability and at the same time maximizing safety practice in the steelmaking operation.

In sensors, the acceleration in digital technology offers an almost unlimited capability for monitioring the process. The development of particular sensors was described in a conference in 1983 (13) where the requirements for improving process performance and maximizing production were outlined.

The requirements of such sensors in order to accomplish improvement in process productivity and control include a direct determination of the state of refining in the process and an ability to indicate processing changes which can be made to assure that the turndown temperature and composition are those desired. Rapid continuous in-situ monitoring of the liquid metal bath composition and temperature would markedly improve process control.

Monitoring of the oxygen steelmaking process for purposes of control will be substantially advanced in the next century. At present, standard practice is to set the oxygen flow rate and the lance position at fixed values for predetermined periods of time. These values are then adjusted, often several times during a heat, on a fixed schedule. The efficiency of the process could be improved by continuously adjusting these parameters based on the monitored status of the operation. The sublance, which is used extensively in Japan and is easier with high hot metal practice where unmelted scrap is not a problem, is the closest technology, but is not entirely satisfactory.

Numerous other sensors which can be brought into digital monitoring systems include those related to maintenance conditions relating to process performance and safety conditions, both in and outside the vessel.

Computer Controlled Production Systems

Closer integration of steelmaking and rolling with continuous casting and implementation of direct hot charging or direct hot rolling requires centralized dynamic production control with casting requirements dominating the system objectives. Steel plant operations can be scheduled to optimize efficiency of production, product quality and/or production costs. Computer integrated manufacturing systems are now being installed with new slab casters, but will be installed in all existing steelplants as production system upgrades. Development of new steelmaking sensors will improve the performance of these production and scheduling control systems.

Raw Materials for Oxygen Steelmaking

The desirability of utilizing a variety of raw materials has been addressed above, particularly from the standpoint of directly utilizing iron ore and coal in particulate form.

In addition, if scrap is to remain as a substantial portion of the charge material, scrap management techniques will have to be employed to minimize residual elements. In fact, maximum effectiveness in scrap management can be achieved if scrap is segregated to limit quality problems associated with recycling of HSLA steels, and at the same time, scrap is segregated to utilize the alloying elements which are high cost and/or scarce.

Continuous Steelmaking

The term "continuous steelmaking" can have many definitions, but may be taken to mean a refining process performed in an oxygen-based steelmaking operation which produces a steady stream of liquid steel based on the uniform continuous addition of raw materials, whatever they may be, to a system of interconnected vessels. The innovative work at IRSID twenty years ago demonstrated the technical viability of this concept (14).

However, current developments have led to vacuum-oxygen-decarburization (VOD) reactors and ladle metallurgy stations where specific refining reactions can be carried out which may not be compatible with the oxygen furnace environment or which would impede the productivity of the steelmaking furnace. Consequently, the concept of continuous steelmaking could better be a series of discrete and non-interconnected reaction vessels. On this basis, implementation of the classic concept of continuous steelmaking may be delayed until even further into the future.

Future Markets

The future demand for steel, domestic and worldwide, and the projections thereof are going to have a substantial influence on process development in the industry. These projections will dramatically affect the rate of advance of development in steelmaking processes. Projections for future demand and revitalized growth of the industry are not particularly encouraging, although extrapolation of steel consumptions offer hope of an expanding market. Recent reviews of the future of the North American industry (15) and future world demand to 1995 (16) indicate at least a stablization. Future growth will depend upon implementation of technology for efficient production of high quality steel. If this is a given, then the future of oxygen steelmaking can be anticipated to be a bright one; with oxygen steelmaking evolving from the BOF, top and/or bottom blown, using a variety of charge materials and supported by sophisticated downstream refining systems.

Conclusions

Oxygen steelmaking continues to evolve and improve. Versatility in raw material usage will be required in the future which could mark the initiation of a decline in BOF steelmaking, at least as it exists today.

In the future, numerous sensors will be developed to improve control of the steelmaking process. Computer integrated manufacturing systems will be universally applied for scheduling and production control.

The outlook for implementation of continuous steelmaking remains uncertain.

REFERENCES

1. Robert D. Pehlke, "Steelmaking - The Jet Age", The 1980 Howe Memorial Lecture, The Iron and Steel Society of AIME, Metallurgical Transactions B, Volume 11B, December, 1980, pp. 539 - 562.

2. J.K. Stone and E.M. Michaelis, L-D Process Newsletter, No. 78, July 1988, VOEST-ALPINE, Ind., Linz, Austria.

3. DIRECT FROM MIDREX, 13, No. 3, pp. 3-5, from "DRI and Scrap As Future Major Raw Materials for Steelmaking" by M. Jellinghaus and W-D Ropke of Krupp Stahltechnik GmbH.

4. Technoeconomic Assessment of Electric Steelmaking Through the Year 2000, Section 3, EM-5445, 1987, CENTER FOR METALS PRODUCTION, Pittsburgh, PA.

5. P.E. Anagbo and J.K. Brimacombe, "The Design and Performance of Porous Plugs in Metal Refining - Part I", Iron and Steelmaker, 15, No. 10, 1988, pp. 38-43.

6. G. Deneir et al, "Industrial Development of Bottom Gas Injection in Top Blown Converters", Iron and Steelmaker, August, 1980, p. 6.

7. H. Takahashi et al., "Metallurgical Characteristics and Operation in Top and Bottom Blown Converter at Kure Works", 5th Int. Iron and Steel Congress Proc., ISS-AIME, Washington, D.C., 69, 1986, pp. 587-590.

8. K. Upadhya, "An Examination of Submerged Injection Processes", Journal of Metals, 35, 1983, pp. 32-36.

9. K. Schafer, et al., "Advances in the K(M)S Process at the Georgsmarienhutte Works of Klockner-Stahl", Iron and Steelmaker, 15, No. 8, 1988, pp. 29-32.

10. S. Sugiura, S. Fujita and N. Demukai, "Development of a New Non-Electric Scrap Melting Process", Transactions ISIJ, 28, 1988, pp. 325-332.

11. T.P. McAloon, "Direct Steelmaking Program Proposed for the United States", Iron and Steelmaker, 15, No. 7, 1988, pp. 30-32.

12. A.D. Mastey and J.J. Innace, "Steelmaking in the Second Millennium", 33 Metal Producing, 18, No. 1, 1980, pp. 37-52.

13. Proceedings of Conference on Sensors in Steelmaking, National Bureau of Standards, Washington, D.C. 1983.

14. B. Trentini: "Comments on Oxygen Steelmaking", Trans. TMS-AIME, 1968, vol. 242, pp. 2377-88.

15. W.A. Tony, "Boom or Bust? A Look at the Future of the North American Steel Industry", Iron and Steelmaker, 14, No. 12, pp. 11-14.

16. T.P. McAloon, "World Steel Demand to Remain Stagnant Through 1995", Iron and Steelmaker, 14, No. 12, pp. 15-17.

THE APPLICATION OF THE SECOND LAW TECHNIQUE FOR THE PREDICTION OF TRENDS IN STEEL MAKING TECHNOLOGY

E.S. Geskin

N.J. Institute of Technology

Abstract

This paper is concerned with the construction of a formal technique for the selection of a metallurgical technology. Exergy consumption is used as the criterion of process efficiency. Different items of the exergy balance are considered and techniques for the reduction of exergy consumption are determined. The selected techniques of exergy saving enable us to identify processes which should be used at different stages of steelmaking. The combination of the selected processes enables us to suggest a technology for conversion of iron ore into steel products. The selection of this technology narrows the range of further search of metallurgical processes. The paper shows that qualitative thermodynamic analysis can be used as the first step in the design of metallurgical processes.

Metallurgical Processes for the Year 2000 and Beyond
Edited by H.Y. Sohn and E.S. Geskin
The Minerals, Metals & Materials Society, 1988

Introduction

The existing processes for making, shaping and treating of steel enable us to convert the available iron bearing ores into marketable steel products. However, the energy cost of metal production is much higher than thermodynamically required, while the metal properties are much lower than thermodynamically available. Increasing cost and declining availability of energy and raw meterials make the creation of new steelmaking technologies imperative. In recent years, the number of novel processes for the making, refining, and shaping of steel have been suggested and tested. Some found a practical application. However, the tested technologies make up a small portion of the physically available routes for the conversion of iron ore into a marketable product. It is also not evident that the optimal routes have been tested and that the selected technologies are the best among those tested.

Although the development of new metallurgical technologies must be based on the experimental and practical information, the number of the alternative routes for the making, shaping and treating of steel, as well as the cost of examining each single route, make necessary the use of analytical techniques for the evaluation of the possible ways of process development.

In the final analysis, the design of a steelmaking process consists in the selection of the optimal trajectory, representing the change of metal composition, shape, and structure in the space of corresponding coordinates. The initial point of this trajectory represents the iron bearing ore, while the final point is a commercial product, for example, a strip. The objective of the design is the minimization of the cost of a product, in the framework of the problem constraints, representing an admissible range of control and process variables.

Although immediate decisions can be made on the basis of the cost in dollars, another objective function is to be used for the long range design. One possible objective function for the technology evaluation is energy consumption. For example, in (1) amount of energy consumed is used to evaluate different ironmaking processes. Process apprisal can be improved if energy consumption will include both energy cost of raw materials as well as material processing (2). Process design can be improved still further, if the exergy function will be used instead of energy (3–9). The application of exergy function for quantitative analysis of a manufacturing process is shown in (10). Amount of information, however, necessary for such analysis is not available at the moment for emerging steelmaking processes. The work (11) suggests the application of exergy function for the qualitative analysis of manufacturing processes. This approach enables us to operate in the framework of available information and to predict trends in technology development.

This paper is concerned with the improvement of the procedure for the design of steelmaking technology. The effect of the process conditions on the exergy consumption at the different stages of the making and shaping of steel is used to estimate the requirements to the processes involved. These requirements are employed for the selection of feasible unit operations.

Exergy Balance

Exergy is the function determining the maximum useful work that can be obtained from a system and, consequently, is a measure of the quality of energy. Using the statistic technique, M. Tribus (7) determines exergy as

$$B = H - T_o S - \sum_i \mu_{oi} N_i \tag{1}$$

where B = exergy, kJ; H = enthalpy, kJ; S = entropy, kJ/K; μ_{oi} = chemical potential of the material "i", N_i = amount of the material "i", kmole. The relationship between exergy and energy is given by the energy grade function R. H. Kevert [12] defines this function as

$$R = \frac{B}{H} \tag{2}$$

The energy grade function characterizes the quality of the energy of a system. For example, for electricity R = 1.0, for natural gas R = 0.913 and for steam at 100° C the value of R is 0.1385. The exergy of electricity, fuel, air, and oxygen consumed in the combustion process determines the cost of energy. Exergy of the ore, fluxes and other used materials characterizes the cost of raw materials. Consequently, the total exergy consumption can be used for the assessment of the overall cost of an operation.

The conditions of the energy utilization in the unit operation are determined by the exergy balance equation:

$$B_i + B_s = B_e + B_r + B_w + B_u + LW \tag{3}$$

where B_i = exergy of a material treated in the unit operation (concentrate, hot metal, molten steel, solid steel, scrap fluxes, etc.), B_s = exergy supplied by electricity, fuel, air, oxygen, fluxes, and other energy careers, B_e = exergy of the product, B_r = exergy of flue gases, slag and other rejected matters, B_w = exergy losses due to heat exchange with surroundings; LW = exergy losses (lost work) due to irreversible processes (chemical reactions, phase transformation, heat exchange, plastic deformation, mixing) occuring in reactors; B_u = exergy output from the energy recovery systems (boilers, vaporizings, cooling systems; gas turbine, etc). For reversible processes LW = 0 and the structure of equation (3) is similar to that of the energy balance equation. The analysis of the exergy utilization at each stage of the steel production can be carried out by the construction of the exergy balances for a stage under study.

Reducing Exergy Consumption

The total exergy lost in process B_t is expressed by the equation:

$$-B_t = B_s + B_i - B_u = B_e + B_r + B_w + LW \tag{4}$$

The objective of the process optimization is the reduction of B_s, B_i, B_e, B_r, B_w and *LW* and the increase of B_u. The lost work is determined by the energy dissipation in the course of the operation. There are several principal ways to reduce *LW*. Practically, all real processes are irreversible and result in energy dissipation. Because if this, *LW* can be reduced by the termination of processes which do not result in the desired change in the thermodynamic state of a system. Among thermodynamically unnecessary processes are coke and pellet production, dissolving of excessive amounts of carbon, silicon and manganese in the metal, etc. Different shapes of the metal obtained in the result of the solidification are almost thermodynamically identical. Because of this electricity (exergy) consumption for plastic deformation, (rolling) is thermodynamically unnecessary; and casting of a liquid steel should be carried out in the form approaching the required metal shape. This, in fact, is quite obvious.

Energy dissipation is also caused by the processes of heat and mass transfer occurring in the reactors. This dissipation can be reduced by the increase in system conductivity. The conductivity can be developed by carrying out transport processes in intensively mixed liquid baths and reducing the size of solid particles supplied into these baths. The objectives above can be achieved by conducting processes of steelmaking in a boiling metal or slag bath supplied by fine coal and concentrate. The use of coke and ore pellets increases exergy losses.

The minimum value of B_e is achieved if flue gases do not contain combustible components and flue gases and slag are rejected at temperatures approaching the surrounding temperature.

The reduction in B_w can be achieved by increasing the rate of process, decrease or termination of the exposure of high temperature metal surfaces to the atmosphere and by the utilization of heat losses, for example, by vaporizing cooling. The rate of process can be increased by the reduction of system resistance, e.g. by carrying out reaction in boiling liquid baths supplied by a fine coal and concentrate.

The making and shaping of the steel involves the use of electricity, oxygen, inert gases, and other components production of which requires significant exergy consumption. The exergy cost of the process can be reduced substantially by the reduction of the use of these components and recycling of the inert gases.

The energy analysis shows that combustion of carbon–oxygen mixture enables us to achieve the temperature required for the steelmaking process; while the energy balance of combustion shows that oxygen in this reaction can be replaced by air without a decrease of the process thermal efficiency, if the air temperature exceeds 1300° C. Consequently, the consumption of oxygen and electricity in steelmaking can be reduced by the use of coal combustion in preheated air as the principle sources of energy.

The comprehensive set of the methods for the reduction of exergy consumption is given in Table I. Analysis of the current practice demonstrates that these methods determine the trend in the development of metallurgical technology.

Development of the Energy Efficiency in the Making and Shaping of Steel

The conditions of the minimization of exergy consumption presented in Table I enable us to outline the following requirements to the processes of the making and shaping of steel:

- making and refining of steel is to be carried out in the sequence of interconnected sealed liquid baths. Process interruption and metal exposure to the atmosphere are to be restricted.

- baths are to be fed by fine concentrate, coal, preheated air, fluxes, inert gases.

- temperature and composition of the bath and the conditions of the supply of concentrate, fluxes and coal are to determine their fast absorption by the bath.

- energy of the process is to be evolved by the coal combustion in air preheated to a temperature more than 1300 ° C.

- use of electricity and oxygen is restricted by processes where the use of air and coal is not technologically feasible.

- strip is to be casted directly from the liquid steel.

- heat of metal is to be extracted by a fluid, for example, an inert gas, which allows controlled heat extraction and heat recovery.

- gases leaving reactors do not contain combustible components.

- heat content of flue gases and slag is to be recovered.

The example of the process implementing the requirements above (13–18) is shown on Fig. 1. According to this figure, fine concentrate, coal, fluxes and preheated air are injected into a liquid bath, similar to the bath of a steelmaking converter at the end of the decarburization process. Carbon content ranges from 0.5 to 1.0 percent. Oxygen of air and iron oxides react with carbon dissolved in metal. The evolved CO gas is burned by the injected air during emerging CO bubbles in the bath or above the bath. The carbon content of the metal is sustained by the dissolving of coal. The rate of coal, ore and air temperature assure the constant composition of the bath and evolving of the amount of heat, which is required by the energy balance of smelting and reducing the concentrate. Silica and other oxides contained in the concetrate are bonded by lime and other injected fluxes and absorbed by the slag. Sulphur is partially removed from coal prior to the coal injection and partially dissolved into the slag. The evolved metal is directed into the next bath where the carbon, silicon, manganese and sulphur contents re reduced down to the required level by the injection of oxygen and fluxes.

Further, metal flow is atomized by the expansion of injected inert gas. Metal drops are purified by submerging in a slag bath and in a flow of an inert gas. The slag and gas treatments result in the removing of nonmetallic inclusion.

TABLE I.

IMPROVEMENT OF EXERGY USE IN STEELMAKING

ITEM OF EXERGY BALANCE	METHOD FOR IMPROVEMENT	MEANS FOR IMPROVEMENT
1	2	3
1. Lost work	Termination of thermodynamically unnecessary processes.	Termination of the use of coke and pellets. Minimization of the contents of iron impurities. Minimization of cross section of a liquid layer during solidification. Casting of composities. Prevention of reheating Integration of thermal and mechanical treatment. Minimization of off–products generation.
	Reduction of entropy production during transport processes.	Use of fine particles of ore, coal and fluxes. Rapid melting of ore particles Use of boiling bath and fluid flow as a reaction medium. Use of membrane for impurities separation. Use of filtration for chemical and physical purification Termination of solid–solid transport processes. Electromagnetic mixing. Direct contact between metal and cooling medium during solidification.
2. Exergy input	Reduction in the use of oxygen and electricity. Termination of the use of coke and pellets. Reduction of energy and material input.	Levitation casting. Preheating of combustion air to temperature above 1300 C. Use of fine ore and coal. Recycling of energy, materials and water. Recycling of inert gases involved in processing.

TABLE I (Cont.)

1	2	3
3. Exergy outflow	Reduction of off–product flow.	Minimization of the off–product generation. Utilization of the off–product and water. Recovery of heat content of metal.
	Reduction of the energy out flow.	Heat recuperation. Cogeneration. Complete combustion.
4. Exergy Losses	Reduction of the residence time of materials in reactors.	Use of fine coal and ore. Rapid smelting. Use of a boiling bath and a gas stream as a reaction medium.
	Recovery of wall losses.	Vaporizing cooling. Gas cooling of reactors and subsequent heat recovery.
	Reduction of metal exposure.	Melting, smelting, refining and solidification in a sequence of sealed interconnected reactors.

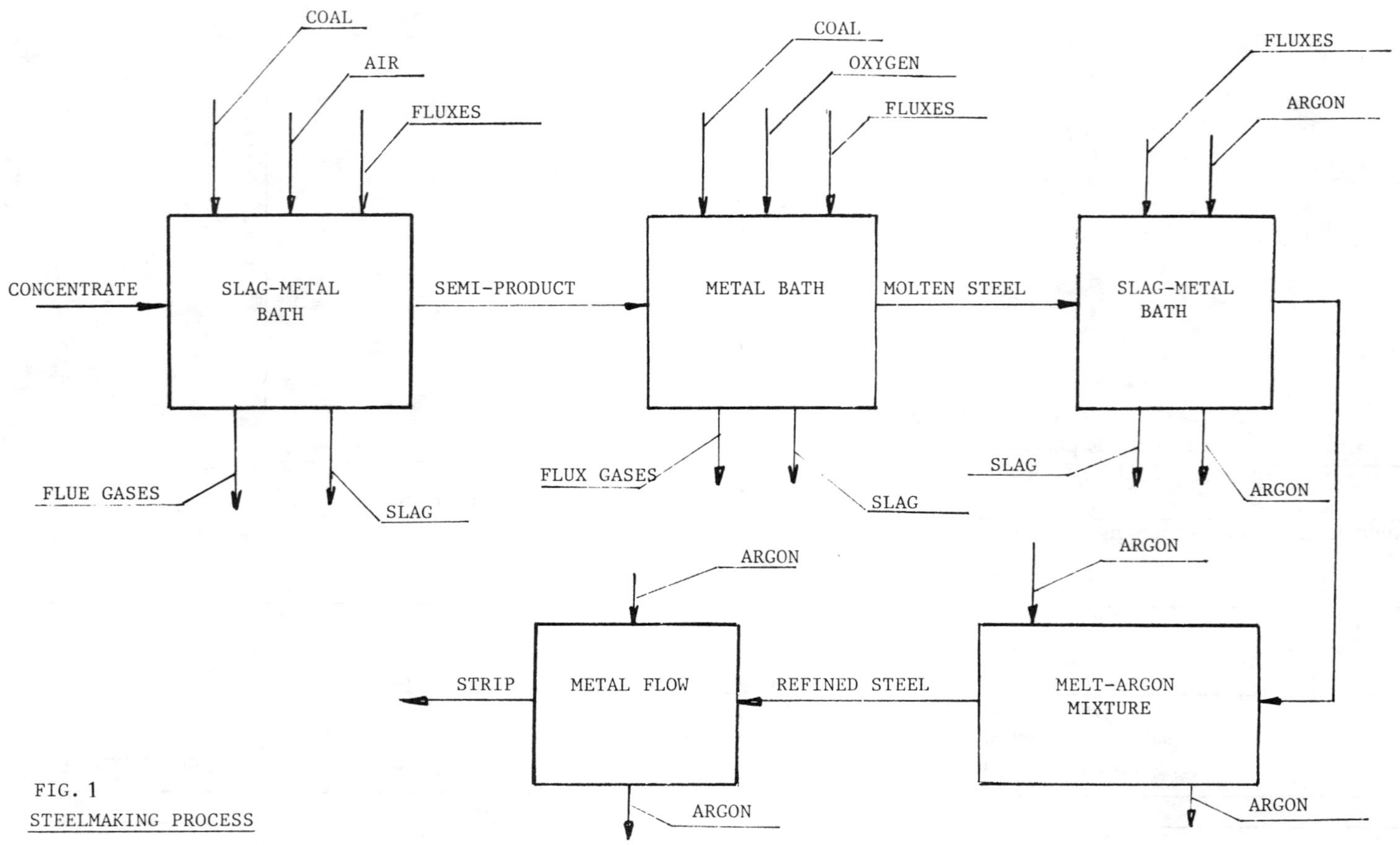

FIG. 1
STEELMAKING PROCESS

Steel solidification is carried out in a thin fluid layer reproducing the shape of a strip. The strip surfaces are blasted by jets of an inert gas. Intensive heat exchange on the impinged solid surface ensures controlled solidification. When solidification is completed, the solid metal can be shaped by rolling. Because the temperature of the metal approaches to the solidus temperature, the work of the deformation is negligible. After the completion of solidification a new portion of the same or different liquid metal can be supplied on the solid surface, and then the processes of the solidification and deformation can be repeated. The jets of inert gas, impinging the strip surface, can be used for the control of metal cooling. The inert gas used for metal treatment is cooled in a turbine of a power plant operating Braton cycles. Cooled and compressed gas is cleaned by membranes or other cleaning devices.

A slag bath is used for extracting heat from slag and flue gases evolved in the course of making and refining of steel. Further cooling of the flue gases is carried out in a conventional recuperator and boiler. The air is preheated firstly in the recuperator and then up to 1300 ° C in the slag bath. Recovered energy is used for preheating of air supplied into reactors and for electricity production.

Conclusions

The exergy analysis is used to outline the requirements to the unit operations involved in the making and shaping of steel. Some of these requirements, for example, the reduction of heat losses or completion of combustion, are obvious and follow straightforward from the energy analysis. Some of the requirements, however, can be defined only in the framework of the exergy analysis, which enables us to outline the comprehensive set of the conditions for the effective energy utilization. For example, only exergy analysis can formally show the effectiveness of the use of fine concentrate and coal, or conducting of the process in a boiling bath. Consequently, exergy analysis shows that direct smelting reduction holds more promise of becoming a primary steelmaking technology.

The discussed procedure is the simplest case of the application of exergy analysis because only tendencies in the exergy change are assessed. However, even this simple assessment provides useful information for the process design. Development of the proposed procedure consists in the quantitative evaluation of the total exergy consumption for unit operation as well as the further process desintegration into more elemental operations. This improvement will bring about the creation of the primary objective procedure for the design of steelmaking technologies.

References

1. N.A. Robbins, "Theoretical Energy Requirements for Ironmaking," Iron and Steelmaker, (1976) 39–61.

2. H.H. Kellog, "Energy Efficiency in the Age of Scarcity," Journal of Metals, June 1974, 39–61.

3. A. Bejan, Entropy Generation Through Heat and Fluid Flows, Chapter 3 (New York, NY: Wiley, 1982).

4. T.J. Kotas, The Energy Method of Thermal Plant Analysis, (Butterworths, 1985), 29–51 and 121–162.

5. G.E. Ahern, The Energy Methods of Energy Systems Analysis, (New York, NY: Wiley, 1980), 73–78.

6. M.J. Moran, Availability Analysis (Prentice Hall, 1982), 146–180.

7. Y.M. El-Saed and R.B. Evans, "Thermoeconomics and the Design of Heat Systems," Journal of Engineering for Power (January 1970), 27–35.

8. G. Tsatsaronis, "A Review of Exergoeconomic Methodologies" (Proceeding of the Fourth International Symposium on Second Law Analysis of Thermal Systems, Rome, 25–29 May 1987), M.J. Moran and E. Scuibo eds., 81–89

9. Y.M. El-Sayed and R.A. Gaggili, "The Integration of Synthesis and Optimization for Conceptual Designs on Energy Systems," Ibid, 43–51.

10. G. Wall, "Exergy Flows in a Pulp and Paper Mills and in a Steel Plant and Rolling Mill" Ibid, 131–141.

11. E.S. Geskin, "The Application of the Second Law Technique for the Prediction of Trends in the Development of a Factory of the Future," Robotics and Factories of the Future '87, R. Radharamanan (Springer-Verlag, 1987), 46–57.

12. H.W. Kevert and S.C. Kevert, "Second Law Analysis: An Alternative Indicator of System Efficiency," Energy, (Aug. 1980), 865–873.

13. E.S. Geskin, "Method of Heating, Melting and Coal Conversion and Apparatus for the Same" U.S. Patent and Trademark Office. Serial Number: 4,422,872. Filing Date: December 27, 1983.

14. E.S. Geskin, "Steelmaking Method" U.S. Patent and Trademark Office. Serial Number: 4,439. Filing Date: February 28, 1984.

15. E.S. Geskin, "Steel Making Method" U.S. Patent and Trademark Office. Serial Number: 4,480,373. Filing Date: November 6, 1984.

16. E.S. Geskin, "Method of Heating, Melting and Coal Conversion" U.S. Patent Office. Serial Number: 4,561,886. Filing Date: December 31, 1985.

AN INTEGRATED STEEL PLANT FOR THE YEAR 2000

Ian F. Hughes

Inland Steel Company
Technology
3210 Watling Street
East Chicago, Indiana 46312

Concepts of an integrated steel plant for sheet products to be built and operating in the year 2000 will be described. The design and operating methodology of this market driven facility will integrate process uniqueness, standardization, manufacturing automation, artificial intelligence and information flow.

Metallurgical Processes for the Year 2000 and Beyond
Edited by H.Y. Sohn and E.S. Geskin
The Minerals, Metals & Materials Society, 1988

An integrated steel plant for the year 2000 is defined for this article as one that will be built on a greenfield site and "started up" in the year 2000 with a capacity of three million tons per year of flat rolled products ranging from cold rolled and electrical steels to high strength formable steels and a full range of coated products.

It may seem that the 21st century and the steel plant we're discussing are far into the future. The startling truth, however, is quite the opposite. As shown in Figure 1, even with accelerated development, a new process on which laboratory development work is begun today might just barely pass through the laboratory phase, pilot plant, and commercial feasibility stages such that it can be considered viable for construction and start up in the year 2000.

On that basis, it is very unlikely that this 21st century steel plant will utilize any technology that isn't at least under investigation in the laboratory in 1988. As a result, we recognize that planning and selecting the facilities for this plant is, in fact, a pragmatic rather than futuristic process, and that its main value is in providing a bridge to longer-term predictions.

The driving force for the flat-rolled steels that this steel plant will produce are its chosen customers' product requirements. In fact, the definition of these requirements could present a bigger challenge than defining the plant technology itself since today's markets are characterized by distinct volatility. Predicting product requirements for the year 2000 requires a knowledge of customers' long-term strategic plans. That, in turn, requires an established relationship between the steel supplier and the customer, since the mutual trust required to share such information needs time to grow.

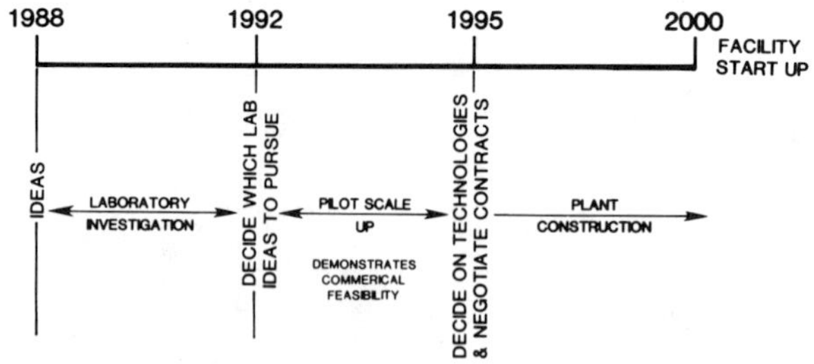

Figure 1 - Timetable for Commercial Process Development

One thing is clear however, that this inherent uncertainty in projecting product trends requires this steel mill to have maximum flexibility in final product production and could see an added dimension of moving to added value parts or systems. Nonetheless, as can be seen in Figure 2 for the anticipated requirements for a zinc iron alloy, significant changes in product uniformity, surface properties, and shape are anticipated for increased consumer demand for the quality that will enhance their ability to be globally competitive.

GAUGE	±2.5%
FLATNESS	Table Flat
WIDTH	-0 to +0.5% depending on width
YIELD STRENGTH	14 Kg/mm^2 Aim ±2.5 Kg/mm^2
TENSILE STRENGTH	30 Kg/mm^2 Aim ±2.5 Kg/mm^2
$\bar{r}$	2.0 Minimum
$\bar{n}$	.25 Minimum
COATING WEIGHT	45 g/m^2±7.5 g/m^2 for Hot Dip
	(±5 g/m^2 for Electrogalvanize
IRON CONTENT	11% ± 1.5% (depends on specific customer
COATING ADHESION	Excellent (minimal powdering in forming)

(Ranges are 6 Sigma)

Figure 2 - Anticipated Product Requirements for DDQ Cold Rolled Zinc-Iron Coated Product

Clearly these requirements will ensure the most fastidious attention to detail in process control and product flow and before analyzing the processes that will best produce the products previously described, let's compare the steel plants of 1980 and 1990. On the one hand, as shown in Figure 3, for solid processing, the Continuous Cold Mill (CCM) has tied together five previously discrete processes into one continuous process. Processing steps that took more than two weeks in 1980 will be carried out in 30 minutes in 1990, with better quality and better matching to customer needs. In contrast, process steps for liquid iron and steel have proliferated.

As the steel plant has changed, so have the criteria by which new processes and facilities are selected. Manufacturing Cost and Productivity were once the main criteria. Currently Capital Cost, Quality, Delivery, Lead Time, and Environmental Concerns are additional factors that will continue to become more important.

It is clear that response to market changes is paramount as the pace of technological change continues to increase. New facilities will also require the process flexibility to use a variety of feed materials. In addition, as the sequence of steelmaking and rolling processes becomes more continuous and more closely tied to short delivery times, processes must be almost completely reliable.

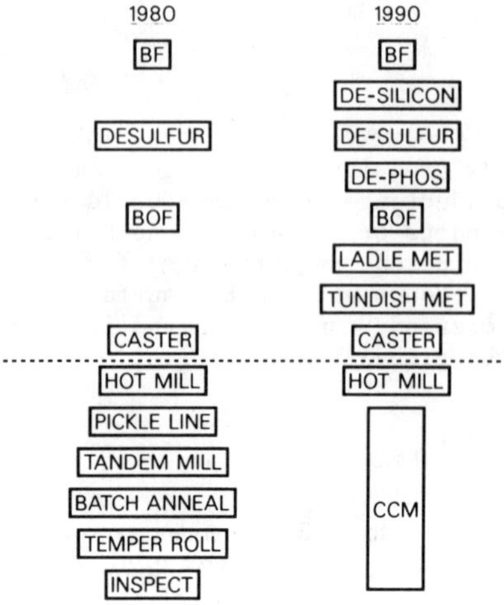

Figure 3 - Process Changes from 1980 to 1990

Steel Plant Overview

The overall facilities and process flow are shown in Figure 4. Iron will be made from ore by a coal-based process. The ore grades and types will be similar to those now in use and not, for example, sulfide ores. Ore has been selected as the basic raw material, rather than scrap, because of the high levels of residuals in most commercially available scrap and the unpredictable price volatility of the scrap market. Recent projections, for example, ascribed a future $20/ton advantage to the Electric Arc Furnace over the BOF, but those assumed a scrap cost of $90/ton. At the time this paper was written, scrap was selling in the Chicago market for approximately $130/ton.

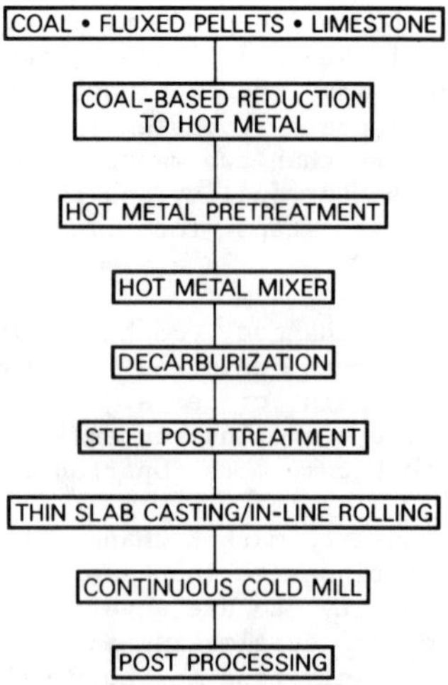

Figure 4 - Facilities and Process Flow

Coal will replace coke as the reductant because environmental restrictions on cokemaking will make coke more costly and less available. Hot metal pre-treatment will continue to be required. A hot metal mixer will smooth out the chemistry and availability of hot metal to the steelmaking process, allowing the remaining liquid and solid steel processing steps to proceed reliably and in close coordination with customer orders.

Steel will be continuously cast as thin slabs, followed by in-line rolling, or cast as thin strip. The hot-rolled coils will then be processed through a continuous cold mill. Post processing will include surface coating followed by a variety of processes which could include part and system manufacturing.

Ironmaking

The results of the 300,000 tpy demonstration COREX unit now in operation in South Africa suggest that "smelt reduction" ironmaking will become and remain the process of choice for new installations for the next ten years. Six 2000 tpd smelt-refining units--five in operation and one under planned maintenance at all times--would provide hot metal to a 3 million tpy plant.

Significant research is currently in progress evaluating other ironmaking processes, including a Japanese National Project, but these efforts will provide the new processes for the year 2010 and beyond.

The Corex process shown in Figure 5 consists of splitting the conventional blast furnace's solid state reduction and smelting steps into two separate steps in two separate vessels means. Metallurgical coke will no longer be needed to support the ore burden and also more available, less expensive non-metallurgical coal can be added to the lower smelting vessel, with the evolved reducing gas routed through the shaft furnace to reduce the ore. If desired, Direct Reduced Iron or even scrap can be added to the smelting vessel. As a result, smelt-reduction units are significantly more flexible in terms of feed materials and operation than a large blast furnace. Unlike a blast furnace, the processing cost per ton does not change significantly as the production rate is increased or decreased.

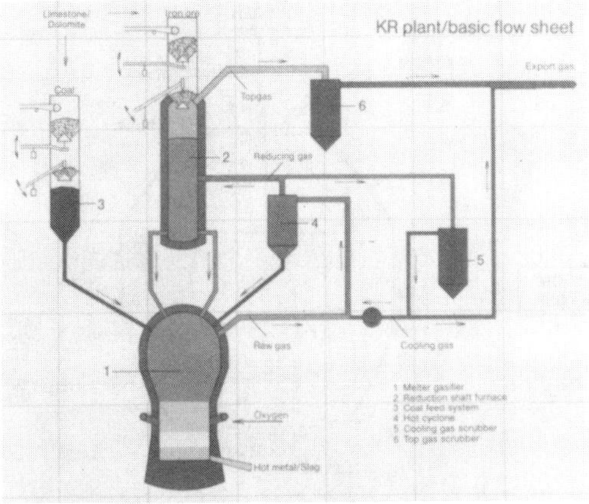

Figure 5 - Corex Process

The advent of smelt-refining does not however, guarantee the demise of the large blast furnace in existing facilities. The development of inexpensive, environmentally sound cokemaking or better utilization of the current worldwide coke production imbalance will allow existing large furnaces to operate effectively for many years.

Steelmaking

There is a considerable amount of process development activity in hot metal treatment and steelmaking, such as the Klockner-Maxhutte Steelmaking, and new AISI direct steelmaking proposal, that have the potential to replace the BOF as the steelmaking process. These processes use various combinations of submerged injection of inert gas, oxygen, and coal; post-combustion of CO; and preheating of scrap or ore to achieve efficient, economical treatment. Despite their potential, it is unlikely they will displace the BOF as the primary steelmaking process by 1995.

Hot metal pretreatment will remain separate from steelmaking to take advantage of the specific slag requirements for removing phosphorus and sulfur. It is possible, however, that with the right ores, smelt-reduction ironmaking will result in a phosphorus level sufficiently low to eliminate the need for dephosphorization. De-siliconization and dephosphorization might still be justified, however, if they permit companies to operate the BOF as a slagless steelmaking facility, with resulting savings in the fluxes needed to absorb silica, the energy needed to melt the fluxes, and iron yield losses to the slag.

The hot metal treatment vessel will have extra freeboard and submerged injection tuyeres. Removal of silicon, phosphorus, sulfur, and carbon, along with addition of aluminum and calcium, could take place in a single vessel, which may itself move from process to process. It will be essential to cast steel without interruption and a totally reliable steelmaking process is required. A channel-inductor hot metal mixer will assure a smooth supply of hot metal to the steelmaking facility and also offer the option of using additional scrap when scrap is inexpensive. Steelmaking most likely will take place in a top and bottom blown vessel, with a slagless (or nearly slagless) operation as shown in Figure 6.

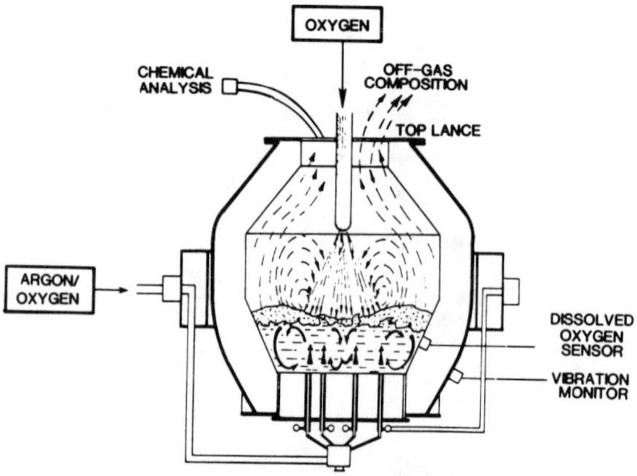

Figure 6 - Combined Blowing BOF Steelmaking

Continuous oxygen, chemistry, vibration, and off-gas sensors will monitor the vessel and process. Computers will dynamically control lance height and composition, and flow of the top and bottom gas streams, based on sensor outputs and on vessel lining volume and thermal history. This dramatic increase in automation and process monitoring will increase predictability in terms of desired end-point carbon and temperature to levels significantly above the 90 percent simultaneous hit ratio currently achieved. The overall refining time for steelmaking will be lower, increasing productivity. Iron loss to the slag will be minimized by reduced slag volume and elimination of reblows.

Ladle and Tundish Metallurgy

As mentioned earlier, this plant will require process flexibility to allow necessary product changes. However, adoption of strip casting removes one opportunity--the hot-rolling step--to control metallurgical properties. This will lead to even greater reliance on changes in chemical composition to produce small lot sizes of unique products. Varying the solid processing parameters as well as steel chemistry will be increasingly necessary. This will require new research to overcome the inherent kinetic limitations of solid state diffusion.

Steel will be vacuum-treated in the ladle to obtain ultra-low carbon when needed, and ladle furnaces will be equipped with plasma-torch heaters to prevent carbon pickup. Large alloy additions, particularly of manganese, will continue to be made to the ladle. To improve chemical control, chemical samplers will provide more representative samples and analysis will be faster and more accurate. Expanded and improved thermodynamic and kinetic models will predict the real time path of reactions between the steel, slag, and refractories. The tundish will have continuous chemistry analysis, a strong mixing region, and a system for making small additions of a large variety of alloys. It will incorporate gas bubbling and electromagnetic flow controls to minimize transition losses, and will include on-line monitoring and filtration of inclusions to meet increasing demands for steel cleanliness. Temperature control will be even more critical for thin slab casting, necessitating induction or plasma heating.

Thin Slab Casting

One or more of the many competing thin slab casting processes, such as the Hazelett twin-belt caster or the Concast-SMS thin slab caster being installed by Nucor at Crawfordsville, Indiana, see Figure 7, will be an established technology with known reliability and process control. The cast thin slabs (25 to 50 mm thick) will be hot-rolled in-line, although it isn't clear yet whether each cast strand will have its own rolling mill or one rolling mill will serve several cast strands.

Ultrasonic sensors to measure the temperature distribution within the cast strand, used in combination with heat transfer models, will control the temperature of the strand during casting and rolling. The process will achieve casting speeds up to 20 m/min, with near perfect surface quality, no internal voids, and minimal internal segregation. In-line hot rolling will save much of the capital cost of the hot strip mill and the energy cost of slab reheating.

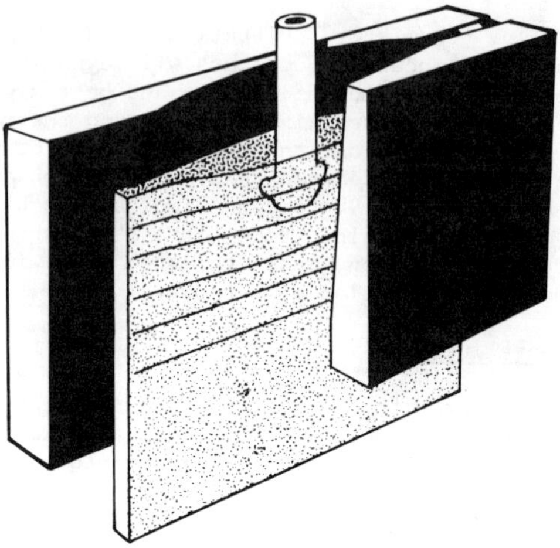

Figure 7 - SMS Thin Slab Mold

Strip Casting

We believe that, by 1995, strip casting will have been demonstrated on a pilot scale and ready for commercialization. If Twin-Roll Casting succeeds among the many competing processes it will mean the successful implementation of a 130-year-old Bessemer patent shown conceptually in Figure 8.

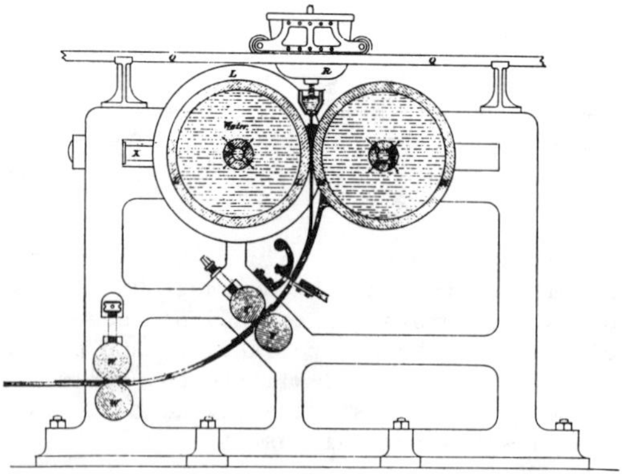

Figure 8 - Bessemer Twin Roll Casting Concept
(Improvement to British Patent No. 221; 1857)

Although the as-cast microstructure is conventional, elimination of hot rolling will result in strip that will respond differently to subsequent processing. Undoubtedly, extensive work, as mentioned previously, will be required to develop chemistries and solid processing techniques that will provide the desired range of final product properties. However, strip casting will allow production of new steel grades that could not previously be rolled to this thickness. It is likely that the steel plant for the year 2000 will have both thin slab and strip casting facilities.

Solid Processing

Hot rolled coils from thin slabs or directly wound coils from thin strip casting will continue to provide a convenient break-point in the processing sequence. The finishing facility for these products will have much in common with the Continuous Cold Mills currently in operation or being built by Inland Steel at New Carlysle, Indiana, incorporating pickling, tandem reduction, heat treating, temper rolling, and inspection into one sequence (Figure 9).

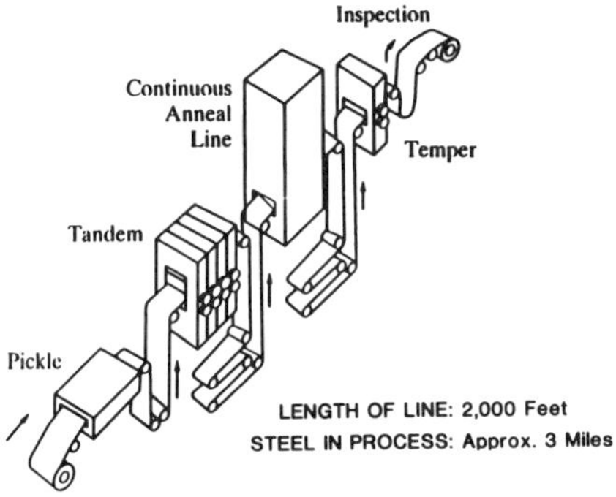

Figure 9 - Schematic of IN/Tek Continuous Cold Mill (CCM)

For the steel plant of the year 2000, the CCM will incorporate improved rolling equipment that will evolve from currently available technology; increased computer-based monitoring and control of temperature, gauge, shape, and atmosphere; and on-line automated inspection. This continuous process will produce products with end-to-end, order-to-order, and month-to-month consistency in mechanical and surface properties and dimensional characteristics and will have process reliability greater than 95%. In comparison to conventional facilities producing the same tonnage, the CCM will reduce manpower requirements by about 35%, reduce energy requirements by 30% and, improve yield by 5%. Lead time will be significantly shortened, and the process reliability will ensure on-time reliability.

Post Processing

The concept of flexible manufacturing will be most definitely seen in post processing steps. In order to supply the customer with a complete, value-added product, surface treatment processes will be placed as satellites to the continuous cold mill, as shown in Figure 10. These processes may include hot-dip galvanizing, electrogalvanizing, vapor deposition, sputtering, laser surface modification, and the application of organic films. These processes will provide materials ranging from preprimed sheet for automotive customers to prepainted products for appliance manufacturers whose only remaining step is stamping the final part. As the varieties of surface coatings increase, it will be more important to design coating lines with the flexibility to change processes and allow innovative combinations of coating processes. The result will be lines with high-quality entry and exit equipment containing a series of modules that can be quickly inserted or removed from the pass line.

Another facet of value added, namely part or system production, could also be incorporated into this steel plant for the year 2000. It is not unikely that iron ore could be the entry end product and a car door be the exiting manufactured system. Clearly, there will be a significant risk element in installation such as these, suggesting that joint ventures will likely become increasingly commonplace.

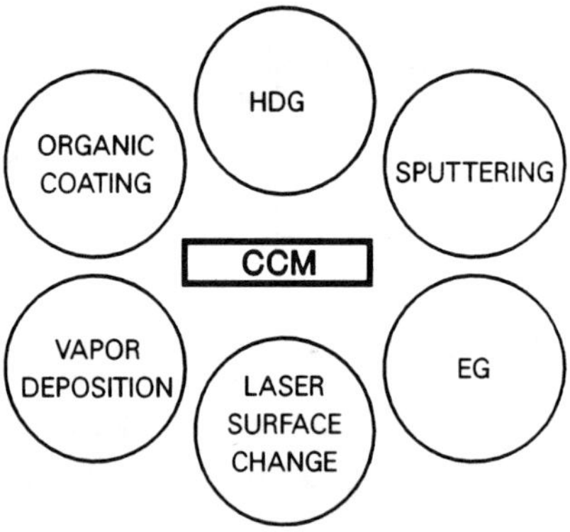

Figure 10 - CCM Post Processing

Operating Methodology

Having now defined the steel plant, let us now consider the area in which most change is likely, namely Operating Methodology. This integrated steel factory will be characterized by the most sophisticated information and process control systems available. It may be virtually paperless. Plantwide, real time information networks will be commonplace, with profound impacts on the bureaucratic organizational structure. Any and all information will be available both vertically and horizontally within and across organizational boundaries. From a user perspective, information will flow like, and be as available as, any other basic utility.

People will be trained to use information in ways suited to their own unique job demands. As a result, manufacturing functions from marketing and sales, through production control and scheduling, and to production and shipping, will be completely integrated, with a built-in capacity to accommodate change. Customers and raw material suppliers will have access to the information loop, as required. Specific applications of information technology include:

- Artificial intelligence
- On-line preventive maintenance systems
- Sensors and instrumentation
- Factory floor workstations
- Robotics

Artificial Intelligence

Artificial intelligence (AI) will be commonly utilized in numerous integrated and stand-along applications. Three applications currently emerging are production scheduling, energy management and maintenance troubleshooting.

In production planning and scheduling, Expert Systems and Model Based Reasoning (MBR) will aggregate discrete customer demands into continuous processing operations, enabling the optimization of multiple, sometimes conflicting, and often changing objectives. These systems will be highly interactive, and will provide the decision maker with ready answers to "what if" questions.

Management of energy consumption, a function incorporating evermore complicated scenarios, will be incorporated in the planning and scheduling systems to achieve optimum economic return. Superimposed over these AI systems will be simulation models of the factory to test production capabilities. The gap between model inception and development and model implementation will become very small, allowing modeling technologies to play a more active partnership role with the planning and scheduling function.

On Line Preventive Maintenance

AI will provide the ability in maintenance management to magnify diagnostic expertise and make it available to a broad spectrum of maintenance personnel. Diagnostic systems currently using isolated medical and mechanical failure applications will be the standard tools for maintaining and repairing steel mill equipment. On-line preventive maintenance system will use real-time computer controlled equipment monitoring. Maintenance diagnosing systems will parallel the integrated business/process control systems referred to previously. Self diagnosing vibration monitors, sonic and ultraviolet temperature devices, and others will continually observe equipment and dictate or recommend required maintenance. Mill motors, turbines, and fans will have continuous monitoring devices measuring amplitude and frequency to monitor the unit's "signature" and predict bearing deterioration prior to failure.

Piezo-electric sensors, also measuring electro-mechanical variations from the norm, will be commonplace for checking caster and mill mechanical accuracy. The reliability and stability of electro-mechanical equipment and machinery will be a prerequisite for cost-effective and delivery-conscious operations in the factory.

Sensor Technology

One of the major developments expected between now and the year 2000 will be in the area of sensor and instrument technologies. Manufacturers are beginning to use Integrated Circuit (IC) and optical technologies to create low cost "smart" instruments, transmitters, and sensors. Much of the labor associated with set-up, calibration, and maintenance will be eliminated. The devices will be more accurate and will have built-in diagnostic capability and sample history.

In-line multiple element chemical analysis will be available to steelmaking operations as the logical extension of sub-lance technology. Developmental work is currently underway utilizing laser vaporization of the molten steel in conjunction with Inductively Coupled Plasma (ICP) for analysis. Vision systems for in-line hot and cold surface inspection will be a practical reality. Vision systems utilizing pattern recognition technology, brightness, and contrast will also be utilized to read product identification codes placed on the steel by the robotisized systems at upstream processes.

In areas where mechanical identification systems are impractical, computer tracking systems will replace identity markings with locations, voiding the need for discrete marking systems. In-line direct measurement of physical properties, such as strength, flatness and grain size, rather than predicting their properties through precise control of process parameters will be available after extensive research has led to new technical breakthrough.

Factory Floor Workstations

The technological changes cited above will significantly change the environment and functionality of operator work stations. The workforce of the year 2000 will be required to be computer literate. The operator will be provided with on-line Statistical Process Control systems, presenting trend reports of product and process capability. Off specification product will be an unnatural event.

The utilization of closed loop process control computers will standardize the processes and require minimal manual intervention. Direct readout of product properties will be coupled directly with information systems, to which customers will have direct access.

Compact Disc Read Only Memory (CD ROM) will make it practical for volumes of standards, practices, procedures, and training materials to reside in readily accessible modes at the worksite. Natural voice input and response systems will greatly improve operator effectiveness by relieving them of the burden of operating via a CRT and keyboard.

Robotics

Classical robotics, which have a human familiarity, will find little application in the actual steel production process. Robots will be used, however, in laboratory sample preparation and testing, as mechanisms for steel identification making systems, for miscellaneous tedious tasks such as lancing of steel scrap and skulls, and possibly in pattern-based tasks such as refractory lining of ladles and tundishes and welding of specific parts.

Summary

One of the most disconcerning conclusions is that this integrated steel plant for the 21st century, as defined here, will not employ radically new process technology. Biochemical extraction of iron from ore or other totally unproven technologies are only feasible for the year 2010 and beyond.

A clear parameter for the success of this venture is the understanding of customers' requirements in a quantitative manner. Thus, if this knowledge base is frail, so will be the ability of the facility to provide flexibility. The massive scale of the continuous processes in themselves required for efficiency of operation, uniformity of properties, and cost effectiveness are the antithesis of manufacturing flexibility.

Producing steel to the exact specification required for the customer's end use, and delivering the product in a Just-In-Time fashion to the customer's receiving point, will demand an in-depth understanding of sales demand forecasting, process capabilities, and interprocess inventories. Managing the factory at the irreducible minimum of raw materials and in-process inventories will contribute significantly to its economic success.

The information revolution is expected to radically change the behavior and character of business and society. This explosion of technology in the areas of computer applications and information gathering and processing, will profoundly impact our "factory of the future" and will require intensive study and planning, in parallel with improved product and process understanding.

I believe the only intrusive flexibility of this steel plant for the 21st century will be in the minds of the men that run it.

THE AISI PROGRAM FOR DIRECT STEELMAKING

Egil Aukrust*

LTV Steel Company, Technology Center
6801 Brecksville Road
Independence, OH 44131

Abstract

An American Iron and Steel Institute Task Force evaluated developing direct steelmaking technology worldwide to seek a replacement for the coke-oven, blast-furnace, basic-oxygen process technology now in use. To remain internationally competitive it is essential to move away from the traditional series of batch processes towards continuous processing and the associated energy savings which are possible with high technology process control systems. The group identified in-bath smelting of iron ore pellets or fines with coal as the process most suitable for domestic raw materials (including abundant scrap) and called for research to study operation with varying degrees of preheat and prereduction of the ore by the gases generated from the coal and various degrees of combustion of these gases above the melt.

A coordinated, 3-year, $27 million, direct steelmaking research program is proposed. Concurrent projects include design, construction, and experimental operation of an in-bath smelting pilot plant; modeling, simulations, and laboratory-scale experiments at universities on slag-phase and metallic-phase reactions and interactions; and large-scale heat transfer and reaction rate trials. Also included are modeling studies and pilot plant trials of on-stream decarbonization and desulfurization; application of circulating fluid bed technology to fine ore reduction and hot gas cleanup; and specific heat transfer and fluid flow research projects.

---

* Dr. Egil Aukrust is Chairman of the AISI Task Force on Direct Steelmaking.

Metallurgical Processes for the Year 2000 and Beyond
Edited by H.Y. Sohn and E.S. Geskin
The Minerals, Metals & Materials Society, 1988

Introduction

The American Iron and Steel Institute (AISI), based on the Task Force studying direct steelmaking, believes strongly that a coordinated program of research on the reduction of domestic raw materials to liquid steel should be undertaken as part of the Steel Initiative program. To specify the steelmaking process for development best suited to the American steel industry and domestic raw materials, it was important to determine the state-of-the-art of direct steelmaking worldwide, both as practiced and under development, and then, by expert analysis of the processes and their potentials, to choose the process most likely to provide a domestic competitive edge.

In August 1987, the AISI assembled a task force of steel industry experts and two university professors eminent in steel process metallurgy to assess practices, developments, and trends in direct steelmaking worldwide, to select the process or processes most likely to provide a domestic competitive edge, and to outline a program of research and development to facilitate rapid implementation of the technology. Members of the task force (Table I) visited 22 organizations in eight countries. Careful compilation and analysis of the information obtained (some under nondisclosure agreements) led to the conclusion that the next generation of processes for the production of steel would be based on in-bath smelting for the production of hot metal.

TABLE I

DIRECT STEELMAKING TASK FORCE

E. AUKRUST (CHAIRMAN)	LTV
G. J. W. KOR	TIMKEN
P. J. KOROS	LTV
D. R. MACRAE	BETHLEHEM
J. C. MYERS	ARMCO
H. R. PRATT	USS
M. G. RANADE	INLAND
J. F. ELLIOTT	MIT
R. J. FRUEHAN	CMU
W. E. DENNIS	AISI

The Task Force issued an AISI internal report in late March 1988, entitled Direct Steelmaking, A Plan for the American Steel Industry.

The Task Force recommended the following concept for development:

1. In-bath smelting of iron from iron ore pellets or iron ore fines with coal in a continuous process to liquid steel.

2. The heat from post combustion of the smelting/reduction gases to be transferred effectively to the bath and the process gases utilized directly for preheating and prereduction of the ore.

3. Flexibility in the use of iron ore and scrap as charge materials to the smelter.

4. The metal will be continuously desulfurized and decarbonized resulting in low-carbon steel that only requires treatment in a ladle station prior to casting.

It was the consensus of the Task Force that the major issues related to process technology should be resolved in a large-scale pilot plant supported by focused laboratory research and coordinated with full-scale experiments of the critical process components. These steps would then lead to a demonstration project.

Technology Overview

Current concepts of in-bath smelting involve the use of a liquid pool of crude iron into which coal, and either raw or partially reduced ore, are introduced. Carbon and carbon monoxide derived from the coal act to reduce the iron oxides to liquid iron. The ore feed is preheated, and prereduced if required, in a separate vessel by means of the off-gases from the smelting unit. A schematic diagram of this type of flowsheet is shown in Figure 1. Several of the smelting concepts under investigation worldwide include post combustion of the effluent gases within the smelting vessel with injected air or oxygen. It is the intent to transfer the heat generated by post combustion back to the slag and metallic bath in the smelter. Control of the post combustion reactions and harnessing of the energy released are essential issues faced by all the organizations working on in-bath smelting methods. There have been many excellent review papers[1-7] on the major processes currently under development.

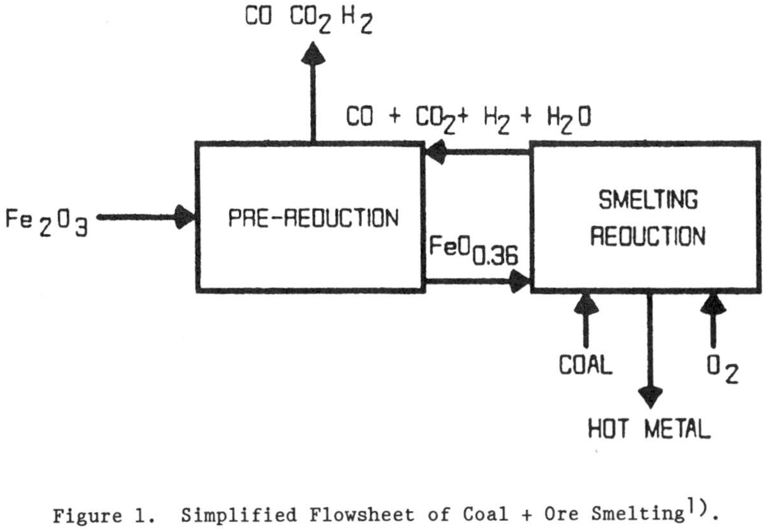

Figure 1. Simplified Flowsheet of Coal + Ore Smelting[1].

In-depth examination of the thinking and of the current development work on coal-ore to liquid metal processes that are in progress in Europe, Japan and Canada was possible because of confidentiality agreements that were made between the process developers and the AISI, and also with individual members of the Task Force. As a result of the exchanges, the breadth of development programs and detailed differences among a number of process concepts became evident.

Critical Process Issues

There are several critical process elements in the proposed smelting technology process including dissolution of the iron oxide in the slag, reduction of the iron oxide, post combustion and heat transfer and slag foaming/gas evolution. Based on discussions in Europe and Japan and our own analysis we have some knowledge of the relative importance of these processes.

Our analysis indicates that post combustion with effective heat transfer is the key to the process. In several processes the foamed slag is used for post combustion and as the heat transfer medium. A high degree of post combustion is necessary for an energy efficient process. Unless heat transfer to the slag and metal is efficient, the off-gas temperature will be too high and the top of the reactor would burn out.

Due to the large gas and slag volumes, slag foaming is significant. The gas evolution is nearly ten times larger per tonne of metal and the slag volume twice that in a BOF. This results in a large volume of foamed slag. In some tests in Japan the rate of production has been limited by the amount of slag foaming which is related to excessive gas evolution.

The rate of solution of the iron oxide in the slag is relatively fast as are the cracking and dissolution of the coal but the gas evolved from cracking may influence the process. The rate of reduction is controlled by energy transfer in the liquid phase and is dependent on the three phase (metal-slag-gas) fluid flow conditions in the reactor. Therefore, if slag foaming is reduced too far, the rate of reduction would be limiting.

In tests to date, heat transfer to the site of the reaction, which is endothermic, has been sufficient; however, at higher production rates, heat transfer may limit the rate. In any case good heat transfer from the post combustion reaction is required to limit the temperature of the gases leaving the reactor and for good energy efficiency.

Smelter. The Task Force's view is that for the U.S., the most likely process is one in which the ore is first charged into a preheating-prereduction vessel and is lightly reduced by the exhaust gas from the smelting reduction furnace. The intermediate hot product is charged continuously into the smelting reduction furnace.

The coal (heat source and reducing agent) is charged directly into the smelting reduction furnace and is burnt by injected oxygen. Work by Oeters[8)] has shown that coal reacts very quickly during in-bath smelting. Upon introduction into the liquid, coal cracks in milliseconds and the carbon dissolves in a fraction of a second. The hydrogen (from the volatile matter in the coal), plus the CO formed by reaction with the unreduced oxides, leave the bath and, if sufficient oxygen is supplied and the space is available, burn partially to CO_2 and H_2O.

The methods for introduction of ore particles, prereduced or not, also raise questions. The effect of sizing is addressed later in this section; an important issue is that fines, not to be "blown-out," must be assimilated by the slag and/or the metal phase -- but reports of "blow through" in bottom injection are of concern.

For a balanced process, e.g., one that does not generate export gas and can operate with a single stage reduction step 50 to 55% of the CO generated would have to undergo secondary combustion to CO_2. This step generates a large amount of heat; the goal is to transfer most of the energy to the iron bath.

1. If the technical upper limit for post combustion is assumed to be about 50%, it is necessary to install only a single stage prereduction facility; under these conditions the energy available in the smelter should be sufficient to complete the reduction step.

2. At $\geq$ 50% post combustion, the prereduction facility may be used only to preheat the ore; an undetermined amount of prereduction would be expected to occur.

3. With conditions (1) and (2), it becomes possible to reduce coal consumption (and the amount of gas produced) to the same or lower levels as in the blast furnace process and thus realize an energy-balanced iron works.

4. If energy must be supplied to the outside, it is possible to achieve this simply by (1) omitting prereduction, and (2) lowering the secondary combustion ratio.

Post Combustion. The effectiveness of the oxygen introduced for post combustion is a significant issue; as there are as many ways to introduce it as there are process developers, this matter requires attention in any pilot program.

Full-scale tests in 110 and 275 NT BOF's by Kawasaki, Nippon Steel and NKK have shown that high post combustion and high heat transfer are possible although demonstration of consistent operation at the $\geq$ 35% post combustion level has been reported only by one group. It is clear that further improvement of heat transfer efficiency in the furnace is the primary subject of current work; at post combustion ratios in excess of 35%, high off-gas temperatures are calculated to occur. Consequently, it is important to develop refractories capable of withstanding exposure to high temperatures. Furthermore, less costly methods for production of oxygen are needed, e.g., less than 99.6% purity, as large quantities are used to burn the coal and the resultant CO.

Slag FeO levels range from 5 to 15%, which affect yields and ultimate metal carbon and sulfur levels. Highly oxidized slag systems appear to allow operation at metal carbon contents in the 2.0-2.5% C range, a tempting target for a steel producing system. Understanding the interplay between slag composition (FeO content and basicity), foaming tendency and ability to complete post combustion and to recover its energy is critical. Our understanding is that carbon addition to the slag is used by some to control slag foaming -- but post combustion oxygen burns this carbon in preference to the CO and H_2 that evolves from coal added to the metal phase. A furnace full of slag foam is less able to allow completion of the post combustion process and to obtain effective heat transfer to the metal below.

Reports show that levels of post combustion between 50 and 60% can be obtained with some forms of carbon as the slag reductant and that lower levels of post combustion were possible with the use of high volatile coals as the slag reductant. The principal problem with this kind of coal is that it causes extensive foaming of the slag. The slag foaming consideration leads to a smelter design that differs from a conventional BOF; perhaps 30% larger for a given nominal holding capacity. Nonetheless, for this type of vessel, oxygen blowing rates more than twice those of a modern BOF are achieved; productivity per volume of furnace exceeds several times that of a modern blast furnace.

Agglomerates Versus Fine Iron Ores. One of the features used to differentiate the process flowsheets is the nature of the iron ore used as a feedstock. The choice is between fine iron ores (either naturally occurring sinter ores or the much finer concentrates resulting from beneficiation) and iron ore agglomerates (pellets). From our North American perspective, the focus is on concentrates and pellets, as sinter ores come from offshore and lump ore availability is low. The situation is different for almost every steel producing location.

Use of unagglomerated ore is attractive because the capital and operating costs associated with agglomeration can be avoided. However, whereas the capital costs associated with maintaining cokemaking capacity are quite high and provide a large economic incentive for using coal instead of coke, the capital costs for maintaining pelletizing plants are relatively low. Operating costs are also low (about $5 to $10 per ton). In addition, the cost of transporting fine ores (specifically, North American concentrates) is not known as the infrastructure is not yet in place. Nevertheless, in taking a long term approach, there is some potential for reducing the delivered cost of iron units to the steel plant by using unagglomerated ore.

A more important issue to be resolved is how the choice of feedstock affects the process in terms of productivity, controllability, and flexibility. These factors are critical to determining the overall process economics. The limitations imposed by the choice of ore type not only restrict the process configuration, but define much of the auxiliary equipment requirements, which affect capital costs. The level of prereduction desired and the means for achieving it also affect, and are affected by, the choice of ore type.

The real attraction of fine iron ores is not so much the avoidance of the agglomeration step as it is the potential for significant increases in reaction rates (a direct result of the small particle size) in both the smelter and the prereduction parts of the process. The resultant high productivity would then be reflected in lower capital costs. Determination of the productivity of the new processes is a critical aspect of any economic assessment or development program.

In Japan, some companies' attention has been focused on the use of fine ores because of the potential for higher productivity in the smelter, as discussed above. In contrast to shaft furnace reduction, it is difficult for fluid beds to achieve high degrees of reduction without multi-staging, due to the lack of counter-current flow. Although some of the processes which utilize fluid beds are aimed at 60-70% reduction, several are limiting prereduction to 33% (FeO). Problems with high temperature sticking can thereby be avoided.

Because of the long-term potential for fluid bed/fine iron ore technology and the development of that technology already underway, it was included in our domestic program.

For the U.S. scene, with ample supply of modern and cost effective pellet plants, the simplicity in using the existing infrastructure is very appealing. Thus, for the short term -- that is, the remaining life of the pellet plants -- we propose to focus on pellet-based processes. This has the advantage of relying on the demonstrated shaft reduction technology. Operation to achieve only about one-third oxygen removal -- to wüstite -- is relatively easy and less prone to sensitivity to swings in smelter post combustion (CO_2/CO ratio) than the fluid bed reducers.

Incentives for the New Steelmaking Process

1. High Productivity. The new processes may achieve several times the production of a modern blast furnace (per inner volume) because of the high reaction rates, and therefore the furnace can be small.

2. Simplification of Facilities. The blast furnace process requires stoves and coke oven batteries in addition to the blast furnace itself, but the new process(es) requires only a converter-type furnace attached to a prereduction vessel, because coal and the prereduced raw materials are best charged directly.

3. Flexibility of Raw Materials. Because coal is added directly without being converted to coke, non-coking coal can be used, i.e., without the need for coking coals. However, high heat value coals are needed for these processes. Substantial quantities of scrap can be introduced to displace ore units.

4. Operational Flexibility. The processes being developed elsewhere are for making pig iron but it appears to us that a "semi-steel" is possible. Further development work will be directed to complementary desulfurization and decarburization processes.

5. Scrap Melting Flexibility. The energy generated in the process can either be used for iron ore smelting or for melting of scrap, a flexibility of great importance in the U.S. An immediate spin-off from involvement in these development programs would be the broadening of conventional BOF capabilities.

AISI Program

The coordinated direct steelmaking research program will have three major components: a pilot plant program, a laboratory research program, and a set of full-scale trials. It will also have several supporting programs including circulating fluid bed studies, decarbonization and desulfurization studies, and mixed-phase heat-transfer studies.

The pilot plant program will entail the engineering and construction of a 10-ton smelting vessel designed for coal injection, oxygen blowing and ore, pellet, or prereduced iron additions and for varying degrees of post combustion of the gases generated by coal addition to the melt. The pilot plant will then be operated for about 18 months to study the various process options. The general flowsheet of the process is shown in Figures 2 and 3, and the key elements of the experimental program is delineated in Table II.

The laboratory research program will be conducted concurrently by Professor Richard J. Fruehan at Carnegie Mellon University and by Professor John F. Elliott at the Massachusetts Institute of Technology. This research

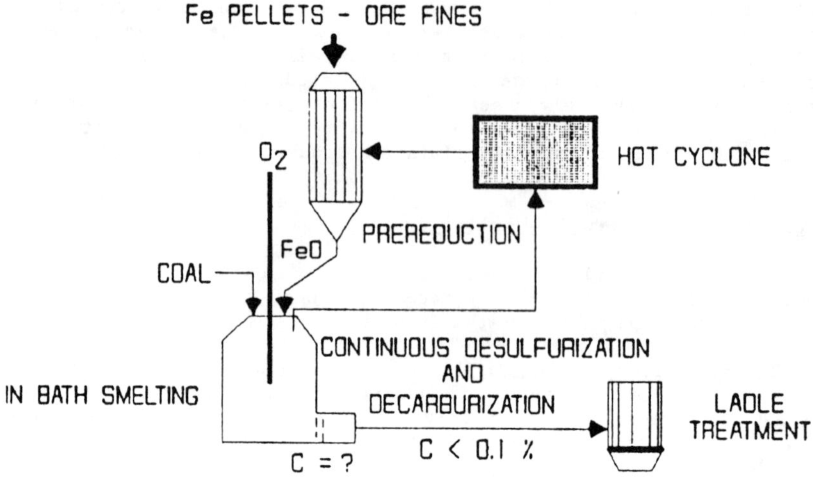

Figure 2. General Flow Sheet of the Direct Steelmaking Process Under Development by AISI.

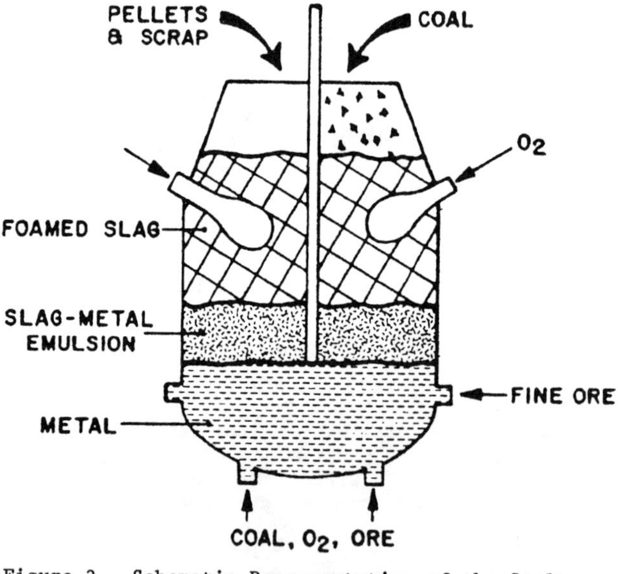

Figure 3. Schematic Representation of the Smelter.

will include modeling, simulations, and laboratory-scale experiments on the various slag-phase and metallic-phase reactions and their interactions.

The full-scale trials will also be conducted concurrently and will be designed to study the critical reactions in production-size basic-oxygen-process furnaces. Coal and oxygen will be injected under various conditions and specialized instrumentation will determine reaction rates and heat transfer rates to identify scale-up problems and to verify potential solutions.

TABLE II

PILOT PLANT IN-BATH SMELTING
10-TON CONVERTER

ISSUES	EXPERIMENTAL PROGRAM
PREREDUCTION	• 0.90% PREREDUCTION TO FeO • COARSE VERSUS FINE ORES
POST COMBUSTION	• SOFT VERSUS HARD BLOWING • FOAM AND EMULSION STABILITY
HEAT TRANSFER	• FLUID BLOW IN CONVERTER AS FUNCTION OF BLOWING AND CHARGING TECHNIQUES • POST COMBUSTION • SCALE-UP CRITERIA
PRODUCTIVITY	• RATE LIMITING MECHANISM: - FOAM STABILITY - HEAT TRANSFER (PR-PC) - SCALE-UP

Studies with circulating fluid beds in copper and lead smelters, cement kilns and a coal gasifier show promise. The application of this technology to hot-gas cleanup, heat extraction, and iron ore reduction will be explored.

Water-flow modeling of continuous decarbonization and desulfurization processes will be conducted. Designs developed in these studies will be tested at the pilot plant site with melts simulating the iron from in-bath smelting.

To assure that experts in heat transfer beyond those in the steel processing community can make contributions, research needs relevant to in-bath smelting will be defined and a work shop will be organized. Researchers and research organizations will be encouraged to submit proposals responsive to the defined research needs.

Summary and Conclusions

A technology based on bath smelting, post combustion methods, and coal injection offers potential for major advances for smelting of ores and electric furnace scrap melting to produce steel.

For successful United States participation in this new era, proposals suitable for the Steel Initiative have been outlined aimed at continuous ore-based steelmaking and more efficient scrap melting.

The development of direct steelmaking will be achieved by an evolutionary process, and thus there is an opportunity for the American industry to play an important role in this development. More urgently, our industry needs to acquire competence in this field if it is to have a stake in the development game in the longer term, and to have a competitive technology a decade from now. Incentives for this technology include, but are not limited to, an estimated $10 to $25 per ton variable operating cost advantage over the coke oven/blast furnace/BOF route, reduced life cycle capital costs, and elimination of coke batteries.

It is the consensus of the Task Force that the major issues related to process technology should be resolved in a large-scale pilot plant supported by focused laboratory research and coordinated with full-scale experiments of the critical process components.

There are also significant short-term benefits from this program through adaptation of the new technology to existing operations. These include increased scrap melting flexibility from post combustion within the BOF process and productivity improvements and electric power savings from post combustion and carbon-oxygen injection into electric furnaces. The development program will foster a revitalization of process metallurgical research within the Industry and the Universities and provide the technical expertise to implement the process in years to come.

This program will provide momentum for the U.S. steel industry along the path toward direct steelmaking. As a first successful step it will lay the foundation for much larger programs to be undertaken jointly and by individual companies to implement this technology during the coming decade.

Acknowledgment

Many thanks are extended to all the members of the AISI Direct Steelmaking Task Force (see Table I) who's dedicated work made this study, and the definition of the AISI program, possible. This paper is based on the internal AISI report.

References

1. Brotzmann, K., "New Concepts and Methods for Iron and Steel Production," Steelmaking Proceedings, ISS-AIME, 70 (1987) 3.

2. Smith, R. D. and Corbett, M. J., "Coal Based Ironmaking," Ironmaking and Steelmaking, The Metals Society, 14 (1987) 49.

3. Tokuda, M. and Kobayashi, S., "Process Fundamentals of New Ironmaking Processes," PTD Proceedings, ISS-AIME, 7 (1988) 3.

4. Oeters, F. and Saatcia, "Some Fundamental Aspects of Iron Ore Reduction With Coal," PTD Proceedings, ISS-AIME, 6 (1986) 1021.

5. Tokuda, M., "Conceptual and Fundamental Problems of Smelting Reduction Process for Ironmaking," Trans. ISIJ, 26 (1986) B192.

6. Hatano, M., et al., "New Ironmaking Process by Use of Pulverized Coal and Oxygen," PTD Proceedings, ISS-AIME, 6 (1986) 1049.

7. Robson, A. L., et al., "A Pilot Plant Study into the Melting Reduction of Iron Ore With Coal and Oxygen," Pyrometallurgy '87, London University.

8. Orsten, S. and Oeters, F., "Dissolution of Carbon in Liquid Iron," PTD Proceedings, ISS-AIME, 6 (1986) 143.

STEEL IN THE YEAR 2000

William T. Hogan, S.J.

Professor of Economics
and
Director, Industrial Economics Research Institute
Fordham University
Bronx, New York

By the year 2000, the steel industry, not only in the United States but in the rest of the world, will undergo significant changes. On the other hand, there will be a number of areas in which change will be minimal.

The current capacity of the industry worldwide is approximately 800-810 million metric tons. This may grow to some extent, as Third World countries install capacity. However, much of it will be neutralized as the industrialized countries shrink their current capacities.

There will be several breakthroughs in terms of technology, some of which deal with the production of iron outside of the blast furnace. However, these processes will not be far enough advanced by that time to replace the blast furnace, rather they will supplement it.

In terms of steelmaking and the subsequent rolling process, there should be a number of basic improvements. The strip mills will roll sheets with a very slight variance in gauge, which will give the customers a better product for automatic stamping machines. Ordinary carbon steel, in a number of instances, will be replaced by high-strength, low-alloy steel, which will reduce the tonnage necessary for present-day applications, such as construction.

The corporate organization will be significantly changed as joint ventures and mergers will reduce the number of companies, as well as facilities in operation. This will enable the industry to reduce its capital investment and improve operating costs, so that it will be able more effectively to compete with other materials, such as plastics, cement, aluminum, etc.

Metallurgical Processes for the Year 2000 and Beyond
Edited by H.Y. Sohn and E.S. Geskin
The Minerals, Metals & Materials Society, 1988

By the year 2000, the steel industry, not only in the United States, but in the rest of the world, will undergo significant changes. On the other hand, there will be a number of areas in which change will be minimal.

The current capacity of the industry worldwide is approximately 800-810 million metric tons. This may grow to some extent, as Third World countries install capacity in the 1990s. However, much of it will be neutralized to some extent as the industrialized countries shrink their current capacities. The principal producers in the Third World are Mainland China, South Korea, Brazil, India, Mexico, and Taiwan. China has increased its output from 20 million metric tons in 1976 to 60 million tons in 1988. Brazil rose from 9.2 million tons in 1976 to 22.7 million in 1987. South Korea rose from 3.5 million tons in 1976 to 16.8 million in 1987, and Taiwan from 1.1 million tons to 5.6 million in the same period.

Most of this increase was achieved through the installation of modern equipment. For example, in South Korea, a completely new integrated, greenfield site mill has been built at Kwangyang which embodies the very latest rolling technologies on its hot-strip mill. The plant currently has a capacity of 5.4 million tons and will ultimately be in excess of 8 million. By contrast, in the industrialized countries, capacity has fallen drastically. In the United States in 1977, raw-steel capacity stood at 160 million net tons. Currently, it is approximately 110 million tons. In Japan, capacity has been reduced by some 30 million metric tons, and the same is true of the European Community. Thus, the growth in the Third World in the past has been more than offset by the shrinkage in the industrialized world.

Investment in the industrialized world has concentrated on improving quality, reducing costs, and increasing productivity. In the Third World, investment has been geared to increasing capacity with the most modern equipment. The Third World will take advantage of new technology as it expands its industry, whereas the industrialized world will use the new technology to replace existing facilities.

In terms of new technology, several major breakthroughs are anticipated, although these will not be sufficiently advanced to replace current facilities completely. They will exist side by side with older equipment by the year 2000, and it will take another 10 years or more before the older equipment is completely eliminated.

At the present time, although attention is directed to virtually every phase of steelmaking and finishing, more emphasis is being placed on the production of iron and steel, particularly that of iron. Current iron and steelmaking technology in the integrated plants is tied to the blast furnace, which produces molten iron that is ultimately turned into steel and steel products. The blast furnace is a huge facility which, if constructed new, would require a staggering investment. A furnace capable of producing 5,000 tons of iron a day would require an investment of more than $300 million. Further, it must be supported by coke ovens, since the operation consists of reducing iron ore to molten iron with coke providing the fuel.

Coke ovens represent another large investment and also involve pollution problems. A battery of coke ovens, capable of producing 1 million tons a year would require an investment of approximately $200 million. Thus, the ironmaking process represents a very large investment, and as a consequence, a great deal of effort and study is being applied to find a substitute that will produce it satisfactorily.

The Japanese are particularly concerned since many of their blast furnaces and coke-oven batteries are to be replaced within the next 10 years, and they would like to avoid the investment if possible through the substitution of another process which would involve smelting iron ore to form molten pig iron through the use of coal, thus bypassing the coke oven and possibly the blast furnace.

Currently, the five major integrated companies in Japan have a joint venture involving the development of a process to make iron outside the blast furnace. Some progress has been made not only by the Japanese but by others in Europe and the United States. However, this has not been sufficient to warrant a projection that the blast furnace will be rendered obsolete by the year 2000. One of these processes, known as the KR Process, has recently been installed in South Africa, and it remains to be seen how successful it will be.

Judging by the rate of progress, these substitute processes will replace a portion of blast-furnace activity, perhaps 15% by the year 2000. An attempt to reduce dependence on coke has been made by injecting coal at the blast furnace tuyeres, as much as 200 pounds per ton of iron produced has been successfully applied, thus reducing the amount of coke required by at least 15%.

A few new coke-oven batteries have been constructed in Western Europe, and at least one is planned for the United States. The new developments in this technology consist principally of widening the individual coke ovens that make up the entire battery. For decades, the slot oven has been 18 inches wide. The new ovens are 24 inches wide and 6 meters high. A number of attempts has been made to reduce the pollution including placing a hood over the car into which the coke is pushed. Whatever coke ovens will be constructed in the future will, most probably, be the wide variety. However, this does not represent a fundamental breakthrough in technology. By the end of the century, coke ovens will still be very much in evidence. However, the possible replacement by a new smelting process could eliminate the need for about 20% of the world capacity.

In regard to steel operations, the basic-oxygen process, which was developed in Austria in the early 1950s, will dominate steel production through the end of the century. It constituted the most dramatic technological breakthrough in the post World War II period. Prior to that time, the open-hearth process was the dominant steel producer throughout the world. In the United States, it accounted for 90% of steel production in the 1940s and early 1950s.

The oxygen process is carried out in a pear-shaped vessel which is charged with molten iron and scrap. In the United States, the proportions are roughly 70% iron and 30% scrap, while in Japan, scrap is less than 10%. An oxygen lance is lowered to within a short distance of the molten metal, and within 25 minutes, the reaction produces steel by reducing the carbon.

The spread of the oxygen process after its introduction in the 1950s was facilitated by the fact that the steel industry in Western Europe and Japan was expanding rapidly in the 1960s. The expansion in the United States had taken place in the 1950s, before the oxygen process was considered fully developed. Thus, a number of steel companies in the United States, in order to meet the rapidly growing demand in the 1950s, installed the familiar open-hearth process. There were some 50 million tons of this capacity put in durir the 1950-60 period.

The principal growth in steel production in Japan took place between 196(and 1975, when output rose form 22 million tons to 119 million. Fortunately for the Japanese, the basic-oxygen steelmaking process was well developed by this time. The same was true of Europe, where many of the companies expanded steelmaking capacity during the 1960-70 period. During the late 1960s, when there was need for steel-capacity replacement in the United States, the basic-oxygen furnace, as well as the electric furnace, expanded considerably.

As of 1986, on a worldwide basis, the oxygen process accounted for 56% of the steel produced, with the electric furnace adding 27%, and the open hearth some 18%. These figures are somewhat distorted by the fact that the Soviet Union produces 53% of its steel by the open-hearth process, while the Eastern European countries produce substantial portions with the same process The United States, Western Europe, and Japan produce less than 2% by this process.

By the year 2000, steel will be produced predominantly by the oxygen method, with the electric furnace accounting for an increasing amount. In terms of the non-Communist Bloc, the oxygen process will account for approximately two thirds, with the electric furnace producing the remainder. In the Soviet Union and the Eastern satellites, the proportion of the oxygen process will increase during the coming decade, as the open hearths are replaced. It is doubtful that there will be a radical change in oxygen or electric steelmaking by the year 2000, although some improvements will take place.

The electric furnace has gained considerable ground in the last two decades. However, its growth is somewhat restricted by the availability and cost of electric power. One of the principal developments has been eccentric bottom pouring, which permits a more efficient emptying of the furnace by a better control of slag. This will grow during the next decade to a point where most of the electric furnaces will be equipped with it.

Currently, electric furnaces are, with few exceptions, fed with iron and steel scrap. This is supplemented in a number of areas by direct-reduced iron, a process which uses gas to remove the oxygen from iron pellets so that a 60-62% iron content is upgraded to 85-90%, and this can be used directly in the electric furnace. The basic difficulty in widespread production of direct-reduced iron is the availability of natural gas.

In a number of areas, such as the Middle East, Venezuela, and Mexico, natural gas is readily available and quite cheap, and it is in these areas that the direct-reduction process has flourished. Venezuela has plans to increase its output by a minimum of 3 million tons in the next few years, providing this material not only for its own use but also for sale throughou the world. It is difficult to estimate the amount of direct-reduced materia that will be produced by the year 2000. To date, progress has been slow. However, the demand of a growing electric-furnace population and the declini availability of high-quality scrap should result in a substantial increase in direct-reduced iron production.

As was stated previously, the basic-oxygen process will remain the dominant facilty and will account for a minimum of 60% of world-steel output. The remainder will be provided, in great part, by electric furnaces, which have grown in number over the past 10 to 15 years. They are particularly popular for small installations, such as those that dominate the Third World countries. One of the serious problems is the cost and availability of electric power in some of the Third World countries.

Another basic change which took place in the second half of the 20th Century was the installation of continuous casting which eliminated four of the steps in the conventional steelmaking process. In continuous casting, the liquid steel from the furnace is cast into a mold, and a semifinished form such as a billet, bloom, or slab emerges. This eliminates the need to put steel into ingot molds, strip them, reheat the ingot in a soaking pit, and roll it down to semifinished form on a slab, billet, or blooming mill. Continous casting produces a superior product in most instances and saves considerable energy.

By 1987, 81% of the steel produced in the European Community was continuously cast. Sixty percent was cast in the United States and 93% in Japan. It is difficult to see how the Japanese percentage can be increased, however, in the United States by the year 2000, the percentage cast will probably be in the range of 90%. The same is true for the European Community. On the other hand, the Soviet Union cast approximately 15% of its steel in 1986 and will probably not reach 50% by the year 2000.

One development in casting which is new is the production of thin slabs, two inches or less in thickness. This has just been introduced by Nucor in the United States, and there is a possibility that it may become widespread. Other experiments are aimed at casting thinner and thinner steel. In Germany, the objective is one-half inch in thickness. In other areas, the attempt is directed at casting strip directly, which, if successful, could reduce and possibly eliminate much of the expensive equipment required for rolling sheet steel. Success in the area of strip casting is, without question, a few years away, and there is no doubt that the continuous hot-strip mill will be very much in evidence by the year 2000.

Both continuous casting and oxygen steelmaking will be operable at the turn of the century. The basic changes in the process of iron and steelmaking that are under development at the present time are much more concerned with the blast furnace and coke ovens than the steelmaking process.

A recent development which will become widespread, if not a standard facility in the steelmaking shops of the world, is the ladle furnace. This provides a buffer between the steel furnace and the continuous caster. It also performs a function of refining steel, so that in an electric furnace, the scrap is merely melted and the remainder of the process takes place in the ladle furnace. It consists of a steel ladle to which a top containing an electrode is added, so that the temperature of the steel can be maintained or increased during the refining process.

The finishing facilities are many and varied. The most active is the hot-strip mill. In 1987, some 43 million tons of steel passed through the hot-strip mills in the United States. When one considers that total shipments of all products amounted to 76 million tons, this is, indeed, a large percentage. The products emerging from the hot-strip mill are further processed and include hot-rolled and cold-rolled sheets, such coated items as tinplate and galvanized sheets, as well as electrical sheets, also known as silicon sheets. These products require cold reduction, annealing, pickling, and plating. Other finishing facilities include structural mills, pipe mills for both seamless and welded items, rod mills, bar mills, rail mills, and plat mills.

In the next decade, radical changes will take place in the production of flat-rolled products. The finishing stands on the strip mills will be equipped with roll bending, side shifting, hydraulic screw-downs, and further computer controls, all of which are directed at improving the quality of the flat-rolled products. A great deal of effort will be expended to produce a uniform gauge, better surface and flatness. This is particularly desirable for consumers that operate repetitive stamping facilities.

The traditional four-high mill, with two work rolls and two back-up rolls, will be replaced, in great part, with a six-high mill, with four back-up rolls and two work rolls. There are two such cold-reduction mills under construction in the United States today. One is at Pittsburg, California, where there is a joint venture between Pohang Iron and Steel Company of South Korea and United States Steel, a subsidiary of USX. The other is a joint venture in northern Indiana between Inland Steel and Nippon Steel of Japan. The quality of the product of these mills should be second to none in the world and better than most and could well force other flat-rolled producers to install like facilities for competitive purposes.

One area which has received considerable attention and requires a substantial capital investment is continuous annealing for sheets. The results are outstanding and may compel the installation of this equipment in a number of companies in order to remain competitive.

Other finishing facilities, particularly for the long products such as bars, rods, and structurals, will undergo improvements. However, at present, there does not seem to be any radical change on the horizon.

Much has been said about the use of lasers in the production of basic iron and steel. Without question, progress will be made, however, it does not seem that the application will be widespread enough by the year 2000.

There are several current developments, some of which are in the experimental stage, others in the pilot-plant stage, and still others in the early production stage. One of these is the Korf Process, referred to as the Energ Optimizing Furnace, for making steel. The process uses scrap which is reduce to molten metal by coal and its gases. It represents a relatively low capita investment.(1)

Another radical change in the steelmaking process, which could achieve some degree of growth in the next 10 years, is spray forming. This process consists of spraying an atomized liquid metal through an inert atmosphere on a metal substrate, such as a roll surface, to form a thin strip. The rotating roll then feeds the metal between two rolls which compact it into a sturdier metal strip. One application of this process is referred to as the Osprey Process.(2)

By 2000, the steel industry, not only in the United States but throughout the world, will be producing fewer tons of crude steel in relation to product shipments. This will be due to a number of developments, including continuous casting and lighter, stronger steel, as well as the increased production of high-strength, low-alloy steels.

Corporate organization will be significantly changed in the next 10 years as more joint ventures are formed. These will be between domestic companies, as well as between domestic and foreign steel producers. The aforementioned Inland and USX international agreements will be multiplied, particularly in the installation of additional facilities for electrolytic continuous-galvanizing operations. There are six of these in existence and more are planned. For example, Nippon Steel and Inland are considering such an installation, as is LTV and Sumitomo.

The joint venture has a number of advantages including a much lower capital investment for each of the parties, as well as a higher operating rate, since one facility will take the place of two.

In the minimill field, there will be much more concentration as firms merge. Already there are three firms operating four or more plants, and this trend will continue.

In summary, changes in the technology for the production of iron and steel, as well as forming it into finished products, will take place by the year 2000. However, many of them will be modifications of existing technology, such as improvement of the hot- and cold-strip mills for producing sheets. A fundamental breakthrough, such as replacing coke ovens or blast furnaces, or for that matter the hot-strip mill itself, will be slow in developing and, by 2000, will not be widespread. It took more than 10 years for the basic-oxygen process to be fully accepted, and it will take at least that much time for fundamental and radical changes in technology to be developed and generally applied. There will be some changes by the turn of the century, but their applications will be limited. Significant changes, however, will take place in the corporate structure and organization of a number of steel companies.

References

1. W.A. Tony, "Connecticut Steel to Build North America's First EOF," Ironmaker and Steelmaker, August 1988, 18-19.

2. Alan W. Cramb, "New Developments in the Continuous Casting of Steel, Part VI: Spray Forming and Other Processes," Ironmaker and Steelmaker, June 1988, 34-37.

Bibliography

Alan W. Cramb, "New Developments in the Continuous Casting of Steel, Part VI: Spray Forming and Other Processes," Ironmaker and Steelmaker, June 1988, 34-3

Richard J. Fruehan, Ladle Refining Furnaces for the Steel Industry(Pittsburgh, PA: Center for Metals Production, March 1985).

_______________, Ladle Metallurgy Principles and Practices(Warrendale, PA: The Iron and Steel Society, Inc., 1985).

Manfred Haissig and Robert A. Heard, "Horizontal Continuous Thin Slab Casting, Iron and Steel Engineer, May 1988, 32-35.

W.A. Tony, "Connecticut Steel to Build North America's First EOF," Ironmaker and Steelmaker, August 1988, 18-19.

THE CONSTEEL PROCESS FOR CONTINUOUS FEEDING-PREHEATING-MELTING AND REFINING STEEL IN THE ELECTRIC FURNACE

John A. Vallomy

Intersteel Technology, Inc.
Charlotte, N.C. 28226

Abstract

The CONSTEEL Process introduces a new concept in electric steelmaking by continuously feeding-preheating-melting and refining ferrous scrap or direct reduced iron. The process uses the furnace off-gas and fuel to preheat the scrap in a specially designed preheater. Thermal incineration of the flue gas reduces the carbon monoxide level to the more stringent requirements. Hot water or steam production from heat of flue gas is optional. The tap-to-tap time is set to match the cycle of the caster. The steel is treated via ladle metallurgy. The environment inside and outside of the steelplant is improved by decreasing levels of noise and hazardous emissions. Energy consumption decreases and disturbance to the electric network is eliminated. A CONSTEEL prototype was tested at Nucor Steel, Darlington, SC. A greenfield project is in the completion stage at the Florida Steel plant of Charlotte, NC, USA.

Metallurgical Processes for the Year 2000 and Beyond
Edited by H.Y. Sohn and E.S. Geskin
The Minerals, Metals & Materials Society, 1988

The Process Philosophy

The philosophy of the CONSTEEL Process is not solely to enhance efficiency; this was just a natural result of the change from a batch to a continuous practice. The primary objective in development was four-fold:

1) 100% process control
2) Rational use of available energy
3) Improved environment inside and outside the plant
4) Stability of energy input from the power system

It was to make electric steelmaking more a science than an art and amenable to modern automation.

The Basic Concept of CONSTEEL

The concept of the process is shown in figure 1 and includes basically a continuous feeding system, a preheater and a melting - refining furnace. A pre-established amount of heat is introduced into the scrap while traveling through the preheater using the sensible and chemical heat contained in the furnace off-gas and additional fuel. Melting-refining is carried out in the electric furnace. Slag formers, slag foaming additions and oxygen are injected to refine the melted steel continuously, to minimize the refractory erosion, to shield the arc, thus optimizing the transfer to the metal bath of the heat generated in the arc, to keep the bath homogeneous in composition and temperature and to produce carbon monoxide (CO) which is fuel for the preheater.

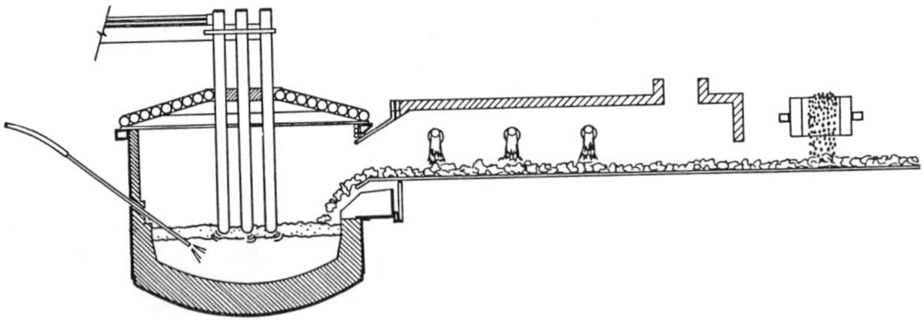

Figure 2 shows the straight-line arrangement of the CONSTEEL process with the continuous feeding system, the preheater and the melting-refining furnace. It has been proven, however, that the feeding system can make a 90 degree turn before entering the preheater.

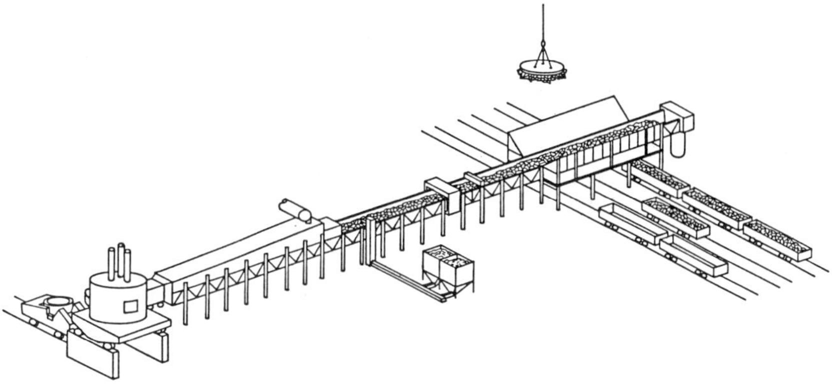

Figure 2 - The straight-line arrangement of the CONSTEEL system

Scrap Feeding

Any commercial scrap grade can be charged in furnaces of about 4.8 m in diameter or more, although some scrap grades may benefit less than others from the preheating. Scrap preparation is not required for smaller furnaces if the dimensions are within specifications and the feeding system has a straight line configuration.

Once the proper scrap mix for the furnace has been established, it is important to keep the scrap bed as homogeneous as possible to increase preheating efficiency.

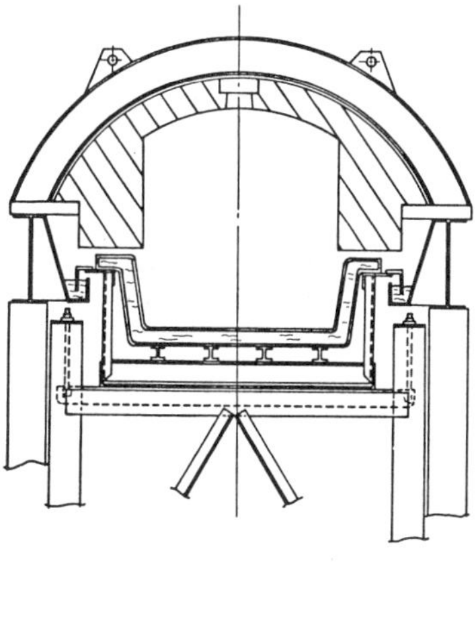

Figure 3 - CONSTEEL preheater

Efficiency of Scrap Preheating

The preheater has a cross-section as shown in figure 3, and is designed to introduce 20 to 40 percent of the heat required to melt the scrap. Preheater efficiency is 30 to 50 percent.

With proper scrap charging, the preheater efficiency increases with the volume of carbon monoxide (CO) generated in the melting furnace and brought to the preheater with minimum combustion to be used as a fuel. Therefore, good control of the melting furnace pressure and air intake from the furnace to the preheater are critical for minimizing the oxidation of the carbon monoxide before the preheater. Figure 4 shows a schematic of the pressure control in the melting furnace which is kept at -0.05 inch of water column at the roof level.

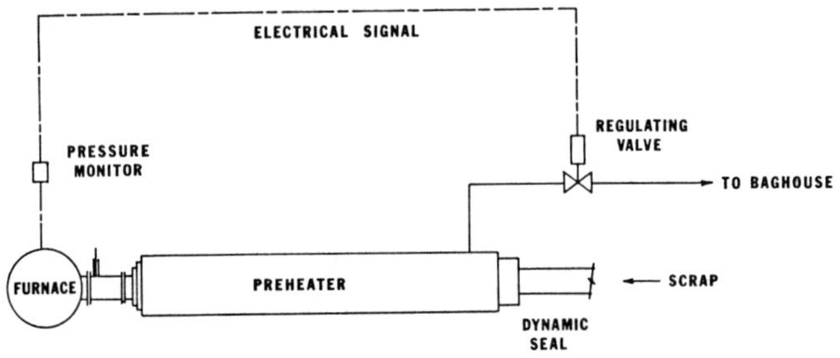

Figure 4 - Schematic of pressure control in melting furnace in CONSTEEL operation

Table I shows that the cost of introducing 1 Kwh of heat in the scrap in the preheater is 50 percent less than introducing it in the form of electricity in the melting furnace. The cost of electrical energy and natural gas assumed in this figure are typical for the USA. The flue gas exiting the preheater is very low in residual carbon monoxide. When required, especially if the steelplant is located in an urban area, the amount of carbon monoxide can be further decreased to meet the local standard by installing an afterburner.

The Melting-Refining Furnace

Continuous melting-refining of the steel is smooth with a hot heel of metal acting as a "thermal flywheel". The slag composition is continuously monitored to minimize refractory erosion and for proper fluidity. In fact, a steady foamy slag is needed at all times to transfer most of the heat generated by the arc to the steel and to protect the furnace walls from radiation (See fig. 5).

Table I - Cost comparison of introducing 1 Kwh of heat into the steel

COMPARISON OF HEATING COST

ELECTRIC FURNACE	PREHEATER
1 Kwh COSTS US$ 0.045	US$ 12.00 PER 1,000,000 Kcal
60% EFFICIENCY	30% EFFICIENCY
1 Kwh IN THE STEEL COSTS US$ 0.075	1 Kwh IN THE STEEL COSTS US$ 0.035

1 Kwh = 860 Kcal = 3413 BTU

FUEL COST = US$ 3.00 PER 1,000,000 BTU
US$ 12.00 PER 1,000,000 Kcal

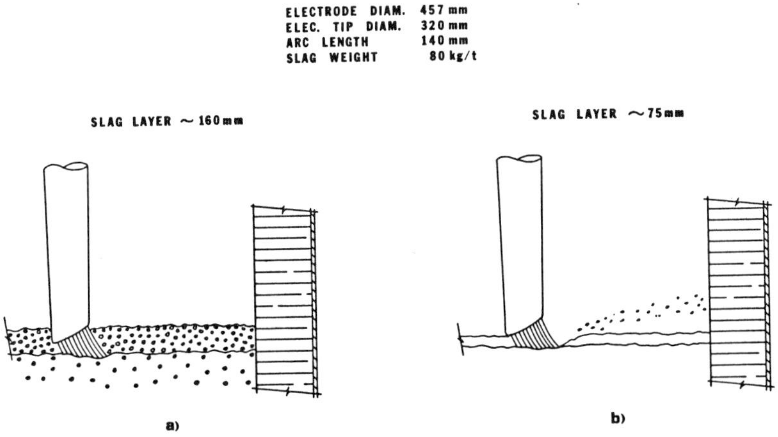

Figure 5 - Transfer of heat generated by the arc is about 80 percent in a) versus about 30 percent in b)

The bath temperature is kept in the range of carbon boil while oxygen injection is used to stir and circulate the bath, (fig. 6) to foam the slag and to generate carbon monoxide to be combusted in the preheater. In the Consteel operation, therefore, each Nm3 of oxygen injected yields a saving of about 5 Kwh versus only 3 Kwh in the conventional EAF operation. Since the furnace needs less power, the distance from the arc to the wall is increased, and the arc radiation is shielded by the foamy slag, there are no water cooling panels in the furnace walls. In effect, the refractory lining design of the Consteel furnace must be more similar to that of an oxygen converter than that of a conventional electric arc furnace. The roof is partially water-cooled for structural stability.

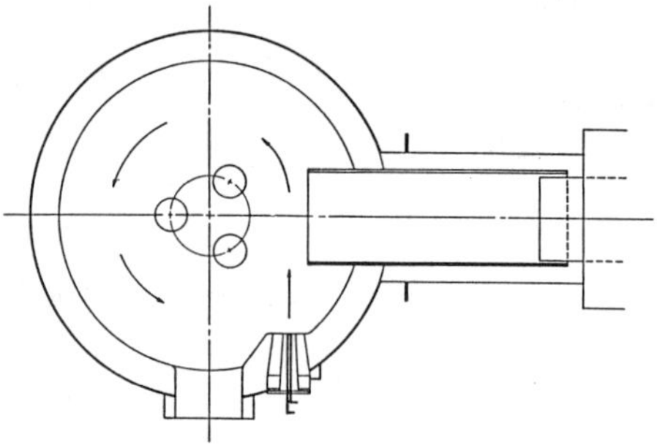

Figure 6 - CONSTEEL submerged tuyere

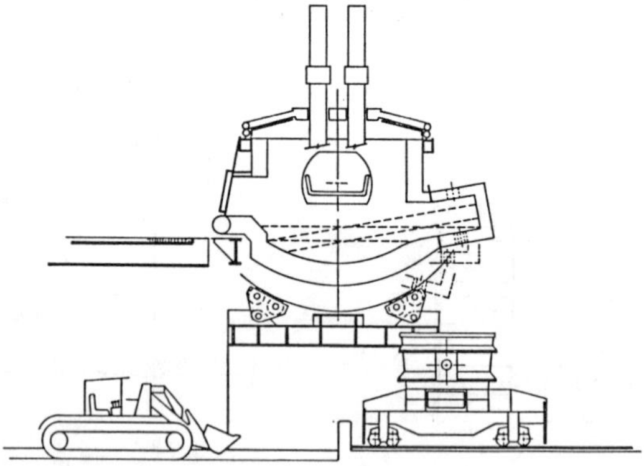

Figure 7 - Cross-section of CONSTEEL melting-refining furnace with feeding opening

Figure 7 shows the electric arc furnace with the scrap feeding opening. The CONSTEEL cycle of 40 min shown in figure 8 can be changed to match the continuous casting machine cycle. The thermal balance of the electric arc furnace is given in figure 9. In this example the oxygen injected is 15 Nm3 per tonne of steel and the scrap is preheated to 540 °C. Note that the sensible and chemical heat of the off-gas made available to the preheater makes up for about 17 percent of the total heat output. Assuming a 50 percent preheater efficiency, the heat available in the EAF off-gas supplies about 50 percent of the sensible heat in the scrap charged.

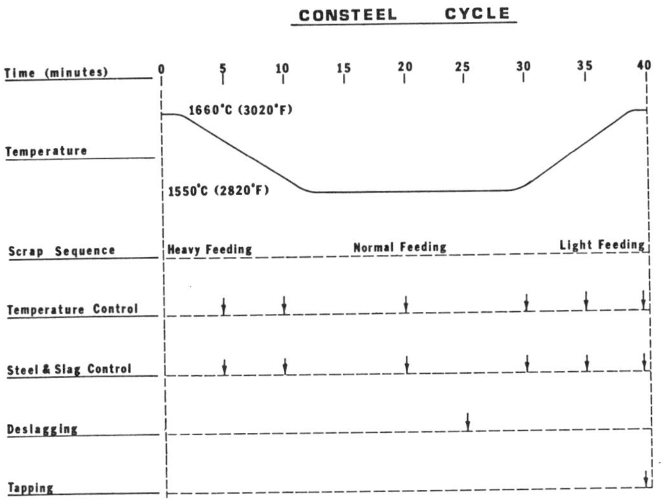

Figure 8 - CONSTEEL cycle

The Economics of the Operation

The energy consumption in the electric arc furnace versus scrap preheating temperature is shown in figure 10 and compared with the electric energy consumption in an all-electric conventional operation. In comparing the conventional and CONSTEEL operation in fig. 10 and 11, the oxygen lancing has been kept at approx. 15 Nm3/tonne in both cases. Figure 11 shows the decrease of energy demand for the same annual capacity and compares the conventional operation with the CONSTEEL operation with an increasing preheating temperature. The decrease in energy demand in the CONSTEEL operation is associated with an increase in the coefficient of utilization of the contracted energy, which results in a decrease of about 10% in the price paid per Kwh by the steelmaker. The productivity increase for an increasing scrap preheating temperature is shown in table II.

Table III shows the savings attainable in processing steel scrap into 1 tonne of billet when preheating the scrap to about 500° C. The savings given in $US show the strong influence of the Kwh cost.

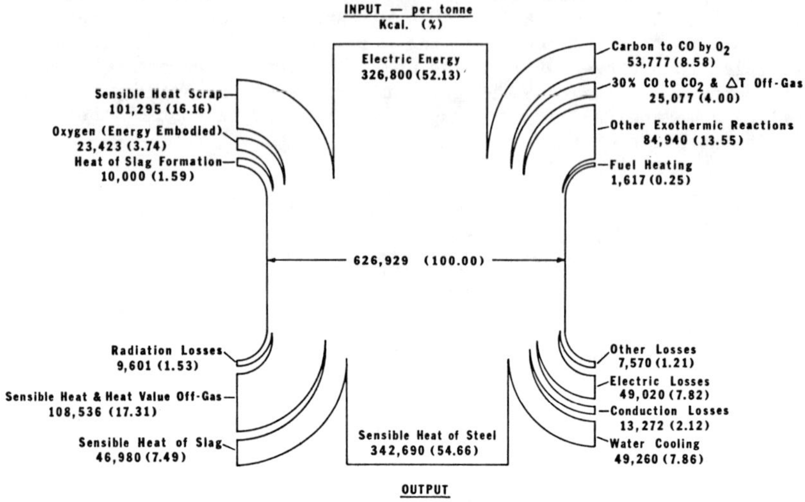

Figure 9 - Thermal balance of CONSTEEL melting-refining EAF - Scrap temperature 540°C, Oxygen used 15 Nm3/tonne

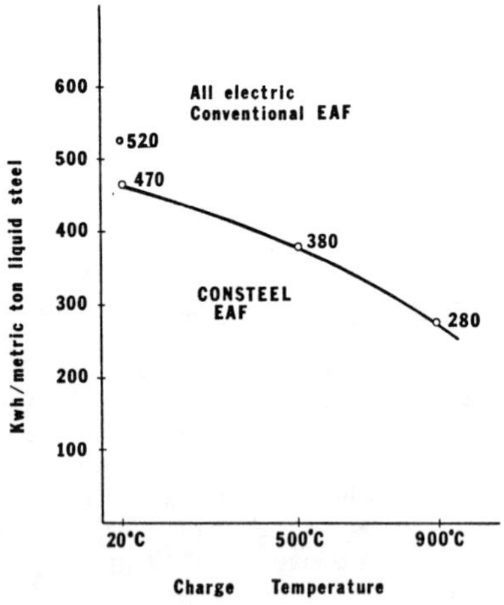

Figure 10 - Relation between scrap temperature and energy consumption in CONSTEEL operation

Table II - Productivity increase of EAF via the CONSTEEL process

Metric ton liquid steel	Metric ton/MVA/h	Short ton/MVA/h
Conventional EAF all electric melting	1.10	1.20
CONSTEEL without charge preheating	1.50	1.65
CONSTEEL with 500°C (930°F) preheating	2.00	2.20
CONSTEEL with 900°C (1650°F) preheating	2.80	3.10

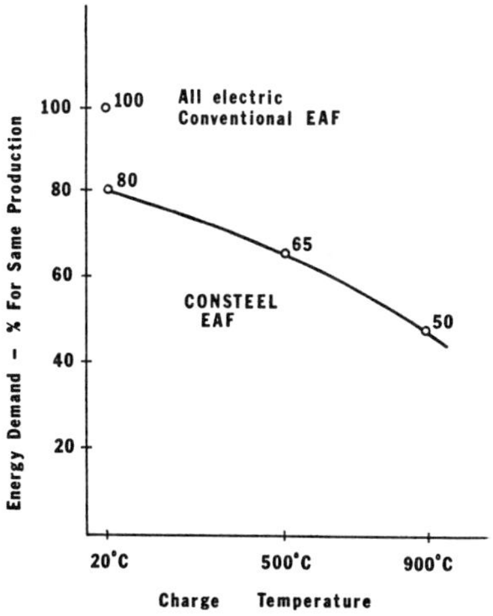

Fig. 11 - Energy demand of conventional operation versus CONSTEEL operation for same annual capacity

Table III - Consteel Savings

Charge Preheating at 500°C

	KWH COST U.S. $		
PER TONNE OF BILLET	0.024	0.045	0.060
Electric Energy	5.00	7.50	10.00
Electrodes	2.50	2.50	4.50
Manpower	4.00	5.00	7.00
Others	1.00	2.50	1.50
Total	12.50	17.50	23.00

The Nucor Steel Experience

The Nucor Steel Corporation was the first to commit the financial and human resources needed for an industrial application of the concept in 1985.

The experience gathered in this industrial prototype (fig. 12) has been invaluable for establishing the feasibility of the process. This is a story of pioneering success considering it was the first application of an untested concept using only the in-plant human resources and in a project lasting only 10 months. Reduction to practice of the concept provided information leading to significant improvements in design and layout of process equipment (Table IV).

On the operational side, the decisive influence of some basic parameters has been confirmed.

- A consistent heel of metal acts as a "thermal flywheel", for quick melting of scrap.
- Keeping the bath temperature within a proper range ensures a constant equilibrium between metal and slag and a continuous bath boil, which results in a bath that is homogeneous in temperature and composition.
- The foaming of the slag can be precisely controlled at any point during the operation and is a must of the process.

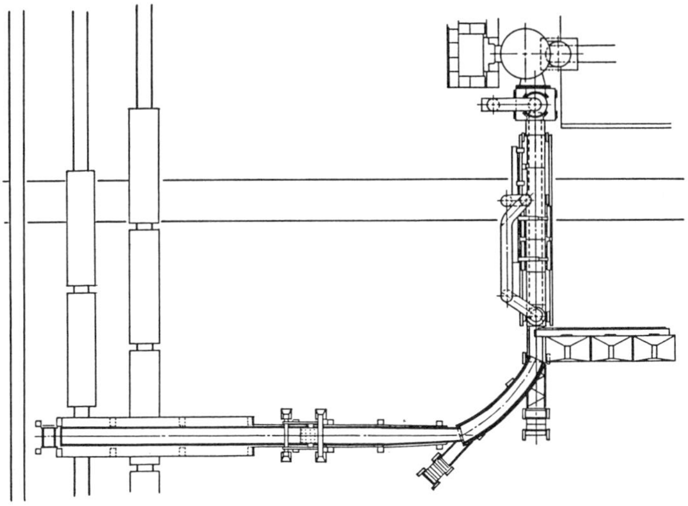

Figure 12 - General layout of Nucor Steel furnace

The Florida Steel Greenfield Project

The project involved a new meltshop built next to an existing one. The layout of the new meltshop is shown in figure 13. It is a straight-lined continuous feeding system which takes the scrap from the rairoad cars to a loading station or charging box, through the preheater, the connecting car, and into the electric arc furnace. This installation features a number of improvements arising from the experience with the first industrial prototype. The human resources and time frame allocated were adequate for a project of this magnitude.

With the meltshop located in the proximity of an urban area, the preheater design includes an afterburner for strict control of the carbon monoxide emissions. The scrap preheating temperature can also be increased up to 700°C. Particular emphasis was on maximizing the air tightness of the system from the electric arc furnace to the preheater for efficient use of the CO generated in the furnace bath as fuel for the preheater. Secondary metallurgy will adjust the chemical composition and temperature of the steel for continuous casting.

An elevation view of the Florida Steel CONSTEEL installation is shown in figure 14. The continuous feeding system shown in figures 13 and 14 includes a deduster to remove dirt from the scrap fed to the preheater and a hopper-feeder to charge cast iron borings. The slag formers are injected into the EAF from the silos shown in the figures.

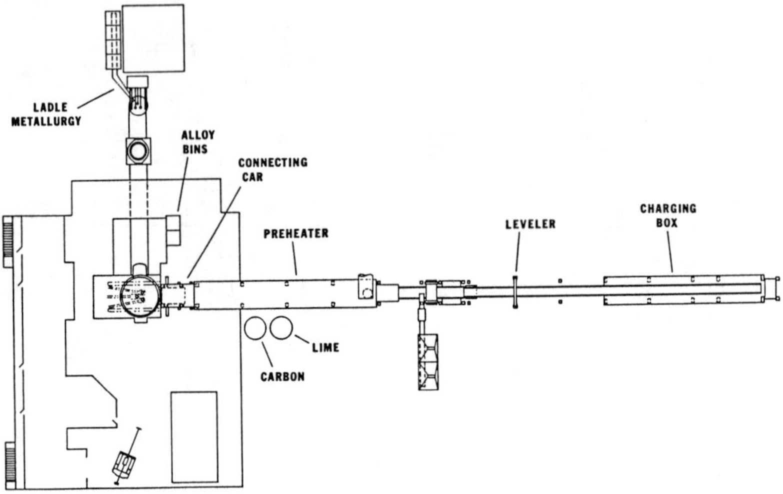

Figure 13 - CONSTEEL layout at Florida Steel

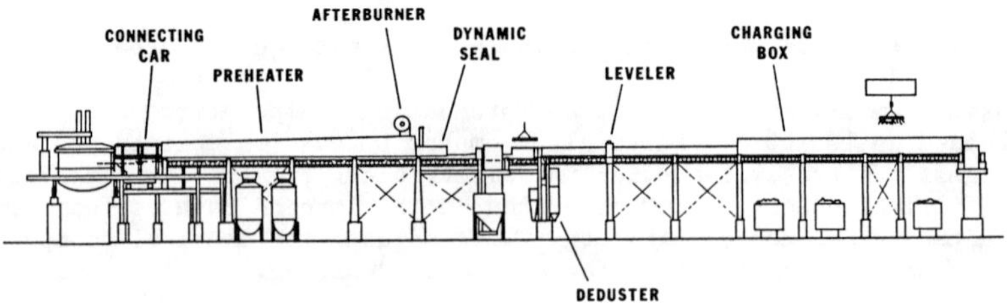

Figure 14 - CONSTEEL installation at Florida Steel

Table IV - Characteristics of CONSTEEL Prototype and Florida Steel Project

NUCOR STEEL CONSTEEL PROTOTYPE

CONTIFEEDING SYSTEM	4 Conveyors in cascade, 1.2m width, 0.3m depth, 75m length, with 90 turn before preheater entry. Scrap loaded from railroad cars.
CONTINUOUS SCRAP PREHEATER	18m in length, N.G. burners, preheating temp. up to 500°C using EAF off-gas and fuel.
ELECTRIC FURNACE	4.85m diameter, 26 MVA transf., 457mm electrodes, heat weight 32 tonnes, tap-to-tap time 35 min. Furnace tilting on rockers. Connecting car.

FLORIDA STEEL CONSTEEL GREENFIELD SYSTEM

CONTIFEEDING SYSTEM	3 Conveyors in cascade, 1.37m width, 0.3m depth, 80m in length, straightline configuration, scrap loaded from railroad cars.
CONTINUOUS SCRAP PREHEATER	24m in length, N.G. burners, preheating temp. up to 700°C using EAF off-gas and fuel.
AFTERBURNER	Afterburner reduces CO emissions to levels required at location of plant in urban area.
ELECTRIC FURNACE	5.0m diameter, 30 MVA transf., 508mm electrodes, heat weight 36 tonnes, tap-to-tap time 45 min. Furnace tilts on rollers, furnace pivoting point in line with preheater and connecting car axis.
LADLE METALLURGY	6 MW transfo, gas stirring and alloy trimming.

The Impact of Advanced Technology

The impact of the transition from conventional electric steelmaking to the CONSTEEL Process is in many ways similar to the shock steelmakers experienced during the transition from ingot casting to continuous casting. The transition will be easier when the changes are not underestimated and the personnel have been fully prepared for the change and trained in all the details of the new technology.

The process offers tremendous advantages but demands that the entire steelmaking procedure be rewritten, from scrap management to furnace operation.

Investment Cost for a New Meltshop and for a Retrofit

For a CONSTEEL meltshop charging scrap preheated from 500° C to 700° C and producing 350,000 to 400,000 tonnes/yr of billets including ladle furnace and induction heating of billets after the casting machine, the cost is in order of US $85/metric tonnes of billets/year.

The investment cost for retrofitting the process in an existing shop with scrap preheating to 500°C is about US $ 2 million in the USA. A study to assess the technical feasiblity of retrofitting the process is required in any case. Table V gives the meltshop characteristics for scrap preheated to 500°C and different annual capacities.
Figure 15 shows an optional arrangement when a straight-line installation is not possible.

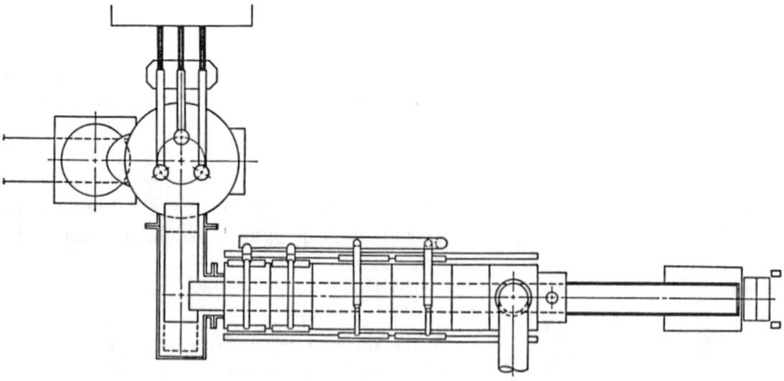

Figure 15 - Optional feeding arrangement for CONSTEEL process

Table V - CONSTEEL Meltshop Characteristics for Scrap Preheated to 500° C

Capacity tonnes/year	Tranformer MVA	Production rate tonnes/hr	Heat size tonnes	Furnace dia. meters	No. of caster strands
70,000	5	9	7	3.6	2
150,000	10	20	16	4.3	3
234,000	15	30	25	4.6	3
300,000	20	40	33	4.9	4
375,000	25	50	42	4.9	4
450,000	30	60	50	5.8	5

Summary

The process represents a logical evolution of the electric steelmaking process. It features

- an efficient use of the energy available with option for co-generation of steam and water;
- an improved efficiency of the casting process which becomes truly continuous, thus favoring direct rolling and increasing the productivity of the rolling mill;
- a lower energy demand, which is attractive where electric emergy is expensive and where the electric network is weak;
- a better human environment in the meltshop and in the surroundings of the steelplant with a decrease in the noise level, fumes and dust emissions;
- the ability to operate in a conventional mode when required, thus preserving the flexibility to produce special steel grades which could not be produced in a continuous feeding mode;
- the possibility to be installed in most existing meltshops with a return of the investment of about one year, and
- the capability to use a DC furnace or a plasma furnace as a melting unit.

Finally, the process establishes the feasibility of an integrated miniplant where hot, direct-reduced iron produced in a contiguous direct reduction plant is continuously charged in an AC or DC furnace with variable percentages of scrap.

COUNTER-CURRENT CONTACTING IN METAL REFINING - A REALITY IN THE YEAR 2000?

D. G. C. Robertson

Generic Mineral Technology Center for Pyrometallurgy
Fulton Hall
University of Missouri-Rolla
Rolla, MO 65401

Abstract

Counter-current staged reactors offer several advantages over batch reactors for mass transfer between two phases. Several metallurgical reactor designs have utilized the counter-current principle, e.g. the WORCA and QSL processes. But before the principle can be widely adopted, it will be necessary to overcome several significant problems. These include the need to minimize longitudinal mixing in a single reactor or to use several separate staged reactors and to carry out different reactions in sequence, e.g. phosphorus oxidation and sulfur reduction. Theoretical and experimental approaches to these problems will be discussed in the paper.

Metallurgical Processes for the Year 2000 and Beyond
Edited by H.Y. Sohn and E.S. Geskin
The Minerals, Metals & Materials Society, 1988

ANALYSIS OF BATH SMELTING PROCESSES

FOR PRODUCING IRON

R.J. Fruehan, K. Ito and B. Ozturk

Department of Metallurgical Engineering and Materials Science
Carnegie Mellon University
Pittsburgh, PA 15213

This paper reprinted with permission from Transactions of the ISS, November 1988

ABSTRACT

A simulation model for bath smelting processes for the production of iron was developed which predicts the coal, flux, ore, and oxygen consumptions and the off gas volume, temperature and composition. The model is comprehensive in that it takes into account all of the important variables including coal composition, metal composition, ore composition, slag basicity, post combustion ratio, (PCR), prereduction degree (PRD), heat transfer coefficient (HTC), flux, scrap charge, and heat losses. Four basic cases were considered: I. 30% PRD-50% PCR; II. 90% PRD-0% PCR; III. 60% PRD-30% PCR; and IV. 0% PRD-50% PCR. Several different coals were considered and a sensitivity analysis of the critical variables was performed. The model also estimates the sulfur content of the metal. The major conclusions included:

1. Post combustion significantly reduces coal consumption but above 20% PCR little reduction of FeO to Fe can be performed with the off gas.
2. Prereducing to FeO (Case I) and having as much post combustion as consistent with good heat transfer is an attractive process. This process only requires a simple prereducer, uses less coal, and is relatively insensitive to the type of coal used.
3. High off gas temperatures may pose a potential problem. The off gas temperature can be reduced by using an O_2-Air mixture for post combustion, limiting post combustion or adding water to the gas.
4. The use of $CaCO_3$ in place of CaO or of supplemental electricity does not appear attractive.
5. The melting unit is theoretically an energy efficient scrap melter. For Case I using 200 kg of scrap as part of the charge the coal consumption decreases by about 80 kg.
6. With PCR > 30% the FeO content of the slag is expected to be 2-5%, and the metal will not be saturated with carbon. These factors and the increased sulfur load since coal is the fuel indicate the sulfur content of the metal may exceed 0.25%

Metallurgical Processes for the Year 2000 and Beyond
Edited by H.Y. Sohn and E.S. Geskin
The Minerals, Metals & Materials Society, 1988

INTRODUCTION

It is generally believed that no new coke plants will be built in the United States and older inefficient blast furnaces will not be replaced. Several processes for producing iron are being developed which use coal in place of coke, have higher productivity per unit volume than a blast furnace, and whose capital cost may be significantly lower than for a coke oven - blast furnace process. Many of these processes have been described recently (1).

The most fully developed process is the Corex (KR) Process. In this process coal is combusted to CO and H_2 to produce the heat to melt iron pellets reduced from oxide pellets reduced using the off gas. However, a family of more promising processes are being developed using in-bath smelting and partial post combustion of the CO and H_2. These processes use less coal and possibly may have a lower capital cost than Corex. An example of such a process is the CRA/Klockner process described by Brotzman (2). Similar processes are being developed in Japan (3).

Briefly, iron ore concentrate or pellets which may be partially reduced are injected into an iron-carbon bath or simply added to the top of the vessel. Coal is injected with a top lance or bottom tuyeres or simply added as lump coal. The iron ore is reduced and the reaction products are CO and H_2. The CO and H_2 are post combusted with oxygen or air above the bath and the heat of post combustion is transferred back to the bath. Lime is added as the flux. The off gas which may be reformed is used for preheating and prereduction of the ore.

In order to analyze and evaluate these processes, precise energy and material balances are required. These balances become complex because of the interaction between the operating variables which include:

Degree of Prereduction (PRD)
Degree of Post Combustion Ratio (PCR)
Heat Transfer Coefficient (HTC) from post combustion
Metal and coal compositions
Air - Oxygen for post combustion

Oeters and Saatic (4) published simplified materials and energy balances for pure carbon and Fe_2O_3 and complete heat transfer (HTC=100%). In a later report (5) they developed complex algebraic equations for using coal and computed the balances for a few simple cases. In these calculations they only considered a HTC of 100%. In any real case it is less than 100% which significantly complicates the calculations and has major impact on the process. Tokuda (6) has also presented some energy and materials balances. However, there has not been a complete evaluation for North American coals and ores showing the effect and sensitivity of the process to various operating variables. The purpose of this work is to evaluate the processes. The effect of varying the PRD, PCR, HTC, metal composition, coal composition, the flux, scrap charge and adding electricity is determined, and an estimate of the sulfur content of the metal is made.

SIMULATION MODEL

Figure 1 shows the flow chart of the computer program. The coal consumption which is the weight (kg) of coal to produce 1 tonne of hot metal is set as an unknown variable. After reading all the initial operating conditions, the program solves multi material balance equations for a given coal consumption. For example, the material balance equation for iron is written in equation (1).

$$Fe_{(flux)} + Fe_{(ore)} + Fe_{(coal)} + Fe_{(scrap)} = Fe_{(hot\ metal)} + Fe_{(slag)} + Fe_{(dust)} \qquad (1)$$

Then, the ore consumption, lime (or lime stone) consumption, and the amount of slag are calculated. The composition of the off-gas is calculated at a given gas temperature in the subroutine using the equilibrium relationship for water shift reaction and the dissociation of CO_2. The heat capacity and the volume of off-gas is then calculated. The temperature of the gas is calculated using its composition, the PCR (post combustion ratio), and the HTC (heat transfer coefficient), where PCR and HTC are defined in equation (2) and (3), respectively,

$$PCR = \frac{\%H_2O + \%CO_2}{\%H_2 + \%H_2O + \%CO + \%CO_2} X\ 100 \qquad (2)$$

$$HTC = \frac{Heat\ transferred\ to\ bath}{Heat\ generated\ by\ post\text{-}combustion} X\ 100 \qquad (3)$$

The consumption of oxygen or air for post combustion is also calculated. The obtained off gas temperature (T_g) is compared with the temperature used to calculate the equilibrium composition (T_g^o). If the absolute value of the difference between T_g and T_g^o is greater than the conversion limit, the new T_g^o is given and the computer repeats the same calculation. The modified two-division method was used for the iteration in this program. At very high temperatures, above 2600°C, the decomposition of CO_2 becomes significant; therefore reaction (4) was considered in calculating the gas composition.

$$CO + 1/2\ O_2 = CO_2 \qquad (4)$$

Based on all the material consumptions and the gas compositions the summation of heat for inputs and outputs are calculated. All the thermochemical data are obtained from the literature (7) If the heat is too much the coal consumption is decreased and the computer repeats the same routine, and vice versa. The two division method was employed for the iteration for the coal consumption. When the difference becomes smaller than the conversion limit, the results are displayed and printed.

The inputs are:

- Metal Temperature
- Metal Composition (%C, %Si)
- PRD (Prereduction Degree)
- PRC (Post Combustion Ratio)
- Slag Basicity
- Slag FeO

Dust
Heat Losses
Scrap Charge
Coal Composition (%C, %H, %N, %O, %S, %H_2O)
Ash Composition (%Al_2O_3 + %SiO_2. %CaO, %Fe_2O_3)
Ore Composition (%Fe, %SiO_2 + %Al_2O_3)
Temperature of inputs (air, ore, flux)

The program offers the option of using $CaCO_3$, scrap and electricity.

The program computes:

- Materials consumption
  - Ore
  - Coal
  - CaO or $CaCO_3$
  - Oxygen
  - Air
- Slag weight
- Gas volume, composition, and temperature

The program also makes a rough estimate of the expected sulfur content of the metal. This calculation is discussed later in detail.

Model Verification

The computer model calculations were verified by manual calculations for a single case. In addition, the model was checked by comparing our results for those cases which have been published (3-6); our results agreed within $\pm 2\%$.

Choice of Input Materials

The most critical choice of inputs is the coal. Coal varies greatly in composition and heat of combustion. Three coals were considered: a typical metallurgical coal (A), a higher volatile coal (B), and a Western coal (C). The composition and heats of combustion are given in Table 1. The ore was taken as a typical Minnesota ore containing 65.72% iron and 5% SiO_2. The dust composition was taken as 50% Fe, 25% Fe_2O_3, 12.5% CaO and 12.5% SiO_2. The coal and ore compositions were supplied by the American Iron and Steel Institute. In some cases all the coals were considered. In doing the sensitivity of the process to several operating conditions only the metallurgical coal was considered.

Cases Considered

Theoretically it is possible to have any amount of prereduction and post combustion. However, in general the highest amount of post combustion is 50%. In addition, in any real process it is difficult to vary PRD and PCR independently. In general as post combustion increases, the degree of prereduction decreases. If the post combustion is high, the gas will not have enough reduction potential to obtain the degree of prereduction desired. For example, for one of the cases considered, 30% PCR, it is not possible to obtain 60% PRD. Of course, it is possible to remove the CO_2 and H_2O or reform the off gas so that it could carry out the prereduction. However, this is expensive in the case of gas scrubbing and requires an additional unit, while for reforming additional coal is required

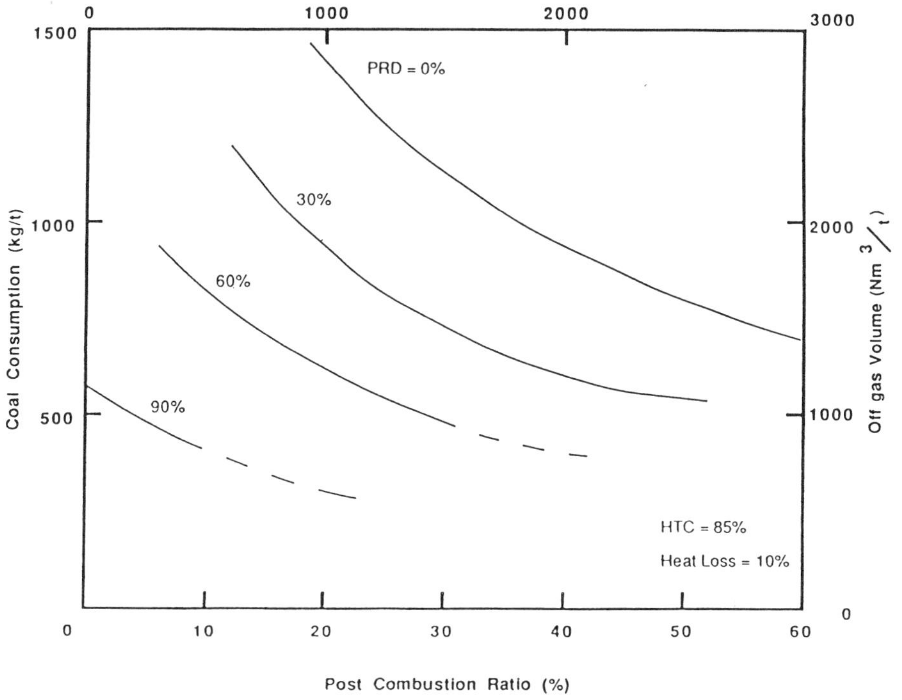

Figure 2: Coal (A) consumption as a function of PCR and PRD, HTC=85% and 10% heat loss.

and produces more gas. Therefore, in the processes being considered the PCR and PRD are chosen so that the amount of gas scrubbing or reforming is small.

Therefore, in the present analysis, four possible cases are considered which are consistent with the above considerations.

TABLE 1: Composition and Heats of Combustion of Coals Used in Simulation

wt%	Coal A	Coal B	Coal C
ash	6.4	5.95	4.9
C	84.7	79.5	73.9
H	4.2	5.2	5.0
N	0.9	1.5	1.5
O	3.0	6.5	14.2
S	0.8	1.32	0.47
ΔH (kcal/kg)	8090	7920	7512

Ash	A and B	C
SiO_2	55	40
Al_2O_3	32	30
CaO	1	20
MgO	1	5
Fe_2O_3	6	4

Case I. 30% PRD and 50% PCR

In this case the ore is reduced only to FeO. Experience indicates that typically 90% of the Fe_2O_3 is reduced to FeO resulting in 30% prereduction. In this process the prereduction can be accomplished in a single stage reducer.

Case II. 90% PRD and 0% PCR

This case is similar to the existing Corex process. The prereduction probably would have to be accomplished in a Midrex type shaft furnace.

Case III. 60% PRD and 30% PCR

This is an intermediate case, and the prereducer could be a shaft or a two stage fluid bed. Calculations indicate the gas off the smelter will not provide 60% PRD and some gas reforming is required.

Case IV. 0% PRD and 50% PCR

This is the simplest possible case in which no prereduction is done. The pellets may be preheated but if it is done with the off gas some prereduction would be expected.

For each case, a sensitivity analysis was performed to determine the effect of important variables such as HTC, silicon content, flux used, effect of electricity, amount of dust, the addition of scrap, etc. The heat loss was arbitrarily set at 10% of the total heat generated in the reactor, which is relatively high. The effect of varying the heat loss is discussed later. The carbon content of the metal was 3.0% for Cases I, III and IV and 4.5% for Case II; varying the carbon has little effect on the energy and material balances.

RESULTS

In general, coal consumption decreases with increasing PRD and PCR as shown in Figure 2 for Coal (A) with HTC = 85% and 10% heat losses. The dashed parts of the curves is when the off gas is not suitable for achieving the prescribed prereduction without significant reforming or scrubbing. The coal consumption for the four cases and for each coal is given in Table 2. For Coal (A) the ore, CaO and oxygen consumptions are given in Table 3 and the off gas composition in Table 4.

Case I

Considering the capital costs, an attractive case is prereduction of the ore only to FeO (Case I) using a single stage prereducer. The simulation results are summarized in Tables 2-4. However, it may be possible to achieve 50% PCR with good heat transfer efficiency. Therefore, the calculations were done for all the coals as a function of PCR and the results are shown in Figure 3. As PCR increases the differences in results obtained for the three coals becomes less. This is because for the H_2 in the coal to be effective it must be oxidized to H_2O. Also even though Coal (B) has a larger heat of combustion than does Coal (C), at a low PCR more of Coal (B) is required. Again this is due to the higher hydrogen content of Coal (B).

Another important operating variable is the heat transfer efficiency. It was assumed for the case above that it is 85%. The coal consumption for Coal (A) and T_g as a function HTC at 30% PRD and 50% PCR are given in Figure 4. If HTC is only 70%, the coal consumption increases to 626 kg/t, the gas temperature is 2262°C, and the gas volume is 1234 Nm^3/t. This high gas temperature would cause excessive heating in the upper part of the vessel and most likely would burn out that area. The gas temperature for the assumed case of 85% HTC, 1881°C, may even be too high. Known systems can generally handle gases up to about 1760°C. The HTC would have to be over 90% to reduce T_g to 1760°C for this case, and this high an HTC may not be possible. There are several ways of lowering off gas temperature in addition to having a higher HTC. Three methods were considered: using air as the post combustion gas, reducing the amount of post combustion and adding water to the gas.

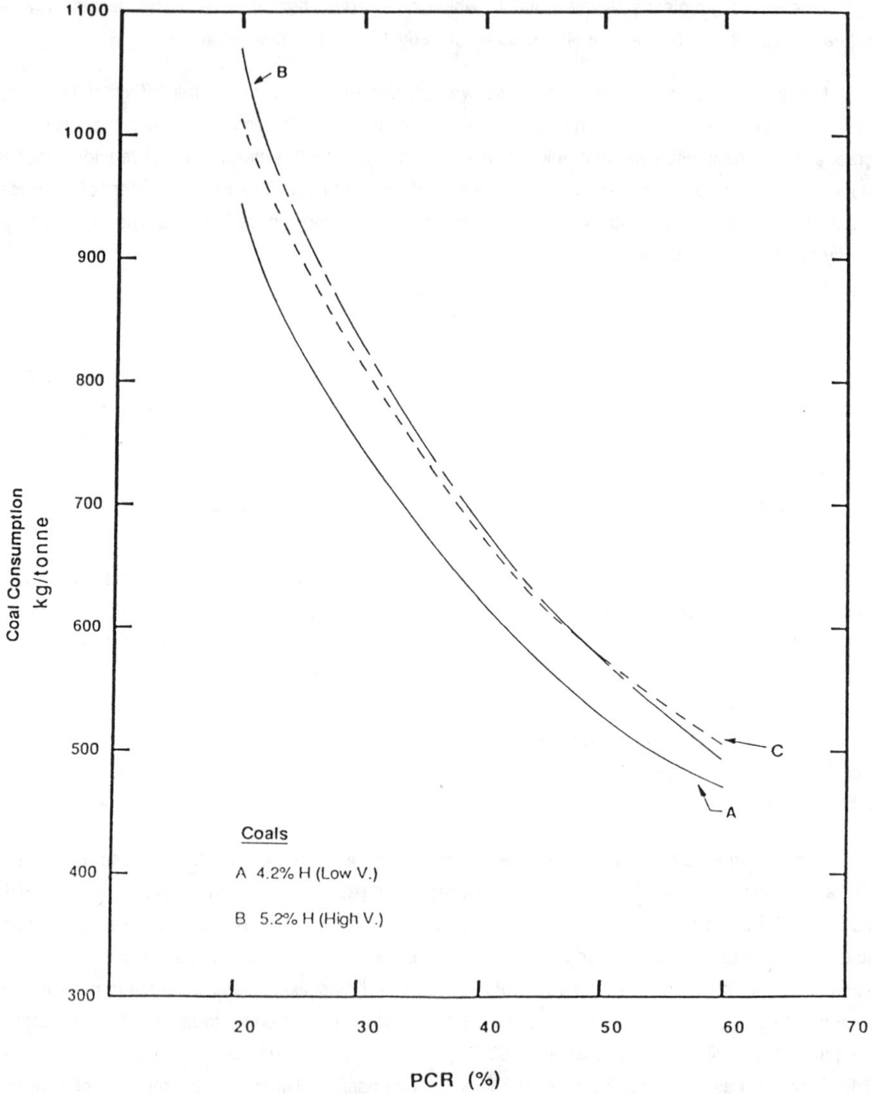

Figure 3: Coal consumption as a function of PCR for 30% PRD, 85% HTC and 10% heat loss.

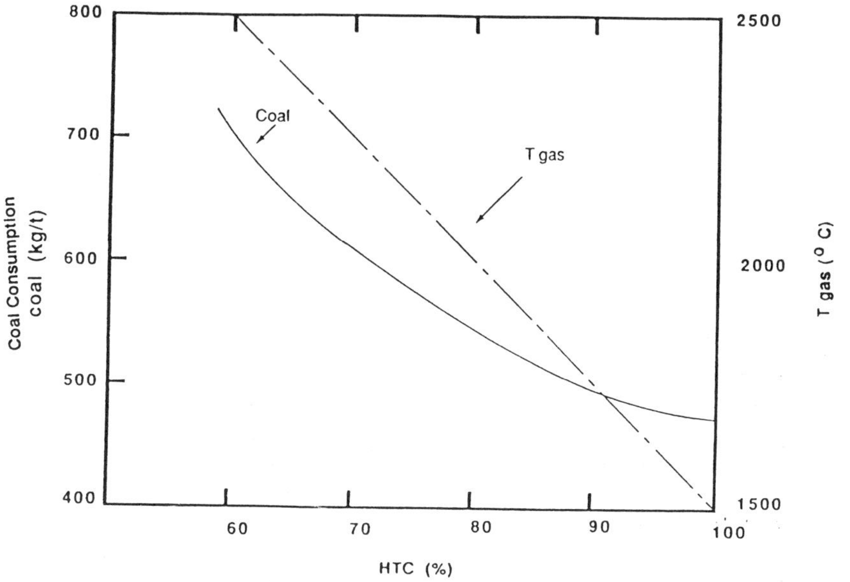

Figure 4: Coal (A) consumption and T_g for 30% PRD, 50% PCR and 10% heat loss using O_2 as a function of HTC.

TABLE 2: Coal Consumption for the Selected Cases of Smelt Reduction and Coals (HTC = 85%)

		Coal, kg		
PRD	PCR	A	B	C
30	50	537	574	576
60(1)	30	495	542	529
90(2)	0	688	881	735
0(3)	50	804	857	859

(1) Additional 53 kg of carbon required in reducer
(2) 1% Si in metal
(3) no ore preheat, all others 800°C

TABLE 3: Oxygen, Ore and CaO Consumptions for Coal A (HTC = 85%)

PRD	PCR	O_2 Nm^3	CaO kg	Ore kg
30	50	444.4	103.8	1492
60	30	381	101	1488
90	0	451	88	1458
0	50	697*	118	1495

*If air 4131 Nm^3

The coal consumption and T_g are shown as a function of HTC in Figure 5 for Case I using 100% air preheated to 1200°C for post combustion. The coal consumption is nearly the same as for un-preheated oxygen but the off gas temperature is lower because of the greater volume of gas due to the nitrogen. For an HTC of 85% the gas temperature is only 1685°C but the gas volume increases to about 2800 Nm^3/t from 1055 Nm^3/t for the base case.

A mixture of O_2 and air could also be used to temper the temperature of the off gas. The gas temperature, gas volume and coal consumption for 30% PRD and 50% PCR with 85% HTC are shown in Figure 6 as a function of the fraction of O_2 in an O_2 - Air mixture. The coal consumption is about the same in all cases but the gas temperature decreases and the gas volume increases as

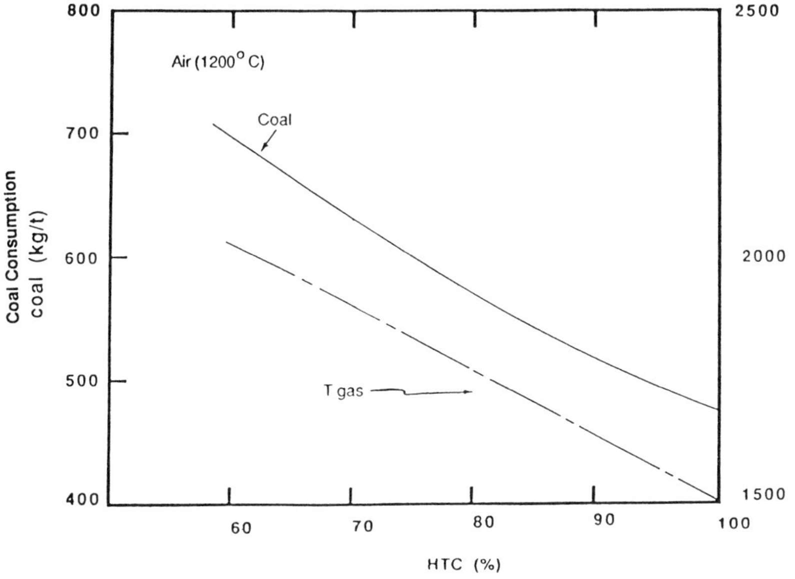

Figure 5: Coal (A) consumption and T_g for 30% PRD, 50% PCR and 10% heat loss for Coal (A) using air as a function of HTC.

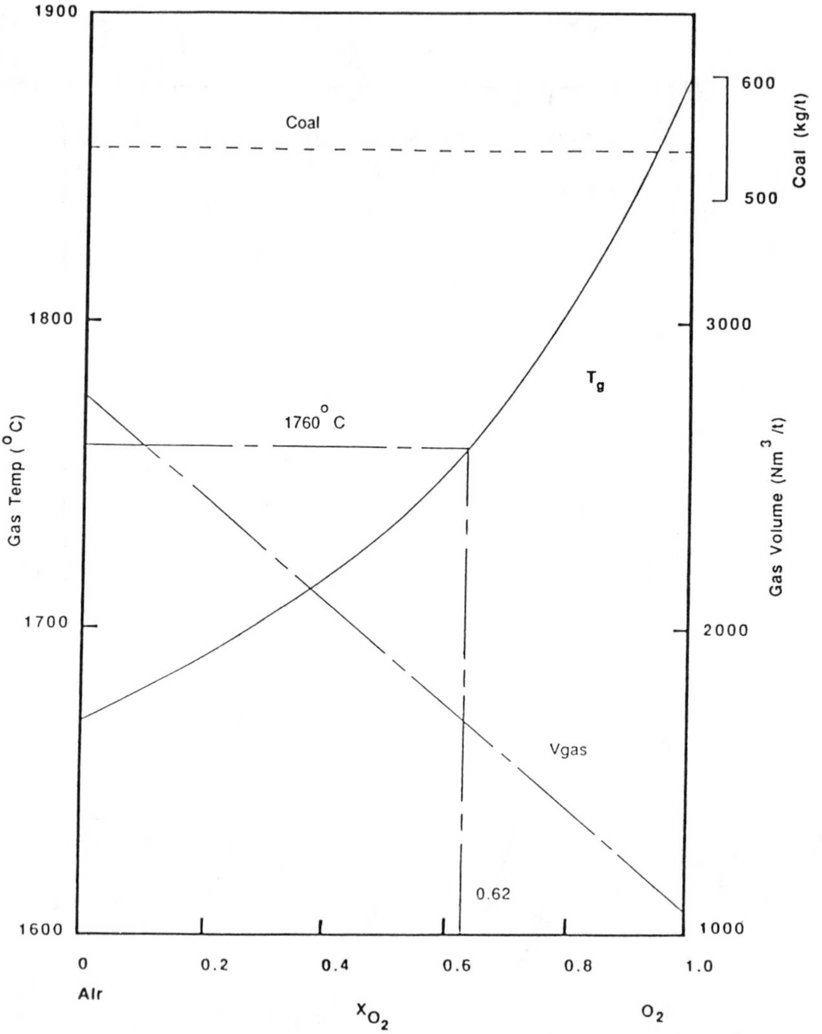

Figure 6: Coal (A) consumption, T_g and gas volume as a function of O_2-Air mixture for PRD = 30%, PCR = 50%, HTC = 85% and 10% heat loss.

TABLE 4: Off Gas Volume, temperature and composition (HTC = 85%)

PRD	PCR	Coal	Vg (Nm^3)	Tg °C	CO %	CO_2 %	H_2 %	H_2O %
30	50	A	1055	1881	44.6	30.9	5.2	18.9
30	50	B	1139	1881	42.9	27.1	6.8	22.6
90	0	A	1334	1500	75.3	0	24.3	0
60*	30	A	964.4	1751	59.2	16.3	10.6	13.6
0	50	A	1600	1881	43.0	27.4	6.6	22.3

* Additional 53 kg of carbon required in pre-reducer which generates an additional 35 Nm^3 of gas.

more air is used. If the maximum allowable temperature is 1760°C, a 62% O_2 - 38% air mixture should be used and the gas volume would be 1690 Nm^3.

Reducing the amount of post combustion will decrease the off gas temperature. In Figure 7 T_g is given as a function of the PCR. For a maximum gas temperature of 1760°C, the post combustion must be reduced to 31.2% in which case the coal consumption increases to 736 kg and the gas volume to 1460 Nm^3.

To reduce the off gas temperature for Case I from 1881°C to 1760°C, 41 liters of water are required. This was calculated based on the latent heat and the water gas shift reaction. With the addition of the water the off gas will still have enough CO and H_2 to reduce Fe_2O_3 to FeO.

Case II

The next general case considered was that for 90% PRD with no post combustion. As indicated in Table 2, the coal consumption is considerably greater than for Case I. For this case the smelter is being run under highly reducing conditions and some silicon reduction is to be expected. The coal consumption is sensitive to the silicon content as shown in Figure 8. Reducing the silicon from 1.0% to 0.5% saves 30 kg of coal.

Also shown in Figure 8 is the effect of 10% post combustion. With as little as 10% PCR, the coal required decreases by about 200 kg. However, the off gas from the smelter will not give 90% PRD as is called for. The gas must be reformed or the CO_2 and H_2 scrubbed from the gas.

Case III

Case III is the intermediate case similar to that proposed by Brotzman (2). In this case the ore is 60% reduced and the off gas is combusted to give a PCR of 30%. For this case the coal consumption and gas generation is less than Case I. However, the 60% PRD cannot be achieved

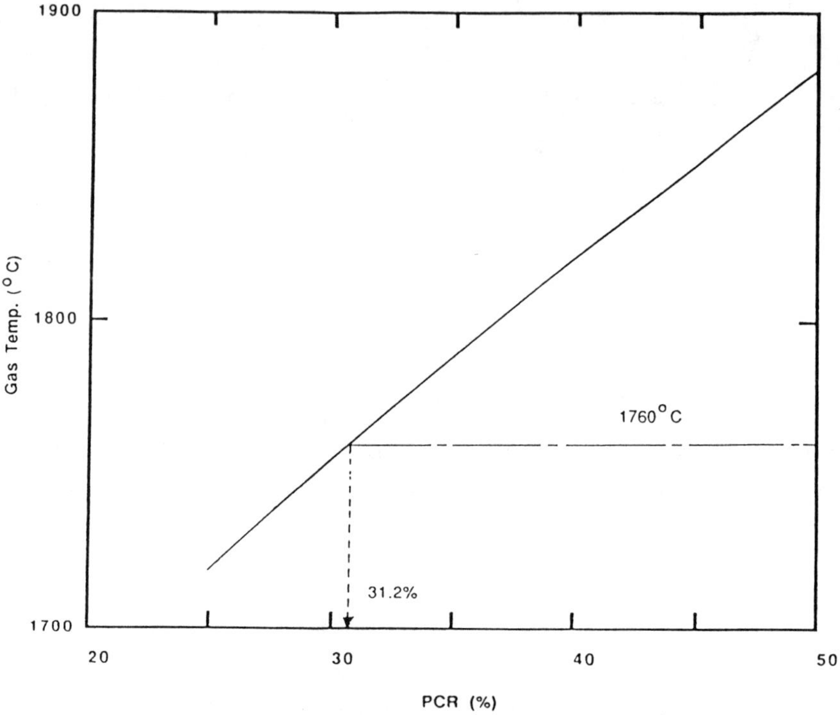

Figure 7: Off gas temperature as a function of PCR for Case I (Coal A) for HTC=85% and 10% heat loss.

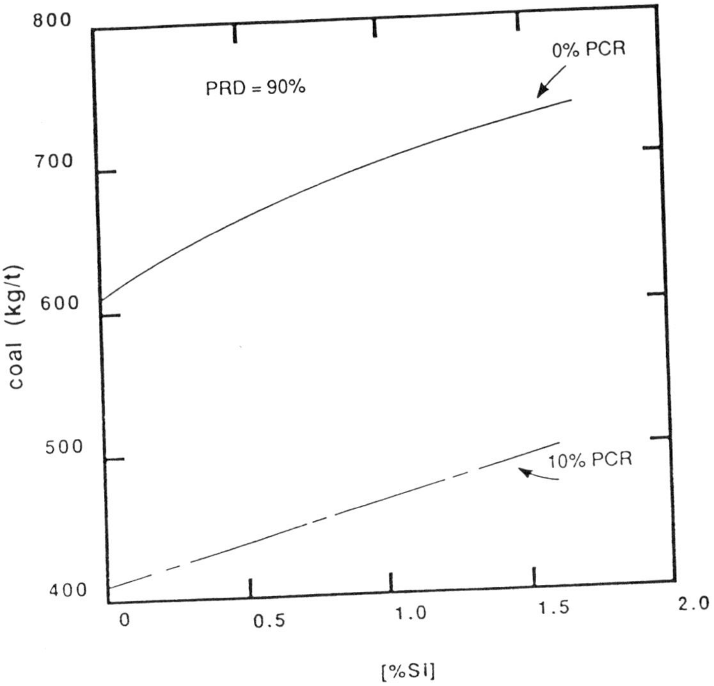

(Figure 8: Coal (A) consumption as a function of assumed silicon content of metal and PCR for 90% PRD and 10% heat loss.

with the off gas and a two stage or counter current reducer must be used. Since the off gas cannot prereduce the ore to 60%, the gas must be reformed or scrubbed. In the case of reforming, an additional 53 kg of carbon, approximately 70 kg of coal depending on the coal type, is required in the final stage of the prereducer.

For the family of intermediate cases with a PCR of 30% and 40%, the coal consumption is given as a function of PRD in Figure 9. As expected coal consumption decreases with increasing PRD and PCR. However, not all the conditions given in Figure 9 are achievable without gas reforming.

Case IV

The simplest possible case is to have no prereduction or preheating of the ore. The coal consumption is very high as indicated in Table 2 but no preheater or prereducer is required. The coal consumption is given as a function of PCR in Figure 10. Also shown in Figure 10 is the coal consumption if the ore is preheated to 800°C. Preheating reduces the coal consumption by about 100 kg. However, thermodynamic calculations indicate that if the off gas from the smelter is used to preheat it will reduce the Fe_2O_3 to FeO and this will be the same as Case I.

Preheated air (1200°C) could be used for post combustion and the coal consumption does not change significantly. However, the gas volumes will be significantly higher. For an HTC of 85% the gas volume is 4216 Nm^3 and the T_g is 1668°C compared to 1600 Nm^3 and 1881°C.

Other Process Variables:

Other process variables that were investigated were the use of $CaCO_3$ as the flux, use of supplemental electricity, coal consumption as affected by the dust generated, heat losses and scrap melting. Another important consideration is the sulfur content of the metal, expected to be higher than in a blast furnace.

Flux: For Case I with HTC = 85% the effect of using $CaCO_3$ as compared to CaO as the flux on coal consumption is shown in Figure 11. At 50% PCR the use of $CaCO_3$ requires about 65 kg more coal. As post combustion decreases the difference increases to over 100 kg. The use of $CaCO_3$ requires heat to calcine the lime stone and introduces CO_2 into the system, both of which have adverse effects on coal consumption. For high temperature reduction processes which are heat limited, the use of $CaCO_3$ in place of CaO is not economical.

Dust: In-bath smelting processes generate a significant quantity of dust. The effect of dust generation on coal consumption for Case I (PRC = 50%, PRD = 30%, HTC = 85%) by using coal (A) is shown in Figure 12. The coal consumption is not very sensitive to the amount of dust. However, the difficulty in cleaning the dusty gases prior to further use may be a significant factor.

Electricity: In several proposed processes, such as the SKF Plasmasmelt process, electricity is used. In Figure 13 the effect of supplemental electricity on coal consumption for Case I and Case II is shown. The use of electricity has little effect on Case I because most of the energy is required

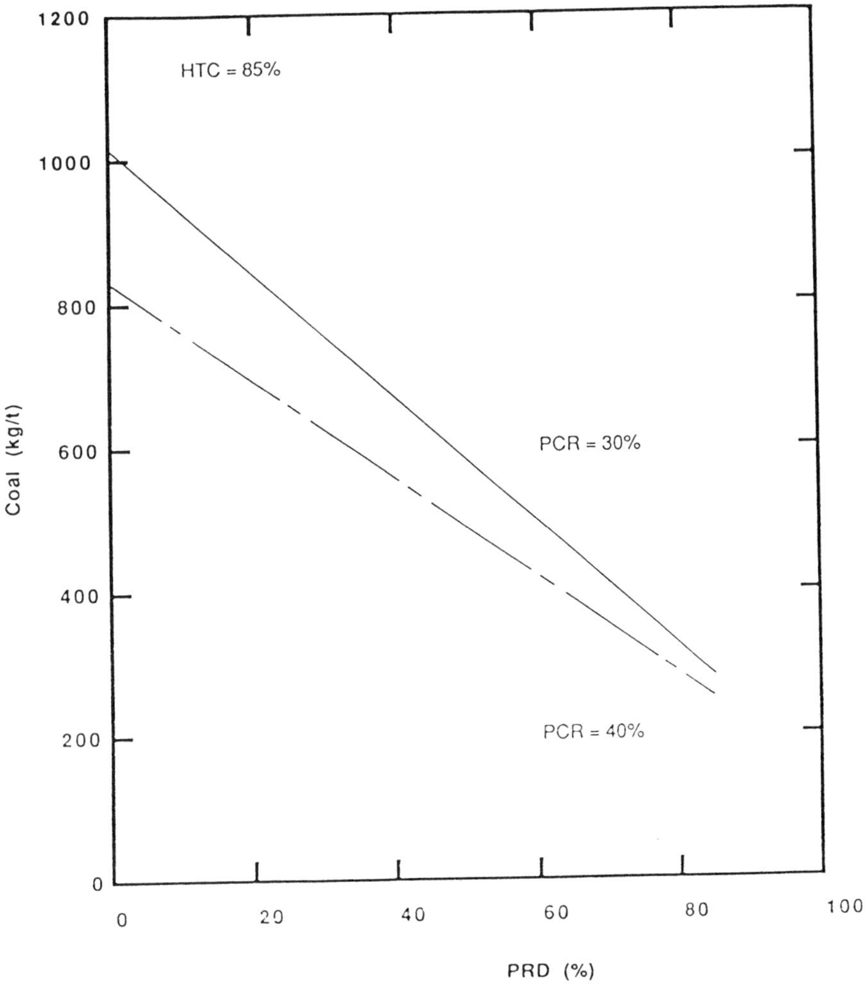

Figure 9: Coal (A) consumption for PCR = 30% and 40% as a function of PRD for HTC=85% and 10% heat loss.

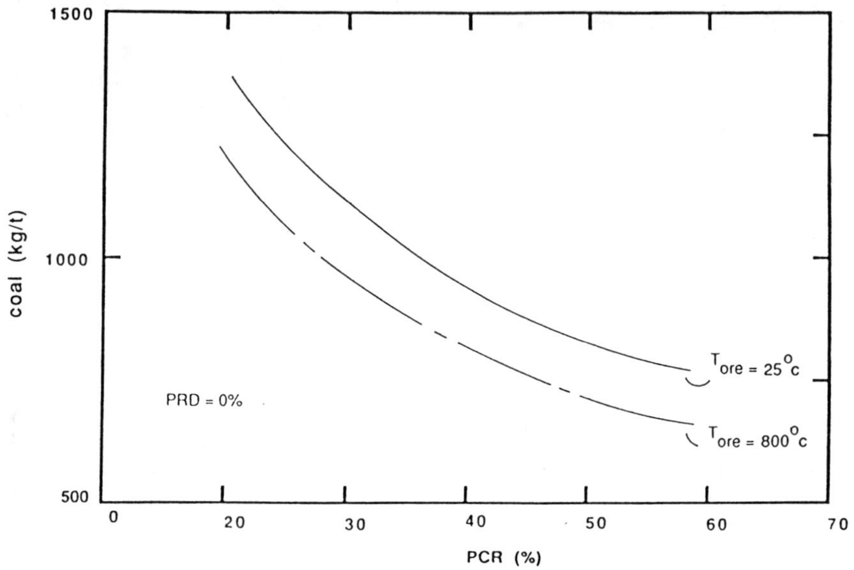

Figure 10: Coal consumption for 0% PRD as a function of PCR for HTC=85% and 10% heat loss.

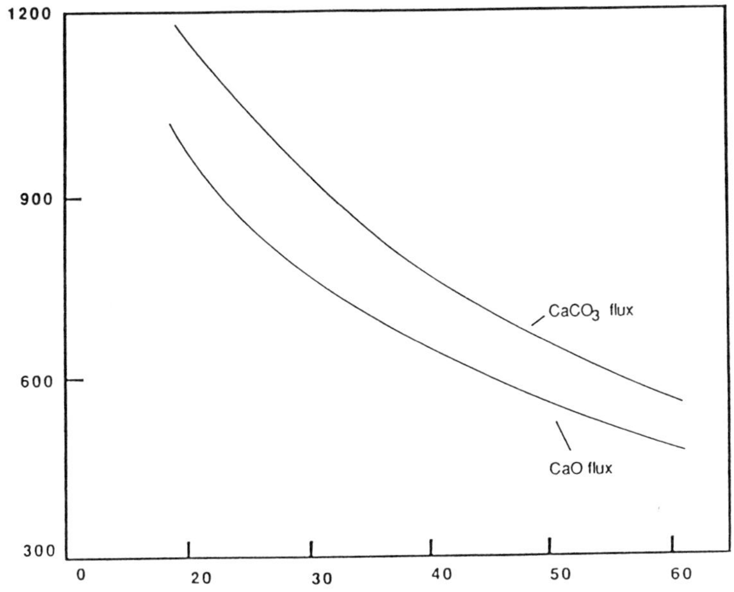

Figure 11: Coal (A) consumption using CaO or $CaCO_3$ as the flux for 30% PRD as a function of PCR for HTC=85% and 10% heat loss.

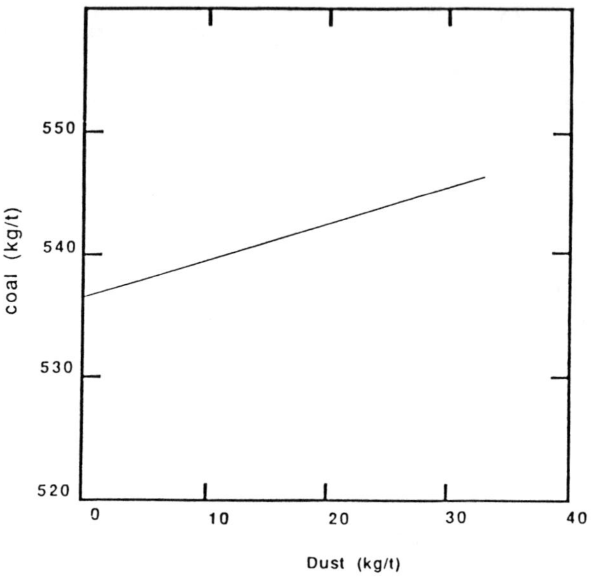

Figure 12: Coal (A) consumption for 30% PRD and 50% PCR as a function of the amount dust.

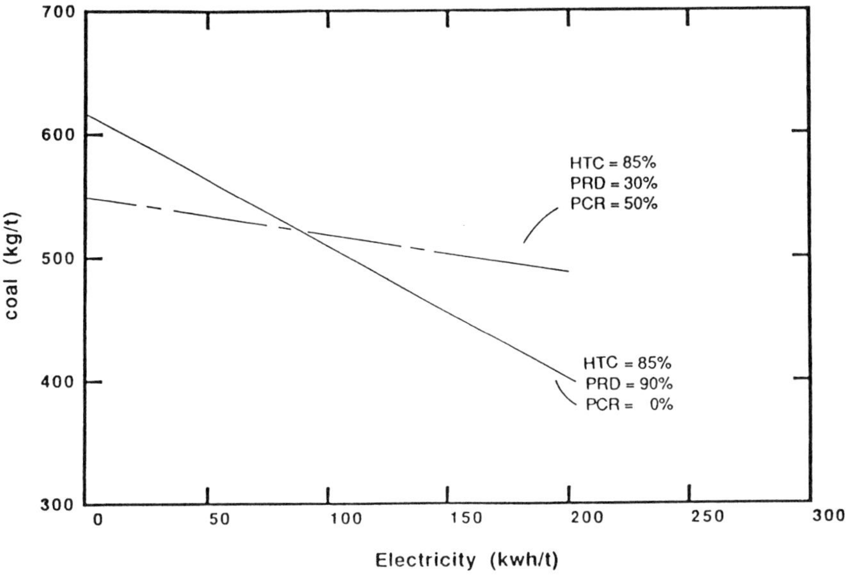

Figure 13: Effect of assumed use of electricity on Coal (A) consumption for 50% PCR - 30% PRD and 0% PCR - 90% PRD with HTC=85% and 10% heat loss.

for reduction and a considerable amount of energy is supplied by post combustion. With 50% PCR the effect of energy supplied by up to 200 kwh of electricity is relatively small. In the case of 90% PRD the energy required is primarily for heating and melting, and with no post combustion less energy is generated in the smelter than for Case I. Consequently, the addition of electricity does effect coal consumption for Case II.

Therefore in processes, such as Plasmasmelt, Elred, and Corex, electricity will effect coal consumption, but with in-bath smelting having high degrees of post combustion the use of electricity is ineffective.

Heat Losses: In all the cases considered, the heat loss was taken as 10% of the total heat required. For a commercial, and possibly even a pilot scale, reactor 10% heat loss may be an over-estimate. If the heat losses are less, the coal consumption and gas volume will decrease accordingly but the off gas composition and temperature will be essentially the same. For example, if the heat losses were 5% for any case considered, the coal consumption and gas volume would decrease by 5% with the gas composition and temperature the same. The flux requirement would decrease very slightly due to the use of less coal which would have a very small effect on the heat required; and, this small effect can be neglected.

Scrap Melting: Since the metal produced in the process is low in carbon and silicon there may be insufficient thermal value in the metal for scrap melting in the decarburization process to follow. However, it may be possible to melt scrap in the smelter. In fact, the smelter is theoretically a very energy efficient method of melting scrap because it uses coal and has significant post combustion.

The simulation model was expanded to consider using scrap at room temperature as part of the charge. For Case I using 100 kg of scrap the coal consumption decreases by nearly 40 kg and for 200 kg by 80 kg of coal. The gas volumes also decrease but the composition and temperature remain the same. There will also be a decrease in CaO consumption. The coal consumption and gas volume is given as a function of the amount of scrap in the charge for Case I (Coal A) in Figure 14.

Sulfur: Since the envisioned processes use coal there is a greater sulfur input than in blast furnace ironmaking. Furthermore since there is a higher oxygen potential in the system, due to post combustion, the sulfur distribution between slag and metal will also be lower. Therefore, the sulfur content of the metal is expected to be relatively high. The computer program was extended to estimate the sulfur content.

In the calculation the published values of the sulfide capacity (C_S) were used (6)

$$C_S = (\%S) \left(\frac{p_{O_2}}{p_{S_2}}\right)^{1/2} \tag{4}$$

where (%S) is sulfur in slag, and p_{O_2}, and p_{S_2} the prevailing oxygen and sulfur pressures. The sulfur

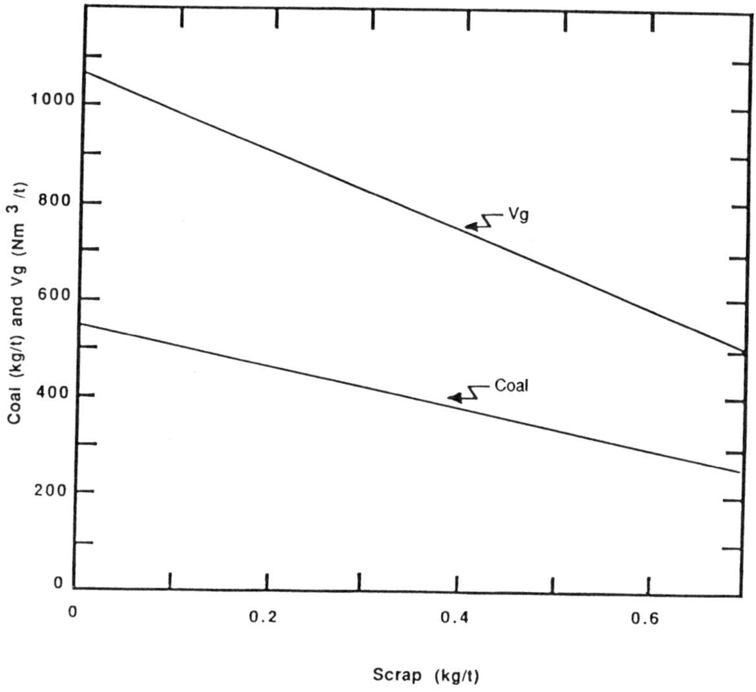

Figure 14: Coal consumption and gas volume as a function of the assumed amount of scrap for Case I (Coal A) with HTC=85% and 10% heat loss.

pressure was calculated from the following equilibrium

$$\underline{S} = 1/2\ S_2 \tag{5}$$

where $\underline{S}$ is the sulfur in the metal. The effect of carbon and silicon on the activity coefficient of sulfur was made using the appropriate interaction coefficients.

An important consideration in the calculation is what is controlling the oxygen pressure. For these reactions it has generally been found that the equilibrium between FeO and Fe determines the oxygen pressure.

$$(FeO) = Fe + 1/2\ O_2 \tag{6}$$

For the slags considered there is little information on the activity of FeO. The limited information (8,9) indicates it is nearly ideal which was assumed in this study.

From equations 4-6 the sulfur partition ratio between slag and metal is calculated. This is then combined with the mass balance, sulfur input, slag weight, and metal weight from the computer program, to calculate the sulfur content of the metal. It should be noted that there are several assumptions and the results greatly depend on the assumed FeO content of the slag and therefore the results give only estimates of the sulfur content. The model was verified by simulating a blast furnace and reasonably good results for sulfur were obtained. The calculated sulfur content in the metal for Case I (Coal A) is given for slag basicities of 1.0 and 1.25 as a function of FeO in the slag in Figure 15.

The results indicate there may be significant sulfur in the metal. For example, for Case I with B=1 5% FeO there will be over 0.35% S in the metal. However, if the FeO is 2% this would be less than 0.25%. In general, the greater PCR, the lower the carbon content of the metal and the higher the FeO content of the slag, both of which contribute to higher sulfur contents. Whereas the present model is only an estimate of the sulfur content, it is clear that the sulfur content will be higher than for blast furnace hot metal.

Conclusions

Mass and energy balances for bath smelting have been carried out. The major conclusions from the process evaluation are:

1. Post combustion greatly reduces coal consumption and the need to market excess gas. However, once post combustion exceeds 20% the gas loses much of its potential to reduce FeO to Fe.
2. An attractive process is prereducing to FeO only and having as much post combustion consistent with good heat transfer. This case requires a simple prereducer, uses less coal, and is insensitive to coal type.

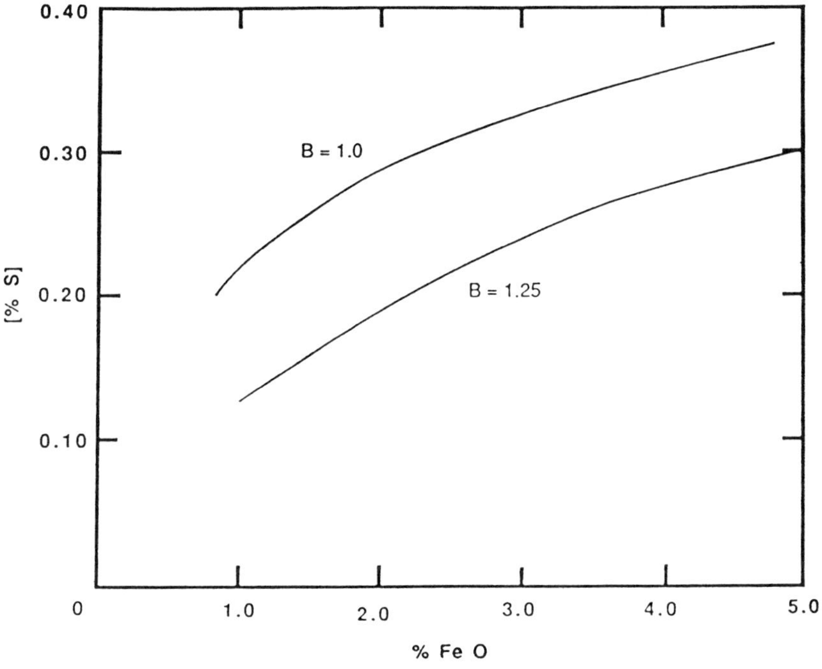

Figure 15: Estimated sulfur content in the metal for Case I (Coal A).

3. One major anticipated problem is that the off gas temperature may be too high to handle with normal techniques. It is possible to reduce the temperature by improving heat transfer, reducing post combustion or using O_2-Air mixtures for post combustion, or spraying water into the off gas. Preheating air to 1200°C is approximately the equivalent with regards to energy required as using oxygen.

4. Gas reforming or CO_2 scrubbing would be necessary in intermediate cases, such as 30% PCR and 60% PRD.

5. $CaCO_3$ should not be used as the flux and supplemental electrical energy is not effective unless the PRD is high.

6. The smelter is theoretically very energy efficient. Using 200 kg of scrap as part of the charge for Case I decreases coal consumption by 100 kg.

7. In the case of PCR >30% the FeO content of the slag will be high (2-5%) and the metal will not be saturated with carbon. These conditions along with the fact that coal is being used could lead to high sulfur levels in the metal, possibly over 0.25%

Acknowledgements: The authors wish to acknowledge the support of the American Iron and Steel Institute (AISI) for this study, the input and discussions with the AISI Task Force on Direct Steelmaking, and AISI for allowing the publication of the results.

REFERENCES

1. R.B. Smith and M.J. Corbett, Ironmaking and Steelmaking, 1987, vol. 14, p. 49.

2. E.K. Brotzman, Steelmaking Proceedings, 1987, vol. 70, p. 3.

3. Y. Hayashi, M. Nakamura and N. Tokumitsu, Process Technology Division Proceedings, Fifth International Iron and Steel Congress, ISS, Washington, D. C., 1986.

4. F. Oeters and A. Saatic, Process Technology Proceeding of ISS, 1986, vol. 6, p. 1021.

5. F. Oeters and A. Saatic, Stahl and Eisen Report, Mass and Heat Balances, 1987, Stahl und Eisen mbH, Dusseldorf.

6. M. Tokuda, Shenyang International Symposium on Smelting Reduction, Chinese Society of Metals, September 1986.

7. O. Kubaschewski and C.B. Alcock, Metallurgical Thermochemistry, fifth edition, 1979, Pergamon Press.

8. F.D. Richardson, Physical Chemistry of Melts in Metallurgy, 1974, Academic Press.

9. T. Arato, M. Tokuda, and M. Ohtani, Testu-to-Hagane, 1982, vol 68, p. 2263.

COAL-BASED IRONMAKING VIA MELT CIRCULATION

N A Warner

Department of Chemical Engineering
University of Birmingham
Birmingham B15 2TT
England

Abstract

The theoretical basis of ideal smelting reduction using coal as reductant in a process employing closed loop forced-circulation of hot metal is outlined. In the proposed process, an extensive area of relatively slag-free hot metal is available as a heat sink for absorbing the large amount of heat produced by complete combustion of CO within the furnace. Any slag that is formed within the oxidizing arm of the furnace loop is quickly swept away as a thin floating layer by the bulk movement of hot metal which is induced positively by forced melt circulation and sustained by causing the hot metal and the thin slag layer to overflow into a lower level hearth in which coal is carbonised and dissolved directly into the circulating melt.

The proposed process is shown to offer very substantial cost savings over all other direct smelting technology. This is mainly because of its inherently low capital cost and its high energy efficiency. Furthermore, the favourable economics are not reliant on fuel gas credits as all the process gas is combusted within the reactor itself.

Metallurgical Processes for the Year 2000 and Beyond
Edited by H.Y. Sohn and E.S. Geskin
The Minerals, Metals & Materials Society, 1988

Introduction

Metallurgical processes in the year 2000 will be compromise solutions to the metallurgical problems that already exist today. It is common knowledge that the mineralogical complexity of certain polymetallic sulphides, for example, precludes high degrees of separation and recovery by conventional mineral beneficiation. Steps are therefore being taken towards the development of a truly polymetallic smelting process which will produce simultaneously zinc, lead and copper as metals from complex ores without prior beneficiation[1]. In the ferrous extraction industry, the shortcomings of the highly efficient modern blast furnace are well known and world-wide there is considerable effort being directed towards coal-based ironmaking without the requirement of metallurgical coke as a process intermediate. The current status of coal-based ironmaking has recently been reviewed by Smith and Corbett[2].

Whatever conditions prevail in the next century, process technology will adapt in an attempt to yield optimal configurations. But conservation will be paramount and in this context it is worth quoting from Professor H.H.Kellogg's Sir Julius Wernher memorial address in London some years ago: "metallurgical processes for the future must be designed to achieve maximum conservation of energy resources, capital resources, and the environment"[3].

In line with the above considerations, this paper will attempt to assess the practicability and competitiveness of certain new proposals made in relation to coal-based ironmaking under the general umbrella title of smelting reduction[4]. Some traditional widely held beliefs may need revision if real progress is to be made in this area because of what are undoubtedly practical problems of considerable magnitude. Perhaps bath smelting based on what has become known as "injection metallurgy" is not the best way forward for smelting reduction but rather forced melt circulation should be in the forefront of future development.

To start with it is taken for granted that fossil fuel will still be the major energy resource in the year 2000. In the absence of cheap hydro-electricity, the choice of fuel versus electrical energy usage, fully discussed by Kellogg[3], requires that wherever possible direct use of fuel is preferred to electricity use at an overall fuel efficiency of only 30%. The author has already explored the prospect of direct fuel usage to replace electricity in smelting processes which have become the preserve of electric furnaces by analysing in detail the pyrometallurgical treatment of nickel laterites by proposed new innovative technology[5].

The key to the new technology is closed-loop melt circulation. An earlier paper described the Mineral Industry Research Organisation (MIRO) project at Birmingham University which is aimed at the development of a new polymetallic smelting furnace[6]. This project, now funded jointly by the Commission of the European Communities and several large industrial companies, employs forced melt circulation. The steel industry's RH vacuum degassing technology is used to circulate molten copper-saturated copper sulphide matte between regions of varying oxygen potential either within a single partitioned furnace or between side-by-side furnaces. The smelting reduction process now to be described brings the basic concept back to its origin, the steel industry, and the melt to be circulated is hot metal comprising mainly iron in association with varying amounts of carbon, silicon and other impurities.

Constraints on New Process Development

With today's expertise in the general area of process engineering, the availability of new advanced materials and the accumulated knowledge of centuries of metallurgical technology, it may appear in principle that ideal solutions can always be implemented. That this is not the case is mainly due to capital charges. There is little point in advocating a replacement for the blast furnace if, for example, the cost of refractories is some five or six times greater in the new process even if it does achieve some marginal reduction in process energy costs. This leads to the concept of process intensity or specific capacity of a furnace expressed in units such as tonne per day throughput per unit volume of furnace enclosure. Over-zealous pursuit of high specific capacity, on the other hand, has its own pitfalls. In any real situation there is an optimum configuration which is a compromise under the conditions currently prevailing.

For environmental reasons it can also be concluded that pyro metallurgical processes of the future will have certain restrictions imposed upon them. Some of these may include:

(1) total filtration or the equivalent of all gaseous emissions to the atmosphere

(2) prohibition of all fugitive in-plant gaseous emission

(3) virtually complete elimination of heat stress exposure to personnel

The ideal metallurgical process of the future will literally resemble a black box. Clearly it will be mandatory to ensure that what is going on inside is conducted safely and all effluents and emissions are contained or discharged to comply with strict codes of practice.

The Float Glass Connection

The foregoing comments on stringent gas emission control have been made in preparation for the introduction of a radically new concept to iron and steelmaking. The origins of the concept can be traced to the commercial float glass process in which molten glass is admitted to the surface of a bath of molten tin to effect solidification and continuous removal of plate glass product. Float glass chambers are typically 5 - 8m in width and 49-61m in length. The tin depth is around 5 to 10cm and the velocity of the glass can be as high as 3.4m/s[7].

The idea of "float steel" has undoubtedly occurred to many and it is also fairly obvious that molten lead has the requisite properties for the bottom liquid layer. Iron and lead are almost completely immiscible as liquids and the question of interface stability when liquid steel is admitted to a surface of molten lead can surely be resolved in principle just as it is achieved in practice in the molten glass/tin system. The extension of this concept to the use of a molten lead pad on which iron and steelmaking reactions are carried out opens up some exciting new horizons for smelting reduction.

Molten Lead Hearth Layer

The maintenance of a lead layer beneath the fast moving hot metal stream of a forced melt circulation process is seen to offer the prospect of chemical attack and erosion limitation whilst providing substitution of cheap refractory materials for otherwise very expensive options. Severe erosion of the linings of the furnace can predictably occur as a result of passage of hot metal at the high circulation rates required for efficient smelting and heat recovery. Whilst replacement of the refractory linings at the sides of such furnaces is possible without completely closing down the plant, it is very much more difficult if not impossible to replace a refractory hearth without a major shutdown.

Since a vacuum lift pump would probably be used as the prime mover for forcibly circulating hot metal, passing the melt through the vacuum lift pump enables dissolved lead to be stripped by vacuum distillation. Because the circuit is a closed loop, the relative efficiencies of vacuum de-leading and the lead pick-up to the circulating hot metal in contact with the lead layer determines the steady-state lead concentration in the hot metal.

Whilst there are adverse health and safety implications associated with the use of molten lead at elevated temperatures, it is believed that close attention to engineering design can reduce the problems to acceptable proportions. The quilibrium solubility in molten iron at 1550°C according to Japanese workers is about 0.16wt%.[8][9]. A less comprehensive but more recent Russian paper[10] indicates a slightly higher solubility of 0.22% Pb at 1550°C. Carbon significantly reduces the solubility of lead in molten iron and for a 2% carbon hot metal at 1400°C, which is typical for the of process to be discussed, the lead saturation level is estimated to be around 0.08 percent. Clearly, the system has to be designed so that the steady state level attained in the circuit is only a fraction of these values.

The Smelting Reduction Route

As defined by Eketorp and co-workers[11], smelting reduction is different from the classical blast furnace method in that reduction of iron oxide at high temperature is carried out by reaction with carbon to CO only. The CO formed is burnt to CO_2 with air or oxygen close to the site of reduction and the heat generated is used for compensation of the strongly endothermic reduction reaction. As pointed out by Eketorp, smelting reduction processes have been tried in the past but so far the difficult problem of heating a reducing bed or bath without at the same time oxidizing it, has not been solved technically despite the considerable energy and capital cost advantages that would accrue.

Fig. 1 is a schematic general arrangement, in plan view, of a proposed plant for the smelting reduction of iron oxide ore. The plant comprises a pair of side-by-side furnaces A and B. At the left hand end of the furnaces A and B as viewed in Fig. 1, a channel C is provided to permit melt flow from furnace B to furnace A. At the opposite end of the furnaces A and B, a lift pump P is provided which operates on the RH vacuum degassing principle in that a reduced pressure is maintained in a reservoir above Legs L_1 and L_2 of the pump to effect forced circulation of hot metal (ie molten iron which is the direct product of the smelting reduction of iron oxide ore and which therefore contains minor amounts of the usual incidental ingredients C, Si, Mn, S etc) in a closed loop path. Hereinafter such hot metal will be referred to simply as molten iron to distinguish it clearly from the molten lead layer described below. A pool of molten lead is provided over the hearths of the furnaces A and B to a depth of 40 - 100mm. The molten iron is provided as a 200 - 500mm thick layer above the pool of liquid lead.

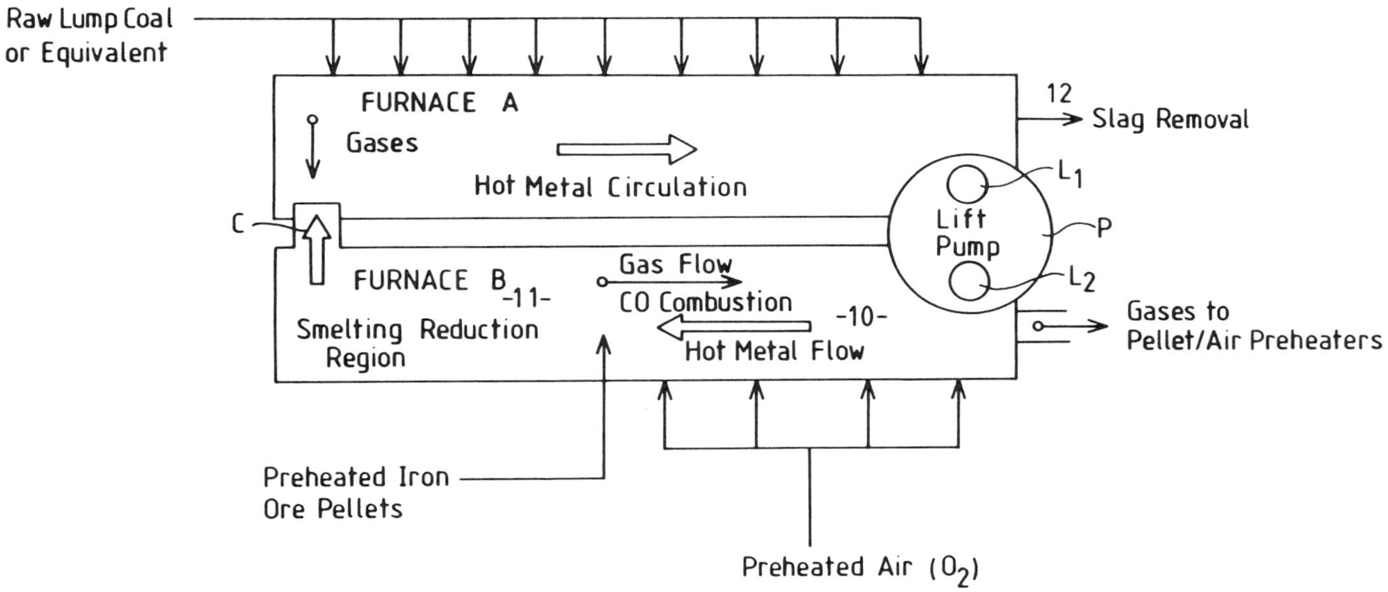

Figure 1 - Schematic plan view of the general arrangement proposed for smelting reduction.

The legs L_1 and L_2 extend into the layer of molten iron in the respective furnaces A and B. When the lift pump P is operated, molten iron is drawn up leg L_1 from furnace A and discharged through leg L_2 into furnace B. At the opposite end of furnace B to lift pump P, molten iron flows through the channel C into furnace A. Thus, a closed loop path of molten iron which circulates in a clockwise direction, as indicated by larger arrows in Fig. 1, is provided. Raw lump coal or the like is supplied as a layer which rests on the upper surface of the molten iron in furnace A. The raw lump coal contacts the surface of the iron over the majority of the length of furnace A. Contact of the coal with the molten iron causes carbon to be dissolved in the iron and for the coal to be gasified. The flow rate of molten iron and the area of contact between the coal and the molten iron are chosen so as to produce a dissolved carbon content of about 2 to 3% in the molten iron at steady state.

As a result of intense heating of the coal in furnace A, coal carbonisation occurs and the coal volatiles produced are passed in countercurrent fashion relative to the flow of molten iron into the left hand end of furnace B through channel C above the molten iron layer. From the left hand end of furnace B, the gases pass to the right hand end of furnace B where a heating zone 10 is provided. The heating zone 10 is supplied with pre-heated air or oxygen to cause combustion of the combustible gases and also combustion of carbon monoxide which is supplied as described later. The heat of combustion is transferred to the molten iron to raise its temperature. The molten iron which is circulated from the heating zone 10 enters a smelting reduction zone 11 in the furnace B. Pre-heated iron ore pellets are supplied to the surface of the molten iron in the smelting reduction zone 11 and reduced by the carbon in the iron to convert the iron oxide to iron and simultaneously oxidise the carbon to carbon monoxide which is passed in the direction of the smaller arrow to the heating zone 10 for combustion with the pre-heated air or oxygen. Hot combustion gases which leave the heating zone 10 are employed for iron ore pellet and air pre-heating.

Any slag which is formed in furnace B is carried by the molten iron over into the lower furnace A and removed together with any raw lump coal residues following carbonisation and carbon dissolution via slag overflow 12. In this way, lift pump P removes the molten iron from under the slag and passes it to the heating zone 10 so that a relatively clean molten iron surface is available for efficient heat transfer from the combustion of gases. The iron oxide pellets fed to the smelting reduction zone 11 can either be prepared with sufficient lime or limestone can be added separately to flux the gangue in the ore as well as the ash in the coal so that a single slag with a low iron oxide content is removed after contacting the full length of the layer of coal in furnace A. This ensures high iron recovery. It is envisaged that the above described plant could sustain 2000 tpd Fe production if furnace A had a size of 40 m length x 10 m wide. Furnace B needs to be of similar size to effect carbon monoxide and coal volatile combustion based on an energy intensity of 350 kW/m^2. This is considered to be possible because only a very thin layer of slag is present on the molten iron in the heating zone 10. To produce the required dissolution of carbon from the raw lump coal or the like, it is considered to be possible to operate with a clearance of say 0.1 m between the layer of coal and the hearth of furnace A, with a local melt velocity of about 0.8 m/s.

Using pre-heated air for combustion without oxygen enrichment gives a gas-diffusion limited rate of slag formation in furnace B of around 2t/h which at the molten iron circulation rates envisaged will yield a maximum thickness of slag less than 0.5 mm thick. The thermal resistance of the thin slag layer removed continuously to a slag reservoir after over-flowing from furnace B is equivalent to a temperature drop across the slag layer of less than 100°C for an energy transfer rate of 400 kW/m^2.

The kinetics and mechanism of the reduction of solid iron oxides in iron-carbon melts at temperatures in the range 1200 - 1500°C have been studied by MacRae[12] and from his work it is immediately apparent that total reduction of pellets at temperatures around 1450°C takes a matter of seconds. For freely floating pellets of average weight 3g, reduction to metallic iron is completed in around 30 seconds, whereas pellets submerged in carbon-saturated iron at the same temperature are fully reduced in about 15 seconds. Process engineering evaluation of smelting reduction based on the present approach indicates that it is not the actual iron oxide reduction reaction itself that is likely to dictate reactor size, but rather dissolution of carbon into the iron melt exerts a much more significant influence. This is particularly the case if one attempts to use raw lump coal as the source of carbon. Forced melt circulation provides the solution to this problem.

Mass and Heat Transfer Considerations

Correlations relevant to the various aspects of heat and mass transfer involved in a high temperature smelting reduction process are readily available in the literature. Most of the thermochemical and transport property data also exist and the process engineering of pyrometallurgy is now developed to the stage where smelting calculations can reasonably be expected to give realistic insights into actual performance.

This does not appear to be the case in the vital area of turbulence structure and transport mechanism at the free surface in an open channel flow in which floating solids are transported along with the mainstream with effectively zero relative velocity. The questions to be resolved range from the effects of a relatively sparse population of solid particles floating along at the stream velocity, to that of a thin continuous slag layer floating on the surface, again moving at substantially the free stream velocity. If there is no boundary layer in the conventional sense, can free surface open channel flow relationships be used? If turbulent eddies from the bulk liquid can reach the underneath of a freely floating thin slag layer as they do at a free surface, surface renewal theory would seem to be applicable but clearly experimental verification would be reassuring. The same considerations apply to the molten iron/liquid lead interface. Diffusional transport across this surface determines the service requirements of vacuum de-leading in the RH vacuum lift system necessary to keep steady state contamination of the hot metal circuit at reasonable levels. Do we assume that the turbulent eddies from the bulk molten iron reach this interface unimpeded so that mass transfer rates are comparable to those expected at a free surface in open channel flow as given by the relationship proposed by Komori et al.[13]?

$$k_L/\sqrt{(D_L\ \bar{U}_{surface})} = 3 \qquad \text{(SI units)}$$

where k_L is the liquid phase mass transfer coefficient, D_L is the molecular diffusivity and $\bar{U}_{surface}$ is the mean value of the velocity at the free liquid surface.

At the other end of the spectrum, interphase mass transfer could take place between the moving hot metal stream and an immobilised liquid lead layer. These are the conditions established beneath a charge column of coal/coke once anchored with respect to the flowing liquid phase and the rate at which solids are assimilated into the melt then reflects smelting intensity for a given physical size of plant. This in turn is determined principally by rates of heat and mass transfer between the solid and liquid phases. For the configuration under discussion, the appropriate dimensionless correlations for fully developed turbulent flow are known as the Sieder-Tate or Colburn equations.

Some attention also needs to be directed towards stability criteria pertaining to the molten lead/iron interface. Classical Kelvin-Helmholtz instability theory has been applied by a number of workers to ideal inviscid fluids. Slugs are pictured to form when the suction pressure generated over a surface wave by the Bernoulli effect is large enough to overcome the stabilizing influence of gravity. In the liquid lead/iron case it is clearly the relative velocity of the interface which is of significance so a means of estimating this is first required, it being recognised of course that the lead pool will recirculate within itself so that the actual interfacial velocity will probably only be a fraction of the iron melt average velocity.

In some respects a similar situation exists in an aluminium reduction cell in which both the molten cryolite and the molten aluminium hearth pad both circulate under the influence of strong electromagnetic forces. Any major instability of the aluminium surface in terms of metal being dragged off into the cryolite layer would seriously reduce the cell efficiency. A stability criterion for the molten aluminium/cryolite interface is referred to by Furman[14] in which the maximum permissible relative velocity is evaluated in terms of the depth and densities of the two immiscible liquids.

A more general treatment relating to two-phase flow modelling is given by Delhaye[15] for the system depicted in Fig. 2. The critical region in the present case appears to be in the area beneath the charge column of coal. For example, if the depth of the hot metal layer is 0.1 m in this region and it resides on top of a lead layer say 0.1 m deep, stability is ensured only at hot metal velocities less than about 0.8 m/s, conservatively based on zero movement in the lead layer. Some numerical computations on the relative velocity expected for the configuration being considered would undoubtedly reveal that much higher circulation rates could be sustained in practice without instability.

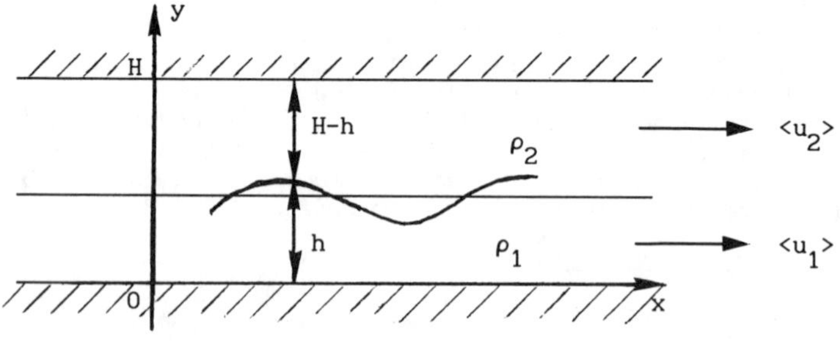

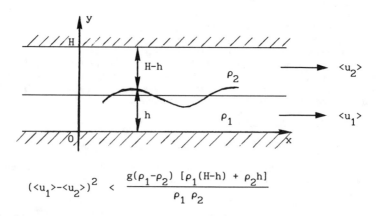

$$(\langle u_1 \rangle - \langle u_2 \rangle)^2 < \frac{g(\rho_1 - \rho_2)\,[\rho_1(H-h) + \rho_2 h]}{\rho_1\,\rho_2}$$

Figure 2 - Stability criteria after Delhaye (15).

Effects of Oxygen Enrichment

Besides the obvious economic factors dictating the level of enrichment of process air with tonnage oxygen, a practical constraint is imposed when refractory temperatures reach an upper limit of around 1800°C. In the melt circulation smelting reduction being discussed, calculation indicates that relatively high gas temperatures can be accommodated without excessive steady-state refractory temperatures being reached. This is the result of two fundamental attributes. First, the smelting reduction reaction is highly endothermic and with bubble agitation on an iron oxide-rich slag layer of relatively thin proportions high heat fluxes are attainable. Secondly, direct heat exchange onto a relatively slag-free metal surface with a very thin film of high emmissivity material immediately established in the oxidizing environment provides an ideal sink for high flux radiative heat transfer. It is the provision of these two thermal sinks that permits high gas temperatures (provided the sinks are freely accessible) and clearly the smelting intensity can be considerably increased by sensible levels of oxygen enrichment.

In what follows it is assumed that the process air is not enriched but clearly the whole question of oxygen enrichment needs evaluation if an optimal configuration is to be reached. Experience with a whole range of non-ferrous smelting operations suggests that at least some modest enrichment up to say 26-32 percent oxygen in the air would be cost effective and highly desirable.

Freeze Lining on Water-Cooled Side Panels

The combination of forced melt circulation and accurate melt level control afforded by the molten lead hearth pad opens up the possibility of a so-called permanent lining for the smelting furnace. Freeze lining is established practice in slag fuming furnaces in the zinc industry but at first sight it may be considered that the high thermal conductivity of molten metals would preclude freeze lining as a viable option. In this regard, highly relevant industrial technology is to be found in Imperial Smelting's Vacuum Dezincing Process (VDZ). In this process, which operated commercially in association with the Imperial Smelting Furnace (ISF) at Swansea Vale in the U.K. for several years, metallic zinc within a vacuum vessel was deposited onto a water-cooled surface from the gaseous state under very high heat flux conditions. The stable zinc layer cycled in thickness according to the zinc production rate but always had the surface at the melting point by self regulation. The heat flux to the water-cooled liquid zinc condenser surface in the Swansea Vale VDZ plant works out at around $0.2 MW/m^2$ on the water side of the condenser, corresponding to a stable metallic zinc layer of about 15 cm thickness.

There is a fundamental question to be resolved in the design of a freeze lining for the iron-carbon system. A truly protective lining would grow on a solid iron surface if the surface was maintained at the solidus temperature. But this could be over conservative and perhaps adequate protection would be afforded by maintaining the surface at the liquidus temperature. This latter approach was taken in the present study.

The following details are presented to give an insight into the likely energy implications of installing a water-cooled strip 30 cm in width along the side walls. The extent of the water-cooled region embodies a true freeze lining in the iron bath over a depth of 20 cm, a 5 cm length immersed in the lead layer and a freeboard distance of 5 cm immediately above the iron melt surface, at equilibrium with a slag coating having a liquidus temperature of 1650K. This is to be seen as the first step towards attainment in principle of an everlasting lining in an ironmaking or steelmaking reactor, the elimination of large volumes of costly refractories and the reduction of routine maintenance to an absolute minimum. The trade-off must be in energy terms but as the following summary hopefully demonstrates for a melt circulation system in which hearth erosion is no longer a major cause for concern, the scheme demands serious consideration.

Consider two side-by-side furnaces each 40 x 10 m. Iron melt depth is 0.20 m on top of a molten lead pad 0.10 m deep. For preliminary design purposes assume the melt velocity is 0.3 m/s in both furnaces, bath composition 2.0 percent carbon at a mean temperature of 1700K and coolant is pressurised water with metal surface temperature on the water side at 370K. Bird et al.[16] recommend the Deissler correlation for Re > 10,000 and Pr above 0.5. Without too much error being introduced this can be used for heat transfer in a hot metal system which has a Prandtl number of about 0.2. The calculated heat fluxes are:

(a)	Above melt total radiative flux to equilibrium slag deposit	0.40 MW/m^2
(b)	In iron bath with ΔT = 30K	0.23 MW/m^2
(c)	In lead layer with equilibrium iron layer deposit	0.30 MW/m^2

The steady state thickness of steel/iron (k_{avg} = 39 $Wm^{-1}K^{-1}$) for the flux in the iron bath = $(39)\ (1525 - 200)/(0.23)10^6 = 0.22m$

and the total heat losses to the water cooled panels are:

(a)	Above melt	=	3.2 MW
(b)	In Iron Melt	=	7.4 MW
(c)	In Lead Bath	=	2.4 MW
	TOTAL		13.0 MW

Assuming that the 13 MW (thermal) is used to generate electricity at 30 percent efficiency, this is equivalent to 3.9 MW (electrical) and at say 340 kWh/t O_2 (99.5% at 20 psig) this is equivalent to 275 tonne per day O_2. Alternatively, this 13 MW thermal energy can be seen to contribute towards the thermal energy requirements of the steam ejectors used in the vacuum pumping system.

Disregarding the fact that the heat loss to the water cooled panels can be usefully employed for electricity or steam generation to satisfy process requirements, a loss of 13 MW thermal energy to the cooling water is equivalent to 156 kWh (thermal) per tonne of iron produced, assuming that a rate of 2000 tpd iron production is attainable. To further put this figure in context, a carbon use of say 0.45 tonne per tonne of iron produced is equivalent to a thermal energy input of some 4100 kWh/t so the freeze lining scheme consumes only four percent of the thermal energy input.

Vacuum Pumping Considerations

Although not absolutely tied to the use of RH vacuum degassing technology, the forced melt circulation requirements of the current proposal would appear to be ideally suited to this technology. Very large tonnages of molten iron are to be moved around the closed loop against a minimal head consistent with ultimate removal of a separated slag layer from the system. Accommodating the upleg and downleg in separate baths or sumps probably means that the normal RH vertical vessel arrangement would need to be replaced by a horizontal vessel with the two nozzles positioned appropriately but in all other respects the systems would be comparable.

The relative merits of steam ejector and mechanical pumps are well documented and for the present discussion it is assumed either option is available. Sickbert[17] describes an installed mechanical pump-set with a total capacity of 230,000 m^3/h and gives the total power consumption of 250 kWh per running hour. These figures can be used to explore the energy implications of moving large tonnages of iron/steel around the system using what must be regarded by all as well established commercial practice.

Scale-up problems would probably be minimised by making the upleg elongated in shape so that gas injection into the upleg penetrates fully. For this reason at this stage it appears logical to keep the width of the upleg at around 0.4m and to inject nitrogen from multiple ports arranged along both faces of the larger dimension. Taking a typical RH degasser capable of achieving steel recirculation of some 45 tonne per minute with gas injection to the upleg at a rate of 750 l/min as a conservative basis, a melt circulation rate of say 320 tonne per minute would require (7.1 x 750) = 5300 l/min nitrogen injection into an upleg with internal dimensions of 0.4 m x 2.8 m. Unpublished research[18] using magnetite/water slurries in which the pumping action of an RH type unit against positive heads around 0.3 m was studied, suggests that the actual gas injection would probably need to be around 1.5 times that indicated to allow for the difference in melt level between each side of the loop so that slag removal can be facilitated. These considerations lead to an estimated 570 kg/h of nitrogen to be pumped from the system.

Besides the initial inert gas injection, the vacuum pumping system may also have to cope with large volumes of $CO_{(g)}$ evolved during degassing. This is particularly the case when considering the circulation of a bath containing a low carbon level. For example, if the melt steady state carbon level was 0.05%C, assuming equilibrium is established between dissolved carbon and oxygen with CO at atmospheric pressure prior to degassing, the inlet dissolved oxygen is calculated to be 380 ppm. Using the procedure reported elsewhere by the author[19] for liquid phase mass transport control within the RH degasser, the estimated pumping requirement at 10 mm Hg pressure is about 250,000 m^3/h, a figure in the same region as the Sickbert plant referred to earlier. The mass flowrate in this case is some ten times greater than the Sickbert example but on the other hand the pressure ratio is one tenth. The estimated electrical power per running hour is thus (10 x 250/ln(10) kWh = 1087 kWh per hr. Thus for the 0.05%C case, the electrical power is 1.09 MW which corresponds to a thermal energy rate of 3.6 MW, a figure which is less than 2 percent of the probable total thermal energy requirement.

At higher carbon levels the equilibrium dissolved oxygen is reduced proportionately and the vacuum pumping volumes become relatively modest, particularly if it is acknowledged that a pressure of 10 mm Hg or even higher is perfectly adequate for melt circulation purposes. However, for lead vaporisation and ultimate recovery a liquid lead condensate for recycle to the bath, there are benefits in operating at lower pressures in the region of 2 to 5 mm Hg. In the present case the carbon level in the bath is taken as 2.0 percent and the estimated CO evolution in the RH upleg is around 90 kg/h giving a total figure of 660 kg/h for the gas to exhausted, assuming zero air ingress. Recent figures published for British Steel's Lackenby vacuum degasser[20] give a pumping speed of 1000 kg/h at a vacuum pressure of 10 mbar with a total steam consumption of 8 t/h for a vacuum system utilizing water ring pumps as the atmospheric stage, backing a three-stage steam ejector, an arrangement specified to produce high performance while minimizing steam consumption. This particular system would appear to well suit the requirements of 2000 tpd iron production.

Gas Phase Mass Transfer

One of the problems normally associated with smelting reduction is the maintenance of strongly oxidizing and reducing zones within the one reactor. Any reaction between carbon dioxide with carbon at the melt surface is counter productive insofar as the carbon is being oxidised without first performing useful reduction. Heat balance considerations indicate just how much of this secondary reaction can be tolerated, but clearly it must be under careful control and must at most only represent a small proportion of the total carbon input.

Using the electro-magnetic levitation technique, the decarburisation of iron-carbon alloys with carbon dioxide has been studied by Baker et al.[21] and it is apparent that the decarburisation rate is controlled by gaseous diffusion until low carbon levels are reached. In the present case, calculations indicate that gaseous diffusion control can reasonably be expected down to about 0.15 percent carbon. Below this carbon level the rate of liquid phase mass transfer of carbon from the bulk liquid iron cannot keep up with the supply of oxidant from the gas phase. In the levitation work this situation was manifested by the appearance of sub-surface growth of carbon monoxide bubbles and the ejection of metal droplets from the surface as the bubbles escape from the gas/liquid interface. The full implications of this behaviour will need careful assessment, if indeed a low carbon level is selected as the best approach on other grounds. For the present paper a 2% carbon level is envisaged so gas phase diffusion control can be anticipated.

If the products of combustion of the smelting reduction and coal devolatilisation are considered to occupy two-thirds of gas space above a 40 m x 10 m furnace with a freeboard distance of say 2.0 m between the melt surface and the underneath side of the refractory roof, the gas phase mass transfer rate on the melt surface can be estimated. At a point mid-way along the combustion zone and a gas temperature of say 2300K, the gas velocity is about 17 m/s, the Reynolds Number is 1.2×10^5 and the mass transfer coefficient is 4.7×10^{-2} m/s. Under these conditions carbon dioxide can diffuse to the melt surface and react with carbon, whether it be as coal particles or dissolved carbon in the melt at a rate equivalent to the consumption of 0.2 kg/s of carbon. For a 2000 tpd Fe production rate this is equivalent to just over 2 percent of the total carbon requirement, or in terms of coal with a fixed carbon of 72 percent is equivalent to about three percent additional coal consumption. Clearly, the configuration of smelting reduction being considered is unlikely to present serious secondary combustion problems.

Thermochemical Aspects of Coal Gasification in Molten Iron

The present proposal is closely related to coal gasification by what is known as the Molten-Iron-Pure Gas (MIP) process but, in the ironmaking mode, captive use is made of all the gas whilst the fixed carbon from the coal is consumed continuously in reduction of iron ore. The thermodynamics and kinetics of coal gasification have been reported recently by Barin et al.[22]. The gas produced has a low sulphur content and is composed primarily of CO and H_2. The sulphur in the coal is transferred together with the ash forming constituents to a liquid slag. Their calculations show that for chemical equilibrium at 1773K, 100 vol ppm S in the gas phase coexists with S contents in the slag and liquid iron of 4.4 percent and 0.03 percent, respectively, when the carbon level is at 3.5 percent in the iron and the weight ratio of CaO/SiO_2 in the slag is 1.81. Increasing the slag basicity from 1.8 to 2 decreases the S content in the gas phase down to 50 ppm at 3.5 percent carbon and 1773K. For the somewhat lower carbon and temperature levels presently being considered in this paper and with a basicity ratio of 1.2, the sulphur in the iron produced will be higher than the 0.03 percent level. Therefore, along with other direct ironmaking routes based on coal, some external desulphurization of the hot metal produced will be necessary.

Another aspect very relevant to the present case has been considered in the Barin paper. They point out that whilst both oxygen and sulphur are surface active and it is known that their segregation at the carbon/liquid iron interface can inhibit carbon dissolution, this effect is important only if the rate of reaction at the interface is less than the diffusion rate. However, all the evidence available suggests that in systems such as that now being considered liquid phase mass transfer will be rate controlling.

Some Process Engineering Considerations

Carbon dissolution from the bottom surface of floating bed of lump coal needs a relatively high melt velocity to obtain reasonable dissolution rates. The active surface of the coked coal is well below the slag layer and there is no reason to doubt the applicability of known chemical engineering correlations for estimating the liquid phase rate of mass transfer. Hence the carbon dissolution rate can be reliably predicted for engineering purposes. But the whole of furnace A is required for this service, meaning that heat transfer from hot gases is restricted only to Furnace B.

Furnace B accommodates intense heat transfer to the melt surface and smelting reduction from freely floating preheated iron oxide pellets can be predicted with confidence from MacRae's published work. Gas combustion would occur over the whole bath surface, the back end say 6 m to 10 m probably being covered with iron oxide-rich slag in the smelting reduction zone, leaving at least 30 m length for high intensity heat transfer without appreciable slag layer. However, it is still necessary to have a high emmissivity surface layer to attain intensive radiant heat transfer.

Without O_2 enrichment, air preheat to 1150K and solids to 1600K, the capacity of a pair of 40 m x 10 m furnaces is around 2000 tpd Fe production. A high iron recirculation ratio is required, typically around 230 to 1. In the carbon dissolution region, the local velocity is 0.8 m/s with an average clearance of 0.1 m. The charge column of coal is assimilated into the melt at a downward rate of about 10 cm per hour.

If the iron melt depth in the coal dissolution furnace was say 50 cm, a single bin covering almost the whole exposed surface area of the furnace would need to have a coal height above the melt surface of somewhere in the region of 3 m. Clearly, lesser depths of molten iron to support the column of coal charge would be needed in Furance A if the coal bunker was partitioned longitundinally with baffles running the whole length and terminating just above the melt level. There would also probably need to be some combustion of volatiles to avoid carbon deposition and accretion problems in this coal dissolution region as well as to facilitate slag removal from the accumulated slag layer.

The depth of melt in furnace B is arranged so that in the smelting reduction region, freely floating pellets have sufficient residence time for complete reduction and assimilation into the bath. For example, a surface melt velocity of 30 cm/s would provide a retention time of 30 seconds over the back 9m length of the surface if pellets (preheated to 1600K) are added as a free falling curtain onto the melt surface. This would necessitate a melt depth of about 20 cm in Furnace B and would leave the front 30 m or so of melt surface virtually slag-free for high intensity heat transfer. However, it is probably reasonable to consider the whole bath surface as receiving heat from the oxidised gases at rates dependent on the surface temperature and condition.

Composite thin Layer Smelting Reduction

If preheated fine ore is blended with crushed raw coal on the surface of the melt so that the coal/ore mixture is in close association under intense heat transfer conditions from above and below, it may be possible to achieve reduction and assimilation of the iron values into the bath accompanied by simultaneous smelting reduction by reaction with dissolved carbon. Sequencing of the addition of coal, then coal/preheated ore together would probably be necessary. Some protection of the refractory walls may be possible by adding coal only to the wall region in an analogous fashion to the provision of a hearth layer in certain sinter plant operations. If all this can be done with a relatively low bath temperature of say 1400°C and carbon level of about 2 percent using conventional refractories without a lead layer, this option may well be the most desirable of all the various schemes proposed for ironmaking via melt circulation[4]. With both Furnaces A and B receiving radiant energy from intense combustion above the whole melt surface, there are indications that the furnace dimensions under discussion could sustain around 5000 tpd iron production. Without experimental verification, however, this particular scheme is highly speculative as the sound process engineering foundation of well-proven heat and mass transfer correlations is lacking for this particular configuration.

Process Thermal and Mass Requirements

Consider the smelting reduction of hematite ore containing 64% Fe, 6.5% SiO_2 with a medium volatile coal of the following analysis (dry basis):-

%C	%H	%N	%S	%O	%Ash	Fixed C	%Volatile
83.7	4.83	1.4	0.99	1.8	7.22	71.7	21.0

Calculated Gross Calorific Value = 33 346 kJ/kg.

The coal requirement to produce hot metal at 1700K containing 3% carbon and 0.6% Si is 0.46 tonne per tonne Fe. Limestone with a CaO equivalent of 0.14 tonne per tonne Fe is used to flux the gangue in the ore and coal and produce a slag with CaO/SiO_2 weight ratio of 1.2. This limestone is incorporated into the feed to a disc pelletiser along with iron ore fines and dust reverts. Green pellets are gently dried and thermally hardened in an annular kiln of the type described by Jennings and Grieve[23]. The annular kiln is fired by the smelting reduction off-gases after temperature moderation by cold air admission and the hardened pellets are assumed to issue forth at 1600K into a shute discharging via a lock hopper or similar arrangement onto the surface of the circulating hot metal in the smelting furnace.

The off-gas from the furnace after complete combustion of all the coal volatiles, hydrogen and carbon monoxide from smelting reduction leaves the furnace via a water-cooled venturi throat in which cold air is admitted to bring the gas temperature down from 2300K to 1650K. Some 65 percent or so of the gas at 1650K is exhausted through the annular kiln to an electrostatic precipitator via an induced draft fan to the stack whilst the remaining 35 percent or thereabouts is passed countercurrent to the incoming combustion air so that the air entering the combustion zone in Furnace B is at 1150K. This relatively modest air preheat enables a high alloy steel recuperative preheater to be used rather than the very much more expensive option of regenerative thermal exchange. The air preheater circuit has its own induced draft fan discharging back into the mainstream electrostatic precipitator/baghouse gas clean-up system.

The whole air and off-gas arrangements resemble in many respects technology now being developed for atmospheric pressure circulating fluidised bed coal combustors (AFBC) for utility electric power generation where latest thinking favours tubular airheaters[24]. Forced and induced draft fans rather than turboblowers, inlet air ports rather than tuyeres and relatively light as opposed to pressure vessel construction, immediately distinguishes the type of plant being considered from the modern iron blast furnace. Table 1 summarises some of the more significant operational parameters.

Table 1

Requirements for 2000 tpd Iron Production

Furnace Dimensions	:	2 side-by-side each 40 m length x 10 m wide
Bath Configuration	:	Iron depth 0.2 m Lead depth 0.1 m
Air Volume	:	178900 Nm^3/h
Gas Effluent	:	321700 Nm^3/h
Melt Circulation Ratio	:	230/1
Raw Materials:		
(a) Coal (dry basis)		922 tonnes
(b) Limestone		540 tonnes
(c) Iron Ore Fines		3190 tonnes
Annular Kiln Pellet Preheater	:	11 m diam with 1.1 annulus width x 2.6 m bed height
Air Preheater	:	196 x 6 inch nominal bore pipes inside 3 m x 3 m square duct 50 m in length. Combustion gases tubeside with air on shellside. High temperature end Type 310 stainless steel or Incoloy 800 with longitudinal fins.

Economic Evaluation

(a) Capital Cost

The capital cost estimates for a 2000 t/day stand alone smelting process were calculated by dividing the plant into five major areas, namely:-

(i) Interconnecting channel furnaces
(ii) Materials handling equipment
(iii) Air preheating equipment and ancillaries
(iv) RH lift pump arrangement
(v) Pellet preheating and ancillaries

A capital cost of £35.1M is estimated which includes an installation and management element. The installation costs have been calculated based on standard estimating practice, whereby the capital costs of the main plant items are multiplied by a factor (generally between 1.5 and 1.9) to arrive at the capital plus installation costs. A 25% mark up has then been applied to include management services costs. Table 2 summarises the major items involved in the capital costs, including installation and management costs. In some cases the cost figures reflect budget quotations from equipment suppliers, in others estimates were based on actual plant costs of similar installations, whilst for the air preheater and its ancillaries, an order of magnitude cost was arrived at based on short-cut methods outlined in the I.Chem.E Guide to Capital Cost Estimating[25].

The cost of the RH lift pump assembly includes three units complete with steam ejectors, standby spares, etc. Also included in the capital cost is the lead bath, plus refractory spares, etc. For an annual production of 660,000 tonnes, the capital requirement represents an investment cost of £53.2 per annual tonne.

As already shown by Smith and Corbett[2], the capital cost per annual tonne of capacity for all the new coal-based ironmaking processes at 0.5 Mt/year is significantly lower than the cost of a conventional works at 0.5 or 1 Mt/year. The range shown by these authors (1985 costs) extends from £110 to £190 per annual tonne of capacity. The estimate provided in Table 2 demonstrates the very considerable cost advantage of the melt circulation process.

(b) Operating Costs

The operating costs are based on the materials requirement listed in Table 3, together with the capital cost estimates for the production of 660,000 t p.a. of liquid iron. The operating cost breakdown listed in Table 3 includes capital charges of 13.15%, based on a 10% IRR, with depreciation over 15 years. The maintenance charges are set at 6% and a further £4.80 per tonne of hot metal is added to cover labour and miscellaneous costs. The raw material and energy requirements, their unit costs and consumptions per tonne of hot metal (thm) containing 3 percent carbon are listed in Table 3. Hot metal costs of £71.21 per thm are indicated. Also included in Table 3 are costs for desulphurisation, provisionally estimated at £5 per thm, bringing the total cost per tonne desulphurised hot metal to £76.21.

(c) Comparison with Other Processes

In Table 3, unit costs for coal and fine iron ore have been taken from data compiled by Smith and Corbett[2] for average European energy and raw material costs (Jan. 1985) with 5% p.a. escalation compounded to 1988. Meaningful comparisons with other new processes or even with existing ironmaking plant is made difficult because of the wide differences in unit costs employed. There is also the complication of the credit value to be given to arising gaseous energy. Smith and Corbett approach this problem by firstly considering the most favourable of circumstances in that the arising gas can be sold at natural gas prices, in their case at average European price levels. Next the impact of fuel gas credits not being available is assessed. Their energy and materials costs in present day terms for the first scenario are £60.2/thm for a conventional ironworks with the new coal-based processes ranging from £54.4/thm to £69.5/thm. By comparison the figure for the melt circulation process is £56.4/thm.

With limited or no gas credits available, Smith and Corbett compared the various new processes then being considered on the basis of the sum of the capital, raw materials and energy costs. With the capital charge being assessed at 20 percent, the lowest cost process in today's terms works out at about £98.4/thm. On the same basis the figure for the proposed melt circulation process is £66.8/thm.

Summary and Conclusions

(1) An entirely new concept for the smelting reduction of iron ore fines with lump coal has been introduced.

(2) The key to the new technology is forced circulation of hot metal in a closed loop around two furnace hearths maintained at slightly different levels using the well established RH vacuum degassing process as the springboard for further development.

(3) Rapid reduction occurs as previously preheated and hardened ore pellets are introduced onto the surface of the moving hot metal stream containing around 2 percent carbon after it has received upstream the thermal energy resulting from total combustion of coal volatiles and CO generated in the smelting reduction reaction.

(4) Without oxygen enrichment of the process air, the temperature above the melt surface is calculated to reach around 2300K at steady-state for 2000 tpd iron production in a 10m wide x 40m long furnace, a situation that can only be sustained without overheating the refractory enclosure if a high quality thermal sink is readily accessible.

(5) Forced circulation of hot metal taken from beneath the slag layer in the lower hearth and then its distribution to the higher hearth by the RH vacuum lift pump followed by its continual overflow back into the lower hearth after its passage along the full length of the upper hearth means that no slag accumulates in the combustion region.

(6) High intensity heat transfer, principally by radiation, occurs between the hot gases and the melt surface in the absence of an appreciable slag layer. Energy transfer rates in the region of 350 kW/m^2 are considered to be attainable.

TABLE 2

CAPITAL ESTIMATES FOR 2000 T/DAY SMELTING REDUCTION PLANT

		£M	£M
(1) Furnaces (Pair)			
	Refractories	1.79	
	Steelwork	0.81	
	Water Cooling	0.41	
	Tuyere + Windbox	1.00	
		4.01	
	Installed Cost (x 1.89)	7.62	
	Management Cost	1.90	
	Total	9.52	9.52
(2) Feeding Equipment			
	Hoppers, Feeders	0.73	
	Installation Cost (x 1.55)	1.14	
	Management Cost	0.28	
	Total	1.42	10.94
(3) Air Preheating, Forced and Induced Draft Fans & Gas Cleaning			
	Equipment	3.15	
	Installed Cost (x 1.60)	5.04	
	Management Cost	1.26	
	Total	6.30	17.24
(4) RH Lift Pump			
	Equipment	4.0	
	Installed Cost (x 1.57)	6.28	
	Management Cost	1.57	
	Total	7.85	25.09
(5) Pellet Preheating System, including Annular Kiln, 2 Disc Pelletisers, I.D. Fans, Electrostatic Precipitator			
	Equipment	4.5	
	Installed Cost (x 1.60)	7.2	
	Management Cost	1.8	
	Total	9.0	34.09
(6) Lead Bath + Spares + Refractories		1.0	
	Total	1.0	35.09
	TOTAL CAPITAL COST		£35.1M

TABLE 3

SMELTING REDUCTION IRONMAKING COSTS

Item	Unit Cost	Unit Consumption per t hot metal	Cost £/t hot metal
Ore (fines)	£20/t	1.55t	31.00
Coal (lump)	£45/t	0.45t	20.25
Limestone	£12/t	0.27t	3.24
Electricity	0.03/kWh	60kWh	1.80
Nitrogen	£20/t	0.007t	0.14
Subtotal raw materials and energy			56.43
Labour and Miscellaneous			4.80
Maintenance 6% of equipment cost/year			3.19
Capital £35.1M 13.15% (10% IRR over 15 years)			6.79
HOT METAL COSTS (ex Runner)			71.21
Desulphurisation:-			
Pre-treatment cost			2.50
Iron Loss (5%)			2.50
DESULPHURISED HOT METAL COST			76.21

(7) The off-gases at 2300K are exhausted from the furnace via a water-cooled venturi throat in which dilution air is added to bring the temperature down to 1650K. Variable speed induced draft fans then draw about 65 percent of the off-gas through a 11m diameter annular kiln, which serves as a pellet hardener and charge preheater, and the remaining 35 percent or so through a recuperative high alloy steel heat exchanger providing air preheat to 1150K for admission to the system by a series of ports arranged along both sides of the combustion furnace.

(8) The furnace itself is maintained under slight negative pressure to ensure that there are no fugitive emissions and all effluent gases are eventually cleaned-up by bag filters or electrostatic precipitators before release to the atmosphere.

(9) Lump coal is assimilated into the melt under forced convection conditions at a rate dependent on the liquid phase mass transfer coefficient, the carbon concentration in the hot metal and the exposed interfacial area. For example, a coal bunker placed above a 10m wide x 40m long furnace can be expected to sustain an iron production rate of at least 2000 tonne per day. Melt circulation ratios in the region

of 230/1 are required for this service.

(10) To protect the furnace hearths from chemical erosion attack, it is proposed that a shallow layer of molten lead is maintained beneath the hot metal. By careful design and due attention to the vacuum de-leading capabilities of the RH vacuum lift pump, an acceptably low steady-state concentration of lead in the hot metal must be maintained. Mathematical modelling and experimental evaluation of the system's ability to cope with the potential lead problem must rank high in research priority.

(11) Given a successful outcome to the above, coal-based ironmaking via melt circulation would appear to offer very substantial cost savings over all other new and emerging technology. This is due to two principal attributes. First, it is inherently a low capital cost solution to ironmaking in the future. Secondly, full combustion of all the coal volatiles and process gases within the reactor itself is energy efficient and makes the process independent of the availability of fuel gas credits to secure its strong economic position.

Acknowledgments

The assistance of British Steel Technical, Teesside Laboratories is gratefully acknowledged. Assistance was also generously provided by Davy McKee Research and Development, Simon-Carves Limited and Vacmetal (U.K.) Limited.

References

1. N.A. Warner, A method of recovering non-ferrous metals from their sulphide ores. British Patent 2 048 309, 1983, U.S. Patent 4 334 918, 1982, European Patent 0 016 595, 1984.

2. R.B. Smith and M. J. Corbett, "Coal-based ironmaking", Ironmaking and Steelmaking, 14, (2), 1987, 49-75.

3. H.H. Kellogg, "Conservation and Metallurgical Process Design", Trans. Inst. Min. & Metall. (Sect. C: Mineral Process Extr. Metall.) 86, 1977, C47-C57.

4. N.A. Warner, Smelting reduction process, U.S. Patent 4, 701, 217, 1987.

5. N.A. Warner, "Innovative smelting with slag energy recovery", in The Reinhardt Schuhmann International Symposium on Innovative Technology and Reactor Design in Extraction Metallurgy, D.R. Gaskell, J.P. Hager, J.E. Hoffmann and P.J. Mackey ed; (T.M.S. AIME Warrendale Pa., 1986), 159-174.

6. N.A. Warner, "Towards polymetallic sulfide smelting", in Complex Sulfides - Processing of Ores Concentrates and By-products, A.D. Zunkel, R.S. Boorman, A.E. Morris and R.J. Wesley, ed; (T.M.S. AIME Warrendale Pa., 1985), 847-865

7. R.W. Serth, T.E. Ctvrtnicek, R.J. McCormick and D.L. Zanders, "Recovery of waste heat from slags via modified float glass process", Energy Communications, 7 (2), 1981, 167-188.

8. T. Asada, T. Sugiyama and S. Inigaki, "Measurement of the solubility of lead in molten steels", Denki Seiko, 34, 1963, 128-132.

9. T. Sugiyama and S. Inigaki, "Study on solubility of lead in steel", Denki Seiko, 34, 1963, 469-475.

10. A. N. Morozov and Yu Ageev, Izvest. Akad. Nauk. SSR, Metals, 4, 1971, 111-114.

11. S. Eketorp and V. Brabie, "Energy considerations in reduction processes for iron- and steelmaking", Scand. J.Met. 3, 1974, 200-204.

12. D.R. MacRae, "Kinetics and mechanism of the reduction of solid iron oxides in iron-carbon melts from 1200°C to 1500°, J.Metals, Dec. 1965, 1391-1395.

13. S. Komori, H.Ueda, F. Ogino and T. Mizushina. "Turbulence structure and transport mechanism at the free surface in an open channel flow", Intl. J.Heat Mass Transfer, 25(4), 1982, 513-521.

14. A. Furman, "Mathematical models applied to aluminum reduction cells" in Heat and Mass Transfer in Metallurgical Systems, D.B. Spalding and N.H. Afgan, eds: Hemisphere Publishing Corp., Washington. 1981,215-234.

15. J.M. Delhaye, "Basic equations for two-phase flow modelling", in Two - Phase Flow and Heat Transfer in the Power and Process Industries, A.E. Bergles, J.G. Collier, J.M. Delhaye, G.F. Hewitt and F. Mayinger, eds: Hemisphere Publishing Corp., Washington, 1981, p86-90.

16. R.B. Bird, W.E. Stewart and E.N. Lightfoot, Transport Phenomena, John Wiley, New York, 1960, p402-404.

17. A. Sickbert, "Discussion on plant and theory" in Vacuum Degassing of Steel, Special Report 92, The Iron and Steel Inst. London, 1965, p55-56

18. C.T. Marsh, "Model studies in vacuum degassing", Dept. Minerals Engineering, Univ. of Birmingham, unpublished report, 1980.

19. N.A. Warner, "Direct smelting of zinc-lead ore", Trans. Instn. Min.Metall. (Sect. C: Mineral Process. Extr. Metall.), 92, 1983, 147-152.

20. G.W. Skinn and S. Knowles, "Design, commissioning, and operation of Lackenby vacuum degasser", Ironmaking and Steelmaking, 14, 1987, 17-21

21. L.A. Baker, N.A. Warner and A.E. Jenkins, "Kinetics of decarburization of liquid iron in an oxidising atmosphere using the levitation technique", Trans. TMS-AIME, 230, 1964, 1228-1235.

22. I. Barin, M. Modigell and F. Sauert, "Thermodynamics and kinetics of coal gasification in a lqiuid iron bath", Met. Trans. B. 18, 1987, 347-354.

23. R.F. Jennings and A. Grieve, "Developments in pellet hardening", Steel Times Annual Review, 1968, 3-10.

24. E.S. Taylor, "Development, design and operational aspects of a 150MW(e) circulating fluidized bed boiler plant", in Circulating Fluidised Bed Technology, Prabir Basu ed; Pergamon Press, 1986, 363-376.

25. Anon. A Guide to Capital Cost Estimating, 3rd Edition, Instn. Chemical Engrs., London, 1988.

PLASMA TECHNOLOGY FOR METAL AND ALLOY PRODUCTION

PRESENT STATUS AND FUTURE POTENTIAL

Sven O. Santén and Jerome Feinman

SCANARC PLASMA TECHNOLOGIES AB
Box 202, 813 00 Hofors, Sweden

Abstract

The principles involved in the generation and application of plasma energy are reviewed briefly and the major developments in commercial and semi-commercial primary metal and alloy production are described and discussed in broad outline, with emphasis on present performance and future potential. New concepts in the application of plasma energy in primary operations are described with emphasis on the expected advantages over conventional technology and the problems that have to be solved to enable effective implementation.

Metallurgical Processes for the Year 2000 and Beyond
Edited by H.Y. Sohn and E.S. Geskin
The Minerals, Metals & Materials Society, 1988

Introduction

The use of plasma energy in industrial processes is generally considered to have started in 1905 with the operation of the Birkeland process for nitrogen fixation for fertilizer production. Although it seems logical to apply plasma energy to metallurgical processes, which also usually require considerable energy at high temperature levels, serious development work along these lines did not begin until the early 1970's. The reasons for this hiatus are associated in part with recognized difficulties in applying plasma energy in processes involving tonnage quantities of solids in severe industrial environments and the delicate nature of the earlier plasma generators. Another, and perhaps overriding factor, is the fairly steady improvement in efficiency and economy of most primary metal producing processes by conventional developments and modifications through the 1950's and 60's. Examples of these evolutionary changes include burden preparation and fuel injection in blast furnace applications and prereduction treatments in ferroalloy operations. By the end of the 1970's most primary metal producing processes had been optimized close to the limit, at least in terms of those operations that were able and willing to implement the developed improvements. In this context, plasma energy offered the possibility to achieve further significant improvements in primary operations where conventional means were exhausted.

Ideas for using plasma energy in metallurgical processes stem in part from a knowledge of how plasma is generated and its physical and chemical properties. We are speaking of thermal plasmas, which are generated by passing an electric current through a gas, whereby the electric energy is transferred to the gas in the form of elevated temperature and dissociation products, as illustrated in Figure 1A and B for so-called non-transferred and transferred arcs. The benefits of plasma energy in primary operations are partly due to its high level (temperature) and density (enthalpy), which distinguish it from energy from combustion of fossil fuels as shown in Table I. Another important difference between plasma and combustion as far as primary metal production is concerned is that plasma gas can have virtually any oxidation potential, from completely oxidized to reducing, while combustion gas is generally limited to neutral to completely oxidized. In this review, we will cover the use of plasma energy in primary metal and alloy production because these processes involve high unit energy consumption and concomitant high efficiency of utilization. Special applications of plasma in metallurgical processes such as spraying, cladding, refining and melting generally use low unit consumptions or are less restrictive about efficiency because of the unique capability of the plasma to produce the desired result. In any event, a complete survey of the use of plasma in metallurgical processes is presented in reference (1).

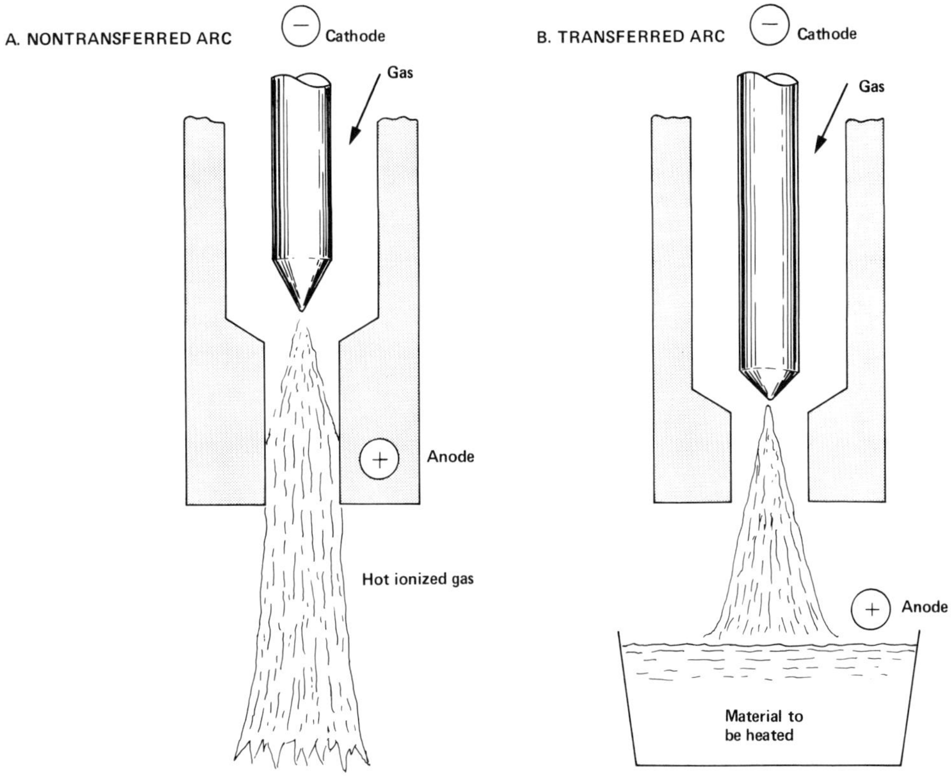

Figure 1 - Schematic representation of nontransferred arc and transferred arc plasmas.

Table I Temperature and Enthalpy for Thermal Plasma and Fossil Combustion

	Temperature, K	Enthaply, kWh/Nm^3
Plasma	up to 10,000	4.0 - 15.0*
Combustion	4,000	1.0 - 1.5

* Lower enthalpies for nontransferred arc, higher enthalpies for transferred arc.

Major Developments

As might be surmised from a consideration of the implications of Figure 1A and B and Table I, the major developments in applying plasma energy to primary metal and alloy production fall in either of the two categories - nontransferred and transferred arc systems. Examples of most of the systems that have been studied to date are presented and discussed in reference (1). Without specifically selecting any of the systems, a general comparison of the two concepts can be made based on the schematic drawings in Figure 2A and B. The nontransferred arc system generally takes the form of a shaft reactor which enables effective contact between the plasmagas, which contains all of the energy input, and the reacting solids/liquids. The transferred arc system generally comprises an empty furnace volume which enables the arc to be maintained between the plasma generator electrode, usually the cathode, and the furnace hearth containing the conducting liquid reaction products, usually forming the anode. In the transferred arc system the reactants are fed at the top or through the generator so that contact between reactants and plasmagas is not as effective as in a countercurrent shaft. However, in the transferred arc system an important compensation for this less efficient contact is obtained by direct transfer of energy to the bath by means of potential drop at the hearth interface (anode fall). Some important characteristics of the two systems are summarized for comparison in Table II. The higher efficiency and the lower plasmagas volume of the transferred arc systems will be partly offset by more difficult gas cleaning and process control requirements. Capacity limitations in scale-up may restrict high tonnage throughput primary operations to nontransferred arc technology. For the primary operations, the plasmagas volume/unit energy is mitigated by the additional gas generated by the reduction reactions, which amounts to about 50 percent of the total gas flow in the nontransferred arc cases and can be as high as 80 to 90 percent in the transferred arc applications.

Table II Comparison of Major Characteristics - Transferred and Nontransferred Arc Systems

Characteristic	Transferred Arc	Nontransferred Arc
Plasmagas volume/ unit energy	low	high
Flexibility wrt plasmagas	high	high
Offgas cleanliness	poor	generally good
Offgas temperature	high	low
Control of energy/feed	must be good	not critical
Efficiency, output/input	90 to 95 %	80 to 85 %
Scale-up	may be limited to about 10 MW	50 MW appears reasonable

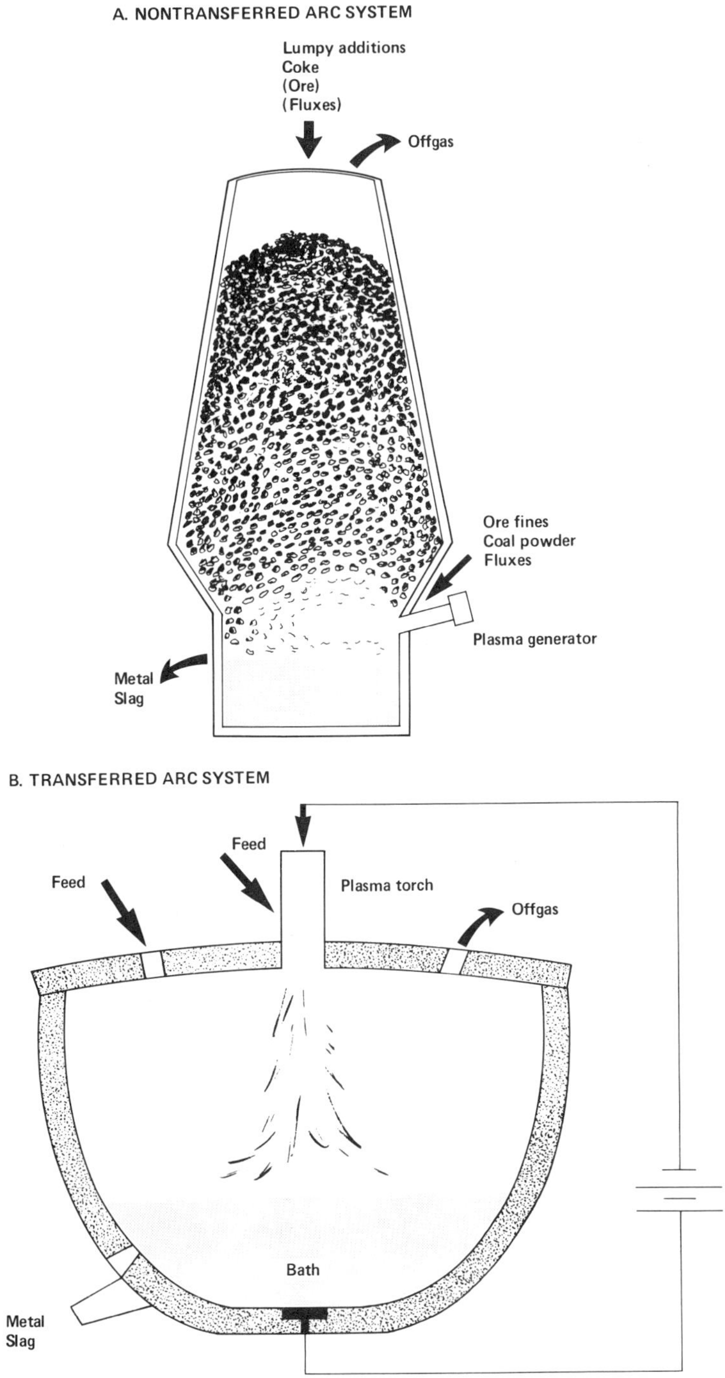

Figure 2 - Schematic representation of nontransferred arc and transferred arc reaction systems.

Status Of Commercial And Semicommercial Plants

The commercial and semicommercial plants in primary metal and alloy production are listed in Table III. The nontransferred arc plasma generators are used in 6 of the 7 cases; the remaining case being based on a transferred DC arc. This distribution is skewed by the fact that four cases involve either plasma-based fuel gasification or gas heating and may not be representative of the future trend of expanded utilization of plasma energy in primary metal and alloy production. A comparison of the operating performance of the SwedeChrome and Middelburg Steel and Alloys ferrochromium plants for the next few years could be decisive in determining which approach will be more successful in competing with conventional submerged arc technology for future new capacity. Both systems have the advantage of being able to process fine chromium ores and fine coal without prior treatment. In other respects, the comparison in Table II may be used as an indication of relative potential.

The performance results of these ferrochromium plants will also provide a good indication about how the two systems might be adapted to the production of ferroalloys such as ferromanganese, ferrosilicon, ferroboron, and other industrial alloys. In the cases of ferromanganese, ferrosilicon and other systems that contain relatively high vapor pressure products and intermediates, the countercurrent shaft reactors with nontransferred arc plasma generators would appear to offer important operating advantages because the lower offgas temperatures attainable will result in lower material losses.

With respect to the processes for treating waste steel plant dusts, it should be emphasized that while these are not primary applications, the operations are equivalent to systems that would be used to produce iron, zinc, phosphorous and other similar metals or alloys. Although tests have been conducted at the 1 to 2 MW level of treatment systems based on the transferred arc generators, no commercial systems have been built. The extensive experience with the ScanDust plant, with continuing improvements and demonstrated reliability and efficiency may make it difficult for alternative plasma-based systems to gain acceptance for future capacity, expected to be considerable in view of the increased environmental restrictions on haphazard disposal of these hazardous wastes.

The direct reduction and blast furnace hot blast heating operations probably represent fairly limited applications of plasma energy in primary operations, particularly in the case of direct reduction. Direct reduction capacity is being utilized at about the 50 percent level (9); new plants are not a high priority item in most locations. The hot blast heating case offers more promise for the future because limitations in coke availability and the potential need for more hot metal on an incremental basis make plasma energy one of the most effective of the alternative solutions.

Table III Summary of Commercial and Semicommercial Plants using Plasma in Primary Metal and Alloy Production

Plant and Location	Description	Type and Capacity	Startup year	Ref.	Status
PLASMARED, Hofors, Sweden	Direct reduction Plasma gasification	Nontransferred arc fuel gasification, shaft reduction-iron ore pellets, 6 MW, 50,000 MTPY	1980	(2)	Decommissioned in 1985, semicommercial development plant
ScanDust AB,* Landskrona, Sweden	Coke-filled shaft smelting of waste steelplant dust	Nontransferred arc 3 6-MW plasma generators 70,000 MTPY dust	1984	(3)	In sustained operation
SwedeChrome AB, Malmö, Sweden	Coke-filled shaft smelting of chromium ore to produce high-carbon ferrochromium	Nontransferred arc 2 shafts, 4 6-MW plasma generators in each, 78,000 MTPY ferrochromium	1986	(4)	In sustained operation
Union Steel Corp, Vaal, South Africa	Direct reduction Plasma gasification	Nontransferred arc fuel gasification, shaft reduction iron ore pellets, 300,000 MTPY, plasma capacity not known.	1986	(5)	In sustained operation
Middelburg Steel & Alloys, Krugersdorp South Africa	Conversion of submerged arc furnace High-carbon ferrochromium	Transferred DC arc furnace, 12-14 MW	1985	(6)	In sustained operation Power Capacity increased in 1988
SFPO Société du Ferro-Manganèse de Paris-Outreau Boulogne-Sur-Mer, France	Plasma heating of hot blast for ferromanganese blast furnace	Nontransferred arc Air preheating, 9 1.5 MW generators	1984	(7)	In sustained operation
HFRSU Hauts Fourneaux Réunis de Saulnes et Uckange Uckange, France	Plasma heating of hot blast for iron blast furnace	Nontransferred arc Air preheating, 6 1.7 MW generators	1987	(8)	In sustained operation

* Although not a primary metal or alloy producing system, included here because shaft smelting operation is equivalent; first commercial application of this technology.

New Concepts

It may seem premature (presumptuous might be a more appropriate expression) to consider new concepts when the existing applications are still not well established and in wide application, but it is the nature of new technology to require fairly long development times. Perhaps the new processes will be developed and applied in less time than the 10 to 15 years that it took to bring in the projects summarized in Table III, but this brings us very close to the 21st century that we are supposed to be describing so it is definitely in order to talk about these concepts now and especially about some of the problems that will be faced. There are basically two areas that we think are relevant - geopolitical and technological. In many respects these areas overlap but it will be easier to consider them separately.

The geopolitical aspects relate to the relationships among the factors that constitute a project - raw materials, energy, manpower, technical infrastructure, logistics, markets, environmental, etc., and are equally applicable to existing plasma processes. Production plants are generally installed where the integration of these factors provides a favorable long-term result compared with alternatives. But not always! Good examples are the MIDREX direct reduction plants at Emden in West Germany and Hunterston in the UK. These plants were located near markets but far from raw materials, energy, and competitive labor and were never operated. On the other hand, the direct reduction capacity near Puerto Ordaz, Venezuela, will always represent an optimum because the raw materials, energy, manpower and logistics are. In the case of Puerto Ordaz, the comparison should eventually be carried through from primary iron to finished steel because the conditions will become overwhelmingly favorable when the experience and concomitant technical infrastructure mature. This will continue into the foreseeable future because the essential resources are essentially inexhaustible.

There are lessons and guidance in this rather unique situation to help us in establishing a sensible approach or philosophy for expanding the use of plasma energy in primary metals production. These can be summarized in a few simple statements.

* Because electric energy is a major cost component, the supply must be fairly priced, consistently available for the foreseeable future, and, except for "acts of God", not subject to whimsical influence. It is clear that abundant hydropower meets these criteria best. The North Countries, Canada, and various locations in South America are the prime candidates. For steel plant waste treatment applications this condition does not apply because logistics surrounding feed supply are overriding.

* Raw material feeds being indispensable, indigenous ores are preferred, otherwise, long-term availability is best assured by establishing sources with properties that make them unsuitable for conventional processing without costly treatment, but which are directly usable with the plasma process, for example, fine chromium and manganese ores in the case of ferrochromium and ferromanganese.

* When "waste" energy in the form of sensible and chemical heat in offgases is a substantial fraction of the total energy input, export uses should be available.

* Environmental concerns are generally less difficult to overcome because plasma-based processes are always closed and do not usually involve dirty ancilliaries, particularly when coke is purchased.

* Plasma processes involve advanced processing and control technology, so skilled manpower is essential - the plants should be considered a community asset, they are clean and represent desirable work opportunities.

* Because many projects will involve international cooperation to satisfy raw material, energy, manpower and market requirements, the concept of joint ownership/operation should be considered to enable the attainment of maximum stability and efficiency with the broadest possible distribution of benefits.

Other considerations will surely become apparent as specific projects appear for evaluation. A thought that may summarize some of these geopolitical aspects better than any detailed statistical or economic evaluation is - if a highly competitive steel industry could be implemented in Japan, without any of the requisites outlined earlier except people and markets, what would be achieved with most, if not all of them? Can we foresee a time when we will rise above local interests to achieve broader benefits?

Some of the proposed new concepts in applying plasma energy to primary metal and alloy production were presented in a very preliminary way in earlier symposia (10,11). Figure 3A and B show schematically the major concept, which we call the slag bath or slag foaming reactor. Based on the nontransferred arc generators, it differs from the nontransferred arc shaft representation in Figure 1, in that the reaction zone is an "agitated" slag bath and freeboard space with slag splashing; there is no coke in the system except that formed from coal injected as a reactant, and metal bearing ore is fed at the top instead of into the plasma gas. In Figure 3A the separation between the reaction zone, which must be highly reducing, and the preheating /melting zone, which must be highly oxidizing, is achieved by locating the coal and oxygen introduction points as far from one another as practical to minimize intermixing of the two. A system as agitated and turbulent as this may have sufficient inherent backmixing to make this objective difficult if not impossible to obtain, therefore the modification shown in Figure 3B is proposed as an alternative that provides a more positive separation.

By splitting the gas leaving the reaction zone, the construction affords a means to allow hot feed to enter through a "purged" bath, with combustion restricted to the space above.

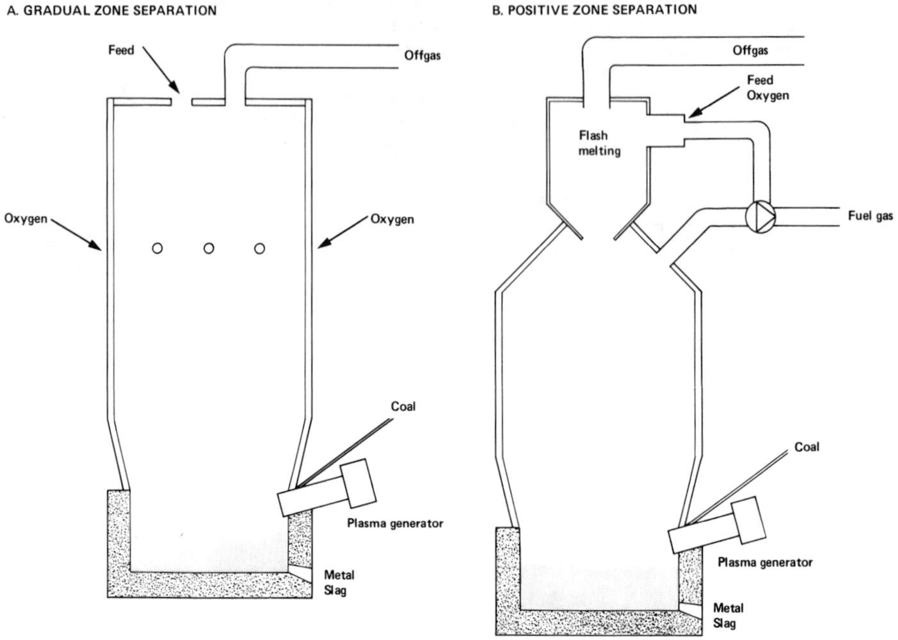

Figure 3 - Schematic representation of slag bath or slag foaming reactor.

The expected advantages of the slag foaming concept include:

* Ability to obtain a range of oxygen potentials in the reaction zone, with concomitant control of carbon content.

* Elimination of dependence on coke.

* Lower total energy consumption by exchange of energy leaving reduction zone with feed solids.

* Possibility to do selective reduction or oxidation.

To illustrate these potential advantages, we have calculated requirements for the production of high-carbon ferrochromium by slag foaming for comparison with the actual requirements for SwedeChrome (11). The results are summarized in Table IV and show that slag foaming uses no coke, requires about 40 percent less electric energy, and produces 65 percent less fuel gas energy for export. Another application that has been considered for treatment using the slag foaming concept is the smelting of ilmenite to produce high titania slag and hot metal. Preliminary tests of this system in a 1.5 MW pilot plant were encouraging (11). The slag foaming concept should also be effective in treating complex oxidic ores where selective reduction conditions are needed to perform the desired separation.

Table IV Comparison of SwedeChrome and Slag Foaming for High-Carbon Ferrochromium per Tonne Product

	SwedeChrome	Slag Foaming
Ore, T	2.3	2.3
Coal, T	0.37	0.51
Coke, T	0.17	-
Oxygen, kmoles	-	12.4
Electricity, kWh	4900	3040
Slag, T	1.2	1.2
Fuel gas, Gcal	2.8	1.0

The major development problems that need to be solved before scale-up to commercial operations include:

* Containment, particularly in the area of vigorous slag agitation.

* Liquid discharge and "level" control in continous operation; the plasma generator outlet should be under the liquid surface at all times.

As we look back over the history and projections that we have assembled, it is a good idea to think about how much of what we have suggested for the future is realistic and how much is wishful thinking. Because we are, in a sense, constrained to be optimistic and enthusiastic about our efforts and goals, it is easy to overestimate the potential and underestimate the problems and obstacles. On reflection, however, we believe that our story is realistic. The geopolitical constraints are not a necessary condition for the expansion we look for, just an extension of present trends in the world. The technical constraints and problems are tractable although they may seem quite challenging even in todays' era of high technology. Nobody with the charter to do development in pyrometallurgy is there because it is easy. If the goals are worthy, the efforts can usually be justified even if the probability of quick solutions is low. We leave it to the audience to make their own judgements.

References

1. "Plasma Technology in Metallurgical Processing," Jerome Feinman, Ed., Iron and Steel Society, Warrendale, PA, 1987.

2. H.G. Herlitz, B. Johansson and S.O. Santén, "A New Family of Reduction Processes based on Plasma Technology," Iron and Steel Engineer, March 1984, pp. 39-44.

3. H.G. Herlitz, B. Johansson, S.O. Santén and J. Feinman, "EAF Dust Decomposition and Metals Recovery at ScanDust," 45th Electric Furnace Conference Proceedings, December 8-11, 1987, Chicago, Illinois.

4. A.B. Wikander, H.G. Herlitz and S.O. Santén, "The First Year of Operation at SwedeChrome," ibid.

5. See Chapter 9 in Reference (1).

6. See Chapter 12 in Reference (1).

7. See Chapter 12 in Reference (1).

8. See Chapter 9 in Reference (1).

9. J. Feinman, "Recent Developments in Direct Reduction and Reduction Smelting," Center for Pyrometallurgy Conference - Gas-Solid Reactions in Pyrometallurgy, Purdue University, Lafayette, Indiana, April 24-25, 1986. Proceedings Published by The Center for Pyrometallurgy, Fulton Hall, University of Missouri-Rolla, MO 65401, pp. 279-304.

10. J. Feinman, Section 3.3, Thermal Plasma Technology Arc Generator Melting and Smelting: Nontransferred Mode, NSF Workshop on Thermal Plasma Systems, August 16-18, 1986, Concord, New Hampshire, jointly sponsored by National Science Foundation, Georgia Institute of Technology and Westinghouse Research and Development Center, pp. 34-39.

11. S.O. Santén and J. Feinman, "Flash Reactions in Conjunction with Shaft and Bath Smelting and Gasification - Theoretical and Experimental Results for Plasma-Based Coal Gasification and Refractory Metal Smelting," Center for Pyrometallurgy Conference on Flash Reaction Processes, June 9-10, 1988, University of Utah, Salt Lake City, Utah, USA, Proceedings Published by Center for Pyrometallurgy, University of Missouri-Rolla, Rolla, MO 65401, pp. 405-419.

STAINLESS STEELMAKING PROCESS BY DIRECT USE

OF THE SMELTING REDUCTION OF CHROMIUM ORE IN BOF

S.Nishioka, K.Yamada, T.Takaoka, Y.Kikuchi,
Y.Kawai, A.Ozeki and M.Yamaga

Steel Research Center, NKK Corporation
1-1 Minamiwatarida-cho, Kawasaki-ku, Kawasaki,
Japan

Abstract

Recently, smelting reduction of chromium ore in BOF has been actively investigated as a key unit process for an innovative future stainless steelmaking. The prominent feature of this new process route is energy and process saving, and production cost saving. The development is already in the midst of application to the commercial production, however there still remain the following key technologies to be established for the full use of the process. One is a speed up of the reduction of chromium ore to match with sequential continuous casting, and the other is efficient combustion of carbonaceous materials in BOF to reduce their consumption, and hence to assure the steel quality. High degree of post combustion technology and new charging technology, i.e., ore-coal-flux-oxygen mixed injection, could be effective for those two subjects. In 2000 or beyond, the new process would have become quite popular owing to the establishment of those key technologies.

Metallurgical Processes for the Year 2000 and Beyond
Edited by H.Y. Sohn and E.S. Geskin
The Minerals, Metals & Materials Society, 1988

Introduction

Since stainless steel was invented, its production has continued to expand because of its attractive combinations of functions of excellent corrosion resistance, good formability, a pleasing appearance, and a wide range of strength levels. New types of stainless steels as well as new applications are continually being developed.

Compared to the production of ordinary steels (1), stainless steel production (2,3) shows steady growth worldwide, while the former production remains on the same level, as Figure 1 shows. It can be estimated that future production would show further growth since materials consumer demands are more fashionable, or more functional, which could be satisfied by stainless steels.

Thus market needs would be met, if it could be produced with reasonable production cost. The emergence of an innovative production technology is then potentially requested as VOD and AOD processes have emerged, which has saved substantial production cost. This paper describes a possibility of a future stainless steelmaking process, which uses the in-bath smelting reduction of chromium ore in BOF, and which would save production as well as energy cost.

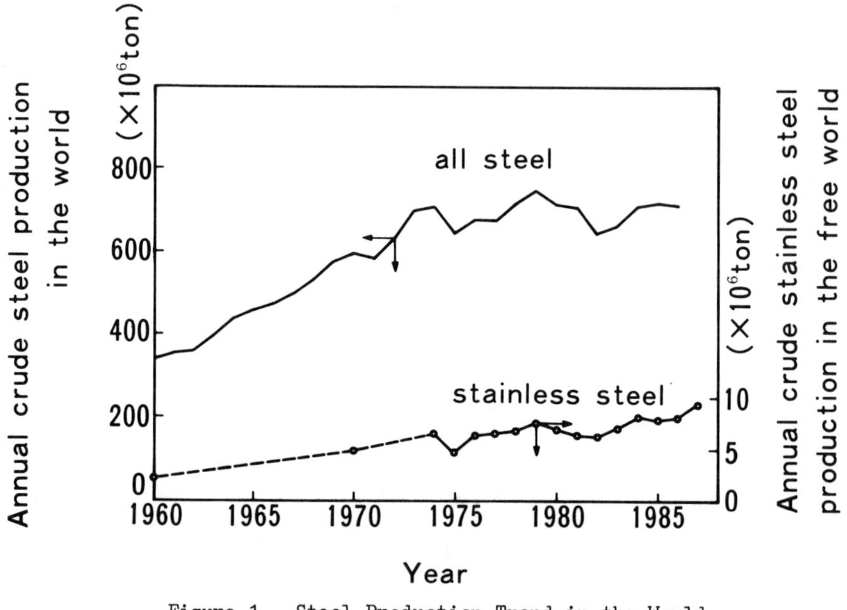

Figure 1 - Steel Production Trend in the World.

Needs of In-Bath Smelting Reduction of Chromium Ore for Stainless Steel Production

Present stainless steel production is carried out mostly by EF-AOD process. AOD process has become a dominant process since it has raw material and processing flexibility as well as high productivity, which eventually yields low production cost. High carbon ferrochromium is a key alloy for the process, taking the place of low carbon ferrochromium, which used to be the essential alloy in the former EF stainless steel production.

In Japan, it has become controversial whether we should depend on the present ferrochromium or not. Ferrochromium is produced in the electric arc furnace, and high in electric energy consumption, and so electricity is rather expensive. In-bath smelting reduction of chromium ore was examined as an alternative route to the EF ferrochromium production by major Japanese steelmakers and ferroalloymakers as a national project (4-10). This process depends on the carbonaceous material as a fuel as well as a reductant for the smelting reduction. Even if the process is successful in being used without electricity, it is still energy inefficient, when the ferrochromium is used as a cold charge. It is therefore believed that in-bath smelting reduction of chromium ore is better used directly for stainless steel mother metal production.

New Process for Future Stainless Steel Production

Smelting reduction of chromium ore in BOF can be combined with blast furnace hot metal production, hot metal pretreatment, decarburization, and continuous casting as shown in Figure 2. Smelting reduction can be done in BOF, charging pretreated (i.e. desulfurized dephosphorized) hot metal, chromium ore, flux, coal or coke, and oxygen, and other process gasses. It is followed by slag removal either by tilting the furnace or by recharging the hot metal to ladle, and followed by decarburization either in the same vessel or in another vessel (AOD, or combined blow BOF).

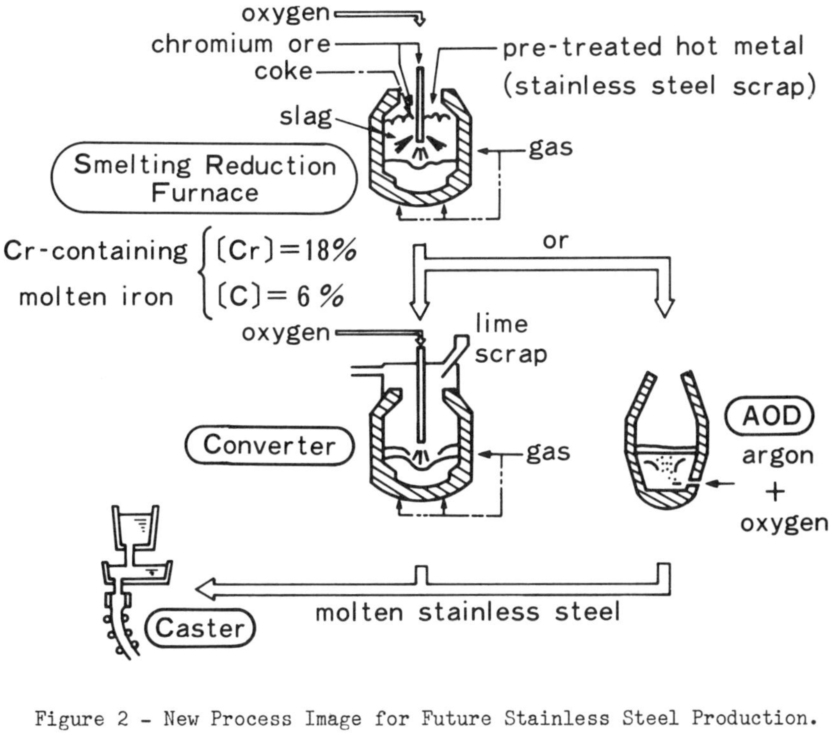

Figure 2 - New Process Image for Future Stainless Steel Production.

Features of New Process

This new process route is characterized by its two prominent features: one is that it is energy, and process saving, and hence it is cost saving; the other is that it is more suitable for establishment in the integrated steelworks, since the process requires BF hot metal, pretreatment, and tonnage oxygen plant.

It is clear that the new route is process saving since ferrochromium production process is not necessary. In order to see how it is energy saving, unit consumptions of energy for making 1% Cr in the metal are compared among three process routes(1: low carbon ferrochromium (FCrL) base; 2: high carbon ferrochromium (FCrH) base; 3: smelting reduction base) as listed in Table I. It is found that Energy consumption of the smelting reduction route is 70 to 90% of the one in FCrH, and 50 to 60% of the one in FCrL, and is the most energy saving. Assumptions have been made for this comparison: 1)Energy transfer efficiency from fossile fuel to electricity is 35%; 2)Oxidation degree, OD(=$CO_2/(CO+CO_2)$), is 0.35 and total heat transfer efficiency is 90%; 3)Unit consumptions for the production of FCrL and FCrH are equal to the figures in the literature (11); 4)Combustion energy of carbon in FCrH(8%), or of hot metal after smelting reduction(7%) is accounted for the energy of subsequent processes; 5)Energy recovery of the process gas in the smelting reduction is 80%, which is the ordinal level in BOF.

Table I. Comparison of Unit Consumptions of Energy for Making 1% Cr in the Metal among Three Process routes

process route	energy consumption (MJ)
FCrL	1370 ~ 1490
FCrH	850 ~ 950
Smelting Reduction	630 ~ 780

Present Stage of Research and development on Chromium Ore Smelting Reduction

Active research and development has been carried out for the new process among Japanese steelmakers (12-25). Some steelshops have already instituted the process in their commercial production (22-25), though the operation is limited to the partial smelting reduction, i.e., some 5 to 10 % Cr out of 13 to 19 % Cr. The reasons why it is limited are that the reduction speed is not sufficient, and the steel quality can not be guaranteed in terms of phosphorus and sulfur.

When d[Cr]/dt is taken as a measure of reduction speed, d[Cr]/dt is desired to be more than 0.3%/min to produce the major stainless steel, 18-8, matching with sequential continuous casting, under assumption of tap-to-tap time of 90 minutes. The presently available reduction speed is, however, about 0.2%/min according to the operational results in 10-to-175ton scale furnace tests (16-25), and it is insufficient for mass-production based on continuous casting.

When smelting reduction is employed, the amount of total charged phosphorus and sulfur inevitably becomes more than the amount which is used in the conventional EF-based process. Because in the former process, a carbonaceous material which contains appreciable amount of phosphorus and sulfur is used mainly for fuel in addition to the reductant. It is reported that phosphorus and sulfur transference to metal is at least 80% and under 20% respectively (23,24). Those impurities could be critical for assuring the product quality.

It is important to reduce the amount of totally charged impurities. Two methods are being considered for such purpose: one is hot metal pretreatment; that is the technology of removing phosphorus and sulfur from hot metal which is established and popular among Japanese Steelmakers. The other method under development is post combustion technology; the latter enables us to decrease the consumption of carbonaceous material by effective carbon combustion and high heat recovery. Phosphorus and sulfur contents in the hot metal after the reduction are estimated as a function of oxidation degreee, OD, as shown in Figures 3 and 4 respectively. In this estimate, coke is used as a cabonaceous material, in which phosphorus and sulfur contents are 0.04% and 0.55% respectively. Two cases have been analyzed depending on the type of hot metal: one is the conventional (P = 0.1%, S = 0.025%), the other is the pretreated (P = 0.01%, S = 0.005%). All chromium source is chromium ore. The product metal is 18 20%[Cr]-6%[C] iron at 1630 C. The heat transfer efficiency is 90%.

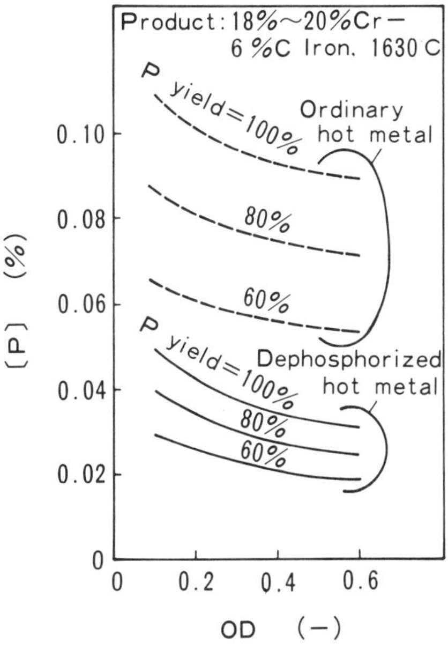

Figure 3 - Estimation of Phosphorus Content in the Product Metal.

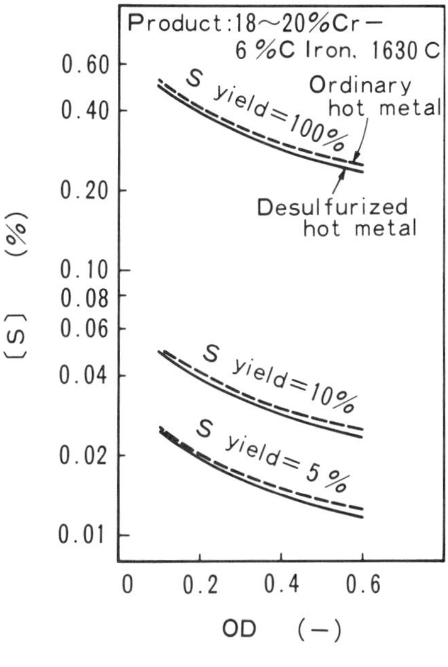

Figure 4 - Estimation of Sulfur Content in the Product Metal.

Phosphorus content is greatly dependent on the hot metal pretreatment(H.M.P.), while sulfur content is not dependent on the H.M.P., on the other hand, it is mostly dependent on its yield of total charged sulfur, since charged sulfur is much greater than the blast furnace hot metal sulfur. In each case, it is evident that OD is definitely influential. When the smelting reduction is done with OD = 0.1, which is the ordinal level in the conventional BOF operation, phosphorus content would be over the level of specification(JIS: under 0.04%, AISI: under 0.045%), even if the pretreated hot metal is employed, since phosphorus yield is reportedly over 80% (23,24). In order to make the smelting reduction feasible in terms of phosphorus specification, it is essential to have both the hot metal dephosphorization and high OD operation in BOF. In terms of sulfur, high OD or the low coke consumption is essential. It is noteworthy that sulfur yield is dependent on OD, and the higher the OD is, the higher the gaseous desulfurization, and the lower the sulfur yield are.

Key Technologies to be developed for Smelting Reduction of Chromium Ore

From the analysis of the previous section, it is clear that in order to apply the smelting reduction of chromium ore to the major stainless steel, 18-8, production, there still remains a further necessity of making a technology breakthrough. Key technologies to be established are 1) speed up of reduction to match with sequential continuous casting, and 2) efficient combustion of carbonaceous materials in BOF to reduce their consumption, and hence to assure the steel melt quality.

Those two subjects can be broken down to the following technical tasks: 1) High degree of post combustion and efficient heat transfer; 2) Intensive mixing or stirring of the bath; 3) Optimum slag composition control; 4) Efficient materials charging method.

Post Combustion and Heat Transfer

A high degree of post combustion and efficient heat recovery is essential to get sufficient heat supply to meet up with high speed charge of chromium ore. This is a key technology in the "iron ore" smelting reduction as well, and a number of research works have been conducted along the line. Oxygen lance and tuyere designs, and side and/or bottom blowing patterns are the subjects to be optimized.

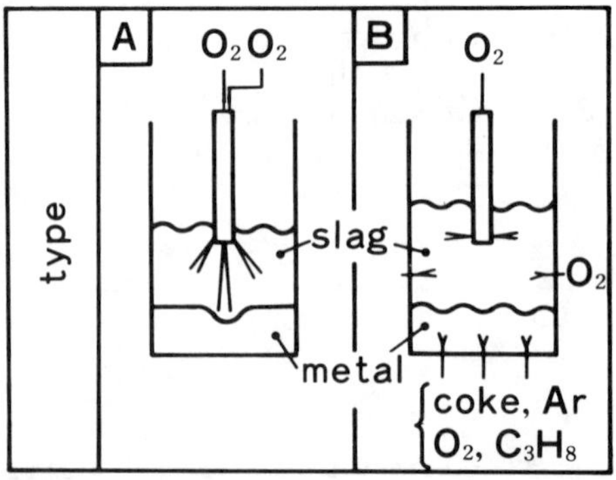

Figure 5 - Example of Future Design Image with High Heat Transfer.

It is pointed out that maximization of oxygen jet trajectory length in the slag zone is one of the most influential items in the heat recovery in the iron ore smelting, and furnace designs are proposed as shown in Figure 5 (26). It is expected that a similar relationship between the furnace design and heat transfer efficiency could be obtained in the chromium ore reduction as in the case of iron ore.

Mixing/Stirring

Intensive mixing of chromium ore, slag, and reductant is essential to an efficient reduction of the ore. For this purpose, slag bath stirring is employed by the side blow tuyere in an experimental converter, as shown in Figure 6 (16). Since hot metal carbon is also a reductant in addition to coke, bottom blow is important to supply carbon containing metal droplets above the metal bath, or in the slag.

Gas stirring either from top lance, or side and bottom tuyeres is effective to the post combustion heat transfer. Thus the design of those blow configurations is important for both the heat and mass transfer in the smelting reduction BOF.

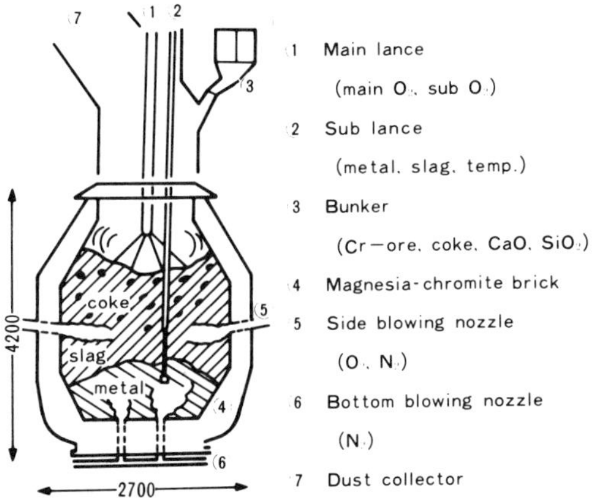

Figure 6 - 10ton Experimental Top and Bottom Blow Converter with Side Blow Tuyere (16).

Control of Slag Chemistry

Chromium ore is known as a highly refractory material, and its melting point is over 1900 C. It is therefore believed that reservoir slag formation to receive the ore, and flux addition to adjust the slag chemistry are necessary so that the ore can be easily molten, and react with the reductant efficiently. Typical composition of chromium ore is listed in Table II (27). A characteristic feature of this composition is that it is high in alumina and magnesia, and those are in the form of spinel, which is believed to have a rigid crystalline structure. Even if the chromium and iron oxide are reduced, the remaining gangue mineral has compositions, which are solid at smelting temperature, as shown in Figure 7. Therefore it is necessary to employ the fluxing agents, such as lime and silica. Alumina and magnesia content should be lowered depending on the smelting temperature. It is known that (Al_2O_3 + MgO) should be less than 29, 35, 45% at 1500, 1550, and 1600 C, respectively (7).

Table II. Examples of Chemical Composition of Chromium Ore in the World (wt%)

Mining Country	Cr_2O_3	SiO_2	FeO	Al_2O_3	MgO	LOI	P	S	Cr/Fe
Turkey	47.49	5.70	15.31	11.38	17.64	1.01	0.003	0.007	2.73
Iran	49.45	5.49	12.67	7.94	19.32	3.20	0.001	0.013	3.43
South Africa	44.79	2.81	25.32	14.19	11.24	0.10	0.003	0.006	1.56
U. S. S. R.	50.49	5.99	12.56	7.44	18.84	1.80	0.004	0.027	3.54
India	53.40	4.48	14.49	11.30	12.14	2.00	0.007	0.010	3.24
Philippine	48.78	3.41	17.86	11.50	15.96	0.93	0.002	0.022	2.40

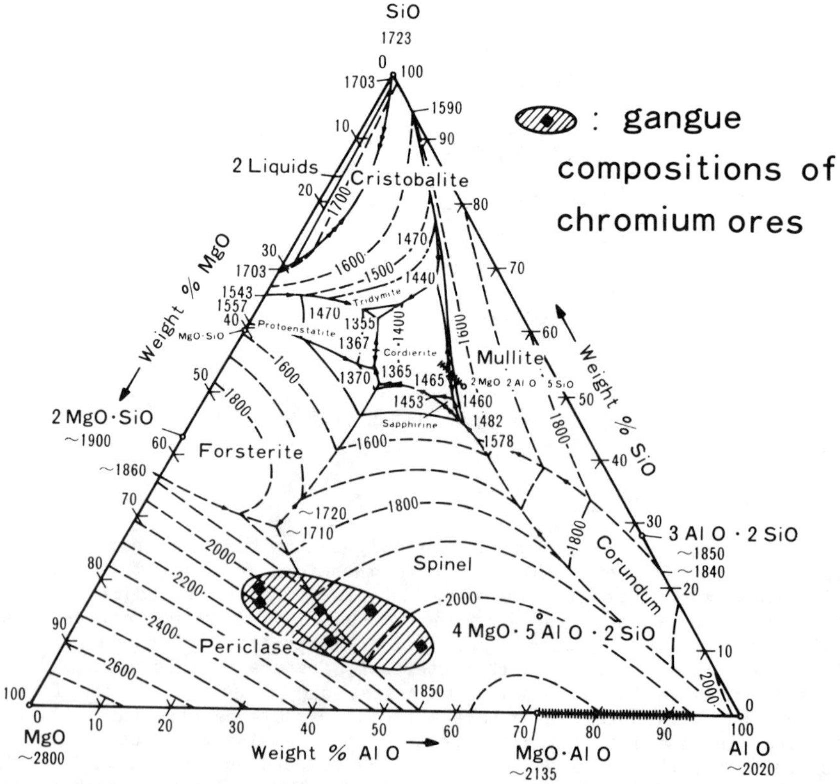

Figure 7 - Gangue Compositions of Chromium Ores.

Material Charging Method

Since available chromium ore is fine particles, pneumatic injection is proposed to get high reduction speed, and to get high yield of the ore, without resorting to pellets (13-15,20,21).

Ore-top-injection with coke and oxygen is found superior to the simple-ore-injection into the metal bath in terms of the metallic recovery yield (20,21). This is attributed to the fact that the ore is heated up by the flame adjacent to the ore stream, so that the ore is thermally abraded to much finer particles, and finally hits the so-called fire spot.

If flux, i.e., lime and silica, is blended to the chromium ore, this top injection performs much better in terms of reduction speed, d[Cr]/dt, than the forementioned injection method, as shown in Figure 8 (13-15).

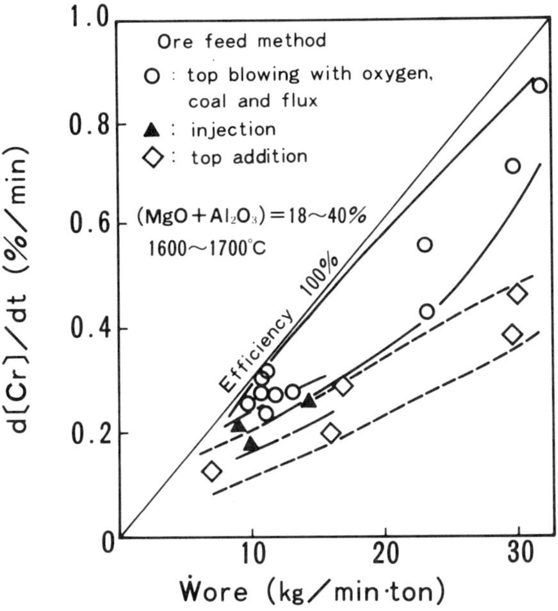

Figure 8 - Influence of Ore Feed Method on Reduction Speed against Ore Feed Speed, $\dot{W}_{ore}$.

Conclusion

In-bath smelting reduction of chromium ore could be a key unit process in order to establish an innovative BOF based stainless steelmaking process, which could take the place of the presently dominant EF-AOD process route. Active research and development has been undertaken, and partial use of the new process has already started in the commercial production. However, for the full use of the smelting reduction for major stainless steel, i.e., 18-8 production, there remains further need of a technology breakthrough. Key technologies to be established are 1) speed up of reduction to meet up with sequential continuous casting, and 2) efficient combustion of carbonaceous materials in BOF to reduce their consumption, and to assure the steel quality.

High heat input and its efficient recovery in BOF is an essential technology for the reduction of both the chromium and iron ores. Research and development projects on the iron ore smelting reduction have been undertaken not only in Japan but also in USA and European countries, and high degree of post combustion technology has been pursued intensively. The technology could be transfered to the chromium ore smelting without difficulty.

In addition to the combustion technology, new charging technology, i.e., ore-coal-flux-oxygen mixed injection could be another key for the high speed and efficient reduction.

In 2000 or beyond, the new smelting reduction based route of stainless steelmaking will have become a quite popular process route for the stainless steelmaking owing to the establishment of the forementioned key technologies, which would be efficient in the number of processing units, in energy, and in production cost.

References

1. Iron Year(Tokyo : The Tekko Shinbun, 1986).

2. "Statistical Data," The Stainless, 32(7)(1988)19.

3. ibid., 26(5)(1982)27.

4. M.Kuwabara et al.,"Influence of Main Factors on the Reaction (Production of Ferro-chrome by the Smelting Reduction in a Stirred Bath-I)," Tetsu-to-Hagané, 70(4)(1984)S116.

5. M.Fujita et al.,"Influence of Slag Compositon and Reaction Mechanism (Production of Ferro-chrome by the Smelting Reduction in a Stirred Bath-II)," Tetsu-to-Hagané, 70(4)(1984)S117

6. M.Kuwabara et al.,"Smelting Reduction of 20% Chromium Molten Iron with 650kg Top and Bottom Blowing Converter(Production of Ferro-Chrome by the Smelting Reduction in a Stirred Bath-III)," Tetsu-to-Haganè, 70(4)(1984)S118.

7. H.Hirata et al.,"Effect of Slag Composition and Temperature on the Reduction Rate of Chrome Oxide," Tetsu-to-Haganè, 72(4)(1986)S114.

8. M.Fujita et al.,"Smelting Reduction of Chrome Ore Pellet in Stirred Bath," Tetsu-to-Haganè, 74(4)(1988)680-687.

9. T.Fukushima et al.,"Smelting Test of Ferrochromium with Miniature Rotary Furnace (Development of a New Ferrochromium Smelting Process-I)," Tetsu-to-Haganè, 69(12)(1983)S834.

10. T.Fukushima et al.,"Smelting Behavior of Ferrochromium in Rotary Furnace (Development of a New Ferrochromium Smelting Process-II)," Tetsu-to-Haganè, 69(12)(1983)S835.

11. Iron & Steel Handbook II, 3rd ed., (Tokyo : The Iron and Steel Institute of Japan, 1979), 412-418.

12.T.Takaoka, Y.Kikuchi, and Y.Kawai, "Rapid Reduction of Chromium Ore in BOF (Fundamental Study on Smelting Reduction Process-III)," Tetsu-to-Hagané, 72(4)(1986)S111.

13.T.Takaoka, Y.Kikuchi, and Y.Kawai, "Effect of Chromium Ore Charging Method on Reduction Rate of Chromium Ore (Fundamental Study on Smelting Reduction Process-VI)," Tetsu-to-Hagané, 72(12)(1986)S969.

14.T.Takaoka, Y.Kikuchi, and Y.Kawai, "Rapid Reduction of Chromium ore by the Top Addition through a Blowing Lance (Fundamental Study on Smelting Reduction Process-VIII)," Tetsu-to-Hagané, 73(12)(1987)S872.

15.Y.Kikuchi et al.,"Fundamental Study on Direct Production of Cr-Containing Iron Through Rapid reduction of Chromite Ore," Proceedings of Shenyang International Symposium on Smelting Reduction,Shenyang, P.R.of China,3-5 September 1986, pp160-172.

16.K.Marukawa et al.,"Operation Results by the Method of Heating with Lumpy Coke and Oxygen Gas in 10t Experimental Converter (Development of a Smelting Reduction Method for Chromium Ore-II)," Tetsu-to-Hagané, 71(12)(1985)S928.

17.K.Marukawa et al.,"Development of Converter Technology Utilizaing Carbon materials for Thermal Source," Tetsu-to-Hagané, 73(3)(1987)A39-A42.

18.K.Marukawa et al.,"Influence of Experimental Conditions on the Reduction Rate of Chromium Ore (Development of Smelting method with High Cr content in 10t Experimental Converter-IV),"Tetsu-to-Hagané, 72(4)(1986)S112.

19.H.Nakamura et al., "Smelting Reduction of Chromium Ore in 5t Test Converter," Tetsu-to-Hagané, 71(4)(1985)S142.

20.S.Takeuchi et al.,"Utilization of the Chromite Sand and the Comparison of Feeding ways of the Chromite Sand (Smelting Reduction of Chromium Ore in 5t Test Converter-II),)" Tetsu-to-Hagané, 72(12)(1986)S968.

21.Y.Kishimoto et al.,"Scrap Remelting and Smelting Reduction of Chromium Ore in 5t Test Converter with a Pulverized Coal Combustion Lance," Tetsu-to-Hagané, 73(3)(1987)A35-A38.

22.H.Baba et al.,"Cr-ore Smelting Reduction by Combined Blowing Converter (Development of Stainless Steel Refining with Cr-ore by Industrial Scale-I)," CAMP-ISIJ, 1(1988)135.

23.H.Baba et al., "Rasult of Continuous Smelting Reduction Operation Using Two Converters for stainless Steel Production (Development of Stainless Steel Refining with Cr-ore by Industrial Scale-II)," CAMP-ISIJ, 1(1988)136.

24.T.Arai et al.,"Result of the Smelting Reduction Test to Make Stainless Steel by Industrial Furnace (Production Test of Stainless Steel by Smelting Reduction Process-II)," Tetsu-to-Hagané, 73(12)(1987)S875.

25.S.Kitamura et al.,"Quantitative Estimation about the Influence of Various Factors on the Smelting Reduction Rate of Cr-ore by Top-and-bottom Blowing Converter," Tetsu-to-Hagané, 74(4)(1988)672-679.

26. Y.Kawai et al.,"Fundamental Study on Post Combustion in Strongly Stirred Iron Bath Reactor for Smelting Reduction," Proceedings of the 7th Process Technology Division Conference of AIME, Toronto, Ontario, Canada, 17-20 April 1988.

27. Iron & Steel Handbook II, 3rd ed., (Tokyo : The Iron and Steel Institute of Japan, 1979),67.

28. Verein Deutscher Eisenhüttenleute, ed., Schlackenatlas (Düsseldorf : Ausschuß für metallurgische Grundlagen,1981),60.

ACOUSTIC SENSORS FOR PROCESS CONTROL IN THE YEAR 2000

N.D.G. Mountford, S. Dawson, I.D. Sommerville and A. McLean
Ferrous Metallurgy Research Group
Department of Metallurgy and Materials Science
University of Toronto
Toronto, Ontario, Canada

Introduction

In order to predict the accepted industrial practice for the year 2000, it may be helpful to first consider the rate of development of scientific knowledge. If the speed with which man can travel on commercial vehicles is taken as a criterion for determining the rate of progression, then it can be seen from Figure 1 that advancement in our present century is very rapid. Likewise, if computer technology is used as a criterion for the measurement of scientific advancement, it becomes evident that technological evolution can be highly exponential. These two examples are sufficient to demonstrate that engineers and managers cannot afford to regard scientific development as being linear with respect to time if they are to avoid underestimating the needs of society.

Within this framework of overall development, the time required for individual ideas to progress from inception to regular application has decreased considerably. For example, X-rays first discovered by Roentgen in the 1890's and now used widely in medicine and engineering, required a developmental period of some eighty years. The motorcar evolved from early beginnings at the turn of the century to a position of great social impact in sixty years. Rectifier materials used in high frequency radar to the present explosion in communications, required forty years. As illustrated in Figure 2, the period is shortening rapidly.

The use of sound waves appears to be following a similar path. There is now a considerable body of knowledge concerning their behaviour and exceptional advances have been made. In medicine and engineering the sonic approach to examination is well established, and acoustic testing forms the basis of several diagnostic and quality control applications. With respect to metals engineering, ultrasonic testing has contributed to the understanding of the relationship between the size and number of internal defects and the probability of premature failure. This has allowed the designer to incorporate appropriate safety factors in his designs; but more importantly, the realization of the deleterious effect of inclusions on the performance of the final product has resulted in a continuous pressure on the material suppliers to reduce the number of defects in their products.

It is not only the designer who is demanding defect free material but also the fabricator, as material breakdown during manufacture can be prohibitively expensive. Examples of this can be found in both the aluminum and steel industries where ultra clean sheet is required to ensure that rejections are minimized during beverage can manufacture and other deep drawing operations.

Metallurgical Processes for the Year 2000 and Beyond
Edited by H.Y. Sohn and E.S. Geskin
The Minerals, Metals & Materials Society, 1988

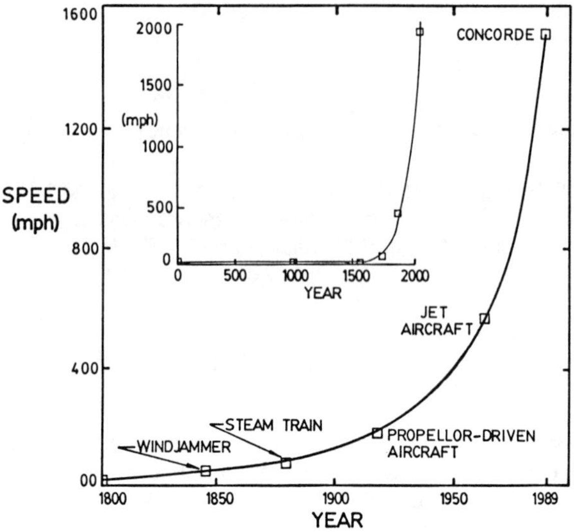

Figure 1. Rate of progress in commercial transportation.

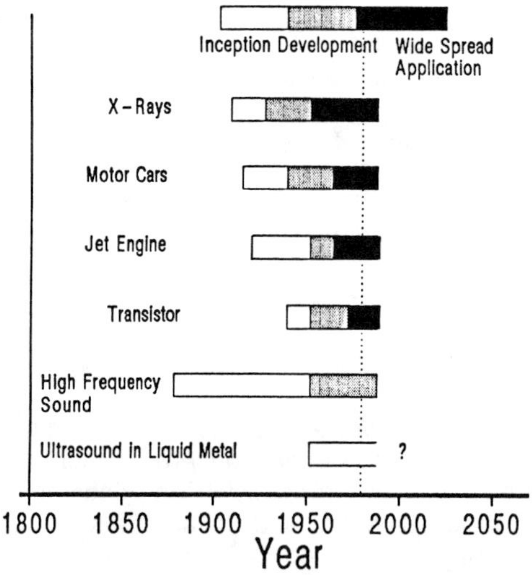

Figure 2. A comparison of developmental rates for various inventions.

Faced with this problem, the supplier can either achieve quality for the customer by close inspection and selection or by controlling the processes of production in order to minimize the number of defects in the final product. In practice this is done by maintaining as rigidly as possible the criteria within their control which will from practical experience yield a suitable product. As a check, the finished product is tested using some form of selective sampling. While this has been a satisfactory approach in the past, it is expensive and often is not capable of meeting the new requirements with respect to sensitivity, sample size, and speed of analysis.

The capability of sound waves to pass through liquid metal[1] opens up a new field for the development of sensors capable of measuring the presence of defect-forming particles and performing other valuable functions at the earliest stages of material manufacturing. Our problem is to predict with our present knowledge whether or not such acoustic devices could be developed and be in practical use by the end of the millenium.

Size and Distribution of Defects

In metal processing industries, quality is frequently associated with the entrainment of non-metallic material within the liquid metal during the primary manufacturing steps. In the case of aluminum, the entrained material can be oxide films together with hard carbide inclusions which are derived from the original smelting process. Aluminum also picks up hydrogen readily which can, on solidification, diffuse to entrained oxide particles, and during the finishing stages, result in blisters on the sheet material. The hard particles, which may be associated with the oxide films, are detrimental in the forming of thin-walled cans. These particles scratch or deform the draw dies and rejections from such a defect can be costly.

The steel industry has similar problems. There are many sources from which particles can be introduced into the liquid metal. Liquid steel can absorb oxygen from the atmosphere during transfer operations, leading to the formation of reoxidation products which may remain within the melt. In addition, other reaction products, such as nitrides and sulphides, may be present. Particle sizes can vary from less than 1 micron to greater than 150 microns, however, above this size natural flotation forces tend to separate the non-metallics to the surface[2]. Reoxidation can be prevented by using slags on exposed surfaces, but here again entrainment of this material can add to the inclusion content of the steel. It is estimated that in a steel with a total oxygen content of 20 ppm present as uniformly sized 10 μm spherical alumina inclusions, there would be approximately 10^{10} inclusions per tonne, with a mean inter-inclusion spacing of 185 μm.

Filtration systems developed by the aluminum industry have been extremely successful as an adjunct to the continuous casting process. However, any handling or movement of the filters can send entrapped oxides or particles down line into the product[4]. Filtration techniques have been applied in the steel industry, but their use is likely to be restricted to final cleaning of the higher value alloys, rather than the high tonnage grades where the presence of large numbers of inclusions leads to premature blockage of the filters.

Detection of Inclusions in Liquid Metals

The principles used in the ultrasonic evaluation of liquid metals are similar to those used in standard NDT ultrasonic flaw detection[4]. High frequency sound pulses are passed into the liquid and since the non metallic particles have different acoustic properties from the liquid[4], they interact with the sound energy pulses. Figure 3 illustrates one arrangement of equipment which has been used for experimental work in the laboratory. A series of electrical pulses are generated by the pulser unit and introduced to the piezoelectric transducer which is attached to the transmitting guide rod. The transducer converts these pulses into sound pressure waves which propagate through the rod and into the melt. The acoustic matching of the solid guide rod and the liquid metal is maximized by employing guide rods which are of the same material as the melt. Thus, when steel guide rods are used for the evaluation of liquid steel, less than 9% of the incident sound pressure is reflected at the interface. Other factors which merit consideration include the melt-back of similar material guide rods, and the development of a highly attenuating austenitic zone at the immersed end of the steel guide rods. While ceramic sleeves have been shown to be sufficient to stabilize the solid/liquid interface, some form of external cooling is required to minimize austenitic growth and allow the sound to enter the melt. This cooling will also contribute to stabilization of the interface, and protection of the heat sensitive transducers.

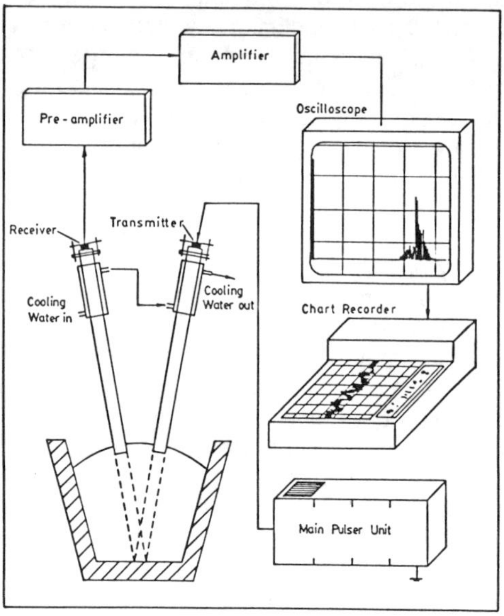

Figure 3. Schematic diagram of the ultrasonic apparatus.

Upon exiting the transmitting guide rod, the sound travels through the melt, reflects off the bottom of the crucible and is collected by the receiving guide rod. The received sound waves are then converted into electrical signals, amplified, and displayed on an oscilloscope. Both a chart recorder and a video camera have been used to record experimental observations.

Within the melt, several attenuating mechanisms are operative and they combine to determine the amount of sound which returns to the receiving guide rod. These include effects which are associated with the temperature, viscosity, and specific heat of the melt; as well as particle interactions. Since the former effects are inherent to the system, it is possible to separate the contribution made by particle interactions, and to use this as a measure of the cleanliness of the melt.

A typical oscilloscope trace which illustrates the pattern of received signals during an ultrasonic evaluation of liquid metal is shown in Figure 4. Essentially, the trace consists of an input pulse which is commonly referred to as the main bang in the field of ultrasonics, a large peak which represents the amount of sound returning from the bottom of the crucible, and a number of smaller signals which result from particle reflections. Based on water modelling studies, it is believed that these discrete signals can be caused by particles which are approximately 100 microns when a detection frequency of 5 MHz is used. By moving the measuring gate to a position above these reflections, the behaviour of large particles in the system can be monitored. Present research is directed toward a device which will be capable of counting the number of reflective echoes. Smaller particles also interact with the sound waves and although they may not produce enough reflective energy to result in a discrete peak on the oscilloscope, they invariably decrease the amount of sound which travels through the melt, and therefore contribute to the overall attenuation. Thus, the general cleanliness of a melt can also be determined by monitoring changes in the magnitude of the bottom echo during processing.

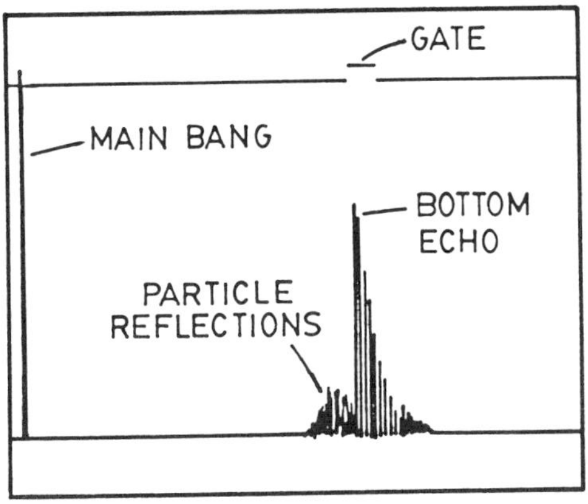

Figure 4. Typical oscilloscope trace from a liquid metal.

These principles were first applied to liquid aluminum in the late 1940's[1]. Since that time there have been a number of developments reported in the literature and there is now a reasonable body of knowledge concerning the application of the principles to the study of both liquid aluminum[4,6,7] and steel[8-10].

Unfortunately the particles which are troublesome in the aluminum industry are hard and very small in diameter, which makes it difficult to employ the discrete counting method. Films of oxide which are also present can be shown to give large reflective echoes but they may or may not contribute to failure. However, they can adhere to the hard particles or act as nucleation sites for the precipitation of deleterious intermetallic compounds[6], thus masking their presence from ultrasonic detectors.

In the steel industry, the size of particle which causes failure in the manufacture of beverage cans is of the order of 50 microns[10], and as mentioned previously, the size of the particle which can be detected using ultrasonics is dependent upon the method of analysis. If overall attenuation is used, then the large number of very small particles can contribute greatly to the degree of sound absorption, but these particles may not affect canmaking operations. However, it may be possible to employ the particle counting technique, as experience would indicate that for frequencies of the order of 2-1/4 MHz, the detectable particle or cluster size may be on the order of 200 microns.

It would appear from recent studies that the intensity of the incident pulse may be a factor to be considered. The use of high powered sound pulses, the minimizing of energy losses at the interface between the guide rod and liquid metal, greater signal amplification and an improvement in the signal/noise ratio may help to provide a better analysis of the range of particles present.

One particular advantage of acoustic sensors is that a relatively large volume of metal can be evaluated. Metal flowing down a runner can be sampled almost at the 30% level as in the method devised by Mansfield[4,7]. Different arrays could cover greater volumes. By developing a more thorough understanding of the way in which sound energy is distributed in the pulsed beam and the relationship between distance from source and the general attenuation in the liquid, it may be possible to record the size of particles and to obtain a measure of the size distribution, as illustrated in Figure 5. Although this technology has not yet been achieved in liquid metal evaluation, it is already embodied in some particle size monitoring applications for mineral processing.

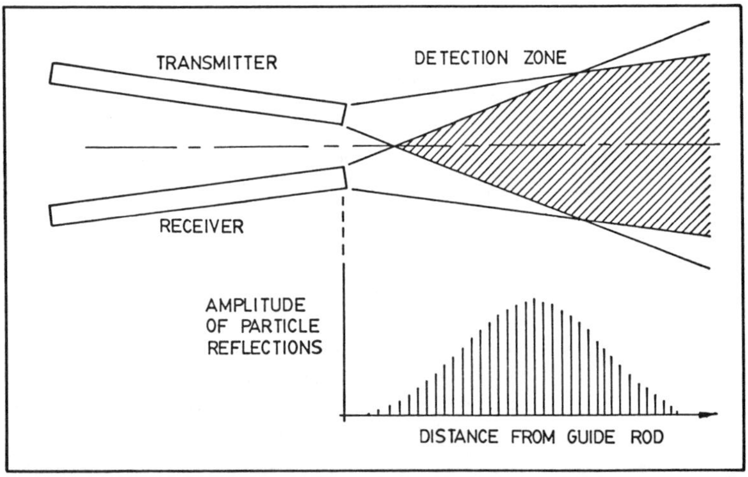

Figure 5. Effect of beam spread on size of received echo.

By careful treatment of large volumes of metal in both the aluminum and steel industries, it is possible to achieve a level of cleanliness such that only a few of the large size particles or clusters remain in the final product. High frequency sound sensors may prove extremely useful for monitoring the quality level in materials of this purity while they are still in the liquid state.

High frequency sound pulses have been passed into liquid aluminum, cast iron and steel and very similar reflections have been observed. Some examples of the application of ultrasonics are provided in the following paragraphs.

In high creep resistant aluminum alloys containing titanium and zirconium, intermetallic compounds may be precipitated in a peritectic reaction some 60°C above the general solidification range. These compounds become entangled with oxide and do not redissolve if the melt temperature is increased. These sporadically occurring clusters of oxide film and intermetallic are entrapped in a random fashion in the castings producing a peculiar type of draw which defies the skill of the foundryman to overcome. In certain circumstances this defect has caused 100% scrap of very expensive castings. Using ultrasonics it was possible to determine that the peritectic shower occurred at a specific temperature, brought about by lowering of the furnace temperature when new ingots were added to restore bath levels. From this knowledge of events taking place in the liquid, new melting procedures were developed and the problem was solved[6].

It is also of interest to note that a similar phenomenon occurs in alloys of 10-15% copper in aluminum, if the copper additions contain dissolved oxygen. When a commercial purity aluminum melt is ultrasonically evaluated, the oscilloscope trace typically consists of an input pulse, and a discrete bottom echo. However, if copper is added to the melt, an abrupt increase in the number and size of particle reflections occurs, and this condition persists regardless of holding time. Again, the presence of these particles may result in unacceptable scrap rates of cast products. Using the ultrasonic technique, it has been shown that if the melt temperature is increased to 950°C, particle reflections are no longer present, and the bottom echo returns to the magnitude observed in pure aluminum, thus indicating the dissolution of the particles.

The ultrasonic measuring technique has also been shown to be capable of responding to aluminum deoxidation reactions in both cast iron and steel. In these experiments, a single rod configuration was employed and the measurements were obtained in a manner similar to that used with oxygen probes. The hand held dip-probe, Figure 6 , is manually immersed in the melt, and because the sound reflects off of the rigid reflector plate, there can be no uncertainty as to the effects of test geometry on the analysis. Despite the high carbon levels in cast iron, and hence the relatively low oxygen levels, the addition of aluminum wire to the melt caused a marked decrease in the height of the bottom echo. Similarly, the addition of aluminum to steel melts resulted in an immediate four-fold increase in the overall attenuation. Perhaps the most significant result of these experiments is that the ultrasonic equipment (at each of 1, 2.25, 5, and 10 MHz) was able to respond to deoxidation products which are less than 2 μm in size[9]. The increased presence of inclusions in this size range was confirmed by metallographic inspection.

Finally, the ultrasonic apparatus has also been employed to determine the efficiency of gas injection through the bottom of a vessel. In this case, the transmitting and receiving guide rods were inserted through the wall of a 225 Kg induction furnace, and cooling water was applied to prevent breakout. With the steel held at 1550°C, a successful rod/melt interface was maintained for more than one hour. The total distance of sound propagation was 60 cm and a discrete echo was obtained from the opposite wall of the furnace. When argon was introduced through a porous insert, reflections were obtained from the bubbles and the echo from the opposite wall was obscured. The signals obtained were responsive to gas flow rates and could be used toindicate blockage of the porous plug.

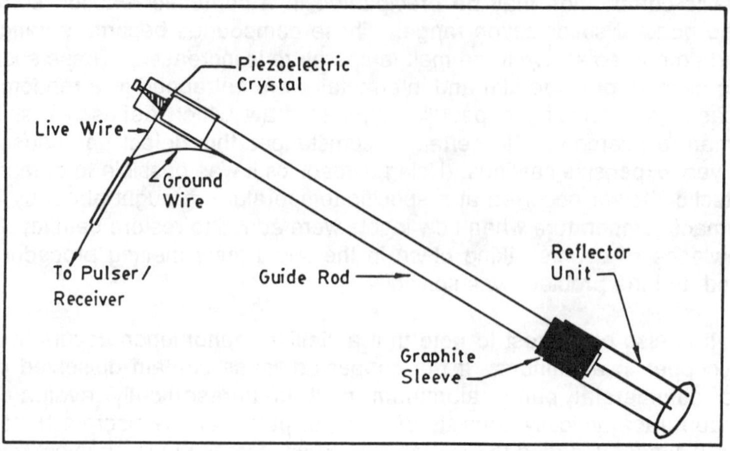

Figure 6. Schematic illustration of the hand-held dip-probe.

The ability of ultrasonics to provide information regarding inclusion precipitation and inclusion populations in liquid metals suggests that it may become a useful sensor for explaining the sporadic incidence of high scrap rejection in foundries, and for detecting the occurrence of metal processing problems such as faulty shrouding during continuous casting.

Depth Monitoring

During the conventional pouring of steel through a submerged nozzle into a receiving vessel, there is no point at which the operator can actually see the location of the slag/metal interface. This can have markedly deleterious effects on the quality of the final product, particularly if slag is entrained.

The continuous monitoring of liquid steel depth in a metallurgical vessel is depicted schematically in Figure 7. The ultrasonic wave energy is generated by an ultrasonic pulser/receiver unit and is introduced to the steel through a transmitting guide rod in the bottom of the ladle. The sound then propagates through the liquid steel, reflects off the slag/metal interface, which is an efficient reflector due to the difference in densities and speed of sound in the two liquids, and returns to the bottom of the ladle where it is collected by a receiving guide rod. The signal from the slag/metal interface appears on the oscilloscope as a discrete peak. The total transit time of the ultrasonic pulse is then represented by the space between the input pulse and the slag/metal interface echo where the horizontal calibration of the oscilloscope is in time units. Knowing the velocity of sound in liquid steel, the transit time can then be used to provide a direct measurement of metal depth. For example, in Figure7 the transit time shown on the oscilloscope is 1.0 ms (0.001 s), and the speed of sound in liquid steel is approximately 4000 mm/μsec[12]. Thus the total distance of liquid steel through which the sound travelled is 4 m, and the depth of steel in the ladle is one-half of that distance or, 2m.

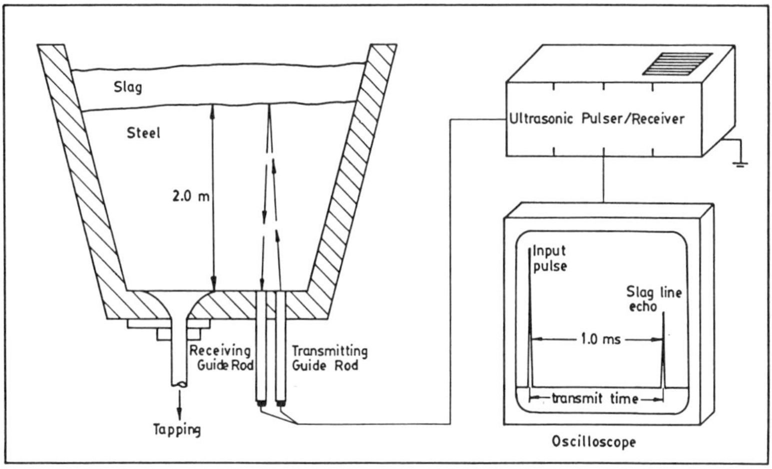

Figure 7. Schematic representation of depth monitoring.

While the realization of such a seemingly simple mechanical design may appear straightforward, success will ultimately depend on the ability to obtain sufficiently high power levels to drive the ultrasonic energy through the appropriate depth of steel. However, this should not be an insurmountable problem as the lower two feet of steel are of greatest interest to the metallurgist with respect to slag carry-over, and penetration through this depth should be relatively easy to obtain. Trials have been successfully conducted with twenty inches of liquid steel[(8)], but attempts to penetrate greater depths have been restricted by furnace size.

<u>Vortex Detection</u>

As a ladle or tundish is emptied, there is a possibility that a vortex may form depending on the fluid flow conditions which develop within the liquid. If a vortex does form, it will prematurely transfer slag into the receiving vessel and adversely affect the quality of the finished product. It would be beneficial to know when a vortex is forming so that the necessary action could be taken to prevent its further development.

The proposed method for ultrasonic vortex detection is schematically illustrated in Figure 8. Under ordinary operating conditions (a), no signal would be obtained by the receiving guide rod. However, if a vortex were to form (b), the metal/slag-air interface would act as a barrier to the propagation of sound and reflect it back toward the receiving guide rod. The presence of the vortex could then be indicated either by the newly formed peak on the oscilloscope, or by an alarm bell. The most advantageous aspect of this detection scheme is that it has the potential to detect the vortex before it matures and entrains slag and air into the tundish or mould.

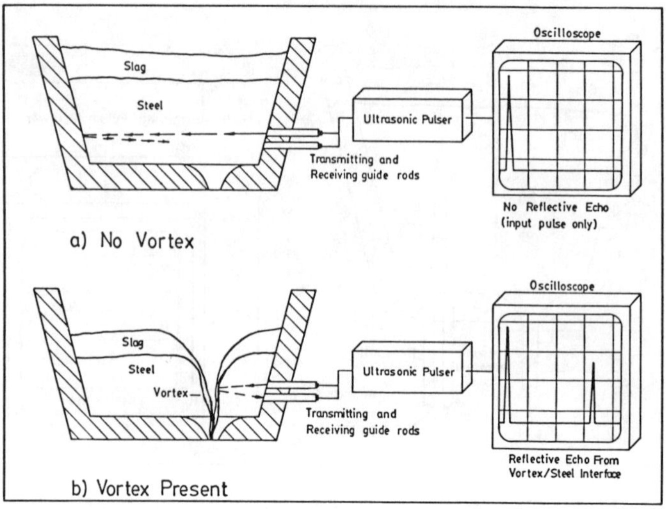

Figure 8. Schematic representation of vortex detection.

Monitoring of Chemical Reactions

Recently Pal et al.[13] have reported their work on sound vibrations of audio frequency emanating from large ladles of steel during desulphurization.

When powdered desulphurizing reagents are injected into the steel ladle using argon gas as a carrier, the gas bubbles vibrate, contracting and expanding and thus originating pressure waves within the liquid metal. This low frequency vibrational energy can pass through the ladle brick work and steel shell, and be monitored by accelerometers attached to the vessel. Sound waves of this kind can also be detected using microphones placed some distance from the metal surface. The basis for this measurement stems from the desulphurization reaction and the resulting increase in the surface tension of the liquid metal as sulphur is removed from solution. The higher surface tension has the effect of changing the natural frequency of vibration of the argon bubbles, and this change in turbulence can be monitored.

Future detection systems for the evaluation of these treatments may provide additional control of the steelmaking process. Additionally, sound pulses could assess the degree of agglomeration and general cleanliness after the level of residual sulphur had been quantified. This, of course, will also require further developmental work.

Enhancement of Inclusion Removal

Reference has been made to the difficulty of using filters to clean large volumes of liquid steel. There is therefore a need for an alternative method of continuous filtration or separation of nonmetallics from molten steel.

Powerful sound waves, if propagated into liquids in such a manner as to form standing waves, can be shown to cause suspended particles to migrate through the liquid to the areas of least movement. Such an effect is demonstrated in Figure 9 where sound waves at 1 MHz frequency were passed into a light oil containing particles of size and distribution similar to those found in steel, and adjusted to create parallel standing waves. The visible bands indicate the alignment of particles at the nodal points.

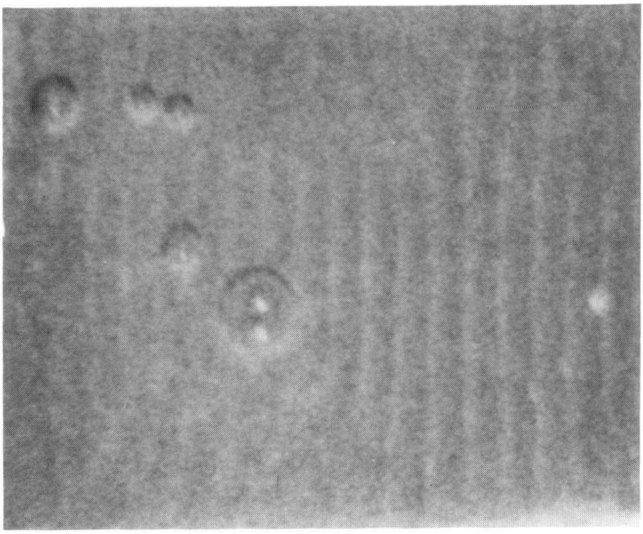

Figure 9. Grouping of particles under the influence of standing waves.

If this behaviour could be obtained in liquid metals then the inclusions may segregate or agglomerate in a similar fashion. Particles would have a greater probability of coalescence and therefore rise faster into the protective slag layer. While this has not yet been achieved in steel, some early experiments with aluminum(1) which used a 500 watt power source suggest that the principle of acoustic agglomeration might be possible. The application of standing waves decreased the hydrogen content by approximately 20% relative to an untreated melt during a 30 minute trial.

The frequencies for these operations will probably be on the order of 250 KHz to 1 MHz with the optimal frequency being a balance between the wave length and the distance particles can be made to move to the nodal points in a given time. The distance between nodes is 1/2 of a wavelength (or approximately 4mm at 0.5 MHz in liquid steel), and the time required for collection of the metal in the sound affected zone will be a function of the driving power of the sound. Furthermore, the required energy of the sound waves for this application will undoubtedly be in excess of the thirty millijoules that are currently used for inclusion detection. This should not be a problem, however, it must be determined if the 'holding strength' of the sound waves is greater than the convective stirring forces within the melt. If the convective forces are too strong, electromagnetic stirring could be used to counteract the inherent circulation and thus provide a quiescent zone adjacent to the ultrasonic transducers. While this technique appears to be quite promising, extensive research is required before the concept can be validated.

Operational Considerations

Increasing the general understanding of ultrasonic evaluation of liquid metals, and translating small scale laboratory experience into industrial practice will occupy the next ten years. During this period, a number of considerations including thermal effects and measuring and recording techniques will need to be addressed.

Perhaps the most significant problem confronting the application of ultrasonics to molten metal systems is the effect of temperature on the propagation of sound and the stability of the apparatus. As mentioned previously the development of an austenitic zone at the base of the immersed steel guide rods can prevent the sound from reaching the end of the guide rod. The inability of sound to travel through the austenitic phase is demonstrated in Figure 10 for 20 cm test pieces, which shows that sound can travel through molten steel, however the amount of received sound falls to zero with the precipitation of an austenitic phase[14]. This condition persists until the austenite is transformed to a ferrite and pearlite matrix at 723°C. Clearly, even a small austenitic zone is undesirable and, during the evaluation of liquid steel, external cooling must be employed to minimize its development[15,16].

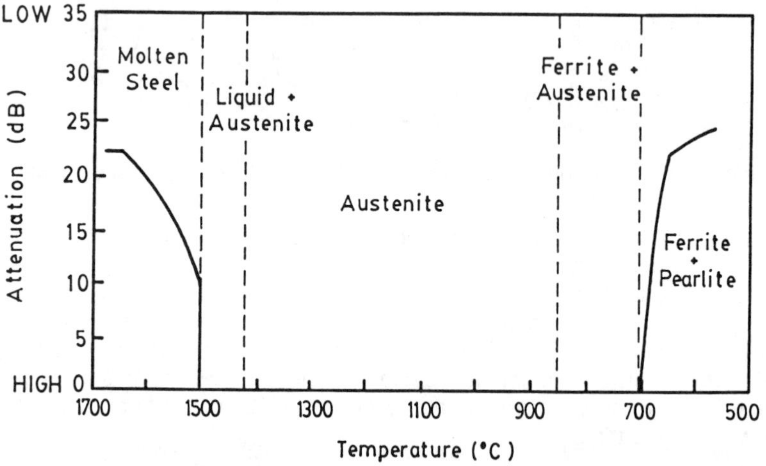

Figure 10. Relationship between attenuation and temperature for a 0.3%C steel.

Three different guide rod configurations can be employed to perform ultrasonic evaluations. These include the pitch-catch assembly shown previously in Figure 3, where the two guide rods can either be immersed in the melt, or introduced through the walls of the vessel; the single rod dip-probe which can be manually immersed through the top surface for quick measurements; and the thru-transmission assembly where the transmitting and receiving guide rods are located on opposite sides of the vessel. This arrangement will be most appropriate for the detection of vortices as metal flows through a tundish. Clearly, the availability of different configurations is beneficial as it will allow ultrasonics to be employed in a variety of situations.

The physical stability of steel guide rods can also be maintained through external cooling, however, experimental evidence has indicated that a simple refractory sleeve is sufficient. These thermal shock resistant sleeves contain the rod/melt interface in an unagitated environment and prevent inward melting at the sides of the guide rods when they are immersed in the molten metal. A computer modelling study on the thermal conditions which develop when the steel guide rods are inserted through the base of the tundish has shown that the existing refractory material provides sufficient insulation to prevent melt-back and breakout. These results are encouraging with respect to depth monitoring and vortex detection.

Finally, with respect to data acquisition, it has been noted that response signals are currently being directly displayed on an oscilloscope, and recorded by a video camera and a chart recorder. The current direction of research will ultimately lead to a computer counting technique which will be capable of counting the number of reflections from discrete inclusions at known locations within the melt. An analysis of this type could potentially provide information regarding inclusion behaviour and it would also help to quantify the relationship between sample volume and inclusion content.

Possible Sensor Applications

With respect to the future, it is interesting to speculate upon the ways in which the unique properties of ultrasonics may be incorporated into industrial practice. While inclusion detection and the ability to evaluate larger quantities of metal are relatively easy to visualize, the following concepts merit further discussion.

In the conventional vertical continuous casting process, there are a number of locations which could benefit from the introduction of continuous monitoring with ultrasonic sensors. Four such locations are illustrated in Figure 11, where the following measurements could be obtained:

1. determination of slag levels and metal depths in the ladle,
2. slag carry-over into the tundish and general melt cleanliness,
3. depth measurement in the tundish, and
4. detection of vortexing in the tundish, and general melt quality above each strand.

Successful monitoring at each of these locations would provide a measure of control which is far greater than that available at present, and would undoubtedly contribute to lower production costs and improved product quality.

If it can be proven that inclusion agglomeration can be accelerated in liquid steel, then Figure 12 demonstrates a possible ladle design which could be employed to utilize this effect. It incorporates vertical troughs at either end of a ladle, and into these are inserted a number of transducers and guide rods which can introduce the powerful sound into the melt and generate standing waves. Electromagnetic stirring devices could be used to gently circulate the metal through the standing wave zones where inclusion coalescence would occur, and the larger inclusion clusters can then rise into the slag phase. While successful implementation of this concept would indeed be advantageous, extensive developmental work will be required to determine its feasibility.

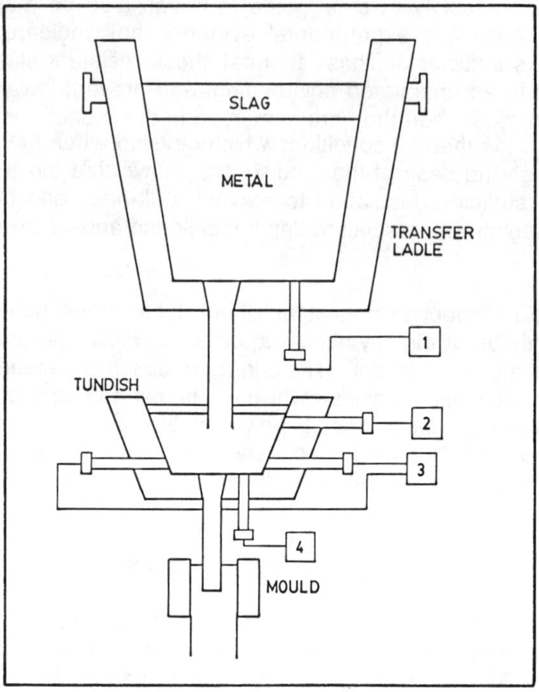

Figure 11. Possible application of sonic sensors in the continuous casting process.

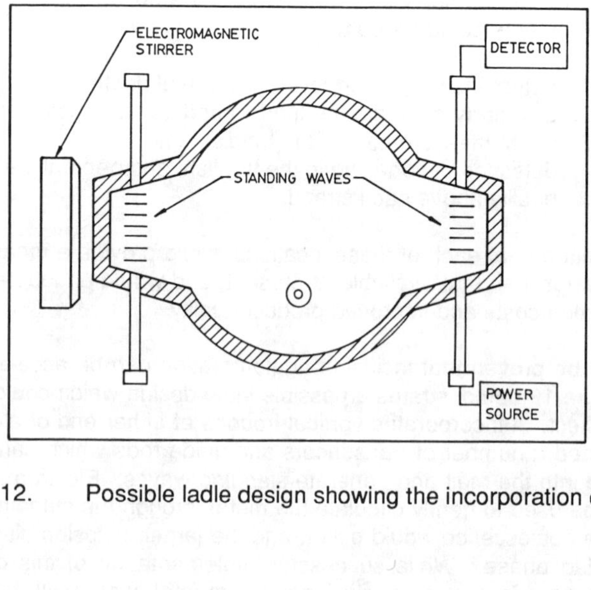

Figure 12. Possible ladle design showing the incorporation of sonic filters.

Summary

The present degree of understanding of the behaviour of ultrasonics in liquid metals makes it possible to predict that sound waves will provide the basis for on-line sensors in the year 2000 and beyond. In particular, it has been shown that sound frequencies as high as 10 MHz can be used to detect inclusions in liquid aluminum and steel, and that systems could be devised which would allow the continuous evaluation of very large quantities of metal. While it is currently possible to detect changes in the number of small inclusions in liquid steel, it is felt that ultrasonics will also be capable of distinguishing the infrequently occurring large inclusions. Other aspects such as depth monitoring and vortex detection, ultrasonic agglomeration and filtering of inclusions, and the sonic monitoring of chemical reactions may also find industrial application in the future.

Ultimately, the real question is whether or not the time span of some ten years is sufficient to solve the problems of translating these principles into industrial practice. From experience in other aspects of scientific development the answer would appear to be yes, and the driving force will probably come from the general pressures on the metal industries to change. Indeed, if these and other novel objectives are to be achieved by the year 2000, a sustained cooperative effort between industry, government, research laboratories, and universities is essential.

Acknowledgements

The authors would like to express their gratitude to Dr. F. J. Fry and the Indianapolis Center for Advanced Research for conducting the initial experimental work on ultrasonic filtering. Special thanks are also due to Mr. M. Brown and Dr. F. Mucciardi of McGill University for performing thermal stability calculations with their Fast-P Computer Model.

Funding for this work has been provided by a Strategic Grant on Tundish Metallurgy from the Natural Sciences and Engineering Research Council of Canada, as well as a grant from the Department of Energy of the Province of Ontario.

References

1. N.D.G. Mountford: "The Application of Ultrasonics to the Detection of Heterogeneous Impurities in Molten Aluminum", M.Sc. Thesis, University of Durham, 1951.
2. B. Hoh, H. Jacubi, H. Wiemer, and K. Wunnenberg: "Improvement of Cleanliness in Continuous Casting,", Fourth International Conference on Continuous Casting, Brussels, May 1988, pp. 211-222.
3. R. Kiessling: "Non-Metallic Inclusions in Steel", The Metals Society, London, 1978.
4. T.L. Mansfield: "Ultrasonic Technology for Measuring Molten Aluminum Quality", Light Metals, The Metallurgaical Society of the AIME, 1982, pp. 969-980.
5. J. Krautkramer and H. Krautkramer: "Ultrasonic Testing of Materials", Third Edition, Springer-Verlag, 1983.
6. N.D.G. Mountford and R. Calvert: "Precipitation Effects in Liquid Aluminum Alloys: Experiments with a Pulsed Ultrasonic Technique", Journal of the Institute of Metals, 1959-60, Vol. 88, pp. 121-127.

7. T.L. Mansfield: "Ultrasonic Technology for Measuring Molten Aluminum Quality", Materials Evaluation, May 1983, Vol. 1, pp. 743-747.
8. N.D.G. Mountford, L.J. Heaslip, E. Bednarek, and A.N. SInclair: "The Development of an Ultrasonic Sensor for Metal Quality in Steel Casting Tundishes", Steelmaking Proceedings, Fifth International Iron and Steel Technology Congress, Washington, April 1986.
9. S. Dawson: "The Application of Ultrasonics to the Evaluation of Liquid Steel Cleanliness", M.A.Sc. Thesis, University of Toronto, 1987.
10. S. Dawson, N.D.G. Mountford, I.D. Sommerville, A. McLean: "Evaluation of Liquid Steel Quality During Metallurgical Processing", Proc. of First International Calcium Treatment Symposium, Scotland, Institute of Metals, 1988.
11. H. Kobayashi, T. Kurokanla, T. Shimomura, K. Matsudo, and S. Miyahara: "Effect of Non-metallic Inclusions of Flange Cracking of Drawn and Ironed Can From Tinplate", Trans. ISIJ, 1983, Vol. 23, pp. 410-416.
12. R.L. Parker, J.R. Manning, and N.C. Peterson: "Application of Pulse-Echo Ultrasonics to Locate the Solid/Liquid Interface During Solidification and Melting of Steel and Other Metals", Applied Physics, Dec. 1985, Vol. 58, pp. 4150-4164.
13. G. Pal, G. Endroczi, G. Hollos, G. Nagy, G. Szonyl, B. Laszlo, Z. Balazs: U.S. Patent No. 4, 617, 830, Oct. 21, 1986.
14. J.D. Lavender: "Transmission of Ultrasound in Liquid Steel and Measurement of Ultrasonic Attenuation During Cooling", NDT, April 1972, Vol. 5, pp. 107-109.
15. N.D.G. Mountford: Canadian Patent No. 1, 235, 476, April 1988.
16. N.D.G. Mountford: United States Patent Application No. 826, 147, July 1988.

DIRECT REDUCED IRON: AN ADVANTAGEOUS CHARGE MATERIAL FOR INDUCTION FURNACE

K. Sadrnezhaad

Dept. of Metallurgical Engineering
Sharif University of Technology
Tehran, Iran

Abstract

Industrial and experimental induction furnaces are used for melting various types of iron ingots, returned scraps and DRI sponge pellets to produce high purity cast-iron and steel heats. The lowest consumption of the electrical energy is determined for continuous feeding operation to be 0.3 KWH/Kg for production of cast-iron in 1.5-ton industrial furnace and 0.45 KWH/Kg for production of steel in 25-Kg experimental furnace. The optimum feeding rate for lowest energy consumption is obtained to be 12.5 grams per second for continuous feeding of DRI in 25-Kg induction furnace. Similar measurements show that the optimum size of the DRI pellets is around 8 millimeter.

Metallurgical Processes for the Year 2000 and Beyond
Edited by H.Y. Sohn and E.S. Geskin
The Minerals, Metals & Materials Society, 1988

Raw Materials

THE INFLUCENCE OF SOLID STATE IMPERFECTIONS

IN MINERAL AND METAL PROCESSING

G. Simkovich and F. F. Aplan

The Pennsylvania State University
University Park, PA 16802

Abstract

A review is given of the effects of solid state point imperfections upon a number of mineral benefication processes and a metal extraction process. It has been found in numerous flotation studies that small amounts of aliovalent impurities (dopants) in binary compounds may show major effects upon flotation behavior. Both semiconducting and ionic conducting compounds are affected. These effects are also seen in hydrometallurgical studies as well as in electrostatic operations. Additionally, de-copperization of lead melts appears to be strongly influenced by similar defects. It is hypothesized that the effects observed in these studies are present in actual operations and should be considered in designing processes. A few speculations concerning future efforts are advanced.

Metallurgical Processes for the Year 2000 and Beyond
Edited by H.Y. Sohn and E.S. Geskin
The Minerals, Metals & Materials Society, 1988

Introduction

Imperfections [point (e.g., vacancies, interstials, quasi-free electrons, electron holes, impurity ions), line (e.g., dislocations), two dimensional (e.g., grain boundaries) and three dimensional (e.g., 2nd phases) defects] are generally present in solid materials and these imperfections may, and often do, have important effects upon a variety of properties and processes. In general, one may, in most cases, attain elimination of three dimensional defects by a separation process of one type or another (e.g., gravity, electrostatic, magnetic, flotation, hydrometallurgical etc.,). Thus, this type of defect has rarely been investigated in a systematic fashion. Two dimensional defects are always present in non-single crystal materials but have also escaped systematic studies in processes primarily because the control of such defects is not easily attained. Line defects are non-equilibrium entities which are present in ill defined quantities so that they present a variable that cannot be easily controlled although through careful systematic experiments one can attain some reproducible relations pertaining to this type of defect.

Contrary to the lack of studies concerned with the effects of 1, 2 and 3 dimensional types of defects there has been some activity concerned with point defects and their effects upon a number of mineral/metallurgical processes, especially flotation processes. Thus, this article will present a review primarily of point defects in relation to their effects upon flotation although other processes and defects will also be mentioned.

Since, as will be illustrated, point defects may have a profound influence upon a number of processes, it is believed that future studies and future applications will include imperfections in materials as variables which aid in controlling processes.

Types of Point Defects

An excellent comprehensive consideration of point defects may be found in the book by Kröger (1) where many topics relating to solid state chemistry are exhaustively covered. In this paper we will be concerned only with binary crystals and the point defects found in such crystals. These defects for a crystal MeX (assume that Me and X are charged primarily at the +2 and -2 levels) are, utilizing the Kröger - Vink(2) notations, as follows.

Cation vacancies (V_{Me}''); cation interstitials ($Me_i^{\cdot\cdot}$); anion vacancies ($V_X^{\cdot\cdot}$); anion interstitials (X_i''); quasi-free electrons, (e'); and electron holes, ($h^{\cdot}$); where the dash and filled dot superscripts indicate excess negative and positive charges in comparison to the defect-free base crystal.

It has been found experimentally, and confirmed theoretically that two of these defects (the major defects) generally appear in the bulk material in equivalent concentrations in order to preserve electrical neutrality, and this "pairing-off" of two types of point defects leads to two major types of electrical conduction, ionic and semi-conductors (metallic conduction will be excluded in this paper, but such conduction does occur in crystals which might initially be considered in the semi-conductor groups, e.g., FeS).

The ionic conductors obtained via the point defect "pairing-off" may be grouped into: Frenkel - equivalent concentrations of cation vacancies and cation interstitials; anti-Frenkel (=

anion Frenkel) - equivalent concentrations of anion vacancies and anion interstitials; Schottky - equivalent concentrations of cation and anion vacancies; and anti-Schottky - equivalent concentrations of cation and anion interstials. The semi-conductors obtainable are as follows: p-type - cation vacancies + electron holes or anion interstitials plus electron holes; n-type - cation interstitials + quasi free electrons or anion vacancies + quasi free electrons; intrinsic semiconductors - quasi free electrons + electron holes; and amphoteric semiconductors which display both n and p-type conduction depending upon the prevailing pressure of $X_{2(g)}$ [or $Me_{(g)}$] above the compound.

The above then is a listing of the type of point defects present in binary compounds. This listing is of the major defects present and it is understood that all crystals contain all types of point defects but other than the major defects the remaining point defects are generally present in minor concentrations.

We now consider some of the studies concerned with defect effects upon various processes.

Flotation Processes

AgCl. The silver halides are Frenkel type defect compounds and a number of studies concerned with flotation as a function of the defects in AgCl have been made (3,4). The defects in this compound are formed in accord with the relation

$$AgCl = V'_{Ag} + Ag^{\cdot}_{i} + Cl^{x}_{Cl} \quad (1)$$

with the equilibrium relation

$$K = [V'_{Ag}] \bullet [Ag^{\cdot}_{i}] \quad (2)$$

where [ ]'s indicate concentration terms and the x superscript indicates no excess charge at the indicated site (the subscript) in comparison to a defect-free, undisturbed crystal.

Doping AgCl with Cd Cl_2 creates cation vacancies as indicated by the following reaction.

$$Cd\,Cl_2 = Cd^{\cdot}_{Ag} + V'_{Ag} + 2\,Cl^{x}_{Cl} \quad (3)$$

while doping with Ag_2S creates silver interstitials in accord with the reaction:

$$Ag_2S = S'_{Cl} + Ag^{\cdot}_{i} + Ag^{x}_{Ag} \quad (4)$$

Unfortunately, the solubility of Ag_2S in AgCl is quite limited so that no firm conclusion as to its effects has been made(4).

The effort carried out by Spearin et al (4) covered a relatively wide range of studies including flotation, zeta potentials and contact angles. The AgCl used in these studies were all synthesized in the laboratory and care was taken to eliminate any light induced silver precipitates.

The zeta potentials studies gave at pH = 6 a zero point of charge (ZPC), which indicates a zero surface charge as detected by the zeta potential measurement, of pAg = 4.5 for pure AgCl and a ZPC of pAg = 3.6 for a 1.5 mole % $CdCl_2$ doped AgCl which is a substantial change in ZPC. Contact angle measurements were also made on these silver chlorides under a wide variety of conditions. The results attained are given in Table I. The values given are the average of six measurements and include the ranges for the measurements.

Table I - Contact Angles (Captive Bubble) on Pure and $CdCl_2$ Doped AgCl

Liquid	pAg	pH	Contact angle average and range (deg) AgCl 'Pure'	0.1% $CdCl_2$	0.5% $CdCl_2$	1.0% $CdCl_2$	1.5% $CdCl_2$
Water	6.0	6.0	25±2	-	-	10±1	-
Water	4.9	6.0	31±3	25±2	14±4	8±2	6±1
Water	2.0	6.0	24±1	-	-	13±3	-
10^{-4} mol l^{-1} Dodecylammonium hydrochloride	6.2*	6.0	68±4	57±3	39±2	21±2	13±3
10^{-4} mol l^{-1} Dodecylammonium acetate	2.0	6.0	13±2	-	-	7±1	-
10^{-4} mol l^{-1} Sodium dodecylsulfonate	2.0	6.0	50±2	-	-	22±2	-
10^{-4} mol l^{-1} Sodium dodecylsulfonate	6.0	6.0	27±2	-	-	15±2	-

*Calculated from pCl measurements made with specific chloride ion electrode.

As may be seen in Table I doping of AgCl with $CdCl_2$ results in a reduction of contact angle under all conditions, i.e., in the absence of any collectors and in the presence of either anionic or cationic collectors at pAg's below, above and near the ZPC. Thus, for all cases the dopant $CdCl_2$ in AgCl renders this compound more hydrophilic.

Flotation tests run on these samples gave results in complete agreement with the contact angle measurements as may be seen from the listing given in Table II.

Table II. Flotation Results on Pure and $CdCl_2$ doped AgCl

Sample and doping level	Sodium dodecyl sulfate, $1x \cdot 10^{-4}$ mol 1^{-1}, pAg 2.0 Reference 4	Dodecylammonium acetate, 5 x 10^{-5} mol 1^{-1}, pAg 6.0 Reference 4	Dodecylammonium hydrochloride 4 x 10^{-4} mol l^{-1}, pAg 6.3 (est.) Reference 3
AgCl - "Pure"	99.5	88.4	99.2
AgCl - 0.1% $CdCl_2$	91.2	27.5	83.6
AgCl - 0.4% $CdCl_2$	-	-	22.1
AgCl - 0.5% $CdCl_2$	68.4	12.5	-
AgCl - 1.0% $CdCl_2$	-	-	-
AgCl - 1.5% $CdCl_2$	-	-	6.8

All of these results indicate a large decrease in the hydrophobicity of Ag Cl due to the introduction of the dopant Cd Cl_2. It appears without question that the presence of the cadium chloride increases the attachment of water to this compound. It would have been desirable to determine the effect of Ag_2S upon the flotation behavior of Ag_2S; however, as noted earlier crystals containing Ag_2S were troublesome in all the studies attempted and therefore these Ag_2S studies were temporarily abandoned.

The ZPC alterations found by Spearin et. al., (4) were somewhat different than those obtained by Honig and Hengst (5) who found that doping AgBr with Pb Br_2 increased the ZPC while doping with Ag_2S decreased the ZPC. Such a difference is as yet not fully understood but may relate to the dopants employed (4).

CaF_2. A number of flotation experiments have been made on the mineral fluorite, CaF_2 (6-10) but only a few (6,7) have considered point defects in the CaF_2 as a variable. This compound has an anti-Frenkel defect structure which is formed in accord with the reaction

$$1/2\ CaF_2 = V_F^{\cdot} + F_i' + 1/2\ Ca_{Ca}^{x} \qquad (5)$$

This compound accommodates relatively large amounts of either NaF or YF_3 in solid solution which may be given as

$$NaF = Na_{Ca}' + V_F^{\cdot} + F_F^{x} \qquad (6)$$

and

$$YF_3 = Y_{Ca}^{\cdot} + F_i' + 2F_F^{x} \qquad (7)$$

In addition to the atomic defects present in CaF_2 this compound exhibits color centers under low flourine (high calcium) pressures (11-12). The formation of these defects may be formulated as:

$$Ca_{(g)} + 2F_F^{x} = CaF_2 + 2e_F^{x} \qquad (8)$$

Furthermore, the electrons at the flourine sites, e_F^{x}, may escape to give some quasi-free electrons (11,12) which is given as

$$e_F^{x} = V_F^{\cdot} + e' \qquad (9)$$

Rao et. al., (6) in their flotation experiments with pure CaF_2 and doped (either NaF or YF_3) CaF_2 found that the presence of the dopants reduced flotation recoveries utilizing either an anionic or a cationic collector. Thus, the behavior was similar to that exhibited by AgCl.

Spearin et., al., (7) found some very interesting results in relation to color centers in CaF_2 when they gamma irradiated an initially white CaF_2 from Rosiclare, Illinois obtained from Wards Natural Science Establishment, Rochester, New York,

Fig. 1 displays the Zeta Potential results obtained on this white fluorite sample as a function of the time of gamma irradiation. It is seen that the radiation, which creates color centers, induces positive surface charges over a wide range of pCa's indicating that the electrons created serve as sites where calcium ions are strongly adsorbed. When the irradiated samples were annealed in a HF atmosphere at 750ºC the samples virtually returned to their original Zeta Potential behavior as may be seen in Fig. 2. Annealing in air at 400ºC or 750ºC which may lead to some oxidation or oxygen doping of the CaF_2, was not nearly as effective as the HF anneal. Thus, it

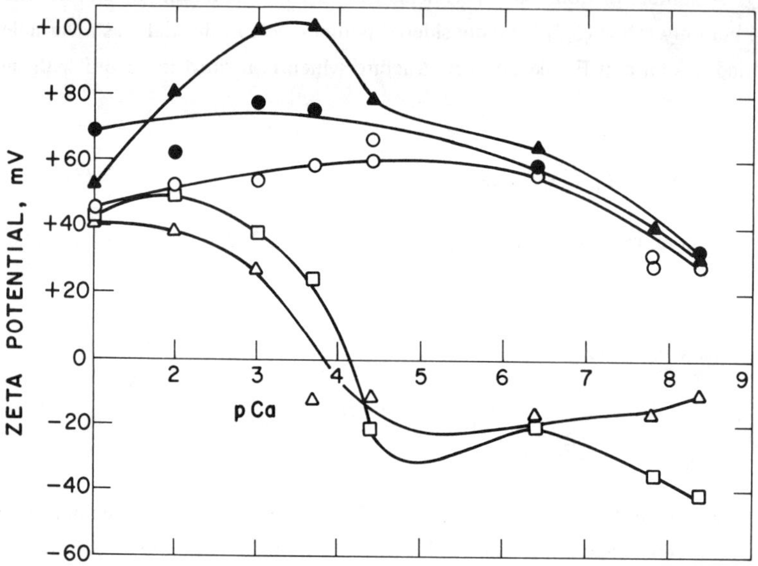

Figure 1. Electrophoretic mobilities of irradiated Rosiclare white fluorite (pH=4.6, Co^{60} ϒ-ray $5x10^{-5}$ R/hr).

Δ - no exposure
❑ - 1 hour
○ - 2 hours
● - 8 hours
▲ - 48 hours

appears that the flourine from the HF enters the CaF_2 lattice and acquires electrons from the color centers and/or the quasi-free electrons and then sits at its normal site or in an interstitial position. This also indicates that mobile defects such as electrons and electron holes may have larger effects at interfaces than do the less mobile ionic defects.

<u>PbI_2</u>. Lead iodide displays a Schottky type defect structure which is given as

$$\text{null} = V_{Pb}'' + 2V_I^{\cdot} \qquad (10)$$

where null indicates an undisturbed defect-free crystal of PbI_2 in this case.

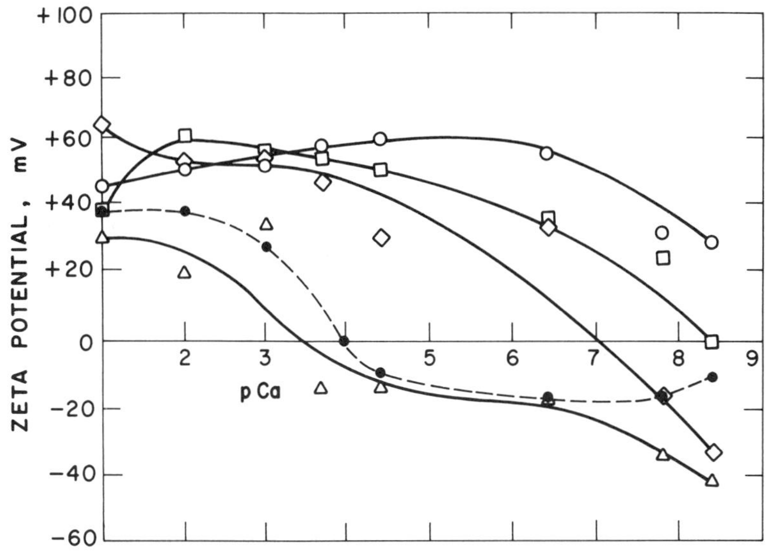

Figure 2. Electrophoretic mobilities of annealed irradiated Rosiclare white fluorite.
● - "as received
○ - Irradiated 2 hours
□ - Irradiated 2 hours, air anneal 1 hour 400°C
◊ - Irradiated 2 hours, air anneal 1 hour 750°C
Δ- Irradiated 2 hours, HF anneal 1 hour 750°C

Doping with AgI or BiI_3 results in the following reactions:

$$AgI = Ag'_{Pb} + V^{\cdot}_{I} + I^{X}_{I} \qquad (11)$$

and

$$2BiI_3 = 2Bi^{\cdot}_{i_{Pb}} + V''_{Pb} + 6I^{X}_{I} \qquad (12)$$

Spearin et. al., (13) have studied this compound in terms of flotation and have found a shift in the ZPC as a function of pPb when doped with AgI to a lower value than pure PbI_2 while doping with BiI_3 increases the ZPC. Flotation follows the electrostatic flotation model i.e., a particle displaying a positively charged surface is floated readily by an anionic collector while particles displaying negatively charged surfaces are readily floated by cationic collectors (14); however, all doping appears to lower the flotation recoveries in comparison to pure PbI_2. This is similar to the ionic conductors AgCl and CaF_2.

<u>NiO</u>. Nickel oxide is a p-type semiconductor whose defects are formed according to the reaction

$$1/2\,O_2 = V_{Ni}'' + 2\,h^{\cdot} + O_O^x \quad (13)$$

with the equilibrium relation

$$K_{13} = [V_{Ni}''][h^{\cdot}]^2/P_{O_2}{}^{1/2} \quad (14)$$

Doping with Li_2O or Al_2O_3 results in the relations

$$1/2O_2 + Li_2O = 2Li_{Ni}' + 2\,h^{\cdot} + 2O_O^x \quad (15)$$

and

$$Al_2O_3 = 2\,Al_{Ni}^{\cdot} + V_{Ni}'' + 3\,O_O^x \quad (16)$$

Hence, one may increase either the cation vacancies or the electron holes by choice of dopant.

The potential determining ions for NiO are hydrogen/hydroxyl ions and it was found that the ZPC = pH of 7.5 for pure NiO, a ZPC = pH of 6.4 for NiO + 1 mole % Li_2O and a ZPC = pH of 9.4 for NiO + 1 mole % Al_2O_3 and for NiO + 1 mole % Cr_2O_3. Thus, dopants distinctly shift the ZPC's and flotation results. Fig. 3 shows this shift quite clearly. Hence, it appears that impurities as dopants play a marked role in the flotation of NiO (15). Note that once again the flotation results are strictly in accord with electrostatic theory (14).

ZnO. Zinc oxide is an n-type semiconductor which apparently has a variety of interstitial cation species, atomic, singly-charged and doubly-charged zinc, compensated by excess electrons. We will assume that doubly-charged zinc interstitials predominate. Hence, the formation of these defects may be given as

$$ZnO = Zn_i^{\cdot\cdot} + 2e' + 1/2O_2 \quad (17)$$

which, at equilibrium gives

$$K_{17} = [Zn_i^{\cdot\cdot}] \bullet [e']^2 \bullet P_{O_2}{}^{1/2} \quad (18)$$

Doping with a lower valent cation, say Li_2O, or a higher valent cation, say Al_2O_3, gives:

$$ZnO + Li_2O = 2\,Li_{Zn}' + Zn_i^{\cdot\cdot} + 2O_O^x \quad (19)$$

and

$$Al_2O_3 = 2Al_{Zn}^{\cdot} + 2\,e' + 2O_O^x + 1/2O_2 \quad (20)$$

Note that doping with an oxide containing the same charge as the matrix metal ion, in this case a +2 ion, should not alter the defect concentration in a major fashion in accord with a reaction such as:

$$CuO = Cu_{Zn}^x + O_O^x \quad (21)$$

Mular (16) investigated ZnO pure and doped with either Al_2O_3 or CuO. The resistivity measurements made by this author indicates that the CuO added to the ZnO may have entered the interstitial sites of ZnO as a divalent or monovalent cation as noted by the large increase in the

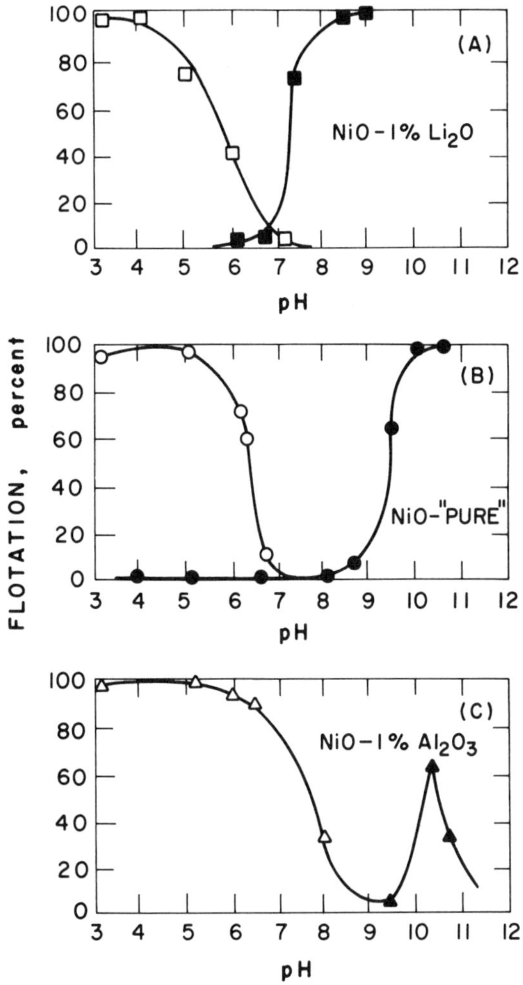

Figure 3. Flotation recovery of lithium-doped nickel oxide, (A) "pure" nickel oxide, (B) and aluminum-doped nickel oxide, (C) Collectors: solid points,, cationic dodecyl amine; open points, anionic dodecyl sulfate.

electrical conductivity of the doped samples. Although the ZPC as a function of pH was not measured Mular (16) estimates the ZPC to be in the range of pH = 9.5 to 10.8.

Fig. 4 is a schematic of Mular flotation results obtained utilizing sodium oleate as a collector and shows a rather pronounced effect of doping with Al_2O_3. Mular discusses these results in terms of the surface electronic states generated by the dopants and indicates that the defects generated take part extensively in the surface reaction occurring in this system. It is noted that the CuO doping had little effect on the flotation results.

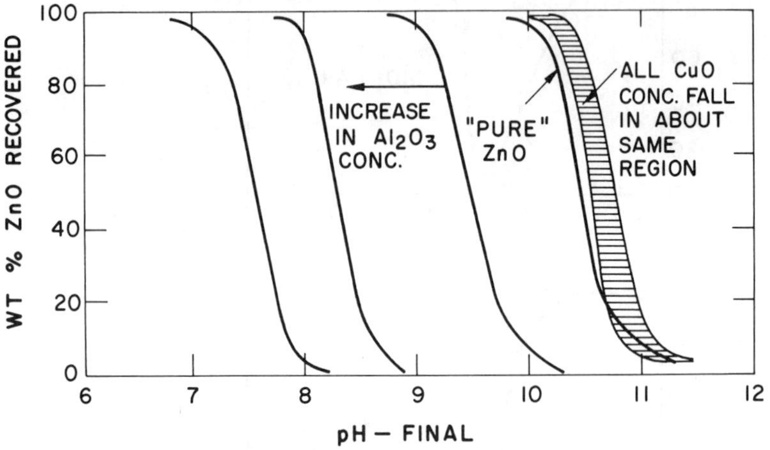

Figure 4. Flotation recovery of pure and doped ZnO - Ref. (16)

$\underline{SnO_2}$. Tin oxide is also an n-type semiconductor whose major defects are generated via the reaction.

$$SnO_2 = 2V_{O}^{\cdot\cdot} + 4e' + Sn_{Sn}^{x} + O_2 \qquad (21)$$

Equilibrium for this reaction is given as:

$$K_{21} = [V_{O}^{\cdot\cdot}]^2 \bullet [e']^4 \bullet P_{O2} \qquad (22)$$

Doping with higher and lower valent cations results in the reactions:

$$Nb_2O_5 = 2Nb_{Sn}^{\cdot} + 2e' + 4O_{O}^{x} + 1/2O_2 \qquad (23)$$

$$Fe_2O_3 = 2Fe_{Sn}' + V_{O}^{\cdot\cdot} + 3O_{O}^{x} \qquad (24)$$

This compound was studied by Senior and Poling (17) and Balachandran et al (18-20) in respect to the effect of impurities upon flotation behavior. The ZPC obtained was found to be a function of both sintering temperature and the dopant. It appears that even at pH's less than the ZPC, where all specimens were charged positively at the surface, that flotation with the anion collector sodium dodecyl sulfate is distinctly a function of dopants. It was found that the recovery of this compound followed as:

Fe_2O_3 doped > undoped > Sb_2O_5 doped

Additional more complex behavior was also determined indicating that one must consider both ionic and electronic defects in this compound's flotation behavior (18-20).

Dissolution Processes

A number of studies concerned with the effects of imperfections upon hydrometallurgical processes have been made and a few of these will be discussed.

Lead sulfide is an amphoteric semiconductor which shows p-type conduction at high sulfur pressures and n-type conduction at low sulfur pressures. The addition of Ag_2S to PbS leads to greater p-type behavior while Bi_2S_3 as a dopant in PbS makes the crystal more n-type.

In an early work on this compound (21) it was found, under conditions where the interface reaction was rate controlling, that the dissolution rate of crystals containing Ag_2S as a dopant was much more rapid in nitric acid than was the rate for any of the other lead sulfides tested (stoichiometric PbS, n-type PbS, p-type PbS and Bi_2S_3 doped PbS). The silver doped material actually proceeded in a fashion similar to an autocatalytic system indicating that the products of reaction were accelerating the reaction.

A number of studies concerning the effect of point defects upon the dissolution rate of NiO in acid solutions have been made (22-27). The point defect structure of NiO was described previously in this paper as well as the defects created by doping with higher or lower valent cations.

Fig. 5 extracted from Lussiez et. al., (27) serves to illustrate the effects observed whenever NiO and NiO containing dopants are dissolved in an H_2SO_4 solution at 60°C. These effects are quite pronounced and leave little doubt of the effect that point defects have upon dissolution kinetics. It was postulated in this case (27) that the electron holes in NiO, increased by the dopant Li_2O, served as sites for SO_4^{2-} ions and hence increased the rate of solution.

The defect structure of zinc oxide and the influence of dopants has already been listed in this paper in respect to flotation behavior. Dissolution behavior of this compound in NaOH solutions was studied specifically by Rogers et. al., (28). Again, the rates were studied on "pure" ZnO, ZnO + Li_2O, ZnO + Al_2O_3 and ZnO + Cr_2O_3. Once again, the rate of dissolution was found to increase with increases in the Li_2O concentration and to decrease with increases of either Cr_2O_3 or Al_2O_3 concentrations. However, in this case, since Li_2O increases the ionic defect, zinc interstitials, it appears that the release of the zinc hydroxo complex is rate determining and that the zinc interstitial facilitates this reaction.

Another dissolution process that appears to be enhanced by imperfections is the ferrie sulphate leaching of $CuFeS_2$(29). For this case the introduction of carbon particles in initial contact with the chalcopyrite enhances considerably the rate of dissolution of the compound. This may result from the increased electronic conductivity of the leached sulfur layer surrounding the partially dissolved particles (30,31) and/or the creation of an electrochemical cell that serves to

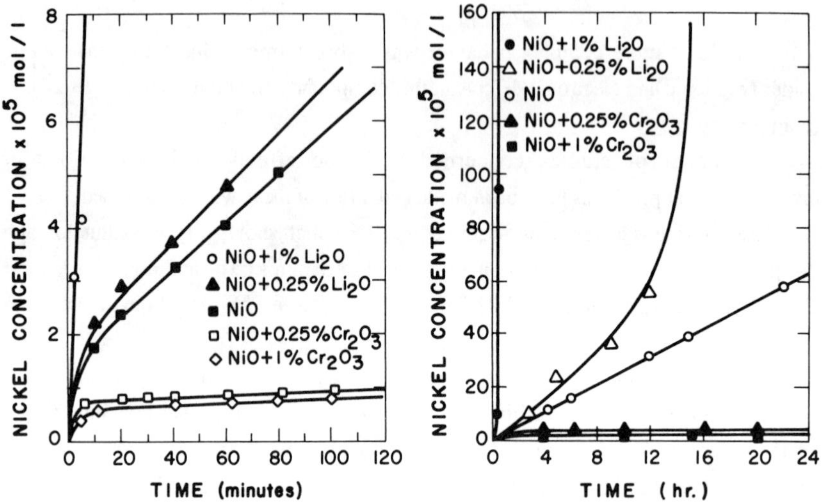

Figure 5. Nickel concentration in solutions as a function of time. Leaching solution was 0.38 mol • dm^3 at 60°C. (A) represents short time and (B) long time results.

facilitate the dissolution reaction (29). In any case, the extended point defect is an aid to this process.

Electrostatic Separation

It is well understood that electrostatic separation processes depend upon differences of the electrical conductivities of the materials being separated. The works cited here are concerned with bulk conductivities and it is assumed that the surface conductivities are in accord with the bulk properties.

Both NiO (32) and ZnO (33), an n and a p type semiconductor, were studied utilizing a Carpco electrostatic separator. Additionally, KCl containing dispersed Ag particles which tend to inject electrons into the KCl (32) and SnO_2 (18) were also included in these studies. A schematic of the bin arrangement is given in Fig. 6.

For NiO doped with Li_2O or Cr_2O_3 it was found that pure and Li_2O doped NiO accumulated in the conductive bins in accord with eqs. (14) and (15) which indicate that electron hole conduction should prevail in these materials. Conversely, doping the NiO with Cr_2O_3 decreased the electron hole concentrations and therefore these particles were found in the non-conductor bins.

ZnO and SnO_2, both n-type semiconductors, showed the opposite effect. When a higher valent cation was added to these compounds it was found, in accord with theory, that the conductivity increased and hence the particles were primarily located in the conductive bins. Again, in accord with theory, lower valent cations decreased the conductivities of these compounds and such particles where located primarily in the non-conductive bins.

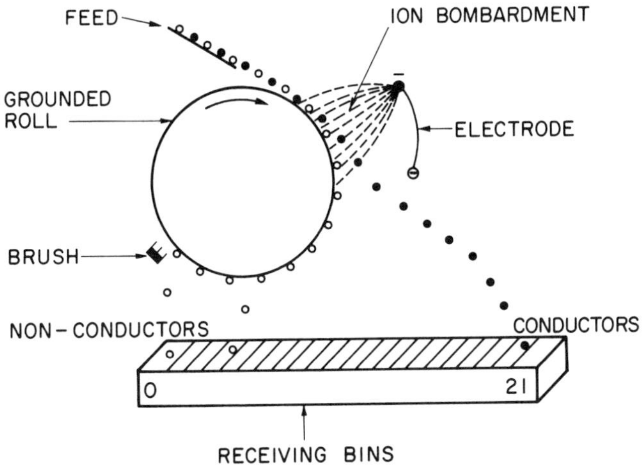

Figure 6. Schematic of Electrostatic Separator

The dispersion of Ag particles in KCl increases the conductivity of this compound in accord with Wagner's dispersed phase theory (30) and such Ag containing particles where found in the conductive bins although the separation in this case was less distinct than in the case of the semiconductors NiO, ZnO and SnO_2.

In addition an attempt was made to utilize the electrostatic separator with NiO or ZnO coated rolls in order to create a p-n junction (34). One may make use of the fact that a p-n junction when biased under a D. C. voltage in one direction will permit current to flow while upon biasing in the opposite direction essentially no current flows. Thus, one may conceivably separate n and p type particles from one another. Some success was attained with the NiO coated drum in attaining an indication of possible separation of p and n type particles but the ZnO coated drums did not show any degree of discrimination between p and n type

Lead De-copperization

Lead is de-copperized by adding sulfur to molten lead to form PbS + Cu_2S + other impurity sulfides. Since PbS in contact with lead is probably an n-type semiconductor it appears that point defect chemistry may affect de-copperization.

Thus, Pin and Wagner (35) found, in tests conducted for long times, that Ag_2S doped PbS decreased the solubility of Cu_2S in the doped PbS while McClincy and Larson (36) in tests conducted for about 100 min, found essentially the opposite effects. Conversely, higher cation valent sulfides, such as Bi_2S_3, increased the solubility of Cu_2S in the lead sulfide according to Pin and Wagner (35) while the results of McClincy and Larson (36) were opposite to those of Pin and Wagner. An answer to this difficulty of opposite experimental findings was provided by Simkovich (37).

It can be shown that when two aliovalent compounds are present as dopants in another compound they will increase each others solubility when they create, individually, opposite type of defects. If however, they create similar type defects they will mutually decrease their solubilities. In PbS it has been shown (38) that cuprous ions are located in interstitial positions for short times, a few hours, while after several days the cuprous ion sits on a normal cation site. This leads to two different types of defects as a function of time in the Cu_2S doped PbS. Thus, since Pin and Wagner's (35) tests were long time tests Ag_2S and Cu_2S should decrease each other's solubilities in PbS while Bi_2S present in PbS should increase the solubility of the Cu_2S. Since McClincy, and Larson's tests were short time tests monovalent copper ions are present as interstitials in PbS (38) then the exact opposite results are to be expected (37).

<u>Summary</u>

Point defects play significant roles in a variety of mineral/metallurgical processes.

In flotation processes it appears that most semi conductor type compounds display a behavior consistent, most of the time, with the electrostatic theory of flotation. Changes in ZPC are generally in the direction anticipated with the changes produced in the mobile electronic defects. Ionic conducting compounds almost always show a reduction in flotation recoveries whenever an aliovalent impurity is present in the matrix compound even though changes of ZPC are taken into account. This may well mean that the ionic conductors are naturally hydrophobic and all disturbances tend to make them more hydrophilic than the pure compound.

Leaching processes are distinctly found to be point-defect sensitive when the interfacial reaction is rate controlling. Such sensitivity appears to be related to preferential sites of reaction provided by the point-defects. Additionally leaching of $FeCuS_2$ was found to be influenced by the presence of carbon suggesting aid from a 3 dimensional defect.

Electrostatic separation was shown to be extremely sensitive to small amounts of dopants in crystals. Furthermore, an analysis of the point defect chemistry allows one to choose the dopants which may enhance or reduce the electrical conductivities of the material under consideration. An attempt to separate p and n type semiconducting particles by use of semiconductor rolls on the electrostatic separator was partially successful but much more work must be performed in this area before a clear delineation between success or failure is realized.

Finally a point defect model applied to decopperization of lead by the formation of PbS + Cu_2S + other impurity sulfides appears to at least partially answer the mode of the reaction model responsible for the effect of the impurities in the lead.

<u>Future Efforts</u>

In accord with the material reviewed in this article these is no question concerning the fact that point defects have major effects upon a variety of processes both metallurgical and mineral. It appears logical to extend these laboratory results to industrial applications by determining the solid solution impurities in mineral deposits and after consideration of the effects that the impurities have upon the point defect properties of the matrix compound to make adjustments in the processes.

It appears worthwhile to study further the possibility of separating electrostatically n and p type semiconductor materials. Such a study would be rather extensive but may well lead to meaningful new types of equipment and separation processes.

The leaching processes in many industrial plants utilize electrochemical considerations to a major extent. Such plants probably could benefit considerably by looking at compounds from the point defect view point.

Finally, it probably is worthy of future studies in a variety of processes to extend imperfection investigations to the 1, 2 and 3 dimensional defects. Such studies may well lead to interesting and practical results.

REFERENCES

1. F. A. Kröger, The Chemistry of Imperfect Crystals (New York, NY: John Wiley and Sons, Inc, 1964.

2. F. A. Kröger and H. J. Vink, "Relations Between the Concentration of Imperfections in Crystalline Solids", in Solid State Physics, eds. F. Seitz and D. Turnbull, (New York, NY: Academic Press Inc., 1956) 307-435.

3. G. Simkovich, "The Influence of Solid State Point Imperfections upon Flotation Processes", Trans. AIME-TMS, 227 (1963) 306-308.

4. E. Y. Spearin, G. Simkovich and F. F. Aplan, "Surface and Flotation Properties of AgCl as a Function of Induced Point Defects", Colloids and Surfaces, 26 (1987) 257-271.

5. E. P. Honig and J. H. Th. Hengst, "The Point of Zero Charge and Solid State Properties of Silver Bromide", J. Colloid and Interface Sci., 31 (1969) 545-556.

6. B. V. P. Rao, H. L. Lovell and G. Simkovich, "The Flotation of Fluorite as a Function of Ionic Point Imperfections", Trans. AIME-SME, 241 (1968) 328-331.

7. E. Y. Spearin, G. Simkovich and F. F. Aplan, unpublished research (1978).

8. J. D. Miller and J. B. Hiskey, "Electrokinetic Behavior of Fluorite as Influenced by Surface Carbonation", J. Colloid and Interface Sci., 41 (1972) 567-573.

9. E. P. Honig and J. H. Th. Hengst, "Point of Zero Charge of Inorganic Precipitates", J. Colloid and Interface Sci., 29 (1969) 510-520.

10. M. C. Fuerstenau, G. Gutierrez and D. A. Elgillani, "The Influence of Sodium Silicate in Nonmetallic Flotation Systems", Trans. AIME-SME, 241 (1968) 319-323.

11. C. Wagner, "Limitations of the Use of CaF_2 in Calvanic Cells for Thermodynamic Measurements Due to the Onset of Electronic Conduction Under Reducing Conditions", J. Electrochem. Soc., 115 (1968) 933-935.

12. J. Delcet, R. J. Heus and J. J. Egan, "Electronic Conductivity in Solid CaF_2 at High Temperature", J. Electrochem. Soc., 125 (1978) 755-758.

13. E. Y. Spearin, "The Influence of Point Defects on Flotation Systems" (PhD. Thesis, The Pennsylvania State University, March, 1979), 166-192.

14. F. F. Aplan and D. W. Fuerstenau, "Principles of Monmetallic Flotation," in Froth Flotation 50th Anniversary Volume (New York, NY: AIME, 1962) 170-214.

15. F. F. Aplan, G. Simkovich, E. Y. Spearin and K. C. Thompson, "Effect of Point Defects on Flotation", in The Physical Chemistry of Mineral-Reagent Interactions in Sulfide Flotation compiled by P. E. Richardson, G. R. Hyde and M. S. Ojalvo (U. S. Bureau of Mines, Information Circular 8818, 1980) 25-34.

16. A. L. Mular, "Effect of Impurities on the Flotation Behavior of Zinc Oxide," Trans. AIME-SME, 232 (1965) 204-211.

17. G. D. Senior and G. W. Poling, "The Chemistry of Cassiterite Flotation", in Advances in Mineral Processing, ed. by P. Somasundaran (Littleton, Colo., AIME-SME, 1986) 229-254.

18. S. B. Balachandran, G. Simkovich and F. F. Aplan, "The Influence of Point Defects on the Floatability of Cassiterite, I. Properties of Synthetic and Natural Cassiterites", International J. Min. Proc., 21 (1987) 157-171.

19. S. B. Balachandran, G. Simkovich and F. F. Aplan, "The Influence of Point Defects on the Floatability of Cassiterite, II. Electrostatic Collector Interactions", International J. Min. Proc., 21 (1987) 173-184.

20. S. B. Balachandran, G. Simkovich and F. F. Aplan, The Influence of Point Defects on the Floatability of Cassiterite, III. The Role of Collector Type", International J. Min. Proc., 21 (1987) 185-203.

21. G. Simkovich and J. B. Wagner, Jr., "The Influence of Point Defects on the Kinetics of Dissolution of Semiconductors", J. Electrochem. Soc., 110 (1963) 513-516.

22. K. Nii, "On the Dissolution Behavior of NiO "Corr. Sci., 10 (1970) 571-583.

23. C. F. Jones, R. L. Segall, R. St. C. Smart and P. S. Turner, "The Effect of Prior Annealing Temperature on Dissolution Kinetics of Nickel Oxide", J. Chem. Soc., Faraday Trans. I., 73 (1978) 1710-1720.

24. C. J. Jones, R. L. Segall, R. St. C. Smart and P. S. Turner, "Semiconducting Oxides: Effects of Electronic and Surface Structure on Dissolution Kinetics of Nickel Oxide", J. Chem. Soc., Faraday Trans. I., 74 (1978) 1615-1623.

25. C. J. Jones, R. L. Segall, R. St. C. Smart and P. S. Turner, "Semiconducting Oxides, Infrared and Rate Studies of the Effects of Surface Blocking by Surfactants in Dissolution Kinetics", J. Chem. Soc., Faraday Trans. I., 74 (1978) 1624-1633.

26. A. Illis, G. C. Nowlan and J. J. Koehler, "Production of Nickel Oxide From Ammonical Process Streams, CIM Bulletin, 63 (1970) 352-361.

27. P. Lussiez, K. Osseo-Asare and G. Simkovich, "Effect of Solid State Impurities on the Dissolution of Nickel Oxide, Met. Trans., 12 B (1981) 651-657.

28. G. Rogers, G. Simkovich and K. Osseo-Asare, "The Effects of Point Defects on the Dissolution of Zinc Oxide in Sodium Hydroxide Solutions", Hydrometallurgy, 10 (1983) 313-328.

29. R. Y. Wan, J. D. Miller and G. Simkovich, "Enhanced Ferric Sulfate Leaching of Copper from $CuFeS_2$ and C. Particulate Aggregates", in International Conf. on Recent Advances in Min. Sci. and Tech., MINITEK 50, (1984).

30. C. Wagner, "The Electrical Conductivity of Semiconductors Involving Inclusion of Another Phase", J. Phys. Chem. Solids, 33 (1972) 1051-1059.

31. J. D. Beckman, A. K. Birchenall and G. Simkovich, "Electrical Conductivity of Dispersed Phase Systems", in First Inter. Conf. on Transport in Non-Stoichiometric Compounds, ed. J. Nowotny (New York, NY: Elsevier Scientific Publ. Co., 1982) 8-28.

32. G. Simkovich, F. F. Aplan and R. A. Fensterer, "Point Defect Effects in the Electrostatic Separation of NiO, I & EC Fundamentals, 11 (1972) 274-276.

33. G. Simkovich and F. F. Aplan, "The Effect of Solid State Dopants upon Electrostatic Separation", in Fine Particle Processing, ed. P. Somasundaran, (New York, NY AIME-SME, 1980) 1342-1355.

34. E. Y. Spearin, G. Simkovich and F. F. Aplan, "Rectifying Junctions in Electrostatic Separation", in Advances in Mineral Processing, ed. P. Somasundaran (Littleton, CO, AIME-SME, 1986) 358-365.

35. C. Pin and J. B. Wagner, Jr., "The Removal of Copper from Liquid Lead by Lead Sulfide Containing Controlled Atomic Defects", Trans. AIME-TMS, 227 (1963) 1275-1281.

36. J. McClincy and A. H. Larson, "The Removal of Copper from Lead with Sulfur", Trans. AIME-TMS, 245 (1969) 193-196.

37. G. Simkovich, "Point Defects in PbS and Their Effects in the Decopperization of Lead with Sulfur", Met. Trans., 9B (1978) 527-529.

38. J. Bloem and F. a. Kröger, "Interstitial Diffusion of Copper in PbS Single Crystals", Philips Research Reports, 12 (1957) 281-302.

Some Milling Practices and Technological

Innovations on Beneficiation of Antimony sulfide ores in China

Hsi-keng Hu

Central South University of Technology

Changsha, Hunan 410012

People's Republic of China

Abstract

This paper deals with the flotability of stibnite and summarizes the major experiences and technological innovations of some typical practices of antimony sulfide ores in China. Experiences of cyanide-free and -less flotation of complex antimony sulfide ores and the application of the heavy media process to the preconcentration are described.

Performance improvements obtained by innovations in operating techniques are also outlined and discussed. Technical innovations involve: (1) ramification roughing and differential-rate cleaning of stibnite-arsenopyrite ore; (2) the application of the new flotation reagent system; (3) process of separation flotation of Sb-Hg bulk concentrate; (4) extraction of gold from Au-Sb bulk concentrate; (5) flotation recovery of coarse stibnite particles.

Introduction

China is rich in natural resources of antimony and is one of the major antimony producers in the world. During the past several decades, a lot of research works have been done and many experiences in milling practices were gained, some of which are summarized briefly as follows.

I. Heavy Media separation of Antimony sulfide Ore at Xikuangshan Mill, Hunan province (1)

HMS as a technique of preconcentration of antimony ore has been performed more than ten years in China with good results.

1. Ore characteristics The antimony ore of Xikuangshan belongs to the epithermal-metasomatic-filled doposit of antimony sulfide. Major minerals are stibnite (92.1%), antimony oxides (7.9%) and a small amount of pyrite and pyrrhotite; the gangue minerals are quartz, silicic limestone, calcite, hydrous kaolin, sericite, barite and others. The specific gravity of the ore is 2.8 and the specific gravity of stibnite and gangue are 4.6 and 2.654, respectively.

When the ore is crushed to 70 mm, a reasonable amount of gangue is liberated from valuable minerals, which is helpful for removing the gangue in coarse sizes by HMS system. As there exists the miarolitic structure in the

Metallurgical Processes for the Year 2000 and Beyond
Edited by H.Y. Sohn and E.S. Geskin
The Minerals, Metals & Materials Society, 1988

ore, the specific gravity of the ore will become lower, thus the stibnite and antimony oxides contained in geodes will be lost in the HMS process.

2. HMS plant The Present flowsheet used at Xikuangshan mill is a combination of handsorting, HMS and flotation processes. The fraction of -150+35mm can be sorted by hands and the fraction of -35+6mm is separated by HMS, the fraction of -12+0mm (including the sinks of HMS) can be treated by flotation. The HMS-stage can discard 40.39% of tailing. The separation result of the plant is that as the feed grade is 2.4% Sb, the total plant concentrate grade is 17.07% Sb (flotation concentrate grade is 48% Sb) and the recovery is 93.11%.

The circuit flowsheet used at the south plant of Xikuangshan is shown in Fig. 1.

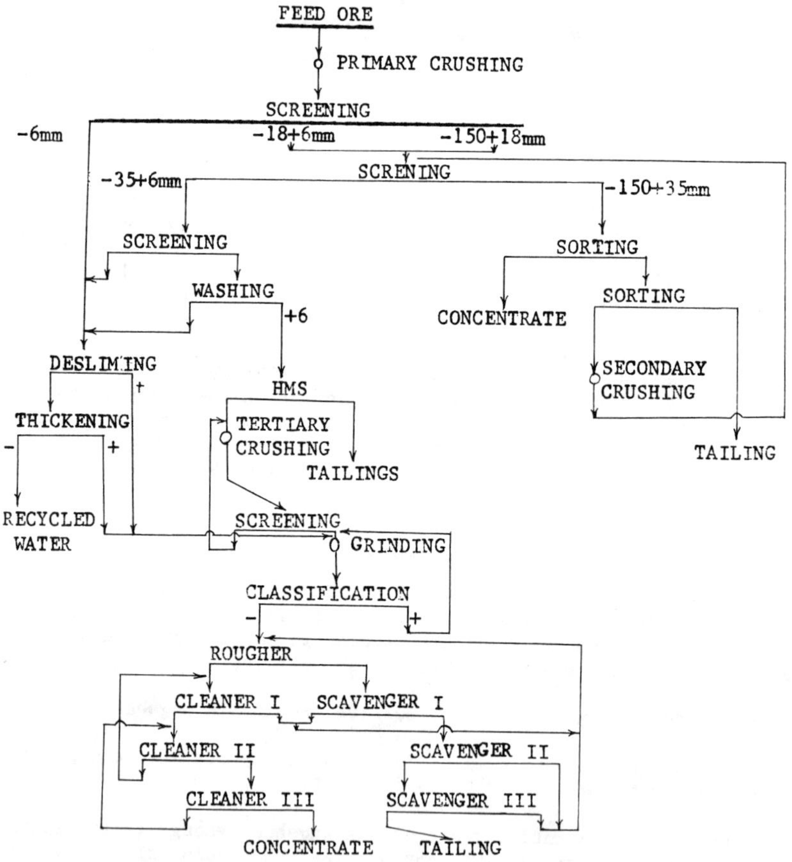

Fig. 1. The Flowsheet of the South Plant of Xikuangshan Mill.

The HMS system consists of four interconnected parts: the ore preparation, the separation of heavies and lights, the media preparation and the media recovery. These are illustrated in Fig.2.

The metallurgicalperformance and media consumption of the recent years of HMS system are given in Table. 1.

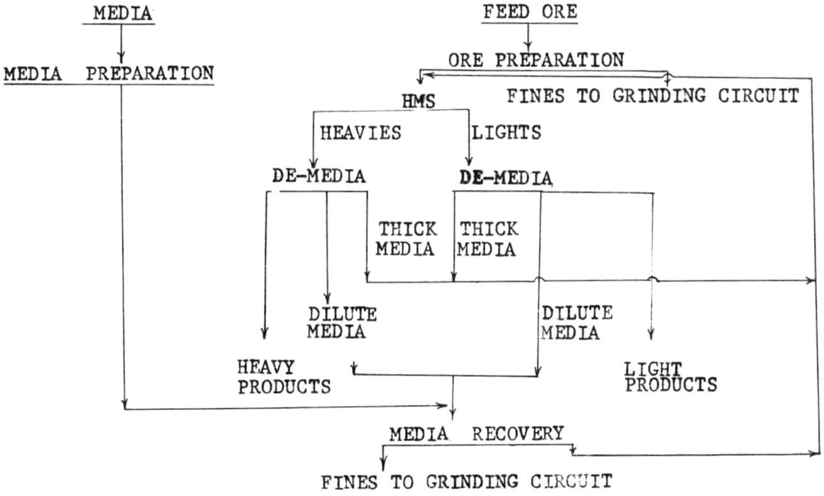

Fig. 2. Typical HMS circuit

Table 1. Metallurgical performance and media consumption of the recent years of HMS system (Average).

Assay,%			Wt. %	Rec. %	Media consumption
Feed	Conc.	Tailing	Tailing	Conc.	kg/t
1.63	2.69	0.20	42.46	94.76	1.84

Because the antimony ore consists of refractory middlings, the plant adopts the Wemco Drum separator. The rotation speed of the drum separator is very important. When the drum rotates too fast, it will disturb the separation due to the media turblence, but when it rotates too slow, the heavies can not be elevated in time and the reduced agitation will increase the settlement of the separator and make the separation process ineffective. The suitable rotation speed is 2 rpm.

According to the milling practices run in the past decade years, it is practicable to use the HMS system for preconcentration of stibium ore. The HMS system with Drum separator is reliable in technology and has great effect on the efficient separation. As the stibium ore does not exhibit marked difference in specific gravity between ore being and waste minerals, it is of critical importance to carefully control the separation specific gravity. The perfect automatic measuring devices for the concentration and density should be researched, on the basis of which the automatic adjustment by the microcomputer can be realized in order to make HMS system more efficient; besides, it is necessary to develop the new type of separators suitable for the -6+3mm fraction separation which can increase the discharging amount of waste minerals so as to increase the economic benefits of the plant.

II. Ramification Roughing and Differential-rate cleaning Flowsheet Used for separating stibnite from Arsenopyrite at Banxi Mill, Hunan province (2)

1. Ore characteristics The antimony ore of Banxi belongs to the epithermal-metasomatic-filled deposit of antimony sulfide. Major minerals are stibnite, arsenopyrite and pyrite, with minor quantities of scheelite, chalcopyrite and native gold. The gangue minerals are quartz and sericite, with minor apatite, feldspar and so forth. Stibnite minerals are unevenly disseminated.

2. Flotation Flowsheet. Before 1982, a selective flotation flowsheet was used with the reagents of CaO, $Pb(NO_3)_2$, BuX and pine oil, but the antimony concentrate was high in arsenic (2.5~2.7% As) and the reagent consumption was higher too.

Since 1982, a new flowsheet of ramification roughing and differential-rate cleaning was adopted instead of the selective sequential flotation, the conventional practice and butyl ammonium aerofloat was used instead of BuX, resulting in good metallurgical results. The present circuit flowsheet is shown in Fig. 3.

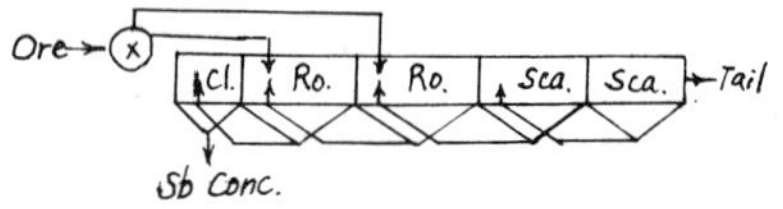

Fig.3. Layout of the Flotation Cells of the Ramification Roughing and Differential-rate Cleaning Flowsheet

The flotation is carried out at pH of 4.8-5.0, with sulfuric acid (1450 g/T) as the pH regulator, $pb(NO_3)_2$ (120g/T) as the activator, butyl ammonium aerofloat (250 g/T) as the collector and pine oil (8-13 g/T) as the frother. The metallurgical results of flotation are as follows:

Ore Grade (%)		Concentrate Grade (%)		Recovery (%)
Sb	As	Sb	As	Sb
6.47	0.43	64.39	0.51	93.52

According to the milling practice of the past few years, it was experienced that the new flowsheet is superior to the conventional selective flowsheet in the following ways: (1) in saving reagent consumption; (2) in upgrading the final concentrate with less arsenic content; (3) in increasing the recovery of stibnite; and (4) by being easier to control the mill operations.

III. Process for the Separation of Stibnite and Cinnabar from the Bulk Concentrate by Flotation with ACMC Reagent (3)

The separation of stibnite and cinnabar from the complex Sb-Hg sulfide ores to get the marketable concentrates is rather difficult. It is owing to the fine dissemination of these minerals in the ore or/and the activation of stibnite by some heavy metal ions. Hence, many of the complex Sb-Hg sulfide ores are treated by flotation to obtain the Sb-Hg bulk concentrate which is then treated by metallurgical process. Some of the complex Sb-Hg sulfide ores can be satisfactorily separated by flotation with $Pb(ND_3)_2$ and $K_2Cr_2O_7$ as the modifiers, but the problem of the environmental pollution will often be happened.

In our work (3), extensive testwork was done on the treating of the separation of stibnitefrom cinnabar. Various activators, including $HgCl_2$, $Pb(NO_3)_2$ and $CuSO_4$, were used in the flotation of stibnite and cinnabar, then various depressants, including ACMC and dichromate were used for the separation of stibnite from cinnabar in a mixture of stibnite and cinnabar. It was found that ACMC is one of the most effective reagent which exibits the strong depressing action on stibnite but no action to cinnabar.

MS flotation cell tests on the Cu-activated mixture of stibnite and cinnabar showed that stibnite canbe more satisfactorily separated from cinnabar at PH 9 with ACMC and Na_2SiO_3 as depressant. Thus, in the flotation of complex Sb-Hg sulfide ores, by the use of ACMC, the separation of stibnite from cinnabar can be achieved.

Detection by XPS and infrared spectra showed that ionic exchange of Cu^{2+} ion was occurred on stibnite surface to form the Cu_2S film, whereas on cinnabar surface it is mainly physical adsorption. At PH 9, ACMC is chemisorbed on stibnite surface, whereas on cinnabar surface is through hydrogen bonding and mast of ACMC adsorbed on the Cu-activated stibnite surface can be removed by xanthate. In the presence of ACMC and butyl xanthate, species adsorbed on the Cu-activated stibnite surface is mainly ACMC, whereas that on the Cu-activated cinnabar surface is mainly xanthate. Therefore, ACMC is essentially the selective depressant for the separation flotation of the Cu-activated mixture of stibnite and cinnabar.

IV. Plant pratice at Xiangxi Concentrator, Hunan province (4,5,6)

1.Ore characteristics The Xiangxi orebody belongs to the epithermal vein-type deposit of complex Sb-W-Au ore. Major minerals are mainly stibnite, scheelite and native gold, with minor quantities of wolframite, pyrite, sphalerite,arsenopyrite, galena, etc. the native gold occurs as free gold and also occurs in pyrite and stibnite with a small amount in quartz and scheelite and a half of the gold are of micro- and submicro-size. The gangue minerals are mainly quartz with some calcite, apatite, dolomite, etc. The economic minerals are unevenly disseminated.

2. Cricuit Flowsheet The flowsheet of the Xiangxi Concentrator is shown in Fig.4. (4) There are three stages of crushing on the surface. The tertiary

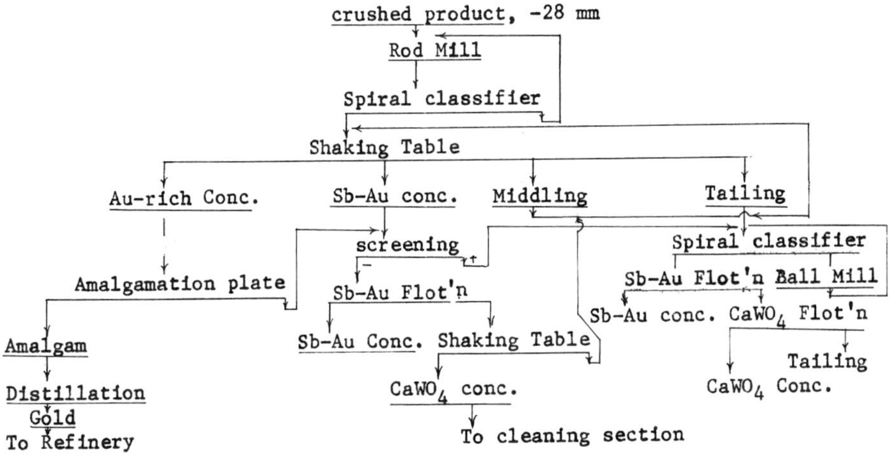

Fig.4. Flowsheet of the Xiangxi concentrator

crusher is in closed circuit with the screen, and its product joins the screen undersize and is conveyed to the fine-ore bin.

The crushed product of 92%-20 mm size is fed to the rod mill which is in closed circuit with spiral classifier. The classifier overflow splits into four streams, each feeding a shaking table. Each table produces four products namely, Au-rich concentrate, Sb-Au bulk concentrate, middling and tailing. The processes for treating these four products are self-explained in Fig.4.

The tailing of the shaking table is ground in the ball mill in closed circuit with spiral classifier, the overflow of which about 75~80% -200 mesh 25% solids is fed to the Sb-Au bulk flotation circuit to float stibnite and gold at PH 7-8 with butyl xanthate as the collector, $Pb(NO_3)_2$ as the activator and pine oil as the frother to get the Sb-Au bulk concentrate which is sent to the gold recovery section and the tailing is the feed to the Scheelite flotation circuit to obtain the crude scheelite concentrate.

3. Cleaning of the crude scheelite concentrate[5] The crude scheelite concentrate generally contains only 5-10% WO_3 which should be cleaned in order to get the final concentrate (60% WO_3). The crude concentrate is fed to the agitation tank is agitated at 50-60% solids, with the addition of Na_2SiO_3 and steaming to 100° C for 30-60 minutes,then it is diluted and decanted twice. The decanted pulp is diluted to 25% solids and is fed to flotation circuit to get the scheelite concentrate grading 65-70% WO_3 with the recovery of 96-97%.This $CaWO_4$ concentrate is further treated by agitation-leaching with hydrochloric acid for de-phosphorization. The simplified flowsheet of Thermal-Agitation-Flotation process is shown in Fig.5.

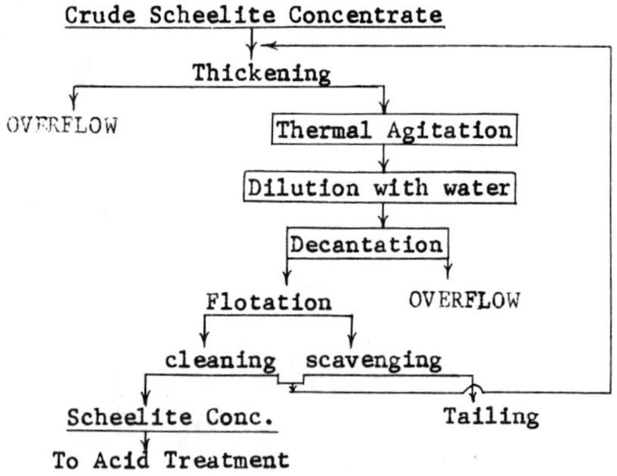

Fig. 5. Simplified Flowsheet of Thermal-Agitation-Flotation Process

4. Gold Recovery from the Sb-Au Bulk Concentrate [6] In the past,an extensive testwork was done to recover the gold from the Sb-Au bulk concentrate and only the Blast-Furnace-Electrolysis Process was successful. The Sb-Au concentrate is pelletized, then roasted in the blast furnace with forehearth, gold and antimony is separated step by step. The simplified flowsheet of the Blast-Furnce-Electrolysis process for gold recovery is shown in Fig.6

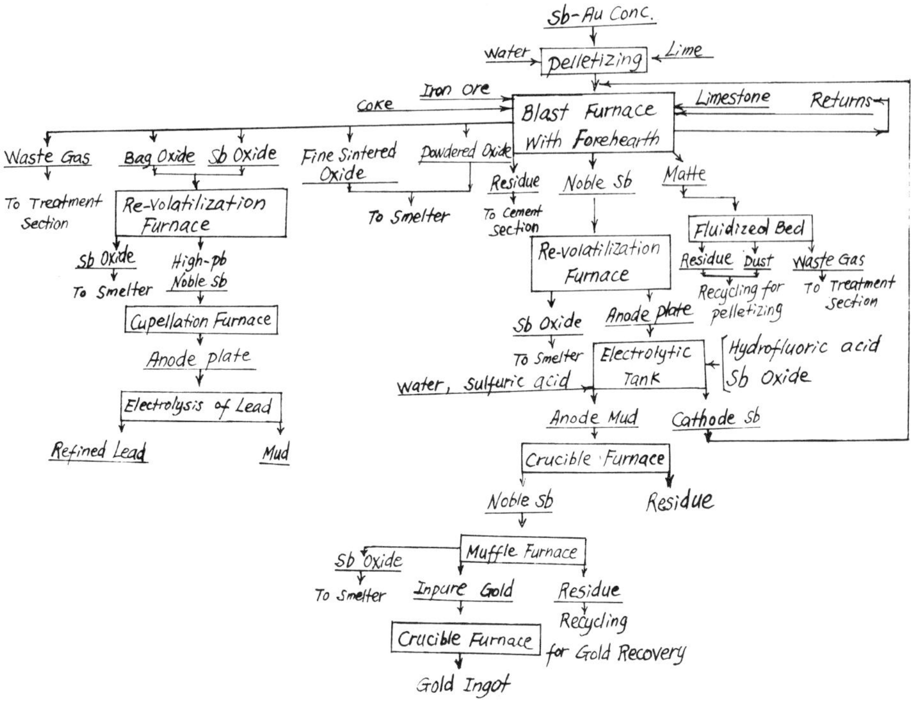

Fig. 6 Simplified Flowsheet of Blast Furnace-Electrolysis process for Gold Recovery

The metallurgical results are shown in Table. 1.

Table 1. Metallurgical Data

Product	Assay			Distribution %		
	Au g/T	Sb %	WO_3 %	Au	Sb	WO_3
Feed	3.75	1.08	0.18			
$CaWO_4$ Conc. of Shaking Table	2.50	0.06	67.91			
$CaWO_4$ Conc. of Flotation	1.64	0.08	6.00			
Sb-Au Bulk Conc.	66.25	41.70	0.02			
Amalgam	99.47	-	-			
Tailing	0.45	0.03	0.03			
Total Recovery				83.3	96.91	83.28

References

1. Sailong Sun, Zhenxun Guo and Songren Li; "Heavy Media Separation of Antimony Sulfide Ore at Xikuangshan, 1988.

2. Hsikeng Hu and Zhenxun Guo; "Benficiation of Antimony Ores," 1987.

3. Xiaojun Xu, Jinhua Liu and Hsikeng Hu; "A process for the separation of stibnite from Cinnabar by Flotation with ACMC Reagent," Extractive Metallurgy and Metals Science, CSUT/TUC, 1987, 9, I-95 ~ 109.

4. Hsikeng Hu; "Mineral Processing of Non-ferrous Sulfide Ores," The Metallurgical Industry press, 1987, pp. 305-312.

5. Hongjun Chang; "Plant Practice of Gold Recovery from the Gold-bearing antimony-Gold Bulk Concentrate," Xiangxi Gold Mine, May, 1981, pp 1-22.

OXYGEN PRODUCTION TECHNOLOGIES

K. J. Murphy, A. P. Odorski and A. R. Smith

Air Products and Chemicals, Inc. 1988
Allentown, PA 18195 U.S.A.

Abstract

The need for large quantities of oxygen for use in metallurgical processes was a major driving force in the commercialization of tonnage air separation plants. Oxygen, nitrogen and argon impact the quality, energy efficiency and environmental aspects of metals production. Cryogenic air separation processes have been used to supply most industrial requirements, however, newer technologies are available or under development to meet the needs of today's and tomorrow's metallurgical processes. This paper reviews the technical features and state of development of adsorption, chemical, cryogenic and membrane technologies. The technologies are compared in terms of plant size, oxygen purity and pressure, nitrogen and argon co-production, product use pattern and back-up requirements. Computer control systems are described that optimize operation over varying load patterns and ambient conditions; substantially lowering energy consumption.

Metallurgical Processes for the Year 2000 and Beyond
Edited by H.Y. Sohn and E.S. Geskin
The Minerals, Metals & Materials Society, 1988

Introduction

Oxygen production totalled 390 billion cubic feet in 1987. Although current production is lower than the peak of 455 billion cubic feet reached in 1979, oxygen remains the fifth leading chemical product manufactured in the U.S. (1) Metals manufacturing and fabrication accounts for sixty percent of the oxygen consumed in the U.S. Oxygen use in the steel industry consists of air enrichment for conversion of iron ore, purification of the iron in basic oxygen or other furnaces, and for cutting and welding the finished steel product. Nonferrous ores are smelted in oxygen enriched atmospheres to reduce energy consumption and increase capture of sulfur compounds to meet environmental requirements. In the future oxygen usage is expected to become even more important in metals production as energy costs increase, environmental standards become more stringent, and as new coal based processes become commercialized. Oxygen usage per ton of metal in the new direct reduction processes can be up to ten times higher than traditional steel making.

The preferred method for oxygen production is the cryogenic air separation process, which accounts for nearly all industrial production. When nitrogen and argon coproducts are required for inerting or can be sold in liquid form to markets near the metals production facility, oxygen purity has traditionally been 99.5% by volume. This purity was established as the economic optimum in the tradeoff between higher argon recovery and increasing power requirements. The traditional metals facility was a large, integrated plant that ran virtually non-stop. Cryogenic oxygen plants were built to operate with higher than 98% on-stream times, and incorporated gaseous and liquid storage capacity to meet fluctuations in product flow requirements and to provide back-up to avoid shutdowns in downstream processes due to unplanned air separation plant outages.

Over the past few years, the size of metals production facilities has decreased and the consumption patterns have become more cyclical. Applications for oxygen purities below 99.5% have increased, and the need for on-site production of nitrogen and argon has decreased. While current trends are to scale down and decentralize metals production, the new direct reduction processes under development in Europe and Japan could be quite large if located near the sources of coal and ore reserves. The result is a need for new technologies to meet the broader range of applications expected in the future.

Adsorption technology has already been employed in metals production where lower purity oxygen is acceptable (2). Membrane systems are currently capable of providing air enrichment purities of below 40% purity, however, substantial research is underway in membrane materials and development. Chemical based systems are also being developed that will offer reduced energy requirements and potential integration opportunities with the metals production facility. All of these processes have advantages and disadvantages depending on the oxygen users specific requirements which include: purity, pressure, pattern of use, co-product needs, energy cost, life of plant and storage or back-up capability.

The following sections describe each technology and a comparison is included which ranks the processes against the major production and operating variables. The processes span the temperature spectrum from cryogenic through ambient (adsorption, membranes) and up to 1200°F (chemical). Similarly, the unit operations employed include heat exchange, distillation, adsorption, diffusion and absorption. All the processes are currently operated sufficiently far from the thermodynamic minimum energy requirement so that future improvements can be expected, although the more mature adsorption and cryogenic processes may have less potential for improvement. As metals processes diversify and change, market niches for each oxygen production technology will develop based on site and project specific considerations.

Adsorption

Adsorption processes are based on the ability of some natural and synthetic materials to preferentially adsorb nitrogen or oxygen molecules as air is passed over them. In the case of zeolites (aluminosilicates), non-uniform electric fields exist in the void spaces of the material, causing polar molecules to be adsorbed more strongly than non-polar molecules. Thus in air separation, nitrogen molecules are more strongly adsorbed than oxygen or argon molecules. As air is passed through a bed of zeolitic material, nitrogen is retained and an oxygen-rich stream exits the bed. Carbon molecular sieves have pore sizes on the same order of magnitude as the size of air molecules. Since oxygen molecules are slightly smaller than nitrogen molecules, they diffuse more quickly into the cavities of the adsorbent. Thus, carbon molecular sieves are selective for oxygen and zeolites are selective for nitrogen.

Zeolites are typically used in adsorption based processes for oxygen production. A typical flowsheet is shown in Figure 1. Pressurized air enters a vessel containing the zeolite adsorbent. Nitrogen is adsorbed and an oxygen-rich effluent stream is produced until the bed has been saturated with nitrogen. At this point, the feed air is switched to a fresh vessel and regeneration of the first bed can begin. Regeneration can be accomplished by heating the bed or by reducing the pressure in the bed, which reduces the equilibrium nitrogen holding capacity of the adsorbent. Heat addition is commonly referred to as Temperature Swing Adsorption (TSA), and pressure reduction as Pressure or Vacuum Swing Adsorption (PSA or VSA). The faster cycle time and simplified operation associated with pressure reduction makes it the process of choice for air separation.

Variations in the process that effect operating efficiency include separate pretreatment of the air to remove water and carbon dioxide, multiple beds to permit pressure energy recovery during bed switching, and vacuum operation during depressurization. Optimization of the system is based on product flowrate, purity and pressure, energy cost and expected operating life. Oxygen purity is limited to a maximum of 93 - 95%. Due to the cyclic nature of the adsorption process, bed size is the controlling factor in capital cost. Since production is directly proportional to bed volume, capital costs increase more rapidly as a function of production rate compared to cryogenic plants. Up to 100 tons per day, the capital cost is relatively low and compensates for the higher energy usage compared to cryogenics (3).

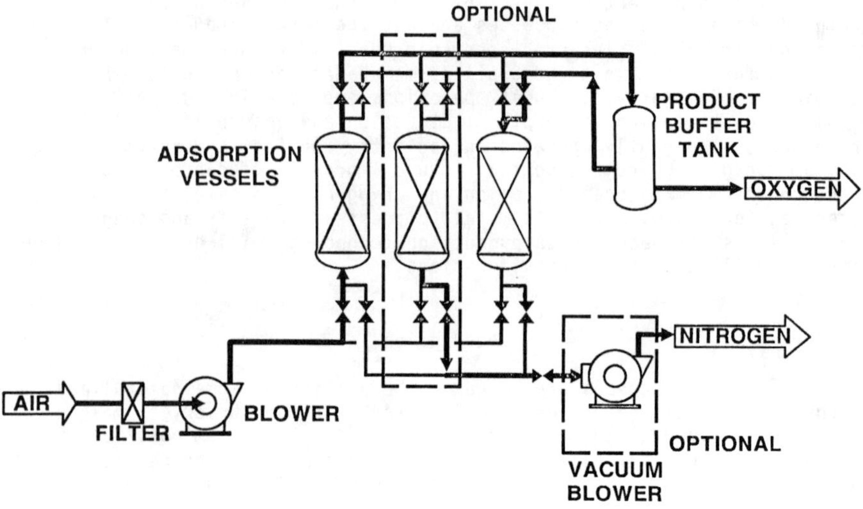

Figure 1 - Adsorption Based Process

Other factors that effect analysis of this process include operability and product storage. Since the system operates at near ambient conditions, start-up and shutdown are relatively fast operations. A unit idled over night or during a weekend can be restarted and reach purity in as little as several minutes. If product storage to meet peak demands, maintenance or other outages is required, a backup liquid oxygen tank may be required. Liquid oxygen would typically be hauled in by an industrial gas supplier. In summary, adsorption systems can deliver 85 to 95% oxygen and are best suited for applications requiring less than 100 tons per day of product.

Membranes

Membrane processes are based on the difference in rates of diffusion of oxygen and nitrogen through a membrane which separates high pressure and low pressure process streams. Flux and selectivity are the two properties that determine the economics of membrane systems, and both are functions of the specific membrane material. Flux determines the membrane surface area, and is a function of the pressure difference divided by the membrane thickness. A constant of proportionality that varies with the type of membrane is called the permeability. Selectivity is the ratio of the permeabilities of the gases to be separated. Due to the smaller size of the oxygen molecule, most membrane materials are more permeable to oxygen than to nitrogen. However, commonly available membrane materials do not have a high enough selectivity to produce medium or high purity oxygen, so that available membrane systems are currently limited to the production of enriched air (25 to 50% oxygen). Active or facilitated transport membranes, which incorporate an oxygen-complexing agent to increase oxygen selectivity, are a potential means to overcome this limitation of present membrane systems.

Figure 2 shows a flowsheet for production of enriched air. A major benefit of membrane separation is the simple, continuous nature of the process and operation at near ambient conditions. An air blower supplies enough head pressure to overcome pressure drop through the filters, membrane tubes and piping. Membrane materials are usually assembled into cylindrical modules that are manifolded together to provide the required production capacity. Oxygen permeates through a fiber (Hollow Fiber) or through sheets (Spiral-Wound) and is withdrawn as product. A vacuum pump maintains the pressure difference across the membrane and delivers oxygen at the required pressure. Carbon dioxide and water usually appear in the enriched air product, since they are more permeable than oxygen for most membrane materials.

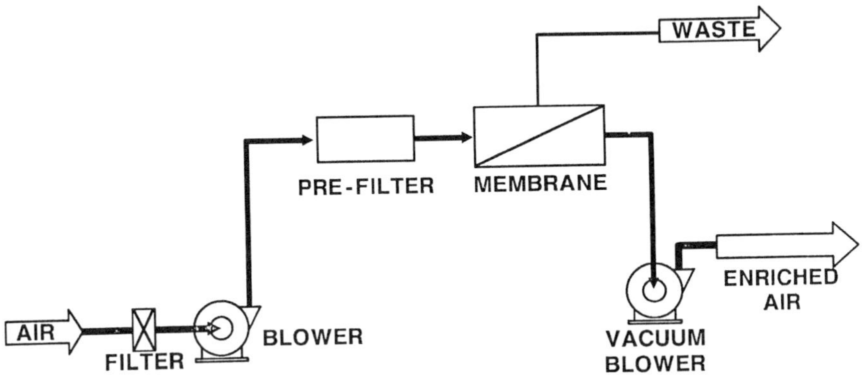

Figure 2 - Membrane Based Process

As with adsorption systems, capital is essentially a linear function of production rate and product backup is not available without a separate liquid oxygen storage tank. Membrane systems will probably fit applications up to 20 tons per day, where air enrichment purities with water and carbon dioxide contaminants can be tolerated. This technology is relatively new, and continued improvements could make membranes attractive for small oxygen requirements that exhibit discontinuous usage patterns.

Chemical

Air Products and Chemicals is currently developing a continuous, liquid phase absorption/desorption chemical air separation process that significantly reduces electric energy consumption compared to cryogenic processes. The MOLTOX™ air separation technology was invented by D. C. Erickson, Energy Concepts Co., Annapolis, MD. Air Products has acquired the exclusive worldwide rights in an agreement with Energy Concepts Co. and began operation of a 1/4 ton per day pilot plant, sponsored in part by the U. S. Department of Energy, in March 1986.

1. PSA - (PRESSURE SWING ABSORPTION)

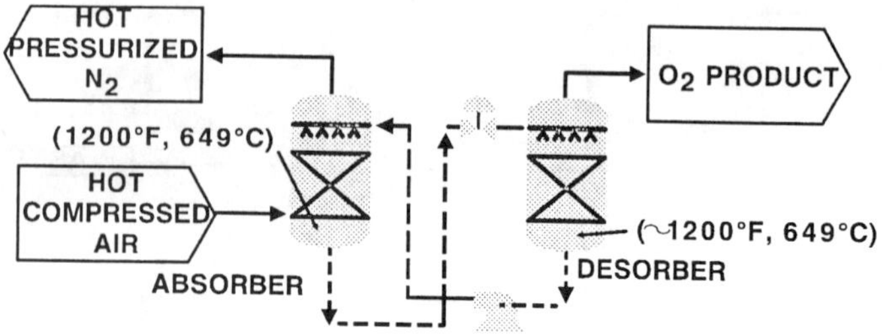

2. TSA - (TEMPERATURE SWING ABSORPTION)

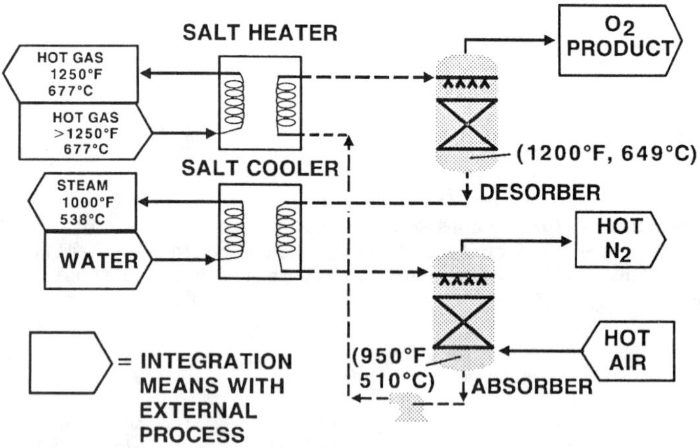

Figure 3 - Chemical Based Process

The two basic MOLTOX oxygen systems, PSA and TSA, are shown in Figure 3. Actual design and operation can be a combination of these processes. Dry, carbon dioxide free air enters the absorber at a temperature of 900 to 1200°F and a pressure of 20 to 186 psia where it contacts molten liquid salt. Oxygen in the air reacts chemically with the salt and is removed with the liquid salt leaving the bottom of the absorber. The oxygen bearing salt flows to the desorber vessel where it is heated and/or reduced in pressure, generating gaseous oxygen. Lean salt is cooled before returning to the absorber to complete the loop.

The MOLTOX systems' major advantage is that the chemical reaction does not consume the compression energy of nitrogen, which is the major portion of the air stream. In the PSA mode, nitrogen leaving the absorber can be expanded to recover its energy; in the TSA mode air/nitrogen is compressed only high enough to overcome the pressure drop through heat exchange and absorption equipment. The TSA process is especially applicable to oxygen users who have waste heat sources, low value fuel streams or boilers that can be integrated with the MOLTOX system.

Pilot plant operation has demonstrated the closed loop, continuous process of absorbing oxygen from air and producing a very high purity oxygen product (99.9%). The next phase of development consists of further laboratory and pilot plant testing to support design and construction of a 5 to 50 ton per day semiworks unit to be located at an oxygen user's facility. Full sized commercial plants are planned for the 1990's, and will be best suited for large oxygen requirements such as 500 tons per day and larger where the operating efficiency offsets the extra capital cost associated with the equipment.

Cryogenics

Cryogenic air separation technology is used to supply nearly all the oxygen consumed by industry today. The process separates the constituents of air by cryogenic distillation. Refrigeration to overcome heat leak and to supply reflux to the distillation columns is obtained by expanding a portion of the compressed gasses. Figure 4 is the flowsheet for a traditional, dual column, low pressure cryogenic process.

Air is compressed to about 90 psia and enters the molecular sieve pretreatment system where water, carbon dioxide and other atmospheric contaminants are removed. The air is cooled to cryogenic temperatures against product and waste streams in the main heat exchanger. Air enters the bottom of the high pressure distillation column where the preliminary separation of a portion of the nitrogen is accomplished. Nitrogen leaving the top of this column is partially warmed against incoming air and then expanded to provide the required refrigeration. A crude oxygen liquid leaves the bottom of the high pressure column, is subcooled and then fed to the low pressure column. The residual nitrogen is removed from the top of this column, mixed with the expanded nitrogen from the high pressure column and warmed against incoming air. This stream is high in purity and is compressed to the required pressure if needed as a coproduct, or vented to the atmosphere. Back in the low pressure column, oxygen and argon purities increase in the lower sections of the column. A sidestream can be withdrawn and sent to a separate distillation column to produce a crude argon product. Residual oxygen from this argon column is returned to the low pressure column. The high and low pressure distillation columns are thermally linked by a reboiler-condenser. High pressure nitrogen gas condenses by boiling liquified low pressure oxygen. The resulting liquefied nitrogen provides reflux to both distillation columns. A portion of the vaporized oxygen flows upward in the low pressure column and the remainder is removed as product, warmed against incoming air and compressed to the required pressure.

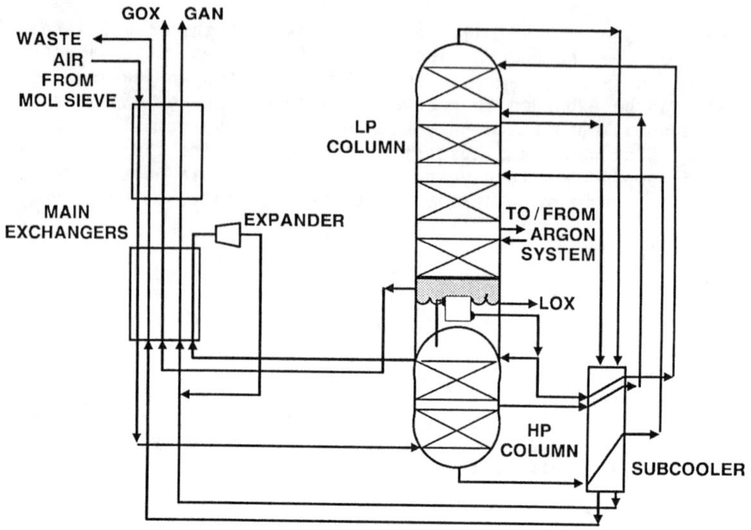

Figure 4 - Cryogenic Based Process

Optimizing the design of the cryogenic process involves the tradeoff between increased equipment size to reduce power consumption and the resulting increase in capital cost. Air Products has developed new process concepts that increase energy efficiency by up to 10%, increase argon recovery by 10% and reduce feed air pressures by nearly 30% compared to the traditional dual column process (4). For oxygen users requiring purities below 96% a new dual reboiler process has been developed that reduces power consumption by up to 10%.

An advantage of cryogenic systems is their ability to produce nitrogen and argon byproducts as well as liquefy a portion of the total output. The liquid is stored in vacuum insulated tanks where it is readily available for use during peak flow requirements, to serve as backup during outages, or for merchant sales. Operation of a cryogenic plant is not as flexible as an adsorption unit for oxygen users who need product for only a portion of the time, shutting the unit down over weekends or at night. Depending on the length of time a cryogenic plant has been shutdown it can take several minutes to several hours to restart, cool down and reach purity. If production schedules are well known, restart times may not adversely effect operations and liquid oxygen is available as backup; however, the unproductive power cost to restart the unit must also be considered in determining the actual cost of supplying oxygen.

Computer Control

Nearly every process or piece of equipment built today is operated via computer control, and many older facilities have been retrofitted to permit computer controls. The four technologies described in this paper can all benefit by application of computer controls. Cryogenic air separation plants are good candidates for process control computers, since the major production cost is electric power for driving the compression equipment. A small savings in power consumption will offset the capital investment necessary to add a computer system, even when performed on a retrofit basis where pneumatic instrumentation must be replaced with computer compatible instruments. Most air separation plants are operating to supply an online customer who has a variable product demand. A computer system can monitor this demand and quickly adjust production accordingly. Air Products has pioneered this technology and today, over 100 plants that Air Products owns and operates have computer systems. Additionally, this technology has been applied to plants that are owned and operated by others.

A computer control system consists of two parts, a baseline control system and an optimizing host computer. The baseline control system provides basic monitoring and regulatory control. This system can take one of two forms: 1) A panel-based system consisting of either pneumatic or electronic instrumentation; or 2) a digital control system which uses microprocessor-based controllers and CRT operator interface units (Figures 5 and 6). The host computer consists of a processing unit and associated peripherals. The optimization applications programs reside in the host computer.

The host computer provides advanced process monitoring, control and optimization. The basis of the monitoring package is a resident database of information. Efficiency calculations such as specific powers and recoveries are performed on-line. Plant logs are automatically generated which contain integrated plant production, consumption, performance and efficiency information. In a panel-based baseline system, the critical loops are actually controlled by the host computer in direct digital control (DDC). In a DCS baseline system, DDC is used for difficult loops, but most of the loops are in supervisory mode where the host computer calculates a new setpoint for each DCS controller.

The primary justification for a host computer is process optimization. Optimization can take several forms. Steady-state optimization is used to calculate the controller setpoints that will minimize the production cost for a given production rate, subject to process and machinery constraints. Load-following optimization is used to change the plant production rate in response to varying customer demand. Dynamic on/off control is used to assist in the start-up or shutdown of a plant.

The objective of steady-state optimization is to simultaneously solve the heat and material balances to yield a given production at minimum cost. Extensive plant testing aided by process simulation is used to establish correlations to predict recovery as a function of several parameters. These correlations are continuously updated via purity feedback to create a dynamic correlation. Air Products has consistently demonstrated a better than five percent reduction in power consumption for a given production rate or five percent increase in

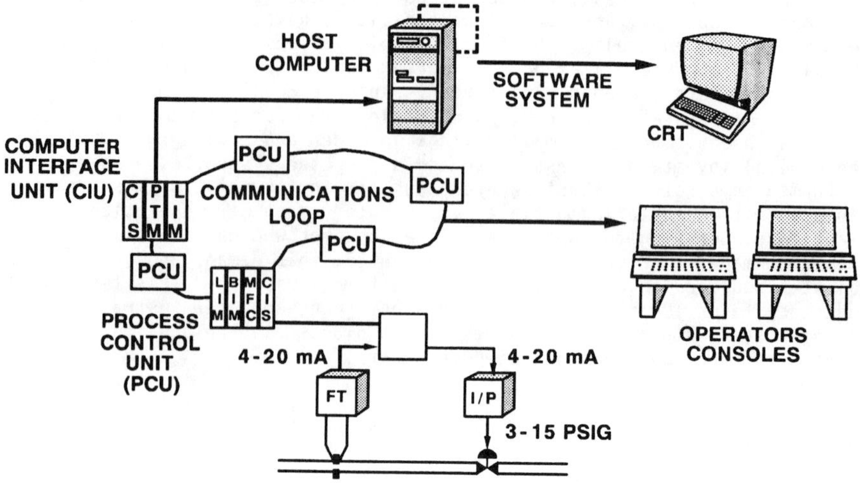

Figure 5 - Supervisory Control With Digital Control System

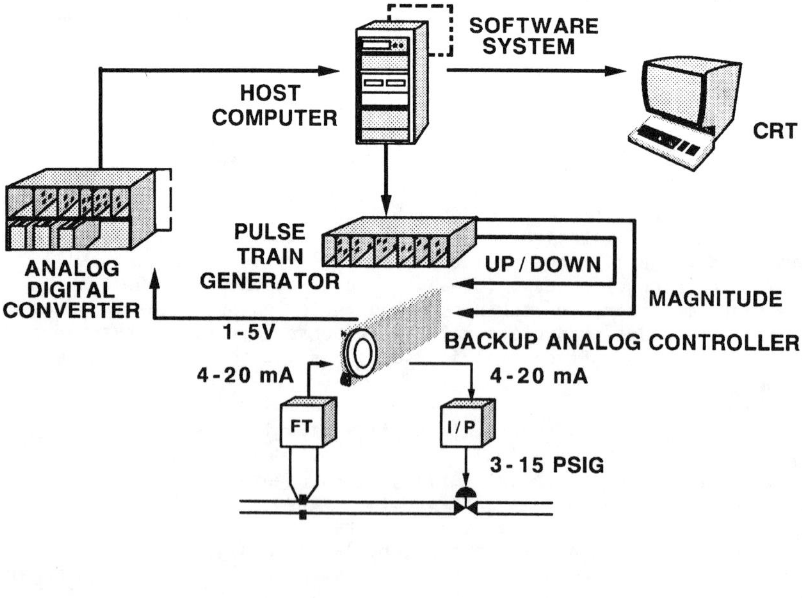

Figure 6 - Direct Digital Control

production for the same power consumption when steady-state optimization is implemented as compared to manual operation. These savings result from mass flow control, cascade purity control, and operating closer to the constraints. The conservatism that is associated with manual operation is eliminated.

As an example, in 1985 Air Products installed an optimizing computer system on an oxygen plant supplying a copper smelter in Arizona. This plant had a DCS baseline control system and 8 percent steady-state savings were documented above the DCS control system performance. Figure 7 is a plot of specific power, defined as power consumption divided by oxygen production, versus production for both the manually controlled and computer controlled cases at the Arizona plant.

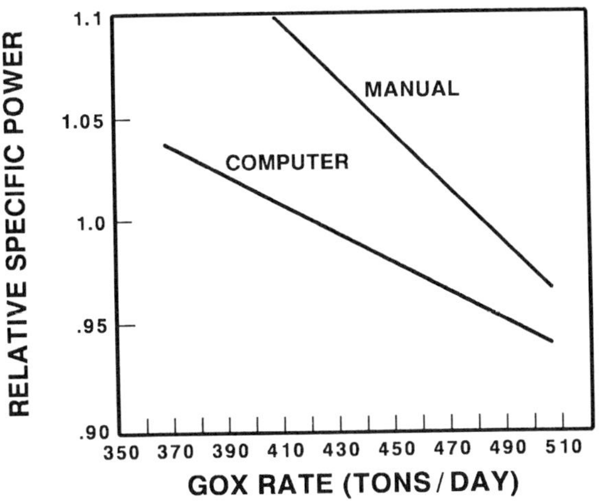

Figure 7 - Optimization Benefits (Computer vs. Manual)

The objective of load-following optimization programs is to automatically match the plant production to the gas demand, subject to process and equipment limitations. The program can be based upon a feedback system, where a process parameter such as pipeline pressure is used to determine the required production rate. This is sufficient for systems where the demand changes slowly, such as the need for oxygen in a wastewater treatment plant. For systems that have much more severe demand changes, a feedforward mechanism is needed. The oxygen needs of many metals production facilities are more variable due to furnace burners being taken in and out of service and due to converters rolling in and out. The flows at the oxygen consumption points were measured on the Arizona smelter so that a feedforward algorithm could be used to determine the oxygen production rate. The performance of a load-following optimization system is dependent upon the maximum ramping rate of the air separation plant, the severity of demand changes, and the capacitance of the system. The objective is to minimize over-production that is vented, while avoiding supplemental liquid vaporization that makes up for production deficiencies.

Computer-assisted plant start-up and shutdown is an example of an application of dynamic on/off control. A system to accomplish this was implemented by Air Products at an oxygen plant owned and operated by the City of Houston (5). Due to electric utility economics, it was advantageous to shutdown the plant daily to avoid on-peak power costs and to be an interruptible electric power consumer. A propane fired boiler was installed in the vaporization system to allow maximum power to be shed while interrupted. Additional valves were automated, and a software package to assist the operator in starting-up and shutting down the plant was written. The operator uses the CRT to monitor the start-up and to enter information into the computer after certain tasks are completed. Using a computer in this application has several advantages. It allows one operator to start-up or shutdown the plant expediently, which minimizes the unproductive power associated with a plant start-up. The increased risk of process equipment failure associated with frequent start-up and shutdowns is minimized.

Once a network of optimized plants is established, the benefits of linking the plant computers to a central location is evident. In 1981 Air Products initiated a program to link all of its plant computers to a central computer in Allentown, PA, where the central engineering and operating departments are located (6). The central computer initiates a daily phone call to each of the sites to retrieve integrated and instantaneous production, consumption, and efficiency information. This data is stored for on-line ad hoc inquiries and for generating exception reports. Trends in data are observed for each plant and degradation in equipment performance is quickly brought to the attention of line management and the plant efficiency engineer. This same system is offered to customers who purchase optimization systems for their own plants.

Implementation of computer monitoring, control and optimization improves operating plant efficiency between 5 and 15 percent, typically resulting in less than a two year payback. Techniques such as linear programming and dynamic optimization will produce additional energy efficiency benefits, while artificial intelligence and similar packages have potential to reduce the programming effort and further improve the profitablity of computer projects. Knowledge gained in computerizing cryogenic air separation plants can also be applied to new oxygen production technologies to increase the efficiency and ultimately reduce the cost of producing oxygen.

Technology Comparison

Oxygen cost is composed of three major parts: power consumption, capital cost and operating costs. Power consumption is the largest portion of total oxygen cost in larger plants, but decreases as plant size decreases. Recovery is one variable that helps to explain the power consumption differences between processes. Recovery is the amount of oxygen produced divided by the amount of oxygen in the feed air stream. It is a relative measure of efficiency and also controls equipment size and cost to a certain extent. Cryogenic processes and the MOLTOX system have the highest recoveries, typically in the mid to high ninety percent range. Adsorption systems recover oxygen at around forty percent, while membranes vary from forty to eighty percent depending on product purity.

Operating pressure is the other variable that combined with recovery determines the overall power consumption of a process. Figure 8 shows the relative power consumption of the four technologies as a function of oxygen purity. The dashed lines show the effect of blending air with high purity oxygen to allow all processes to be compared on an equal purity basis. Liquid oxygen has not been addressed separately since it is a form of cryogenic production, but has been included for comparison purposes. The transportability and the existing network to supply liquid oxygen to industrial customers brings the economies of scale of large plants to small users. Its flexibility in terms of variable flowrate, pressure and instantaneous supply are major advantages.

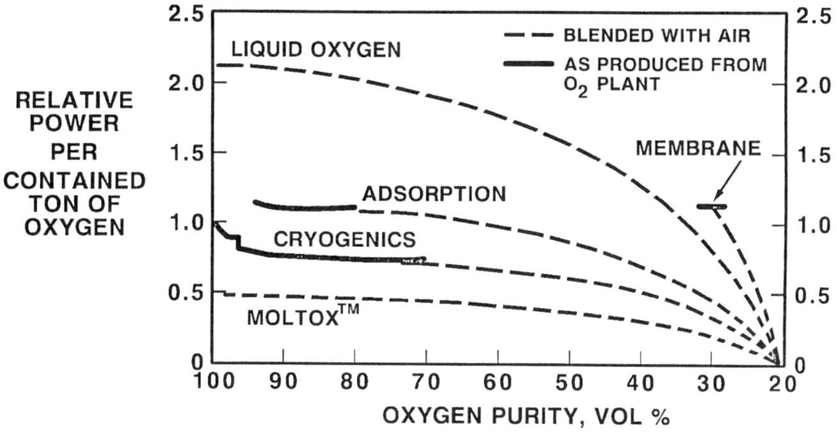

Figure 8 - Oxygen Power Comparison at 0 PSIG

In addition to power cost, operating criteria vary for the different processes as shown in Table I. This table compares the four technologies and liquid oxygen based on the following categories:

Status is the degree to which the technology has been commercialized and varies from mature for cryogenics, through developing for membranes, to pilot plant for chemical methods.

Economic Range is the typical production range where the technology is economically viable. As expected, the less efficient but less complex technologies are favored at the lower production rates; while the more power efficient but more complex technologies take over above several hundred tons per day.

Specific Power Ranking lists the technologies in order of power consumption from lowest (least costly to operate) to highest.

Table I. Characteristics of Oxygen Technologies

	OXYGEN PRODUCTION TECHNOLOGY				
	ADSORPTION	CHEMICAL	CRYOGENIC	LOX	MEMBRANES
STATUS	SEMI-MATURE	PILOT PLT	MATURE	MATURE	DEVELOPING
ECONOMIC RANGE	<100T/D	>500T/D	>20T/D	<50T/D	<20T/D
SPECIFIC POWER RANKING (1 = LOWEST)	3	1	2	5	4
PURITY LIMIT	95%	99.9%+	99.8%	99.5%+	ca. 40%
STORAGE/BACKUP	ADD-ON OR HAUL IN	ADD-ON OR HAUL IN	STD	STD	ADD-ON OR HAUL IN
STARTUP TIME	MINUTES	HOURS	HOURS	MINUTES	MINUTES
FLEXIBLE PRODUCTION RANKING (1 = HIGHEST)	3	5	4	1	2

Purity Limit is the maximum purity that can be economically produced by the technology. All the technologies can provide lower than maximum purity by changes in cycle design or by blending air with the higher purity product stream.

Storage/Backup is the ability of the technology to provide continuous product flow in the event of a plant shutdown. Liquid oxygen is readily made and stored in cryogenic cycles, providing any level of backup from manual start to automated, instantaneous flow. The other technologies require liquid oxygen to be hauled to the site by an industrial gas company or require the addition of small cryogenic liquefiers.

Start-up Time is the period required to restart the process and reach specified purity after a shutdown. Again, the non-cryogenic processes have shorter restart times than the more complex processes.

Flexible Production Ranking is a measure of how well the technology can perform at off design rates. Liquid oxygen is the most flexible with nearly infinite variation in flow. As the processes become more complex, they typically become less flexible in operating range.

Conclusion

Figure 9 shows the size and purity ranges where the technologies are typically economically viable. All the processes will continue to improve in the future, however some shifts in the ranges in Figure 9 are expected. Although membrane systems currently have limited uses, future development could increase their range of application. As with all new processes, there is more potential for improvement compared to more mature technologies. Adsorption systems have been applied in some metals production applications, and may benefit from future improvements is adsorbent materials research. Cryogenics, often considered mature technology, has improved substantially over the past years in terms of power consumption and argon recovery. A number of companies have found it profitable to replace older, smaller oxygen plants with newer, more efficient plants with enhanced argon recovery and operating characteristics. Chemical air separation processes are expected to compete with cryogenics in the 1990's, based on the extremely low power consumption and the ability to be integrated with high temperature oxygen using processes.

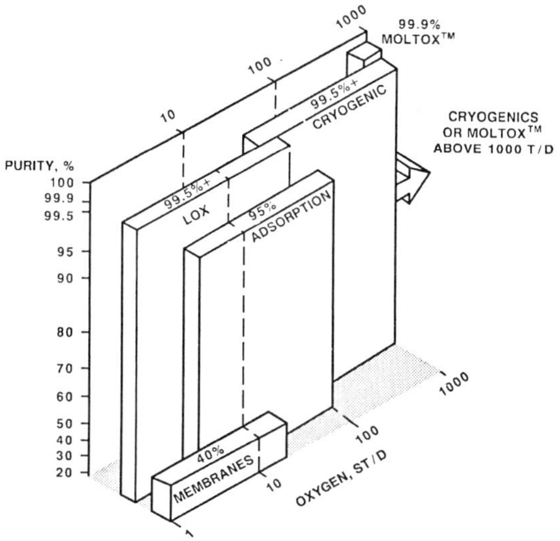

Figure 9 - Typical Application Ranges

As discussed in this paper, the cost of oxygen is dependent on the following key parameters:

1. Size
2. Purity
3. Nitrogen and/or Argon Requirements
4. Pressure
5. Pattern of Use
6. Storage and Backup Requirements

Each of the technologies has characteristics that makes it amenable to certain applications. It is important for the oxygen user to understand the cost impact of each parameter as well as the limits for each oxygen technology.

References

1. Anon. "Top 50 Chemical Production." Chemical and Engineering News. April 11, 1988.

2. S. V. Schaedel. "Low-Cost Oxygen Breathes Life Into Industrial Furnace." GRID, 8, 10-17, 42,43.

3. J. Paffenbarger. "Review of Oxygen Production Technologies for IGCC Power Plants." EPRI Report: RP8005-5.

4. K. B. Wilson, D. W. Woodward and D. C. Erickson. "New, Low-Energy Processes for Cryogenic Air Separation." (International Cryogenic Engineering Conference). July 1988. Southampton, U.K.

5. S. L. Russek, T. M. Beckowski and D. R. Vinson. "Computer Automated Start-Up and Shutdown of an Air Separation Plant." (1985 Industrial Energy Technology Conference). Houston, TX.

6. N. Chatterjee, D. J. Hersh and J. R. Couch . "Total Energy Management With Mini-Computer Systems." (Eighth Energy Technology Conference. Washington, D.C. 1981).

RECENT "ADVANCES" IN SULFURIC ACID TECHNOLOGY-REVIEW

AND ANALYSIS

Leonard J. Friedman

Acid Engineering & Consulting, Inc.
4619 Kings Point Court
Lakeland, FL 33813

Abstract

There have been considerable changes in systems, equipment design and materials used in the modern Sulfuric Acid process. Many of these will shape the metallurgical sulfuric acid plant for the year 2000 and beyond. Advances have come from all over the world, and include new scrubbing area designs, catalyst contacting devices, silicon stainless alloys for unlined acid systems, energy recovery from acid heat, stainless converters and heat exchangers, etc. This paper describes each of the process "advances" and presents a review and analysis indicating advantages, operating experience and limitations.

Metallurgical Processes for the Year 2000 and Beyond
Edited by H.Y. Sohn and E.S. Geskin
The Minerals, Metals & Materials Society, 1988

PROCESSING OF COMPLEX SULPHIDES

J. M. Figueiredo*, M. C. Coelho*, A. R. Silva* and R. M. Guedes**

* - LNETI, Department of Materials Technology
Azinhaga dos Lameiros, 1600 Lisbon, Portugal
** - QUIMIGAL, Development Division
Av. Infante Santo, 2, 1300 Lisbon, Portugal

Abstract

With the continuous depletion of conventional sources of non ferrous metals, low grade complex sulphide deposits have been regarded as an alternative source of supply of zinc, copper, lead and also silver and gold. In spite of very important efforts in research in this domain it is still a long way to go in order to achieve reliable technological solutions both in ore dressing and metallurgical treatments which enable these complex materials to be regarded as commodities. The RLE process to treat bulk concentrates such as those obtained from portuguese ores, under development at LNETI and QUIMIGAL premises, combines experienced unit operations and innovative configurations and is an effective contribution for the future use of these materials. The results of three years of experimental work and main conclusions on the overall feasibility of the process are presented.

Metallurgical Processes for the Year 2000 and Beyond
Edited by H.Y. Sohn and E.S. Geskin
The Minerals, Metals & Materials Society, 1988

Introduction

In spite of the very well known problems associated to the development of complex sulphide deposits, the economical importance of the enormous reserves of these materials, widespread all over the world, is so evident that its future use as source of supply of basic and precious metals and other minor elements is almost unavoidable.

Critical steps in that direction are mainly concerned with favourable economical conditions and sound metallurgical practice for the treatment of non ferrous metals concentrates, which are nowadays still considered as non conventional raw materials.

Since Portugal is one of the countries where these orebodies occur more extensively, Portuguese authorities have dedicated since the beginning of the seventies, a big effort on finding economical and reliable processes for the recovery of copper, lead, zinc and also gold and silver, and for this purpose, several alternatives on ore dressing and metallurgical processing were studied, however without full sucess.

Five years ago LNETI, a research laboratory, and QUIMIGAL, a chemical company, both state owned, decided to launch a research programme for the development of a metallurgical process for the treatment of NFM concentrates, which received financial support from NATO under its programme "Science for Stability".

The basic idea of this stepped programme was to develop a process whereby the simultaneous use of experienced technologies and new technical solutions could result in a reliable and economical way to treat such materials and also applicable to other similar products, namely dirty zinc concentrates, more and more frequent in industrial practice.

The main big challenges antecipated were: (i) to roast in a fluidized bed a very fine sulphide concentrate with a copper plus lead content of about 15%, (ii) to handle high copper concentrations in a zinc circuit and, (iii) to remove iron as jarosite in presence of calcine with very high levels of lead and silver.

The bulk concentrate used in the experimental work was expressly produced from local ores in an ore dressing pilot plant existing in Aljustrel, south of Portugal, and averaged the following composition:

Table I - Bulk concentrate composition

S(%)	Fe(%)	Cu(%)	Pb(%)	Zn(%)	Ag(g/t)	Au(g/t)	As(%)
36.0	23.0	8.5	6.0	20.0	140	0.5	0.3

With this work we are joining those trying to overcome several problems through research activities, and to make possible that in a near future this concentrate or similar ones would not be mistrustfully regarded.

Process Description

The simplified diagram which is shown in Figure 1, is mainly based on a roasting operation, and a hydrometallurgical circuit to treat the calcine in order to recover the main non ferrous metals mostly in upgraded forms.

The NFM concentrate is roasted in oxidizing conditions, at temperatures ranging 700 to 720°C, to produce calcine containing copper and zinc partially as sulphates, oxides and ferrites, in optimal proportions for further equilibrium of the balance of sulphates and sulphuric acid in the hydrometallurgical circuit.

This calcine is partially fed to a neutral leaching step where at 60-80°C and with a residence time of 1-1.5 hours it is possible to solubilize all copper and zinc in weak acid solube forms and simultaneously to precipitate iron as iron hydroxide at a pH of about 3.0 to avoid a co-precipitation of copper basic sulphates.

The unleached residues from the neutral leaching and preneutralization sections are fed at the HAL. This operation is conducted at 90 and 95°C, in two stages, with different levels of acidity and iron concentrations to get high dissolution rates for copper and zinc ferrites (> 90%), and to avoid lead and silver jarosite formation. The HAL residue, mainly constituted by lead sulphate, silver, hematite and gold, is solubilized in a brine leaching step at 85°C with control of redox potential for further recovery of gold and silver as a cement and lead as a metal.

The iron is removed as sodium jarosite in absence of calcine, which are only used in a preneutralization section, to neutralize the residual acidity from the HAL step as well as the acidity from part of the iron depleted solution which is also recycled back to this operation.

The clarified preneutralized solution containing 15 to 20 g/l of iron and an acidity not higher than 3 g/l is fed to iron precipitation step, with a residence time of 6 hours at 98°C. A small amount of the acid generated during jarosite precipitation is neutralized with milk of lime to get a final iron concentration less than 5 g/l.

The pregnant liquor outcoming from the neutral leaching operation, which is the main copper and zinc stream, is fed to a copper solvent extraction circuit to remove copper from 20 g/l down to less than 100 mg/l. This is achieved using a 22.5% ACORGA extractant in Escaid and an intermediate pH control step with milk of lime in order to get values in the range of 1.5 to 2.0. The loaded organic is stripped with a spent copper electrolyte and copper is then electrowon.

The raffinate, which is an impure electrolyte with 130 g/l of zinc, is neutralized to pH 4.5 with milk of lime and after a conventional purification step the zinc is electrowon and the spent zinc electrolyte recycled back to the neutral and hot acid leaching operations.

Process Development

In the course of the experimental development of the process diagram a stepwise approach was considered, including:

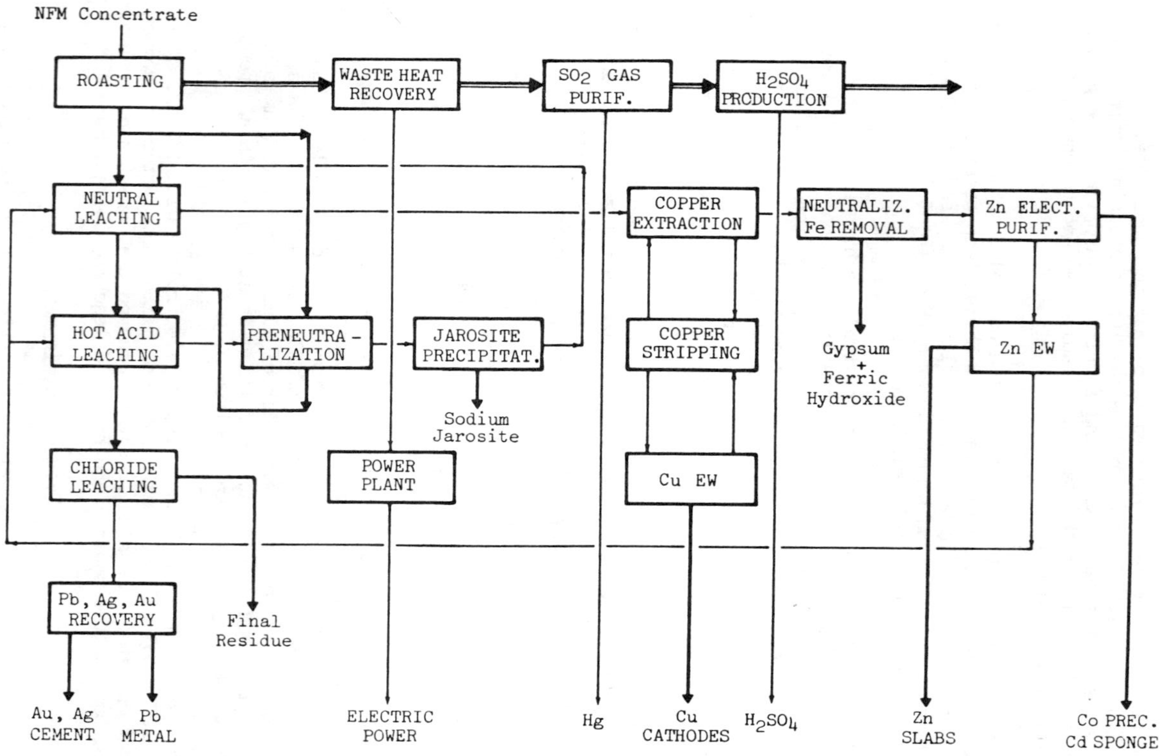

Figure 1 - Schematic diagram of LNETI - QUIMIGAL process.

- Separate experimental work for roasting, neutral leaching, hot acid leaching, iron removal, copper solvent extraction and brine leaching, according to preselected process inflow conditions;

- Optimization of overall treatment process diagram; and,

- Simulation of the process under open and close regimes.

Whilst the nature of the hydrometallurgical operations, namely their reliability, reproductiveness of their results and the possibility of utilization of wide multiplicative factors made possible to limit to a laboratorial continuous pilot set-up the work to be done, the roasting tests are also to be performed in a semi-industrial roasting plant with a nominal capacity of 2.4 tpd of concentrate, already assembled at QUIMIGAL premises.

The remaining operations, namely the waste heat recovery, SO2 conversion, zinc and copper electrowinnings and zinc electrolyte purification are very common in industrial practice and for the purpose of this project, were considered to be set in the standard conditions of their normal application, and no experimental work was done on them.

Experimental Equipment

Roasting pilot plant

The equipment was essentially constituted by a 3m tall jacketed fluidized bed roaster in 1.4742 steel tube with 15 cm of id. The bed could be discharged by a grate level underflow with a star valve and the carryover was collected into two cyclones in series. The dedusted gases were cooled and washed in a vertical tower prior to be fanned to the stack. The concentrate was loaded in a hopper and fed to the roaster by a screw feeder at an average feed rate of 4 kg/h. Monitoring and control of the plant was made via 11 thermocouples, 3 pressure sensors, 1 flow sensor and 1 pH sensor associated to the respective controllers. The data were automatically registered in a data logger.

In figure 2 is shown the process flowsheet for this operation.

Hydro pilot plant

This pilot plant was flexible enough to be assembled with different configurations according to the unit operations to be studied. The major components were mechanically agitated FGRP reactors with variable capacities up to 10 liters, peristaltic pumps automatically controlled via current or voltage signals, thickners, screw feeders for solids, filters, and silica red rods for heating. The basic module for control and monitoring included 4 temperature controllers, 3 pH and 2 redox potential controllers with respective transmitters and sensors and a data logger system for automatic registration of data for further scaling-up purposes.

The equipment for copper solvent extraction included mainly 2 water jackted batteries, with 5 mixer-settlers each, in transparent PVC. The capacities of mixers (pump-mix type) and settlers were 0.5 and 1.7 liters respectively allowing a total flow rate of 10 l/h approximately for a residence time of 3 minutes in the mixers. These batteries were also

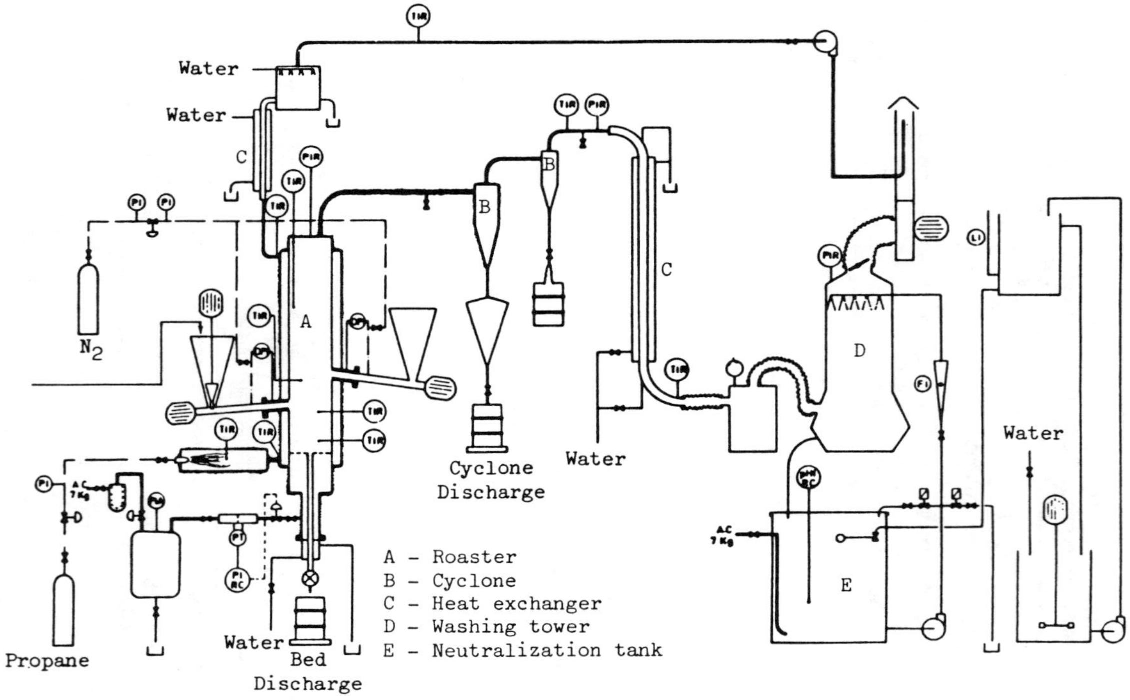

Figure 2 - Process flowsheet for roasting

provided with internal recirculation of the aqueous phase and level devices to control the height of interphases in the settlers.

Results and Discussion

In Tables II and III are shown the average feed compositions for solids and solutions corresponding to the unit operations which were tested in a continuous way, but separately.

It has to be stressed that the composition of the bulk concentrate used in tests is somewhat far from the expected one since the production of 9 tonnes of concentrate in a 1.5 t/h ore dressing pilot plant did not enable a fully representative composition.

Therefore some corrections had to be done in order to obtain more realistic calcine for subsquent hydrometallurgical testing.

Table II- Average Composition of Solids Feed in Different Unit Operations

Element (%)	Roasting	Neutral* Leaching	Hot Acid Leaching	Brine Leaching	Final Residue
Zn	20.0	30.8	14.0	9.0	1.2
Cu	8.5	4.9	4.3	0.23	0.14
Pb	6.0	3.4	6.6	24.5	0.12
Ag	140(g/t)	85(g/t)	150(g/t)	507(g/t)	36(g/t)
Fe	23.0	23.0	34.3	25.4	37.7
S	36.0	5.4	4.5		
S(SO4=)			1.7		

* Blended with calcine obtained from roasting of an impure zinc concentrate to get a composition closer to the expected one.

Table III- Average Composition of Aqueous Feeds in Different Unit Operations

Elements (g/l)	Neutral Leaching	HAL	Brine Leaching	Jarosite Precipitation	Copper Solvent Extraction	Neutraliz. and Iron Removal
Zn	110-115	50	-	70	119	119
Cu	16-17	-	-	2.0	20	60(mg/l)
Fe	2.5-3.0	-	-	30.0	0.9	0.9
H2SO4	32	150	-	3	pH=3	6
Cl	-	0.1	170	-	-	-
Ca	-	-	36	-	-	-
Na	-	-	71	4.5	-	-

Roasting

After the commissioning stage made with pyritic and zinc concentrates, and some preliminar orientative testwork with bulk concentrate , several weektests were performed in a total amount of 650 effective hours, during which 3 tonnes of bulk concentrate were fed into the roaster.

Temperature and excess air were varied respectively between 700 and 720°C and 18 and 46%, causing air velocity to range 0.5 - 0.6 m/s.

Bed height was maintained in most tests in the range of 40 - 50 cm but sometimes reached 60 - 70 cm.

Average particle size in the bed was of 641 ± 123μ and in cyclone discharge 44.3 ± 7.9 μ.

From the point of view of leachibility the cinders were sistematically analysed not only for metal contents but also for their water and acid soluble forms, as it is shown in Table IV where some average results are presented.

Table IV - Average Results on Roasting

Temperature (°C)	Excess Air (%)	Water Soluble* (%)				Acid Soluble* (%)				Sulphide Content (%)	
		Cyclone		Bed		Cyclone		Bed		Cyclone	Bed
		Cu	Zn	Cu	Zn	Cu	Zn	Cu	Zn		
700	24	33	36	38	32	89	89	70	73	2.8	0.4
700	33	52	49	51	43	90	86	77	80	2.9	0.5
710	24	42	42	40	34	87	90	74	76	1.8	0.4
720	22	21	30	25	27	95	89	84	73	1.2	0.5
720	18	38	38	34	36	87	89	76	76	1.3	0.6

* Copper and zinc in water soluble and acid soluble forms must be understood as the percentage of these elements which are solubilized respectively with a 0.025N H_2SO_4 solution at room temperature and with a 60 g/l H_2SO_4 solution at 90°C during 1 hour.

Excess air of about 18 - 20% at a temperature of 720°C were the most favourable conditions, in which the best compromise was reached between residual sulphide, sulphate sulphur and acid solubility of non ferrous metals.

Residual sulphide could probably be much more reduced if the furnace would have had an expanded freeboard.

Neutral Leaching

The main handicap for this operation when incorporated in a treatment diagram for bulk concentrates with high copper content is the necessity of using lower final pH values in order to minimize the copper precipitation as copper basic sulphate with potential negative consequences on both final iron concentrations and kinetics of copper solvent extraction.

The tests were performed under different configurations but the most extensively worked out is presented in Figure 3.

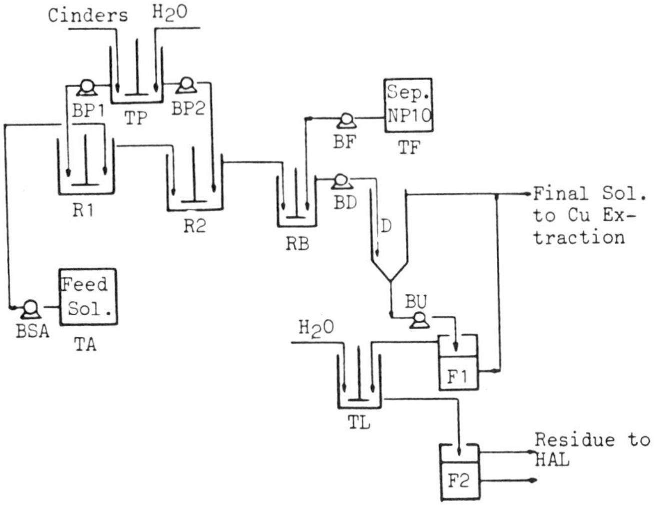

Figure 3 - Process flowsheet for neutral leaching

Some of the average results concerning copper and iron precipitation and zinc solubilization yields are shown in Table V for different levels of pH in reactors R1 and R2, assuming fixed operating conditions for temperature (60°C) and residence time (0.5h in each reactor).

Table V - Average Results on Neutral Leaching

pH		Precipitation yields (%) Fe		Cu*		Leaching yields (%)	Final concentrations (g/l)			
R1	R2	R1	R2	R1	R2	Zn	Zn	Cu	Fe	As
1.5	3.0	0	53.1	0	6.5	71.2	119.0	17.1	0.82	≈5mg/l
2.0	3.5	8.6	65.7	0	9.4	73.0	117.4	16.3	0.60	"
2.0	4.2	35.6	73.1	1.4	19.3	73.9	117.1	15.3	0.61	"
3.0	4.3	60.9	82.0	10.0	28.6	73.2	115.4	12.0	0.41	"

* Calculated on basis of copper fed in initial solution plus copper solubilized.

These data point out very clearly to a strong influence of pH on copper precipitation and consequently an upper limit of about 3.0 has to be fixed.

Despite the final iron concentrations are according to the target values, those were negatively affected by the presence of some iron (II) resulting from the presence of non roasted sulphides in the calcine.

The behaviour of zinc and arsenic was quite normal and according to the expected figures.

Hot Acid Leaching

These tests were carried out in two stages according to the flowsheet shown in Figure 4, working countercurrently, each one with two reactors in series. The two main purposes were to obtain very high leaching yields for zinc and copper ferrites and to keep lead and silver jarosite formation at a very low level.

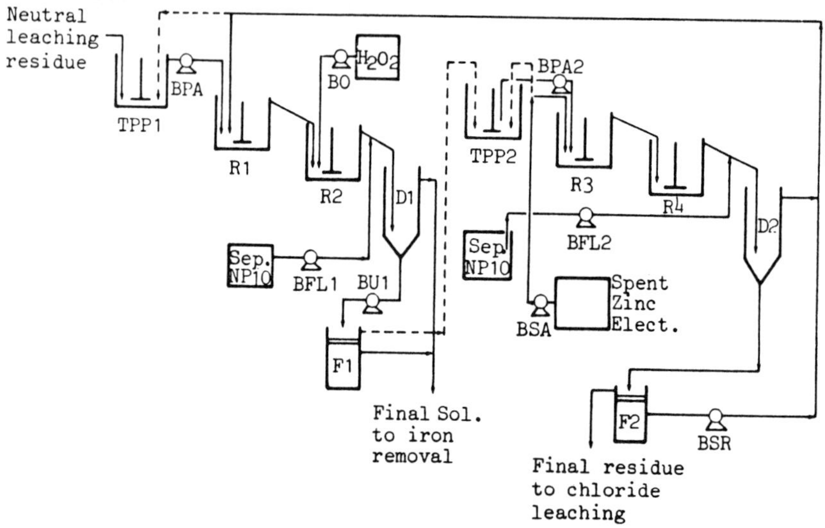

Figure 4 - Process flowsheet for HAL

In Table VI are presented the average results corresponding to the following set of operating conditions:

- Temperature - 90° C (R1 and R2)
  - 95° C (R3 and R4)
- Residence time - 6 h (R1 + R2)
  - 3 h (R3 + R4)
- Final adicity - 40 g/l (H2SO4)

Table VI - Average Results on Hot Acid Leaching

Element	Solution Composition (g/l)		Leaching yields (%)				Decomposition* yields (%)	
	R2	R4	R1	R2	R3	R4	R2	R4
Zn	66.5	53.3	66.1	79.5	85.9	85.4	-	-
Cu	5.2	0.4	75.5	89.4	97.7	98.4	-	-
Fe	31.4	4.2	58.3	67.8	80.6	82.0	-	-
Pb	-	-	-	-	-	-	88.0	99.8
Ag	0	-	-	-	-	0	56.5	98.1
H2SO4	49.3	145.3	-	-	-	-	-	-

* Decomposition yields for lead and silver were determined through a brine leaching (180 g/l NaCl and 110 g/l $CaCl_2$) at 85° C during 2 h. The levels of acidity and redox potential were kept constant and equivalent to a pH 1.5 and 700mV respectively.

From the results it can be concluded that, in these operating conditions, very high leaching yields for copper and zinc ferrites can be achieved keeping almost all lead and silver in their chloride solube forms.

The lower figures for zinc are due to the presence in the feed material of an important amount of zinc sulphide, which was proved and quantified through chemical testing in the HAL residue. If the zinc leaching yields were recalculated with basis only in the zinc ferrites content in the feed, 95% instead of 85.4% is reached.

The presence of as little as 100 mg/l of chloride ions in solution completely inhibits silver solubilization.

Iron Removal

Precipitation of jarosite using calcine as neutralizing agent is a very well known operation. In case of calcine resulting from the roasting a bulk concentrate serious losses of silver and lead would occur if such materials were to be used in the conventional way.

Several alternatives for iron removal in absence of calcine and other neutralizing agents were investigated.

In Figure 5 is shown one of the tested process flowsheets including a preneutralization step (RN reactor) conducted at 90°C during 0.5 h, and the iron precipitation step conducted at 98°C during 6 hours (R1, R2, R3, R4 reactors).

Main potential problems were: (i) to neutralize with calcine the acidity down to 3 g/l at 90°C without jarosite precipitation; and, (ii) to remove iron from 30 g/l to 5 g/l or less, minimizing the use of lime slurry to keep the acid balance close to equilibrium.

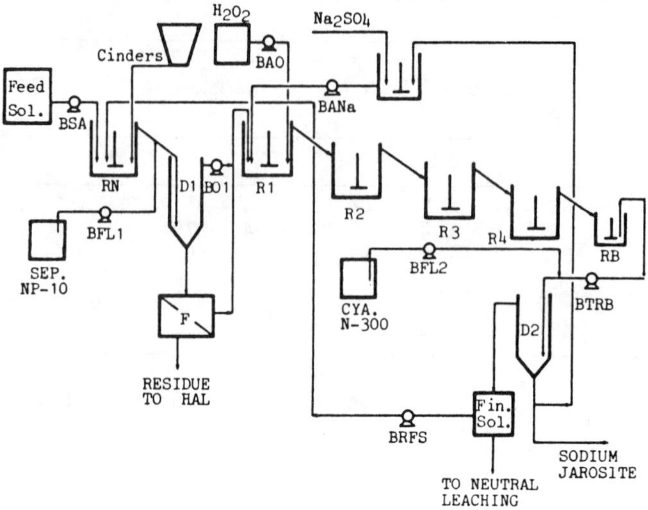

Figure 5 - Process flowsheet for iron removal

Average results with and without lime slurry addition are shown in Table VII.

Table VII - Average results on Iron Removal

Element (g/l)	Lime Addition	Feed Solution (g/l)	RN	R1	R2	R3	R4
Fe	No	30	20.3	13.4	11.2	10.3	9.1
H2SO4		40	2.1	15.9	19.5	22.4	25.8
Fe	Yes	30	16.9	10.4	8.2	5.3	4.5
H2SO4		40	3.5	13.8	16.5	11.0	15.0

In both cases part of the final solution was recycled back to the preneutralization step in a ratio 1:1 regarding feed solution, aiming to dilute the initial iron concentration from 30 to 20 g/l or less, and to neutralize with calcine a surplus amount of acid generated in jarosite precipitation.

Besides the existence of other alternatives for iron removal in absence of neutralizing agents, this one is very promising since the target values for iron concentration in the final solution (less than 5 g/l) was obtained using lime addition corresponding to only 30% of the acid generated.

The levels of jarosite precipitation in the preneutralization step were non significant even in presence of a sodium concentration of 1 to 1.5 g/l, and so the residue can be recycled back to HAL for further leaching without any special inconvenient.

Chloride Leaching

Additionally to the batch chloride tests for determination of the levels of silver and lead jarosite formation in different stages of HAL, continuous test runs were also performed with the bulk of tha HAL residue according to the flowsheet shown in Figure 6.

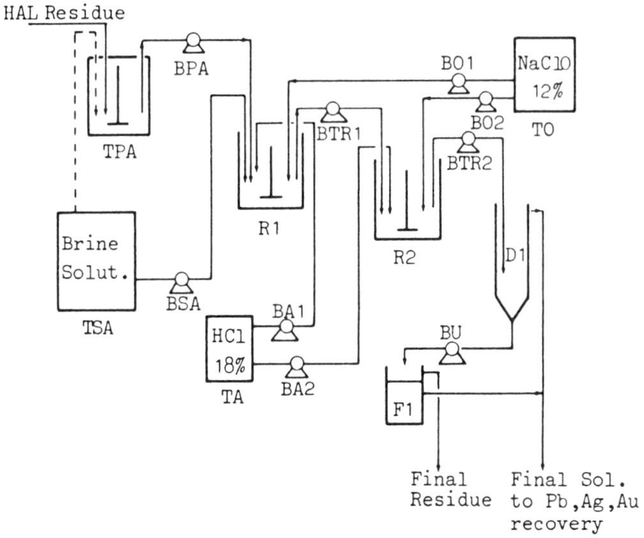

Figure 6 - Process flowsheet for chloride leaching

This configuration was mainly based on two leaching reactors R1 and R2, working cocurrently corresponding to a total residence time of 1.5 hours. Parameters like pH, redox potential were also mesured and controlled, reaching values of respectively 1.5 and 300 or 700 mV in reactor R1 and 2.0 and 1020 mV in reactor R2. Temperature was kept constant at 85°C.

In Table VIII are shown some average results concerning lead, silver and gold solubilization. The presence of zinc sulphide in the feed is clearly identified by the very high zinc leaching yields reached at high levels of redox potential.

Leaching yields higher than 99 and 94% can be easily reached respectively for lead and silver, which means that their jarosite forms are pratically absent in the feed material. Values for gold solubilization were not completely cleared up due to analytical problems, but some figures point out that levels of at least 70% can be obtained.

Table VIII - Average Results on Chloride Leaching

Residence time (min.)		Redox Potencial (mv)		Leaching yields (%)								
				R1				R2				
R1	R2	R1	R2	Pb	Ag	Zn	Fe	Pb	Ag	Zn	Fe	Au
45	45	300	1020	99.8	81.8	0	≈0	99.8	94.0	79.9	≈0	-
30	60	300	1020	99.7	80.4	0	≈0	99.8	94.4	86.8	≈0	77.7
30	60	700	1020	99.7	90.2	38.9	≈0	99.7	95.2	91.0	≈0	-

Copper Solvent Extraction and Neutralization

The main antecipated problem was to control the acidity generated during copper removal from 20 g/l to less than 100 mg/l without using conventional neutralizing agents like NH3, NaOH, etc., which would lead to the build-up of undesirable species in the circuit. Alternatives like dilution of initial copper concentration with a copper depleted and neutral solution, acid removal with a tertiary amine, and an intermediate neutralization with lime slurry were studied, this last one presenting better overall results.

In Figure 7, is shown one of the tested process flowsheets.

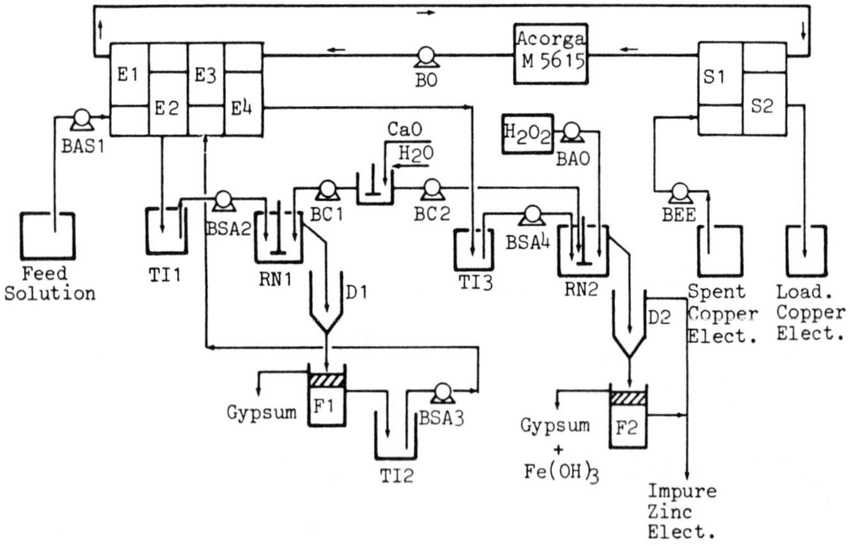

Figure 7 - Process flowsheet for copper extraction and acid neutralization

This configuration is mainly constituted by 4 mixer-settlers for extraction (E1 to E4) and 2 mixer-settlers for stripping (S1 and S2). The intermediate neutralization on the flow outcoming from stage E2 was performed in the reactor RN1 at a pH level of 1.5 during 1 hour at 50°C. The final neutralization and precipitation of residual iron in the copper depleted solution was performed in the reactor RN2 at a pH level of 4.5 during 1 hour at 80°C.

Two extractants P-5100 and M-5615 from ACORGA were tested. In Table IX are presented the average results for ACORGA M-5615 in Escaid 102, with a reagent strength of 22.5%.

Table IX - Average Results on Copper Solvent Extraction and Neutralization

Stages	Feed Solutions (g/l)				Aqueous Phase (g/l)				Organic Phase (g/l)
	Cu	H2SO4	Fe	pH	Cu	H2SO4	Fe	pH	Cu
E1	20.2	-	0.9	3.0	8.4	18.3	-	-	11.0
E2	-	-	-	-	2.0	27.2	-	-	7.0
RN1	2.0	27.4	0.9	-	1.9	-	0.9	1.5	-
E3	1.9	3.8	0.9	1.5	61(mg/l)	6.1	-	-	4.8
E4	-	-	-	-	28(mg/l)	6.3	-	-	4.1
RN2	32(mg/l)	6.1	0.9	-	32(mg/l)	-	0.4(mg/l)	4.5	-
S1	30	180	-	-	33.3	177	-	-	4.1
S2	-	-	-	-	44.8	159	-	-	5.7

According to these results, it is possible to accomplish the target value of less than 100 mg/l of copper in the raffinate, provided that an intermediate stage of neutralization be introduced in the extraction circuit. The residual copper sulphate was expressly left in this flow to act as catalyzing agent in the zinc electrolyte purification step.

The expected copper depletion was reached in only three stages, making the fourth useless. No crude formation was observed during the testwork. The efficiencies in every extraction and stripping stages were in the range of 97 to 100%.

A further advantage of the intermediate neutralization step (RN1) is to render allowable lower pH values in the neutral leaching solution without negatively affecting overall copper extraction kinetics and lowering the copper content in the NL residue fed to the hot acid leaching step.

Finally, these results are in full accordance with the simulation model developed by ACORGA, extensively used during the continuous test runs.

Conclusions

Starting from a bulk concentrate obtained from Aljustrel complex sulphide ore, those operations of the proposed diagram which were considered raw material dependent were studied separately, in continuous pilot and bench scales.

The main anticipated challenges were successfully overcame, namely:

- The roasting of very fine bulk concentrates with a Cu + Pb content of about 15% proceeded steadily in a fluidized bed roaster for periods in excess of 100 hours, with consistent particle size in the bed, negligible build-up of accretions in the roaster, and targeted cinders quality;

- The removal of high concentration of copper (15-20 g/l) in the zinc circuit was successfully carried out by solvent extraction with ACORGA M-5615 in Escaid 102 down to less than 100 mg/l, leading to the production of a neutral zinc electrolyte with a similar composition to those of conventional zinc electrowinning plants; and,

- The use of cinders with high content of lead and silver as a neutralizing agent prior to jarosite precipitation could be done with negligible formation of lead and silver jarosites, which would negatively affect the yield of the chloride leaching.

Implementation of the proposed diagram, which has been already patented, is dependent on results of further testing in a semi-industrial roaster already assembled and on favourable market conditions.

SOME PROMISING TECHNIQUES FOR

COMPLEX IRON ORE METALLURGICAL PROCESSING

L.I.Leontjev, N.A.Vatolin and S.V.Shavrin

Institute of Metallurgy

Ural Division USSR Academy of Sciences

101 Amundsen St., 620219, Sverdlovsk GSP-312, USSR

Abstract

The ever growing interest to the complex iron ores may be attributed to possible production of the naturally alloyed metals and the efficient recovery of the associated elements. However, the conventional technique for complex ore processing is complicated and in some cases even impossible. A pyrometallurgical beneficiation of such ores without melted slag formation has been advanced. A version of the technique assumes pellet production with metal coagulated within a slag shell; on grinding the pellets, the metal phase is mechanically separated from the slag shell. The technique has been run up with ferrochromonickel, high-titanium, high-magnesia, aluminaphosphorus ores and concentrates. Another version of the pyrometallurgical beneficiation without melted slag formation in a rotary kiln has been developed from the schemes for processing red mud of alumina manufacture and high-titanium concentrates. Means have been put up to improve schemes available for complex iron ore processing with primary emphasis on material pretreatment before blast furnace and steelmaking processes. Vanadium-bearing titanomagnetite concentrate oxidized pellets possess low strength at reduction. The technique suggested for pelletizing in the controlled atmosphere improves the aforesaid pellet metallurgical properties. Production of metallized pellets in a conveyor roasting machine holds considerable promise for blast furnace and steelmaking processes. In processing the high-sulphur iron ores still another technique is proposed to roast the fluxed pellets with 95% and more desulphuration.

Metallurgical Processes for the Year 2000 and Beyond
Edited by H.Y. Sohn and E.S. Geskin
The Minerals, Metals & Materials Society, 1988

Introduction

The ever growing interest to the complex iron ores may be attributed to possible production of naturally alloyed metals and efficient recovery of the associated elements. However, the conventional technique for complex ore processing is complicated and in some cases even impossible.

With regard to this some new processes and technical approaches for the complex ore treatment have been developed, definite stages of the conventional technological schemes improved.

The main trends in this work may be stated as:

pyrometallurgical concentration without melted slag formation;

improvements in preparation to metallurgical conversion through the formation of partly oxidized wustitomagnetite and metallized structures in pellets on roasting in the controlled atmosphere;

metallized pellets production from complex ores by high-temperature carbothermal processes at grate machines;

increasing the degree of desulphuration of fluxed pellets made from pyrotinbearing concentrates on roasting in the controlled atmosphere.

Complex Ore Pyrometallurgical Benefication

Benefication of titanomagnetite, ferrochromonickel, alumina-phosphorous and other complex ores by mechanical methods is not efficient, being complicated in realization, it does not provide quality concentrates fit for subsequent conventional processing. Thus, from titanomagnetites of a series of ore deposits it has not been possible to obtain low TiO_2 ferrovanadium concentrate which would permit coventional blast furnace melting. Yield of concentrates of sufficiently high(over 60%) iron content from ferrochromonickel and alumina-phosphorous ores is not justified, it being practically impossible to control the alloying elements in concentrates and tails. The methods of pyrometallurgical beneficiation without slag melt formation offer wide possibilities. The most known among them, based on iron reduction and coagulation in rotary kilns, has a limited scope of application because of a relatively low

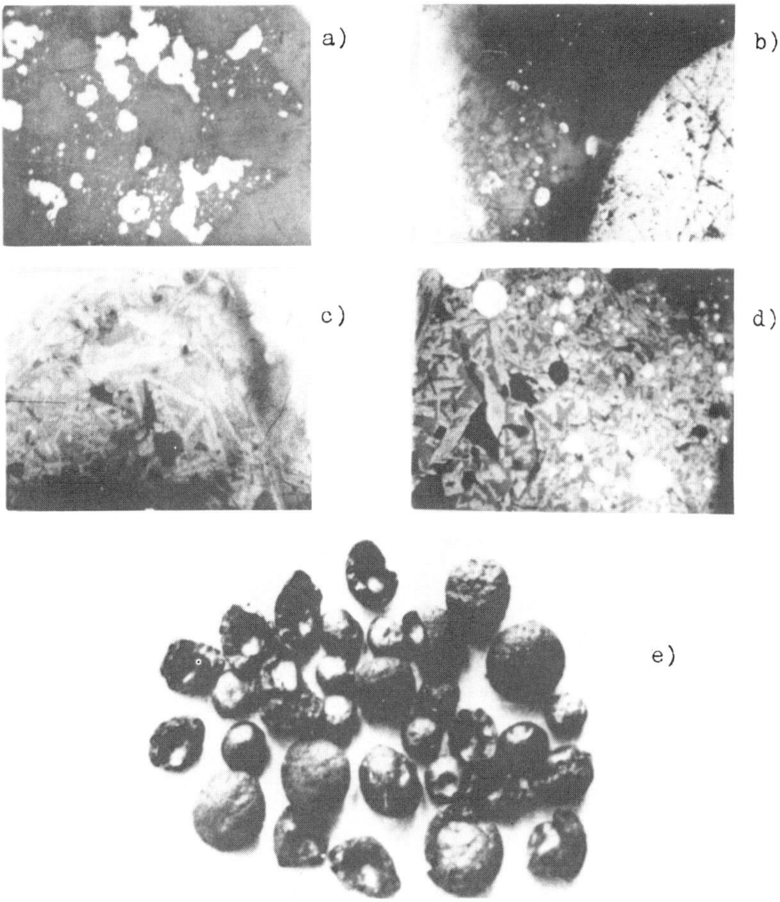

Fig.1. a).Pellet from titanomagnetite concentrate reduced at 1450°C. Metal coagulates in metallic shots,uniformly distributed through a pellet.The main phase of slag is anosovite. Anosovite needle crystals are seen in glass.Reflected light, x50; b).Pellet from titanomagnetite concentrate reduced at 1500°C.Slag shell entrapping a metallic shot consists of anosovite and glass; c).Slag shell microstructure (outside zone).Reflected light,x 100; e).General view of pellets with metal coagulated within slag shell. d).Slag shell microstructure (shot adjoining zone). Reflected light, x 100.

temperature level allowed, particular requirements for physico-chemical properties of slag and difficulties arising from lining and burden interaction. The pyrometallurgical benefication technique developed by the Institute of Metallurgy (Ural Divis., Acad.Sci., USSR) is based on high metallization and coagulation of coal-bearing iron ore pellets by carbothermal reduction in the grate-type machine [1-4], the "nut" structure pellets being produced with metal coagulated within slag shell (Fig.1).

Formation of such structure requires strict observation of technological parameters peculiar for each type of materials. More simple is the process of obtaining grains of coagulated material scattered within pellet.

In both cases upon suitable crushing the metallized concentrate is magnetically separated from the reduced product. This method enables pyrometallurgical benefication of various materials irrespective of physico-chemical properties of slag-forming agents. The absence of refractories in the grate-type machines eliminates the difficulties arising from their interaction with the burden. Any type of solid fuel is acceptable as a reductant. The technique provides yield of metallized iron concentrate and naturally alloyed (e.g. with nickel, chrom, vanadium, phosphorus) concentrates. At the same time conditions of reduction permit to control distribution of alloying elements in the concentrates and tails. Thus, 89% Fe metallized concentrate with traces of phosphorus has been obtained from brown iron ore containing 44-49% Fe, 9-17% SiO_2, about 5% Al_2O_3 and 1% P_2O_5. Simultaneous yield of low (0.01-0.2%) phosphorous and high (upto 3.6%) phosphorous metallized concentrates is also possible [5].

The pyrometallurgical beneficiation method holds considerable promise for ferrotitanium vanadium-containing materials. Separation of iron from titanium without melting is possible with 95-97% Fe in the metallized concentrate at 90-94% extraction and with >50% TiO_2 in the non-magnetic product; V_2O_5 remains in the non-magnetic product, its concentration depends on the initial content in the beneficiated material.

The hydrochemical treatment of a nonmagnetic product containing 59.4% TiO_2 and *.85% V_2O_5 results in 85% and 80%

Table 1. Data on Pyrometallurgical Beneficiation with Metallized Concentrate Production

Initial materials	Components in concentrate,%							Extraction of Iron into concentrate, %
	Fe_{total}	$Fe_{met.}$	C	P		(Ti) Ni	(V) Cr	
Brown iron ore ($Fe=44-49\%, SiO_2=9-17\%, Al_2O_3=5\%, P_2O_5=1,0\%$)	80,4-88,8	77,0-84,5	0,1-0,2	0,01	traces	-	-	88,4-97,5
Ultimate titanomagnetite concentrate ($Fe=53-55\%, TiO_2=13-20\%, V_2O_5=0,9-1,0\%$)	95,0-97,0	03,0-94,0	0,1-0,2	-	(0,2÷0,3)	(0,01)	(0,06)	90,0-94,0
Red mud of alumina production($SiO_2=8-12\%, Fe=30-35\%, Al_2O_3=12-15\%, P_2O_5=1,0\%$)	69,7-78,6	66,5-74,5	0,1-0,2	0,04 0,01	0,04 traces	-	-	79,0-95,5
Siderite($Fe=34-39\%, MgO=10-13\%$)	90-92	85,5-87,5	0,1-0,2	traces	traces	-	-	93,0-95,0
Oxidized nickel Ore ($Fe=18-20\%, SiO_2=30-35\%, Ni=0,5-0,8\%, Cr_2O_3=1,2-1,5\%$)	80,0-85,0	76,0-81,0	0,1-0,2	traces	traces	1,5-2,5	2,0-2,5	85,0-90,0

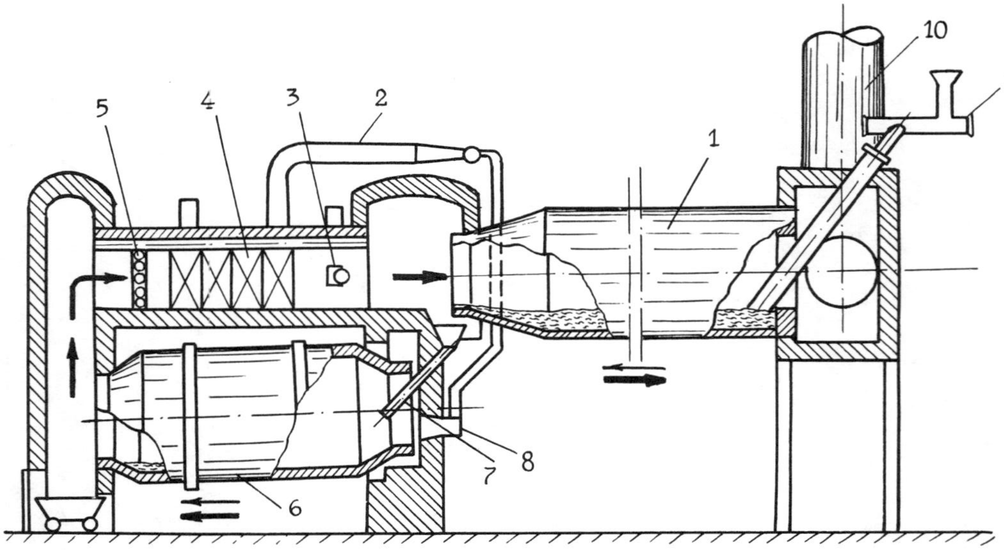

Fig.2.Scheme of Pilot Grate-Type Machine

1.Reduction kiln 2.Hot air pipe 3.Emulsifying agent 4.Pipe type heat exchanger 5.Gas temperature decreasing grid 6.Melting kiln 7. Tube 8.Burner 9.Feeding screw 10. Tube

← material

⬅ flue gas

extraction of TiO_2 and V_2O_5, respectively.

The pyrometallurgical beneficiation applicability has been assessed for red muds of alumina manufacture, siderite and ferrochromonickel ores /Table 1/. The metallized concentrates obtained may be used in the electric furnace practice and in powder metallurgy.

Red muds (by-products of alumina production) and high-titanium concentrates have furnished an example to illustrate another version of pyrometallurgical beneficiation in rotary kilns without slag melt formation.

The expediency of this method for such materials arises from the specific properties of the high-titanium and high-alumina slags obtained: high melting point, rapid change of viscosity at temperature alterations, heterogeneity within wide temperature range.

Among various metallurgical processes it is from rotary kilns only, that the discharge of slags of any aggregate state, including solid state, is possible. The process may be realized in the rotary kilns at relatively low temperature and energy consumption. Among the advantages of the kiln process are possibility to adopt any readily available cheap fuels and reductants, yield of metals free of noxious impurities(P,S).

The disadvantages of the prototype processes by Bosse[6] and Stulzerberg[7] involving formation of skull have been successfully avoided by separate reduction and melting, by diminishing possible contact of burden with refractories at softening points (charging the prereduced material on melt layer).

In the case of separate reduction and melting in the successive rotary kilns (Fig.2) the softening point range has been shortened by abrupt temperature change of the material when transferring it from reduction to melting kilns, the first having countercurrent heat flow and the second concurrent one [8]. Such heat exchange scheme permits the combustion product heat content to be best used, its excess utilized in heating air for burning, the temperature of the heat-transfer agent lowered to the level allowed for the rotary kiln process.

The products of red mud processing (Fe=33-35%, FeO=5-12%, SiO_2=5.9-7.8%, Al_2O_3 =11.0-14.5%, TiO_2=2.5-4.5%, P_2O_5=0.7-0.8%,

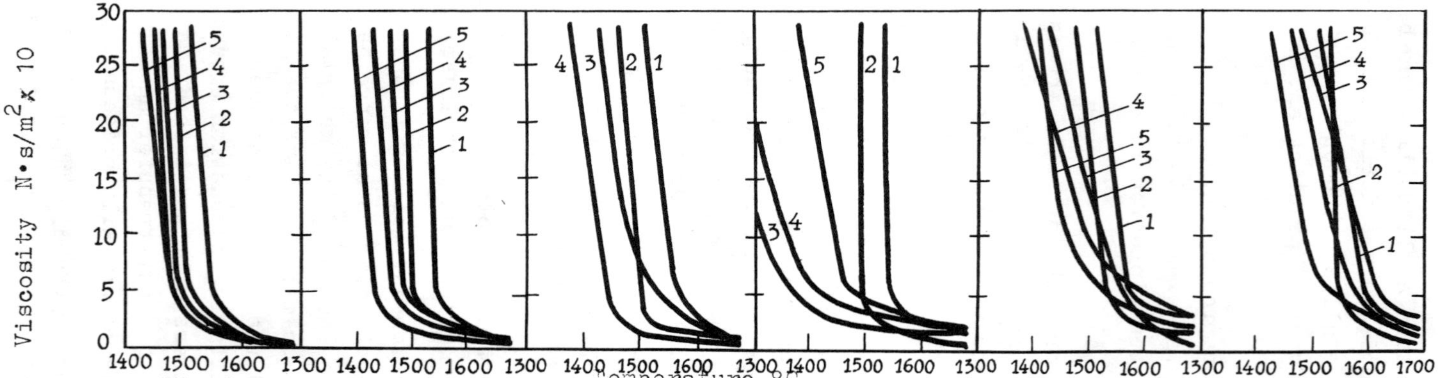

Fig.3.1 Fig.3.2 Fig.3.3. Fig.3.4. Fig.3.5. Fig.3.6.

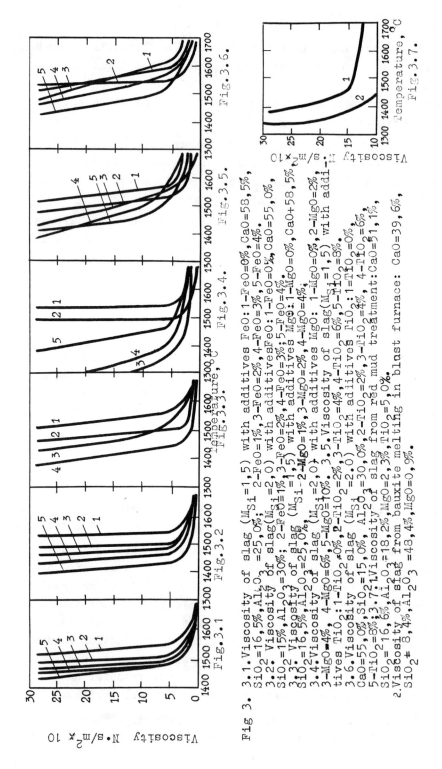

Fig 3. 3.1.Viscosity of slag (M_{Si}=1,5) with additives FeO: 1-FeO=0%, CaO=58,5%, SiO_2=16,5%, Al_2O_3=25,0%; 2-FeO=1%, 3-FeO=2%, 4-FeO=3%, 5-FeO=4%.
3.2. Viscosity of slag(M_{Si}=2,0) with additives FeO: 1-FeO=0%, CaO=55,0%, SiO_2=15%, Al_2O_3=30%; 2-FeO=1%, 3-FeO=2%, 4-FeO=3%; 5-FeO=4%.
3.3. Viscosity of slag (M_{Si}=1,5) with additives Mg0: 1-MgO=0%, Ca0+58,5%, SiO_2=16,5%, Al_2O_3=25,0%, 2-MgO=1%, 3-MgO=2%, 4-MgO=4%;
3.4. Viscosity of slag (M_{Si}=2,0) with additives MgO: 1-MgO=0%, 2-MgO=2%, 3-MgO=4%, 4-MgO=6%, 5-MgO=10%. 3.5. Viscosity of slag(M_{Si}=1,5) with additives TiO_2: 1-TiO_2=0%, 2-TiO_2=2%, 3-TiO_2=4%, 4-TiO_2=6%, 5-TiO_2=8%.
3.6. Viscosity of slag (M_{Si}=2.0) with additives TiO_2: 1=TiO_2=0%, CaO=55,0%, SiO_2=15,0%, Al_2O_3=30,0%, 2-TiO_2=2%, 3-TiO_2=4%, 4-TiO_2=6%, 5-TiO_2=8%; 3.7. 1.Viscosity of slag from red mud treatment: CaO=51,1%, SiO_2=16,6%, Al_2O_3=18,2%, MgO=2,3%, TiO_2=5,0%.
2.Viscosity of slag from bauxite melting in blast furnace: CaO=39,6%, SiO_2= 8,4%, Al_2O_3 =48,4%, MgO=0,9%.

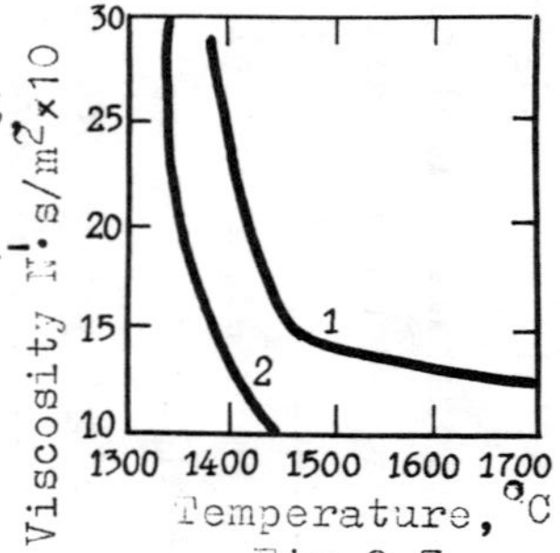

Fig.3.7.

$MgO=1.3-1.6\%, S=1.6\%, R_2O=2.7\%$)were 2-4.5% C ferrocarbide melt and alumocalcium slag with $(CaO)_2SiO_2$ and $(CaO)_{12}(Al_2O_3)_7$ (Fig3) phases. Representative change in the viscosity of such slag evidences its solid state at 1350°C. Indeed, the form of the slag discharged on separate melting of the prereduced burden in the semi-scale commercial rotary kiln at 1320-1370°C has been that of solid balls and ellipsoids of ~100 mm diam[9]. Production in a rotary kiln of high-titanium (TiO_2 > 40%)slag in ductile state is also possible.

Improvements in Complex Ore Preparation for Metallurgical Conversion

It is known[10,11],that oxidized pellets from iron rich concentrates possess high strength when cold, but lose strength and even disintegrate under reduction. Pellets from concentrates with low silica content, i.e. with less pronounced silica bond are found to disintegrate more drastically[17].

Loss of strength is more marked in pellets from titanomagnetite concentrates, processing of these in a blast furnace being especially sensitive to burden quality.

In order to improve the quality of pellets from titanomagnetite concentrates ($Fe=62.15\%, TiO_2=2,30\%, V_2O_5=0.56\%, CaO=1.48\%, MgO=2.14\%, SiO_2=3.64\%, Al_2O_3=2.49\%, Mn=0.22\%, S=0.003\%$) new techniques have been developed and run up for roasting pellets in a controlled atmosphere. The techniques ensure substantial drop in pellet disintegration under reduction due to decrease in the degree to which the material is oxidized and formation of wustitomagnetite or metallized structures[13,14,15].

Laboratory and commercial scale investigations have evidenced that swelling decreases by a factor of 1.3, when the bivalent iron content of pellet increases from 1.5 to 4%. Substitution of such pellets for oxidized ones yields 1% decrease in coal consumption with the same rise in output.

Commercial production of 6-10% FeO pellets, their use in the blast furnace burden have shown 10-13 kg decrease in coke consumption per ton of iron.

More crucial improvement of metallurgical properties is observed in heat-treatment under the controlled atmosphere of

Table 2. Data on Pellet Quality at Roasting under Controlled Atmosphere

Pellet type	Content of solid reductant in burden, %	Content in finished pellets, %		Degree of metallization $\frac{Fe_{met.}}{Fe_{total}} \times 100$	Crushing Strength (mean) kgf/pellet	Strength and abrasion under reduction, % as per GOST (mean)	
		Fe_{total}	$Fe_{met.}$ ($\overline{FeO}$)			+100mm	-0,5m
From titanomagnetite concentrate Fe=62,15%, SiO_2=3,7%, TiO_2=2,8%							
with raised FeO content	-	57,3-60,2	(16,2-25,2)	-	220	64,3	10,7
wustitomagnetite	3,0	58,3-63,1	(41,1-56,8)	-	124	31,2	3,2
metallized	10,0	60,1-63,5	9,2-13,5	14,9-29,1	161	91,8	3,0
oxidized(datum point)	-	58,6	(2,4)	-	179	0,7	55,7
From magnetite concentrate Fe=61,6%, CaO=0,6%, SiO_2=6,2%, Al_2O_3=5,3%							
wustitomagnetite	3,0	53,5-58,6	0,4-4,94 (15,9-29,6)	0-3,0	39	31,6	1,0
metallized with basicity 1,0	8,0	56,7-58,8	11,6-16,8	20,0-29,6	178	96,0	1,8
	11,5	55,3-61,4	11,7-24,3	21,4-39,6	129	93,5	3,5
oxidized (datum point)	-	58,0	(2,0-5,0)	-	200	27,5	6,0

coal-bearing iron ore pellets with wustitomagnetite or metallized structures formation. Utilization of metallized pellets in the blast furnace burden may yield 5-7% decrease in coke consumption and output rise per every 10% metallization[16].

Table 2 gives an insight into metallurgical properties of pellets of different phase structure.

The new technique has been developed to fit the type of the conveyer grate machines available. Means have been put up to convert these machines to the new production process.

Metallized Pellets for Direct Production of Iron from Complex Ores

Low temperature reduction processes with gaseous reductant are not efficient for the production of highly metallized (over 80%) pellets from complex ores. Some modifications have been worked out for high metallization on grate machines with solid reductant introduced in the granulated burden.

High metallization in oxidizing atmosphere has been reached through some technological approaches including combination of reductants of different activity[17], which extends reductant temperature range, decreases the degree of its burnout. Pellet sintering allowable for the grate-type machines and relatively high temperature (1200-1300°C), which furnish high rates in simultaneous metallized structure formation through pellet cross section are also the factors promoting sufficient metallization.

The metallized pellets obtained under these conditions are not pyrophorous. Table 2 lists data on metallized pellets from different concentrates.

Increasing the Degree of Desulfuration of Fluxed Pellets Made from Sulphuric Materials

Increasing the basicity of the oxidized pellets is known to hinder the removal of sulphur because of calcium sulphate formation. The technique developed[18,19] ensures 95-98% desulphuration at the basicities of $\frac{CaO}{O_2}$ upto 1.6 without the temperature level rise as compared to the conventional practice, i.e. at temperature below 1300°C. It also provides for roasting in the controlled gas atmosphere with and without

Table 3. Data on Pellet Quality at Roasting Sulphurous Magnetite Concentrates under Controlled Atmosphere

Pellet Type	Content of solid Reductant in Burden,%	Content in finished pellets,%		Crushing strength (mean) kgf/pel.	Strength and abrasion under Reduction,% as per GOST (mean)		Sulphur content, %
		Fe_{total}	FeO		+10 mm	-0,5 mm	
Magnetite Concentrate Fe=66,5%, SiO=4,5% S =0,41%							
1. No reductant in Burden	-	61,5-62,8	8,2-19,0	156	33,9	8,6	0,0096 0,044
2. With 1% reductant in Burden	1,0	62,4-63,4	17,2-25,3	124	48,5	3,8	0,014 0,048
3. Oxidized (datum point)	-	62,8	1,06	199	23,1	6,5	0,07

introduction of solid fuel into the granulated burden. The mean solphur content of the pellets from concentrates with 64-66% iron,0,4-0,59% sulphur and at a maximum 1250-1280°C roasting temperature with 1.2-1.6 basicity is 0,025% with 8.2-25.3% FeO.Pellets made from the burden with upto 1% solid fuel reveal less disintegration on reduction than those oxidized and roasted in oxygen-free atmosphere /Table 3/.

REFERENCES

1. V.B.Fetisov, L.I.Leontjev, B.S.Kudinov. "Process of Metallized Pellets Production". USSR Inventor's Certificate N396368, Filed 22.09.70, Ser.N1472771, Int.Cl.C21B 13/00.

2. V.A.Kobelev, L.I.Leontjev, B.S.Kudinov et al. "Process for Metallized Concentrate Production". USSR Inventor's Certificate N787483, Filed 03.11.78, Ser.N2681159, Int.Cl.C22B1/02.

3. V.A.Kiselev, L.I.Leontjev, S.V.Shavrin et al. "Metallized Concentate Production from Ural Titanomagnetite Ores" (Paper presented at the 2-nd All-Union Conference on Complex Utilisation of Ores and Concentrates, Moscow (Part II) (1983) 17-18.

4. A.P.Kasakov, L.I.Leontjev, B.S.Kudinov "Beneficiation of Lisakovsk Ore with Metallized Concentrate Production and Mechanism of Its Reduction by Carbon". Tr. Ural Nauchno-Issled. Inst. Chern. Met. (34) (1978).

5. V.A.Kobelev, A.P.Kasakov, L.I.Leontjev et al. "Sulphur and Phosphorus Removal in Direct Reduction". Sb. "Teorija i praktika poluch. Zeleza", Moscow, Izd. "Nauka", (1986).

6. A.N.Pokhvistnev, I.Ju.Kozevnikov, L.N.Spektor et al. "Direct Reduction Abroad". Moscow. Izd. Metallurgya (1964), 244-246.

7. Eulenstein, A.Krus. "Eisengewinnung im Trommel Ofen", Stahl und Eisen, 57(H1) (1937) 6-7.

8. B.S.Kudinov, L.I.Leontjev, A.I.Chernogolov et al."Heat Transfer Scheme of Rotary Kiln for Red Mud Processing". Tr.Inst.Metallurgii UFAN USSR 22 (1970) 50-55.

9. B.S.Kudinov, A.I.Bychin, L.I.Leontjev."Semi-scale Commercial Run for Red Mud Processing in Rotary Kiln." Tsvet.Met. 1 (1967), 63-65.

10. I.E.Rushkin, B.V.Kachula, S.N.Antonova et al."Ferro-Vanadium Pellets from Kachkanar GOK."Tr.Ural Nauchno-Issled.Inst.Chern. Met., Sverdlovsk (1973), 32-37.

11. F.G.Khokhlov, "Agglomerate and Pellet Disintegration in Reduction, Metallurgy, 1 (1971), 13-15.

12. Ju.S.Jusfin, G.N.Basilevich."Optimum Degree of Iron Ore Beneficiation", Stal' 8 (1971) 681-683.

13. S.V.Shavrin, L.I.Leontjev, V.M.Antonov et al."Oxidation and Metallurgical Properties of Kachkanar Pellets", Bul.Tsentr.Nauchno-Issled.Inst.Chern.Met., 10 (1973).

14. A.P.Efimov, L.I.Leontjev, S.V.Shavrin."The Technique for Wustitomagnetite Pellets Production from Kachkanar Concentrates with Solid Fuel". Sb.: Okuskov.Zelezn.Rud.i Kontsentratov, Sverdlovsk, (1975).

15. L.I.Leontjev, V.A.Kobelev, A.P.Kasakov et al."Properties of Metallized Pellets Produced in Semi-Scale Commercial Grate Machine OK- 5,2 SSGOK." Deposited in Vsesojusn. Inst.Nauchn.i Techn. Inform. N4935 (1985).

16. V.F.Knjasev, A.I.Gimmelforb, A.M.Nemenov, "Non-Coke Metallurgy of Iron." Moscow, Izd. Metallurgiya (1972) 272.

17. A.P.Kasakov, L.I.Leontjev, B.S.Kudinov et al. "Process for Coal-Bearing Iron Ore Pellet Production." USSR Inventor's Certificate N 755370, Filed 28.07.78, Ser.N2649867, Int.Cl.C22B 1/14

18. V.A.Kobelev, L.I.Leontjev, B.S.Kudinov et al., "Phosphorus Removal from Coal-Bearing Iron Ore Pellets." Deposited in Vsesojusn. Inst.Nauchn.i Techn.Inform. N4911(1980).

19. V.A.Kobelev, L.I.Leontjev, B.S.Kudinov et al., "High-Iron-Monoxide Pellets Production in Semi-Scale Commercial Grate Machine OK-5.2 SSGOK." Deposited in Vsesojusn.Inst. Nauchn.i Techn.Inform. N4936(1985).

Author Index

Apelian, D., 183
Aplan, F.F., 765
Asai, S., 17, 59
Aukrust, E., 637
Aw, C.Y., 465

Barin, I., 361
Bartlett, R.W., 3
Brinckman, F.E., 427
Brown, D.J., 465
Brown, S.B., 9
Brupbacher, J.M., 71
Byrne, J.G., 103

Cao, R.-J., 383
Chen, W.L., 203
Cheung, P.W., 465
Christodoulou, L., 71
Coelho, M.C., 809
Cooper, P.G., 273

Dawson, S., 745

Eisele, T.C., 411
El-Kaddah, N., 169
Engl, H.W., 139
Eriksson, G., 361

Fanning, R.E., 225
Feinman, J., 721
Figueiredo, J.M., 809
Flemings, M.C., 9
Flinn, J.E., 103
Fray, D.J., 363, 493
Friedman, L.J., 807
Fruehan, R.J., 673
Fujikawa, M., 301

Geskin, E.S., 203, 613
Goto, S., 301
Grjotheim, K., 535
Guangwen, Z., 583
Guedes, R.M., 809
Guo, B., 323

Hanniala, P., 285
Hogan, W.T., 647
Holcomb, G., 69
Holl, H., 139
Hu, H.-K., 783
Hughes, I.F., 623

Im, S.J., 57
Ito, K., 673

Jiang, H., 323
Johnsonbaugh, D., 427

Kamio, S., 253
Kanamori, K., 253
Kaneko, K., 347
Kar, A., 101
Kawai, Y., 733
Kawatra, S.K., 411
Keller, R., 551
Kikuchi, Y., 733
Kim, D., 57
Kim, J.-J., 103
Kim, W., 103
Kimura, T., 347
Kiuchi, M., 83
Klein, A.S., 129
Kozuka, T., 59

Langthaler, T., 139
Leontjev, L.I., 825
Liao, E., 465
Lowry, H.R., 129

Maeda, Y., 287
Mazumder, J., 101
McCarter, M.K., 451
McKelliget, J.W., 169
McLean, A., 745
McMinn, C.J., 517
Meystel, A., 183
Min, I.S., 57
Moinpour, M., 465
Mountford, N.D.G., 745
Murphy, K.J., 791

Nagano, T., 37B
Nagle, D.C., 71
Nishikawa, M., 301
Nishioka, S., 733

Odorski, A.P., 791
Ohba, H., 347

Ohno, A., 155
Okunev, A.I., 391
Olson, G.J., 427
Ozeki, A., 733
Ozturk, B., 673

Pardee, W.J., 73
Paul, B.C., 451
Pehlke, R.D., 115, 603
Pirzada, S., 167
Progelhof, R.C., 215

Qiuzhan, T., 583

Ranson, W.F., 215
Rao, G.M., 571
Rapp, R.A., 69
Richards, N.E., 517
Robertson, D.G.C., 671

Saavedra, A.F., 517
Sadrnezhaad, K., 761
Santén, S.O., 721
Sassa, K., 59
Sauert, F., 361
Schwaha, K.L., 139
Shaff, M.A., 73
Shavrin, S.V., 825
Shibasaki, T., 253
Shim, I.-0., 103
Shuzhen, D., 583
Silva, A.R., 809
Simkovich, G., 765
Smith, A.R., 791
Smith, R.W., 395
Sohn, H.Y., 451
Sokolowski, R.S., 37A
Sommerville, I.D., 745
St.Pierre, G.R., 69
Sugiyama, S., 83
Sulanto, J., 285
Suzuki, S., 287
Szekely, J., 165

Takaoka, T., 733
Tan, W., 583
Tang, X.-H., 383
Taylor, P.R., 167
Thompson, L.C., 439
Trout, T.K., 427
Tunold, R., 563

Vallomy, J.A., 655
Vatolin, N.A., 825
Vora, A., 203

Wakamatsu, N., 287
Warner, N.A., 699
Welch, B., 535
Wharton, Jr., R.A., 395

Yamada, K., 733
Yamaga, M., 733
Yang, S., 323
Yang, Z., 395
Yazawa, A., 241
Yongnian, D., 335

Zhang, Y.J., 563